Formulas for Computing Volumes and Surface Areas

Rectangular Solid: The length is l, the width w, and the height h.
Volume: $V = lwh$
Surface Area: $S = 2(lw + lh + wh)$

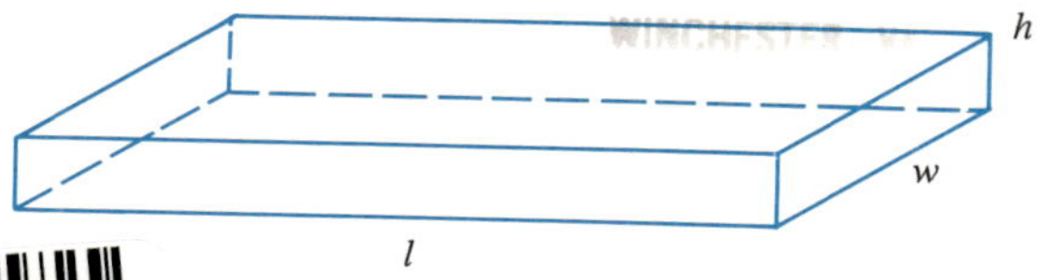

Right Circular Cylinder: The height is h. The radius of each circular base is r.
Volume: $V = \pi r^2 h$
Total Surface Area: $S = 2\pi r^2 + 2\pi rh$
Lateral Surface Area: $L = 2\pi rh$

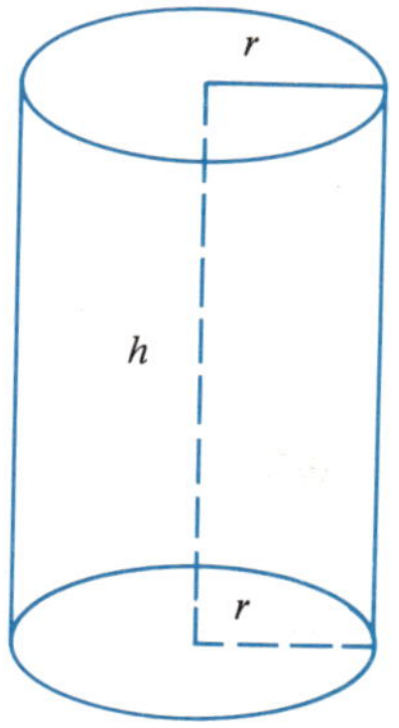

Right Circular Cone: The height is h. The radius of the circular base is r.
Volume: $V = \frac{1}{3}\pi r^2 h$

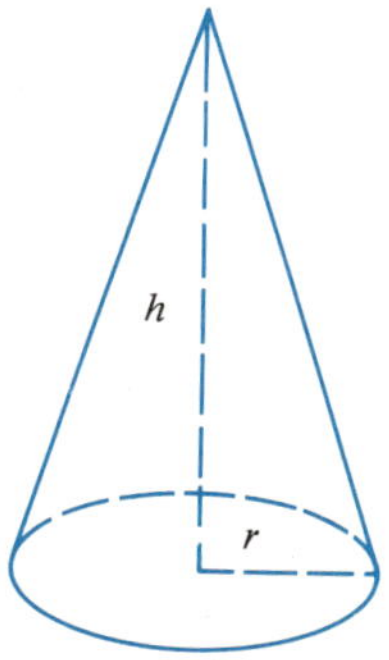

Right Pyramid: The height is h. The area of the base is B.
Volume: $V = \frac{1}{3}Bh$

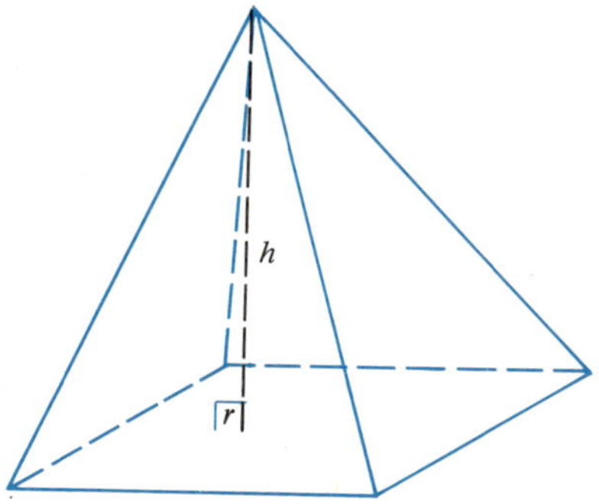

Other Formulas

Distance: If t is time, r is speed, and d is distance, then

$$d = rt$$

Simple Interest: If t is time, r is the rate of interest (usually expressed as a percent), p is the principal, and i is the interest earned, then

$$i = prt$$

INTERMEDIATE ALGEBRA

INTERMEDIATE ALGEBRA

Andrew Demetropoulos
Montclair State College

Kenneth C. Wolff
Montclair State College

MACMILLAN PUBLISHING COMPANY
NEW YORK

Collier Macmillan Publishers
London

Cover photo: Detail, eroded butte, Caineville, California. Photo by William Garnett.

Printed in the United States of America.

Macmillan Publishing Company
866 Third Avenue, New York, New York 10022

Collier Macmillan Canada, Inc.

Library of Congress Cataloging in Publication Data

Demetropoulos, Andrew.
Intermediate algebra.

Includes index.
1. Algebra. I. Wolff, Kenneth C. II. Ti
QA154.2.D45 1985 512.9 83
ISBN 0-02-328530-3

Printing: 1 2 3 4 5 6 7 8 Year 5 6 7 8 9 0 1 2

ISBN 0-02-328530-3

PREFACE

In *Intermediate Algebra* we offer a topic selection and presentation that is a complete college-level intermediate algebra text written for the student. Although we assume that students have had a previous course in algebra, we review all the basic topics of elementary algebra.

Our primary goal in the design of the text is to enable students to acquire the necessary skills and understanding required for subsequent courses such as college algebra, precalculus, a short course in calculus/analytic geometry, elementary finite mathematics, mathematics for business, and so forth.

In writing the text, we have aimed at achieving a balance between conceptual and technical competence appropriate for the intermediate algebra student. Without a doubt, technical or mechanical competence is a first requisite for success in mathematics. However, the authors contend that it is an error to ignore stating mathematical results in their proper form such as axioms, definitions or theorems, which are the basis of the algebraic techniques being used. Thus, by stating general results in their proper form, fundamental concepts are disentangled from the mechanical procedures that depend on these fundamental ideas. Accordingly, the student is not put in the position of struggling to acquire a mass of details that are not illuminated by any general unifying concepts.

In accordance with the purposes outlined above, we have written a textbook that will accomplish the following:

1. First and foremost, develop the student's technical competence in using algebraic techniques and procedures. Therefore, exercises are primarily geared to developing algebraic skills appropriate to intermediate algebra.
2. Develop the student's ability to deal with applied problems (word problems) that require algebra for their solution. To this end, where appropriate, numerous ''word problems'' are provided in the Exercise Sets.
3. Develop conceptual competence by stating important mathematical consequences as theorems. Although most theorems are not proved, a few simple proofs are provided for use as the instructors judge appropriate. Exercise sets are not dependent on proofs.
4. Develop a coherent and mathematically complete text consistent with the level of the intended audience.

Specific features of this book that are the result of the general aims and purposes stated above follow.

Pedagogy

- Each chapter contains a brief introduction to the topics that are presented in that chapter.
- Examples have step-by-step explanations, usually with supporting reasons for each step.
- Verbal (word) problems are introduced early in the text.
- Applied problems of a real-life nature are included.
- Attention is given to carefully wording general mathematical statements. This provides a bridge to the type of statements that a student may encounter in future mathematics courses.
- The meaning of a general mathematical statement is illustrated by relevant and specific worked-out examples.
- Most general mathematical statements are followed by English statements to aid student comprehension.
- Exercises are keyed to worked-out examples.
- Selected exercises whose solution is aided by the use of a *calculator* are marked accordingly with a calculator symbol (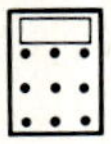).
- Extensive Chapter Reviews are provided at the end of each chapter.
- Chapter Tests are provided at the end of each chapter.

Exercises

- The text contains over 4500 exercises.
- Exercise sets are graduated in difficulty with some exercises at the very end to challenge the more proficient student.
- In general, an even-numbered exercise corresponds to the preceding odd-numbered exercise.
- Inclusion of real life problems such as profit and revenue, exponential growth and decay, variation problems involving applications from the physical sciences, and so forth.
- Answers to odd-numbered exercises in the Exercise Sets and all answers to Chapter Reviews and Chapter Tests are given at the back of the book.

Other Features

- Important rules and procedures are highlighted or boxed in color.
- Sets are reviewed briefly in Chapter 1 but are only used in the text when the result is economy of thought and language.
- Examples that contain fractions, decimals, and percents are included, thus enabling the student to review the related arithmetic skills. These examples are usually found at the end of the Exercise Sets.

Support Material

An Instructor's Manual is available, which contains answers to all even-numbered exercises; multiple forms of Chapter Tests; and final examinations. Answers to all these tests and examinations are included. A Student Solutions Manual is also available.

We wish to thank the following reviewers for their helpful suggestions for improving the text: James Arnold, University of Wisconsin—Milwaukee; R. Paul Beem, Indiana University at South Bend; Arlene Dowshen, Peirce Junior College; Warren B. Gordon, Baruch College, City University of New York; Irving H. Hart, Oregon Institute of Technology; Michael Karelius, American River College; Richard Marshall, Eastern Michigan University; Joyce Saxon, Morehead State University; and Ann Watkins, Pierce College.

We wish to thank the editors at Macmillan for their confidence and support—in particular, Susan Saltrick, our Mathematics Editor, for her kindly and thoughtful assistance; and Elaine Wetterau, our Production Supervisor, for her gracious and expert guidance of the manuscript through the production process. We also wish to thank Linda Lodzinski, who checked all the solutions. Our families were also supportive despite long periods of inattentiveness. Finally, we thank Pat Demetropoulos for an excellent job of typing the manuscript and the Instructor's Manual, many times at short notice.

Andrew Demetropoulos
Kenneth C. Wolff

CONTENTS

10 Functions 377

11 Exponential and Logarithmic Functions 423

12 Sequences and Series; Counting Principles 459

A Synthetic Division 497

B The Remainder Theorem and the Factor Theorem 503

INTERMEDIATE ALGEBRA

CHAPTER 1

Fundamentals

Positive and negative numbers are used extensively in mathematics, but they also have other practical applications. For example, these numbers are used in connection with counting time during the launch of a space vehicle. The time before launch is negative (−), time at blast off is 0, and after launch is positive (+).

National Aeronautics and Space Administration

In this chapter we discuss fundamental algebraic concepts and rules that we will use in our study of intermediate algebra. In addition, we familiarize ourselves with the symbolism or notation used to refer to these basic algebraic notions. Although some of the material presented may be familiar to you from earlier courses, do not omit it. It is good practice to review and reinforce the basic algebraic ideas that you will need in order to understand the remainder of the text.

1.1 Sets

An acquaintance with the basic concepts, operations, and notation involving sets is important for the study of a great deal of modern mathematics. We shall use sets in a variety of ways in our study of algebra.

Intuitively,

> A *set* is a collection or group of objects. The objects making up the set are called *elements* or *members* of the sets.

A set must be specified in a *well-defined* manner.

A set is *well-defined* when it is possible to determine whether a particular object does or does not belong to the set. The *listing* or *roster* method is one way of specifying sets so that they are well-defined. We specify a set with the listing method by writing the names of the members of the set between braces { }. For example, the set of odd numbers from 1 to 5 inclusive is written as

$$\{1, 3, 5\}.$$

Commas are used to separate the names of members of the set and the order in which they are listed does not matter. Thus {1, 3, 5}, {5, 3, 1}, and {3, 1, 5} all represent the same set of numbers. The set { } containing *no* elements is a special set called the *null* or *empty* set. It is usually represented by the symbol ∅.

We use capital letters of the alphabet—A, B, C, and so forth—to represent sets. Thus, using the symbol for equality, =, we write

$$A = \{1, 2, 3\}$$

to indicate that we are calling the set {1, 2, 3} by the name A.

Example 1. Use the listing or roster method to specify the set of odd numbers from 1 to 9 inclusive.

Solution: We write

$$\{1, 3, 5, 7, 9\}.$$

Example 2. Let V represent the set of letters of the alphabet that represent the vowels (assume that y is not a vowel). List the members of the set V.

Solution

$$V = \{a, e, i, o, u\}.$$

For sets containing many members, it is not necessary or sometimes not possible to list every member of the set. For example, the set of all *counting* numbers or all *natural* numbers is designated by

$$\{1, 2, 3, 4, \ldots\},$$

where the three dots mean continue according to the same pattern. An alternative and sometimes more efficient way for specifying a set is called the *rule* or *statement* method. This method utilizes *set-builder notation,* which has the following form: If p denotes a certain property, the notation

$$\{x \mid x \text{ has the property } p\}$$

↑ the bar is read "such that"

represents the set of *all* objects x for which the statement "x has the property p" is *true*. Thus, in set-builder notation,

$$\{x \mid x \text{ is a natural number}\}$$

is read as "the set of all x such that x is a natural number" (remember, the vertical bar means "such that") and denotes the same set as

$$\{1, 2, 3, 4, \ldots\}.$$

The letter "x" is called a "variable." We usually use the letters at the end of the alphabet, such as x, y, and z as variables.

Example 3. Use the rule method to specify the following sets.
(a) $B = \{1, 3, 5, 7, 9, 11, \ldots\}$. (b) $C = \{2, 4, 6, \ldots, 96, 98, 100\}$.

Solution

(a) $B = \{x \mid x \text{ is an odd natural number}\}$.
(b) $C = \{y \mid y \text{ is an even natural number from 2 to 100 inclusive}\}$.

We use the symbol $\in$ to denote that an object *belongs to* or *is a member* (element) of a set. We negate most mathematical symbols by placing a slash, /, through the symbol. Consequently $\notin$ means "is *not* a member of the set." Thus, given the set $A = \{1, 2, 3\}$, we may write

$3 \in A$. read as "3 is an element of set A"

To show, for example, that 4 does not belong to set A, we write

$$4 \notin A. \qquad \text{read as ``4 is not an element of set } A\text{''}$$

If A and B are sets that contain exactly the same elements, the sets are said to be *equal*. To symbolize that two sets A and B are equal, we use the equality symbol, $=$, and write

$$A = B.$$

For example, suppose that C and D are the sets

$$C = \{x \mid x \text{ is a natural number}\} \qquad \text{and} \qquad D = \{1, 2, 3, 4, 5, \ldots\}.$$

Since C and D contain the same elements, we may write

$$C = D.$$

If two given sets A and B are not equal, we write $A \neq B$.

A set A is a *subset* of B if every element of A is also an element of B. We express this relationship symbolically by writing

$$A \subseteq B. \qquad \text{read ``}A \text{ is a subset of } B\text{''}$$

To denote that A is *not a subset* of B, we write

$$A \not\subseteq B. \qquad \text{read ``}A \text{ is } not \text{ a subset of } B\text{''}$$

It can be shown that, for *any* given set A, the following relations are valid:

1. $\varnothing \subseteq A$. The null set is a subset of any set.
2. $A \subseteq A$. Any set is a subset of itself.

Example 4. List all the subsets of the set $A = \{1, 2, 3\}$.

Solution: Every element of a subset of A must be a member of set A.

$$\varnothing, \quad \{1\}, \quad \{2\}, \quad \{3\}, \quad \{1, 2\}, \quad \{1, 3\}, \quad \{2, 3\}, \quad \{1, 2, 3\}.$$

Example 5. How many possible committees consisting of two people can be formed from the set of people {Julio, Jane, Marie}?

Solution: The number of two-person committees is equal to the number of subsets consisting of two people selected from the set {Julio, Jane, Marie}. They are {Julio, Jane}, {Julio, Marie}, {Jane, Marie}. The number of two-member committees is three.

We may combine sets to form new sets. The two basic operations performed on sets to form new sets are called *union* and *intersection*. These set operations are defined as follows:

The *union* of sets, A and B, is the set of elements belonging to *either A or*

B or possibly both. We denote this new set by

$$A \cup B. \qquad \text{read ``}A \textit{ union } B\text{''}$$

The definition of the union of sets A and B may be written symbolically as

$$A \cup B = \{x \mid x \in A \text{ or } x \in B\}.$$

The *intersection* of sets A and B is the set of elements *common to both A and B*. We denote this new set by

$$A \cap B. \qquad \text{read ``}A \textit{ intersection } B\text{''}$$

Symbolically, the definition of intersection of sets A and B is given by

$$A \cap B = \{x \mid x \in A \text{ and } x \in B\}.$$

Example 6. Given the sets $A = \{1, 3, 5, 7\}$ and $B = \{5, 7, 9\}$, find
(a) $A \cup B$. (b) $A \cap B$.

Solution

(a) To find $A \cup B$, we combine the elements of A and B into one set. Thus

$$A \cup B = \{1, 3, 5, 7, 9\}.$$

Note that we do not list an element common to both sets twice when forming their union.

(b) $A \cap B = \{5, 7\}$.

If two given sets A and B have *no* elements in common, then by definition $A \cap B = \{\ \} = \varnothing$. If $A \cap B = \varnothing$, we say that sets A and B are *disjoint*.

Sets may be characterized as *finite* or *infinite*. *Finite* sets consist of a specific number of elements. Sets that are *not* finite are called *infinite* sets.

Example 7. Given $D = \{1, 3, 5, 7, \ldots, 99\}$, $E = \{2, 4, 6, 8, 10, 12, \ldots\}$
(a) Determine if D and E are finite or infinite sets.
(b) Find $D \cap E$.

Solution

(a) Set D is finite because it contains 50 elements. It consists of the first 50 odd natural numbers. Set E is an infinite set because we cannot specify the number of elements it contains.

(b) $D \cap E = \varnothing$. D and E are *disjoint* sets.

EXERCISES 1.1

For Exercises 1–10, use the roster method to specify the set. See Examples 1 and 2.

1. The set of even natural numbers between 5 and 13.

2. The set of odd natural numbers between 2 and 10.

3. The set of vowels in the word "college."

4. The set of vowels in the word "algebra."

5. The set of natural numbers that divide evenly into 7.

6. The set of natural numbers that divide evenly into 11.

7. The set of odd natural numbers from 2 to 90.

8. The set of even natural numbers from 1 to 71.

9. The set of even counting numbers larger than 10.

10. The set of odd counting numbers larger than 6.

For Exercises 11–16, use the rule method to specify the set. See Example 3.

11. $\{2, 4, 6, 8, \ldots, 396, 398, 400\}$ **12.** $\{1, 3, 5, 7, \ldots, 997, 999\}$

13. $\{a, b, c\}$ **14.** $\{x, y, z\}$

15. $\{5, 6, 7, 8, 9, \ldots\}$ **16.** $\{2, 4, 6, 8, 10, \ldots\}$

For Exercises 17–28, given the sets

$$A = \{2, 3, 5, 7, 11\}, \qquad B = \{1, 4, 6, 8\}, \qquad C = \{1, 6\},$$

fill in the blank with the correct symbol: $\in$, $\notin$, $\subseteq$, or $\not\subseteq$.

17. 2 ____ A **18.** 3 ____ B **19.** 3 ____ C **20.** 5 ____ A

21. C ____ B **22.** C ____ A **23.** B ____ C **24.** B ____ A

25. $\varnothing$ ____ A **26.** $\varnothing$ ____ C **27.** B ____ B **28.** A ____ A

For Exercises 29–40, list all the subsets of the given set. See Example 4. [HINT: *Don't forget about* $\varnothing$.]

29. $\{a\}$ **30.** $\{b\}$ **31.** $\{0\}$ **32.** $\{1\}$

33. $\{x, y\}$ **34.** $\{a, b\}$ **35.** $\{2, 4\}$ **36.** $\{1, 3\}$

37. $\{a, b, c\}$ **38.** $\{x, y, z\}$ **39.** $\{2, 3, 4, 5\}$ **40.** $\{4, 6, 8, 10\}$

For Exercises 41–52, given $A = \{2, 3, 5, 7, 8\}$, $B = \{4, 6, 7, 8, 10\}$, $C = \{2, 11, 13\}$, *find the indicated union or intersection. See Example 6.*

41. $A \cup B$ **42.** $A \cap B$ **43.** $A \cap C$ **44.** $A \cup C$

45. $B \cup C$ **46.** $C \cap B$ **47.** $A \cap \varnothing$ **48.** $A \cup \varnothing$

49. $B \cup B$ **50.** $A \cap A$ **51.** $C \cap C$ **52.** $A \cup A$

For Exercises 53–60, determine if the set is finite *or* infinite. *See Example 7.*

53. {1, 2, 3, 4, 5, . . . , 9999, 10000} **54.** {2, 4, 6, 8, . . . , 998, 1000}

55. {3, 6, 9, 12, 15, . . . , 96, 99} **56.** {4, 8, 12, 16, 20, . . . , 100, 104}

57. $\{x \mid x$ is a counting number smaller than 1 million$\}$

58. $\{x \mid x$ is a counting number smaller than 10 billion$\}$

59. $\{x \mid x$ is a counting number larger than 1 billion$\}$

60. $\{x \mid x$ is a counting number larger than 1 million$\}$

For Exercises 61–68, determine whether the statement is true or false.

61. $\varnothing \subseteq \varnothing$ **62.** $\{4, 6, 8\} = \{8, 6, 4\}$ **63.** $\{123\} = \{1, 2, 3\}$

64. $\varnothing = \{0\}$ **65.** If $A = B$ and $8 \in B$, then $8 \in A$.

66. If $C \subseteq D$ and $3 \in C$, then $3 \in D$.

67. If $A \subseteq B$ and $5 \notin A$, then $5 \notin B$.

68. If $A \subseteq B$ and $7 \notin B$, then $7 \notin A$.

Solve Exercises 69–72. See Example 5.

69. How many possible committees consisting of three people can be selected from the following set of people? {Rupert, Juanita, Bob, Mary}.

70. How many possible committees consisting of four people can be selected from the following set of people? {Maria, Alice, Juan, Bob, Joe}.

71. How many different tool kits consisting of four tools can one assemble from the following set of tools? {hammer, screwdriver, pliers, wrench, file}.

72. A person purchasing a new car has the following set of audio options
{stereo radio, cassette player, eight-track tape deck}.
If the person's budget allows for exactly two options, how many possible options does the person have?

Real Numbers and Subsets of Real Numbers 1.2

In this section we review the real numbers and subsets of real numbers. We will start with the more basic subsets of the real numbers and gradually build up to a description of the real numbers themselves.

Natural Numbers (N): $N = \{1, 2, 3, 4, 5, \ldots\}$.

The set of natural numbers N are also called the *counting numbers*.

Whole Numbers (W): $W = \{0, 1, 2, 3, 4, 5, \ldots\}$.

Notice that the set W consists of the natural numbers together with zero.

Integers (Z): $Z = \{\ldots, -3, -2, -1, 0, +1, +2, +3, \ldots\}$.

The numbers $+1, +2, +3, \ldots$ are called the *positive* integers, while $-1, -2, -3, \ldots$ are called the *negative* integers. The set of natural numbers $\{1, 2, 3, \ldots\}$ and positive integers $\{+1, +2, +3, \ldots\}$ are thought of as representing the same set of numbers. Thus the set of integers Z can be specified as follows:

$$Z = \{\ldots, -3, -2, -1, 0, 1, 2, 3, \ldots\}.$$

Notice that the sets N, W, and Z are related as follows:

$N \subseteq W \subseteq Z$. N is a *subset* of W, and W is a *subset* of Z

Rational Numbers (Q): $Q = \left\{\frac{p}{q} \;\middle|\; p \text{ and } q \text{ are integers, provided that } q \neq 0\right\}$.

Note that $q \neq 0$ is read as "q is not equal to 0."

We can describe a rational number more precisely as any number that can be *put into the form* $\frac{p}{q}$, where p and q are integers, $q \neq 0$. Notice that any integer is also a rational number, because it can be put into the form of a fraction. The denominator of the fraction could be 1 and the integer itself would be the numerator of the fraction. For example, the integers

$$-2, \qquad 0, \qquad 2$$

are by definition also rational numbers, since they can be expressed in the following form:

$$-2 = \frac{-2}{1}, \qquad 0 = \frac{0}{1}, \qquad 2 = \frac{2}{1}.$$

Therefore, we see that the set of integers is a *subset* of the rational numbers. Using set notation, we express this relationship between the sets by writing $Z \subseteq Q$.

It can be shown that the *set of rational numbers* is equal to the set of numbers that can be represented by either *a decimal that terminates* or *a nonterminating repeating decimal*. The following examples illustrate these two representations of some rational numbers.

$$\frac{1}{50} = .02 \qquad -\frac{5}{6} = -.833333\ldots = -.8\overline{3}$$

$$\frac{3}{8} = .375 \qquad \frac{8}{3} = 2.66666\ldots = 2.\overline{6}$$

$$\frac{5}{11} = .454545\ldots = .\overline{45} \qquad -\frac{29}{9} = -3.22222\ldots = -3.\overline{2}$$

Notice that for convenience we indicate a repeating decimal by the use of a bar over the repeating part.

Irrational Numbers (K): $K = \{x \mid x \text{ is a nonterminating, nonrepeating decimal}\}$.

The number represented by the decimal

$$.01001000100001\ldots$$

is an irrational number, since it does not repeat, because as the pattern continues there is *one more* 0 after each 1.

It can be shown that the following numbers are irrational numbers:

$$\sqrt{2} = 1.414213\ldots$$
$$\pi = 3.141592\ldots$$
$$-\sqrt{7} = -2.645751\ldots.$$

Observe that a number cannot be both rational and irrational. In set notation this means that

$Q \cap K = \varnothing$. Q and K are disjoint sets; or Q and K have no elements in common

Real Numbers ($\mathbb{R}$): $\mathbb{R} = \{x \mid x \text{ can be represented by a decimal expansion}\}$.

The set of real numbers, $\mathbb{R}$, consists of all rational numbers Q and irrational numbers K. In set notation, we have

$$\mathbb{R} = Q \cup K, \qquad \text{where } Q \cap K = \varnothing.$$

An outline of the relationship between the set of real numbers and some of its important subsets is given in Figure 1.1.

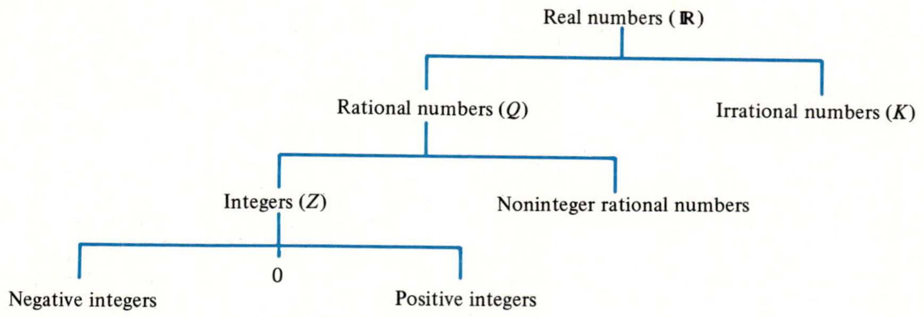

Figure 1.1

Example 1. Use set notation to list the numbers in the set

$$\left\{-5, \frac{-7}{2}, 0, \frac{1}{2}, 6, \frac{9}{1}\right\}$$

that are
(a) Integers.
(b) Rational numbers that are not integers.
(c) Rational numbers.

Solution

(a) $\left\{-5, 0, 6, \frac{9}{1}\right\}$. note that $\frac{9}{1} = 9$

(b) $\left\{\frac{-7}{2}, \frac{1}{2}\right\}$.

(c) $\left\{-5, \frac{-7}{2}, 0, \frac{1}{2}, 6, \frac{9}{1}\right\}$.

Example 2. Use set notation to list the numbers in the set

$$\left\{\frac{2}{-3}, .4\overline{6}, .020020002\ldots, \sqrt{2}, \frac{22}{7}\right\}$$

that are (a) rational numbers, (b) irrational numbers.

Solution

(a) $\left\{\frac{2}{-3}, .4\overline{6}, \frac{22}{7}\right\}$.

(b) $\{.020020002\ldots, \sqrt{2}\}$.

The set of real numbers can be pictured geometrically on a straight line. This is done by taking any two points on the line and labeling the points 0 and 1. The distance between these two points is called the *unit length* or simply a *unit*. We usually locate the point associated with 1 to the right of the point associated with 0. The point associated with 0 is called the *origin*. A positive integer such as 3 is associated with a point 3 units to the *right* of the origin and the corresponding negative integer, -3, with a point 3 units to the *left* of the origin. All the integers are located on the line in a similar way. The location of a portion of the integers is shown in Figure 1.2. Rational

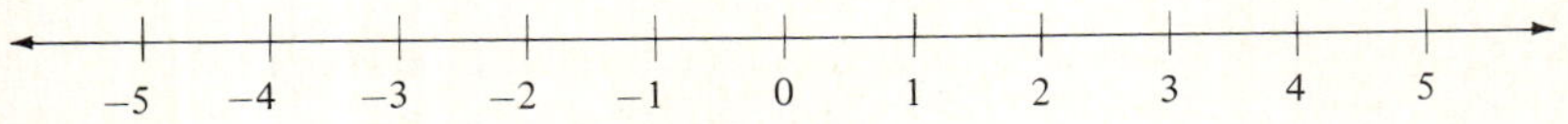

Figure 1.2

numbers such as $\frac{1}{2}, \frac{5}{3}, -\frac{5}{3}$ are located in relation to the unit length as shown in

Figure 1.3. Locating irrational numbers on the number line is a little more complicated, but each irrational number is also associated with a point on the line. Points to the right of the origin are labeled with positive real numbers and points to the left of the origin with the negative real numbers. The line with a real number attached to each point as outlined above is called the *real number line,* or simply the *number line*.

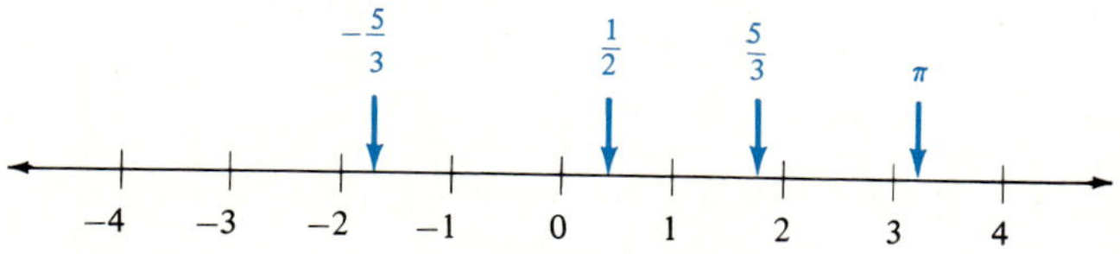

Figure 1.3

We note that to each point on the number line there corresponds exactly one real number and to each real number there is associated exactly one point. As a result of this association of geometrical points and numbers, the words *point* and *number* are often used interchangeably. Thus, with the number line in mind, we use phrases such as "the point 2" and "the number 2" to mean the same thing. The number associated with a point on the number line is called the *coordinate* of the point and the point is called the *graph* of the number. The graph of a number is indicated by a dot on the number line, as illustrated in the following example.

Example 3. Graph the set of numbers $\left\{-2, -\frac{1}{2}, 0, \frac{2}{3}, 2\right\}$.

Solution: Figure 1.4 shows the solution.

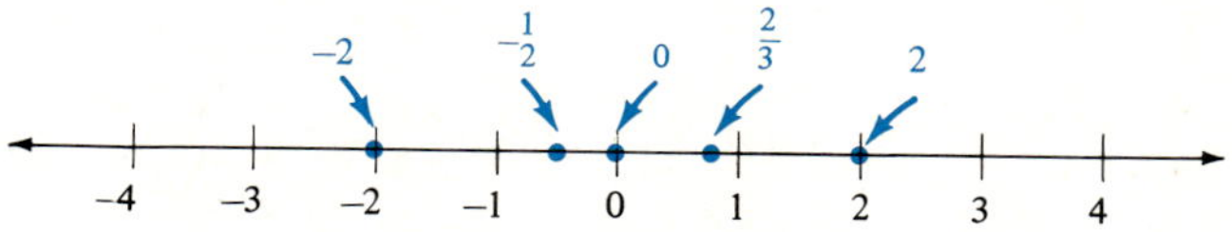

Figure 1.4

We now introduce two concepts that will be useful when we discuss addition and subtraction of real numbers. They are the *additive inverse,* and the *absolute value* of a real number.

Real numbers that are the same distance from the origin but on opposite sides of the origin are called *additive inverses* of each other. Thus $+3$ is the additive inverse of -3, and -3 is the additive inverse of $+3$. The term *opposite* is often used in place of additive inverse. In general, if we use x to represent a real number, we designate the additive inverse of x by

$-x$. read "additive inverse of x" or "opposite of x"

Before giving an example on finding the additive inverse of a number, it is appropriate at this point to make a few comments on the different uses of the symbol "$-$" in algebra. We observe that "$-$" can be used to denote

$7 - 5$	subtraction; the symbol "$-$" is often read as "minus"
-3	a negative number; the symbol "$-$" is often called "the negative sign"
$-(-3)$.	additive inverse or opposite of -3

The reader should not be concerned about any possible confusion that might be caused by the multiple uses of "$-$" in algebra. Algebra is so structured that the three meanings of "$-$" are interrelated and as a result we automatically shift from one use to the other without difficulty.

Example 4. Find $-x$ (additive inverse, opposite of x) when x has the following values:

(a) $+2$. (b) -7. (c) $-\dfrac{1}{3}$. (d) .25

Solution

x	$-x$
$+2$	$-(+2) = -2$
-7	$-(-7) = +7$
$-\dfrac{1}{3}$	$-\left(-\dfrac{1}{3}\right) = +\dfrac{1}{3}$
.25	$-(.25) = -.25$

Example 4 illustrates that the additive inverse of a positive real number is a negative number and conversely the additive inverse of a negative number is positive. We note that $-0 = 0$.

If we located the numbers $+3$ and -3 on the number line, we would see that $+3$ is located a distance of 3 units to the right of the origin, and -3 a distance of 3 units to the left of the origin, as shown in Figure 1.5.

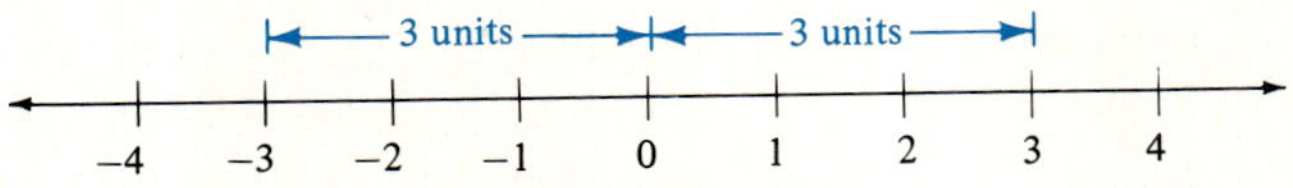

Figure 1.5

The *distance* of a real number from the origin on the number line, without reference to whether it is located to the left or right of the origin, is called the *absolute value* of a number. We denote the absolute value of a number with vertical bars as follows:

$|+3| = 3$ read "absolute value of +3 is 3"
$|-3| = 3.$ read "absolute value of −3 is 3"

The absolute value of a number is never negative, since it measures distance without reference to direction. However, since +3 and 3 are used interchangeably, we may write $|+3| = +3$. Similarly, we often write $|-3| = +3$.

In general, the *absolute value* of a real number x is given by

$$|x| = \begin{cases} x & \text{if } x \text{ is positive} \\ 0 & \text{if } x \text{ is zero} \\ -x & \text{if } x \text{ is negative.} \end{cases}$$

Note carefully that if *x is a negative number,* then *−x is a positive number*. For example, if $x = -2$, we have by definition

$$|\underbrace{-2}_{x}| = -(\underbrace{-2}_{x}) = +2.$$

Example 5. Find the values of the following expressions.
(a) $|+8|$. (b) $|-5|$. (c) $-|-6|$. (d) $-(|-3| + |-2|)$.

Solution

(a) $|+8| = 8$.

(b) $|-5| = 5$.

(c) $-|-6| = -(6) = -6$.

(d) $-(|-3| + |-2|) = -(3 + 2) = -(5) = -5$.

EXERCISES 1.2

For Exercises 1–4, see Examples 1 and 2 and the preceding discussion.

1. Let $A = \left\{-6, \frac{-5}{2}, -2, 0, \frac{4}{3}, 3, 7\right\}$. List the members in set A that are

(a) Natural numbers. **(b)** Whole numbers.
(c) Negative integers. **(d)** Positive integers.
(e) Integers. **(f)** Rational numbers that are not integers.

2. Let $A = \left\{-8, \frac{-7}{3}, -3, 0, \frac{5}{2}, 4, 6\right\}$. List the members of set A that are

(a) Natural numbers. **(b)** Whole numbers.
(c) Negative integers. **(d)** Positive integers.
(e) Integers. **(f)** Rational numbers that are not integers.

3. Use set notation to list the members in the set

$$\left\{-4, -.3\overline{5}, \sqrt{2}, 1.414, \pi, \tfrac{9}{2}\right\}$$

that are
(a) Rational numbers. **(b)** Irrational numbers. **(c)** Real numbers.

4. Use set notation to list the members in the set

$$\left\{-9, -\sqrt{2}, .4\overline{6}, \tfrac{22}{7}, \pi\right\}$$

that are
(a) Rational numbers. **(b)** Irrational numbers. **(c)** Real numbers.

For Exercises 5–14, identify the number as rational or irrational. See the discussion preceding Example 1.

5. $.3\overline{7}$ **6.** .070070007 . . . **7.** .12112111211112 . . . **8.** $\frac{11}{12}$ **9.** $\frac{19}{27}$

10. $.627\overline{25}$ **11.** $\dfrac{-24}{25}$ **12.** .31331333133331 . . . **13.** $.6281\overline{72}$ **14.** .825

For Exercises 15–26, determine whether the relationship between the following sets is true or false.
natural numbers (N), whole numbers (W), integers (Z),
rational numbers (Q), irrational numbers (K), real numbers (ℝ)

15. $W \subseteq Z$ **16.** $Z \subseteq Q$ **17.** $Z \subseteq N$ **18.** $Q \cap K \neq \varnothing$

19. $Q \cap K = \varnothing$ **20.** $K \not\subseteq Q$ **21.** $Q \subseteq K$ **22.** $Z \cup Q = K$

23. $Q \subseteq \mathbb{R}$ **24.** $K \subseteq \mathbb{R}$ **25.** $Q \cup K = \mathbb{R}$ **26.** $Z \cup N = Z$

For Exercises 27–30, graph the set on the number line. See Example 3.

27. $\left\{-2, -\frac{3}{2}, -1, 0, \frac{5}{2}\right\}$ **28.** $\left\{\dfrac{-4}{3}, -\dfrac{1}{2}, 0, \dfrac{2}{3}, 2\right\}$

29. $\left\{-3, 0, 2, \frac{5}{2}, 3\right\}$ **30.** $\left\{-\frac{6}{2}, -\frac{4}{2}, 0, \frac{4}{2}, \frac{5}{2}\right\}$

For Exercises 31–38, find the indicated additive inverse. See Example 4.

31. $-(-10)$ **32.** $-(+8)$ **33.** $-(+9.2)$ **34.** $-(-17)$

35. $-(-\frac{1}{5})$ **36.** $-(-\frac{1}{6})$ **37.** $-(+\frac{1}{8})$ **38.** $-(.\overline{3})$

For Exercises 39–48, find the value of the expression. Express your answer without absolute value bars. See Example 5.

39. $|-10|$ **40.** $|+7|$ **41.** $-|-4|$ **42.** $-|-20|$

43. $-|12 - 2|$ **44.** $-|15 - 3|$ **45.** $|-5| + |-3|$ **46.** $|-7| - |-2|$

47. $-(|-5| + |-4|)$ **48.** $-(|-6| + |-2|)$

For Exercises 49–52, determine whether the expression is true or false. If true, the statement must be valid for all *values of x.*

49. $|-x| = x$ **50.** $-|x| = -x$

51. $|20-x| = |20| - |x|$ **52.** $|5 + x| = |5| + |x|$

Evaluating Algebraic Expressions; Order of Operations 1.3

In arithmetic, the basic operations of addition, subtraction, multiplication, and division are performed on definite numbers. In algebra these basic operations are also performed on symbols—usually letters of the alphabet—which stand for numbers. These letters standing for numbers are sometimes called variables and sometimes constants. We have already mentioned the term *variable* briefly in the preceding two sections. We now give a more explicit definition of a variable and also explain how the term *constant* is used.

In general, a *variable* is a symbol appearing in a mathematical expression that may be replaced by *any* member of a specified set of two or more numbers called the *domain* or *replacement set* of the variable. A *constant* is a symbol that stands for a fixed number in a mathematical expression.

It is customary to use letters at the end of the alphabet x, y, z, t, s for variables, and letters at the beginning of the alphabet a, b, c, d for constants.

Below, as a review, we list some *basic algebraic expressions* involving variables together with a verbal description of their meaning.

Let x and y represent real numbers.

	Basic Algebraic Expression	Literal Meaning
Addition	$x + y$	The *sum* of x and y
Subtraction	$x - y$	The *difference* of x and y
Multiplication	xy, $x(y)$, $(x)y$, $x \cdot y$	The *product* of x and y; x and y are called *factors* of the product
Division	$x \div y$, $\frac{x}{y}$, x/y	The *quotient* of x by y or x *divided* by y

In general, by an *algebraic expression* we mean a mathematical expression consisting of a definite number of symbols of *operation* (addition, subtraction, etc.), variables, constants, and grouping symbols (parentheses, brackets, braces). Below we list illustrations of such expressions:

$$7xy, \qquad 2x + 3y, \qquad 5(y + z), \qquad y/(x + z),$$
$$2s[r - (t + 4)], \qquad s\{w + [r + (t + s)]\}.$$

In the following example, we give an English phrase that is equivalent to the meaning of some simple algebraic symbolism.

Example 1. Translate the following algebraic expressions into verbal statements. (a) $5x + 2$. (b) $5(x - y)$. (c) $(ab)/c$. (d) $(x + y)(x - y)$.

Solution

(a) $5x + 2$.	sum of the product $5x$ and 2
(b) $5(x - y)$.	five times the difference of x and y
(c) $(ab)/c$.	quotient of the product ab by c
(d) $(x + y)(x - y)$.	product of the sum and difference of x and y

Translating words into algebraic symbolism, and vice versa, is important for solving problems using algebraic methods. Notice in the previous discussion how we used the words "sum," "difference," "product," and "quotient" to indicate operations. Thus we *add* two numbers x and y to form their *sum;* when we *multiply* them, we form their *product*. To *subtract* y from x, we form the *difference* $x - y$; when we *divide* x by y, the result is the *quotient* x/y.

We now illustrate the algebraic translation of two English phrases that occur quite often in algebra: "more than" and "less than." Thus

"4 *more than* twice x" is written symbolically as $2x + 4$

and

"6 *less than* the product of 3 and y" is symbolized by $3y - 6$.

Note also that "minus" often indicates "subtraction." As a result,

"7 *minus* the product of 3 and z" is written as $7 - 3z$.

Example 2. Translate the following verbal statements into algebraic expressions.
(a) Two less than the product of x and y.
(b) Two more than the product of x and y.
(c) x minus the quotient of y by z.

Solution

(a) $xy - 2$. (b) $xy + 2$. (c) $x - \dfrac{y}{z}$.

We now discuss what it means to evaluate an algebraic expression for values of the variables appearing in the expression. The individual *numbers* in the replacement set of a variable are called *values* of the variable. When we replace the variables appearing in an algebraic expression with some particular values, that is, with definite numbers, and simplify the resulting expression, we say that we have *evaluated* the expression for those values. Usually, when we wish to replace a variable, such as x, with a particular value, such as 2, in an algebraic expression, we will write "let $x = 2$," "if $x = 2$," or "for $x = 2$." For example, to evaluate

$$5x + y$$

for $x = 2$ and $y = 6$, we write

$$5(2) + 6 = 5 \cdot 2 + 6 = 10 + 6 = 16.$$

We say that $5x + y$ *has value* 16 for $x = 2$ and $y = 6$.

Note that in evaluating the expression $5 \cdot 2 + 6$ we multiplied first and then added to obtain the answer 16. This is the correct procedure according to rules agreed upon for evaluating an expression with more than one operation. We state the rules governing the *order* in which operations are to be performed below. We should note, however, that we can use *grouping symbols,* such as parentheses ( ), brackets [ ], and braces { }, to *change* the conventional order of operations. Thus, as we mentioned above, to evaluate $5 \cdot 2 + 6$, we multiply first, then add. However, the expression $5 \cdot (2 + 6)$ means add 2 and 6 first, then multiply by 5 to obtain the answer 40. To avoid confusion when evaluating expressions, the following rules of procedure are observed.

Order of Operations Rule

1. *Without grouping symbols*—Do the operations in the following order.
   (a) Proceed from left to right doing multiplication and division in the order that they occur.
   (b) Next, again proceed from left to right doing addition and subtraction in the order in which they occur.
2. *With grouping symbols*—Work from the innermost grouping symbols to the outermost. Within each grouping symbol, use the order of operations rule (1) given above.

Example 3. Evaluate the following expressions.
(a) $5 \cdot 6 - 3 \cdot 2$. (b) $12 \div 4 \cdot 3 + 8 \cdot 2$.
(c) $3 \cdot (4 + 7 \cdot 2)$. (d) $5[16 - (15/3 - 2)]$.

Solution

(a) $5 \cdot 6 - 3 \cdot 2 = 30 - 6$ multiply first
$= 24.$ subtraction

(b) $12 \div 4 \cdot 3 + 8 \cdot 2 = 3 \cdot 3 + 8 \cdot 2$	starting left to right division first, next
$= 9 + 8 \cdot 2$	multiplication
$= 9 + 16$	multiplication continuing left to right
$= 25.$	addition
(c) $3 \cdot (4 + 7 \cdot 2) = 3(4 + 14)$	do operations in the parentheses first
$= 3(18)$	continuation of previous step
$= 54.$	multiplication
(d) $5[16 - (15/3 - 2)] = 5[16 - (5 - 2)]$	first do the work in the innermost grouping symbols ( ) using Rule 1
$= 5[16 - (3)]$	continue as in the previous step
$= 5[13]$	do work inside next grouping symbols [ ]
$= 65.$	multiplication

Example 4. Evaluate the following algebraic expressions when $x = 3$, $y = 5$, and $z = 6$.
(a) $7x + yz$. (b) $yz - x$. (c) $y(z - x)$.

Solution

(a) Replacing x with 3, y with 5, and z with 6 in the expression, we obtain

$$7(3) + (5)(6) = 21 + 30 = 51.$$

(b) Similarly, as in part (a), we obtain

$$5(6) - 3 = 30 - 3 = 27.$$

(c) $5(6 - 3) = 5(3) = 15.$

The following example shows how to evaluate complicated expressions involving horizontal fraction bars. For such expressions, we do all the operations in the numerator and denominator first, and then simplify as far as possible. Thus the *fraction bar* acts as a *grouping symbol*.

Example 5. Evaluate the following expressions when $s = 8$, $t = 4$.

(a) $\dfrac{s + t}{t}$. (b) $\dfrac{3s + t}{t + 10}$.

Solution

(a) $\dfrac{8 + 4}{4} = \dfrac{12}{4}$ do addition in the numerator first

$= 3.$ simplify

(b) $\dfrac{3(8) + 4}{4 + 10} = \dfrac{24 + 4}{14} = \dfrac{28}{14} = 2.$

EXERCISES 1.3

For Exercises 1–18, translate the algebraic expression into a verbal statement. See Example 1.

1. $x + 5$
2. $3 + y$
3. $x - 2$
4. $y - 3$
5. $2(x + 4)$
6. $8(3 + y)$
7. $9(x - 2)$
8. $10(y - 3)$
9. $3x + 1$
10. $6y + 3$
11. $7z - 3$
12. $5z - 2$
13. $(xy)/z$
14. $(ab)/2$
15. $y/(x + 2)$
16. $z/(y + 3)$
17. $(x + 2)(x - 2)$
18. $(x - 1)(x + 1)$

For Exercises 19–28, translate the verbal statement into an algebraic expression. See Example 2.

19. Three less than the product of 2 and x.
20. One less than the product of 5 and y.
21. Four less than twice y.
22. Six less than z.
23. Ten more than twice z.
24. Five more than the product of 5 and y.
25. One more than the difference of x and y.
26. Seven more than the difference of x and z.
27. y minus the quotient of x by two.
28. z minus the quotient of three by y.

For Exercises 29–48, evaluate the expression. See Example 3.

29. $3 \cdot 2 + 4 \cdot 5$
30. $5 \cdot 2 + 7 \cdot 3$
31. $8 \cdot 6 - 6 \cdot 4$
32. $7 \cdot 5 - 5 \cdot 2$
33. $24 \div 4 \cdot 2$
34. $36 \div 12 \cdot 3$
35. $48 \div 6 \cdot 2$
36. $27 \div 3 \cdot 3$
37. $16 \div 4 \cdot 2 + 2 \cdot 3$
38. $25 \div 5 \cdot 5 + 10 \cdot 2$
39. $4(2 + 4 \cdot 3)$
40. $2(1 + 7 \cdot 3)$
41. $50/10 - 5 + 7$
42. $15/5 - 2 + 3$
43. $6[(7 - 1) - (4 - 3)]$

44. $4[(6 + 3) - (5 - 2)]$ **45.** $2[20 - (18/3 - 2)]$ **46.** $3[10 - (28/7 - 3)]$

47. $3\{12 - [15 - 2(7 - 4)]\}$ **48.** $4\{24 - 2[16 - (6 + 2)]\}$

For Exercises 49–62, evaluate the expression for $x = 8$, $y = 6$, $z = 2$. See Example 4.

49. $2x + 4y$ **50.** $3y + 5z$ **51.** $5y - 2x$ **52.** $8x - 10z$

53. $x(x + 2)$ **54.** $y(y + 4)$ **55.** $3xz - 6y$ **56.** $2yz - 3x$

57. $x + yz - z$ **58.** $xz - y + z$ **59.** $(x + y)(x - y)$ **60.** $(x + y)/(y - 2z)$

61. $3x - (y - z)/z$ **62.** $2x - 8z/(y - 2)$

For Exercises 63–72, evaluate the expression for $a = 6$, $b = 3$. See Example 5.

63. $\dfrac{a + b}{b}$ **64.** $\dfrac{a - b}{b}$ **65.** $\dfrac{a + b}{a - b}$ **66.** $\dfrac{a - b}{a + b}$

67. $\dfrac{4a - 2b}{2a - 2b}$ **68.** $\dfrac{3a + 15b}{3a + b}$ **69.** $\dfrac{a + 3(a - b)}{2a + b}$ **70.** $\dfrac{b + 2(a - b)}{4a - 7b}$

71. $\dfrac{b[(3a - b) - 5]}{(a - 1)[(5b - a)/b]}$ **72.** $\dfrac{(3b + 1)[(8b - a) - 6]}{a[(6a - 2b) - 20]}$

1.4 Properties of Real Numbers and Equality

In mathematics we generally have to begin with certain terms whose meaning are intuitively understood. These are called *undefined* terms. Some examples of undefined terms used so far in our study of algebra are: set, natural number, integers, and straight line. Using these particular undefined terms, we are then able to define the meaning of other terms, such as subset, rational number, real number line, and so forth. In general, *definitions* of terms are stated in terms of undefined terms and previously defined terms. Definitions are usually stated in the form "p means q" or "p if and only if q," where the letter p represents a term being defined in terms of known concepts denoted by the letter q.

In addition to *undefined terms* and *definitions,* the logical development of a mathematical discipline such as algebra requires that we *assume* that certain basic statements are *true* about the properties of the undefined or defined concepts to be studied. These basic statements are called *axioms* or *postulates*. The axioms or postulates form the starting point from which we proceed to logically prove or deduce other mathematical statements that we call *theorems*. In general, theorems are proven using undefined terms, definitions, axioms, previously proven theorems, and the standard rules of logic. By rules of logic we mean, for example, that we accept as a logical rule that a statement *cannot both* be true and false at the same time.

In our study of algebra *we shall not stress proofs of theorems*. Our main object will be to understand and use the theorems *to develop algebraic skills*. However, when we wish to outline the proof of a theorem, we shall do so in an informal manner. When we omit the proof of a theorem, as will usually be the case, we shall indicate this in our discussion by using such phrases as "it can be shown that" or "it follows from."

In our study of algebra we assume the following properties for addition and multiplication of real numbers. These properties are important because they form the basis from which we derive many of the rules used in algebra.

Axioms for the Set of Real Numbers, $\mathbb{R}$

If x, y, and z are any real numbers in $\mathbb{R}$:

Closure: The sum $x + y$ is a unique real number in $\mathbb{R}$; the product xy is a unique real number in $\mathbb{R}$. (The set of real numbers is said to be *closed* under the operations of *addition* and *multiplication*.)

Commutative: $x + y = y + x$, $xy = yx$

Associative: $(x + y) + z = x + (y + z)$, $(xy)z = x(yz)$

Distributive: $x(y + z) = xy + xz$, $(y + z)x = yx + zx$

Identity Element for Addition: There is exactly one real number, 0, such that

$$x + 0 = x \quad \text{and} \quad 0 + x = x;$$

0 is called the *additive identity*.

Identity Element for Multiplication: There is exactly one real number, 1, such that

$$x \cdot 1 = x \quad \text{and} \quad 1 \cdot x = x;$$

1 is called the *multiplicative identity*.

Additive Inverse: For each $x \in \mathbb{R}$ there is a unique real number, denoted $-x$, such that

$$x + (-x) = 0 \quad \text{and} \quad (-x) + x = 0;$$

$-x$ is called the *additive inverse* of x; we note that $-x$ is also called the "*negative*" or "opposite" of x.

Multiplicative Inverse: For each $x \in \mathbb{R}$, provided that $x \neq 0$, there is a unique real number, denoted $\frac{1}{x}$, such that

$$x \cdot \frac{1}{x} = 1 \quad \text{and} \quad \frac{1}{x} \cdot x = 1;$$

$\frac{1}{x}$ is called the *multiplicative inverse* or *reciprocal* of x.

The following examples illustrate in more concrete terms the properties mentioned above.

Example 1. State the properties that justify the following statements of equality.
(a) $(-7) + 3$ is a real number. (b) $x + 1 = 1 + x$.
(c) $(y + 2) + 6 = y + (2 + 6)$. (d) $(+2) + (-2) = 0$.

Solution

(a) The closure property for addition. We note that $(-7) + 3 = -4$. In Section 1.6 we shall review addition of signed numbers.
(b) $x + 1 = 1 + x$ is an example of the commutative law for addition. It asserts that adding 1 to x yields the same result as adding x to 1.
(c) The associative law for addition enables us to write

$$(y + 2) + 6 = y + (2 + 6).$$

The associative law says that we get the same result whether we add the sum of y and 2 to 6 or y to the sum of 2 and 6.
(d) Additive inverse for addition. We note that -2 is the additive inverse of $+2 = 2$.

Example 2. State the properties that justify the following statements of equality.
(a) $(-5)(2)$ is a real number. (b) $2x = x2$.
(c) $2(3x) = (2 \cdot 3)x$. (d) $5 \cdot \frac{1}{5} = 1$.

Solution

(a) $(-5)(2)$ is a real number by the closure property for multiplication. We note that $(-5)(2) = -10$. Rules for evaluating products of signed numbers will be reviewed in Section 1.7.
(b) Commutative property for multiplication.
(c) Associative property for multiplication.
(d) $\frac{1}{5}$ denotes the reciprocal or multiplicative inverse of 5. Thus the product of 5 times $\frac{1}{5}$ is 1.

The significance of the commutative properties and associative properties can best be understood by considering operations for which these properties are not valid. Notice that for division and subtraction we have

$$10 \div 5 \neq 5 \div 10 \quad \text{and} \quad 6 - 2 \neq 2 - 6.$$

Note that $6 - 2 = 4$ and $2 - 6 = -4$. We will discuss evaluating expressions such as $2 - 6$ when we review subtraction of real numbers in Section 1.6. Also, it is easy to check that subtraction and division are not associative by considering the following expressions:

$$\begin{aligned}(10 - 5) - 2 &\neq 10 - (5 - 2) \\ (5) - 2 &\neq 10 - (3) \\ 3 &\neq 7\end{aligned} \qquad \begin{aligned}(12 \div 6) \div 2 &\neq 12 \div (6 \div 2) \\ (2) \div 2 &\neq 12 \div (3) \\ 1 &\neq 4.\end{aligned}$$

Example 3. Use the distributive property to rewrite the following expressions. (a) $2(x + 3)$. (b) $5a + 5b$.

Solution

(a) $2(x + 3) = 2x + 2 \cdot 3$
$= 2x + 6.$

(b) $5a + 5b = 5(a + b)$.

Much of our discussion on the logical justification of the rules of algebra (the simplification of algebraic expressions and solving equations) depends on a proper understanding of the use of the symbol of equality, =. Note that $a = b$, read "a equals b," means that a and b stand for the same object. Thus $\frac{1}{4} = .25$ means that "$\frac{1}{4}$" and ".25" represent the same number. Below we list the basic properties of equality.

Properties of Equality

Let a, b, and c represent real numbers:

Reflexive Property: $a = a$.

Symmetric Property: If $a = b$, then $b = a$.

Transitive Property: If $a = b$ and $b = c$, then $a = c$.

We also state for completeness the following principle:

Substitution Principle: If $a = b$, then a may be substituted for b, or b for a, in any statement without changing the truth of the statement.

Example 4. State the property of equality justifying the following statements.
(a) $x + 5 = x + 5$.
(b) If $2 = x$, then $x = 2$.
(c) If $x + y = z + 2$ and $z + 2 = 3$, then $x + y = 3$.
(d) If $x + y = 10$ and $y = z$, then $x + z = 10$.

Solution

(a) Reflexive property.
(b) Symmetric property.
(c) Transitive property, or Substitution principle
(d) Substitution principle.

The following theorems are the basis for manipulating many algebraic expressions. Theorems are usually written in the form "If p, then q," where p and q represent statements. The p statement is called the *hypothesis, given,* or *assumption* of the theorem, and q is called the *conclusion* of the theorem. "If p, then q" is often read as "p implies q." Some theorems are stated in the form "p if and only if q." This form is a concise way of stating that "p implies q" *and* "q implies p."

Theorem 1

If $a = b$, then

$$\text{(a) } a + c = b + c \qquad \text{(b) } ac = bc.$$

From the assumption of Theorem 1 that $a = b$, we wish to show that we can proceed logically to the conclusion $a + c = b + c$. To this end we write the following steps:

Statement	Reason(s)
1. $a = b$	Hypothesis or given
2. $a + c = a + c$	Reflexive property of equality
3. $a + c = b + c$	Steps 1 and 2 and substitution principle; substitute b for a on the right side of the equality in Step 2 to obtain Step 3

The statements listed above together with the reasons on the right justifying them constitute a proof of part (a) of Theorem 1. The proof of part (b) of Theorem 1 is very similar to that for part (a).

The next theorem, stated without proof, deals with products involving zero.

Theorem 2: Zero Property

If a is a real number, then

$$a \cdot 0 = 0.$$

Examples 5 and 6 illustrate how we may simplify algebraic expressions using the axioms and theorems discussed thus far in this section.

Example 5. Simplify the following expressions.
(a) $(x + 2) + 5$. (b) $5(6x)$.

Solution

(a) $(x + 2) + 5 = x + (2 + 5)$ associative property of addition

$= x + 7$ $7 = 2 + 5$; substitution principle

Therefore,

$(x + 2) + 5 = x + 7.$ Steps 1 and 2 and the transitive property of equality

(b) $5(6x) = (5 \cdot 6)x$ associative property of multiplication
$= 30x$ $30 = 5 \cdot 6$; substitution principle

Thus

$5(6x) = 30x.$ Steps 1 and 2 and transitive property of equality

Sometimes as a shortcut we also write a sequence of equalities on one line. Thus, in Example 5(b), we could write the indicated steps as follows:

$$5(6x) = (5 \cdot 6)x = 30x.$$

As a result we conclude that $5(6x) = 30x$. Of course, the reasons for each step remain the same. However, we often omit explicitly mentioning the substitution principle and transitive property of equality. We illustrate the preceding discussion in parts (c) and (d) of the next example.

Example 6. Simplify the following expressions.
(a) $2x + 6x$. (b) $3x + 5 + 7x + 2$. (c) $5 \cdot 0 \cdot x$. (d) $8(3x + 2)$.

Solution

(a) $2x + 6x = (2 + 6)x$ distributive property
$= 8x.$ $2 + 6 = 8$

Thus we see that $2x + 6x = 8x$.

(b) $3x + 5 + 7x + 2 = 3x + 7x + 5 + 2$ commutative property (addition)
$= (3x + 7x) + 5 + 2$ associative property (addition)
$= (3 + 7)x + 5 + 2$ distributive property
$= 10x + 5 + 2$ $3 + 7 = 10$
$= 10x + 7.$ associative property (addition)

Therefore, we conclude that

$$3x + 5 + 7x + 2 = 10x + 7.$$

(c) $5 \cdot 0 \cdot x = 0 \cdot x = 0$. Thus $5 \cdot 0 \cdot x = 0$. Notice that we used the *zero property* twice in the first line of equalities.

(d) $8(3x + 2) = 8(3x) + 8 \cdot 2 = (8 \cdot 3)x + 16 = 24x + 16$. The reader is invited to supply the reasons for each step in the chain of equalities reading from left to right—also, the reason why we can conclude from the sequence of equalities that $8(3x + 2) = 24x + 16$.

EXERCISES 1.4

For Exercises 1–26, state a property of the real numbers that justifies the statement. All variables represent real numbers. See Examples 1–3.

1. $x + 4 = 4 + x$

2. $z + 5 = 5 + z$

3. $2 + \pi$ is a real number.

4. $\pi + (-2)$ is a real number.

5. $5y = y5$

6. $9t = t9$

7. $(+3) + (-3) = 0$

8. $(+4) + (-4) = 0$

9. $(x + 6) + 5 = x + (6 + 5)$

10. $(y + 7) + 2 = y + (7 + 2)$

11. $3(4y) = (3 \cdot 4)y$

12. $7(2y) = (7 \cdot 2)y$

13. $(-5) + 0 = -5$

14. $(+3) + 0 = +3$

15. $\pi(\sqrt{2})$ is a real number.

16. 3π is a real number.

17. $1 \cdot x = x \cdot 1$

18. $0 + x = x + 0$

19. $1 \cdot z = z$

20. $y + 0 = y$

21. $(-4) \cdot \dfrac{1}{-4} = 1$

22. $(-7) \cdot \dfrac{1}{-7} = 1$

23. $1 \cdot (x + y) = 1 \cdot x + 1 \cdot y$

24. $1 \cdot (x + 3) = x + 3$

25. $y \cdot 7 + y \cdot 2 = y(7 + 2)$

26. $z \cdot 5 + z \cdot 3 = z \cdot (5 + 3)$

For Exercises 27–38, state the name of the property or properties that justify the statement. See Example 4 and Theorems 1 and 2.

27. If $x = y$ and $y = 2$, then $x = 2$.

28. If $y + 2 = z$ and $z = 3$, then $y + 2 = 3$.

29. If $2 = z$, then $z = 2$.

30. If $t = -5$, then $-5 = t$.

31. $x + 2 = x + 2$.

32. $y + z = y + z$.

33. If $x + 1 = 4$, then $(x + 1) + (-1) = 4 + (-1)$.

34. If $3x = 6$, then $\frac{1}{3}(3x) = \frac{1}{3}(6)$.

35. $0 \cdot \pi = 0$

36. $0 \cdot \sqrt{2} = 0$

37. If $x + z = z$ and $z = 5$, then $x + 5 = 5$.

38. If $x + 5 = z$ and $z = -5$, then $x + 5 = -5$.

For Exercises 39–68, use the properties discussed in this section to simplify the expression. See Examples 5 and 6.

39. $(y + 2) + 8$

40. $5 + (6 + x)$

41. $5x + 3x$

42. $8x + 2x$

43. $\frac{1}{7}(7x)$

44. $\frac{1}{9}(9x)$

45. $7(3x)$
46. $4(2x)$
47. $(3 + z) + 7$
48. $(4 + y) + 10$
49. $7 \cdot 0 \cdot y$
50. $8 \cdot 2 \cdot 0 \cdot y$
51. $3(2x + 4)$
52. $7(2x + 1)$
53. $(x + 2) \cdot 6$
54. $(y + 3) \cdot 5$
55. $[(2 + 5x) + 3x] + 3$
56. $[(3 + 7x) + 2x] + 2$
57. $[(1 + 2x) + 6x] + 2$
58. $[(1 + 8x) + 2x] + 3$
59. $[(8x + 2) + 4x] + 7$
60. $[(6x + 10) + 7x] + 2$
61. $9x + 8 + 3x + 4$
62. $10x + 2 + 6x + 5$
63. $5(2x + 4) + 3x + 2$
64. $3(3x + 2) + x + 4$
65. $(3x + 7x) + 8x$
66. $(10x + 5x) + 6x$
67. $12x + 2x + 6x$
68. $7x + 3x + 5x$

For Exercises 69–72, determine if the given set of numbers is closed under the specified operation. [HINT: *Remember that each member of a set may be used more than once.*]

69. {0, 1}; multiplication
70. {0, 1}; addition
71. {2, 4, 6, 8, 16, 32, 64, 128, . . .}; addition
72. {2, 4, 8, 16, 32, 64, 128, . . .}; multiplication

Inequality Properties of Real Numbers 1.5

In the preceding section we discussed equality of numbers and the properties of the equality symbol, =. In this section we discuss the relation of inequality between two real numbers x and y. We use the symbol $<$ to represent the relation "less than" between real numbers x and y.

We define

$$x < y \qquad \text{read} \quad \text{"}x \text{ is less than } y\text{"}$$

to mean that there exists a positive real number p such that

$$x + p = y.$$

Furthermore, we may express this relationship between x and y in a different way using the symbol $>$ and writing

$$y > x. \qquad \text{read} \quad \text{"}y \text{ is greater than } x\text{"}$$

Thus $x < y$ (x is less than y) and $y > x$ (y is greater than x) are equivalent statements. Note that the symbols $<$ and $>$ always point to the smaller quantity. Also,

$$x \le y \qquad \text{means "either } x < y \text{ or } x = y\text{"}$$

and

$$x \ge y. \qquad \text{means "either } x > y \text{ or } x = y\text{"}$$

As long as we compare positive real numbers, there is usually no problem determining whether one real number is less than or greater than another real number. For example, we know intuitively that $2 < 5$. Formally, by the definition of "less than" given above, since 3 is a positive number such that $2 + 3 = 5$, we know that $2 < 5$. But how do we compare numbers such as -2 and -5, or 2 and -3? When we review how to add signed numbers, we will use the definition of "less than" given above to answer these questions. But for now, the real number line provides us with a geometric guideline for comparing signed numbers, which we now state.

Let x and y represent real numbers:

$x < y$ if x lies to the *left* of y on the real number line

$y > x$. if y lies to the *right* of x on the real number line

We illustrate this in Figure 1.6.

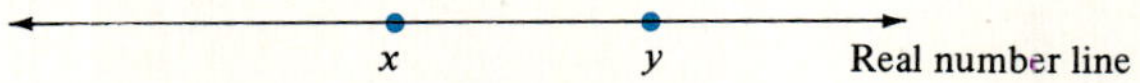

Figure 1.6

Example 1. Fill in each blank with $<$ or $>$.
(a) -5 ____ -3. (b) 2 ____ -4. (c) -4 ____ 0. (d) 2 ____ 0.

Solution: First, as shown in Figure 1.7, we graph the set of numbers to be compared:

$$\{-5, -3, 2, -4, 0\}.$$

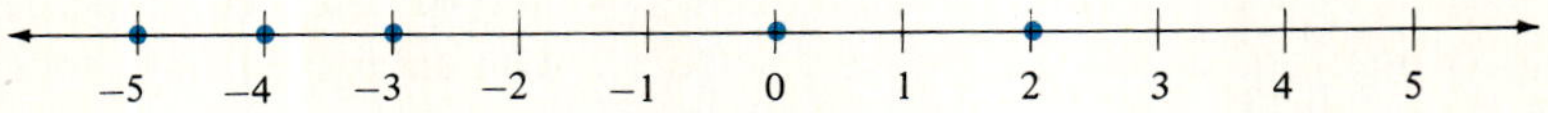

Figure 1.7

(a) $-5 < -3$. (b) $2 > -4$. (c) $-4 < 0$. (d) $2 > 0$.

Notice that any positive number x lies to the right of 0 on the number line, as indicated in Figure 1.8.

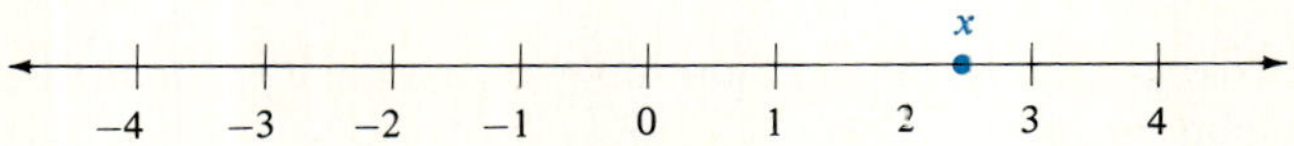

Figure 1.8

Thus the inequality

$x > 0$ is often read "x is a positive number."

Similarly, since any negative number x lies to the left of 0 on the number line, we have

$x < 0$ which is often read "x is a negative number."

Also, we symbolize the *nonnegative* numbers (positive numbers and 0) by writing $x \geq 0$.

Example 2. Write a verbal statement describing each of the following inequalities.
(a) $x \leq -4$. (b) $x \leq 0$. (c) $x \geq 3$. (d) $x > 0$.

Solution

(a) $x \leq -4$. x is less than or equal to -4
(b) $x \leq 0$. x is less than or equal to 0
(c) $x \geq 3$. x is greater than or equal to 3
(d) $x > 0$. x is greater than 0 or x is positive

If x is a real number, the statement of inequality

$$-2 < x < 3$$

means that

x is greater than -2 *and* x is less than 3.

On the number line this means that x lies to the *right* of -2 but to the *left* of 3. Thus the graph of the set of numbers $\{x \mid -2 < x < 3\}$ is represented by a solid line between -2 and 3 as shown in Figure 1.9. Notice the small circles around -2 and 3 to show that these numbers are not part of the graph. When either of the endpoints is included, we shade in the circle.

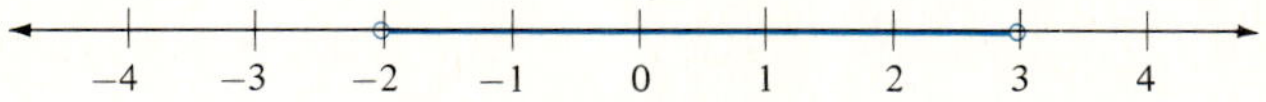

Figure 1.9

Example 3. Graph the set of real numbers $\{x \mid -3 \leq x < 1\}$.

Solution: The given set consists of the real numbers between -3 and 1, including -3 but excluding 1. Thus the graph is as shown in Figure 1.10.

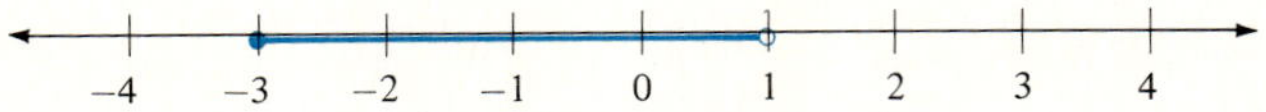

Figure 1.10

Example 4. Graph the following sets of real numbers.
(a) $\{x \mid x > 2\}$. (b) $\{x \mid x \leq -1\}$.

Solution

(a) The graph of the given set consists of all points to the right of 2, but not including 2. The graph of this set is indicated in Figure 1.11 by a solid line extending to the right of 2.

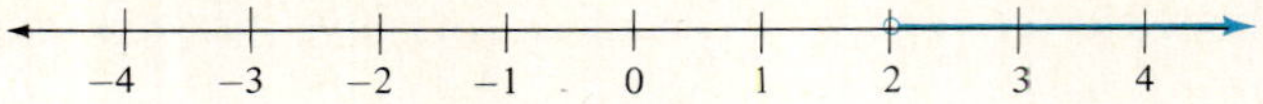

Figure 1.11

(b) The graph of the given set is shown in Figure 1.12.

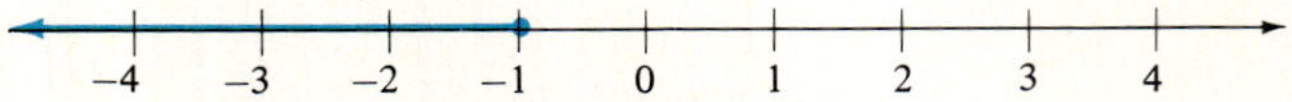

Figure 1.12

Example 5. If x represents an integer, graph the set $\{x \mid -4 \leq x \leq 2\}$.

Solution: The replacement values of x is taken from the set of integers. Thus the given set consists of the integers to the right of -4 and to the left of 2, including -4 and 2. The graph of this set is shown in Figure 1.13.

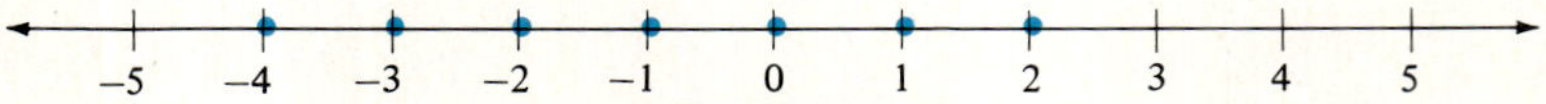

Figure 1.13

We conclude this section with the following two assumptions on the properties of the inequality relation.

Properties of Inequality

Transitive Property: If $x < y$ and $y < z$, then $x < z$.

Trichotomy Property: If x, y, and z are real numbers, then exactly one of the following is true:

$$x < y, \qquad x = y, \qquad x > y.$$

EXERCISES 1.5

For Exercises 1–12, fill in each blank with < or >. See Example 1.

1. 2 ____ 5 **2.** 3 ____ 2 **3.** 3 ____ −4 **4.** 3 ____ −2

5. −2 ____ −4 **6.** −3 ____ −2 **7.** −2 ____ 0 **8.** 5 ____ 0

9. −4 ____ −3 **10.** −5 ____ −4 **11.** −5 ____ 5 **12.** 2 ____ −2

For Exercises 13–20, write a verbal statement describing the inequality. See Example 2.

13. $x \leq 4$ **14.** $x \geq 5$ **15.** $z \geq -2$ **16.** $t \leq -6$

17. $y > 0$ **18.** $z \leq 0$ **19.** $x \leq -3$ **20.** $y \leq a$

For Exercises 21–32, graph the set. Assume that the domain (replacement set) of the variables is the set of real numbers. See Examples 3 and 4.

21. $\{x \mid -1 < x \le 3\}$ **22.** $\{t \mid -2 \le t < 1\}$ **23.** $\{y \mid -1 \le y \le 2\}$

24. $\{z \mid 0 < z < 3\}$ **25.** $\{x \mid -3 < x < -1\}$ **26.** $\{z \mid -4 < z < 2\}$

27. $\{x \mid x \le 3\}$ **28.** $\{y \mid y \ge 2\}$ **29.** $\{z \mid z > -2\}$

30. $\{x \mid x < 1\}$ **31.** $\{x \mid x < 0\}$ **32.** $\{y \mid y \ge 0\}$

For Exercises 33–38, graph the set. Assume that the domain (replacement set) is the set of integers. See Example 5.

33. $\{x \mid -3 \le x \le 2\}$ **34.** $\{y \mid -2 \le y \le 3\}$ **35.** $\{z \mid -2 < z < 4\}$

36. $\{x \mid -1 < x < 3\}$ **37.** $\{z \mid 0 \le z < 4\}$ **38.** $\{y \mid -3 < y \le 2\}$

For Exercises 39–42, fill in the blank so that the resulting statement is an example of the given property. Assume that the variables represent real numbers.

39. $x < 0$ or $x = 0$ or x ____ 0; trichotomy property for inequality

40. $z < -1$ or z ____ -1 or $z > -1$; trichotomy property for inequality

41. If $x < y$ and $y < 5$ then x ____ 5; transitive property

42. If $y < z$ and $z < 0$ then y ____ 0; transitive property

Addition and Subtraction of Real Numbers 1.6

In this section we review the rules for adding and subtracting real numbers. In our development of these rules we will state formally some of the properties of addition and subtraction that are the basis for these rules. We will also illustrate the application of these rules with concrete examples. We begin by assuming the following two properties of real numbers.

Addition Property for Positive Numbers

> If $x > 0$, $y > 0$, then
>
> $$x + y > 0.$$

The *addition property* for positive real numbers is the assumption that

The sum of two positive numbers is positive.

The *trichotomy property for real numbers* is the assumption that

Trichotomy Property for Real Numbers

> For a given real number x, *exactly one* of the following alternatives holds:
>
> Either x *is positive*, or x *is zero*, or $-x$ *is positive*.

The content of the following theorem enables us to compute the sum of the additive inverses of two real numbers.

Theorem 3

> If x and y are real numbers, then
>
> $$(-x) + (-y) = -(x + y).$$

We will not prove this theorem but note that a proof can be based essentially on the property that a real number has *one and only one* additive inverse (or opposite).

We observe by the *trichotomy* property for real numbers that *if* x and y are *positive* numbers, then $-x$ and $-y$ are *negative* numbers. Notice that in this case x and y represent the *absolute values* of $-x$ and $-y$, respectively. Thus, from Theorem 3, we obtain the following rule.

Addition of Two Negative Numbers

> The sum of two *negative* numbers is obtained by adding their absolute values and then taking the negative of this sum.

We illustrate the application of the rule above in Example 1.

Example 1. Find the indicated sums.
(a) $(-3) + (-7)$. (b) $(-10) + (-4)$. (c) $[(-5) + (-3)] + (-1)$.

Solution

(a) $(-3) + (-7) = -(3 + 7) = -10$. Note that $|-3| = 3$, $|-7| = 7$.

(b) $(-10) + (-4) = -(10 + 4) = -14$.

(c) $[(-5) + (-3)] + (-1) = [-(5 + 3)] + (-1) = (-8) + (-1) = -(8 + 1) = -9$.

Because real numbers are positive, zero, or negative, we often refer to them as signed numbers. Two numbers have *like signs* when both are positive or both are negative. Two numbers are said to have *unlike signs* when one number is positive and the other number is negative. Theorems similar to Theorem 3 involving positive and negative numbers are the basis for the following rule for adding numbers with *unlike* signs.

Addition of Numbers with Unlike Signs

> To add two numbers with *unlike signs* (one positive, the other negative), subtract the smaller absolute value from the larger absolute value. To this difference attach the sign of the number that has the larger absolute value.

We illustrate the application of the rule above in Example 2.

Example 2. Find the sums.
(a) $(+10) + (-4)$. (b) $(-7) + 5$. (c) $(-2) + 4$. (d) $8 + (-15)$.

Solution: We apply the rule stated above for adding numbers with unlike signs.

(a) Notice that $|+10| = 10$, $|-4| = 4$, and that $10 > 4$. Subtracting the smaller absolute value from the larger absolute value, we obtain $10 - 4 = 6$. To this difference we attach the sign of the number that had the larger absolute value, namely, $+10$. Thus

$$(+10) + (-4) = +6.$$

(b) We see that $|-7| = 7$ and that $7 > 5$. Next we take the difference $7 - 5 = 2$. To this difference we attach the sign associated with -7. Thus we have

$$(-7) +5 = -2.$$

(c) $(-2) + 4 = |4| - |-2| = 4 - 2 = 2.$

(d) $(-15) + 8 = -(|-15| - |8|) = -(15 - 8) = -(7) = -7.$

When adding several real numbers, we may use the associative and commutative properties to group and add numbers in any order that we wish. We illustrate this in the next example.

Example 3. Find the sums.
(a) $(-7) + 4 + (-8)$. (b) $10 + (-6) + 5 + (-4) + (-2)$.

Solution

(a) We may add the numbers proceeding from left to right.

$$(-7) + (4) + (-8) = (-3) + (-8) = -11.$$

(b) To evaluate the expression, it is simpler first to group the negative and positive numbers together, then add to find the desired sum. Thus we proceed as follows:

$10 + (-6) + 5 + (-4) + (-2)$

$= 10 + 5 + (-6) + (-4) + (-2)$	commutative property
$= 15 + (-12)$	addition of numbers with *like* signs
$= 3.$	addition of numbers with *unlike* signs

When we did subtraction problems in elementary arithmetic, we subtracted a smaller number from a larger number. For example, we evaluated differences such as $7 - 5 = 2$. However, we did not encounter differences such as $5 - 7$. The difference $5 - 7$ did not "make sense" until we were introduced to negative numbers. We now recall the definition of subtraction in algebra.

Definition of Subtraction

Let x and y represent real numbers; then

$$x - y = x + (-y).$$

This definition states that *to subtract y from x,* add to x the additive inverse (opposite) of y. In less formal language this definition can be stated as follows:

To subtract y from x, change the sign of y to the opposite sign and proceed as in addition.

As an illustration of this procedure we evaluate the following difference using the definition.

$$\underset{\text{subtraction}}{5 \underset{\uparrow}{-} 7} = 5 + (\underset{\substack{\uparrow \\ \text{additive inverse}}}{-}7)$$

$$= -2. \qquad \text{addition rule for adding numbers with unlike signs}$$

The following theorem can be established using the uniqueness of the additive inverse of a real number. The theorem is useful for evaluating the difference of two signed numbers.

Theorem 4: Double-Negative Theorem

If x is a real number, then

$$-(-x) = x.$$

The following example illustrates the application of the double-negative theorem.

Example 4. Find
(a) $-(-3)$. (b) $-[-(-3)]$.

Solution

(a) If we replace x with 3 in the double-negative theorem, we have

$$-(-3) = 3.$$

(b) Similarly, for $x = -3$, we see that

$$-[-(-3)] = -3.$$

Example 5. Find the differences.
(a) $4 - 12$. (b) $8 - (-4)$. (c) $(-10) - (-15)$. (d) $-15 - 10$.

Solution

(a) $4 - 12 = 4 + (-12) = -8$. definition of subtraction; addition

(b) $8 - (-4) = 8 + [-(-4)]$ definition of subtraction
$= 8 + 4$ double-negative theorem
$= 12$.

When evaluating differences such as the one given above, we often skip the step involving the double-negative property. Thus, with this understanding, we write more briefly,

$8 - (-4) = 8 + 4$ subtraction; double-negative theorem
$= 12$.

(c) $(-10) - (-15) = -10 + 15 = 5$.

(d) $-15 - 10 = -15 + (-10) = -25$.

The next example illustrates how to deal with expressions involving both subtraction and addition.

Example 6. Rewrite each expression in terms of addition; then evaluate the sum.
(a) $-9 - 6 + 18$. (b) $-5 - 3 - 4$.

Solution

(a) $-9 - 6 + 18 = -9 + (-6) + 18$ definition of subtraction
$= -15 + 18$ addition
$= 3$.

(b) $-5 - 3 - 4 = -5 + (-3) + (-4)$ definition of subtraction
$= -8 + (-4)$ addition
$= -12$.

Problems involving subtraction such as those given in Example 6 can be done quickly by supplying the addition operation "mentally." This is illustrated in the following example.

Example 7. Evaluate the following expressions.
(a) $-4 - 8 - 3$. (b) $-10 - 3 - (-6)$.

Solution

(a) By definition, $-4 - 8 - 3$ means $-4 + (-8) + (-3)$. Thus, to evaluate $-4 - 8 - 3$, we add -4, -8, and -3 using the rules for adding numbers with like signs to obtain -15. As a result,

$$-4 - 8 - 3 = -15.$$

(b) $-10 - 3 - (-6) = -10 - 3 + 6$ $\quad -(-6) = 6$
$= -13 + 6$
$= -7.$

Example 8. Evaluate the following expressions when $x = -4$, $y = 5$, and $z = -12$.
(a) $x - y + z$. (b) $x - y - z$.

Solution

(a) We replace x with -4, y with 5, and z with -12 in the given algebraic expression. As a result, we obtain

$$(-4) - (5) + (-12) = (-4) + (-5) + (-12) = -21.$$

(b) The value of $x - y - z$ for $x = -4$, $y = 5$, and $z = -12$ is

$$(-4) - (5) - (-12) = -4 - 5 + 12 \quad \text{double-negative theorem}$$
$$= -9 + 12$$
$$= 3.$$

EXERCISES 1.6

For Exercises 1–18, find the sum. See Examples 1 and 2.

1. $(+7) + (+3)$
2. $(+8) + (+2)$
3. $(-4) + (-6)$
4. $(-10) + (-2)$
5. $(-2.5) + (-1.5)$
6. $(-3.6) + (-1.3)$
7. $[(-6) + (-1)] + (-3)$
8. $[(-5) + (-3)] + (-4)$
9. $(+12) + (-4)$
10. $(+13) + (-3)$
11. $(-15) + 10$
12. $(-20) + 5$
13. $(-6) + 10$
14. $(-7) + 14$
15. $9 + (-12)$
16. $(-\frac{2}{3}) + (\frac{1}{3})$
17. $(\frac{5}{8}) + (-\frac{2}{8})$
18. $(-\frac{1}{2}) + (-\frac{3}{4})$

For Exercises 19–30, find the sum. See Example 3.

19. $(-9) + 3 + (-4)$
20. $(-12) + 6 + (-3)$
21. $13 + (-3) + 2$
22. $14 + (-4) + 5$
23. $10 + (-6) + (-4) + 5$
24. $8 + (-2) + (-6) + 5$
25. $(-7) + 3 + (-7) + 12$
26. $(-8) + 5 + (-8) + 15$
27. $12 + (-3) + 4 + (-7) + (-2)$
28. $15 + (-4) + 10 + (-8) + (-6)$
29. $(-9) + 5 + (-8) + 7 + 3$
30. $(-7) + 2 + (-5) + 4 + 4$

For Exercises 31–38, use the Double-Negative Theorem to simplify. See Example 4.

31. $-(-9)$ **32.** $-(-5)$ **33.** $-[-(-6)]$ **34.** $-[-(-3.2)]$
35. $-[-(+8)]$ **36.** $-[-(+2)]$ **37.** $-[-(-\frac{1}{4})]$ **38.** $-[-(+\frac{2}{3})]$

For Exercises 39–56, find the difference. See Example 5.

39. $5 - 10$ **40.** $6 - 8$ **41.** $9 - (-3)$
42. $10 - (-4)$ **43.** $15 - (-5)$ **44.** $7 - (-7)$
45. $(-5) - (-3)$ **46.** $(-10) - (-6)$ **47.** $-8 - (-12)$
48. $(-17) - (-20)$ **49.** $-14 - 10$ **50.** $-12 - 6$
51. $-20 - 5$ **52.** $-13 - 12$ **53.** $\frac{5}{6} - \frac{7}{6}$
54. $\frac{3}{4} - (-\frac{2}{4})$ **55.** $(-\frac{1}{3}) - (-\frac{2}{3})$ **56.** $-\frac{3}{7} - \frac{2}{7}$

For Exercises 57–74, evaluate the expression. See Examples 6 and 7.

57. $-3 - 4 + 2$ **58.** $-5 - 10 + 3$ **59.** $-7 - 4 + 12$
60. $-12 - 3 + 20$ **61.** $6 - 8 - 2$ **62.** $18 - 5 - 3$
63. $-9 - 6 - 5$ **64.** $-5 - 5 - 4$ **65.** $-7 - 3 - 2$
66. $-11 - 9 - 2$ **67.** $4 - 5 - (-3)$ **68.** $16 - 9 - (-2)$
69. $10 - (-8) - 20$ **70.** $14 - (-6) - 30$ **71.** $14 - (-2) - (-5)$
72. $-\frac{1}{3} - (-\frac{1}{3}) + \frac{4}{3}$ **73.** $-\frac{2}{5} - (-\frac{2}{5}) + \frac{4}{5}$ **74.** $\frac{7}{3} - (-\frac{4}{3}) - (-\frac{5}{3})$

For Exercises 75–83, evaluate the expression for $x = -3$, $y = 4$, and $z = -10$. See Example 8.

75. $x - y + z$ **76.** $x + y - z$ **77.** $x - (y + z)$
78. $x - (y - z)$ **79.** $x - y - z$ **80.** $y - x - z$
81. $|x| + y + |z|$ **82.** $x + y + |z|$ **83.** $x - (|y| + |z|)$

For Exercises 84–89, determine whether the statement is true or false.

84. The set of positive real numbers is closed under addition.

85. The set of negative real numbers is closed under addition.

86. The set of positive real numbers is closed under subtraction.

87. The set of negative real numbers is closed under subtraction.

88. The double-negative theorem states that $-(-x)$ is always positive.

89. The set of real numbers is closed under subtraction.

1.7 Multiplication and Division of Real Numbers

We begin our discussion on multiplication of *real* numbers with the following axiom.

Multiplication Property for Positive Numbers

> If $x > 0$, $y > 0$, then
>
> $$xy > 0.$$

The *multiplication property* stated above is the assumption that

The product of two positive real numbers is positive.

For example, $4(6) = 24$, $3(.1414) = .4242$, and $(+2)(+5) = +10 = 10$.

The following theorem forms the basis for the rules governing the sign of the product of two real numbers.

Theorem 5

> If x and y are real numbers, then
>
> (a) $$-x(y) = -(xy).$$
>
> (b) $$x(-y) = -(xy).$$
>
> (c) $$(-x)(-y) = (xy).$$

We observe that if x is positive, then $-x$ is negative. In this case x would also represent the absolute value of $-x$. For example, if x were 3, then the opposite of x, namely, $-x$, would be -3. Note also that $3 = |-3|$. As a consequence of the multiplication property for positive numbers and Theorem 5, we have the following rule for multiplying signed numbers.

Multiplication of Signed Numbers

> The product of two numbers having
>
> 1. *Unlike signs* (one positive, the other negative) is the product of their absolute value and the sign of this product is *negative*.
> 2. *Like signs* (both positive or both negative) is the product of their absolute values and the sign of this product is *positive*.

We apply the above rule in Examples 1 and 2.

Example 1. Evaluate the indicated products.

(a) $(-7)(3)$. (b) $(5)(-6)$. (c) $3\left(-\frac{1}{3}\right)$. (d) $(4)(-2)(5)$.

Solution

(a) $(-7)(3) = -(7 \cdot 3) = -21$.

(b) $5(-6) = -(5 \cdot 6) = -30$.

(c) $3(-\frac{1}{3}) = -(3 \cdot \frac{1}{3}) = -1$.

(d) $4(-2)(5) = -(4 \cdot 2)(5) = -8(5) = -40$.

Note that $|-7| = 7$

$|-6| = 6$

Example 2. Evaluate the indicated products.

(a) $(-3)(-6)$. (b) $(-2)(-5)$. (c) $(-2)(-\frac{1}{2})$. (d) $(-2)(-3)(4)(-1)$.

Solution

(a) $(-3)(-6) = +(3 \cdot 6) = 18$.

(b) $(-2)(-5) = +(2 \cdot 5) = 10$.

(c) $(-2)(-\frac{1}{2}) = +(2 \cdot \frac{1}{2}) = 1$.

(d) $(-2)(-3)(4)(-1) = 6 \cdot 4(-1) = 24 \cdot (-1) = -24$.

$|-3| = 3$, $|-6| = 6$

Using the definition of subtraction and the axioms and theorems discussed so far, we can establish that we can distribute multiplication over subtraction. Thus, if x, y, and z are real numbers, we have

$$x(y - z) = (xy) - (xz)$$

and

$$(y - z)x = (yx) - (zx).$$

We shall also refer to these properties as *distributive properties*. We also note that the distributive properties are valid for more than two variables within the parentheses. For example,

$$r(s + t + u + \cdots + z) = rs + rt + ru + \cdots + rz.$$

The following theorem stated without proof proves useful for removing parentheses preceded by a negative sign.

Theorem 6

Let x be a real number, then

$$-1 \cdot x = -x.$$

We illustrate the application of the additional distributive properties mentioned above and Theorem 6 in the following example.

Example 3. Rewrite the following expressions without parentheses.
(a) $-5(x-2)$. (b) $(y-3)(2)$ (c) $-(x-y+z)$.

Solution

(a) $-5(x-2)$	$= -5x-(-5)2$	distributive property
	$= -5x-(-10)$	multiplication of numbers with unlike signs
	$= -5x+10.$	definition of subtraction; double-negative theorem
(b) $(y-3)(2)$	$= y(2)-3(2)$	distributive property
	$= 2y-6.$	commutative property
(c) $-(x-y+z)$	$= -1\cdot(x-y+z)$	Theorem 6
	$= -1\cdot x-(-1)y+(-1)z$	distributive property
	$= -x-(-y)+(-z)$	Theorem 6
	$= -x+y-z.$	definition of subtraction

We observe from Example 3(c) that

$$-(x-y+z) = -x+y-z.$$

We call x, $-y$, and z the terms within parentheses. This example illustrates the following rule:

> When *removing* parentheses immediately preceded by a negative sign, change the sign of each term within parentheses to the opposite sign.

We now turn our attention to the rules regulating division of signed numbers. In the preceding section we defined subtraction in terms of addition. In this section we define division in terms of multiplication using the concept of the multiplicative inverse (or reciprocal) of a real number.

Definition of Division

> If x and y are real numbers, then
>
> $$x \div y = x\cdot\frac{1}{y} \qquad \text{provided that } y \neq 0.$$

Recall that dividing x by y is also symbolized

$$\frac{x}{y} \qquad \text{or} \qquad x/y$$

and is called the *quotient* of x divided by y. Division by 0 is not defined, since 0 does not have a reciprocal. Observe that the reciprocal of a number y is a number $\frac{1}{y}$ such that $y \cdot \frac{1}{y} = 1$. But if $y = 0$ by the zero-property, we have

$$0 \cdot \frac{1}{y} = 0.$$

Therefore, for 0 there *does not exist* a real number $\frac{1}{y}$ such that $0 \cdot \frac{1}{y} = 1$.

The following theorem can be proven using the definition of division, uniqueness of the reciprocal of a number, and Theorem 5.

Theorem 7

> If x, y are real numbers and $y \neq 0$, then
>
> $$\frac{x}{-y} = \frac{-x}{y} = -\frac{x}{y}$$
>
> and
>
> $$\frac{-x}{-y} = \frac{x}{y}.$$

As a result of Theorem 7 we obtain the following rule of signs for division.

Rule of Signs for Division

> The sign of the *quotient* of two real numbers with
> 1. *Like* signs (both positive or both negative) is positive.
> 2. *Unlike* signs (one positive, the other negative) is negative.

This rule is illustrated in the following examples.

Example 4. Evaluate the following quotients.

(a) $\frac{15}{5}$. (b) $\frac{-8}{4}$. (c) $\frac{9}{-3}$. (d) $\frac{-30}{-5}$. (e) $\frac{0}{-7}$.

Solution

(a) $\frac{15}{5} = 3.$

(b) $\frac{-8}{4} = -\left(\frac{8}{4}\right) = -2.$

(c) $\frac{9}{-3} = -\frac{9}{3} = -3.$

(d) $\dfrac{-30}{-5} = \dfrac{30}{5} = 6.$

(e) $\dfrac{0}{-7} = -\dfrac{0}{7} = -\left(0 \cdot \dfrac{1}{7}\right) = -(0) = 0.$

We note in part (e) of Example 4 that the quotient of 0 by a nonzero number such as -7 is valid. It is division *by* 0 that is not valid for reasons already discussed above.

Example 5. Evaluate the following expressions.

(a) $\dfrac{-20}{-5} + \dfrac{-12}{6}.$ (b) $\dfrac{(-4)(-5) + (-2)(3)}{(-2)(3) + 4}.$

Solution

(a) $\dfrac{-20}{-5} + \dfrac{-12}{6} = 4 + (-2) = 2.$

(b) If we view the horizontal bar as indicating division, we simplify the "numerator" and "denominator" separately and then divide. Thus

$$\frac{(-4)(-5) + (-2)(3)}{(-2)(3) + 4} = \frac{20 - 6}{-6 + 4} = \frac{14}{-2} = -7.$$

In concluding this section, we note that the multiplicative inverse of $\dfrac{a}{b}$ is $\dfrac{b}{a}$, since $\dfrac{a}{b} \cdot \dfrac{b}{a} = \dfrac{ab}{ba} = \dfrac{ab}{ab} = 1$. Thus the multiplicative inverse (or reciprocal) of $\dfrac{-2}{3}$ is $\dfrac{3}{-2}$. We use this fact in the following example.

Example 6. Evaluate the quotients.

(a) $\dfrac{-4}{3} \div \dfrac{6}{-9}.$ (b) $-\dfrac{7}{2} \div (-3).$

Solution

(a) $\dfrac{-4}{3} \div \dfrac{6}{-9} = \dfrac{-4}{3} \cdot \dfrac{-9}{6} = \left(-\dfrac{4}{3}\right)\left(-\dfrac{9}{6}\right) = +\left(\dfrac{36}{18}\right) = 2.$

Another and equivalent way of evaluating this quotient is as follows:

$$\frac{-4}{3} \div \frac{6}{-9} = \frac{-4}{3} \cdot \frac{-9}{6} = \frac{(-4)(-9)}{3 \cdot 6} = \frac{36}{18} = 2.$$

(b) $-\dfrac{7}{2} \div (-3) = -\dfrac{7}{2} \cdot \dfrac{1}{-3} = -\dfrac{7}{2}\left(-\dfrac{1}{3}\right) = +\left(\dfrac{7}{2} \cdot \dfrac{1}{3}\right) = \dfrac{7}{6}.$

EXERCISES 1.7

For Exercises 1–24, evaluate the product. See Examples 1 and 2.

1. $(-8)(2)$
2. $(-9)(3)$
3. $7(-5)$
4. $6(-4)$
5. $4(-\frac{1}{4})$
6. $-5(\frac{1}{5})$
7. $-9(\frac{1}{9})$
8. $-10(\frac{1}{10})$
9. $3(-2)(5)$
10. $4(-3)(2)$
11. $5(-4)(3)(2)$
12. $6(-2)(2)(3)$
13. $(-4)(-5)$
14. $(-7)(-2)$
15. $(-3)(-\frac{1}{3})$
16. $(-7)(-\frac{1}{7})$
17. $(-3)(-2)(-4)$
18. $(-4)(-2)(-1)$
19. $3(0)(-3)(5)$
20. $(-2)(0)(-2)(-3)$
21. $(-3)(-1)(-3)(-2)$
22. $(-4)(-2)(-1)(-2)$
23. $(-\frac{2}{3})(-\frac{3}{2})(-\frac{1}{5})$
24. $(-\frac{2}{7})(-\frac{7}{2})(-\frac{1}{3})$

For Exercises 25–36, rewrite the expression without parentheses. See Example 3.

25. $-3(x - 2)$
26. $-2(y - 4)$
27. $(z - 3)5$
28. $(t - 1)(-2)$
29. $-(x - y - z)$
30. $-(3x - y - 2z)$
31. $-2(s + t - u)$
32. $-4(s - t - 3u)$
33. $-3(4t - 2u - 5v)$
34. $-5(2x - 4y - 3z)$
35. $(3s - 2t + v)(3)$
36. $(5s - t + 3v)(-4)$

For Exercises 37–48, evaluate the quotient. See Example 4.

37. $\frac{18}{3}$
38. $\frac{49}{7}$
39. $\dfrac{-8}{2}$
40. $\dfrac{-10}{5}$
41. $\dfrac{27}{-3}$
42. $\dfrac{32}{-8}$
43. $\dfrac{-36}{-9}$
44. $\dfrac{-24}{-6}$
45. $\dfrac{-4(3)}{2}$
46. $\dfrac{-5(6)}{2}$
47. $\dfrac{(-4)(-7)}{-2}$
48. $\dfrac{(-3)(-8)}{-6}$

For Exercises 49–60, evaluate the expression. See Example 5.

49. $\dfrac{-16}{4} + \dfrac{36}{-6}$
50. $\dfrac{-30}{15} + \dfrac{20}{-10}$
51. $\dfrac{-20}{-4} + \dfrac{-40}{-8}$
52. $\dfrac{-10}{-2} + \dfrac{-25}{-5}$
53. $\dfrac{15}{-3} + \dfrac{-21}{7}$
54. $\dfrac{-28}{4} + \dfrac{36}{-9}$
55. $\dfrac{(-2)(-3) + (-4)(3)}{(-2)(-3)}$
56. $\dfrac{(-3)(3) + (-4)(4)}{(-5)(-1)}$
57. $\dfrac{(-5)(2) + (-2)(7)}{(-3)(-4)}$
58. $\dfrac{(-2)(-5) + (-3)(-10)}{(-2)(-5)}$
59. $\dfrac{(-4)(5) + (-2)(10)}{(-3)(-2) + 4}$
60. $\dfrac{(-8)(3) + 4(-6)}{6 + (-2)(-5)}$

For Exercises 61–68, write the quotient (or fraction) in two equivalent ways using the result,

$$\frac{x}{-y} = \frac{-x}{y} = -\frac{x}{y} \qquad (y \neq 0).$$

61. $-\frac{3}{4}$ **62.** $\frac{-3}{7}$ **63.** $\frac{4}{-9}$ **64.** $\frac{5}{-8}$

65. $\frac{-5}{11}$ **66.** $-\frac{2}{5}$ **67.** $\frac{-7}{4}$ **68.** $\frac{10}{-3}$

For Exercises 69–76, evaluate the quotient. See Example 6.

69. $\frac{-9}{2} \div \frac{3}{-6}$ **70.** $\frac{-8}{5} \div \frac{2}{-5}$ **71.** $-\frac{5}{2} \div (-7)$ **72.** $-\frac{3}{2} \div (-3)$

73. $\frac{11}{3} \div \frac{11}{-3}$ **74.** $\frac{-13}{2} \div \frac{1}{2}$ **75.** $\frac{15}{-3} \div \frac{5}{3}$ **76.** $\frac{20}{4} \div \frac{5}{-4}$

For Exercises 77–82, determine whether the statement is true or false.

77. If x is a real number, then $x \cdot x$ must be positive.

78. If x is a real number, $x \neq 0$, then $\frac{0}{x} = 0$.

79. If x is a real number, then $\frac{x}{0}$ is not defined.

80. The reciprocal of $\frac{0}{1}$ is $\frac{1}{0}$.

81. The set of negative real numbers is closed under multiplication.

82. The set of negative real numbers is closed under division.

Key Words

Absolute value
Algebraic expression
Constant
Equality (=)
  Reflexive
  Symmetric
  Transitive
  Substitution principle
Inequality (<, >)
  Transitive
  Trichotomy
Numbers
  Natural (N)
  Whole (W)
  Integers (Z)
  Rational (Q)
  Irrational (K)
  Real ($\mathbb{R}$)
Operations
  Addition
  Subtraction
  Multiplication
  Division
Order of operations
Properties of real numbers
  Closure

- Commutative
- Associative
- Distributive
- Identity
  - Additive
  - Multiplicative
- Inverse
  - Additive (negative, opposite)
  - Multiplicative (reciprocal)

Real number line
- Coordinate
- Graph

Set
- Listing method
- Rule method
- Subset ($\subseteq$)
- Null set ($\varnothing$)
- Member of a set ($\in$)
- Union ($\cup$)
- Intersection ($\cap$)

Variable
- Domain
- Replacement set

CHAPTER 1 REVIEW

SECTION 1.1

Use the roster method to specify the following set.

1. The set of odd natural numbers between 4 and 12.

Use the rule method to specify the following set.

2. $\{6, 8, 10, 12, \ldots, 86, 88, 90\}$.

3. List all the subsets of the set $\{p, q, r\}$.

Given the sets $A = \{-5, -3, 2, 4\}$ and $B = \{-5, 0, 2, 4\}$, *find*

4. $A \cup B$ **5.** $A \cap B$

Determine whether the statements are true or false.

6. If $A \subseteq B$ and $7 \in B$, then $7 \in A$.

7. If $A \subseteq B$ and $B \subseteq C$, then $A \subseteq C$.

SECTION 1.2

Given the set $A = \{-3, -1, 0, \frac{2}{3}, \frac{3}{1}, 6, \frac{9}{4}\}$, *list the members in the set A that are*

8. Positive integers.

9. Negative integers.

10. Integers.

11. Rational numbers that are not integers.

Identify the numbers as rational or irrational.

12. $\frac{-7}{8}$ **13.** .050050005 . . . **14.** $.5\overline{6}$

Find the indicated inverses.

15. $-(-6)$ **16.** $-(+9)$ **17.** $-(7.2)$ **18.** $-(-\frac{1}{4})$

Express without absolute value bars.

19. $|-8|$ **20.** $-|-9|$ **21.** $|+3|$

SECTION 1.3

Translate the verbal statements into algebraic expressions.

22. Seven less than twice x.

23. Five more than the product of x and y.

Evaluate the expressions.

24. $4 \cdot 5 + 5 \cdot 3$ **25.** $36 \div 6 \cdot 3$ **26.** $3[6 - (15/5 - 2)]$

Evaluate the expressions for $x = 6$, $y = 2$.

27. $4x + 3y$ **28.** $xy - (x - y)/y$ **29.** $\frac{x + y}{y}$

SECTION 1.4

State the property of real numbers that justifies each statement.

30. $7 + x = x + 7$ **31.** $(y + 3) + 5 = y + (3 + 5)$

32. $2(a + b) = 2a + 2b$ **33.** $x + 0 = x$

State the properties or theorems that justify each statement.

34. If $3 = x$, then $x = 3$.

35. If $x + y = z$ and $z = a + b$, then $x + y = a + \mathrm{b}$.

36. If $x = 4$ and $y + 5 = x + 5$, then $y + 5 = 4 + 5$.

37. $z \cdot 0 = 0$

SECTION 1.5

Fill in each blank with $<$ or $>$.

38. -3 ____ 2 **39.** 5 ____ -7 **40.** -6 ____ -5 **41.** 3 ____ -3

Graph the set on the real number line.

42. $\{x \mid -2 < x \leq 4\}$, where x is a real number

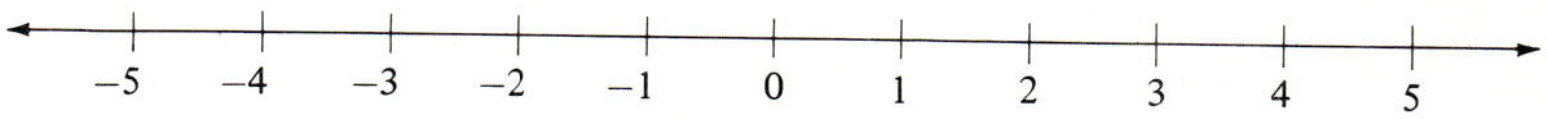

SECTION 1.6

Evaluate the expressions.

43. $-14 + 10$ **44.** $-8 - (-10)$ **45.** $-5 + (-9) + 4$

46. $-\frac{1}{4} - \frac{3}{4} + 14$

Evaluate the expressions for $x = -2$, $y = 5$, *and* $z = -7$.

47. $x + y - z$ **48.** $y - z - |x|$

SECTION 1.7

Evaluate the products.

49. $(-6)(2)$ **50.** $(-4)(-3)$ **51.** $(-5)(-2)(-4)$

Rewrite the expressions without parentheses.

52. $-5(y - 3)$ **53.** $-(x - y - 4)$

Evaluate the quotients.

54. $\dfrac{-45}{5}$ **55.** $\dfrac{-30}{-6}$ **56.** $\dfrac{12}{-3}$ **57.** $\dfrac{0}{-5}$

Evaluate the expressions.

58. $\dfrac{(-3)(4) + (6)(4)}{(-2)(2)}$ **59.** $\dfrac{(-2)(-5) - (-6)(5)}{(-8)(2) + 4(-1)}$

Evaluate the quotients.

60. $\dfrac{-8}{6} \div \dfrac{2}{-3}$ **61.** $\dfrac{-10}{3} \div \dfrac{-2}{6}$

CHAPTER 1 TEST

1. Given the sets $A = \{-2, -1, 0, 3, 4\}$ and $B = \{0, 3, 5, 6\}$, find
(a) $A \cup B$ **(b)** $A \cap B$

2. Given the set $A = \left\{-\pi, -3, \frac{-5}{2}, 0, \sqrt{2}, 3, \frac{22}{7}, 3.5, 4.\bar{2}, 4\right\}$, find the members of A that are

(a) Natural numbers. **(b)** Whole numbers. **(c)** Integers.
(d) Rational numbers that are not integers.
(e) Irrational numbers. **(f)** Real numbers.

3. Find the values of
(a) $|-5|$ **(b)** $|+5|$ **(c)** $-|-7|$ **(d)** $|-(-3)|$

4. Translate the verbal statements into algebraic expressions.
(a) Ten less than the sum of x and y.
(b) Five more than 10 times z.

5. Evaluate the expressions.
(a) $5 \cdot 6 + 2 \cdot 3$ **(b)** $45 \div 15 \cdot 3$ **(c)** $4[8 - (28/7 - 3)]$.

6. State the property of real numbers that justifies each statement.
(a) $(x + 2) + 3 = (2 + x) + 3$ **(b)** $5(6y) = (5 \cdot 6)y$
(c) $-4 + 0 = -4$ **(d)** $1 \cdot x = x$ **(e)** $2(y + x) = 2y + 2x$.

7. State the property or theorem that justifies each statement.
(a) If $x + 2 = y$ and $y = z + 3$, then $x + 2 = z + 3$.
(b) If $3 = y + 2$, then $y + 2 = 3$.

8. Graph the set on the real number line.

$$\{x \mid -3 < x \le 2\}, \quad \text{where } x \text{ is a real number}$$

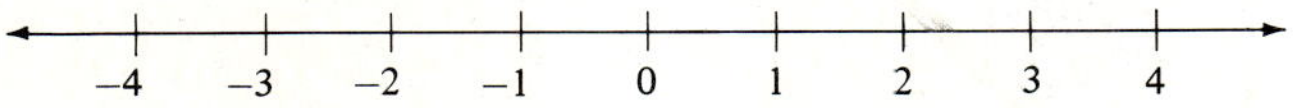

9. Evaluate.
(a) $-3 + (-7)$ **(b)** $(-10) + 3$ **(c)** $15 + (-3) + 10 + (-2)$
(d) $-18 - (-20)$ **(e)** $7 - (-8) - (-2)$ **(f)** $7 - 5 - 5$

10. Evaluate.
(a) $(-5)(-20)$ **(b)** $-3(8)$ **(c)** $(-3)(-1)(-4)$
(d) $\frac{36}{-9}$ **(e)** $\frac{-50}{-10}$ **(f)** $\frac{0}{-10}$

11. Evaluate.
(a) $\frac{(-5)(-3) + 7(-1)}{(-1)(2)}$ **(b)** $\frac{(2 - 6)(5 - 2) + (6)(-2)}{(4 - 2)(2 - 4)}$

12. Evaluate the expressions for $x = -5$, $y = 2$, $z = -6$.
(a) $2x - 3y - 2z$ **(b)** $\frac{3x - 5y + 4z}{z - 1}$

13. Fill in the blanks in each statement to obtain a true statement.
(a) x ____ -3 or $x = -3$ or $x > -3$
(b) If $y < -6$ and $-6 < -2$, then y ____ -2.

CHAPTER 2

First-Degree Equations and Inequalities

First-degree equations can be used to solve practical problems. For example, in order to winterize his car, an economically minded college student needs to drain off just enough liquid and replace it with antifreeze so as to increase the antifreeze content of his car's radiator from 30% to 50%. If the radiator has a capacity of 22 quarts, then by solving the equation $.30(22 - x) + x = .50(22)$ he will know how much liquid should be drained and replaced with pure antifreeze.

The ability to solve equations and inequalities efficiently is very important to students of mathematics. In this chapter we are concerned primarily with statements of equality or equations such as

$$3x = 51, \qquad 4y - 7 = 23, \qquad -8 = 3z - 2.$$

Each of these equations is called a *first-degree equation in one variable*. Any first-degree equation in one variable can be written in the form

$$ax + b = 0,$$

where x is the variable and a and b are real numbers with $a \neq 0$. A statement of inequality such as $4w + 7 \geq 35$ is a *first-degree inequality in one variable*. In this chapter we discuss first-degree inequalities and applications that involve first-degree equations. We also discuss equations and inequalities involving absolute value. We begin by discussing how to simplify first-degree expressions.

2.1 Simplifying First-Degree Expressions

By carefully applying the properties of real numbers that were discussed in Chapter 1, it is possible to write a first-degree expression such as

$$3x - 17 + 4x + x + 2$$

more compactly. The preceding expression has five *terms*. They are $3x$, -17, $4x$, x, and 2. A *term* is either a number or a number multiplied by at least one variable. The numerical part of a term is called the *numerical coefficient* of the term. For example, the numerical coefficients of the terms $5x$, $-3w$, $17xyz$, and -2 are 5, -3, 17, and -2, respectively. The numerical coefficient of the term x is understood to be 1. Terms that contain the same variables with the same exponents are called *like terms*. Thus each of the following sets consists of elements that are like terms.

$$\{-2x, 7x, 11x\}$$
$$\{-9xy^2, -xy^2, 3xy^2\}$$
$$\{2abc, 7abc, -23abc\}.$$

We can use the distributive property to add two or more like terms. For example, to add $-2x$, $7x$, and $11x$, we proceed as follows:

$$\begin{aligned} -2x + 7x + 11x &= (-2 + 7 + 11)x && \text{distributive property} \\ &= 16x. && \text{addition} \end{aligned}$$

This procedure is called *combining like terms*. Notice that the original expression, $-2x + 7x + 11x$, had three terms, but the final expression, $16x$, has only one term.

Example 1. Combine the like terms in the expression

$$5y + 16y - 13y + y.$$

Solution

$$\begin{aligned} 5y + 16y - 13y + y &= (5 + 16 - 13 + 1)y && \text{distributive property} \\ &= 9y. && \text{addition} \end{aligned}$$

Often, an algebraic expression contains different types of like terms, as shown in the next example.

Example 2. Combine the like terms in the expression

$$23x + 7 + x - 12 - 17x + 1.$$

Solution

$$\begin{aligned} &23x + 7 + x - 12 - 17x + 1 \\ &= (23x + x - 17x) + (7 - 12 + 1) && \text{commutative and associative properties} \\ &= (23 + 1 - 17)x + (7 - 12 + 1) && \text{distributive property} \\ &= 7x - 4. && \text{addition} \end{aligned}$$

When we rewrite an algebraic expression more compactly by combining like terms, we say that we have *simplified* the expression. Note that when we combine like terms we simply add the numerical coefficients of the like terms. Once we understand how to combine like terms by adding their numerical coefficients, we do not show all the steps that involve the commutative, associative, and distributive properties.

Example 3. Simplify each of the following expressions.
(a) $4w + 3 - 8 + 12w - 1$.
(b) $7(z + 1) + 2z - 5(2z + 3)$.
(c) $9ax - (ax + 6) + 4ax - 10$.

Solution

(a) $4w + 3 - 8 + 12w - 1$

$$\begin{aligned} &= \overbrace{4w - \underbrace{5 + 12w - 1}_{-6}}^{16w} && 3 - 8 = -5;\ \text{combining like terms} \\ &= 16w - 6. \end{aligned}$$

(b) $7(z + 1) + 2z - 5(2z + 3)$

$$\begin{aligned} &= 7z + 7 + 2z - 10z - 15 && \text{distributive property} \\ &= -z - 8. && \text{combining like terms} \end{aligned}$$

(c) $9ax - (ax + 6) + 4ax - 10$

$= 9ax - ax - 6 + 4ax - 10$ $\qquad -(ax + 6) = -ax - 6$

$= 12ax - 16.$

EXERCISES 2.1

For Exercises 1–10, combine the like terms in the expression. See Example 1.

1. $7x + 3x - 5x$
2. $14y + 2y - 9y$
3. $6z + 4z - 8z$
4. $12w + 2w - 9w$
5. $3p - 7p + p + 6p$
6. $4q - 9q - q + 8q$
7. $8r + 5r - 4r - r$
8. $11s - 7s + 2s - s$
9. $3t + 6t - t - 4t$
10. $2u + 9u - 3u - u$

For Exercises 11–20, combine the like terms in the expression. See Example 2.

11. $5t + 4 - 2t + 1 + 3t - 3$
12. $4u + 3 - 3u + 8 + u - 1$
13. $9v - 6 + v + 4 + 2v - 3$
14. $10w + 11 - 3w - 4 + w - 2$
15. $17x + 10 + 3x - 2 - 14x + 1$
16. $19y + 4 - 11y + 2 + y - 5$
17. $20 + 31z + 6 - 12z - 15 + 2z$
18. $15 - 13p - 8 + 2p + 4 + 17p$
19. $17 + q - 14 + 7q - 20 - 10q$
20. $18 + 6r - 13 - 8r - 11 + r$

For Exercises 21–40, combine the like terms in the expression. See Example 3.

21. $4(x + 3) + 3x - 2(x - 1)$
22. $5(y + 2) + 4y - 3(y + 2)$
23. $2(w + 5) - 3w + 2(w + 4)$
24. $3(z + 4) - 2z + 5(z + 6)$
25. $4(p + 6) - (p + 2) + 3p$
26. $5(q + 1) - (q + 5) + 1$
27. $10(2r + 3) - (5r - 4) + 3r$
28. $7(3r + 5) - (7r - 2) + r$
29. $-11(3r - 2) + 2(3r - 5) + 9$
30. $-15(2s - 1) + 3(6s - 2) + 20$
31. $-14(2x - 3) + 5(3x - 2) + 11$
32. $-17(3y - 2) + 4(12y - 15) + 6$
33. $.45(z + 2) + .30(z + 1) + .25z$
34. $.25(w + 3) + .40(w + 2) + .35w$
35. $-.5(x - 3) + .7(x + 2) + .10$
36. $-.6(y - 2) + .9(y + 4) + .2$
37. $\frac{1}{5}(w + 4) + \frac{3}{5}(3w + 1) + \frac{3}{5}$
38. $\frac{1}{3}(x + 2) + \frac{2}{3}(4x + 1) + \frac{2}{3}$
39. $\frac{1}{6}(3x + 2) + \frac{1}{2}\left(x - \frac{1}{3}\right) + \frac{5}{6}$
40. $\frac{1}{10}(7y + 3) + \frac{3}{10}(y + 2) + \frac{1}{10}$

Solving First-Degree Equations 2.2

The statement $7x + 4 = 25$ is either true or false, depending on the value given to the variable x. If the number that replaces the variable results in a true statement, that number is a *solution* of the equation. The number 3 is a solution of the equation $7x + 4 = 25$ because when we replace x with 3 we get a true statement. The set that consists of all the solutions of an equation is called the *solution set* of the equation. It may be shown that any first-degree equation in one variable of the form $ax + b = 0$ with $a \neq 0$ has exactly one solution. The solution set of $7x + 4 = 25$ is $\{3\}$. When we have found the solution to an equation, we have *solved* the equation.

We will use our ability to simplify expressions to solve first-degree equations in one variable. In addition, we will use the following theorem similar to Theorem 1 of Section 1.4 to obtain simpler equations with the same solution set.

Theorem 1

If A, B, and C are algebraic expressions, then $A = B$ is equivalent to

(a) $A + C = B + C$. *Addition property of equality*
(b) $AC = BC$ $(C \neq 0)$. *Multiplication property of equality*

The *addition property of equality* implies that adding the same expression to both sides of an equation produces another equation with the same solution set. Recall that subtracting a number is the same as adding the opposite of the number. Since these expressions represent numbers, *subtracting* the same expression from both sides of an equation produces another equation with the same solution set.

The *multiplication property of equality* may be used to show that multiplying both sides of an equation by the same expression representing a *nonzero* number produces another equation with the same solution set. Recall that dividing by a nonzero number is the same as multiplying by the reciprocal of the number. Thus, *dividing* both sides of an equation by the same *nonzero* expression representing a number produces another equation with the same solution set.

Equations with the same solution set are called *equivalent equations*. Note that

$$7x + 4 = 25, \qquad 7x = 21, \qquad x = 3$$

are all equivalent equations because $\{3\}$ is the solution set of each equation. Whereas the solution to the equation $7x + 4 = 25$ is not obvious, the solution to the equation $x = 3$ is obvious. The goal in solving a first-degree equation is to obtain an equivalent equation of the form $x = a$. To do this, we start with the original equation and use the addition and multiplication properties of equality, together with our ability to simplify algebraic expressions. We thus obtain a sequence of simpler equations that are all equivalent to the original equation until we arrive at an equation whose solution is obvious. The following examples illustrate this procedure.

Example 1. Find the solution set of $4y - 11 = 21$.

Solution

$$\begin{aligned} 4y - 11 &= 21 \\ 4y - 11 + 11 &= 21 + 11 && \text{add 11 to both sides; Theorem 1(a)} \\ 4y &= 32 && \text{addition} \\ \left(\frac{1}{4}\right)4y &= \left(\frac{1}{4}\right)32 && \text{multiply both sides by } \frac{1}{4}\text{; Theorem 1(b)} \\ y &= 8. && \text{multiplication} \end{aligned}$$

Provided that we did not make any mistakes, all five of the equations above are equivalent. The only solution to the last equation is clearly 8. If we replace y with 8 in the original equation, the result is

$$\begin{aligned} 4(8) - 11 &= 21 \\ 32 - 11 &= 21 \\ 21 &= 21. && \text{a true statement} \end{aligned}$$

Thus $\{8\}$ is the solution set of $4y - 11 = 21$.

Example 2. Solve the equation $3z + 2 + 4z = 13 + z + 7$.

Solution: First, combine any like terms on each side of the equation. The result of this is

$$7z + 2 = z + 20.$$

Next, we add -2 to each side of the equation.

$$\begin{aligned} 7z + 2 + (-2) &= z + 20 + (-2) \\ 7z &= z + 18. \end{aligned}$$

To get all the terms with z on one side of the equation, we add $-z$ to each side.

$$\begin{aligned} 7z - z &= z + 18 - z && \text{add } -z \text{ to each side} \\ 6z &= 18 \\ \left(\frac{1}{6}\right)6z &= \left(\frac{1}{6}\right)18 && \text{multiply each side by } \frac{1}{6} \\ z &= 3. \end{aligned}$$

The solution to the last equation is 3. We now check to see if 3 is also the solution to the original equation.

CHECK

$$\begin{aligned} 3z + 2 + 4z &= 13 + z + 7 \\ 3(3) + 2 + 4(3) &= 13 + 3 + 7 && \text{let } z = 3 \\ 9 + 2 + 12 &= 13 + 3 + 7 \\ 23 &= 23. && \text{a true statement} \end{aligned}$$

Since 3 produced a true statement, the solution set of the original equation is $\{3\}$.

Example 3. Solve $\frac{2}{3}x - 11 = 1$.

Solution: First we add 11 to each side of the equation. After that we multiply each side by $\frac{3}{2}$, which is the reciprocal of $\frac{2}{3}$.

$$\frac{2}{3}x - 11 = 1$$

$$\frac{2}{3}x - 11 + 11 = 1 + 11 \qquad \text{add 11 to each side}$$

$$\frac{2}{3}x = 12$$

$$\left(\frac{3}{2}\right)\frac{2}{3}x = \frac{3}{2}(12) \qquad \text{multiply each side by } \frac{3}{2}$$

$$x = 18.$$

We now check that 18 is the solution to the original equation.

CHECK

$$\frac{2}{3}x - 11 = 1$$

$$\frac{2}{3}(18) - 11 = 1 \qquad \text{let } x = 18$$

$$12 - 11 = 1$$

$$1 = 1. \qquad \text{true}$$

Thus the solution set is $\{18\}$.

Example 4. Solve $5(w - 2) + 3w = 2(w + 4) + 12$.

Solution

$$5(w - 2) + 3w = 2(w + 4) + 12$$

$$5w - 10 + 3w = 2w + 8 + 12 \qquad \text{distributive property}$$

$$8w - 10 = 2w + 20 \qquad \text{combining like terms}$$

$$8w - 10 + 10 = 2w + 20 + 10 \qquad \text{add 10 to each side}$$

$$8w = 2w + 30$$

$$8w - 2w = 2w + 30 - 2w \qquad \text{subtract } 2w \text{ from each side}$$

$$6w = 30$$

$$\frac{6w}{6} = \frac{30}{6} \qquad \text{divide each side by 6}$$

$$w = 5.$$

The reader should check that 5 is the solution to the original equation. Thus the solution set for the original equation is $\{5\}$.

When an equation contains fractions it is often easier first to *clear fractions* by multiplying both sides of the equation by a common denominator of all the fractions in the equation. The next example shows how this is done.

Example 5. Find the solution set of

$$\frac{3}{5}x - \frac{1}{3} = \frac{1}{3}x + 1.$$

Solution: To clear fractions, we first multiply each side of the equation by the common denominator 15.

$$15\left(\frac{3}{5}x - \frac{1}{3}\right) = 15\left(\frac{1}{3}x + 1\right) \quad \text{multiply by 15}$$

$$15\left(\frac{3}{5}x\right) - 15\left(\frac{1}{3}\right) = 15\left(\frac{1}{3}x\right) + 15(1) \quad \text{distributive property}$$

$$9x - 5 = 5x + 15$$

$$9x - 5 + 5 = 5x + 15 + 5 \quad \text{add 5 to each side}$$

$$9x = 5x + 20$$

$$9x - 5x = 5x + 20 - 5x \quad \text{subtract } 5x \text{ from each side}$$

$$4x = 20$$

$$\frac{4x}{4} = \frac{20}{4} \quad \text{divide each side by 4}$$

$$x = 5.$$

Check that $\{5\}$ is the solution set by substituting 5 for x in the original equation.

As the next example demonstrates, there are equations composed of first-degree expressions that have *no* solutions, and there are also equations composed of first-degree expressions that have an *infinite number* of solutions.

Example 6. Solve each of the following equations.
(a) $z + 3 = 5 + z$. (b) $y + 7 = 7 + y$.

Solution

(a)

$$z + 3 = 5 + z$$

$$z + 3 - 3 = 5 + z - 3 \quad \text{subtract 3 from each side}$$

$$z = z + 2$$

$$z - z = z + 2 - z \quad \text{subtract } z \text{ from each side}$$

$$0 \cdot z = 2$$

or

$$0 = 2. \quad \text{false}$$

The equations $z + 3 = 5 + z$ and $0z = 2$ are equivalent. That is, they have the same solution set. But there is no real number that is a solution to the equation $0z = 2$. We conclude that the original equation also has no solution. Careful inspection of the original equation would also indicate that it has no solution. This is because there is no real number z with the property that $z + 3$ equals $z + 5$. Note that $z + 3$ and $z + 5$ will always differ by 2. If an equation that is composed of first-degree expressions has no solution, it can be simplified to a false statement such as $0 = 2$. Since there are no solutions, the solution set is the empty set, $\{\ \}$.

(b) Careful inspection of the equation $y + 7 = 7 + y$ reveals that it is a true statement for any value of y. Any equation that is true for all possible values of the variable is called an *identity*. Thus $y + 7 = 7 + y$ is an identity. If we had not noticed that $y + 7 = 7 + y$ was an identity and proceeded to solve for y, the result would be a true statement such as $0 = 0$, as shown below.

$$
\begin{aligned}
y + 7 &= 7 + y \\
y + 7 - 7 &= 7 + y - 7 && \text{subtract 7 from each side} \\
y &= y \\
y - y &= y - y && \text{subtract } y \text{ from each side} \\
0 &= 0. \quad \text{true}
\end{aligned}
$$

Whenever an equation that is composed of first-degree expressions can be simplified to a true statement such as $0 = 0$, we conclude that it is an identity. Since *all* real numbers are solutions to the identity $y + 7 = 7 + y$, the solution set is $\mathbb{R}$.

The following steps are a summary of the techniques for solving a first-degree equation in one variable. The order in which they are applied may not be the order in which they are listed. It may not be necessary to apply all of the steps to a particular problem. It may also be necessary to apply some of the steps more than once. The steps used and the order in which they are applied will depend on the particular equation that is being solved.

Steps for Solving a First-Degree Equation in One Variable

Step 1. Use the multiplication property of equality to clear fractions.

Step 2. Combine like terms on both sides of the equation.

Step 3. Use the addition property of equality to get all the variable terms on one side of the equation and all the constant terms on the other side.

Step 4. Use the multiplication property of equality to obtain a coefficient of 1 for the variable term, thus obtaining an equation of the form $x = a$.

Step 5. Check your answer by substituting it in the original equation.

EXERCISES 2.2

For Exercises 1–20, solve the equation. See Example 1.

1. $5x - 2 = 13$
2. $3q - 4 = 14$
3. $11w - 10 = 34$
4. $12r - 5 = 31$
5. $6y + 7 = 25$
6. $7s + 2 = 30$
7. $13p + 5 = 57$
8. $15t + 4 = 49$
9. $17 + 6z = 5$
10. $19 + 4u = 3$
11. $23 + 4w = 3$
12. $31 + 7q = 3$
13. $6 - 5x = 21$
14. $11 - 7y = 25$
15. $17 - 8z = 57$
16. $13 - 12v = 49$
17. $47 - 9q = 11$
18. $38 - 7r = 3$
19. $89 - 10t = 19$
20. $30 - 13s = 4$

For Exercises 21–30, solve the equation. See Example 2.

21. $5x + 4 + 2x = 11 + 3x + 5$
22. $9w + 8 + 4w = 20 + 5w + 4$
23. $11y + 3 + y = 22 + 5y + 9$
24. $14z + 5 + z = 40 + 2z + 4$
25. $3p + 4 - 11p = 19 + 2p + 15$
26. $2q + 3 - 13q = 40 - 2q + 8$
27. $r + 5 - 10r = 43 + 3r + 10$
28. $s + 7 - 8s = 50 - 3s + 1$
29. $8t + 7 - 12t = 5 + 2t - 10$
30. $7w + 13 - w = 2 + w - 9$

For Exercises 31–40, solve the equation. See Example 3.

31. $\frac{1}{2}x = 10$
32. $\frac{1}{3}y = 4$
33. $\frac{3}{5}w - 4 = 2$
34. $\frac{2}{5}z - 3 = 1$
35. $\frac{2}{7}p + 1 = 3$
36. $\frac{4}{5}q + 1 = 9$
37. $\frac{2}{3}r + 7 = 1$
38. $\frac{2}{5}s + 3 = 1$
39. $\frac{3}{4}t - 1 = -7$
40. $\frac{2}{9}v - 1 = -7$

For Exercises 41–50, solve the equation. See Example 4.

41. $2(x + 1) + 3x = 3(x + 2) + 4$
42. $4(y + 2) + 3y = 2(y + 3) + 12$
43. $4(w + 3) + 6w = 5(w + 8) + 2$
44. $5(z + 4) + 7z = 3(z + 9) + 2$
45. $3(u - 6) + 2u = 4(u + 1) - 5$
46. $7(v - 2) + 3v = 9(v + 2) - 11$
47. $8(2t + 3) + t = 3(t + 1) - 7$
48. $6(5s + 6) + s = 7(s - 1) - 5$
49. $6 - 5(2x - 3) = x - 2(3x + 2)$
50. $16 - 7(2y - 5) = 2y + 3(2y - 5)$

For Exercises 51–60, solve the equation. [HINT: *Clear fractions first.*] *See Example 5.*

51. $\frac{1}{4}p - 5 = p - \frac{1}{2}$
52. $\frac{1}{6}q - 7 = q - \frac{1}{3}$
53. $\frac{3}{5}w + \frac{1}{4} = \frac{2}{5}w + \frac{5}{4}$
54. $\frac{2}{5}x + \frac{1}{3} = \frac{1}{5}x + \frac{4}{3}$
55. $\frac{3}{10}r + \frac{1}{2} = \frac{1}{5}r + \frac{9}{10}$
56. $\frac{1}{2}s + \frac{3}{5} = \frac{2}{5}s + \frac{7}{10}$
57. $\frac{7}{15}t + \frac{4}{15} = \frac{3}{5}t + \frac{2}{3}$
58. $\frac{11}{12}u - \frac{1}{2} = \frac{1}{2}u + \frac{3}{4}$
59. $\frac{1}{2}y - \frac{1}{7} = \frac{2}{7}y + \frac{1}{2}$
60. $\frac{1}{3}z - \frac{2}{5} = \frac{1}{5}z + \frac{2}{3}$

For Exercises 61–70, determine the solution set of the equation. If the equation is an identity, the solution set is $\mathbb{R}$. If the equation has no solution, the solution set is the empty set, { }. See Example 6.

61. $w + 3 = 3 + w$

62. $w + 5 = 5 + w$

63. $-y + 11 = 11 - y$

64. $-p + 13 = 13 - p$

65. $x + 12 = 12 + x + 2$

66. $24 - x = 10 - x + 11$

67. $x - 13 = 5 + x - 18$

68. $z + 2 = -6 + z + 8$

69. $w + 3 = w$

70. $q = q - 2$

For Exercises 71–80, use the techniques of this section to solve the exercise.

71. $.7z + 2.4 = .3$

72. $.3p + 1.7 = .5$

73. $-.4q + .37 = .01$

74. $-1.1r + .229 = .108$

75. $.3(4x - 10) = 4(1.5 - .2x) + 1$

76. $.7(8y - 20) = .3(2y + 10) - 2$

77. $1.9(6z + 11) = .6(4z + 5) - 10$

78. $2.3(4t - 5) = .7(8t + 5) + 3$

79. $\frac{2}{3}(6y - 1.5) = \frac{1}{4}(4y - .4)$

80. $\frac{4}{5}(.2z - 1.5) = \frac{1}{3}(.03z + 5.4)$

Formulas 2.3

A *formula* is an equation that relates two or more variables. Some examples of formulas that you may be familiar with are given next.

$A = \pi r^2$	area (A) of a circle with radius (r)
$d = rt$	distance (d) equals the product of the rate (r) and the time (t)
$i = Prt$.	simple interest (i) equals the product of the amount invested (P), the annual interest rate (r), and time in years (t); the amount invested is called the principal

A list of the formulas used in this text may be found inside the front cover of the book.

The ability to use formulas is important for people who are training for careers in agriculture, biology, business, chemistry, computer science, engineering, physics, and many other areas. Usually, when we work with a formula, it is either

1. To find the value of one variable when we are given specific numerical values for the other variables, or
2. To solve the formula for one of the variables in terms of the other variables.

Example 1. Use the formulas $A = \pi r^2$ and $C = 2\pi r$ to find the (a) area A and (b) circumference C of a circle that has a radius of 4 inches. Use 3.14 as an approximation for π.

Solution: Direct substitution of 4 for r and 3.14 for π results in

(a) $A = \pi r^2 = (3.14)4^2 = (3.14)16 = 50.24$. Thus the area A is 50.24 square inches.

(b) $C = 2\pi r = 2(3.14)4 = (6.28)4 = 25.12$. Thus the circumference C is 25.12 inches.

Example 2. Use the formula $C = \frac{5}{9}(F - 32)$ to find the temperature in degrees Celsius (C) if the Fahrenheit (F) temperature is 77°.

Solution

$$\begin{aligned} C &= \frac{5}{9}(F - 32) & \\ &= \frac{5}{9}(77 - 32) & F = 77 \\ &= \frac{5}{9}(45) & \text{subtraction} \\ &= 25. & \text{multiplication} \end{aligned}$$

Thus the temperature is 25° Celsius when it is 77° Fahrenheit.

In Example 2 we used the formula $C = \frac{5}{9}(F - 32)$ to convert from the Fahrenheit temperature scale to the Celsius temperature scale. We can solve this formula for the variable F to obtain a formula that will give us the temperature in degrees Fahrenheit if we know the temperature in degrees Celsius. To do this, we first clear fractions by multiplying each side of $C = \frac{5}{9}(F - 32)$ by 9.

$$\begin{aligned} C &= \frac{5}{9}(F - 32) & \\ 9C &= 5(F - 32) & \text{clearing fractions} \\ 9C &= 5F - 160 & \text{distributive property} \\ 9C + 160 &= 5F & \text{add 160 to each side of the equation} \\ 5F &= 9C + 160 & \text{symmetric property of equality} \\ F &= \frac{9}{5}C + 32. & \text{divide each side of the equation by 5} \end{aligned}$$

We can use this last formula to convert from the Celsius to the Fahrenheit scale. Thus, to find the Fahrenheit temperature that corresponds to 30° Celsius, we substitute into the formula as shown.

$$\begin{aligned} F &= \frac{9}{5}C + 32 & \\ &= \frac{9}{5}(30) + 32 & C = 30 \end{aligned}$$

$$= 54 + 32 \qquad \text{multiplication}$$
$$= 86. \qquad \text{addition}$$

Thus 30° Celsius is 86° Fahrenheit.

Example 3

(a) Solve the formula $i = Prt$ for P.

(b) Use the result of part (a) to determine the amount of principal (P) that must be invested for 3 years (t) at an annual interest rate of 12% (r) if the goal is to earn \$900.00 in simple interest (i).

Solution

(a) $i = Prt$

$$\frac{1}{rt}(i) = \frac{1}{rt}(Prt) \qquad \text{multiply each side by } \frac{1}{rt}$$

$$\frac{i}{rt} = P \qquad \text{multiplication}$$

or

$$P = \frac{i}{rt}. \qquad \text{symmetric property of equality}$$

(b) Note that 12% means 12 one hundredths; thus 12% denotes $\frac{12}{100}$ or .12. If we replace i with 900, r with .12, and t with 3, the result is

$$P = \frac{i}{rt}$$
$$= \frac{900}{(.12)(3)} = \frac{900}{.36}$$
$$= 2500.$$

Thus the necessary principal is \$2500.

Example 4. The surface area (S) of a rectangular solid (Figure 2.1) with length (L), width (W), and height (H) is given by the formula

$$S = 2(LW + LH + HW).$$

Solve this formula for the variable H.

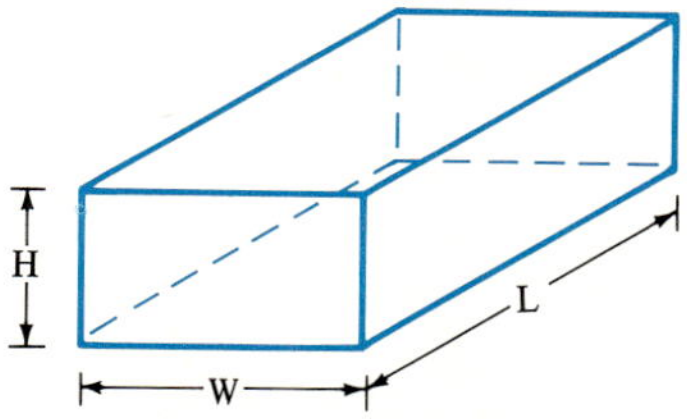

Figure 2.1

Solution

$$\begin{aligned} S &= 2(LW + LH + HW) && \\ &= 2LW + 2LH + 2HW && \text{distributive property} \\ S - 2LW &= 2LH + 2HW && \text{subtract } 2LW \text{ from each side} \\ S - 2LW &= H(2L + 2W) && \text{distributive property} \\ H(2L + 2W) &= S - 2LW && \text{symmetric property of equality} \\ H &= \frac{S - 2LW}{2L + 2W}. && \text{divide each side by } 2L + 2W \end{aligned}$$

EXERCISE 2.3

For Exercises 1–6, for the circle with radius (r), find (a) the area and (b) the circumference. Use 3.14 as an approximation for π. See Example 1.

1. $r = 5$ feet **2.** $r = 6$ feet **3.** $r = \frac{2}{5}$ yard

4. $r = \frac{3}{5}$ yard **5.** $r = 2.3$ feet **6.** $r = 3.4$ feet

For Exercises 7–10, change the temperature given in degrees Fahrenheit to degrees Celsius. See Example 2.

7. 68°F **8.** 14°F **9.** 212°F **10.** 32°F

For Exercises 11–14, change the given Celsius temperature to a Fahrenheit temperature. See the discussion following Example 2.

11. −5°C **12.** 15°C **13.** 35°C **14.** 5°C

For Exercises 15–30, refer to the formulas on the inside of the front cover for the appropriate formula and then compute the value of the indicated variable.

15. How far will you travel in $3\frac{1}{2}$ hours if you maintain a steady speed of 40 miles per hour?

16. If you drive at 45 miles per hour for $2\frac{1}{3}$ hours, how far will you travel?

17. If you maintain a steady speed of 80 kilometers per hour, how far will you travel in $2\frac{1}{2}$ hours?

18. How many kilometers will you travel in $3\frac{2}{3}$ hours if you drive at 90 kilometers per hour?

19. If you invest $2000 for 2 years in an account that pays simple yearly interest at the rate of 11%, what will be the total interest earned?

20. If $5000 is invested in an account that pays simple annual interest at the rate of 9%, how much interest will be earned after 3 years?

21. Find the area and perimeter of each triangle shown. The lengths indicated are measured in inches.

(a)

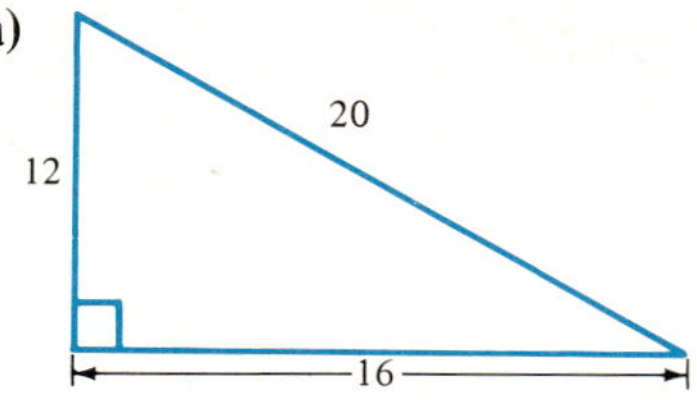

(b)

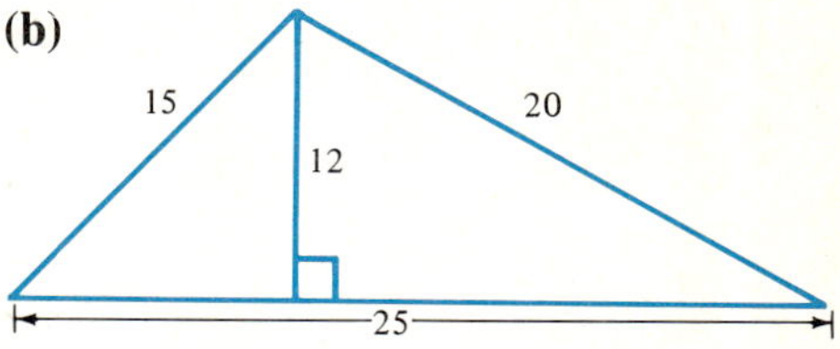

22. Find the area and perimeter of each triangle shown. The lengths indicated are measured in feet.

(a)

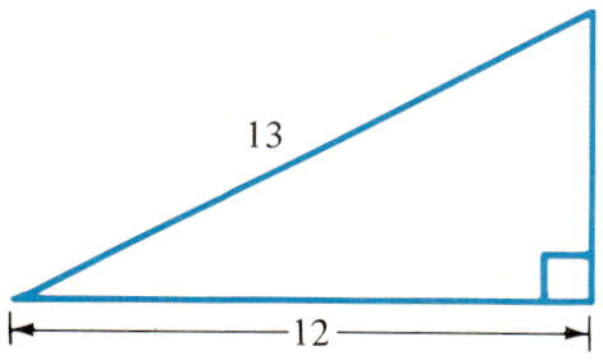

(b)

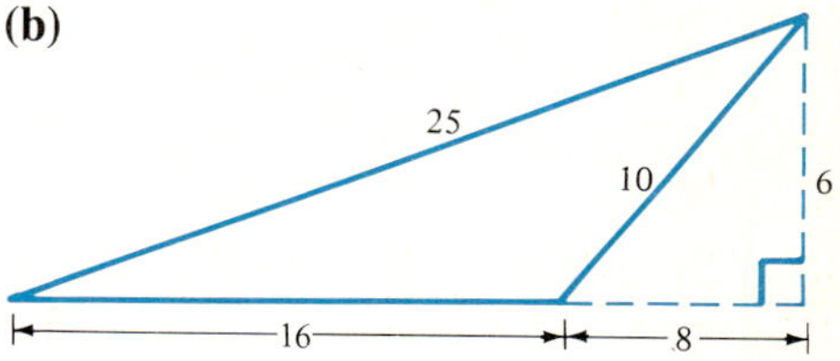

23. Find the area of the trapezoid shown.

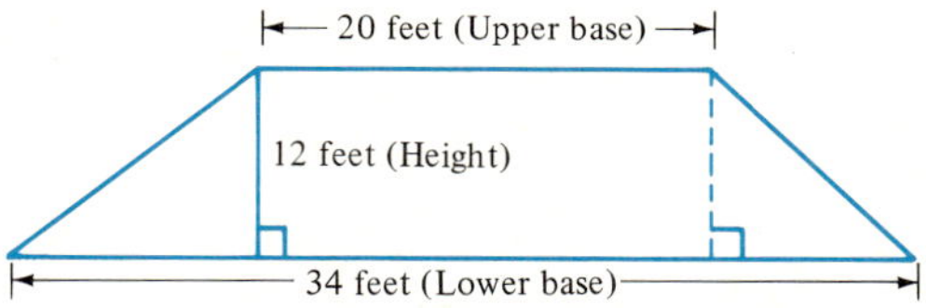

24. Find the area of the trapezoid shown.

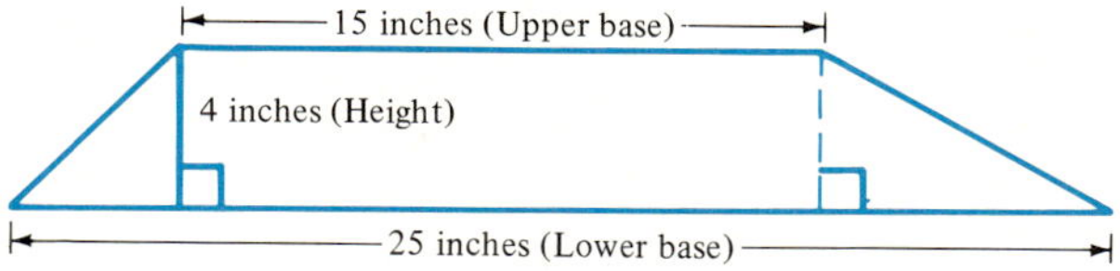

25. Find the volume and surface area of a rectangular solid that has a length of 8 feet, a width of 5 feet, and a height of 3 feet.

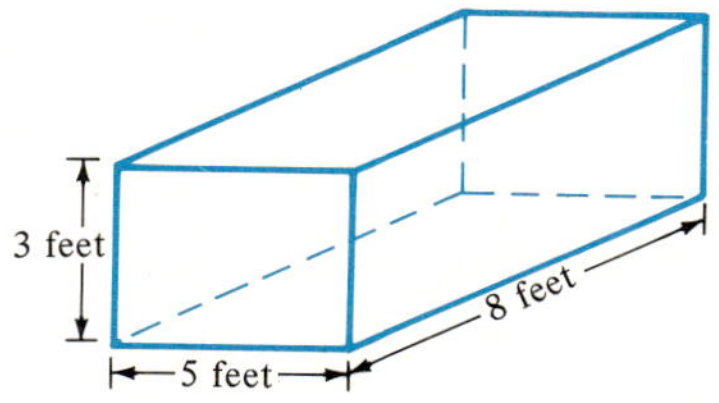

26. Find the volume and surface area of a rectangular solid that has length equal to 9 feet, width equal to 4 feet, and a height of 2 feet.

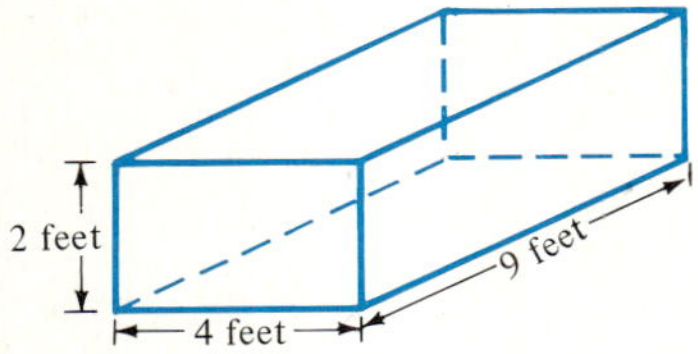

27. Find the volume and surface area of a right circular cylinder with radius 2 inches and height 7 inches. Use 3.14 as an approximation for π.

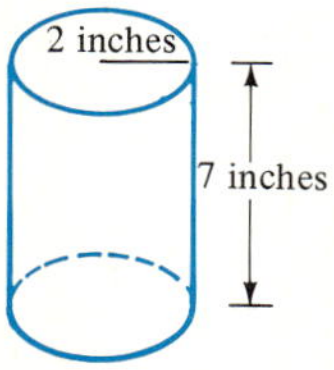

28. Find the volume and surface area of a right circular cylinder with radius 3 feet and height 5 feet. Use 3.14 as an approximation for π.

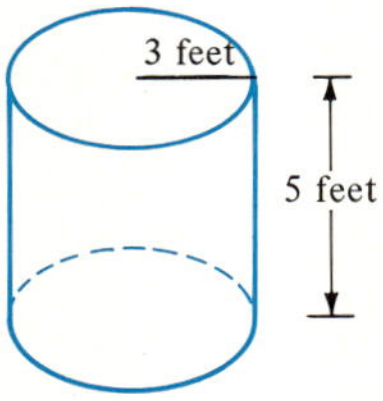

29. Find the volume of a right pyramid that has a height of 6 inches and a square base that is 4 inches on each side.

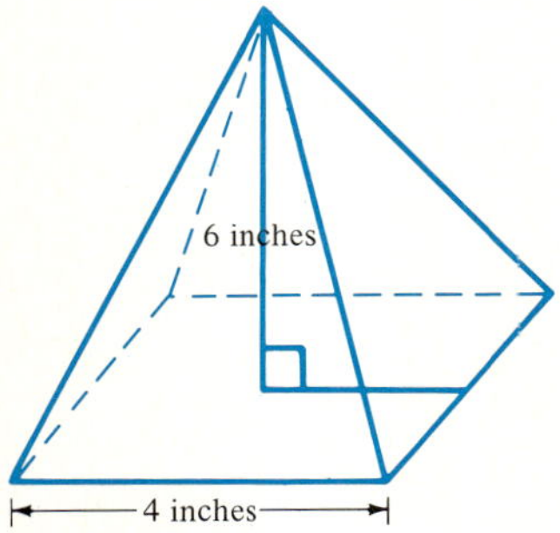

30. Find the volume of a right pyramid that has a height of 2 feet and a square base that is 3 feet on each side.

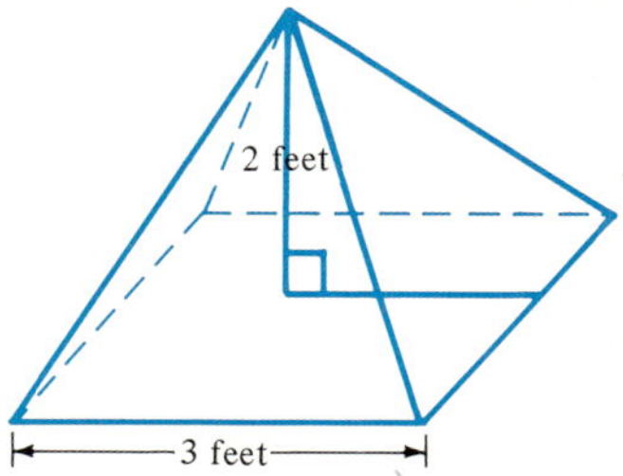

Most stores allow customers to purchase items such as furniture or large appliances and then to pay for these items with equal monthly installments over a period of many months. Because the store is actually providing you with a loan, they add a finance charge *to the purchase price. The formula*

$$r = \frac{24f}{b(n + 1)}$$

gives an approximation to the true annual interest rate *that the customer is being charged. In the formula,*

r = approximate annual interest rate
f = finance charge
b = purchase price (amount borrowed)
n = number of equal payments.

Use this formula to answer Exercises 31–34.

31. Furniture that was purchased for $2100 with an additional finance charge of $420 is to be paid for in 24 equal monthly payments.
(a) What is the amount of each payment?
(b) What is the approximate annual interest rate?

32. A home entertainment unit that sells for $1600 with an additional finance charge of $140 may be paid for in 12 equal monthly payments.
(a) What is the amount of each payment?
(b) What is the approximate annual interest rate?

33. A central air conditioning unit that costs $6500 may be paid for over a 3-year period by equal monthly payments. If this method of payment is selected, there is an additional finance charge of $610.
(a) What is the amount of each payment?
(b) What is the approximate annual interest rate?

34. A multifuel furnace that burns wood, coal, or oil sells for $3700. For an additional finance charge of $620, the furnace may be paid for by equal monthly installments over a 2-year period.
(a) What will be the amount of each monthly payment?
(b) What is the approximate annual interest rate?

For Exercises 35–44, solve the formula for the variable indicated. See Examples 3 and 4.

35. $d = rt$ for t

36. $i = Prt$ for r

37. $C = 2\pi r$ for r

38. $E = IR$ for R

39. $A = \frac{1}{2}bh$ for b

40. $V = LWH$ for H

41. $P = 2L + 2W$ for W

42. $S = 2\pi rh + 2\pi r^2$ for h

43. $A = \frac{1}{2}h(B + b)$ for B

44. $S = \frac{1}{2}(a + b + c)$ for b

For Exercises 45–60, decide which formula is appropriate, solve the formula for the required variable, and then find the value of the variable. See Examples 3 and 4 and the inside of the front cover if necessary.

45. A truck driver is 260 miles from a warehouse that he wants to arrive at in 5 hours. What average speed must he maintain to accomplish his goal?

46. A train is 175 miles from its next stop. At what average speed must it travel to arrive at the stop in exactly $2\frac{1}{2}$ hours?

47. How long does it take to travel 175 miles at a speed of 50 miles per hour?

48. How long does it take to travel 220 kilometers at a speed of 80 kilometers per hour?

49. A bank is offering 1-year certificates of deposit (CD's) that have an annual interest rate of 9%. How much principal is required to earn $135.00 interest?

50. A bank is offering 1-year-certificates of deposit (CD's) that have an annual interest rate of 8.5%. How much principal is required to earn $170.00 interest?

51. The area of a triangle is 90 square feet. If the base is 18 feet, what is the height?

52. The area of a triangle is 132 square meters. If the base is 22 meters, what is the height?

53. The perimeter of a rectangle is 36 feet. If the length is 11 feet, what is the width?

54. The perimeter of a rectangle is 56 feet. If the length is 15 feet, what is the width?

55. The area of a trapezoid is 119 square inches. If the upper base is 13 inches and the height is 7 inches, what is the length of the lower base?

56. The area of a trapezoid is 270 square yards. If the lower base is 42 yards and the height is 9 yards, what is the length of the upper base?

57. The radius of a can is 4 inches. Find the height of the can if the total surface area is 56π square inches.

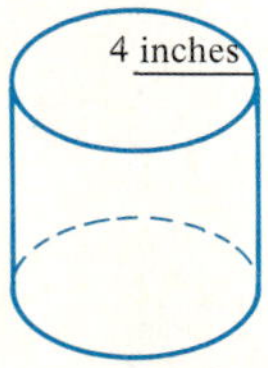

58. The total surface area of a can is 154π square centimeters. If the radius of the can is 7 centimeters, what is the height?

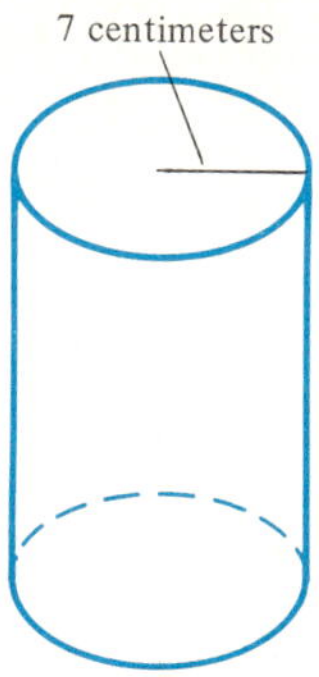

59. A store determines its finance charges by using an annual rate of 18%. What is the finance charge on an $800 television set that is to be paid for in 12 equal monthly installments? See the formula preceding Exercise 31.

60. A furniture store uses an annual rate of 16% to determine its finance charges. What is the finance charge on $2400 worth of furniture that is going to be paid for in 24 equal monthly payments? See the formula preceding Exercise 31.

Equations Involving Absolute Value 2.4

In this section we study equations involving *absolute value*. From our study in Chapter 1 we know that if x is a point on the number line, then $|x|$ is the distance between x and 0. Note also that $|x| \geq 0$ for all real x. Furthermore, for any two points x and y on the number line, we define $|x - y|$ as the distance between x and y.

The equation $|x| = 7$ may be read "find all x such that the distance between x and 0 is 7." As Figure 2.2 illustrates, there are two solutions to this equation, -7 and 7. This observation is generalized by noting that, for $b > 0$, the solution set of the equation $|x| = b$ is $\{-b, b\}$. If $b = 0$, the solution set is $\{0\}$.

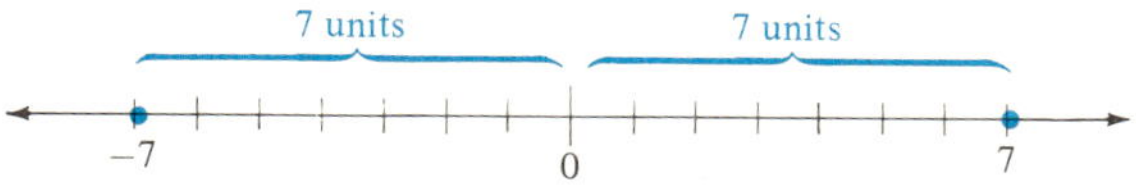

Figure 2.2

Example 1. Find the solution set of $|y| = 15$.

Solution: Since $15 > 0$, the solution set is $\{-15, 15\}$.

Another method of solving equations that involve absolute value is to recall that the formal definition of $|x|$ is

$$|x| = \begin{cases} x & \text{if } x \geq 0 \\ -x & \text{if } x < 0. \end{cases}$$

This definition is often more convenient to apply when solving absolute value equations. Using this definition we see that, for $k \geq 0$, the equation $|y| = k$ may be replaced with the two equations

$$y = k \quad \text{if } y \geq 0 \qquad \text{or} \qquad -y = k \quad \text{if } y < 0.$$

Multiplying both sides of the second equation by -1, we see that $|y| = k$, is equivalent to $y = k$ or $y = -k$. The preceding discussion leads to the following theorem.

Theorem 2

If $k \geq 0$, the equation $|y| = k$ is equivalent to the pair of equations

$$y = k \qquad \text{or} \qquad y = -k.$$

Example 2. Use Theorem 2 to find the solution set for $|4x| = 20$.

Solution: Theorem 2 tells us that we may replace $|4x| = 20$ with the two equations

$$4x = 20 \qquad \text{or} \qquad 4x = -20.$$

Thus

$$x = 5 \qquad \text{or} \qquad x = -5.$$

It is easy to check that both -5 and 5 are solutions to $|4x| = 20$. Thus the desired solution set is $\{-5, 5\}$.

Example 3. Solve $|3z + 5| = 4$.

Solution: This equation will be true if either

$$3z + 5 = 4 \qquad \text{or} \qquad 3z + 5 = -4.$$

Solving these two equations, we see that

$$3z = -1 \qquad \text{or} \qquad 3z = -9.$$

Thus

$$z = -\frac{1}{3} \qquad \text{or} \qquad z = -3.$$

CHECK

When $z = -\dfrac{1}{3}$: When $z = -3$:

We see that

$$|3z + 5| = 4 \qquad\qquad |3z + 5| = 4$$

$$\left|3\left(-\frac{1}{3}\right) + 5\right| = 4 \qquad\qquad |3(-3) + 5| = 4$$

$$|-1 + 5| = 4 \qquad\qquad |-9 + 5| = 4$$

$$|4| = 4 \quad \text{true} \qquad\qquad |-4| = 4 \quad \text{true}$$

Thus the solution set is $\left\{-\frac{1}{3}, -3\right\}$.

Example 4. Find the solution of $|4x + 3| = -8$.

Solution: Since $-8 < 0$ and $|4x + 3|$ is never negative (<0), we see that there is no solution. Therefore, the solution set is the empty set, { }.

Example 5. Solve $|2x - 7| + 3 = 8$.

Solution: In order to use Theorem 2, we will isolate $|2x - 7|$ on one side of the equals sign. If we subtract 3 from each side, we get

$$|2x - 7| = 5.$$

Applying Theorem 2 results in

$$2x - 7 = 5 \quad \text{or} \quad 2x - 7 = -5.$$

Thus

$$2x = 12 \quad \text{or} \quad 2x = 2$$
$$x = 6 \qquad\qquad x = 1.$$

The reader should check that $\{1, 6\}$ is the solution set by substituting 1 and 6 in the original equation.

The next situation we consider involves equations with more than one set of absolute values. To this end we consider the statement $|a| = |b|$. This statement is equivalent to a and b being equal or being additive inverses (opposites) of each other. Thus we have the following theorem.

Theorem 3

$$|a| = |b| \quad \text{if and only if} \quad a = b \quad \text{or} \quad a = -b.$$

Example 6. Solve $|w - 7| = |3w + 5|$.

Solution: Applying Theorem 3, we see that either

$$w - 7 = 3w + 5 \quad \text{or} \quad w - 7 = -(3w + 5).$$

Solving each of these equations, we get

$$-2w = 12 \qquad\qquad w - 7 = -3w - 5$$
$$w = -6 \qquad\qquad 4w = 2$$
$$w = \frac{1}{2}$$

To check these, notice that

when $w = -6$:

$$|w - 7| = |3w + 5|$$
$$|-6 - 7| = |3(-6) + 5|$$
$$|-13| = |-13|$$
$$13 = 13 \quad \text{true}$$

when $w = \frac{1}{2}$:

$$|w - 7| = |3w + 5|$$
$$\left|\frac{1}{2} - 7\right| = \left|3\left(\frac{1}{2}\right) + 5\right|$$
$$\left|-\frac{13}{2}\right| = \left|\frac{13}{2}\right|$$
$$\frac{13}{2} = \frac{13}{2} \quad \text{true}$$

Thus the solution set is $\left\{\frac{1}{2}, -6\right\}$.

Example 7. Solve $|x - 7| = |x + 3|$.

Solution: Using Theorem 3, we get the following two equations:

$$x - 7 = x + 3 \qquad \text{or} \qquad x - 7 = -(x + 3)$$
$$0 = 10 \quad \text{false} \qquad\qquad x - 7 = -x - 3$$
$$2x = 4$$
$$x = 2.$$

Note that the first equation led to a false statement, therefore the only possible solution is 2. It is easy to check that 2 is indeed a solution and thus the solution set is $\{2\}$.

Example 8. Solve $|x + 5| = x + 5$.

Solution: First consider an equation such as $|y| = y$. From the definition of absolute value we know that this is true only when $y \geq 0$. Replacing y with $x + 5$, we see that $|x + 5| = x + 5$ only when $x + 5 \geq 0$ or when $x \geq -5$. Thus the solution set is $\{x \mid x \geq -5\}$.

EXERCISES 2.4

For Exercises 1–8, solve the equation. See Example 1.

1. $|x| = 4$ **2.** $|y| = 9$ **3.** $|w| = 13$ **4.** $|z| = 17$

5. $|p| = 100$ **6.** $|q| = 96$ **7.** $|r| = -10$ **8.** $|s| = -1$

For Exercises 9–16, find the solution set for the equation. See Example 2.

9. $|3x| = 15$ **10.** $|2y| = 10$ **11.** $|4y| = 2$ **12.** $|3y| = 6$

13. $|2q| = 4.6$ **14.** $|3q| = 6.6$ **15.** $|\frac{1}{3}z| = 2$ **16.** $|\frac{1}{4}w| = 5$

For Exercises 17–52, find the solution set for the equation. See Examples 3–5.

17. $|x + 1| = 5$ **18.** $|y + 3| = 4$ **19.** $|w + 2| = 6$

20. $|z + 5| = 11$ **21.** $|p - 2| = 1$ **22.** $|q - 1| = 5$

23. $|2 - r| = 5$ **24.** $|1 - s| = 3$ **25.** $|4 + t| = 5$

26. $|5 + v| = 4$ **27.** $|2x + 3| = 7$ **28.** $|2y + 5| = 11$

29. $|3p + 4| = 8$ **30.** $|3q + 2| = 7$ **31.** $|7w + 1| = 13$

32. $|6z + 5| = 7$ **33.** $|3w + 1| = -4$ **34.** $|5w + 2| = -3$

35. $|\frac{1}{2}p + 3| = 5$ **36.** $|\frac{1}{2}q + 5| = 1$ **37.** $|\frac{2}{3}x - 4| = 2$

38. $|\frac{2}{3}y - 2| = 4$ **39.** $|\frac{4}{5}w + 7| = 1$ **40.** $|\frac{3}{5}z + 2| = 1$

41. $|3(x + 2) + 4| = 2$ **42.** $|4(y + 1) + 3| = 1$ **43.** $|11(2y + 3) - 11| = 22$

44. $|13(y + 5) - 52| = 13$ **45.** $|x + 4| + 3 = 5$ **46.** $|y + 2| + 7 = 9$

47. $|w - 2| + 3 = 9$ **48.** $|z - 5| + 2 = 11$ **49.** $|2p + 1| + 1 = 5$

50. $|2q + 3| + 1 = 13$ **51.** $|5r - 2| + 6 = 8$ **52.** $|5s - 3| + 7 = 9$

For Exercises 53 – 70, find the solution set for the equation. See Examples 6–8.

53. $|x + 2| = |2x + 7|$ **54.** $|y - 15| = |3y + 7|$ **55.** $|3w - 1| = |2w + 11|$

56. $|2z - 5| = |3w + 10|$ **57.** $|6q - 1| = |3q - 5|$ **58.** $|5p - 3| = |2p - 7|$

59. $|7u + 11| = |2u - 4|$ **60.** $|8v + 13| = |3v - 2|$ **61.** $|\frac{2}{3}x + 5| = |\frac{4}{3}x - 7|$

62. $|\frac{3}{4}y + 7| = |\frac{5}{4}y - 3|$ **63.** $|w - 5| = |w + 9|$ **64.** $|z - 3| = |z + 11|$

65. $|3x - 8| = |3x - 4|$ **66.** $|5y + 11| = |5y + 9|$ **67.** $|p + 3| = p + 3$

68. $|q - 2| = q - 2$ **69.** $|3 - v| = v - 3$ **70.** $|10 - u| = u - 10$

2.5 Solving First-Degree Inequalities

In many practical situations we are often interested in finding the solution set of an inequality. In this section we consider *first-degree inequalities in one variable* such as $7x + 2 \leq 10$. Some other examples of first-degree inequalities in one variable are

$$2y - 4 > 1, \qquad -\frac{1}{2}z + \frac{1}{4} < \frac{3}{4}, \qquad 4.3w + 1.2 \geq -7.4.$$

A statement of the form

$$ax + b \leq c \qquad a,\ b,\ \text{and } c \text{ constants, } a \neq 0$$

is called *a first-degree inequality in one variable*. In this section we state the rules in terms of the $\leq$ symbol. However, similar rules are also valid when we replace $\leq$ with either $>$, $<$, or $\geq$.

A *solution* to a first-degree inequality is any value of the variable that makes the inequality a *true* statement. The set of all solutions is called the *solution set* of the inequality. When we have identified the solution set, we say that we have solved the inequality. Inequalities that have the same solution set are called *equivalent inequalities*. As we did with first-degree equations, we will produce a sequence of equivalent inequalities until we get an inequality such as $x \leq 2$, for which the solution set is obvious.

There are two basic properties that are used when solving an inequality. The first of these is the *addition property of inequality,* which states that the same quantity may be added to each side of an inequality to obtain an equivalent inequality. We note that the direction (or sense) of the inequality symbol remains the same.

Theorem 4. Addition Property of Inequality

If a, b, and c are real numbers, then $a \leq b$ is equivalent to

$$a + c \leq b + c.$$

Example 1. Solve each inequality and graph the solution set.
(a) $x - 4 \leq 1$. (b) $y + 3 > 2$.

Solution

(a) If we add 4 to each side of the equality, the result is

$$x - 4 + 4 \leq 1 + 4$$
$$x \leq 5.$$

Thus the solution set is $\{x \mid x \leq 5\}$. The graph of this set is shown in Figure 2.3.

Figure 2.3

(b) If we add -3 to each side of the inequality, we get

$$y + 3 - 3 > 2 - 3$$
$$y > -1.$$

Thus the solution set is $\{y \mid y > -1\}$. Its graph is shown in Figure 2.4.

Figure 2.4

The coefficients of the unknowns in Example 1 were both equal to 1. If we want to solve an inequality such as $4x \leq 16$, it seems reasonable to multiply each side by $\frac{1}{4}$. Before doing this, we investigate what happens to the symbol of inequality when we multiply each side of an inequality by a number. Thus consider the inequality $-1 < 2$. If we multiply both sides of the inequality by 3, we get the products -3 and 6. Notice that $-3 < 6$. Notice that the direction of the inequality symbol remains the same. What happens when we multiply both sides of an inequality by a negative number? Consider again the statement $-1 < 2$. If we multiply each side of this inequality by -3, the resulting products are $(-1)(-3) = 3$ and $(2)(-3) = -6$. But $3 > -6$. That is, the direction (or sense) of the inequality symbol has changed. The previous discussion about inequalities involved particular numbers. In general, it can be shown that the following theorem is valid.

Theorem 5. Multiplication Properties of Inequality

If a, b, and c are real numbers and $a \leq b$ *then*

(a) $ac \leq bc$ if c is positive ($c > 0$).
(b) $ac \geq bc$ if c is negative ($c < 0$).

Example 2. Solve each of the following inequalities and graph the solution set.
(a) $4x \leq 16$. (b) $-\frac{1}{3}y > 2$.

Solution

(a) Multiply each side of the inequality by $\frac{1}{4}$. Since $\frac{1}{4}$ is positive, Theorem 5(a) tells us that we do not reverse the inequality symbol:

$$4x \leq 16$$
$$\left(\frac{1}{4}\right)4x \leq \left(\frac{1}{4}\right)16$$
$$x \leq 4.$$

The graph of the solution set $\{x \mid x \leq 4\}$ of the given inequality is shown in Figure 2.5.

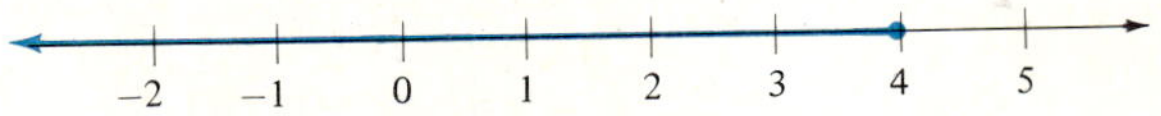

Figure 2.5

(b) Multiply each side by -3. Since -3 is negative, Theorem 5(b) tells us that we reverse the direction of the inequality symbol.

$$-\frac{1}{3}y > 2$$

$$(-3)\left(-\frac{1}{3}y\right) < (-3)(2) \qquad \text{multiply by } -3 \text{ and change } > \text{ to } <$$

$$y < -6.$$

The graph of the solution set $\{y \mid y < -6\}$ of the inequality is shown in Figure 2.6.

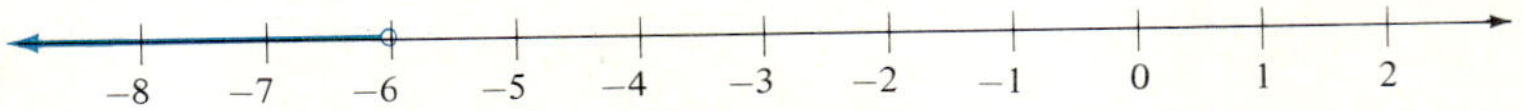

Figure 2.6

Example 3. Solve $-4x + 3 \geq 11$ and graph the solution set.

Solution

$$-4x + 3 \geq 11$$

$$-4x + 3 - 3 \geq 11 - 3 \qquad \text{add } -3 \text{ to each side}$$

$$-4x \geq 8$$

$$\left(-\frac{1}{4}\right)(-4x) \leq \left(-\frac{1}{4}\right)(8) \qquad \text{multiply each side by } -\frac{1}{4} \text{ and reverse the direction of the inequality symbol}$$

$$x \leq -2.$$

The graph of the solution set $\{x \mid x \leq -2\}$ is shown in Figure 2.7.

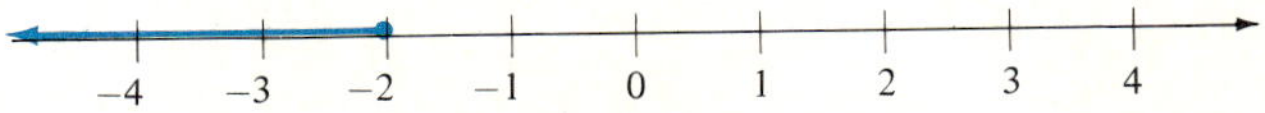

Figure 2.7

Example 4. Solve $19 + 8(z - 3) \geq 3(2z + 1) - 2$ and graph the solution set.

Solution: We begin by using the distributive property and then combine like terms on each side of the inequality.

$$19 + 8(z - 3) \geq 3(2z + 1) - 2$$

$$19 + 8z - 24 \geq 6z + 3 - 2 \qquad \text{distributive property}$$

$$8z - 5 \geq 6z + 1$$
$$8z - 5 + 5 \geq 6z + 1 + 5 \qquad \text{add 5 to each side}$$
$$8z \geq 6z + 6$$
$$8z - 6z \geq 6z + 6 - 6z \qquad \text{add } -6z \text{ to each side}$$
$$2z \geq 6$$
$$\left(\frac{1}{2}\right)2z \geq \frac{1}{2}(6) \qquad \text{multiply each side by } \frac{1}{2}$$
$$z \geq 3.$$

The solution set is $\{z \mid z \geq 3\}$. Its graph is shown in Figure 2.8.

Figure 2.8

In some areas of mathematics it is necessary to solve *compound inequalities* such as

$$5x + 3 \leq 8 \quad \textit{and} \quad 2x + 7 > 3$$

or perhaps

$$2x + 3 \leq 1 \quad \textit{or} \quad -x + 1 > -2.$$

A compound inequality is a mathematical statement with two or more inequality statements joined with connective words such as *and* or *or*. Recall from Chapter 1 that if A and B are sets and the element x has the property $x \in A$ and $x \in B$, then $x \in A \cap B$. In fact, $A \cap B$ is composed of precisely those elements that belong to *both A and B*. Therefore, we see that the solution set of a compound inequality that uses the connective *and* is the *intersection* of the solution sets of each inequality. Using similar reasoning, we conclude that when the connective is *or*, the solution set is the *union* of the solution sets of each inequality.

Example 5. Graph the solution set of each compound inequality.
(a) $5x + 3 \leq 8$ and $2x + 7 > 3$.
(b) $2x + 3 \leq 1$ or $-x + 1 < -2$.
(c) $4x + 1 \geq 9$ or $7 - 2x < 5$.

Solution

(a) First we solve each inequality separately.

$$5x + 3 \leq 8 \quad \text{and} \quad 2x + 7 > 3$$
$$5x \leq 5 \quad \text{and} \quad 2x > -4$$
$$x \leq 1 \quad \text{and} \quad x > -2.$$

The graphs of the solution sets $\{x \mid x \leq 1\}$ and $\{x \mid x > -2\}$ are shown in Figure 2.9. Because the connecting word is *and,* we want the *intersection* of these two solution sets. The intersection of these two solution sets consists of those numbers that are common to each of the sets. Thus the intersection is

$\{x \mid -2 < x \leq 1\}$. This is the solution set to the given compound inequality. The graph of the solution set is shown in Figure 2.10.

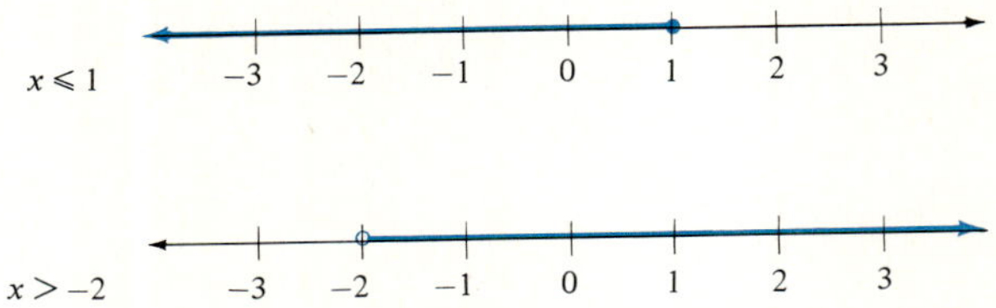

Figure 2.9

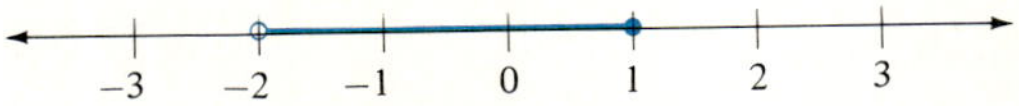

Figure 2.10

(b) Again, we begin by solving each inequality separately.

$$\begin{aligned} 2x + 3 &\leq 1 & \text{or} && -x + 1 &< -2 \\ 2x &\leq -2 & \text{or} && -x &< -3 \\ x &\leq -1 & \text{or} && x &> 3. \end{aligned}$$

The graphs of the two solution sets are shown in Figure 2.11. Since the connect-

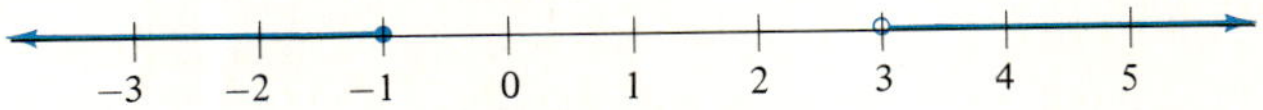

Figure 2.11

ing word is *or,* the solution set of the compound inequality is the union of these two solution sets. The union of two sets consists of all numbers that are in either one or both of the two sets. In this case, the graph of the union of the two solution sets is precisely the graph shown in Figure 2.11. This union may be described as $\{x \mid x \leq -1 \text{ or } x > 3\}$.

(c) Solving each inequality, we get

$$\begin{aligned} 4x + 1 &\geq 9 & \text{or} && 7 - 2x &< 5 \\ 4x &\geq 8 & \text{or} && -2x &< -2 \\ x &\geq 2 & \text{or} && x &> 1. \end{aligned}$$

The graphs of the two solution sets are shown in Figure 2.12. Since the connecting word is *or,* we want the union of these two sets, which is the set $\{x \mid x > 1\}$.

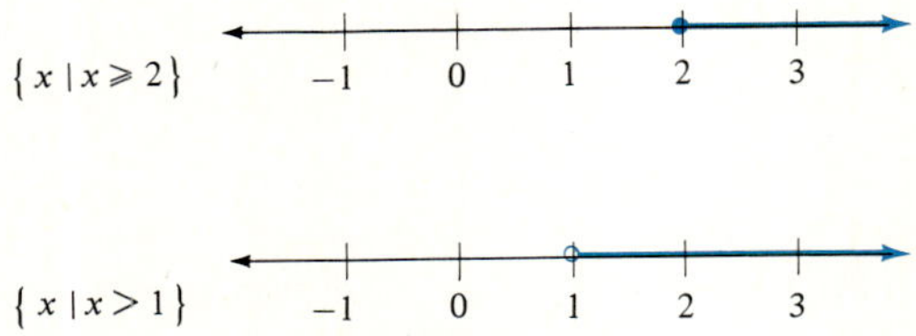

Figure 2.12

This is because every number in the set $\{x \mid x \geq 2\}$ is also in the set $\{x \mid x > 1\}$. Thus the graph of the given compound inequality is the graph of the set $\{x \mid x > 1\}$.

Example 6. Graph the solution set of the inequality

$$-1 < 5w + 4 < 19.$$

Solution: The given inequality is equivalent to the compound inequality

$$-1 < 5w + 4 \quad \text{and} \quad 5w + 4 < 19.$$

We will solve the two inequalities at the same time.

$$-1 < 5w + 4 < 19$$
$$-1 - 4 < 5w + 4 - 4 < 19 - 4 \qquad \text{add } -4 \text{ to each inequality}$$
$$-5 < 5w < 15$$
$$\left(\frac{1}{5}\right)(-5) < \left(\frac{1}{5}\right)5w < \left(\frac{1}{5}\right)15 \qquad \text{multiply by } \frac{1}{5}$$
$$-1 < w < 3.$$

The graph of the solution set is shown in Figure 2.13.

Figure 2.13

EXERCISES 2.5

For Exercises 1–16, solve the inequality and graph the solution set. See Examples 1 and 2.

1. $x - 1 < 2$ **2.** $y - 3 > 1$ **3.** $w + 3 \geq 5$ **4.** $z + 7 \leq 8$

5. $5p \leq 20$ **6.** $4q \geq 12$ **7.** $2r > -6$ **8.** $7s < -14$

9. $-\frac{2}{5}x \leq 4$ **10.** $-\frac{3}{4}y \geq 6$ **11.** $\frac{4}{3}u < \frac{8}{9}$ **12.** $\frac{2}{5}v > \frac{4}{15}$

13. $3.1p < 12.4$ **14.** $2.3q > 9.2$ **15.** $-2.1z \geq -6.3$ **16.** $-1.2w \leq -2.4$

For Exercises 17–50, solve the inequality and graph the solution set. See Examples 3 and 4.

17. $2t - 2 \leq 6$ **18.** $3t - 5 \geq 4$ **19.** $5u - 1 \geq 2$

20. $4v - 3 \leq 2$ **21.** $4w + 3 < 11$ **22.** $5z + 2 > 12$

23. $-7x + 2 > 16$ **24.** $-6y + 5 < 17$ **25.** $8 - 3p \leq -1$

26. $11 - 4q \geq -1$ **27.** $-4 + 5s \geq 6$ **28.** $-3 + 7r \leq 4$

29. $-5 - 3u < -20$ **30.** $-4 - 5v > 11$ **31.** $4(3p + 1) \leq 16$

32. $6(5q + 1) \geq 36$ **33.** $2(3u - 1) \geq 4$ **34.** $3(2v - 3) \leq 9$

35. $5(2 - 3w) > 10$ **36.** $9(1 - z) < 18$ **37.** $\frac{1}{2}(r - 1) < 3$

38. $\frac{2}{3}(s + 1) > 1$ **39.** $1.5(3x + 2) \leq 12$ **40.** $2.5(y + 2) \geq 10$

41. $2.3(10 - 4u) > -4.6$ **42.** $3.7(10 - 6v) < -7.4$

43. $5 + 4(x - 3) \geq 2(x + 5) - 3$ **44.** $3 + 5(y - 2) \leq 3(y + 4) - 3$

45. $7 + 5(y - 1) < 2(y + 3) + 5$ **46.** $1 + 7(z - 2) > 4(z + 2) + 3$

47. $3 + 6(t + 1) \leq 2(t - 3) - 1$ **48.** $2 + 9(u + 2) \geq 4(u - 3) + 12$

49. $2 + 4(v + 5) \geq 5(v + 2) + 9$ **50.** $5 + 6(w + 3) \leq 7(w + 3) + 1$

For Exercises 51 –62, graph the solution set of the compound inequality. See Example 5.

51. $x - 3 \leq 1$ and $x + 1 > 2$ **52.** $y - 2 > 1$ and $y + 1 \leq 7$

53. $2u + 5 > 1$ and $3u + 1 \leq 10$ **54.** $3z + 7 \geq 1$ and $2z + 3 < 9$

55. $11 - 4w \leq 3$ and $17 - 5w \geq -3$ **56.** $2 - 5v < 17$ and $7 - 4v > -1$

57. $x + 4 < 7$ or $x - 3 \geq 2$ **58.** $y + 6 \leq 10$ or $y - 2 > 5$

59. $w + 2 \geq 5$ or $2 - w \leq 1$ **60.** $z + 5 \leq 11$ or $7 - z \geq 5$

61. $2(p + 3) > 9$ or $7(1 - p) < 7$ **62.** $3(q - 5) \leq 6$ or $4(2 - q) \geq 8$

For Exercises 63–70, graph the solution set of the inequality. See Example 6.

63. $5 < 2x + 3 \leq 11$ **64.** $3 \leq 2y + 1 < 7$ **65.** $2 \leq 3w + 5 \leq 14$

66. $3 \leq 4z + 7 \leq 15$ **67.** $-12 < 5p + 3 < -2$ **68.** $-10 < 3q + 2 < -4.$

69. $-2 < 4 - 3x \leq 13$ **70.** $-1 \leq 5 - 2y < 2$

2.6 Inequalities Involving Absolute Value

From our work in Section 2.4 we know that if x is a point on the number line, the absolute value equation $|x| = 4$ has two solutions, -4 and 4. In this section we consider *absolute value inequalities* such as $|x| < 4$, $|y| \geq 3$, and $|2z - 3| \leq 5$.

Recall that the absolute value equation $|x| = 4$ may be read "the distance between the point x and the origin 0 is *exactly* 4 units." Similarly, the absolute value inequality $|x| < 4$ may be read "the distance between the point x and the origin 0 is *less than* 4 units." The graph of all points satisfying this condition is shown in Figure 2.14. Thus the solution set of the inequality $|x| < 4$ is $\{x | -4 < x < 4\}$.

Figure 2.14

Example 1. Find the solution sets of the following absolute value inequalities. (a) $|x| \leq 3$. (b) $|x| > 3$.

Solution

(a) The inequality $|x| \leq 3$ denotes all numbers that are at most 3 units from 0. The graph of these numbers is shown in Figure 2.15. Thus the solution set is $\{x | -3 \leq x \leq 3\}$.

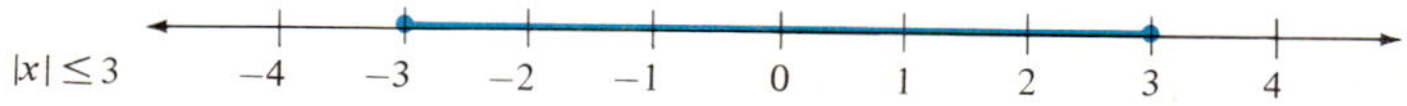

Figure 2.15

(b) The inequality $|x| > 3$ denotes all numbers that are more than 3 units from 0. The graph of all such numbers is shown in Figure 2.16. Here the solution set is $\{x | x < -3 \text{ or } x > 3\}$.

Figure 2.16

The results of Example 1 suggest the following theorem.

Theorem 6

If $c > 0$, then

(a) $|x| \leq c$ is equivalent to $-c \leq x \leq c$.

(b) $|x| \geq c$ is equivalent to $x \leq -c$ or $x \geq c$.

We now turn our attention to inequalities of the form $|ax + b| < c$. In particular, we will graph the solution set of $|x - 1| < 4$. To simplify the problem, we will replace $x - 1$ with y. With this change the absolute value inequality $|x - 1| < 4$ becomes $|y| < 4$. But Theorem 6(a) tells us that $|y| < 4$ is equivalent to $-4 < y < 4$. Since $y = x - 1$, we see that $-4 < y < 4$ is the same as $-4 < x - 1 < 4$. Thus $|x - 1| < 4$ is equivalent to $-4 < x - 1 < 4$. Using the methods described in Section 2.5, we solve this inequality by adding 1 to each part of the inequality as follows.

$$-4 < x - 1 < 4$$

$$-4 + 1 < (x - 1) + 1 < 4 + 1 \qquad \text{add 1 to each part of the inequality}$$

$$-3 < x < 5. \qquad \text{addition}$$

Thus the solution set to $|x - 1| < 4$ is $\{x | -3 < x < 5\}$. The graph of this solution set is shown in Figure 2.17.

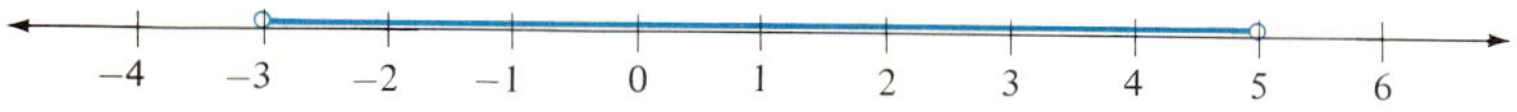

Figure 2.17

Example 2. Graph the solution set of $|2z - 3| \le 5$.

Solution: Proceeding as we outlined in the preceding discussion, we have

$$-5 \le 2z - 3 \le 5$$

$$-5 + 3 \le (2z - 3) + 3 \le 5 + 3 \qquad \text{add 3 to each part of the inequality}$$

$$-2 \le 2z \le 8$$

$$-1 \le z \le 4. \qquad \text{multiply by } \tfrac{1}{2}$$

Thus the solution set of $|2z - 3| \le 5$ is $\{z | -1 \le z \le 4\}$. The graph of this set is shown in Figure 2.18.

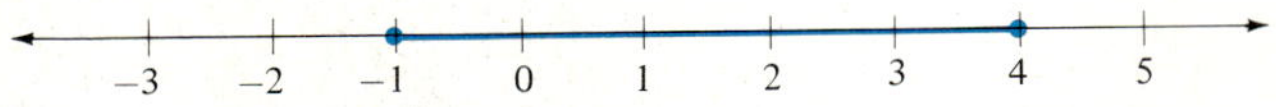

Figure 2.18

Example 3. Solve the inequality $|3q + 2| > 8$ and graph the solution set.

Solution: First we let $x = 3q + 2$ and get $|x| > 8$, which is equivalent to the compound inequality $x < -8$ or $x > 8$. Replacing x with $3q + 2$, we have

$$3q + 2 < -8 \quad \text{or} \quad 3q + 2 > 8.$$

We solve these as follows:

$$3q + 2 < -8 \quad \text{or} \quad 3q + 2 > 8$$

$$(3q + 2) - 2 < -8 - 2 \quad \text{or} \quad (3q + 2) - 2 > 8 - 2 \qquad \text{subtract 2 from each side of each inequality}$$

$$3q < -10 \quad \text{or} \quad 3q > 6 \qquad \text{simplify}$$

$$q < -\frac{10}{3} \quad \text{or} \quad q > 2. \qquad \text{divide by 3}$$

The graph of the solution set $\{q \mid q < -\frac{10}{3} \text{ or } q > 2\}$ is shown in Figure 2.19.

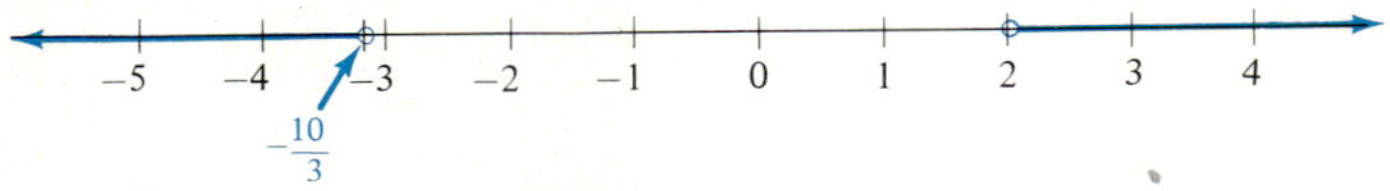

Figure 2.19

Examples 2 and 3 suggest the following theorem which is equivalent to Theorem 6.

Theorem 7

If $a \ne 0$ and c is positive, then

(a) $|ax + b| \le c$ is equivalent to $-c \le ax + b \le c$.

(b) $|ax + b| \ge c$ is equivalent to $ax + b \le -c$ or $ax + b \ge c$.

It is often helpful to note that using the geometric intrepretation of absolute value, we may read $|x - b| \le c$ as "the values of x that are *at most* c units from b on the number line." Using this interpretation, we can graph the solution set immediately as

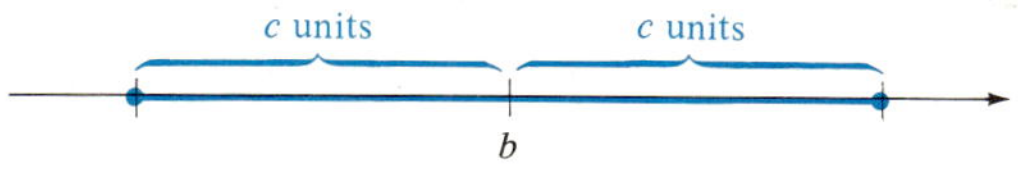

Figure 2.20

shown in Figure 2.20. In a similar manner, the inequality $|ax - b| > c$ may be read "the values of x such that the distance between ax and b on the number line is *greater than* c units."

Table 2.1 summarizes the results of this section. In the table, $a \neq 0$ and $c > 0$.

Table 2.1

Absolute Value Inequality	Equivalent Compound Inequality	English Interpretation
$\|x\| \le c$	$-c \le x \le c$	All numbers x that are less than or equal to c units from 0
$\|x\| \ge c$	$x \le -c$ or $x \ge c$	All numbers x that are c or more than c units from 0
$\|ax - b\| \le c$	$-c \le ax - b \le c$	All values of x such that the distance between ax and b is less than or equal to c
$\|ax - b\| \ge c$	$ax - b \le -c$ or $ax - b \ge c$	All values of x such that the distance between ax and b is greater than or equal to c

Example 4. Give an English interpretation for each of the following absolute value inequalities.
(a) $|7x - 5| < 2$. (b) $|3x + 2| > 5$.

Solution

(a) "All numbers x such that the distance between $7x$ and 5 is less than 2 units."
(b) First note that the given inequality may be written as $|3x - (-2)| > 5$. Thus an interpretation is: "all numbers x such that the distance between $3x$ and -2 is greater than 5 units."

Sometimes an absolute value inequality has no solution. It is also possible that every real number is a solution to an absolute value inequality. The following example illustrates these possibilities.

Example 5. Solve the following inequalities.
(a) $|4p + 2| \le -3$. (b) $|3q - 4| \ge -5$.

Solution

(a) The left-hand side of $|4p + 2| \le -3$ is always positive or zero. Thus no value assigned to p will result in a number that is less than or equal to -3. Thus the solution set is the empty set, { }.

(b) The expression $|3q - 4|$ is always nonnegative. That is, $|3q - 4| \ge 0$ for all values of q. But -5 is less than 0 or equivalently $0 > -5$. We conclude that $|3q - 4| > -5$ for all values of q and therefore the solution set is $\mathbb{R}$.

EXERCISES 2.6

For Exercises 1–12, solve the absolute value inequality and graph the solution set. See Example 1.

1. $|x| \le 5$ **2.** $|y| \le 2$ **3.** $|w| < 1$ **4.** $|z| < 6$

5. $|p| \ge 2$ **6.** $|q| \ge 1$ **7.** $|r| > 5$ **8.** $|s| > 4$

9. $|x| + 2 \le 3$ **10.** $|y| + 1 < 4$ **11.** $|w| - 1 > 7$ **12.** $|z| - 2 \ge 4$

For Exercises 13–36, solve the inequality and graph the solution set. See Examples 2 and 3.

13. $|x - 2| < 1$ **14.** $|y - 1| < 1$ **15.** $|z + 1| < 5$ **16.** $|w + 2| < 3$

17. $|p + 3| \le 4$ **18.** $|q + 4| \le 2$ **19.** $|r - 5| \le 6$ **20.** $|s - 3| \le 7$

21. $|t + 1| > 2$ **22.** $|u + 2| \ge 3$ **23.** $|v - 3| \ge 5$ **24.** $|w - 1| \ge 2$

25. $|2x + 1| \le 3$ **26.** $|2y + 3| \le 5$ **27.** $|3x - 1| < 7$ **28.** $|3z - 1| < 4$

29. $|5p + 1| < 1$ **30.** $|4q + 3| < 1$ **31.** $|6w - 5| \le 7$ **32.** $|7z - 4| \le 3$

33. $|4x + 3| \ge 9$ **34.** $|4y + 5| \ge 7$ **35.** $|7w - 5| > 9$ **36.** $|5z - 3| > 2$

For Exercises 37–44, write an English interpretation for the absolute value inequality. See Example 4 and Table 2.1.

37. $|x - 2| \le 4$ **38.** $|y - 1| < 3$ **39.** $|w + 1| > 2$ **40.** $|z + 3| \ge 5$

41. $|2w - 5| < 1$ **42.** $|3z - 5| \le 4$ **43.** $|4p + 5| \ge 7$ **44.** $|4q + 3| > 5$

For Exercises 45–48, write an absolute value inequality that corresponds to the statement. See Example 4 and Table 2.1.

45. All numbers y such that the distance between y and 4 is less than 3 units.

46. All numbers z such that the distance between z and 3 is less than or equal to 11 units.

47. All numbers x such that the distance between $3x$ and 1 is greater than or equal to 2 units.

48. All numbers y such that the distance between $5y$ and 2 is greater than 7 units.

For Exercises 49–52, determine the solution set. See Example 5.

49. $|x| > -1$ **50.** $|y + 1| \ge -2$ **51.** $|3w + 1| \le -4$ **52.** $|4z - 3| < -1$

For Exercises 53–60, use the methods of this section to solve the inequality and then graph the solution set. Remember that if both sides of an inequality are multiplied by a negative number, the direction of the inequality is reversed. It is also helpful to note that $|a - b| = |b - a|$.

53. $|2 - x| \le 3$ **54.** $|3 - y| < 5$ **55.** $|5 - 2y| < 1$ **56.** $|3 - 2z| \le 5$

57. $|1 - w| > 1$ **58.** $|5 - z| \ge 2$ **59.** $|4 - 5p| \ge 1$ **60.** $|3 - 4q| > 5$

Applications 2.7

In this section we illustrate how to translate word problems into mathematical expressions. The ability to do this is very important, particularly in management science and in the traditional sciences. As discussed in Chapter 1, the following table contains some common English phrases, together with their corresponding mathematical expression.

Table 2.2

English Phrase	Mathematical Expression
The sum of a number and 3	$x + 3$
5 more than twice a number	$2x + 5$
2 less than three times a number	$3y - 2$
The difference of x and y	$x - y$
The product of two numbers	xy
The product of 6 and y is 12	$6y = 12$
The quotient of x by y	$\frac{x}{y}$
Two-thirds a number less 4	$\frac{2}{3}x - 4$
The sum of twice x and 11	$2x + 11$
Twice the sum of x and 11	$2(x + 11)$

Although there is no specific method for translating verbal problems into mathematical expressions, the following steps provide a useful guide for dealing with word problems.

Step 1. Read the problem carefully in order to understand the situation that is being described. You may have to read the problem several times.

Step 2. If appropriate, use sketches, figures, charts, or tables.

Step 3. Identify what values are known and what values are to be found (unknown). Represent the unknown value by an appropriate symbol (variable).

Step 4. If there is more than one unknown, try to express them all in terms of one variable.

Step 5. Write an equation involving the unknown that represents the information given in the problem.

Step 6. Solve for the unknown value and check your answer in the original problem.

The following examples illustrate the general procedures outlined above.

Example 1. The length of a rectangle is 3 feet more than twice its width. The perimeter of the rectangle is 42 feet. What are the dimensions of the rectangle?

Solution: Because the problem involves a geometric figure, it is helpful to sketch and label a figure. From the problem as given we know that we have to find the length and width of the rectangle. If we let W denote the width, then $2W + 3$ represents the length, L, as shown in Figure 2.21. The formula for the perimeter of

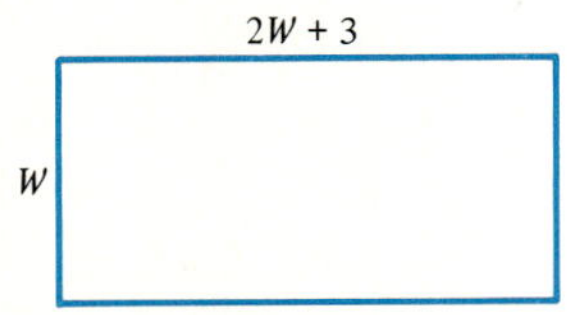

Figure 2.21

a rectangle is $P = 2W + 2L$. But we are given that $P = 42$ and $L = 2W + 3$. If we replace P with 42 and L with $2W + 3$ in

$$P = 2W + 2L,$$

we obtain

$$\begin{aligned} 42 &= 2W + 2(2W + 3) \\ 42 &= 2W + 4W + 6 \\ 42 &= 6W + 6 \\ 36 &= 6W && \text{subtract 6 from each side} \\ 6 &= W. && \text{divide each side by 6} \end{aligned}$$

The width W is 6 feet and the length $L = 2W + 3 = 2(6) + 3 = 15$ feet. We can check this by noting that $P = 2(6) + 2(15) = 42$, as required.

Example 2. After 7 is subtracted from 8 times a number, the difference is divided by 3. If the resulting quotient is 11, what was the original number?

Solution: Let x denote the number. Then "7 subtracted from 8 times the number" is represented by $8x - 7$. "The difference is divided by 3" is $\frac{8x - 7}{3}$.

Finally, the statement "the resulting quotient is 11" gives us the equation

$$\frac{8x - 7}{3} = 11.$$

Solving this equation, we get

$$\begin{aligned} \frac{8x - 7}{3} &= 11 & \\ 8x - 7 &= 33 & \text{multiply each side by 3} \\ 8x &= 40 & \text{add 7 to each side} \\ x &= 5. & \text{divide each side by 8} \end{aligned}$$

Thus the number is 5. Checking 5 in the original problem, we see that

$$\frac{8(5) - 7}{3} = \frac{33}{3} = 11,$$

as required.

Example 3. Find three consecutive positive integers such that the second is 12 times the difference of the third and first.

Solution: Let x denote the first integer. Consecutive integers such as 3, 4, 5 or 18, 19, 20 differ by 1. Thus we have $x + 1$ and $x + 2$ as the second and third consecutive integers. We may then write

$$\underbrace{x + 1}_{\text{second integer}} \underbrace{=}_{\text{is}} \underbrace{12 \cdot}_{\text{12 times}} \underbrace{[(x + 2) - x]}_{\text{difference of third and first integers}}.$$

Solving this equation, we get

$$\begin{aligned} x + 1 &= 12[x + 2 - x] \\ x + 1 &= 12(2) \\ x + 1 &= 24 \\ x &= 23. \end{aligned}$$

If 23 is the first number, 23, 24, and 25 are the three numbers in question.

CHECK: $24 = 12(25 - 23) = 12 \cdot 2 = 24$.

Example 4. Pat has invested a total of \$30,000 in two different accounts. Part of the money went into a long-term CD (certificate of deposit) that pays interest at the rate of 12% per year. The remaining money was invested in a utility stock fund. Last year the stock fund paid a 10% dividend. If Pat's total income from these two investments was \$3320, how much did she invest in each account?

Solution: Let x represent the amount invested at 12%. The remaining amount is represented by $30{,}000 - x$. This is invested at 10%. Using the formula $i = Prt$ with $t = 1$ (one year), we compute the interest earned in each account. Note that in the formula we replace P with x. Also 12% means $\frac{12}{100} = .12$ and 10% means $\frac{10}{100} = .10$. Thus we have

$$\begin{aligned} x(.12)(1) &= \text{interest earned at 12\% in 1 year} \\ (30{,}000 - x)(.10)(1) &= \text{interest earned at 10\% in 1 year} \end{aligned}$$

Since the total interest income was \$3320, we may write the following equation:

$$\underbrace{\text{total income}}_{3320} \;\overset{\text{is}}{=}\; \underbrace{\text{interest at 12\%}}_{x(.12)} \;\overset{\text{plus}}{+}\; \underbrace{\text{interest at 10\%}}_{(30{,}000 - x)(.10).}$$

Solving this equation, we get

$$\begin{aligned} 3320 &= .12x + (30{,}000)(.10) - .10x \\ 3320 &= .02x + 3000 \\ 320 &= .02x \\ 16{,}000 &= x. \end{aligned}$$

Thus Pat invested \$16,000 at 12% and \$14,000 at 10%. It is easy to check that these amounts produce a total investment income of \$3320.

The formula $d = rt$ is used to compute distance (d), given a rate (r) and time (t). We stated this formula in Section 2.3. We could solve this formula for time t and get $t = \frac{d}{r}$, or for rate r and get $r = \frac{d}{t}$. Problems that involve distance, rate, and time are often encountered in applied situations. The next example illustrates a typical method of solution for such problems.

Example 5. A group of tourists rents a steamboat for a 4-hour river party. The speed of the boat in still water is 16 miles per hour and the speed of the river's current is 4 miles per hour. How far upstream should the boat go if the round trip is to take exactly 4 hours?

Solution: Let d represent the distance that the boat will travel upstream. Because the speed of the river is 4 miles per hour, we see that the upstream speed of the boat is decreased by 4 miles per hour. Thus the upstream speed of the boat is $16 - 4 = 12$ miles per hour. Similarly, when the boat is going downstream, its speed is increased by the speed of the current. Thus the downstream speed of the boat is $16 + 4 = 20$ miles per hour. Recall that time $= \frac{\text{distance}}{\text{rate}}$ or $t = \frac{d}{r}$. Since the total time for the trip is 4 hours, we may write the following equation:

$$\underbrace{\text{time upstream}}_{\frac{d}{12}} \ \underset{+}{\overset{\text{plus}}{\downarrow}} \ \underbrace{\text{time downstream}}_{\frac{d}{20}} \ \underset{=}{\overset{\text{equals}}{\downarrow}} \ \underbrace{\text{total time}}_{4.}$$

To solve this equation, we first clear fractions by multiplying each side of the equation by 60. Thus we have

$$\begin{aligned} 60\left(\frac{d}{12} + \frac{d}{20}\right) &= 60(4) \\ 5d + 3d &= 240 \\ 8d &= 240 \\ d &= 30. \end{aligned}$$

Therefore, the boat should go 30 miles upstream.

CHECK

$$\text{time upstream} = \frac{\text{distance}}{\text{rate}} = \frac{30}{12} = 2\frac{1}{2} \text{ hours}$$

$$\text{time downstream} = \frac{\text{distance}}{\text{rate}} = \frac{30}{20} = 1\frac{1}{2} \text{ hours}$$

$$\text{total time} = 2\frac{1}{2} + 1\frac{1}{2} = 4 \text{ hours.}$$

Example 6. A mechanic has 16 quarts of a 50% antifreeze solution. How much of a 90% antifreeze solution must he add to get an 80% solution?

Solution: Recognizing that the total amount of pure antifreeze in the final solution is the sum of the amounts of antifreeze in the 50% solution and the 90% solution, we have the following diagram. In the diagram, x denotes the number of quarts of the 90% solution that are required to make an 80% solution.

quarts of solution	16 quarts of 50% solution	+	x quarts of 90% solution	=	$(x + 16)$ quarts of 80% solution
quarts of antifreeze	$(.50(16) = 8)$	+	$.90x$	=	$.80(x + 16)$

It follows that

$$8 + .90x = .80(x + 16).$$

Thus

$$8 + .90x = .80x + 12.8$$
$$.10x = 4.8$$
$$x = 48 \text{ quarts of the 90\% solution.}$$

CHECK: $(.50)16 + (.90)(48) = 8 + 43.2 = 51.2$ quarts of antifreeze in $16 + 48 = 64$ quarts of solution. Note that 80% of 64 is 51.2.

EXERCISES 2.7

Exercises 1–12 involve geometric figures. Draw a figure and if necessary refer to the inside of the front cover for an appropriate formula. See Example 1.

1. The length of a rectangle is 5 feet more than four times the width. If the perimeter is 100 feet, what are the dimensions of the rectangle?
2. The length of a rectangle is 1 inch more than five times the width. If the perimeter is 110 inches, what are the dimensions of the rectangle?
3. The perimeter of a rectangle is 400 meters. The width is 1 meter less than one-half the length. What are the dimensions of the rectangle?
4. The perimeter of a rectangle is 80 centimeters. The width is 5 centimeters less than one-half the length. What are the dimensions of the rectangle?
5. The longest side of a triangle is 2 inches shorter than three times the length of the shortest side. The third side is 1 inch more than twice the length of the shortest side. If the perimeter is 29 inches, what are the lengths of the three sides?
6. The longest side of a triangle is 5 feet longer than twice the length of the shortest side. The third side is 1 foot longer than twice the shortest side. If the perimeter is 56 feet, what are the lengths of the three sides?
7. The sides of a triangle are labeled a, b, and c. The length of side a is two-thirds the length of the longest side b. The length of the third side c is 4 centimeters less than the length of side a. Find the lengths of the three sides if the perimeter of the triangle is 45 centimeters.
8. The length of the shortest side of a triangle is one-half the length of the longest side. The length of the third side is 5 millimeters more than the length of the shortest side. Find the lengths of the three sides if the perimeter is 65 millimeters.
9. The longest side of a triangle is twice the length of the shortest side. The third side of the triangle is 3 feet longer than the shortest side. Find the length of the shortest side if the perimeter of the triangle is 31 feet.
10. The perimeter of a triangle is 78 feet. The shortest side is half as long as the longest side. The length of the other side is two-thirds the length of the longest side. What is the length of the longest side of the triangle?
11. The perimeter of the isosceles trapezoid shown is 36 feet. Each of the two nonparallel sides has length 5

feet as shown. The length of the upper base is 2 feet longer than twice the height. The length of the lower base is four times the height. What is the height?

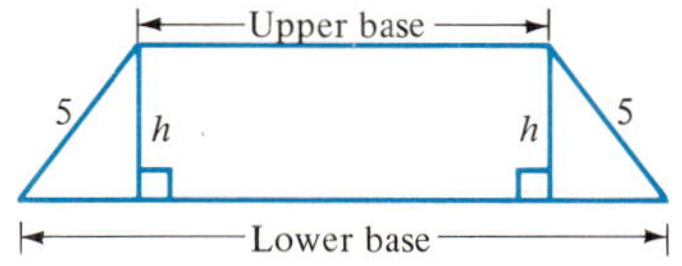

12. The perimeter of the isosceles trapezoid shown is 84 millimeters. The length of each of the two nonparallel sides is 13 millimeters as shown. The length of the upper base is twice the height. The length of the lower base is 2 millimeters less than three times the height. What is the height?

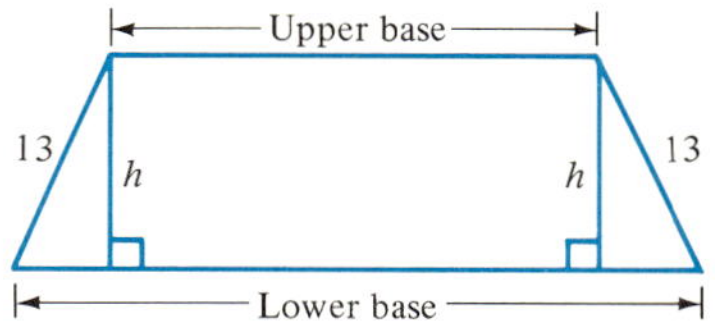

Solve Exercises 13–18. See Example 2.

13. After 5 is subtracted from a number, the difference is multiplied by 7. The resulting product is 42. Find the number.

14. After 9 is added to a number the sum is multiplied by 5. The resulting product is 110. Find the number.

15. The sum of twice a number and 5 is 2 less than three times the number. Find the number.

16. The sum of four times a number and 2 is the same as 11 more than the number. Find the number.

17. After 9 is subtracted from 6 times a number, the difference is divided by 3. If the resulting quotient is -17, what was the original number?

18. After 20 is added to seven times a number, the sum is divided by 3. If the resulting quotient is 2, what was the original number?

Solve Exercises 19–24. See Example 3.

19. Find two consecutive integers with a sum of 47.

20. Find two consecutive integers with a sum of 55.

21. Find three consecutive integers with a sum of 186.

22. Find three consecutive integers with a sum of 264.

23. Find three consecutive positive integers such that the second is nine times the difference of the third and first.

24. Find three consecutive positive integers such that the second is 26 times the difference of the third and first.

Solve Exercises 25–30. See Example 4.

25. Don has invested \$12,000 that he won in a state lottery in two accounts. One account, a short-term CD, pays interest at the annual rate of 9%. The other account is a long-term CD that has an annual interest rate of 11%. If Don's total income from these two accounts is \$1160, how much is invested in each account?

26. Juan has invested \$15,000 in two accounts. One account is an ordinary savings account with an annual interest rate of 5%. The other is a long-term CD that pays interest at the rate of 12% per year. If Juan's total income from these two accounts was \$1590, how much did he invest in each account?

27. Yvonne has invested a certain amount of money at 10% and \$4000 more than this amount at 8%. If her total income from these two investments is \$1400, how much did she invest at each rate?

28. Carla invested a certain amount of money at 12% and \$5000 less than this amount at 8%. If her total interest income from these two investments is \$3200, how much did she invest at each rate?

29. The amount of \$8000 is invested in an account that pays interest at 9%. How much more money must be invested at 12% in order to have a 10% rate of return on the total amount of money invested?

30. The amount of \$5000 is invested at 8%. How much more money must be invested at 11% in order to have a 10% rate of return on the total amount of money invested?

Solve Exercises 31–36. See Example 5.

31. Jose has rented a canoe for 5 hours. He and his friend can paddle upstream at a rate of 2 miles per hour and downstream at a rate of 8 miles per hour. How far upstream should they paddle if they want to be back in exactly 5 hours?

32. Tina has rented a rowboat for 3 hours. There is a steady current in the river with a rate of 1 mile per hour. In still water Tina can row at a rate of 3 miles per hour. How far should she row against the current before turning around if she wants to be back in 3 hours?

33. A south-bound freight train crosses the Newark city line at 9 A.M. and maintains a constant speed of 40 miles per hour. At 11 A.M. a south-bound passenger train crosses the same city line on a track that parallels the track of the freight train. If the passenger train maintains a constant speed of 60 miles per hour, how long will it take it to catch the freight train?

34. A tortoise and a hare are having a race. The tortoise can travel at a constant rate of $\frac{1}{4}$ kilometer per hour (kph). The hare can run at a constant rate of $2\frac{1}{4}$ kph. If the tortoise is given an 8-hour head start, how long will it take the hare to catch up?

35. Julianne drove 200 miles in 4 hours last weekend. For the first 3 hours of her trip she maintained a steady speed. However, the last hour she had to reduce her car's speed by 20 miles per hour because of construction. What was her car's speed during the first 3 hours?

36. Jennifer piloted her small plane a distance of 400 miles in 3 hours. The first 2 hours she maintained a steady cruising speed. However, because of head winds, the plane's speed was reduced by 20 miles per hour during the last hour. What was the plane's cruising speed during the first 2 hours?

Exercises 37 and 38 involve levers. Relative to a fulcrum, a lever is balanced when the sum of all the forces times their corresponding distances from the fulcrum on one side of the fulcrum equals the sum of all the forces times their corresponding distances from the fulcrum on the other side of the fulcrum as shown.

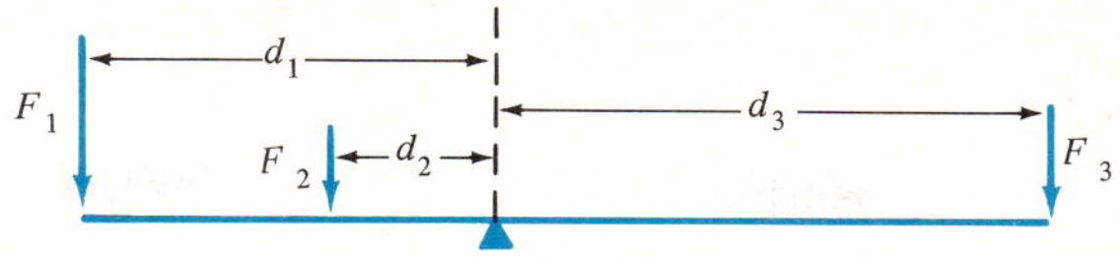

$$F_1d_1 + F_2d_2 = F_3d_3$$

37. A fulcrum is placed 40 inches from one end of a 100-inch steel rod. An 80-pound weight is placed at the short end of the rod and a 20-pound weight is placed at the long end of the rod. How far from the fulcrum should a weight of 40 pounds be placed in order to balance the rod?

38. A weight of 120 pounds is placed at one end of a 70-inch lever and a weight of 90 pounds is placed at the other end. To balance the lever, where should the fulcrum be placed?

Solve Exercises 39–44. See Example 6.

39. A chemist has 250 milliliters of a 10% acid solution and 300 milliliters of a 50% solution of the same acid. How much of the 50% solution must be added to the 10% solution to obtain a 30% solution?

40. A laboratory technician has 200 milliliters of a 5% alcohol solution and 750 milliliters of a 60% solution of the same alcohol. How much of the 60% solution must be added to the 5% solution to obtain a 20% solution.

41. A 20-quart radiator is filled with a 50% solution of antifreeze. How much must be drained and then replaced with pure antifreeze to get a 60% solution?

42. A 30-quart container is filled with a 40% solution of antifreeze. How much of the solution must be removed and then replaced with pure antifreeze to get a 60% solution?

43. A biologist has a container with 500 milliliters of a 5% saline solution and 100 milliliters of a 15% saline solution. How much of the 5% solution must be added to the 15% solution to obtain a 10% solution?

44. A young botanist has a 500-milliliter beaker with 100 milliliters of a 20% acid solution. She also has available 600 milliliters of a 5% solution of the same acid. How much of the weaker solution must be added to the 20% solution to get a 10% solution?

For Exercises 45–60, read the question carefully and, if appropriate, draw a figure or a diagram. Next write and solve an appropriate equation.

45. Hector is 3 years older than his sister Rosita. In 5 years the sum of their ages will be 29. How old is Hector?

46. Valerie is 4 years younger than her cousin Marilyn. In 10 years the sum of their ages will be 42. How old is Valerie?

47. Norman's age is $\frac{1}{6}$ the age of his mother. In 20 years he will be $\frac{1}{2}$ as old as his mother. How old is Norman?

48. Loretta is $\frac{1}{4}$ as old as her aunt. In 10 years she will be $\frac{2}{5}$ as old as her aunt. How old is Loretta?

49. Bob's age is $\frac{8}{7}$ of his wife's age. When they were married 10 years earlier, his age was $\frac{6}{5}$ her age. How old is Bob?

50. Mary's age is $\frac{7}{8}$ of her husband's age. When they were married 8 years earlier, her age was $\frac{5}{6}$ his age. How old is Mary?

51. A computer company charges \$6.50 per hour for connect time and \$.70 per CPU (central processing unit) second. If Jim's program cost \$13.25 and used 5 seconds of CPU time, how much connect time was used?

52. A computer company charges \$4.50 per hour for connect time and \$.80 per CPU second. Janet's program used 5 CPU seconds and had a total cost of \$15.25. How much connect time did she use?

53. The City Taxi Company has an initial charge of \$1.00 plus \$.20 for each $\frac{1}{9}$ mile. If you are charged \$4.60 for a taxi ride, how many miles was the ride?

54. The Yellow Cab taxi company has an initial charge of \$1.25 plus \$.20 for each $\frac{1}{6}$ mile. If you are charged \$3.05 for a ride, how long was the ride?

55. A collection of dimes and nickels contains five fewer nickels than twice the number of dimes. If the total face value of the coins is \$8.75, how many dimes are in the collection?

56. A jar contains \$7.00 worth of pennies and nickels. If the number of nickels is 20 more than the number of pennies, how many pennies are in the jar?

57. Bob weighs 50 pounds more than Ted, who weighs 30 pounds more than Alice. If their combined weight is 455 pounds, how much does Ted weigh?

58. Matthew's weight is $\frac{1}{4}$ that of his father's weight. His mother's weight is 15 pounds more than twice his weight. The total of their three weights is 400 pounds. How much does Matthew weigh?

59. Instead of adding enough vinegar to a 50-ounce mixture of salad dressing to obtain a mixture that is 20% vinegar, a student chef added 20 ounces of vinegar. How much more of the salad dressing mixture should be added to obtain a 20% solution of vinegar, as intended?

60. A gravy recipe calls for 25% skim milk. By mistake, Ken added 25 ounces of skim milk to 32 ounces of beef stock. How much more of the beef stock is needed to get a 25% skim milk content, as intended?

Key Words

Absolute value
Combining like terms
Equivalent equations
First-degree equation
First-degree inequality
Formula
Like terms
Numerical coefficient
Simplified
Solution
Solution set
Term

CHAPTER 2 REVIEW

SECTION 2.1

Combine the like term in each expression.

1. $13w - 8w + 7w$

2. $7x + 3 - x + 2 + 4x - 1$

3. $3(y + 5) - 2(y + 1) + 2$

4. $\frac{3}{4}(z + 2) + \frac{1}{4}(z - 2)$

SECTION 2.2

Solve the equations.

5. $9x - 11 = 7$ **6.** $\frac{4}{5}y = 8$ **7.** $3w - 2 + 4w = 2w + 13$

8. $4(z - 2) + 3(z + 1) = 2$ **9.** $2u - 3(2u - 5) = 3(u - 2)$ **10.** $\frac{2}{3}v - \frac{2}{5} = \frac{2}{5}v + \frac{2}{3}$

SECTION 2.3

Refer to the inside of the front cover for the appropriate formula and answer the questions.

11. Find the area and perimeter of a rectangle that is 18 inches long and 7 inches wide.

12. Find the area and circumference of a circle with radius 7 feet. Use 3.14 as an approximation for π.

13. How many kilometers will you travel in $4\frac{1}{3}$ hours if you drive at 72 kilometers per hour?

14. Find the area and perimeter of the triangle shown. The lengths are measured in inches.

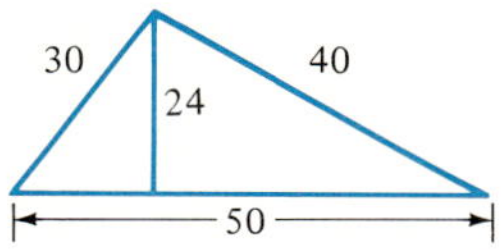

Solve each formula for the variable indicated.

15. $i = Prt$ for t **16.** $S = \frac{n}{2}(f + e)$ for e

SECTION 2.4

Solve the equations.

17. $|x| = 11$ **18.** $|y| = -2$ **19.** $|w| = 0$ **20.** $|z + 3| = 7$

21. $|2z + 5| = 13$ **22.** $|7 - 4t| = 9$ **23.** $|u + 3| = |u - 7|$ **24.** $|v - 5| = |2v + 7|$

SECTION 2.5

Solve the inequalities and graph the solution set.

25. $x + 7 \geq 11$ **26.** $3y < -12$ **27.** $-2w < 6$

28. $3(4z + 1) \leq 15$ **29.** $2p + 3 \geq -5$ and $p + 1 < 3$ **30.** $q + 3 \leq 8$ or $q - 1 \geq 7$

SECTION 2.6

Solve each absolute value inequality and graph the solution set.

31. $|x + 2| < 5$ **32.** $|3y + 2| \leq 2$ **33.** $|z + 5| \geq 5$ **34.** $|3w - 1| > 4$

SECTION 2.7

Solve each word problem.

35. The perimeter of a rectangle is 126 inches. The width is three-fourths of the length. What are the dimensions of the rectangle?

36. Find three consecutive integers with a sum of 222.

37. Raul has invested \$4000 at 7%. How much more money must he invest at 10% to have an 8% rate of return on the total amount of money invested?

38. Donna drove her van 150 miles in 3 hours. During the first 2 hours she maintained a steady speed. However, because of construction, she had to reduce her speed by 15 miles per hour during the last hour. What was the speed of the van during the first 2 hours?

39. A 16-quart radiator is filled with a 25% solution of antifreeze. How much of the solution must be drained and then replaced with pure antifreeze in order to get a 40% solution?

40. A computer company charges \$5.50 per hour for connect time and \$.70 per CPU second. Tina's program cost \$8.60 and used $\frac{1}{2}$ second of CPU time. How much connect time did she use?

CHAPTER 2 TEST

For Problems 1–4, simplify and solve the equation.

1. $5x - 2 - x = x + 16$

2. $\frac{3}{7}y = 9$

3. $4(z + 1) + 2z = 7(z - 1) + 10$

4. $\frac{1}{2}w - \frac{2}{3} = \frac{1}{3}w + \frac{2}{3}$

Solve Problems 5 and 6.

5. Find the area and perimeter of a rectangle that is 24 inches long and 10 inches wide.

6. How long does it take to travel 140 miles at 40 miles per hour?

For Problems 7 and 8, solve the equation.

7. $|z| = 11$

8. $|2x - 5| = 7$

For Problems 9–14, solve the inequality and graph the solution set.

9. $4x - 5 \leq 7$

10. $-3y + 1 > 7$

11. $z + 2 < 5$ and $z + 1 \geq 2$

12. $3(w + 1) \leq 6$ or $5 - w < 1$

13. $|4x + 3| \leq 1$

14. $|5q - 1| > 2$

Solve Problems 15–18.

15. The perimeter of a rectangle is 500 inches. The width is two-thirds the length. What are the dimensions of the rectangle?

16. Find three consecutive integers with a sum of 276.

17. Fred has rented a rowboat for 2 hours. There is a steady current of 1 mile per hour on the river and Fred can row at a rate of 3 miles per hour in still water. If he wants to return in 2 hours, how far should he row upstream before turning around?

18. A chemist has 400 milliliters of a 15% saline solution and 700 milliliters of a 40% saline solution. How much of the 40% solution should be added to the 15% solution to obtain a 20% solution?

CHAPTER 3

Exponents and Polynomials

Exponents are a useful notational device in mathematics and science. For example, in order to measure the immense distances of stars from the Earth, astronomers use different units of measure such as light years and parsecs. Approximations for these units in miles are

1 light year = 6000000000000 miles
1 parsec = 19200000000000 miles.

If we use exponents, we can write these units of measure in miles in a compact form as

1 light year $= 6 \times 10^{12}$ **miles**
1 parsec $= 1.92 \times 10^{13}$ **miles.**

Astronomer adjusts the spectrograph entrance slit at the coude focus of the 2.1-meter (84-inch) telescope in preparation for beginning an observation. This instrument is used to obtain high resolution spectra of starlight.

Mathematical symbolism should be concise so that it is visible at one glance of the eye and can be written quickly. For example, we may quickly write down the formula

$$s = 16t^2.$$

This formula, written down so compactly, contains a great deal of information. For objects near the surface of the earth, it enables us to compute the distance, s, that a freely falling object has fallen from rest in t seconds if we neglect air resistance. Thus, if we drop a steel ball from a tower, we will be able to compute, with the aid of this formula, the distance it has fallen after t seconds. The integer 2 in the expression $16t^2$ is called an exponent. We will see that exponents are an excellent example of mathematical symbolism. In this chapter we will develop rules for working with integer exponents. The expression $16t^2$ by itself is an example of a monomial. Monomials are a particular type of algebraic expression called a polynomial. We discuss polynomials and use the rules for exponents in order to add, subtract, and multiply polynomials.

3.1 Fundamentals of Exponents

When we write a product such as $2 \cdot 3 \cdot 5 = 30$, we say that 2, 3, and 5 are *factors* of the product 30. A product with repeated factors such as $3 \cdot 3 \cdot 3 \cdot 3$ can be written quickly and compactly by representing it symbolically by the expression 3^4. Thus

$$3^4 = \underbrace{3 \cdot 3 \cdot 3 \cdot 3.}_{\text{four factors}}$$

Similarly, $3 \cdot 3 \cdot 3 \cdot 3 \cdot 3 \cdot 3 \cdot 3$ can be conveniently represented by 3^7. The positive integers 4 and 7 appearing in the expressions 3^4 and 3^7 are called *exponents*. The number 3 is called the *base*. The complete expression, such as 3^4, is called a *power*. We read 3^4 as "3 raised to the fourth power" or simply "3 to the fourth." Similarly, 3^7 is read as "3 to the seventh power." The expressions 3^4 and 3^7 are said to be in *exponent form*. When we multiply the repeated factors symbolized by 3^4, that is, as follows:

$$3^4 = 3 \cdot 3 \cdot 3 \cdot 3 = 81,$$

we say that we have *evaluated* 3^4. We also say that 3 *raised to the fourth power* is 81 or *the fourth power* of 3 is 81. Other examples of powers are

$$2^5 \quad \text{fifth power of 2} \qquad \text{and} \qquad (-4)^3 \quad \text{third power of } -4.$$

Note that $2^5 = 2 \cdot 2 \cdot 2 \cdot 2 \cdot 2 = 32$ and $(-4)^3 = (-4)(-4)(-4) = -64$. We note that variables may appear in exponent form. Thus we write x^3 to represent the product $x \cdot x \cdot x$. The discussion above illustrates the following definition.

nth Power of a

For any real number a and any positive integer n,

$$a^n = a \cdot a \cdot \cdots \cdot a \qquad n \text{ factors of } a$$

n is the *exponent*, a the *base*, and a^n is called the nth *power of a*.

We note that a^1 is the same as a. Special powers such as a^2 and a^3 are read as "a squared" and "a cubed," respectively.

Example 1. For each of the following exponent forms, identify the exponent and the base, and evaluate the expression.
(a) 6^3. (b) $(-4)^2$. (c) -4^2.

Solution

(a) The exponent is 3, the base is 6, and

$$6^3 = 6 \cdot 6 \cdot 6 = 216.$$

(b) The exponent is 2, the base is -4, and

$$(-4)^2 = (-4)(-4) = 16.$$

(c) The exponent is 2, the base is 4, and

$$-4^2 = -(4 \cdot 4) = -16.$$

In Example 1, note the difference between the values of

$$(-4)^2 = 16 \qquad \text{and} \qquad -4^2 = -16.$$

The reason for this difference in values is that in the first exponent form, with parentheses, the base is -4 and the exponent refers to -4. In the second exponent form without parentheses, -4^2, the base is 4 and the exponent 2 refers to the base 4. The negative sign is applied *after* performing the multiplication. In general, an exponent refers to the *one number* or *symbol* that immediately precedes it. Thus in the expression $3x^5$ the exponent 5 refers only to x. In other words, $3x^5 = 3 \cdot x \cdot x \cdot x \cdot x \cdot x$.

Example 2. Write each of the products using exponents.
(a) $2 \cdot 2 \cdot 2$. (b) $xxxyyyy$. (c) $(-a)(a)bbb$. (d) $aa - bb$.

Solution

(a) $2 \cdot 2 \cdot 2 = 2^3$.
(b) $xxxyyyy = x^3y^4$.
(c) $(-a)(a)bbb = -(a \cdot a)bbb = -a^2b^3$.
(d) $aa - bb = a^2 - b^2$.

We now obtain a rule for writing the product of two exponent forms with the same base as a single exponent form. For example, we see that

$$2^3 \cdot 2^4 = \underbrace{(2 \cdot 2 \cdot 2)}_{3\text{ factors}} \cdot \underbrace{(2 \cdot 2 \cdot 2 \cdot 2)}_{4\text{ factors}} = \underbrace{2 \cdot 2 \cdot 2 \cdot 2 \cdot 2 \cdot 2 \cdot 2}_{3+4=7\text{ factors}}$$
$$= 2^7.$$

Thus

$$2^3 \cdot 2^4 = 2^{3+4} = 2^7.$$

In general, if a is a real number, and m and n are positive integers,

$$a^m \cdot a^n = \underbrace{(a \cdot a \cdot a \cdot \cdots \cdot a)}_{m\text{ factors}} \cdot \underbrace{(a \cdot a \cdot a \cdot \cdots \cdot a)}_{n\text{ factors}}$$
$$= \underbrace{a \cdot a \cdot a \cdot \cdots \cdot a}_{m+n\text{ factors}}.$$

The discussion above implies the following property of exponents.

Property 1

> If a is any real number and m and n are positive integers, then
> $$a^m \cdot a^n = a^{m+n}.$$

We shall see at the end of this section that Property 1 is also valid for zero and negative integer exponents.

Example 3. Use Property 1 to simplify the following products.
(a) $5^3 \cdot 5^6$. (b) $(-7)^2(-7)^3$. (c) $(x^4y)(x^2y^3)$. (d) $(3a^2bc^3)(2a^3b^2c^4)$.

Solution

(a) $5^3 \cdot 5^6 = 5^{3+6} = 5^9$.

(b) $(-7)^2(-7)^3 = (-7)^{2+3} = (-7)^5$.

(c) Using the commutative and associative properties of multiplication as well as Property 1 for exponents, we have

$$\begin{aligned}(x^4y)(x^2y^3) &= (x^4x^2)(yy^3)\\ &= x^{4+2}y^{1+3} \qquad \text{note that } y = y^1\\ &= x^6y^4.\end{aligned}$$

(d) $$\begin{aligned}(3a^2bc^3)(2a^3b^2c^4) &= (3 \cdot 2)(a^2a^3)(bb^2)(c^3c^4)\\ &= 6a^{2+3}b^{1+2}c^{3+4}\\ &= 6a^5b^3c^7.\end{aligned}$$

Thus far we have considered only expressions with positive integer exponents. We will now show how to interpret an exponent form with an exponent equal to zero and exponent forms with negative integer exponents.

In the following discussion we assume that Property 1 for exponents holds for zero and negative integers. For example, consider

$$3^2 \cdot 3^0 = 3^{2+0} = 3^2$$

or

$$3^2 \cdot 3^0 = 3^2. \qquad \text{transitive property of equality}$$

Multiplying both sides of the last equation by $\frac{1}{3^2}$, we see that

$$\frac{1}{3^2} \cdot 3^2 \cdot 3^0 = \frac{1}{3^2} \cdot 3^2$$

$$1 \cdot 3^0 = 1 \qquad \text{since} \qquad \frac{1}{3^2} \cdot 3^2 = 1$$

or

$$3^0 = 1.$$

Thus it appears that 3 with a zero exponent is 1. More generally, if a is a real number, $a \neq 0$ and n is positive integer, we have

$$a^n \cdot a^0 = a^{n+0} = a^n$$

or

$$a^n \cdot a^0 = a^n \qquad \text{transitive property of equality}$$

$$\frac{1}{a^n} \cdot a^n \cdot a^0 = \frac{1}{a^n} \cdot a^n$$

$$a^0 = 1. \qquad \text{since} \qquad \frac{1}{a^n} \cdot a^n = 1$$

The previous discussion is the basis for the following definition.

Property 2. Zero Exponent

If a is any nonzero real number, then

$$a^0 = 1 \qquad (a \neq 0).$$

The expression 0^0 is undefined. Thus, in this text, all properties for exponents are stated with the understanding that the undefined expression 0^0 is excluded.

We now obtain an interpretation for expressions with negative exponents such as 3^{-4}. Our original definition of an exponent as a symbol indicating repeated factors does not appear to make sense in this situation. For how can we repeat 3 as a factor -4 times? To give a meaning to negative exponents, let us consider the following sequence of equalities in which we assume that Property 1 holds for negative exponents.

$$3^2 \cdot 3^{-2} = 3^{2+(-2)} = 3^0 = 1$$

Thus we have

$$3^2 \cdot 3^{-2} = 1.$$

Multiplying both sides of the last equation by the reciprocal of 3^2, $\frac{1}{3^2}$, we obtain

$$\left(\frac{1}{3^2} \cdot 3^2\right)3^{-2} = \frac{1}{3^2} \cdot 1$$

or

$$3^{-2} = \frac{1}{3^2}. \qquad \text{since} \quad \frac{1}{3^2} \cdot 3^2 = 1$$

Thus we see that 3^{-2} can be interpreted as the reciprocal of 3^2. In general, if a is a real number, $a \neq 0$, and n is a positive integer,

$$a^n \cdot a^{-n} = a^{n+(-n)} = a^0 = 1. \qquad \text{note that since } n \text{ is a positive integer, } -n \text{ is a negative integer}$$

Thus

$$a^n \cdot a^{-n} = 1. \qquad \text{transitive property of equality}$$

Multiplying both sides of the equation by the reciprocal of a^n, $\frac{1}{a^n}$, we obtain

$$\left(\frac{1}{a^n} \cdot a^n\right)a^{-n} = \frac{1}{a^n} \cdot 1$$

$$a^{-n} = \frac{1}{a^n}.$$

This discussion motivates the following definition.

Property 3. Negative Integer Exponents

If a is any nonzero real number and n is a positive integer, then

$$a^{-n} = \frac{1}{a^n} \qquad (a \neq 0).$$

With Properties 2 and 3 we see that Property 1,

$$a^m \cdot a^n = a^{m+n},$$

may be extended to *all* integer exponents (except 0^0). Following is a summary of the properties of exponents developed in this section.

Property 1:	$a^m \cdot a^n = a^{m+n}$	m and n are integers; a is a real number
Property 2:	$a^0 = 1$	a is real, $a \neq 0$
Property 3:	$a^{-n} = \dfrac{1}{a^n}$	n positive integer; a is real, $a \neq 0$

Example 4. Write the following expressions with positive exponents and evaluate when possible.

(a) 2^{-4}. (b) $(-5)^{-2}$.

(c) x^3x^{-5}. (d) $(y^{-2}z^{-3})(y^{-4}z^6)$.

Solution

(a) $2^{-4} = \dfrac{1}{2^4} = \dfrac{1}{16}$. note that $2^4 = 2 \cdot 2 \cdot 2 \cdot 2 = 16$

(b) $(-5)^{-2} = \dfrac{1}{(-5)^2} = \dfrac{1}{25}$.

(c) $x^3 \cdot x^{-5} = x^{3+(-5)} = x^{-2} = \dfrac{1}{x^2}$.

(d) $(y^{-2}z^{-3})(y^{-4}z^6) = y^{-6}z^3 = \dfrac{1}{y^6}z^3 = \dfrac{z^3}{y^6}$.

EXERCISES 3.1

For Exercises 1–10, (a) identify the base; (b) identify the exponent; and (c) evaluate the expression. See Example 1.

1. 2^3 **2.** 3^2 **3.** $(-5)^2$ **4.** $(-3)^3$ **5.** -10^4

6. -10^5 **7.** $(\frac{1}{3})^2$ **8.** $(\frac{1}{2})^3$ **9.** $-(-1)^4$ **10.** $-(1)^5$

For Exercises 11–30, rewrite the expression using exponents. See Example 2.

11. $9 \cdot 9 \cdot 9$ **12.** $8 \cdot 8 \cdot 8 \cdot 8$ **13.** $(-7)(-7)(-7)$

14. $(-10)(-10)(-10)(-10)$ **15.** xx **16.** yyy
17. $zzzz$ **18.** $aaaaa$ **19.** $rrsss$
20. $ppppqq$ **21.** $-2aabb$ **22.** $-2 \cdot 2yyyzz$
23. $(-b)(-b)ccc$ **24.** $(-r)(r)ss$ **25.** $m(-m)mnn$
26. $ww(-y)(y)$ **27.** $aaa + aab$ **28.** $ppp + pqq$
29. $5mm - mm \cdot nnn$ **30.** $6ppq - 6pqq$

For Exercises 31–50, simplify the product as far as possible as one exponent form. See Example 3.

31. $7^2 \cdot 7^8$ **32.** $5^4 \cdot 5^3$ **33.** $5^2 \cdot 5^3 \cdot 5^4$
34. $3^4 \cdot 3^3 \cdot 3^2$ **35.** $(-8)^2(-8)^3$ **36.** $(-11)^4(-11)^4$
37. $(\frac{1}{4})^3(\frac{1}{4})^5$ **38.** $(\frac{1}{3})^6(\frac{1}{3})^5$ **39.** $x^3 \cdot x^2$
40. $y^4 \cdot y^3$ **41.** $x^2 \cdot x^4 \cdot x$ **42.** $y^3 \cdot y^5 \cdot y$
43. $(x^5y^2)(x^3y^4)$ **44.** $(p^2q^3)(p^5q^2)$ **45.** $(m^3n^2)(m^4n^3)$
46. $(r^7s^3)(r^2s^5)$ **47.** $(-2x^3y^5)(3xy^2)$ **48.** $(-3a^4b^3)(4ab^3)$
49. $(a^3b^4c^5)(a^2b^3c^4)$ **50.** $(p^2q^4r^6)(pq^2r^3)$

For Exercises 51–58, write the expression with positive exponents and evaluate. See Example 4.

51. 7^0 **52.** 5^0 **53.** 5^{-3} **54.** 6^{-2}
55. $(-11)^0$ **56.** $(-13)^0$ **57.** $(-2)^{-5}$ **58.** $(-3)^{-3}$

For Exercises 59–74, simplify as much as possible. Write all answers with positive exponents when possible. See Example 4.

59. y^5y^{-2} **60.** $z^{-4}z^7$ **61.** t^4t^{-9} **62.** q^5q^{-7}
63. x^4x^{-6} **64.** $y^{-2}y$ **65.** $r^{-2}r^{-4}$ **66.** $s^{-3}s^{-5}$
67. t^2t^{-2} **68.** $x^{-3}x^3$ **69.** $(x^{-2}y^3)(x^3y^{-2})$ **70.** $(a^{-1}b^{-3})(a^2b^4)$
71. $(r^3s^{-2})(r^{-4}s^3)$ **72.** $(p^5q^{-3})(p^{-6}q^4)$ **73.** $(x^2y^{-1})(x^3y^2)^0$ **74.** $(m^{-1}n^2)^0(m^2n^3)$

3.2 Additional Properties of Exponents

In this section we discuss additional properties of exponents related to powers of products and quotients.

We see from Property 1 for exponents, and the commutative and associative properties for multiplication, that

$$(2 \cdot 3)^5 = \underbrace{(2 \cdot 3)(2 \cdot 3)(2 \cdot 3)(2 \cdot 3)(2 \cdot 3)}_{\text{5 factors of } (2 \cdot 3)} \qquad \text{definition of exponents}$$

$$= \underbrace{(2 \cdot 2 \cdot 2 \cdot 2 \cdot 2)}_{\text{5 factors of 2}}\underbrace{(3 \cdot 3 \cdot 3 \cdot 3 \cdot 3)}_{\text{5 factors of 3}} \qquad \text{commutative, associative properties}$$

$$= 2^5 \cdot 3^5. \qquad \text{Property 1 for exponents}$$

Thus we see that

$$(2 \cdot 3)^5 = 2^5 \cdot 3^5.$$

For negative integer exponents we observe that

$$(2 \cdot 3)^{-5} = \frac{1}{(2 \cdot 3)^5} \qquad \text{Property 3 for exponents}$$

$$= \frac{1}{2^5 3^5} \qquad \text{since } (2 \cdot 3)^5 = 2^5 \cdot 3^5$$

$$= \frac{1}{2^5} \cdot \frac{1}{3^5}$$

$$= 2^{-5} \cdot 3^{-5}. \qquad \text{Property 3 for exponents}$$

Therefore, we have

$$(2 \cdot 3)^{-5} = 2^{-5} \cdot 3^{-5}.$$

The previous discussion suggests Property 4 for exponents.

Property 4

> If a and b are real numbers and n is an integer, then
>
> $$(ab)^n = a^n b^n.$$

We note that by the symmetric property of equality, Property 4 may also be stated as follows:

$$a^n b^n = (ab)^n.$$

Example 1. Use Property 4 to evaluate or simplify.

(a) $5^4 \cdot 2^4$. (b) $2^5\left(\frac{1}{2}\right)^5$. (c) $(2x)^3$. (d) $(3xyz)^2$.

Solution

(a) We use the following form of Property 4: $a^n b^n = (ab)^n$. Applying this form to the given expression, $5^4 \cdot 2^4$, we see that

$$5^4 \cdot 2^4 = (5 \cdot 2)^4 = 10^4 = 10{,}000.$$

(b) Similarly, we see that

$$2^5\left(\frac{1}{2}\right)^5 = \left(2 \cdot \frac{1}{2}\right)^5 = 1^5 = 1.$$

(c) Using Property 4 in the form $(a \cdot b)^n = a^n \cdot b^n$, we have

$$(2x)^3 = 2^3x^3 = 8x^3. \qquad \text{note that } 2^3 = 8$$

(d) Although Property 4 is stated for two factors, it may be extended to expressions with more than two factors. Thus

$$(3xyz)^2 = 3^2 \cdot x^2y^2z^2 = 9x^2y^2z^2.$$

Next we consider a property of exponents that will enable us to write expressions such as $(a^2)^3$ with a single exponent. We note that

$$\begin{aligned}(a^2)^3 &= a^2 \cdot a^2 \cdot a^2 && \text{definition of exponent}\\ &= a^{2+2+2} && \text{Property 1}\\ &= a^{3(2)} && 2 + 2 + 2 = 3(2)\\ &= a^6.\end{aligned}$$

As a result, we see that

$$(a^2)^3 = a^{2(3)} = a^6.$$

We now state Property 5 for exponents, which is a generalization of the last result.

Property 5

If a is a real number and m and n are integers, then

$$(a^m)^n = a^{mn}.$$

Notice that Property 5 is valid for m and n if either one or both exponents are negative integers. We illustrate the application of Property 5 in the following example.

Example 2. Simplify the following expressions. Write all answers with positive exponents and when possible evaluate.

(a) $(2^4)^2$. (b) $(3^{-2})^2$. (c) $(x^2)^4$. (d) $(x^{-3})^5(y^{-4})^{-2}$.

Solution

(a) $(2^4)^2 = 2^{4(2)} = 2^8$. note that $2^8 = 256$

(b) $(3^{-2})^2 = 3^{(-2)(2)} = 3^{-4} = \dfrac{1}{3^4}$. note that $\dfrac{1}{3^4} = \dfrac{1}{81}$

(c) $(x^2)^4 = x^{2(4)} = x^8$.

(d) $(x^{-3})^5(y^{-4})^{-2} = x^{(-3)(5)}y^{(-4)(-2)} = x^{-15}y^8$

$$= \frac{y^8}{x^{15}}. \qquad \text{by Property 3, } x^{-15} = \frac{1}{x^{15}}$$

In the next example we use several of the properties of exponents discussed thus far to write expressions in a simpler form.

Example 3. Use the properties of exponents 1–5 to simplify the following expressions. Write all answers with positive exponents.

(a) $(a^3b^4)^3$. (b) $(2^2y^3)^2$. (c) $(3y^2z^{-1})^3$. (d) $(r^2s)^4(rs^4)^2$.

Solution

(a) $(a^3b^4)^3 = (a^3)^3(b^4)^3$ Property 4

$= a^9b^{12}$. Property 5

(b) $(2^2y^3)^2 = (2^2)^2(y^3)^2$ Property 4

$= 2^4y^6 = 16y^6$. Property 5

(c) $(3y^2z^{-1})^3 = 3^3y^6z^{-3}$ Property 4

$= 27y^6\frac{1}{z^3}$ Property 3

$= \frac{27y^6}{z^3}$.

(d) $(r^2s)^4(rs^4)^2 = [(r^2)^4(s^4)][r^2(s^4)^2]$ Property 4

$= r^8s^4r^2s^8$ Property 5

$= r^8r^2s^4s^8 = r^{10}s^{12}$. Property 1

The final two properties of exponents that we develop in this section involve quotients. We first discuss the quotient of two exponent forms with the same base a $(a \neq 0)$, such as $\frac{a^5}{a^3}$.

Observe that

$$\frac{a^5}{a^3} = \frac{a \cdot a \cdot a \cdot a \cdot a}{a \cdot a \cdot a} \quad \begin{array}{l}\longleftarrow \text{ five factors of } a \\ \longleftarrow \text{ three factors of } a\end{array}$$

$$= \frac{a \cdot a \cdot \cancel{a} \cdot \cancel{a} \cdot \cancel{a}}{\cancel{a} \cdot \cancel{a} \cdot \cancel{a}}$$

$$= a \cdot a \quad \longleftarrow (5 - 3) = 2 \text{ factors of } a$$

$$= a^2.$$

Thus we see that the simplified form of $\frac{a^5}{a^3}$ has the same base a, and a single exponent of 2. Furthermore, the exponent appearing in the simplified form is obtained by subtracting the exponent appearing in the denominator from the exponent in the numerator. As a result of this observation we see that the expression $\frac{a^5}{a^3}$ can be easily simplified as follows:

$$\frac{a^5}{a^3} = a^{5-3} = a^2.$$

The procedure described above is also valid when the exponent appearing in the denominator is larger than the exponent in the numerator. Consider, for example,

$$\frac{a^3}{a^5} = \frac{a \cdot a \cdot a}{a \cdot a \cdot a \cdot a \cdot a} = \frac{\not{a} \cdot \not{a} \cdot \not{a}}{\not{a} \cdot \not{a} \cdot \not{a} \cdot a \cdot a} = \frac{1}{a^2}.$$

However, if we subtract exponents we obtain the same result; that is,

$$\frac{a^3}{a^5} = a^{3-5} = a^{-2} = \frac{1}{a^2}.$$

These observations motivate Property 6 for exponents.

Property 6

If a is a nonzero real number and m and n are integers, then

$$\frac{a^m}{a^n} = a^{m-n} \qquad (a \neq 0).$$

Finally, recall that $\frac{a}{b}$ is defined for all real numbers a, b provided that $b \neq 0$. Let us assume that n is a positive integer. Note that

$$\left(\frac{a}{b}\right)^n = \frac{a}{b} \cdot \frac{a}{b} \cdot \frac{a}{b} \cdot \cdots \cdot \frac{a}{b} \quad \longleftarrow n \text{ ``factors'' of } \frac{a}{b}$$

$$= \frac{a \cdot a \cdot a \cdot \cdots \cdot a}{b \cdot b \cdot b \cdot \cdots \cdot b} \quad \begin{array}{l} \longleftarrow n \text{ factors of } a \\ \longleftarrow n \text{ factors of } b \end{array}$$

$$= \frac{a^n}{b^n}.$$

This is Property 7 for exponents. This property holds for positive, negative integers, and zero exponents.

Property 7

If a and b are real numbers ($b \neq 0$) and n is an integer, then

$$\left(\frac{a}{b}\right)^n = \frac{a^n}{b^n} \qquad (b \neq 0).$$

We shall exclude any application of the foregoing rule that results in division by zero. It can be shown that the following useful result is valid. If a and b are nonzero real numbers and n is a positive integer, then

$$\left(\frac{a}{b}\right)^{-n} = \left(\frac{b}{a}\right)^n. \qquad \text{note that } -n \text{ is a negative integer, since } n > 0$$

Example 4. Use Properties 6 and 7 to simplify the following exponent forms. Write all answers with positive exponents and evaluate.

(a) $\dfrac{7^{10}}{7^8}$. (b) $\dfrac{4^3}{4^5}$. (c) $\left(\dfrac{3}{4}\right)^2$. (d) $\left(\dfrac{5}{3}\right)^{-2}$.

Solution

(a) $\dfrac{7^{10}}{7^8} = 7^{10-8} = 7^2 = 49.$

(b) $\dfrac{4^3}{4^5} = 4^{3-5} = 4^{-2} = \dfrac{1}{4^2} = \dfrac{1}{16}.$

(c) $\left(\dfrac{3}{4}\right)^2 = \dfrac{3^2}{4^2} = \dfrac{9}{16}.$

(d) $\left(\dfrac{5}{3}\right)^{-2} = \left(\dfrac{3}{5}\right)^2 = \dfrac{3^2}{5^2} = \dfrac{9}{25}.$ $\qquad \left(\dfrac{a}{b}\right)^{-n} = \left(\dfrac{b}{a}\right)^n$

Example 5. Use the properties of exponents to simplify the following expressions. Write all answers with positive exponents.

(a) $\dfrac{x^7}{x^3}$. (b) $\left(\dfrac{x^2}{y^3}\right)^4$. (c) $\dfrac{a^{-2}b^3}{a^{-5}b^4}$. (d) $\left(\dfrac{x}{y}\right)^2 \dfrac{y^7}{y^4}$.

Solution

(a) $\dfrac{x^7}{x^3} = x^{7-3} = x^4.$ Property 6

(b) $\left(\dfrac{x^2}{y^3}\right)^4 = \dfrac{(x^2)^4}{(y^3)^4} = \dfrac{x^8}{y^{12}}.$ Properties 7 and 5

(c) $\dfrac{a^{-2}b^3}{a^{-5}b^4} = \dfrac{a^{-2}}{a^{-5}} \cdot \dfrac{b^3}{b^4} = a^{-2-(-5)} \cdot b^{3-4}$ Property 6

$= a^3 \cdot b^{-1}$ note that $-2 - (-5) = 3$ and $3 - 4 = -1$

$= \dfrac{a^3}{b}.$ Property 3; $b^{-1} = \dfrac{1}{b}$

(d) $\left(\dfrac{x}{y}\right)^2 \dfrac{y^7}{y^4} = \dfrac{x^2}{y^2} \cdot y^{7-4} = \dfrac{x^2}{y^2} \cdot \dfrac{y^3}{1}$ Properties 7 and 6

$= \dfrac{x^2}{1} \cdot \dfrac{y^3}{y^2}$

$= x^2y^{3-2} = x^2y.$ Property 6

We collect together all the properties of exponents discussed in the preceding section and this section and list them below for convenient reference.

Properties of Exponents

If a and b are real numbers and m and n are integers:

Property 1: $a^m \cdot a^n = a^{m+n}$

Property 2: $a^0 = 1 \quad (a \neq 0)$

Property 3: $a^{-n} = \dfrac{1}{a^n} \quad (a \neq 0)$

Property 4: $(a \cdot b)^n = a^n \cdot b^n$

Property 5: $(a^m)^n = a^{mn}$

Property 6: $\dfrac{a^m}{a^n} = a^{m-n} \quad (a \neq 0)$

Property 7: $\left(\dfrac{a}{b}\right)^n = \dfrac{a^n}{b^n} \quad (b \neq 0)$

NOTE. 0^0 is not defined.

EXERCISES 3.2

For Exercises 1–16, use Property 4 to evaluate or simplify the expression. See Example 1.

1. $4^2 \cdot 5^2$ **2.** $5^3 \cdot 2^3$ **3.** $6^2 \cdot 5^2$ **4.** $8^2 \cdot 5^2$

5. $(-3)^7(\frac{1}{3})^7$ **6.** $(-5)^4(\frac{1}{5})^4$ **7.** $12^4(\frac{1}{6})^4$ **8.** $27^3(\frac{1}{9})^3$

9. $(3x)^2$ **10.** $(2y)^4$ **11.** $(3r)^3$ **12.** $(4s)^2$

13. $(2xyz)^3$ **14.** $(4rst)^2$ **15.** $(xyzt)^4$ **16.** $(rstv)^5$

For Exercises 17–32, simplify the expression. Write all answers with positive exponents. See Example 2.

17. $(3^2)^2$ **18.** $(2^3)^3$ **19.** $(5^{-2})^2$ **20.** $(7^{-1})^5$ **21.** $(y^3)^4$

22. $(z^5)^2$ **23.** $(t^{-3})^{-4}$ **24.** $(r^{-2})^{-5}$ **25.** $(x^0)^{10}$ **26.** $(y^0)^5$

27. $(s^{-2})^3$ **28.** $(a^{-2})^4$ **29.** $(x^2)^3(y^3)^4$ **30.** $(t^5)^2(s^4)^3$ **31.** $(x^2)^{-3}(y^3)^2$

32. $(t^3)^4(s^{-1})^9$

For Exercises 33–44, simplify the expression. Write all answers with positive exponents. See Example 3.

33. $(x^2y^3)^4$ **34.** $(y^3z^4)^3$ **35.** $(3^2y^4)^2$ **36.** $(2^3x^4)^2$

37. $(2t^3z^{-2})^4$ **38.** $(3x^3y^{-4})^2$ **39.** $(5x^{-2}y)^3$ **40.** $(3y^{-1}z^{-1})^{-1}$

41. $(x^3y^2)^2(xy)^3$ **42.** $(st^2)^4(st)^2$ **43.** $(3r^3s^2)^2(rs)^5$ **44.** $(xy)^5(2x^2y^2)^2$

For Exercises 45–60, simplify the expression. Write all answers with positive exponents and then evaluate. See Example 4.

45. $\frac{5^6}{5^4}$ **46.** $\frac{2^9}{2^7}$ **47.** $\frac{3^{10}}{3^9}$ **48.** $\frac{9^5}{9^4}$ **49.** $\frac{4^2}{4^3}$

50. $\frac{7^5}{7^6}$ **51.** $\frac{3^2}{3^4}$ **52.** $\frac{6^7}{6^9}$ **53.** $(\frac{2}{3})^3$ **54.** $(\frac{2}{5})^2$

55. $(\frac{3}{5})^{-2}$ **56.** $(\frac{2}{3})^{-3}$ **57.** $(\frac{1}{6})^{-1}$ **58.** $(\frac{1}{8})^{-1}$ **59.** $\left(\frac{1}{2^3}\right)^{-1}$ **60.** $\left(\frac{1}{3^2}\right)^{-1}$

For Exercises 61–76, simplify the expression. Write all answers with positive exponents. See Example 5.

61. $\frac{z^9}{z^3}$ **62.** $\frac{x^6}{x^2}$ **63.** $\left(\frac{y^3}{z^2}\right)^3$ **64.** $\left(\frac{t^4}{s^2}\right)^3$

65. $\frac{b^7b^3b^2}{(b^3)^2}$ **66.** $\frac{(x^3)^2x^4}{x^3x^5}$ **67.** $\frac{a^4 \cdot a^6(3a^2)^4}{(3a^2)^3}$ **68.** $\frac{(2w^2)^3w^3 \cdot w}{8w^{12}}$

69. $\left(\frac{a^5}{b^4}\right)^2$ **70.** $\left(\frac{x^3}{y^2}\right)^3$ **71.** $\frac{x^{-3}y^3}{x^{-4}y^2}$ **72.** $\frac{y^5z^{-4}}{y^3z^{-6}}$

73. $\left(\frac{a}{b}\right)^3\left(\frac{a}{c}\right)^2$ **74.** $\left(\frac{t}{s}\right)^4\left(\frac{r}{s}\right)^2$ **75.** $\left(\frac{y^{-1}}{z^{-1}}\right)^{-2}$ **76.** $\left(\frac{a^{-1}}{b^{-1}}\right)^{-1}$

3.3 Scientific Notation

In this section we introduce a notation that uses integral powers of 10 to represent very large or very small numbers in a compact form. Numbers of very large or very small magnitude are usually encountered in the sciences. For example, scientists have determined that the speed at which light travels to be approximately

30,000,000,000 centimeters per second.

Notice that

$$30{,}000{,}000{,}000 = 3 \times 10{,}000{,}000{,}000.$$

↑ times symbol

Scientists would write this number as

3×10^{10} read "three times 10 raised to the tenth power"

In doing this, we make use of the fact that

$$10^{10} = 10{,}000{,}000{,}000.$$

Notice that 10^{10} is 1 followed by 10 zeros. As an example of a very small number we note that scientists who pioneered in atomic theory estimated the radius of the nucleus of a copper atom to be less than

.0000000000017 centimeter.

Using integer powers of 10, a scientist might write this number as

1.7×10^{-12} read "1.7 times 10 to the negative 12

Thus we see that the advantage of powers of 10 notation is that we can use it to represent concisely very small as well as very large numbers. The previous discussion suggests a way of representing numbers, called scientific notation, which is defined as follows:

Scientific Notation

> A number is said to be in *scientific notation* if it is written in the form
>
> $$m \times 10^n,$$
>
> where *m is a number between* 1 *and* 10 $(1 \le m < 10)$ and *n is an integer*.

The following numbers are written in scientific notation:

$$3.2 \times 10^{14}, \qquad 6.02 \times 10^{26}, \qquad 1.67 \times 10^{-27}.$$

If a number is written in decimal form (positional notation) such as 3.1415, we shall say that it is in *ordinary form*. To write a number that is in ordinary form in scientific notation, we use the following procedure.

Steps for Writing a Number in Decimal Form in Scientific Notation

$$m \times 10^n;\ 1 \le m < 10,\ n \text{ an integer}$$

Step 1. To find the exponent, n, count the number of decimal places, p, that the decimal point has to be moved to obtain a number m between 1 and 10.

Step 2. The exponent n

(a) Equals p and is *positive* if the decimal point must be moved to the *left*.

(b) Equals $-p$ and is *negative* if the decimal point has to be moved to the *right*.

We will apply the steps listed above to write the numbers 423,000 and .000218 in scientific notation. First, for 423,000 = 423000., we move the decimal point five places to the left as shown below, to obtain a number between 1 and 10:

4 2 3 0 0 0 . note that $1 \le 4.23 < 10$

five places to the left

Thus

$$423{,}000 = 4.23 \times 10^5.$$

Following a similar procedure for .000218, we see that

. 0 0 0 2 1 8 note that $1 \le 2.18 < 10$

four places to the right

As a result, we see that

$$.000218 = 2.18 \times 10^{-4}.$$

Example 1. Write the following numbers in scientific notation.
(a) 43,700 (b) .000375 (c) 2,000,000 (d) .000001

Solution

(a) $43{,}700 = 4.37 \times 10^4$. (b) $.000375 = 3.75 \times 10^{-4}$.
(c) $2{,}000{,}000 = 2 \times 10^6$. (d) 1×10^{-6}.

We note that multiplying a number in decimal notation by 10 shifts the decimal point to the right. For example, $3.78 \times 10 = 37.8$. We also observe that multiplying a number by 10^{-1} or $\frac{1}{10}$ shifts the decimal point one place to the left. Hence

$3.78 \times 10^{-1} = .378$. Thus, if we wish *to convert a number from scientific notation to ordinary form,* we do the following.

If the exponent is

1. *Positive:* move the decimal point that many decimal places to the *right*.
2. *Negative:* move the decimal point that many decimal places to the *left*.

We illustrate this procedure in the next example.

Example 2. Convert the following numbers from scientific notation to ordinary form.
(a) 7.182×10^5. (b) 3.15×10^{-3}.

Solution

(a) $7.182 \times 10^5 = 718{,}200$
(b) $3.15 \times 10^{-3} = .00315$

We have seen that scientific notation enables us to write very large or very small numbers in a compact and comprehensible way. Another important advantage of scientific notation is that it enables us to do computations with these numbers in an efficient way. For example, suppose that we wish to multiply

$$14{,}000{,}000{,}000 \qquad \text{and} \qquad .000000025$$

It would be tedious and difficult to multiply these two numbers. However, if we write each number in scientific notation and use the properties of exponents and real numbers, we can quickly arrive at the answer. Thus

$$\begin{aligned}(14{,}000{,}000{,}000)(.000000025) &= (1.4 \times 10^{10})(2.5 \times 10^{-8}) \\ &= (1.4)(2.5) \times 10^{10+(-8)} \\ &= 3.5 \times 10^2 \\ &= 350.\end{aligned}$$

We note that if we used a calculator to calculate the product (14,000,000,000)(.000000025), the numbers would have to be entered in scientific notation.

Example 3. Calculate and express the result in scientific notation or ordinary form.

(a) $(5 \times 10^3)(2 \times 10^6)$. (b) $\dfrac{(2.5 \times 10^5)(8.0 \times 10^{-3})}{2.0 \times 10^8}$.

Solution

$$\begin{aligned}\text{(a)}\ (5 \times 10^3)(2 \times 10^6) &= (5)(2) \times 10^3 \times 10^6 && \\ &= 10 \times 10^9 && \text{not in scientific form} \\ &= 1 \times 10 \times 10^9 && 10 = 1 \times 10 \\ &= 1 \times 10^{10}. && 10 \times 10^9 = 10^{10}\end{aligned}$$

(b) $\dfrac{(2.5 \times 10^5)(8.0 \times 10^{-3})}{2.0 \times 10^8} = \dfrac{(2.5)(8.0) \times 10^{5+(-3)}}{2.0 \times 10^8}$

$= \dfrac{20 \times 10^2}{2.0 \times 10^8}$

$= 10 \times 10^{-6} = .000010$ in ordinary form.

As a concluding remark we note that on scientific hand-held calculators, calculations that result in very large or very small numbers are displayed by means of scientific notation. For example, the display

1.551121 25

means 1.551121×10^{25}.

EXERCISES 3.3

For Exercises 1–16, write the number in scientific notation. See Example 1.

1. 370 **2.** 570 **3.** 1000 **4.** 10,000

5. .002 **6.** .004 **7.** 673 **8.** 972

9. 600,000 **10.** 300,000 **11.** .000071 **12.** .0000082

13. 48,700,000 **14.** 762,000,000 **15.** .00000972 **16.** .00006721

For Exercises 17–28, convert the number from scientific notation to ordinary decimal form. See Example 2.

17. 8.6×10^2 **18.** 5.6×10^3 **19.** 6.81×10^3 **20.** 4.51×10^3

21. 7×10^{-3} **22.** 2×10^{-4} **23.** 3.1×10^{-5} **24.** 3.2×10^{-5}

25. 5.72×10^{-6} **26.** 2.51×10^{-6} **27.** 7.531×10^{-3} **28.** 8.671×10^{-3}

For Exercises 29–42, calculate and express the answer in scientific notation or ordinary decimal form. See Example 3.

29. $(2.3 \times 10^4)(3.0 \times 10^5)$ **30.** $(4.1 \times 10^6)(2.0 \times 10^5)$ **31.** $(2.1 \times 10^3)(4.0 \times 10^4)$

32. $(3.2 \times 10^7)(3.0 \times 10^6)$ **33.** $(1.5 \times 10^8)(4.0 \times 10^{-5})$ **34.** $(2.3 \times 10^{-4})(3.0 \times 10^8)$

35. $(1.2 \times 10^{-3})(4.0 \times 10^{-2})$ **36.** $(4.1 \times 10^{-3})(3.0 \times 10^{-2})$ **37.** $(7.1 \times 10^3)(1.0 \times 10^{-7})$

38. $\dfrac{(8.0 \times 10^9)(3.0 \times 10^{-4})}{2.0 \times 10^4}$ **39.** $\dfrac{(6.0 \times 10^4)(8.0 \times 10^{-4})}{1.2 \times 10^{-3}}$ **40.** $\dfrac{(4.0 \times 10^5)(6.0 \times 10^{-4})}{2.0 \times 10^{-2}}$

41. $\dfrac{(1.2 \times 10^6)(3.2 \times 10^3)}{6.0 \times 10^5}$ **42.** $\dfrac{(4 \times 10^3)(6 \times 10^{-6})}{8 \times 10^{-3}}$

For Exercises 43–50, use scientific notation to compute, and express the answer in scientific notation.

43. (160)(20,000,000)(.00025) **44.** (150)(32,000,000)(.00002) **45.** $\dfrac{(20{,}000)(180{,}000)}{.000009}$

46. $\dfrac{(.006)(.000008)}{.000012}$

47. If 1 gram equals .002205 pound and the mass of the earth is approximately 6×10^{27} grams, find the mass of the earth in pounds.

48. The mass of a neutron is approximately 1.6×10^{-24} gram. Find the total mass of 4,000,000 neutrons.

49. One meter is equal to 100 centimeters. The speed of light is 3×10^{10} centimeters per second. Express the speed of light in meters per second.

50. There are 5280 feet in a mile. Find the number of inches in 500 miles.

3.4 Polynomials

In this section we discuss special types of algebraic expressions called polynomials. Polynomials play a very important role in a great deal of mathematics. We will learn how to identify particular polynomials and how to add and subtract polynomials.

We begin with a description of a monomial. A *monomial* is a constant or the product of a constant and *positive integer* powers of variables. The following algebraic expressions are examples of monomials:

$$7, \quad x, \quad 6y, \quad 3x^2y, \quad 9x^3yz^2, \quad 18t^2s.$$

We note that the following expressions are *not* examples of monomials:

$$\frac{1}{x} = x^{-1}, \quad \frac{x^2}{y^2}, \quad \frac{xy}{z}, \quad \frac{3}{x^2 + 3x + 2}.$$

The numerical factor of a monomial, such as $8x^2y^3$, is called the *numerical coefficient* or simply the *coefficient* of the monomial. Thus 8 is the coefficient of $8x^2y^3$. The numerical coefficient of monomials such as x^2 or yz^3 is understood to be 1. That is, $x^2 = 1 \cdot x^2$.

A *polynomial* is a monomial or the indicated sum of a finite number of monomials. Examples of polynomials are

$$x + 10, \quad 16y^2 - x^2, \quad x^2 + 3x + 2, \quad x + 3y + z - w, \quad \frac{1}{3}x^2 + x.$$

Example 1. Determine which of the following expressions are polynomials.
(a) $x^2 - \dfrac{1}{x} + 2$. (b) $x^3 - 2xy + \dfrac{1}{3}$. (c) $\dfrac{1}{x + 1}$. (d) $\dfrac{1}{2}x^2 + 3x + 1$.

Solution

(a) Since $-\dfrac{1}{x}$ is not a monomial, the given expression is not a polynomial.

(b) Polynomial. Note that $\dfrac{1}{3}$ is a constant, and therefore is a monomial.

(c) The expression is *not* a polynomial, since $\dfrac{1}{x+1}$ is not a monomial.

(d) Polynomial.

Each of the monomials appearing in a polynomial is called a *term* of the polynomial. Since we have defined a polynomial as the *sum* of monomials, the signs appearing in a polynomial indicate positive or negative numerical coefficients. Recall from the definition of subtraction that $x - y = x + (-y)$. Thus, since

$$2x^2y - 3x - 2xy + 5 = 2x^2y + (-3x) + (-2xy) + 5,$$

we see that the terms of the polynomial given above are $2x^2y$, $-3x$, $-2xy$, and 5.

As we have seen, a polynomial containing one term is called a monomial. Those with exactly *two* terms, such as $2x^2 + y$, are called *binomials;* and those with exactly *three* terms, such as $5x^2y - 2xy + 1$, are called *trinomials*. No special names are attached to polynomials containing more than three terms. They are referred to generally as polynomials.

Example 2. Identify each of the following expressions as a monomial, a binomial, a trinomial, or a polynomial. Choose the word that gives the best description. (a) $3x^5 - 6$. (b) $5x$. (c) $6x^2 - 3xy + 1$. (d) $5xy + y + x + 6$.

Solution

(a) Binomial. (b) Monomial. (c) Trinomial. (d) Polynomial.

The *degree of a monomial,* with a nonzero coefficient, is the *sum* of the exponents on the variables appearing in the monomial. Thus the degree of $7x^3y$ is $3 + 1 = 4$. A monomial, with a nonzero coefficient, which does not involve variables, such as 5, is said to have degree 0 since $5x^0 = 5(1) = 5$. Monomials with 0 coefficients do not have any degree. Thus 0 has no degree. The *degree of a polynomial* is the largest degree of the nonzero terms or monomials of the polynomial. Thus in the polynomial

$$7x^3yz^2 - 6xy^2 + 5xy$$

the term with the largest degree is $7x^3yz^2$. The degree of $7x^3yz^2$ is $3 + 1 + 2 = 6$. Thus the polynomial above is of degree 6.

Example 3. Determine the degree of the following polynomials. (a) $9x + 2$. (b) $x^2 + 3x - 4$. (c) $5xy^2 + xy + y$. (d) $5x^7 + 6x^2y^3 + 9y^3$.

Solution

(a) The degree of $9x$ is 1, and the degree of 2 is 0. Thus the degree of $9x + 2$ is 1.
(b) The monomial with the largest degree in the given polynomial is x^2. Thus the degree of the polynomial is 2.
(c) The term with the largest degree is $5xy^2$. The degree of $5xy^2$ is $1 + 2 = 3$. Thus the degree of the polynomial is 3.
(d) The degree of the polynomial is 7.

Two monomials that *differ only* in their numerical coefficients are called *like terms*. Thus $5xy^2$ and $2xy^2$ are like terms. Similarly, $7x^3$ and $-3x^3$ are like terms. The terms $3x^2$ and $2y^2$ are *not* like terms; similarly, $3x^3$ and $-2a^3$ are *not* like terms. It is customary to combine like terms of a polynomial by use of the distributive property. For example, to combine $5xy^2$ and $2xy^2$, we proceed as follows:

$$\begin{aligned} 5xy^2 + 7xy^2 &= (5 + 2)xy^2 && \text{distributive property} \\ &= 7xy^2. && \text{addition} \end{aligned}$$

Note that to combine two like terms, we simply add the corresponding numerical coefficients. Thus, to combine $2x^2z$ and $-5x^2z$, we add 2 and -5 to get -3 and our answer is $-3x^2z$. In writing a polynomial, it is customary to combine like terms: for example,

$$3x^2 + 2xy^2 - 6xy^2 = 3x^2 - 4xy^2.$$

The result is a polynomial expressed as a finite sum of unlike terms. We then say that the polynomial has been *simplified*. In general, to simplify a polynomial we use the commutative and associative properties of addition and then combine like terms. We illustrate this procedure in the next example.

Example 4. Simplify the following polynomials.
(a) $3x - 8y + 7x + 5y$. (b) $6x^2y - xy - x^2y + 2x^2y$.

Solution

(a)
$$\begin{aligned} 3x - 8y + 7x + 5y &= 3x + 7x - 8y + 5y && \text{commutative property} \\ &= 10x - 3y. && \text{combining like terms} \end{aligned}$$

(b)
$$\begin{aligned} &6x^2y - xy - x^2y + 2x^2y \\ &\quad = 6x^2y - x^2y + 2x^2y - xy && \text{commutative property} \\ &\quad = (6 - 1 + 2)x^2y - xy && \text{combining like terms} \\ &\quad = 7x^2y - xy. \end{aligned}$$

For most of this chapter we will be concerned primarily with polynomials that contain only one variable. For example, $5x^4 - 4x^3 + 3x^2 - 6x + 2$ is called a polynomial in x. Note that we have written the polynomial in *decreasing powers* of x. That is, as we read from left to right, the exponents are decreasing. When a polynomial in one variable is written in this form, we say that it is in *standard form*. We also note that in this form the first coefficient, the coefficient of the highest-degree term, is called the *leading coefficient*. Thus the leading coefficient of $6x^3 - 2x^2 + x + 3$ is 6.

To add two polynomials, we simply combine like terms. To do this, we add the numerical coefficients of like terms. Two formats are usually employed when adding (or subtracting) polynomials. One is referred to as the *vertical* format, the other one as the *horizontal* format for adding polynomials. The following example illustrates how to add two polynomials using both the vertical and horizontal formats.

Example 5. Add $4x^3 + 6x^2 - 3x + 2$ and $5x^3 - 2x^2 - 3 + x$. Express your answer in standard form.

Solution: To use the vertical format, we write each polynomial in standard form, line up the like terms vertically, and then add the numerical coefficients.

$$\begin{array}{rrrrr} & 4x^3 + & 6x^2 - & 3x + & 2 \\ \text{(add)} & 5x^3 - & 2x^2 + & x - & 3 \\ \hline & 9x^3 + & 4x^2 - & 2x - & 1 \end{array} \qquad \textit{vertical format}$$

To add the polynomials using the horizontal format, we write

$$(4x^3 + 6x^2 - 3x + 2) + (5x^3 - 2x^2 - 3 + x).$$

Note how we used parentheses to distinguish the polynomials that are being added. Next, we combine like terms. Thus we see that

$$\begin{aligned} &(4x^3 + 6x^2 - 3x + 2) + (5x^3 - 2x^2 - 3 + x) \\ &\qquad = (4x^3 + 5x^3) + (6x^2 - 2x^2) + (-3x + x) + (2 - 3) \\ &\qquad = 9x^3 + 4x^2 - 2x - 1. \end{aligned}$$

The format for adding polynomials that should be employed is the one that the reader finds most comfortable and efficient. Usually, when adding more than two polynomials, the vertical format is easier to use, as shown in the following example.

Example 6. Add $3x^5 - x^3 + x + 2$, $5x^4 + 2x^3 + 3$, and $2x^5 + x^4 - 5x^2 + 1$.

Solution: We use the vertical arrangement to add the given polynomials, lining up like terms vertically.

$$\begin{array}{rrrrrrr} & 3x^5 & & - x^3 & & + x & + 2 \\ & & 5x^4 & + 2x^3 & & & + 3 \\ \text{(add)} & 2x^5 & + x^4 & & - 5x^2 & & + 1 \\ \hline & 5x^5 & + 6x^4 & + x^3 & - 5x^2 & + x & + 6 \end{array}$$

As mentioned before, subtraction of real numbers is defined in terms of addition; that is, $x - y = x + (-y)$. Since the variables appearing in a polynomial represent real numbers, we use the same definition to subtract polynomials. Thus, to subtract $3x^2 - x$ from $5x^2$, we proceed as follows:

$$5x^2 - (3x^2 - x) = 5x^2 + [-(3x^2 - x)]. \qquad \text{definition of subtraction}$$

From Chapter 1, recall that to remove parentheses preceded by a negative sign we change the sign of each term within parentheses. Thus we have

$$= 5x^2 + [-3x^2 + x]. \qquad -(3x^2 - x) = -3x^2 + x$$

Combining like terms, we have

$$\begin{aligned} &= 5x^2 + (-3x^2) + x \\ &= 2x^2 + x. \end{aligned}$$

As a result, we have

$$5x^2 - (3x^2 - x) = 2x^2 + x.$$

We can shorten this procedure by simply writing

$$\begin{aligned} 5x^2 - (3x^2 - x) &= 5x^2 - \underbrace{3x^2 + x}_{\text{signs changed}} && \text{remove parentheses} \\ &= 2x^2 + x. && \text{combine like terms} \end{aligned}$$

The shortened procedure used above for subtracting polynomials can also be accomplished using a vertical format.

$$\begin{array}{rr} & 5x^2 \\ \text{(subtract)} & \underline{3x^2 - x} \end{array} \quad \text{rewrite as} \quad \begin{array}{rr} & 5x^2 \\ \text{(add)} & \underline{-3x^2 + x} \\ & 2x^2 + x \end{array} \longleftarrow \text{change each sign}$$

The vertical format for subtracting polynomials will be used later when we divide polynomials.

In the following example, we use the shortened procedures shown above for finding the difference of two polynomials containing more than two terms.

Example 7. Subtract $3x^2 - x + 5$ from $7x^2 + x - 3$.

Solution: We first use the horizontal form and parentheses to indicate the difference of the two given polynomials. Then we evaluate the difference. Thus we write

$$\begin{aligned} (7x^2 + x - 3) - (3x^2 - x + 5) &= 7x^2 + x - 3 - \underbrace{3x^2 + x - 5}_{\text{signs changed}} \\ &= 4x^2 + 2x - 8. \qquad \text{combine like terms} \end{aligned}$$

Now if we use the vertical form, we proceed in a similar manner.

	$7x^2 + x - 3$			$7x^2 + x - 3$	
(subtract)	$3x^2 - x + 5$	rewrite as	(add)	$-3x^2 + x - 5$	⟵ change each sign
				$4x^2 + 2x - 8$	

EXERCISES 3.4

For Exercises 1–16, state whether the expression is or is not a polynomial. See Example 1.

1. $x + 1$ **2.** $y - 1$ **3.** $x^2 - \frac{1}{5}$ **4.** $y^3 + \frac{1}{2}$

5. $3x^2 + y - 3$ **6.** $5t^{-4} + s^2$ **7.** $z^3 - z + 1$ **8.** $y^4 - y^2 + 7$

9. $\frac{7}{x^3} + 5x + 2$ **10.** $\frac{2}{xy} + 3xy$ **11.** $xy^2 - x^3y$ **12.** $yx^5 - xy^5$

13. $\frac{xy}{9} + \frac{y^2}{3}$ **14.** $\frac{t^2}{5} + \frac{t}{2}$ **15.** $\frac{1}{x^2 - 1}$ **16.** $\frac{1}{z^3 - 1}$

For Exercises 17–32, use one of the following terms that best describes the given expression: monomial, binomial, trinomial, polynomial. Also determine the degree of the polynomial. See Examples 2 and 3.

17. $8x + 3$ **18.** $10x + 7$ **19.** $6n^2 - n + 1$ **20.** $7x^2 - x + 1$

21. $-3a^4b$ **22.** $7x^3yz^2$ **23.** $r^2 - t^2$ **24.** $x^3 - y^3$

25. $q^4 - q^3 + 3q + 8$ **26.** $2p^7 + p^5 - p + 5$ **27.** $xyz + yz$ **28.** $abc^2 + bc^2$

29. $-8xy^2 - xy + x$ **30.** $-9x^2yz + 3y^5$ **31.** $x^2 - y^2 + xy + 1$ **32.** $tx^3 - t^2x^2 + t - 1$

For Exercises 33–50, simplify the polynomial. See Example 4.

33. $5x - 3x$ **34.** $-8t + 5t$ **35.** $9x - 6x + 7x$

36. $-10y + 2y - 6y$ **37.** $7x^2 - x^2 + 5x + 2x$ **38.** $10z^2 + 2z^2 + 6z - 4z$

39. $9t^2 - t + t^2 - 3t$ **40.** $12m^2 - 5m + 3m^2 + 2m$ **41.** $5x - 7y + 3x + 4y$

42. $12p - 9q - 3p + 6q$ **43.** $-8r^2 + 5t - 3r^2 - 2t$ **44.** $-6s + t^2 - 5t^2 - 4s$

45. $5xy^3 - 2xy + 7xy^3 - 3xy$ **46.** $6st^4 - 4t - 5st^4 - t$ **47.** $5p^3q - 9p^2q - p^3q + 2p^2q$

48. $3(2x - y) + 2y - 4x$ **49.** $5(3t - 2s) - 7t + 3s$ **50.** $6(4r - 2t) - t - 20r$

For Exercises 51–64, perform the indicated operation. See Examples 5–7.

51. $(x^2 - 2x + 1) + (3x^2 - 5x + 4)$ **52.** $(3x^3 - 6x^2 + 5) + (2x^3 - 3x^2 - 5)$

53. $(-5y^4 + y^2 + 6y) + (y^4 - 4y^3 - 6y)$ **54.** $(10z^5 - 3z^3 + 4z) + (-6z^5 + 3z^3 + 1)$

55. $(6x^3 + 3x - 1) - (5x^3 - x - 1)$ **56.** $(4y^4 + 3y^3 - 1) - (y^4 + 2y^3 - 3)$

57. $(7x^3 - x^2 + 5x + 3) - (x^3 - 2x^2 - 3x + 5)$ **58.** $(9y^4 - 2y^2 + 3y + 5) - (y^4 - 3y^2 - 3y + 1)$

59. $(5xy^2 + 3xy) + (xy^2 - 2xy)$ **60.** $(7y^2z - 5yz) + (2y^2z + 4yz)$

61. $(5x^2 + x) - (2x^2 - x) + (x^2 + 2x)$

62. $(8y^2 - y) - (2y^2 + 3y) + (-y^2 + 5y)$

63. $(7t^2 - 5t + 1) - (2t - 3) - (t^2 + 2t + 3)$

64. $(s^2 - 5t) - (4s^2 + t - 1) - (7t + 3)$

For Exercises 65–72, perform the operation. See Examples 6 and 7.

65. Add $3x^4 - x^3 + 1$, $2x^4 + x^2 - x + 2$, and $2x^3 + x^2 + 3$.

66. Add $5y^5 + 2y^3 - y^2 + y - 3$, $-y^5 - 4y^3 + 3y$, and $7y^5 + 3y^3 + y^2 - 4$.

67. Subtract $4x^3 - x^2 + 2x - 1$ from $-2x^3 + x^2 + 2x + 5$.

68. Subtract $7y^3 + 2y^2 - 3y + 5$ from $10y^3 + 3y^2 - 3y - 4$.

69. Add $3x^2 - 2x + 5$ and $2x^2 - x + 2$, then subtract $x^2 + 2x + 1$.

70. Add $4z^2 - 7z - 3$ and $3z^2 - 2z + 9$, then subtract $5z^2 - 4z - 2$.

71. Add $-2(5t^2 - t + 6)$ and $6(t^2 - t + 1)$.

72. Add $3(4s^2 - 3s + 4)$ and $-4(2s^2 - 3s + 3)$.

3.5 Multiplying Polynomials

In this section we learn how to multiply two polynomials. The product of any two polynomials can be obtained by repeated uses of the commutative, associative, and distributive laws together with Property 1 for exponents. We begin our discussion by considering the product of two monomials.

Recall from our discussion on exponents in Section 3.1 that the product of two monomials, such as $5x^3y^2$ and $3xy^4$, is obtained as follows:

$$(5x^3y^2)(3xy^4) = 5 \cdot 3 \cdot x^3 \cdot x \cdot y^2 \cdot y^4 \qquad \text{commutative property of multiplication}$$

$$= 15x^4y^6. \qquad \text{associative property; Property 1 for exponents}$$

We use this result to discuss the product of a polynomial and a monomial. To multiply a polynomial by a monomial, we use the distributive law. For example, to multiply $m + n$ by $3m$, we write

$$3m(m + n) = 3m(m) + 3m(n) \qquad \text{distributive property}$$

$$= 3m^2 + 3mn. \qquad \text{Property 1 for exponents}$$

If the polynomial has more than two terms, we use the extended form of the distributive property:

$$r(s + t + u + \cdots + z) = r(s) + r(t) + r(u) + \cdots + r(z).$$

Thus the product of $3x^2$ and $x^2 - 3xy + y^2$ is

$$\begin{aligned}3x^2(x^2 - 3xy + y^2) &= 3x^2(x^2) + 3x^2(-3xy) + 3x^2(y^2)\\ &= 3x^4 - 9x^3y + 3x^2y^2.\end{aligned}$$

Example 1. Find the products.
(a) $2x^2y(3y^2 - xy)$. (b) $ab^2(a^3 - 3b^2 + ab - 2)$.

Solution

(a)
$$\begin{aligned}2x^2y(3y^2 - xy) &= 2x^2y(3y^2) + 2x^2y(-xy)\\ &= 2 \cdot 3x^2 \cdot y \cdot y^2 - 2x^2 \cdot x \cdot y \cdot y\\ &= 6x^2y^3 - 2x^3y^2.\end{aligned}$$

(b)
$$\begin{aligned}ab^2(a^3 - 3b^2 + ab - 2) &= ab^2(a^3) + ab^2(-3b^2) + ab^2(ab) + ab^2(-2)\\ &= a \cdot a^3 \cdot b^2 - 3ab^2b^2 + a \cdot a \cdot b^2 \cdot b - 2ab^2\\ &= a^4b^2 - 3ab^4 + a^2b^3 - 2ab^2.\end{aligned}$$

Until now we have discussed multiplying two monomials, and a polynomial by a monomial. We now discuss the more general case of the product of two polynomials where each polynomial has two or more terms. For example, let us find the product of $2x - 3$ and $x^2 - 4x + 5$. When we multiply two such polynomials, we treat one polynomial as a single term and make repeated use of the distributive property, as shown below. We treat $2x - 3$ as a single term by introducing parentheses and use the distributive property. Thus

$$\begin{aligned}&(2x - 3)(x^2 - 4x + 5)\\ &\qquad = (2x - 3)x^2 + (2x - 3)(-4x) + (2x - 3)(5)\\ &\qquad = 2x(x^2) - 3(x^2) + 2x(-4x) + (-3)(-4x) + 2x(5) + (-3)(5)\\ &\qquad = 2x^3 - 3x^2 - 8x^2 + 12x + 10x - 15\\ &\qquad = 2x^3 - 11x^2 + 22x - 15.\end{aligned}$$

Note that in multiplying the two polynomials above, we used the distributive property four times and in the last line we combined like terms.

When multiplying two polynomials containing many terms, it is sometimes more efficient to use a vertical format rather than the horizontal format we have been using. The vertical format is similar to the vertical format used for multiplying large numbers in arithmetic. We illustrate this form of multiplying polynomials by multiplying

$$2x - 3 \qquad \text{and} \qquad x^2 - 4x + 5.$$

Notice that we multiplied these two polynomials above using the horizontal format. First the two polynomials should be in standard form (descending powers of the variable) with like terms written in the same column as follows.

$$\begin{array}{lrl} & x^2 - 4x + 5 & \\ \text{(multiply)} & \underline{2x - 3} & \\ & 2x^3 - 8x^2 + 10x & \longleftarrow 2x(x^2 - 4x + 5) \\ \text{(add)} & \underline{- 3x^2 + 12x - 15} & \longleftarrow -3(x^2 - 4x + 5) \\ & 2x^3 - 11x^2 + 22x - 15. & \end{array}$$

The previous discussion illustrates that whether we multiply two polynomials horizontally or vertically, the resulting product is the same. These two procedures illustrate the following rule for multiplying two polynomials, which is merely an abbreviation for the repeated use of the distributive property.

> To multiply two polynomials, multiply each term of one polynomial by each term of the other polynomial and then combine like terms.

We illustrate the application of the rule above in the following examples.

Example 2. Multiply the polynomials $a + b$ and $a^2 + 2ab + b^2$ using (a) The horizontal format. (b) The vertical format.

Solution

(a) $(a + b)(a^2 + 2ab + b^2)$

$$= a(a^2) + a(2ab) + a(b^2) + b(a^2) + b(2ab) + b(b^2)$$
$$= a^3 + 2a^2b + ab^2 + a^2b + 2ab^2 + b^3$$
$$= a^3 + 3a^2b + 3ab^2 + b^3.$$

(b)

$$\begin{array}{lrl} & a^2 + 2ab + b^2 & \\ \text{(multiply)} & \underline{a + b} & \\ & a^3 + 2a^2b + ab^2 & \longleftarrow a(a^2 + 2ab + b^2) \\ \text{(add)} & \underline{a^2b + 2ab^2 + b^3} & \longleftarrow b(a^2 + 2ab + b^2) \\ & a^3 + 3a^2b + 3ab^2 + b^3. & \end{array}$$

Example 3. Use the vertical format to multiply the polynomials $3y^2 - 4 + 2y$ and $y - 2y^2 + 1$.

Solution: We write each polynomial in standard form (descending powers of the variable) with like terms in the same column.

$$\begin{array}{lrl} & 3y^2 + 2y - 4 & \\ \text{(multiply)} & \underline{-2y^2 + y + 1} & \\ & -6y^4 - 4y^3 + 8y^2 & \longleftarrow -2y^2(3y^2 + 2y - 4) \\ & + 3y^3 + 2y^2 - 4y & \longleftarrow y(3y^2 + 2y - 4) \\ \text{(add)} & \underline{3y^2 + 2y - 4} & \longleftarrow 1 \cdot (3y^2 + 2y - 4) \\ & -6y^4 - y^3 + 13y^2 - 2y - 4. & \end{array}$$

The product of *two binomials* is a special product that often comes up in algebra. We will develop a memory device called *FOIL,* which will enable us to calculate quickly the product of two binomials.

Consider the product $(2x + 5)(3x - 4)$. Using the horizontal format for multiplying polynomials, we obtain

$$
\begin{aligned}
&(1) \qquad & (2x + 5)(3x - 4) &= 2x(3x) + 2x(-4) + 5(3x) + 5(-4) \\
&(2) \qquad & &= 6x^2 - 8x + 15x - 20 \\
&(3) \qquad & &= 6x^2 + 7x - 20.
\end{aligned}
$$

Observe that on line (2), before we combined like terms, the first term, $6x^2$, of the polynomial $6x^2 - 8x + 15x - 20$ is the product of the *First* terms in each set of parentheses. That is,

$$\overbrace{(2x + 5)(3x}^{6x^2} - 4). \qquad 2x \text{ and } 3x \textit{ first} \text{ terms}$$

Similarly, the terms $-8x$ and $15x$ are the products of the *Outermost* and *Innermost* terms in each set of parentheses. That is,

$$\overbrace{(2x + 5)(3x - 4)}^{-8x}. \qquad 2x \text{ and } -4 \textit{ outermost} \text{ terms}$$

$$(2x + \underbrace{5)(3x}_{15x} - 4). \qquad 5 \text{ and } 3x \textit{ innermost} \text{ terms}$$

Finally, -20 is the product of the *Last* term in each set of parentheses:

$$(2x + \underbrace{5)(3x - 4}_{-20}). \qquad 5 \text{ and } -4 \textit{ last} \text{ terms}$$

In summary, we have

$$
\begin{aligned}
&\text{F}: 6x^2 \quad \text{O}: -8x \quad \text{I}: 15x \quad \text{L}: -20 \\
&\qquad\qquad\qquad\qquad\quad \text{F} \;+\; \text{O} \;+\; \text{I} \;+\; \text{L} \\
(2x + 5)(3x - 4) &= 6x^2 + (-8x) + 15x + (-20) \\
&= 6x^2 + 7x - 20.
\end{aligned}
$$

The "word" FOIL is easy to remember and proves very useful for multiplying two binomials.

Example 4. Find the product $(5x + y)(2x + 3y)$.

Solution

$$\begin{aligned} & \qquad\qquad\qquad\quad \text{F} \;\; + \;\; \text{O} \;\; + \;\; \text{I} \;\; + \;\; \text{L} \\ (5x + y)(2x + 3y) &= 10x^2 + 15xy + 2xy + 3y^2 \\ &= 10x^2 + 17xy + 3y^2. \end{aligned}$$

Several special products involving binomials occur very frequently in algebra. We list below formulas for evaluating these special products together with the names commonly used to refer to them. The reader can verify them by performing the multiplication.

Special Product Formulas

Square of a Binomial:	$(x + y)^2 = x^2 + 2xy + y^2$	(1)
	$(x - y)^2 = x^2 - 2xy + y^2$	(2)
Product of the Sum and Difference of Two Terms:	$(x + y)(x - y) = x^2 - y^2$.	(3)

We note that the variables x and y appearing in the special product formulas above may be replaced by an algebraic expression. We also note that the polynomial $x^2 - y^2$ is referred to as the *difference of two squares*. We illustrate the application of the special product formulas in the following example.

Example 5. Find the indicated products.
(a) $(2a + 3b)^2$. (b) $(3t^2 - s)^2$. (c) $(5p - 2q)(5p + 2q)$.

Solution

(a) To find the square of $(2a + 3b)$, we use formula (1) listed under "Special Product Formulas": $(x + y)^2 = x^2 + 2xy + y^2$. To this end we let $x = 2a$ and $y = 3b$. Substituting in the formula, we obtain

$$\begin{aligned} (2a + 3b)^2 &= (2a)^2 + 2(2a)(3b) + (3b)^2 \\ &= 2^2a^2 + 2 \cdot 2 \cdot 3a \cdot b + 3^2b^2 \\ &= 4a^2 + 12ab + 9b^2. \end{aligned}$$

(b) To evaluate the indicated product, we use Special Product Formula (2): $(x - y)^2 = x^2 - 2xy + y$. We let $x = 3t^2$ and $y = s$. Thus

$$(3t^2 - s)^2 = (3t^2)^2 - 2(3t^2)(s) + s^2$$
$$= 9t^4 - 6t^2s + s^2. \qquad \text{note: } (3t^2)^2 = 3^2(t^2)^2 = 9t^4$$

(c) We will use Special Product Formula (3): $(x + y)(x - y) = x^2 - y^2$. Letting $x = 5p$ and $y = 2q$, we see that

$$(5p - 2q)(5p + 2q) = (5p)^2 - (2q)^2$$
$$= 5^2p^2 - 2^2q^2 \qquad \text{Property 4 of exponents}$$
$$= 25p^2 - 4q^2.$$

EXERCISES 3.5

For Exercises 1–12, multiply the polynomial by the monomial. See Example 1 and the discussion preceding it.

1. $5x(x + y)$ **2.** $7p(p + q)$ **3.** $st(s - t)$

4. $2mn(m - n)$ **5.** $3y(y^3 - 5y + 2)$ **6.** $-4x(x^2 - 3x + 2)$

7. $-6z^2(z^2 - 2zy + y)$ **8.** $-5a^2(a^3 - 2ab - b)$ **9.** $3x^2y(2y^2 - xy)$

10. $4rs^2(r^2s - 2rs)$ **11.** $xy^3(x^4 - 5y^2 + xy - 3)$ **12.** $t^2s(t^3s^2 + 4t - 5s - 7)$

For Exercises 13–30, use either the horizontal or vertical format to multiply the polynomials. See Example 2 and the discussion preceding it.

13. $(2a + b)(a + 3b)$ **14.** $(y + 3z)(4y + z)$ **15.** $(3x - 4y)(2x + 5y)$

16. $(6t + 5s)(2t - 3s)$ **17.** $(w + 4)(w^2 - w + 2)$ **18.** $(z - 2)(z^2 - 3z + 4)$

19. $(3x - 2)(x^2 + 2x + 3)$ **20.** $(5y - 1)(y^2 + 4y + 2)$ **21.** $(2t + 3)(t^2 - 5t + 4)$

22. $(4t + 5)(t^2 - 2t + 1)$ **23.** $(x + y)(x^2 + 3xy + y^2)$ **24.** $(u + v)(u^2 + 5uv + v^2)$

25. $(2x + y)(3x^2 + xy + 2y^2)$ **26.** $(r + 3s)(2r^2 + 3rs + s^2)$ **27.** $(y + 2)(y - 3)(y - 1)$

28. $(z - 2)(z - 3)(z + 4)$ **29.** $(3x + y)(2x + y)(x + y)$ **30.** $(2t + s)(4t + s)(t + s)$

For Exercises 31–36, use the vertical format to multiply the polynomials. See Example 3.

31. $(x^2 + x + 1)(x^2 - x + 2)$ **32.** $(y^2 - y + 2)(y^2 + 2y - 1)$

33. $(2z^2 + z + 3)(z^2 - 2z + 3)$ **34.** $(3t^2 - t + 1)(t^2 + 4t - 2)$

35. $(t - 3t^2 + 2)(2t^2 - 3t + 1)$ **36.** $(w - 2w^2 + 3)(2w^2 - 4w + 1)$

For Exercises 37–56, use the word FOIL as a memory aid to find the product of binomials. See Example 4 and the discussion preceding it.

37. $(x + 3)(x + 2)$ **38.** $(y + 4)(y + 5)$ **39.** $(z - 1)(z - 2)$

40. $(t - 5)(t - 2)$ **41.** $(s + 1)(s - 1)$ **42.** $(w + 2)(w - 2)$

43. $(2x + 3)(3x + 1)$ **44.** $(5x + 2)(2x + 3)$ **45.** $(3y - 2)(3y - 4)$

46. $(5w - 3)(2w - 3)$ **47.** $(2x + 3)(2x - 3)$ **48.** $(5t - 2)(5t + 2)$

49. $(1 - x)(2 + x)$
50. $(3 - y)(4 + y)$
51. $(2 - x)(3 - x)$
52. $(5 - x)(6 - x)$
53. $(3r + s)(2r + s)$
54. $(5a + b)(3a + b)$
55. $(2x - 3y)(3x - 2y)$
56. $(3a + 4b)(4a - 2b)$

For Exercises 57–74, find the special product. See Example 5.

57. $(2x + 4)^2$
58. $(5y + 2)^2$
59. $(3x - 2)^2$
60. $(4z - 1)^2$
61. $(2r + s)(2r - s)$
62. $(3u + v)(3u - v)$
63. $(5x + 3y)^2$
64. $(4y + 2z)^2$
65. $(2r - 3s)^2$
66. $(3u - 4v)^2$
67. $(2p + 3q)(2p - 3q)$
68. $(5a - 4b)(5a + 4b)$
69. $(2x^2 - 1)^2$
70. $(4t^2 + 2)^2$
71. $(3x^2 - y)(3x^2 + y)$
72. $(2t^2 - s)(2t^2 + s)$
73. $(p^3 + q^3)(p^3 - q^3)$
74. $(p^4 + q^4)(p^4 - q^4)$

For Exercises 75–86, find the product. See the equations below and the Special Product Formulas. Assume that all variable exponents represent positive integers.

$$x^n(x^n + x) = x^{n+n} + x^{n+1} = x^{2n} + x^{n+1}$$
$$(a^m + b^n)(a^m - b^n) = (a^m)^2 - (b^n)^2 = a^{2m} - b^{2n}$$

75. $x^{2n}(x^n - x)$
76. $y^{3n}(y^{2n} + y)$
77. $z^{n-2}(z^{n+2} + z^2)$
78. $t^{n-1}(t + t^{n+1})$
79. $(x^m + 3)(x^m - 3)$
80. $(y^{2n} + 4)(y^{2n} - 4)$
81. $(y^n + 1)^2$
82. $(z^n - 1)^2$
83. $(x^n + y^n)(x^n - y^n)$
84. $(y^{2n} + z^{3m})(y^{2n} - z^{3m})$
85. $(x^n + y^n)^2$
86. $(x^n - y^n)^2$

3.6 Introduction to Factoring

When the integers 2 and 3 are multiplied, the result is the product 6. That is, $2 \cdot 3 = 6$. We call 2 and 3 integer *factors* of 6, or simply *factors* of 6. Some other factors of 6 are 1 and 6, since $1 \cdot 6 = 6$.

A positive integer greater than 1 that has no positive integer factors except itself and 1 is called a *prime number*. There are an infinite number of prime numbers. The first eight prime numbers are

$$2, 3, 5, 7, 11, 13, 17, \text{ and } 19.$$

A positive integer, other than 1, that is not prime is called a *composite number*. Examples of composite numbers are 4, 6, 8, 9, 10, 12, and so on. When a composite number is written as the product of powers of all its prime factors, it is said to be *completely factored*. For example, $60 = 6 \cdot 10$. But $6 = 2 \cdot 3$ and $10 = 2 \cdot 5$. Thus

$$60 = (2 \cdot 3)(2 \cdot 5) = 2^2 \cdot 3 \cdot 5.$$

Since 2, 3, and 5 are all prime numbers, we see that 60 has been completely factored into $2^2 \cdot 3 \cdot 5$.

We note that a positive integer has *one* and *only one* set of prime factors to specific powers.

We now define the terms "prime" and "composite" for negative integers. If n is a positive integer, then $-n$ is a negative integer. In this case $-n$ (a negative integer) is prime if n is prime. For example, -3 is prime because 3 is prime. Also, we will adopt the convention that a negative integer, $-n$, is *composite* if the corresponding positive integer n is composite. Thus, to *factor* $-n$ *completely,* we will express $-n$ as the product of -1 and the prime factors of the positive integer n. For example, to express -6 in completely factored form, we write $-6 = (-1)(2)(3)$.

Example 1. Factor the following numbers completely.
(a) 90. (b) 23. (c) 54. (d) -72.

Solution
(a) $90 = 9 \cdot 10 = (3 \cdot 3)(2 \cdot 5) = 2 \cdot 3^2 \cdot 5$.
(b) Prime number.
(c) $54 = 2 \cdot 27 = 2 \cdot (3 \cdot 3 \cdot 3) = 2 \cdot 3^3$.
(d) $-72 = (-1)(2 \cdot 36) = (-1)2 \cdot (2 \cdot 18) = (-1)2^2(2 \cdot 3 \cdot 3) = (-1)2^3 \cdot 3^2$.

To *factor a polynomial* means to express it as a product of other polynomials. For example, to factor $3xy + 3x$, we use the distributive property to write

$$3xy + 3x = (3x)y + (3x) \cdot 1 = 3x(y + 1).$$

We call $3x$ and $(y + 1)$ *factors* of $3xy + 3x$. We say variously that we have *factored out,* or *removed, the common factor* $3x$ from the polynomial $3xy + 3x$. The process of expressing a number or polynomial in terms of its factors is called *factoring*. Notice that factoring is the "reverse" of multiplying.

In this chapter we study factoring of polynomials with *integer coefficients*. Furthermore, we require that the factors of a polynomial be *polynomials* with *integer coefficients*. Using the guidelines above, we see that $x + 1$ *cannot* be factored as shown below, for the reasons given.

(1) $$x + 1 = 2\left(\frac{1}{2}x + \frac{1}{2}\right)$$ *fractions are not allowed as coefficients of the variables*

and

(2) $$x + 1 = x\left(1 + \frac{1}{x}\right)$$ $\left(1 + \frac{1}{x}\right) = (1 + x^{-1})$ *is not a polynomial*

The simplest factorization of a polynomial involves factoring out a common monomial factor from each term of the polynomial. This type of factorization is based on the extended distributive property:

$$ax + ay + az + aw = a(x + y + z + w).$$

Recall that a polynomial is the sum of monomials. Each monomial is called a term of the polynomial. Thus we define

> the *greatest common monomial factor* (g.c.m.f.) of a polynomial as the monomial of highest degree (power) and with the largest numerical coefficient (in absolute value) that is a factor of each term of the polynomial.

To aid our understanding of the definition given above, we will identify the greatest common monomial factor (g.c.m.f.) of the polynomial

$$18x^2y^3 + 27xy^2z.$$

We observe that some of the monomials that are factors *of each term* of the polynomial above are 3, $9x$, y^2, and $6xy$. But the monomial of highest degree with the largest numerical coefficient that *is a factor of each term* of $18x^2y^3 + 27xy^2z$ is $9xy^2$. Thus $9xy^2$ is by definition the greatest common monomial factor (g.c.m.f.) of the given polynomial. We emphasize this fact by isolating the g.c.m.f. in each term of the polynomial, as follows:

$$18x^2y^3 + 27xy^2z = (9xy^2)2xy + (9xy^2)3z.$$

Note that $2xy$ and $3z$ on the right-hand side of the equation have no factors in common.

It is necessary to make one additional comment on the g.c.m.f. of a polynomial. If we are asked to factor out the g.c.m.f. from a polynomial such as

$$-2xy + 10x,$$

there are two possible answers. We could factor the polynomial as

$$2x(-y + 5) \qquad \text{or} \qquad -2x(y - 5).$$

Thus we see that the g.c.m.f. could be either $2x$ or $-2x$. In either case 2 or -2 is the largest numerical coefficient in *absolute value* common to each term of the polynomial. Note that $|2| = 2$ and $|-2| = 2$. The g.c.m.f. we choose as far as sign is concerned depends on the form of the factorization that we prefer.

In the following example we illustrate a technique for determining the greatest common monomial factor of a polynomial. We should note that a polynomial does not necessarily have a g.c.m.f. (other than 1 or -1).

Example 2. Factor out the g.c.m.f. from $60x^3y + 100x^2y^2$.

Solution: First, factor each *numerical coefficient* of the given polynomial completely into prime factors.

$$2^2 \cdot 3 \cdot 5x^3y + 2^2 \cdot 5^2x^2y^2.$$

Then group together in each term, the *highest power* of each prime factor *common to all coefficients*. Therefore, as a first approach to determining the g.c.m.f. of $60x^3y + 100x^2y^2$, we rewrite it in the following form:

$$(2^2 \cdot 5)3x^3y + (2^2 \cdot 5)5x^2y^2.$$

Next, for each variable (if possible) group together the *highest power* of the variable that is *a factor of each term*. As a result, we have

$$(2^2 \cdot 5x^2y)3x + (2^2 \cdot 5x^2y)5y.$$

The g.c.m.f. of the given polynomial is $2^2 \cdot 5 \cdot x^2y = 20x^2y$. Finally, using the distributive property, we factor out the greatest common monomial factor (g.c.m.f.):

$$\begin{aligned} 60x^3y + 100x^2y^2 &= (2^2 \cdot 5 \cdot x^2y)(3x + 5y) \\ &= 20x^2y(3x + 5y). \end{aligned}$$

Example 2 illustrated a method that will produce the greatest common monomial factor of a polynomial. However, it may be difficult to apply primarily because factoring numbers completely into primes may at times prove difficult. As we gain experience from multiplying and factoring polynomials, many of the steps outlined in Example 2 may be omitted. In actual practice, when seeking the g.c.m.f. of a polynomial we keep factoring out common monomial factors from the terms of the polynomial until there are none left. We illustrate this approach in the next example.

Example 3. Factor the g.c.m.f. from the following polynomials.
(a) $16x^2y + 24xy^2 - 32xy$. (b) $56a^3b - 20a^2b^2 + 24a^2b^2$.

Solution

(a) First, we usually factor out the largest numerical factor (in absolute value) common to each term of the polynomial. Thus we write

$$16x^2y + 24xy^2 - 32xy = 8(2x^2y + 3xy^2 - 4xy).$$

Next, we identify those variables appearing *in each term* of the polynomial within parentheses. For each such variable we factor out the largest power of the variable that is a factor of *all* the terms of the polynomial. As a result, we have

$$= 8xy(2x + 3y - 4).$$

The g.c.m.f. of $16x^2y + 24xy^2 - 32xy$ is $8xy$. Thus to express the given polynomial with the g.c.m.f. factored out, we write

$$16x^2y + 24xy^2 - 32xy = 8xy(2x + 3y - 4).$$

As a check that we factored correctly, we can multiply $8xy(2x + 3y - 4)$ to see if we obtain the original polynomial $16x^2y + 24xy^2 - 32xy$. *Readers are encouraged to check their work by doing this whenever factoring a polynomial.*

(b) Following the procedure outlined in part (a), we write

$$\begin{aligned} 56a^3b - 20a^2b^2 + 24a^2b &= 4(14a^3b - 5a^2b^2 + 6a^2b) \\ &= 4a^2b(14a - 5b + 6). \end{aligned}$$

We note that $4a^2b$ is the g.c.m.f. of the given polynomial.

We have used the distributive property to remove common monomial factors from a polynomial. It is sometimes possible to remove a *common binomial factor* from certain polynomials, as illustrated in the following examples.

Example 4. Factor $x^2 + 3x + ax + 3a$.

Solution

1. Group the terms of the polynomial into *two* groups each containing two terms.

$$(x^2 + 3x) + (ax + 3a).$$

2. Remove the greatest common monomial factor from each group.

$$x \cdot (x + 3) + a \cdot (x + 3).$$

The expression above is a two-term expression consisting of the terms $x(x + 3)$ and $a(x + 3)$.

3. Since there is a common *binomial factor,* use the distributive property to remove it.

$$x \cdot \underline{(x + 3)} + a \cdot \underline{(x + 3)} = \underline{(x + 3)}(x + \text{a}).$$

common binomial factor

Thus

$$x^2 + 3x + ax + 3a = (x + 3)(x + a).$$

Example 5. Factor $2x^3 - 3x^2 + 6x - 9$.

Solution

1. Group the terms.

$$(2x^3 - 3x^2) + (6x - 9).$$

2. Remove the common factors from each group.

$$x^2(2x - 3) + 3(2x - 3).$$

3. Remove the common binomial factor.

$$x^2(2x - 3) + 3(2x - 3) = (2x - 3)(x^2 + 3).$$

Thus

$$2x^3 - 3x^2 + 6x - 9 = (2x - 3)(x^2 + 3).$$

EXERCISES 3.6

For Exercises 1–20, factor the integer completely into prime numbers. If the given number is prime, state that it is prime. See Example 1.

1. 6 **2.** 4 **3.** 24 **4.** 11 **5.** 22 **6.** 12 **7.** 25
8. 18 **9.** 100 **10.** 27 **11.** −36 **12.** −50 **13.** 35 **14.** 150
15. 70 **16.** 14 **17.** 28 **18.** 31 **19.** 17 **20.** 51

For Exercises 21–38, factor the g.c.m.f. from the polynomial. If the polynomial does not have a g.c.m.f. other than 1 or −1, so state. See Examples 2 and 3 and the discussion preceding them.

21. $5x + 10$ **22.** $3x + 12$ **23.** $6x^2 + 18x$
24. $5y^2 - 30y$ **25.** $-4xy + 16x$ **26.** $-7ab + 14b$
27. $2x^2y - 18xy$ **28.** $2uv^2 + 16uv$ **29.** $9x^4 - 27x^3 + 18x^2$
30. $8z^5 + 16z^3 - 24z^2$ **31.** $16x^3y + 32x^2y$ **32.** $25x^2y^3 - 75x^2y^2$
33. $18a^2b^3 + 27ab^2c$ **34.** $15x^2y^2 - 30xy^2z$ **35.** $12a^3 - 6a^2b - 12ab^2$
36. $30s^4 - 5s^2t^2 + 15s^2t$ **37.** $21a^3b^2 - 14a^2b + 7a^2$ **38.** $20u^4v + 5u^3v^2 - 35u^3v$

For Exercises 39–52, factor by removing a common binomial factor. See Examples 4 and 5.

39. $x(z - 3) + 3t(z - 3)$ **40.** $m(2n + 5) - 4(2n + 5)$ **41.** $7s(a - b) - 3t(a - b)$
42. $x^2(y + 2) + 4(y + 2)$ **43.** $2a - 2b + 5ac - 5bc$ **44.** $x - xy + y - y^2$
45. $m^2 + 5m + bm + 5b$ **46.** $p^2 + 2p + 5p + 10$ **47.** $2xy - 3y + 10x - 15$
48. $3xz - z + 6x - 2$ **49.** $q^3 + 2q^2 - 3q - 6$ **50.** $x^3 + 3x^2 + 4x + 12$
51. $3y^3 - 2y - 6y^2 + 4$ **52.** $5t^3 - t - 10t^2 + 2$

Factoring Trinomials 3.7

In the preceding section we factored a common monomial from the terms of a polynomial. In this section we examine factoring trinomials (a polynomial with three terms) that have at least two binomial factors. We require that the binomial factors have integer coefficients. To be more specific, we observe that in *multiplying* polynomials, we are given two factors and asked to find the product. For example, the product of the binomials $(x + 3)$ and $(x + 4)$ is a polynomial:

$$(x + 3)(x + 4) = x^2 + 7x + 12.$$

In *factoring* we are given a polynomial, such as $x^2 + 7x + 12$, and asked for the factors from which it can be obtained. Thus, to factor $x^2 + 7x + 12$, we write

$$x^2 + 7x + 12 = (x + 3)(x + 4).$$

Therefore, $(x + 3)(x + 4)$ is the *factored form* of $x^2 + 7x + 12$. The factors $(x + 3)$ and $(x + 4)$ are the *binomial factors* of $x^2 + 7x + 12$.

In this section we study primarily trinomials of second degree that are factorable, such as

$$x^2 + 4x - 21, \qquad 3x^2 - 5x - 2, \qquad x^2 + xy - 6y^2.$$

As we observed before, in multiplication we have a routine method for determining a product. However, in factoring, our task is more difficult because we are given a polynomial and asked to find the factors of which it is composed. The factoring approach used in this section relies on an intelligent "trial-and-error method." It requires a good understanding of how to multiply binomials and how to factor integers.

We will first discuss factoring second-degree trinomials that have 1 as a coefficient of the second-degree term, such as

$$x^2 + 12x + 35.$$

The following formula will provide us with a guide for factoring such trinomials into the product of binomials:

product of two binomials $\longrightarrow$ $(x + b)(x + d) = x^2 + dx + bx + bd$

$$= x^2 + (b + d)x + bd.$$ $\longleftarrow$ *trinomial*

In this formula, b and d represent integer constants.

When seeking the binomial factors $(x + b)$ and $(x + d)$ of the trinomial, we observe from the formula given above that

1. The product of the integers b and d is the constant term of the trinomial, bd.
2. The sum of b and d, $b + d$, is the coefficient of the first-degree term, also called *the middle term*.

The following examples illustrate how to factor some trinomials.

Example 1. Factor $x^2 + 9x + 14$.

Solution: We would like to find integers b and d such that

(1) $$x^2 + 9x + 14 = (x + b)(x + d)$$

or

$$x^2 + 9x + 14 = x^2 + (b + d)x + bd.$$

Thus we need a pair of factors, b and d, of the constant term 14 whose product $bd = 14$ and whose sum $b + d = 9$. The possible pairs of positive factors of 14 are as follows: 1, 14 and 2, 7. Since $2 \cdot 7 = 14$ and $2 + 7 = 9$, we substitute $b = 2$ and $d = 7$ in line (1).

Thus

$$x^2 + 9x + 14 = (x + 2)(x + 7).$$

Note that if we were to let $b = 7$ and $d = 2$, we would obtain

$$x^2 + 9x + 14 = (x + 7)(x + 2).$$

By the commutative property this result is the same as $(x + 2)(x + 7)$.

Example 2. Factor $x^2 - 3x - 10$.

Solution: We seek integers b and d such that

(1) $$x^2 - 3x - 10 = (x + b)(x + d)$$

or

$$x^2 - 3x - 10 = x^2 + (b + d)x + bd.$$

Thus we need to find integers b and d so that

$$bd = -10 \qquad \text{and} \qquad b + d = -3.$$

Since the product of b and d is a negative integer, they must be opposite in sign. We list possible pairs of factors of -10 and their corresponding sum.

Factors b, d	Sum $b + d$
-1, 10	9
1, -10	-9
-2, 5	3
2, -5	-3

We see that the last pair of factors listed above are the required integers. Thus we let $b = 2$, $d = -5$ in line (1) and write

$$x^2 - 3x - 10 = (x + 2)(x - 5).$$

We now make some observations on the trial-and-error process used in Examples 1 and 2 that may prove helpful when factoring trinomials.

We observe that in Example 1, the constant term of the trinomial to be factored, $x^2 + 9x + 14$, is positive $(+14)$. For such trinomials the two factors of the constant term that we seek will have *like signs* and we will add their absolute values. Thus, in Example 1, to add algebraically the correct factors $+2$ and $+7$ we *add* the absolute values, 2 and 7, of the factors. Then we attach the common positive sign of the factors to this sum to obtain the coefficient of the middle term, $+9$.

In Example 2, the trinomial to be factored, $x^2 - 3x - 10$, has a negative constant term (-10). The two factors of -10 that we need will have *unlike signs* and we take the *difference* of their absolute values. Thus, in Example 2, to add algebraically the correct factors, -5 and 2, we take the *difference* of their absolute values, 5 and 2. Then we attach the appropriate sign to obtain the numerical coefficient of the middle term, which is -3.

The next example illustrates the use of the observations made above. It also illustrates that we should be alert to see if there is a greatest common monomial factor that can be removed from the trinomial so that the factoring process can be made easier.

Example 3. Factor $2x^2y + 8xy - 42y$.

Solution: We observe that each term has a common monomial factor of $2y$, and we begin by removing this g.c.m.f. Thus we have

$$2x^2y + 8xy - 42y = 2y(x^2 + 4x - 21).$$

To factor $x^2 + 4x - 21$, we seek integer factors of -21 whose algebraic *sum* is 4. We ignore 1 and -21 and -1 and 21 because the difference of their absolute values (20) is too large. If we try 3 and -7, we get a sum of -4. However, if we switch signs, we obtain the correct factors -3 and 7. Thus

$$x^2 + 4x - 21 = (x - 3)(x + 7)$$

and

$$\begin{aligned} 2x^2y + 8xy - 42y &= 2y(x^2 + 4x - 21) \\ &= 2y(x - 3)(x + 7). \end{aligned}$$

Example 4. Factor $x^2 - 7xy + 12y^2$.

Solution: To factor the given trinomial, we need to find integers b and d such that

$$x^2 - 7xy + 12y^2 = (x + by)(x + dy)$$

or

$$x^2 - 7xy + 12y^2 = x^2 + (b + d)xy + (bd)y^2.$$

Thus we require factors b and d of 12, since $bd = 12$, so that $b + d = -7$. Because the product of the factors is positive and their sum is negative, both factors must be negative. We can ignore -1 and -12, since their sum is too large. The correct factors to choose are -3 and -4. Thus the desired factorization is

$$x^2 - 7xy + 12y^2 = (x - 3y)(x - 4y).$$

At this point we should note that not every polynomial with integer coefficients can be factored into polynomials with integer coefficients, as we have required. We exclude $+1$ and -1 as polynomial factors. For example,

$$x^2 + x + 1, \qquad x^2 - 2x + 4, \qquad x^2 + 4x + 1$$

cannot be factored as $(x + b)(x + d)$, where b and d are integers. Such polynomials are said to be irreducible over the integers.

We now turn our attention to factoring second-degree trinomials having coefficients other than 1 or -1 as coefficients of the second-degree term. An example of such a trinomial is

$$2x^2 + 7x + 6.$$

To express the general quadratic trinomial $Ax^2 + Bx + C$ as the product of two binomials, $(ax + b)$ and $(cx + d)$, we need to find integers a, b, c, and d such that

(1) $$Ax^2 + Bx + C = (ax + b)(cx + d).$$

If we multiply out the right-hand side of (1) using the FOIL method, we obtain

(2) $$\begin{aligned} & \quad\;\; \text{F} \quad\; + \quad \text{O} \quad + \quad \text{I} \quad + \text{L} \\ Ax^2 + Bx + C &= (ac)x^2 + (ad)x + (bc)x + bd \\ &= \underset{\text{F}}{\underset{\uparrow}{(ac)x^2}} + \underset{\text{(O + I)}}{\underset{\uparrow}{(ad + bc)x}} + \underset{L}{\underset{\uparrow}{bd}}. \end{aligned}$$

Equations (1) and (2) suggest the following procedure for factoring a general quadratic trinomial $Ax^2 + Bx + C$.

Write down possible binomial products of the form $(ax + b)(cx + d)$ with the aim of finding values of a, b, c, and d such that

1. The product of a and c equals the coefficient of x^2. That is, $ac = A$, the coefficient of the *second-degree term* or *first term* of the trinomial.
2. The product of b and d equals the *constant term* C, or $bd = C$.
3. The coefficient of x or the middle term is $ad + bc$, that is, $ad + bc = B$.

The following examples illustrate the procedure outlined above for factoring quadratic trinomials of the form $Ax^2 + Bx + C$ when $A \neq 1$.

Example 5. Factor $2x^2 + 7x + 3$.

Solution: To factor $2x^2 + 7x + 3$, if it can be factored, we realize that its factors must be of the form $(ax + b)(cx + d)$. Thus we write

(1) $$2x^2 + 7x + 3 = (ax + b)(cx + d)$$

or

$$2x^2 + 7x + 3 = (ac)x^2 + (ad + bc)x + bd.$$

Therefore, we need to find integers a, b, c, and d such that $ac = 2$, $bd = 3$, and $ad + bc = 7$. We usually start by choosing the positive factors of the coefficient of the x^2 term. Thus we let $a = 2$ and $c = 1$, and as a result we have

(2) $$2x^2 + 7x + 3 = (2x + b)(x + d) = \underset{\text{F}}{\underset{\uparrow}{2x^2}} + \underset{\text{(O + I)}}{\underset{\uparrow}{(2d + b)x}} + \underset{\text{L}}{\underset{\uparrow}{bd}}.$$

Next, we have to choose factors of 3, b and d, such that $bd = 3$, which will give us the coefficient of the middle term, $2d + b = 7$. We see that $d = 3$ and $b = 1$ gives us the desired result. Thus, substituting in line (2), we get

$$2x^2 + 7x + 3 = (2x + 1)(x + 3).$$

Example 6. Factor $3y^2 - 11y + 10$.

Solution: We wish to write $3y^2 - 11y + 10$ as the product of two binomials $(ay + b)(cy + d)$. Thus we set

(1) $$3y^2 - 11y + 10 = (ay + b)(cy + d)$$

or

$$3y^2 - 11y + 10 = (ac)y^2 + (ad + bc)y + bd.$$

Next, we must find integer values a, b, c, and d such that $ac = 3$, $bd = 10$, and $ad + bc = -11$. After trying several possibilities in line (1), we see that $a = 3$, $c = 1$ and $b = -5$, $d = -2$ are the required values. As a result, we find that

$$3y^2 - 11y + 10 = (3y - 5)(y - 2).$$

Example 7. Factor $6x^2 + 5xy - 25y^2$.

Solution: To factor the given trinomial into binomial factors, we have to express $6x^2 + 5xy - 25y^2$ in the following form:

(1) $$6x^2 + 5xy - 25y^2 = (ax + by)(cx + dy).$$

Note that a, b, c, and d represent integers. The positive pairs of factors a and c of 6 to be considered are as follows: 1 and 6 or 2 and 3. The factors b, d of -25 could be 1 and -25, -1 and 25, or -5 and 5. After trying combinations of the factors of 6 with those of -25 in the binomial factors $(ax + by)(cx + dy)$ in line (1), we find that

$$6x^2 + 5xy - 25y^2 = (2x + 5y)(3x - 5y).$$

As a check we multiply the binomial factors to make sure that we obtained the correct factorization of $6x^2 + 5xy - 25y^2$. As a result, we see that we have factored correctly, since

$$\begin{aligned}(2x + 5y)(3x - 5y) &= 6x^2 - 10xy + 15xy - 25y^2\\ &= 6x^2 + 5xy - 25y^2.\end{aligned}$$

EXERCISES 3.7

For Exercises 1–32, factor the trinomial. First check to see if there is a g.c.m.f. See Examples 1–4.

1. $x^2 + 4x + 3$ **2.** $x^2 + 3x + 2$ **3.** $x^2 + 8x + 7$

4. $x^2 + 6x + 5$
5. $x^2 + 5x + 4$
6. $x^2 + 5x + 6$
7. $x^2 - 4x - 5$
8. $x^2 - 6x - 7$
9. $y^2 - 2y - 8$
10. $z^2 - z - 12$
11. $x^2 - 7x - 18$
12. $x^2 - 4x - 21$
13. $x^2 - 6x + 8$
14. $x^2 - 7x + 12$
15. $t^2 - 9t + 14$
16. $s^2 - 8s + 15$
17. $x^2 - 9x + 18$
18. $x^2 - 11x + 24$
19. $x^2 + 8x - 9$
20. $x^2 + 6x - 7$
21. $w^2 + 10w - 11$
22. $v^2 + 12v - 13$
23. $m^2 + 4m - 12$
24. $n^2 + 6n - 16$
25. $x^2 + 9xy + 18y^2$
26. $x^2 + 12xy + 27y^2$
27. $u^2 - 13uv + 36v^2$
28. $s^2 - 4st - 45t^2$
29. $3x^2 - 39xy + 66y^2$
30. $4t^2s - 40ts + 64s$
31. $2xy^2 - 2xy - 40x$
32. $5x^2z - 35xz + 50z$

For Exercises 33–59, factor the trinomial. Check to see if there is a g.c.m.f. and then factor. See Examples 5–7.

33. $2x^2 + 15x + 7$
34. $2x^2 + 5x + 3$
35. $2x^2 + 11x + 5$
36. $2x^2 + 5x + 2$
37. $3y^2 - 8y + 5$
38. $3y^2 - 10y + 7$
39. $3x^2y - 16xy + 5y$
40. $3x^2z - 14xz + 8z$
41. $7t^2 + 2t - 5$
42. $7s^2 + 4s - 3$
43. $5t^2 + 12t - 9$
44. $50y^2 + 20y - 70$
45. $60t^2 - 10t - 150$
46. $11u^2 - 4u - 7$
47. $6x^2 - 5x - 6$
48. $4m^2 - 5m - 6$
49. $6x^2y + 11xy - 10y$
50. $8y^2x - 2yx - 3x$
51. $9x^2 + 18xy + 8y^2$
52. $10t^2 + 29ts + 10s^2$
53. $6x^2 - 11x - 10$
54. $4y^2 - 20y + 21$
55. $9x^2 - 17x + 8$
56. $12x^2 - x - 20$
57. $20y^2 - 50y - 30$
58. $20x^2 - 130x - 70$
59. $10t^2 - 19t + 6$

For Exercises 60–70, factor out the g.c.m.f. and then factor the remaining trinomial.

60. $2x^2y^2 + 9x^2y - 5x^2$
61. $2x^2z^2 - 5xz^2 + 2z^2$
62. $a^5 + 3a^4 + 2a^3$
63. $b^5 + 5b^4 + 6b^3$
64. $9x^3 + 33x^2 - 60x$
65. $8x^3y + 24x^2y + 16xy$
66. $5xy^3 + 5xy^2 - 30xy$
67. $24t^4 - 4t^3 - 48t^2$
68. $16x^2 + 32xy + 16y^2$
69. $18x^3 + 3x^2y - 105xy^2$
70. $15m^3 - 18m^2n - 24mn^2$

Factoring Special Polynomials 3.8

We conclude our discussion of factoring in this chapter with a discussion on factoring polynomials that have a special pattern or form. They occur frequently in mathematics and therefore it is important that we become acquainted with them.

We begin by recalling that in Section 3.5 we presented the following formulas for finding special binomial products:

(a) $(x+y)^2 = x^2 + 2xy + y^2$
(b) $(x-y)^2 = x^2 - 2xy + y^2$
(c) $(x+y)(x-y) = x^2 - y^2.$

Since factoring is the "reverse" of multiplying, these formulas, rewritten by interchanging right and left sides of the equations, provide us with the following *factoring formulas:*

Factored Form of Some Special Polynomials

(1) $x^2 + 2xy + y^2 = (x+y)^2$
(2) $x^2 - 2xy + y^2 = (x-y)^2$
(3) $x^2 - y^2 = (x+y)(x-y).$

We see from lines (1) and (2) that trinomials of the form

$$x^2 + 2xy + y^2 \qquad \text{and} \qquad x^2 - 2xy + y^2$$

can be expressed as the squares of the binomials $(x + y)$ and $(x - y)$, respectively. As a result, they are called *perfect square trinomials*. We note that a trinomial will have the same *pattern* or *form* as $x^2 + 2xy + y^2$ if it has the sum of *two* terms that are perfect squares and the remaining term is *twice the product* of the terms being squared.

A polynomial having the same pattern or form as $x^2 - y^2$, appearing in line (3) is called the *difference of two squares*. We observe that generally the sum of two squares, such as $x^2 + y^2$, is *not* factorable using integers. For example, $x^2 + 1$ is not factorable using integers.

Learning to recognize which polynomials "fit" the special polynomial patterns discussed above is important for correctly applying formulas (1)–(3). In Example 1 we examine some examples of such special polynomials.

Example 1. Determine whether the following polynomials are perfect square trinomials, the difference of two squares, or neither.
(a) $9z^2 + 24z + 16$. (b) $4a^2 - 25b^2$. (c) $49a^2 - 21ab + 9b^2$.

Solution

(a) If we look at the patterns of the perfect square trinomials appearing in formulas (1) and (2) and compare them to $9z^2 + 24z + 16$, we see that

perfect square
trinomial pattern: $x^2 \quad + \quad 2xy \quad + \quad y^2$

trinomial: $(3z)^2 + 2(4)(3z) + (4)^2 = 9z^2 + 24z + 16.$

Therefore, $9z^2 + 24z + 16$ is a perfect square trinomial.

(b) We see that

$$4a^2 - 25b^2 = \underset{\underset{x^2}{\uparrow}}{(2a)^2} - \underset{\underset{y^2.}{\uparrow}}{(5b)^2} \qquad \text{difference of two squares}$$

Thus $4a^2 - 25b^2$ is the difference of two squares.

(c) For the trinomial $49a^2 - 21ab + 9b^2$, we see that $49a^2 = (7a)^2$ and also $9b^2 = (3b)^2$ are perfect squares. We compare its pattern to the perfect square trinomial $x^2 - 2xy + y^2$ appearing in formula (2). We notice, however, that the middle term of the given polynomial, $21ab \neq 2(7a)(3b) = 42ab$. Therefore, $49a^2 - 21ab + 9b^2$ is *not* a perfect square trinomial.

Whenever we encounter and recognize polynomials that are *perfect square trinomials* or the *difference of two squares* we can factor them using Formulas (1)–(3) for Special Polynomials. However, if we do not recognize these perfect square forms, the previous techniques developed in this chapter can be used to factor the polynomial.

Example 2. Factor $x^2 + 8x + 16$.

Solution: We see that $x^2 + 8x + 16$ is a perfect square trinomial, since

$$x^2 + 8x + 16 = x^2 + 2(x \cdot 4) + 4^2.$$

Thus, using formula (1), $x^2 + 2xy + y^2 = (x + y)^2$, we factor $x^2 + 8x + 16$ as follows:

$$x^2 + 8x + 16 = (x + 4)^2.$$

Example 3. Factor $9a^2 - 25b^2$.

Solution: We observe that the given polynomial is the difference of two squares, since

$$9a^2 - 25b^2 = (3a)^2 - (5b)^2.$$

Thus, using factoring formula (3), $x^2 - y^2 = (x + y)(x - y)$, with $x = 3a$ and $y = 5b$, we write

$$(3a)^2 - (5b)^2 = (3a + 5b)(3a - 5b)$$

or

$$9a^2 - 25b^2 = (3a + 5b)(3a - 5b).$$

Example 4. Factor as far as possible the trinomial $x^4 + x^2 - 2$.

Solution: The given trinomial is *not* a perfect square trinomial. However, we can factor it as we did the general trinomial.

$$x^4 + x^2 - 2 = (x^2 - 1)(x^2 + 2)$$

But note that $x^2 - 1 = x^2 - 1^2$ is the difference of two squares and can be factored as

$$x^2 - 1 = (x + 1)(x - 1).$$

As a result, we have

$$\begin{aligned} x^4 + x^2 - 2 &= (x^2 - 1)(x^2 + 2) \\ &= (x + 1)(x - 1)(x^2 + 2). \end{aligned}$$

Observe that $x^2 + 2$ cannot be factored using integer coefficients. Thus

$$x^4 + x^2 - 2 = (x + 1)(x - 1)(x^2 + 2).$$

Example 4 illustrates how to factor a polynomial as far as possible with integers. The next example illustrates again that we should always remove the g.c.m.f. from a polynomial before attempting to factor it.

Example 5. Factor $2xy^2 - 24xy + 72x$.

Solution: First we factor out from the given polynomial the g.c.m.f. $2x$. Thus we have

$$2xy^2 - 24xy + 72x = 2x(y^2 - 12y + 36).$$

The other factor $y^2 - 12y + 36$ is a perfect square trinomial and is the square of $(y - 6)$. Thus

$$\begin{aligned} 2xy^2 - 24xy + 72x &= 2x(y^2 - 12y + 36) \\ &= 2x(y - 6)^2. \end{aligned}$$

Example 6. Factor $4a^2 + 12ab + 9b^2$.

Solution: The polynomial $4a^2 + 12ab + 9b^2$ is a perfect square trinomial of the same pattern as $x^2 + 2xy + y^2$. Note that $4a^2 = (2a)^2$, $9b^2 = (3b)^2$, and that the middle term of the trinomial $12ab = 2(2a)(3b)$. Consequently, we factor the given polynomial by setting $x = 2a$ and $y = 3b$ in the factoring formula

$$x^2 + 2xy + y^2 = (x + y)^2.$$

As a result, we have

$$\begin{aligned} 4a^2 + 12ab + 9b^2 &= (2a)^2 + 2(2a)(3b) + (3b)^2 \\ &= (2a + 3b)(2a + 3b) \\ &= (2a + 3b)^2. \end{aligned}$$

To conclude this section, we present factorizations of two additional special polynomials: the sum of two cubes, $x^3 + y^3$, and the difference of two cubes, $x^3 - y^3$.

Sum and Difference of Two Cubes

$$x^3 + y^3 = (x + y)(x^2 - xy + y^2).$$
$$x^3 - y^3 = (x - y)(x^2 + xy + y^2).$$

We note that the trinomial factors $(x^2 - xy + y^2)$ and $(x^2 + xy + y^2)$ in these formulas cannot be factored further using integer coefficients.

The formulas given above can be verified by multiplying the products on the right side of each equation. Note that although we cannot in general factor the sum of two squares, we can factor the sum of two cubes.

In the following examples, we illustrate the application of the formulas for factoring polynomials that are the sum of two cubes or the difference of two cubes.

Example 7. Factor the following polynomials.

(a) $x^3 + 1$. (b) $27m^3 + 8$.

Solution

(a) $x^3 + 1 = (x)^3 + (1)^3 = (x + 1)(x^2 - x \cdot 1 + 1^2)$
$= (x + 1)(x^2 - x + 1)$.

Thus

$$x^3 + 1 = (x + 1)(x^2 - x + 1).$$

(b) $27m^3 + 8 = (3m)^3 + (2)^3 = (3m + 2)[(3m)^2 - (3m)(2) + (2)^2]$
$= (3m + 2)(9m^2 - 6m + 4)$.

Example 8. Factor the following polynomials.

(a) $a^3 - 64$. (b) $8x^3 - 125y^3$.

Solution

(a) $a^3 - 64 = (a)^3 - (4)^3 = (a - 4)(a^2 + a(4) + 4^2)$
$= (a - 4)(a^2 + 4a + 16)$.

Thus

$$a^3 - 64 = (a - 4)(a^2 + 4a + 16).$$

(b) $8x^3 - 125y^3 = (2x)^3 - (5y)^3 = (2x - 5y)[(2x)^2 + (2x)(5y) + (5y)^2]$
$= (2x - 5y)(4x^2 + 10xy + 25y^2)$.

EXERCISES 3.8

For Exercises 1–12, identify the polynomial as a perfect square trinomial, the difference of two squares, or neither. See Example 1.

1. $x^2 - 10x + 25$ **2.** $y^2 + 14y + 49$ **3.** $t^2 + 4t + 16$

4. $9z^2 - 15z + 25$ **5.** $16x^2 - y^2$ **6.** $a^2 - 49b^2$

7. $16x^2 - 40xy + 25y^2$ **8.** $9u^2 - 42uv + 49v^2$ **9.** $x^2 + x + 1$

10. $y^2 - 2y + 1$ **11.** $9x^2 - 18y^2$ **12.** $4x^3 - 12y^2$

For Exercises 13–50, factor the polynomial. See Examples 2–6.

13. $x^2 - 4x + 4$ **14.** $x^2 + 6x + 9$ **15.** $y^2 + 14y + 49$

16. $z^2 - 16z + 64$ **17.** $y^2 - 14y + 49$ **18.** $x^2 + 18x + 81$

19. $t^2 - 2t + 1$ **20.** $u^2 + 2u + 1$ **21.** $4u^2 - v^2$

22. $s^2 - 16t^2$ **23.** $9x^2 - 49y^2$ **24.** $81a^2 - 100b^2$

25. $4x^2 - 64y^2$ **26.** $3s^2 - 27y^2$ **27.** $x^4 - 3x^2 - 4$

28. $x^4 - 8x^2 - 9$ **29.** $x^4 - 7x^2 - 18$ **30.** $x^4 - 14x^2 - 32$

31. $y^4 - 5y^2 + 4$ **32.** $z^4 - 13z^2 + 36$ **33.** $2x^2y - 44xy + 242y$

34. $2a^2b + 48ab + 288b$ **35.** $a^3 + 20a^2 + 100a$ **36.** $b^3 - 26b^2 + 169b$

37. $9y^2 - 48y + 64$ **38.** $4x^2 - 20x + 25$ **39.** $4a^2 + 24ab + 36b^2$

40. $49u^2 + 28uv + 4v^2$ **41.** $9x^2 - 24xy + 16y^2$ **42.** $4t^2 + 20ts + 25s^2$

43. $25x^2 - 40xy + 16y^2$ **44.** $4a^2 - 28ab + 49b^2$ **45.** $(x + 2)^2 - 9$

46. $(y + 1)^2 - 4$ **47.** $16 - (a + b)^2$ **48.** $25 - (x + y)^2$

49. $(z - 1)^2 - 1$ **50.** $(a - 2)^2 - 4$

For Exercises 51–74, factor the sum or difference of the two cubes. See Examples 7 and 8.

51. $x^3 - 1$ **52.** $a^3 - 8$ **53.** $m^3 - 27$ **54.** $r^3 - s^3$

55. $u^3 - v^3$ **56.** $27n^3 + 1$ **57.** $x^3 + 64$ **58.** $y^3 - 125$

59. $a^3 - 216$ **60.** $64 - x^3$ **61.** $8a^3 - b^3$ **62.** $125y^3 - z^3$

63. $64m^3 + 8n^3$ **64.** $216x^3 + 27y^3$ **65.** $z^3 - 343$ **66.** $t^3 + 343$

67. $y^6 + 1$ **68.** $b^6 - 64$ **69.** $2m^3 + 16$ **70.** $4x^3 - 108$

71. $(x - 1)^3 + 1$ **72.** $(x + 1)^3 - 1$ **73.** $(x + 2)^3 - 8$ **74.** $(x - 3)^3 + 27$

Key Words

Base
Binomial
Coefficient
Composite number
Cubes
  Sum of two cubes
  Difference of two cubes
Degree of a polynomial

Factor
FOIL
Greatest common monomial factor
Leading coefficient
Like terms
Monomial
Polynomials
  Standard form
  Term
  Irreducible
Power
Prime number
Scientific notation
Squares
  Difference of two squares
Trinomial
  Perfect square trinomial

CHAPTER 3 REVIEW

SECTION 3.1

Evaluate the exponent forms.

1. $(-7)^2$ **2.** -9^2 **3.** $(-4)^3$ **4.** $(-3)^{-2}$ **5.** 8^0

Write the expressions in exponent form.

6. $3 \cdot 3 \cdot 3 \cdot 3$ **7.** bbb **8.** $aaabb$ **9.** $(-r)(r)ssss$

SECTION 3.2

Simplify the expressions as much as possible. Write all answers with positive exponents, when possible.

10. x^7x^{-2} **11.** $y^{-4}y^4$ **12.** $(a^{-3}b^5)(a^4b^{-6})$

Simplify the expressions. Write all answers with positive exponents.

13. $(4x)^3$ **14.** $(z^2)^4$ **15.** $(r^3s^2)^4$ **16.** $(t^{-2})^3$ **17.** $(x^2)^{-3}(y^{-3})$

18. $(x^3y^2)(x^5y)$ **19.** $\left(\frac{t^2}{s^3}\right)^3$ **20.** $\left(\frac{a}{b}\right)^{-1}$ **21.** $\frac{r^5s^4}{r^2s^6}$ **22.** $\frac{a^{-5}b^4}{a^{-6}b^3}$

SECTION 3.3

Express the numbers in scientific notation.

23. 5,300,000,000 **24.** .00000023

Convert the numbers from scientific notation to ordinary decimal form.

25. 7.2×10^{-3} **26.** 3.28×10^4

Calculate and express the answer in scientific notation or ordinary decimal form.

27. $\dfrac{(4.0 \times 10^{6})(8.0 \times 10^{-6})}{1.6 \times 10^{-4}}$

SECTION 3.4

Simplify the polynomials.

28. $8x^2 - 2x + 4x^2 - x$ **29.** $7a^3b - 3a - 2a^3b - 5a$

Perform the operations indicated.

30. $(5t^2 - 3t + 2) + (7t^2 + t - 5)$ **31.** $(6y^4 + y^3 - 1) - (2y^4 - y^3 - 1)$

SECTION 3.5

Find the indicated products.

32. $xy^2(x^2y^3 - y)$ **33.** $(2x + 3y)(4x - 5y)$ **34.** $(3r + s)^2$

35. $(y - 1)(y^2 + y + 1)$ **36.** $(3r + 2s)(3r - 2s)$

SECTION 3.6

Factor the integers completely into prime numbers.

37. 14 **38.** 90

Factor the polynomials as far as possible.

39. $7x^2 + 28x$ **40.** $3x^2y - 27x^2y^2$

SECTION 3.7

Factor the trinomials.

41. $x^2 - 5x - 6$ **42.** $3x^2 + 7x + 2$ **43.** $xy^2 + 3xy + 2x$

SECTION 3.8

Factor the polynomials.

44. $4a^2 - 4ab + b^2$ **45.** $4x^2 - 25y^2$ **46.** $9t^2 + 24st + 16s^2$ **47.** $s^3 - 8$

48. $8t^3 + 27$

CHAPTER 3 TEST

For Problems 1–4, evaluate the exponent form.

1. $(-10)^2$ **2.** -5^3 **3.** 7^0 **4.** $(-2)^{-3}$

For Problems 5–9, simplify the expression. Write all answers with positive exponents.

5. $(x^7y^3)(x^2y)$ **6.** $(r^3s^4t^2)(r^{-1}s^{-2}t^3)$ **7.** $(t^3)^4(t^2)^{-2}$ **8.** $\dfrac{x^6y^7z^2}{x^3y^5z^4}$ **9.** $\left(\dfrac{a^3}{b^2}\right)^{-2}$

For Problems 10 and 11, express the number in scientific notation.

10. .0000089 **11.** 93,000,000

For Problems 12–19, perform the indicated operation.

12. $(2x^2 - 3x + 5) + (6x^2 + 7x - 1)$

13. $(10y^5 - 3z^2 + 4) - (3y^5 + 4z^2 - 2)$

14. $xy(x^2y - xy^2)$

15. $(5x + 1)(2x + 3)$

16. $(3x + 2y)(2x - 9y)$

17. $(4t + 3)^2$

18. $(a^m - b^n)(a^m + b^n)$

19. $(x + 1)(x^2 - x + 1)$

For Problems 20–26, factor the polynomial as far as possible.

20. $x^2(y + 1) - (y + 1)$ **21.** $x^3 + x^2 - x - 1$ **22.** $6x^2 + 11x + 4$

23. $y^2 - 14y + 49$ **24.** $8x^2 - 50y^2$ **25.** $64a^3 - 125b^3$

26. $27y^6 + 8$

CHAPTER 4

Basic Operations with Rational Expressions

Equations with rational expressions can be used to solve applied problems. For example, the spray tank on a tree surgeon's truck can be filled by a large hose in 50 minutes. It takes 75 minutes for a smaller hose to fill the same tank. The solution to the equation

$$\frac{1}{50} + \frac{1}{75} = \frac{1}{x}$$

will tell how long it takes to fill the tank using both hoses simultaneously.

U.S. Department of Agriculture

CHAPTER 4
CHAPTER 4
CHAPTER 4
CHAPTER 4
CHAPTER 4
CHAPTER 4
CHAPTER 4
CHAPTER 4
CHAPTER 4
CHAPTER 4
CHAPTER 4
CHAPTER 4
CHAPTER 4
CHAPTER 4
CHAPTER 4
CHAPTER 4
CHAPTER 4
CHAPTER 4
CHAPTER 4
CHAPTER 4
CHAPTER 4
CHAPTER 4
CHAPTER 4
CHAPTER 4
CHAPTER 4
CHAPTER 4
CHAPTER 4
CHAPTER 4
CHAPTER 4
CHAPTER 4
CHAPTER 4
CHAPTER 4
CHAPTER 4
CHAPTER 4

In this chapter we study rational expressions. We learned in Chapter 1 that a rational number is any number that may be written as the quotient of two integers, with the denominator not zero. A *rational expression* is a fraction whose numerator and denominator are both polynomials, with the denominator not zero. Each of the following is an example of a rational expression:

$$\frac{x^2 - 1}{x^2 + 1}, \qquad \frac{7x^3 - 2x + 5}{4x + 7}, \qquad \frac{3xy}{2}, \qquad \frac{11}{5}.$$

Since any polynomial such as $x^2 + x + 1$ can be written as a fraction with a denominator of 1, $\frac{x^2 + x + 1}{1}$, we see that all polynomials are rational expressions.

4.1 Simplifying Rational Expressions

The variables in a rational expression may be replaced by any real number *except those numbers that make the denominator equal to zero*. We say that a fraction is *undefined* for those values that make the denominator zero. Thus $\frac{3x}{x + 2}$ is undefined when $x + 2 = 0$, that is, when $x = -2$.

Example 1. For what values of the variables are the following rational expressions undefined?

(a) $\frac{4y - 9}{y - 1}$. (b) $\frac{3x}{x + 10}$. (c) $\frac{5z + 2}{11}$.

Solution: Each of these fractions is defined for any value except for those values that make its denominator equal zero. Thus, to determine for what numbers each fraction is undefined, we need to know for what values each denominator is zero.

(a) For $\frac{4y - 9}{y - 1}$ we solve the equation $y - 1 = 0$. Thus, when $y = 1$, $\frac{4y - 9}{y - 1}$ is undefined.

(b) For $\frac{3x}{x + 10}$ we solve the equation $x + 10 = 0$. Thus $\frac{3x}{x + 10}$ is defined except when $x = -10$.

(c) The denominator of $\frac{5z + 2}{11}$ is the constant 11. Thus this denominator is never zero. We conclude that $\frac{5z + 2}{11}$ is defined for all real numbers.

It is often necessary to rewrite a rational number with a different denominator. For example, $\frac{6}{9}$ may be rewritten as $\frac{2}{3}$ or $\frac{12}{18}$ because

$$\frac{6}{9} = \frac{2 \cdot 3}{3 \cdot 3} = \frac{2}{3} \quad \text{or} \quad \frac{6}{9} = \frac{6 \cdot 2}{9 \cdot 2} = \frac{12}{18}.$$

Similarly, rational expressions may be rewritten by applying the *fundamental property of rational expressions.*

Fundamental Property of Rational Expressions

If P, Q, and D are polynomials with $Q \neq 0$ and $D \neq 0$, then

$$\frac{P \cdot D}{Q \cdot D} = \frac{P}{Q}.$$

This property tells us that we may multiply or remove from both the numerator and denominator of a rational expression the same nonzero polynomial. If we read the above equation from right to left, we see that we may multiply both the numerator and denominator of a rational expression by the same nonzero polynomial without changing the value of the rational expression. Similarly, reading the equation from left to right, we may remove from both the numerator and denominator the same nonzero polynomial. When the numerator and denominator of a rational expression have no common factor (other than ± 1), we say the rational expression has been *reduced to lowest terms*. We often use the fundamental property of rational expressions to reduce rational expressions to lowest terms.

Example 2. Reduce $\frac{28x^2}{35xy}$ to lowest terms.

Solution: First note that $7x$ is a common factor of both the numerator and denominator. Thus $\frac{28x^2}{35xy} = \frac{(7x)4x}{(7x)5y}$. Next, we apply the fundamental property of rational expressions. Thus $\frac{28x^2}{35xy} = \frac{(7x)4x}{(7x)5y} = \frac{4x}{5y}$. Since $D = 7x$ and $Q = 5y$ are zero when either $x = 0$ or $y = 0$, this result assumes that $x \neq 0$ and $y \neq 0$.

The process of removing the same factor from the numerator and denominator is usually indicated by *placing slash marks through the common factors*. Thus, in Example 2, we would write

$$\frac{28x^2}{35xy} = \frac{\cancel{(7x)}4x}{\cancel{(7x)}5y} = \frac{4x}{5y}.$$

Similarly, to reduce $\frac{5z - 15}{z^2 - 6z + 9}$, we would write

$$\frac{5z - 15}{z^2 - 6z + 9} = \frac{5\cancel{(z - 3)}}{\cancel{(z - 3)}(z - 3)} = \frac{5}{z - 3},$$

provided that $z - 3 \neq 0$.

For the rest of this chapter we assume that the variables in the denominator of a rational expression do not assume values that make the denominator zero. Thus we will usually not mention the restrictions that are implicit in the examples and problems.

Example 3. Reduce each rational expression to lowest terms.

(a) $\frac{y^3 - 2y^2 - 15y}{y^3 - 9y^2 + 20y}$. (b) $\frac{a - 4}{24 - 6a}$.

Solution

(a) Factoring numerator and denominator, we see that

$$\frac{y^3 - 2y^2 - 15y}{y^3 - 9y^2 + 20y} = \frac{\cancel{y}(y + 3)\cancel{(y - 5)}}{\cancel{y}(y - 4)\cancel{(y - 5)}} = \frac{y + 3}{y - 4}.$$

(b) The denominator of $\frac{a - 4}{24 - 6a}$ may be factored as $24 - 6a = 6(4 - a)$. However, since the factor $(4 - a) = -(a - 4)$ we could also factor the denominator as $-6(a - 4)$. Thus

$$\frac{a - 4}{24 - 6a} = \frac{1\cancel{(a - 4)}}{-6\cancel{(a - 4)}} = \frac{1}{-6} = -\frac{1}{6}.$$

Sometimes it is necessary to rewrite a fraction such as $\frac{6}{9}$ with a larger denominator. For example, to rewrite $\frac{6}{9}$ with a new denominator of 18, we use the fundamental property of rational expressions and multiply the numerator and denominator by 2. We use 2 because we notice that $18 = 9 \cdot 2$. Thus we see that

$$\frac{6}{9} = \frac{6 \cdot 2}{9 \cdot 2} = \frac{12}{18}.$$

We extend this procedure to rational expressions as shown in the next example.

Example 4. Rewrite each rational expression with the indicated denominator.

(a) $\frac{6w}{11z}$; with denominator $44wz^3$.

(b) $\dfrac{x-1}{2x-3}$; with denominator $2x^2 - x - 3$.

Solution: The polynomial that we multiply numerator and denominator of each given rational expression by is the polynomial that we need to multiply the old denominator by in order to obtain the new denominator.

(a) $44wz^3 = (11z)4wz^2$. Thus we get

$$\frac{6w}{11z} = \frac{6w(4wz^2)}{11z(4wz^2)} = \frac{24w^2z^2}{44wz^3}.$$

(b) Since $2x^2 - x - 3 = (2x-3)(x+1)$, we multiply numerator and denominator of the given rational expression by $(x+1)$. Thus we get

$$\frac{x-1}{2x-3} = \frac{(x-1)(x+1)}{(2x-3)(x+1)} = \frac{x^2-1}{2x^2-x-3}.$$

EXERCISES 4.1

For Exercises 1–12, find the values of the variable, if any, for which the rational expression is not defined. See Example 1.

1. $\dfrac{4x}{7x}$ **2.** $\dfrac{3z^3}{2z}$ **3.** $\dfrac{5x}{4}$ **4.** $\dfrac{11w}{2}$ **5.** $\dfrac{7}{2p}$ **6.** $\dfrac{3}{5q}$

7. $\dfrac{1}{u-2}$ **8.** $\dfrac{1}{v-5}$ **9.** $\dfrac{3}{p+6}$ **10.** $\dfrac{5}{q+10}$ **11.** $\dfrac{3x+2}{x-5}$ **12.** $\dfrac{5y-1}{y-9}$

For Exercises 13–36, reduce the expression to lowest terms. See Examples 2 and 3.

13. $\frac{33}{55}$ **14.** $\frac{60}{84}$ **15.** $\dfrac{15x^2}{10x}$ **16.** $\dfrac{20y}{15y^2}$

17. $\dfrac{42x^2y}{60xy^2}$ **18.** $\dfrac{35wz^3}{49w^2z}$ **19.** $\dfrac{84a^2bc^3}{126ab^2c}$ **20.** $\dfrac{39pq^3r^2}{26p^2qr^3}$

21. $\dfrac{125r^3st^4}{(5rst^2)^2}$ **22.** $\dfrac{16w^7x^3y^5}{(2wx^2y^2)^2}$ **23.** $\dfrac{6x(x+3)}{24x^2(x+3)}$ **24.** $\dfrac{14y(y-5)}{21y^2(y-5)}$

25. $\dfrac{w^2-1}{w^2-2w+1}$ **26.** $\dfrac{z^2-4}{z^2-4z+4}$ **27.** $\dfrac{4z+8}{z^2+5z+6}$ **28.** $\dfrac{5y-10}{y^2-8y+12}$

29. $\dfrac{p^2-3p}{p^2+2p-15}$ **30.** $\dfrac{q^2+5q}{q^2+2q-15}$ **31.** $\dfrac{6y^2-24}{3y+6}$ **32.** $\dfrac{5z^2-45}{10z-30}$

33. $\dfrac{w-2}{2-w}$ **34.** $\dfrac{7-t}{t-7}$ **35.** $\dfrac{3p-21}{42-6p}$ **36.** $\dfrac{8+4q}{6q+12}$

For Exercises 37–48, rewrite the rational expression with the indicated denominator. See Example 4.

37. $\frac{3}{4}$; with denominator 32

38. $\frac{2}{7}$; with denominator 35

39. $\dfrac{4x}{5}$; with denominator $15x^2$

40. $\dfrac{3y}{8}$; with denominator $16y^3$

41. $\dfrac{7p}{3q}$; with denominator $12pq^2$

42. $\dfrac{5w}{2z}$; with denominator $10w^2z$

43. $\dfrac{x-5}{x+1}$; with denominator $x^2 - 1$

44. $\dfrac{y+3}{y-4}$; with denominator $y^2 - 16$

45. $\dfrac{w+7}{w+2}$; with denominator $w^2 + 4w + 4$

46. $\dfrac{z-5}{z+3}$; with denominator $z^2 + 6z + 9$

47. $\dfrac{p-1}{p+7}$; with denominator $2p^2 + 17p + 21$

48. $\dfrac{q+4}{q-3}$; with denominator $2q^2 - 4q - 6$

For Exercises 49–52, determine whether the statement is true or false.

49. Every rational number is a rational expression.

50. Every polynomial is a rational expression.

51. Every rational expression is a polynomial.

52. If a value for a variable makes the numerator of a rational expression equal to zero, the rational expression is undefined at that number.

4.2 Polynomial Division

In the preceding chapter we learned how to add, subtract, and multiply two polynomials. We now show how to divide a polynomial by another polynomial and express the result using only polynomials. This can be done using the idea of a remainder. Thus, to divide a polynomial P (the dividend) by a polynomial S (the divisor) *means* to find a polynomial Q (the quotient) and a polynomial R (the remainder) such that

$$P = Q \cdot S + R,$$

where the degree of R is less than the degree of S.

We will consider division by a monomial first.

Example 1. Divide $10x^4 - 5x^3 + 15x^2$ by $5x$.

Solution: Rewrite the division problem as shown, factor the numerator, and apply the Fundamental Property of Rational Expressions.

Thus

$$\begin{aligned}
&(10x^4 - 5x^3 + 15x^2) \div 5x \\
&= \frac{10x^4 - 5x^3 + 15x^2}{5x} \\
&= \frac{\cancel{5x}(2x^3 - x^2 + 3x)}{\cancel{5x}} \\
&= 2x^3 - x^2 + 3x .
\end{aligned}$$

To check, we multiply the quotient $2x^3 - x^2 + 3x$ by the divisor $5x$. $5x(2x^3 - x^2 + 3x) = 10x^4 - 5x^3 + 15x^2$. Since this product equals the original dividend, the answer is correct.

We now turn our attention to division problems in which the divisor is a binomial. In this case we proceed as we did with the long division of two whole numbers. For example, to divide $8x^3 + 16x^2 + 8x + 1$ by $2x + 1$, we proceed as follows.

$$\begin{array}{rl}
& 4x^2 \\
2x + 1 \!\!\!& \big)\overline{8x^3 + 16x^2 + 8x + 1}
\end{array}$$

divide the first term of the divisor into the first term of the dividend

$$\begin{array}{rl}
& 4x^2 \\
2x + 1 \!\!\!& \big)\overline{8x^3 + 16x^2 + 8x + 1} \\
& 8x^3 + 4x^2
\end{array}$$

multiply the entire divisor $2x + 1$ by $4x^2$

$$\begin{array}{rl}
& 4x^2 \\
2x + 1 \!\!\!& \big)\overline{8x^3 + 16x^2 + 8x + 1} \\
& \underline{8x^3 + 4x^2} \\
& \phantom{8x^3 + {}} 12x^2 + 8x
\end{array}$$

subtract $8x^3 + 4x^2$ from $8x^3 + 16x^2$ and bring down the next term in the dividend

$$\begin{array}{rl}
& 4x^2 + 6x \\
2x + 1 \!\!\!& \big)\overline{8x^3 + 16x^2 + 8x + 1} \\
& \underline{8x^3 + 4x^2} \\
& \phantom{8x^3 + {}} 12x^2 + 8x \\
& \phantom{8x^3 + {}} \underline{12x^2 + 6x} \\
& \phantom{8x^3 + 16x^2 + {}} 2x + 1
\end{array}$$

divide $12x^2$ by $2x$ and get $6x$; multiply $6x$ and $(2x + 1)$ and obtain $12x^2 + 6x$; subtract $12x^2 + 6x$ from $12x^2 + 8x$ and bring down the next term in the dividend

$$\begin{array}{rl}
& 4x^2 + 6x + 1 \\
2x + 1 \!\!\!& \big)\overline{8x^3 + 16x^2 + 8x + 1} \\
& \underline{8x^3 + 4x^2} \\
& \phantom{8x^3 + {}} 12x^2 + 8x \\
& \phantom{8x^3 + {}} \underline{12x^2 + 6x} \\
& \phantom{8x^3 + 16x^2 + {}} 2x + 1 \\
& \phantom{8x^3 + 16x^2 + {}} \underline{2x + 1}
\end{array}$$

repeating the procedure, we see that $2x$ divides $2x$ once (1) and that the result of the subtraction is 0

Since the result of the subtraction is 0, the remainder is 0. To check, we multiply the divisor $2x + 1$ and the quotient $4x^2 + 6x + 1$.

$$\begin{aligned}(2x + 1)(4x^2 + 6x + 1) &= 8x^3 + 12x^2 + 2x + 4x^2 + 6x + 1 \\ &= 8x^3 + 16x^2 + 8x + 1.\end{aligned}$$

Since this product is the original dividend, the answer is correct.

When using long division to divide two polynomials, it is important to write them both in standard form. It is also important to note that we continue the division process until the result of the subtraction is a polynomial *with degree less than the degree of the divisor*. We illustrate these ideas in the next example.

Example 2. Divide $2x + 9x^2 + 3 + 14x^3$ by $x + 7x^2$.

Solution: Rewriting each polynomial in standard form and proceeding as before, we get

$$\begin{array}{r} 2x + 1 \\ 7x^2 + x \overline{)14x^3 + 9x^2 + 2x + 3.} \\ \underline{14x^3 + 2x^2} \\ 7x^2 + 2x \\ \underline{7x^2 + x} \\ x + 3 \end{array}$$

Since the degree of $x + 3$ is less than the degree of the divisor, the remainder is $x + 3$ and the quotient is $2x + 1$. We now check to see if the product of the quotient and divisor, plus the remainder, equals the dividend.

CHECK

$$\begin{aligned}&\overbrace{(2x + 1)}^{\text{quotient}}\ \overbrace{(7x^2 + x)}^{\text{divisor}} + \overbrace{(x + 3)}^{\text{remainder}} \\ &= 14x^3 + 2x^2 + 7x^2 + x + x + 3 \\ &= 14x^3 + 9x^2 + 2x + 3.\end{aligned}$$

The result of this division is sometimes written in the form

$$2x + 1 + \frac{x + 3}{7x^2 + x}$$

If the dividend is written in standard form and it does not contain a power of the variable that would follow the leading term, it is helpful to write this missing power with a coefficient of zero. The next example illustrates this idea.

Example 3. Divide $18x^3 - 41x + 15$ by $3x + 5$.

Solution: First, rewrite the dividend as $18x^3 + 0x^2 - 41x + 15$. Thus

$$
\begin{array}{r}
6x^2 - 10x + 3 \\
3x + 5\overline{)18x^3 + 0x^2 - 41x + 15.} \\
\underline{18x^3 + 30x^2} \\
-30x^2 - 41x \\
\underline{-30x^2 - 50x} \\
9x + 15 \\
\underline{9x + 15}
\end{array}
$$

Note that the result of the division process is the quotient $6x^2 - 10x + 3$ with a remainder of 0.

EXERCISES 4.2

For Exercises 1–20, perform the division and check your answer. See Example 1.

1. $(18x^2 + 6x) \div 3x$

2. $(25y^2 + 10y) \div 5y$

3. $(40w^2 + 16w) \div 8w$

4. $(45z^2 + 30z) \div 15z$

5. $(72p^4 + 54p^3) \div 9p^2$

6. $(64q^4 + 48q^3) \div 16q^2$

7. $(12u^5 + 8u^3 - 4u^2) \div 4u$

8. $(15v^5 + 10v^3 - 5v^2) \div 5v$

9. $(39y^6 - 26y^5 + 13y^4) \div 13y^3$

10. $(60z^6 - 48z^5 + 24z^4) \div 12z^2$

11. $(150w^8 + 100w^7 - 75w^6 - 50w^5) \div 25w^3$

12. $(120z^8 + 80z^7 - 60z^6 - 40z^5) \div 20z^3$

13. $(30x^4 + 15x^3 + 45x^2) \div 15x^2$

14. $(100y^4 + 50y^3 + 30y^2) \div 10y^2$

15. $(56v^5 + 28v^4 - 14v^2) \div 7v^2$

16. $(72u^4 - 36u^3 + 24u^2) \div 12u^2$

17. $(75y^9 - 100y^5 + 25y^2) \div 25y^2$

18. $(12x^5 + 6x^4 + 6x^3) \div 3x^2$

19. $(-21a^4 + 42a^3 - 14a^2) \div 7a^2$

20. $(-12b^{10} + 24b^7 - 4b^6) \div 4b^6$

For Exercises 21–40, perform the division and check your answer. See Examples 2 and 3.

21. $(8x^2 + 14x + 5) \div (4x + 5)$

22. $(6x^2 + 25x + 24) \div (2x + 3)$

23. $(12y^2 + 17y + 6) \div (4y + 3)$

24. $(6y^2 + 19y + 10) \div (3y + 2)$

25. $(8w^2 + 8w - 6) \div (2w - 1)$

26. $(15w^2 - w - 6) \div (3w - 2)$

27. $(15z^2 + 4z - 4) \div (5z - 2)$

28. $(8z^2 - 2z - 3) \div (4z - 3)$

29. $(x - x^2 + x^3 - 1) \div (x - 1)$

30. $(2y - y^2 + y^3 - 2) \div (y - 1)$

31. $(8p^3 + 16p^2 + 8p + 3) \div (2p + 3)$

32. $(6p^3 + 7p^2 + 5p + 2) \div (3p + 2)$

33. $(10q^3 - q^2 - 17q - 4) \div (5q + 2)$

34. $(6q^3 - 17q^2 + 18q - 6) \div (3q - 4)$

35. $(3x^3 - 4x + x^4) \div (2x + x^2)$

36. $(3x^3 + 16x + x^4) \div (4x + x^2)$

37. $(6y^4 + 4y^3 + 7y^2 + 6y + 1) \div (2y^2 + 3)$

38. $(8z^4 + 4z^3 + 10z^2 + 10z + 5) \div (2z^2 + 5)$

39. $(2u^3 - 2u^2 + 3u + 6) \div (u^2 - 1)$

40. $(3u^3 + 12u^2 + 5u + 6) \div (3u^2 + 1)$

4.3 Products and Quotients of Rational Expressions

We note that the product of two rational numbers is found according to the following rule. If a, b, c, and d are real numbers with $b \neq 0$ and $d \neq 0$, then

$$\frac{a}{b} \cdot \frac{c}{d} = \frac{a \cdot c}{b \cdot d}.$$

The product of two rational expressions is found in the same manner. Thus the numerator of the product of two rational expressions is the product of the two numerators and the denominator of the product is the product of the two denominators.

Product of Two Rational Expressions

If P, Q, R, and S are polynomials with $Q \neq 0$ and $S \neq 0$, then

$$\frac{P}{Q} \cdot \frac{R}{S} = \frac{P \cdot R}{Q \cdot S}.$$

The following examples illustrate how to find the product of two rational expressions. It is common practice to reduce the product of two rational expressions to lowest terms.

Example 1. Compute the following products. Reduce the answer to lowest terms.

(a) $\dfrac{15x^2}{14y^2} \cdot \dfrac{7y}{30x}$. (b) $\dfrac{(x-3)^2}{(x+5)} \cdot \dfrac{(x+5)^2}{(x-3)^3}$.

Solution

(a) $\dfrac{15x^2}{14y^2} \cdot \dfrac{7y}{30x} = \dfrac{15x^2 \cdot 7y}{14y^2 \cdot 30x}$ numerator *times* numerator denominator *times* denominator

$= \dfrac{105x^2y}{420xy^2}$ multiplication

$= \dfrac{x\cancel{(105xy)}}{4y\cancel{(105xy)}}$ factoring and reducing

$= \dfrac{x}{4y}.$

(b) $\dfrac{(x-3)^2}{(x+5)} \cdot \dfrac{(x+5)^2}{(x-3)^3} = \dfrac{(x-3)^2 \cdot (x+5)^2}{(x+5) \cdot (x-3)^3}$ multiplication of rational expressions

$= \dfrac{\cancel{(x-3)^2}\cancel{(x+5)}(x+5)}{\cancel{(x+5)}\cancel{(x-3)^2}(x-3)}$ factoring and reducing

$= \dfrac{x+5}{x-3}.$

Notice in part (b) of Example 1 that the numerators and denominators were in factored form. When this is not the case, it is easier to factor numerators and denominators of the rational expressions and to remove common factors from both rational expressions before multiplying. We illustrate this procedure in the following examples.

Example 2. Compute the product $\dfrac{4x-12}{x^2+6x+5} \cdot \dfrac{x+5}{2x^2-18}$ and reduce to lowest terms.

Solution: Factor as much as possible and remove common factors before multiplying.

$$\frac{2 \cdot \cancel{2(x-3)}}{(x+1)\cancel{(x+5)}} \cdot \frac{\cancel{(x+5)}}{\cancel{2(x-3)}(x+3)} = \frac{2}{(x+1)(x+3)}.$$

Example 3. Compute the product $\dfrac{x^2-3x-10}{x^2+2x-35} \cdot \dfrac{x^2+7x+12}{x^2+5x+6}$ and reduce to lowest terms.

Solution: Factor, reduce, and then multiply.

$$\frac{\cancel{(x+2)}\cancel{(x-5)}}{(x+7)\cancel{(x-5)}} \cdot \frac{\cancel{(x+3)}(x+4)}{\cancel{(x+2)}\cancel{(x+3)}} = \frac{x+4}{x+7}.$$

The reciprocal of $\dfrac{3}{5}$ is $\dfrac{5}{3}$ because $\dfrac{3}{5} \cdot \dfrac{5}{3} = 1$. More generally, if P and Q are nonzero polynomials, *the reciprocal of* $\dfrac{P}{Q}$ *is* $\dfrac{Q}{P}$ because $\dfrac{P}{Q} \cdot \dfrac{Q}{P} = \dfrac{P \cdot Q}{Q \cdot P} = 1$.

The rule for dividing one real number by another was stated in Chapter 1. When this rule is applied to real numbers, we obtain the following result. If a, b, c, and d are real numbers with $b \neq 0$, $c \neq 0$, and $d \neq 0$, then

$$\frac{a}{b} \div \frac{c}{d} = \frac{a}{b} \cdot \frac{d}{c}.$$

The quotient of two rational expressions is found in the same manner.

Quotient of Two Rational Expressions

> If P, Q, R, and S are polynomials with $Q \neq 0$, $R \neq 0$, and $S \neq 0$, then
>
> $$\frac{P}{Q} \div \frac{R}{S} = \frac{P}{Q} \cdot \frac{S}{R}.$$

Thus, to find the quotient of two rational expressions, multiply the dividend by the reciprocal of the divisor.

Example 4. Compute the following quotients. Reduce the answer to lowest terms.

(a) $\dfrac{9y}{7x} \div \dfrac{15y^2}{14x^2}$. (b) $\dfrac{7q - 14}{6q} \div \dfrac{q^2 - 4}{2q^2}$.

Solution

(a) $$\frac{9y}{7x} \div \frac{15y^2}{14x^2} = \frac{\overset{3}{\cancel{9y}}}{\underset{1}{\cancel{7x}}} \cdot \frac{\overset{2x}{\cancel{14x^2}}}{\underset{5y}{\cancel{15y^2}}}$$ multiply by the reciprocal and reduce

$$= \frac{3}{1} \cdot \frac{2x}{5y}$$

$$= \frac{6x}{5y}.$$

(b) Take the reciprocal of the divisor, factor, reduce, and multiply.

$$\frac{7q - 14}{6q} \div \frac{q^2 - 4}{2q^2} = \frac{7\cancel{(q - 2)}}{\cancel{2} \cdot 3 \cdot \cancel{q}} \cdot \frac{\cancel{2} \cdot q \cdot \cancel{q}}{\cancel{(q - 2)}(q + 2)}$$

$$= \frac{7q}{3(q + 2)}.$$

When an expression contains a string of products and quotients, we can avoid difficulties with the order of operations if we first change all quotients to products by using reciprocals. This approach is illustrated in the next example.

Example 5. Perform the operations indicated.

$$\frac{a^2 - 2a - 3}{3a + 21} \div (a^2 - 1) \cdot \frac{5a - 5}{a^2 - 9}.$$

Solution: First change the division by $(a^2 - 1)$ to multiplication by its reciprocal. Then factor, reduce, and multiply.

$$\frac{\cancel{(a-3)}\cancel{(a+1)}}{3(a+7)} \cdot \frac{1}{\cancel{(a+1)}\cancel{(a-1)}} \cdot \frac{5\cancel{(a-1)}}{\cancel{(a-3)}(a+3)} = \frac{5}{3(a+7)(a+3)}.$$

EXERCISES 4.3

For Exercises 1–30, compute the indicated product. Reduce your answer to lowest terms. For Exercises 1–30, see Examples 1–3.

1. $\frac{7}{12} \cdot \frac{4}{3}$

2. $\frac{15}{4} \cdot \frac{12}{10}$

3. $\dfrac{x}{2} \cdot \dfrac{6}{x^3}$

4. $\dfrac{y^2}{5} \cdot \dfrac{15}{y^3}$

5. $\dfrac{w}{4z^2} \cdot \dfrac{12z^3}{w^3}$

6. $\dfrac{8p^2}{q} \cdot \dfrac{q^2}{4p^3}$

7. $\dfrac{50ab^3}{20a^2b} \cdot \dfrac{16ab^2}{5ab}$

8. $\dfrac{45cd^3}{5c^2d} \cdot \dfrac{4cd}{3c^2d^2}$

9. $\dfrac{(p-1)^2}{p+3} \cdot \dfrac{(p+3)^2}{(p-1)^3}$

10. $\dfrac{(q+4)^3}{(q-5)^2} \cdot \dfrac{q-5}{(q+4)^2}$

11. $\dfrac{(r+4)^5}{(r-3)^3} \cdot \dfrac{(r-3)^2}{(r+4)^2}$

12. $\dfrac{(t+11)^3}{(t+7)^6} \cdot \dfrac{(t+7)^4}{(t+11)^2}$

13. $\dfrac{(2w-5)^4}{(3w+1)^3} \cdot \dfrac{3w+1}{(2w-5)^2}$

14. $\dfrac{7z+1}{(4z-5)^2} \cdot \dfrac{(4z-5)^4}{(7z+1)^3}$

15. $\dfrac{x+3}{3} \cdot \dfrac{6}{(x+3)^2}$

16. $\dfrac{y+4}{50} \cdot \dfrac{100}{(y+4)^2}$

17. $\dfrac{(a+2)^2}{30a^4} \cdot \dfrac{82a^3}{(a+2)^2}$

18. $\dfrac{(b+1)^3}{42b^2} \cdot \dfrac{93b}{(b+1)^3}$

19. $\dfrac{(x-4)^2}{7} \cdot \dfrac{14}{4-x}$

20. $\dfrac{(y-2)^2}{15} \cdot \dfrac{5}{2-y}$

21. $\dfrac{(w-11)^2}{3w-12} \cdot \dfrac{8-2w}{(11-w)^3}$

22. $\dfrac{(z-8)^3}{3w-21} \cdot \dfrac{28-4w}{(8-z)^2}$

23. $\dfrac{5p-15}{p^2+2p+1} \cdot \dfrac{p^2-1}{6-2p}$

24. $\dfrac{4q-20}{q^2-2q+1} \cdot \dfrac{q^2-1}{10-2q}$

25. $\dfrac{2r-6}{r^2-6r+9} \cdot \dfrac{r^2-9}{9-3r}$

26. $\dfrac{15-3s}{s^2-25} \cdot \dfrac{s^2+10s+25}{7s-35}$

27. $\dfrac{x^2-1}{x+3}\cdot\dfrac{x+2}{x+1}\cdot\dfrac{x^2-9}{x^2+x-2}$

28. $\dfrac{y+5}{y^2-4}\cdot\dfrac{y+2}{y+3}\cdot\dfrac{y^2+y-6}{y^2-25}$

29. $\dfrac{a^2+5a+6}{a^2-1}\cdot\dfrac{a^2-4a+3}{a^2+8a+12}\cdot\dfrac{a^2+6a+5}{a^2-9}$

30. $\dfrac{b^2-b-6}{b^2+8b+15}\cdot\dfrac{b^2-1}{b^2-4}\cdot\dfrac{b^2+b-6}{b^2-4b+3}$

For Exercises 31–54, compute the indicated quotient. Reduce your answer to lowest terms. See Example 4.

31. $\dfrac{6x^2}{15y^2}\div\dfrac{3x}{10y}$

32. $\dfrac{4w}{21z}\div\dfrac{2w^2}{14z^2}$

33. $\dfrac{5p^3q^2}{6rs^2}\div\dfrac{25pq^3}{18r^2s}$

34. $\dfrac{9a^2b}{2c^2d^3}\div\dfrac{3ab^2}{4c^3d^2}$

35. $\dfrac{3(w^2v)^3}{(5tw^2)^2}\div\dfrac{(6wv)^2}{10(t^2w)^2}$

36. $\dfrac{(2w^2x)^3}{3(y^2z)^2}\div\dfrac{4w^3x^4}{9y^3z^3}$

37. $\dfrac{(-2x^2y)^2}{15p^3q^2}\div\dfrac{(-xy^2)^3}{(3pq)^2}$

38. $\dfrac{(-a^2b)^3}{20c^5d^3}\div\dfrac{(-a^2b^2)^2}{(4c^3d^2)^2}$

39. $\dfrac{5a}{(a+3)^2}\div\dfrac{25}{a+3}$

40. $\dfrac{7b}{(b-4)^3}\div\dfrac{21}{b-4}$

41. $\dfrac{5(c+3)^2}{14(c-2)}\div\dfrac{10(c+3)^3}{21(c-2)^2}$

42. $\dfrac{6(d-5)^2}{25(d+5)^3}\div\dfrac{9(d-5)}{15(d+5)^2}$

43. $\dfrac{p+7}{5p+15}\div\dfrac{4p+28}{p+3}$

44. $\dfrac{2q+10}{q-3}\div\dfrac{q+5}{3q-9}$

45. $\dfrac{7r+14}{2r-14}\div\dfrac{3r+6}{3r-21}$

46. $\dfrac{3s+12}{2s-10}\div\dfrac{5s+20}{3s-15}$

47. $\dfrac{t^2-16}{t^2-1}\div\dfrac{6t+24}{5t-5}$

48. $\dfrac{5u+25}{3u-6}\div\dfrac{u^2-25}{u^2-4}$

49. $\dfrac{p^2+8p+7}{p^2-2p-3}\div\dfrac{p^2+3p+2}{p^2-p-6}$

50. $\dfrac{x^2-3x+2}{x^2+x-6}\div\dfrac{x^2+2x-3}{x^2+4x+3}$

51. $\dfrac{2b^2-b-3}{b^2-3b+2}\div\dfrac{10b^2-5b-15}{3b^2+3b-6}$

52. $\dfrac{18-2c^2}{c^2+4c+3}\div\dfrac{15-20c+5c^2}{c^2-1}$

53. $\dfrac{5a^2-8a+3}{6a^2+13a+6}\div\dfrac{5a^2+2a-3}{2a^2+5a+3}$

54. $\dfrac{3w^2-4w+1}{2w^2-w-1}\div\dfrac{3w^2-16w+5}{2w^2+3w+1}$

For Exercises 55–60, perform the indicated operation. Reduce your answer to lowest terms. See Example 5.

55. $\dfrac{6a^4}{5b^2}\div\dfrac{3a^2}{10b}\cdot\dfrac{5b^3}{2a^3}$

56. $\dfrac{6a^4}{5b^2}\cdot\dfrac{3a^2}{10b}\div\dfrac{5b^3}{2a^2}$

57. $\dfrac{a^2+7a+10}{7a^3-21a^2}\cdot\dfrac{a^3}{a^2+a-2}\div\dfrac{a+5}{a-1}$

58. $\dfrac{x^2-x-2}{x^2+5x}\cdot\dfrac{2x^2}{x^2-3x+2}\div\dfrac{x+1}{x+5}$

59. $\dfrac{4q^2 - 4}{q^2 + 5q + 6} \div \dfrac{6q^3 + 6q^2}{q^2 + q - 2} \cdot \dfrac{3q^3}{q^2 - 2q + 1}$

60. $\dfrac{2z^2 + 4z - 30}{3z + 3} \div \dfrac{z^2 - 4z + 3}{z^2 + 2z + 1} \cdot \dfrac{3z - 3}{7z + 35}$

4.4 Addition and Subtraction of Rational Expressions

If two fractions have the same denominator, their sum or difference is found by adding or subtracting the two numerators and placing it over the *common denominator*. Thus if a, b, and c are real numbers, $c \neq 0$, then

$$\frac{a}{c} + \frac{b}{c} = \frac{a+b}{c} \quad \text{and} \quad \frac{a}{c} - \frac{b}{c} = \frac{a-b}{c}.$$

We use the same procedure to add two rational expressions that have a common denominator.

Addition and Subtraction of Two Rational Expressions

If P, Q, and R are polynomials with $Q \neq 0$, then

$$\frac{P}{Q} + \frac{R}{Q} = \frac{P+R}{Q} \quad \text{and} \quad \frac{P}{Q} - \frac{R}{Q} = \frac{P-R}{Q}.$$

The next two examples illustrate the application of this rule for adding two rational expressions.

Example 1

(a) Add $\dfrac{7x}{3}$ and $\dfrac{2y}{3}$.

(b) Find the sum of $\dfrac{2x}{x+5}$ and $\dfrac{3-x}{x+5}$.

(c) Subtract $\dfrac{2x}{4x-1}$ from $\dfrac{3x+1}{4x-1}$.

Solution: In each case, the denominators of the two rational expressions are the same. Thus, in parts (a) and (b), we add the numerators and use the common denominator. In part (c) we subtract the numerators and use the common denominator.

(a) $\dfrac{7x}{3} + \dfrac{2y}{3} = \dfrac{7x + 2y}{3}$.

(b) $\dfrac{2x}{x+5} + \dfrac{3-x}{x+5} = \dfrac{2x + 3 - x}{x+5} = \dfrac{x+3}{x+5}$.

(c) $\dfrac{3x + 1}{4x - 1} - \dfrac{2x}{4x - 1} = \dfrac{3x + 1 - 2x}{4x - 1} = \dfrac{x + 1}{4x - 1}.$

As the next example illustrates, after adding or subtracting two rational expressions, it may be possible to reduce the sum to lowest terms.

Example 2

(a) Add $\dfrac{3y + 5}{6y + 9}$ and $\dfrac{5y + 7}{6y + 9}$. (b) Subtract $\dfrac{2z + 5}{z^2 - z - 6}$ from $\dfrac{3z + 2}{z^2 - z - 6}$.

Solution

(a)
$$\frac{3y + 5}{6y + 9} + \frac{5y + 7}{6y + 9} = \frac{8y + 12}{6y + 9}$$
$$= \frac{4\cancel{(2y + 3)}}{3\cancel{(2y + 3)}}$$
$$= \frac{4}{3}.$$

(b)
$$\frac{3z + 2}{z^2 - z - 6} - \frac{2z + 5}{z^2 - z - 6} = \frac{(3z + 2) - (2z + 5)}{z^2 - z - 6}$$
$$= \frac{\cancel{(z - 3)}}{\cancel{(z - 3)}(z + 2)}$$
$$= \frac{1}{z + 2}.$$

If two rational expressions have different denominators, before they can be added or subtracted they must be rewritten with a common denominator. We usually find the *least common denominator* using the following procedure:

Steps for Finding the Least Common Denominator

Step 1. Factor each denominator completely.

Step 2. Form the product of all the distinct factors from Step 1.

Step 3. Write each factor in Step 2 with the largest exponent that it has in Step 1.

It is customary to use LCD to denote the least common denominator.

We will demonstrate the procedure by adding $\dfrac{5}{12w}$ and $\dfrac{2}{15w^3}$. Before we can add these two rational expressions, we must rewrite them with a common denominator. Applying the three steps of the procedure to $12w$ and $15w^3$, we get

Step 1. $12w = 2^2 \cdot 3 \cdot w$ and $15w^3 = 3 \cdot 5 \cdot w^3$

Step 2. $2 \cdot 3 \cdot 5 \cdot w$

Step 3. $2^2 \cdot 3 \cdot 5 \cdot w^3 = 60w^3$

Thus the least common denominator (LCD) is $60w^3$. We now want to write each fraction with a denominator of $60w^3$. To do this, we multiply the numerator and denominator of each fraction by the result of dividing the old denominator into the least common denominator. First we consider $\frac{5}{12w}$. Note that $5w^2$ is the result of dividing $60w^3$ by $12w$. Thus we multiply the numerator and denominator of $\frac{5}{12w}$ by $5w^2$. Similarly, for $\frac{2}{15w^3}$, we multiply the numerator and denominator by 4. Thus we get

$$\begin{aligned}\frac{5}{12w} + \frac{2}{15w^3} &= \frac{5w^2}{5w^2} \cdot \frac{5}{12w} + \frac{4}{4} \cdot \frac{2}{15w^3} \\ &= \frac{25w^2}{60w^3} + \frac{8}{60w^3} \\ &= \frac{25w^2 + 8}{60w^3}.\end{aligned}$$

Example 3. Find the following sums.

(a) $\frac{4}{75xy^2} + \frac{3}{20x^2y}$. (b) $\frac{7}{6x - 4} + \frac{2x}{3x - 2}$.

Solution

(a) The two denominators in factored form are

$$75xy^2 = 3 \cdot 5^2 \cdot x \cdot y^2 \quad \text{and} \quad 20x^2y = 2^2 \cdot 5 \cdot x^2 \cdot y.$$

From this we see that the LCD is

$$2^2 \cdot 3 \cdot 5^2 \cdot x^2 \cdot y^2 = 300x^2y^2.$$

Next we rewrite each rational expression with this LCD. First note that

$$\frac{300x^2y^2}{75xy^2} = 4x \quad \text{and} \quad \frac{300x^2y^2}{20x^2y} = 15y.$$

Thus, to add the fractions, we proceed as follows.

$$\frac{4}{75xy^2} + \frac{3}{20x^2y} = \frac{4x}{4x} \cdot \frac{4}{75xy^2} + \frac{15y}{15y} \cdot \frac{3}{20x^2y}$$
$$= \frac{16x}{300x^2y^2} + \frac{45y}{300x^2y^2}$$
$$= \frac{16x + 45y}{300x^2y^2}.$$

(b) The first denominator of the expression may be factored as $6x - 4 = 2(3x - 2)$. The second denominator $(3x - 2)$ cannot be factored. Thus the LCD is $2(3x - 2)$. Rewriting the second term with this denominator, we get

$$\frac{7}{6x - 4} + \frac{2x}{3x - 2} = \frac{7}{6x - 4} + \frac{2}{2} \cdot \frac{2x}{3x - 2}$$
$$= \frac{7}{6x - 4} + \frac{4x}{6x - 4}$$
$$= \frac{4x + 7}{6x - 4}.$$

Recall that in Example 2, after adding two fractions with common denominators, we were able to reduce the sum to lowest terms. The next two examples illustrate that when working with fractions with *different* denominators, after finding the LCD and adding or subtracting the rational expressions, it is sometimes possible to reduce the answer.

Example 4. Subtract $\frac{3}{w - 4}$ from $\frac{7w - 4}{w^2 - 16}$ and then reduce tne difference to lowest terms.

Solution: First we factor each denominator. The denominator $w^2 - 16 = (w - 4)(w + 4)$. The other denominator, $w - 4$, is not factorable. Thus the LCD is $(w - 4)(w + 4)$. Next we rewrite the problem as

$$\frac{7w - 4}{w^2 - 16} - \frac{3}{w - 4} = \frac{7w - 4}{(w - 4)(w + 4)} - \frac{3(w + 4)}{(w - 4)(w + 4)}$$
$$= \frac{4w - 16}{(w - 4)(w + 4)}$$
$$= \frac{4\cancel{(w - 4)}}{\cancel{(w - 4)}(w + 4)}$$
$$= \frac{4}{w + 4}.$$

Example 5. Find $\frac{2}{3} + \frac{y-1}{y+1} - \frac{y-1}{y^2+3y+2}$.

Solution: First we rewrite the problem with all the denominators factored.

$$\frac{2}{3} + \frac{y-1}{y+1} - \frac{y-1}{(y+1)(y+2)}.$$

The LCD is $3(y+1)(y+2)$. Next, we multiply the numerator and denominator of each term by an appropriate polynomial so that all the terms have the same denominator. Thus we get

$$\frac{2}{3}\cdot\frac{(y+1)(y+2)}{(y+1)(y+2)} + \frac{y-1}{y+1}\cdot\frac{3(y+2)}{3(y+2)} - \frac{y-1}{(y+1)(y+2)}\cdot\frac{3}{3}$$

$$= \frac{2(y^2+3y+2)}{3(y+1)(y+2)} + \frac{3(y^2+y-2)}{3(y+1)(y+2)} - \frac{3(y-1)}{3(y+1)(y+2)}$$

$$= \frac{2y^2+6y+4+3y^2+3y-6-3y+3}{3(y+1)(y+2)} \qquad \text{add numerators}$$

$$= \frac{5y^2+6y+1}{3(y+1)(y+2)} = \frac{(5y+1)\cancel{(y+1)}}{3\cancel{(y+1)}(y+2)} \qquad \text{combine like terms and factor}$$

$$= \frac{5y+1}{3(y+2)} = \frac{5y+1}{3y+6}.$$

Thus

$$\frac{2}{3} + \frac{y-1}{y+1} - \frac{y-1}{y^2+3y+2} = \frac{5y+1}{3y+6}.$$

EXERCISES 4.4

For Exercises 1–6, find the indicated sum or difference. When possible, reduce your answer to lowest terms.

1. $\frac{3}{5} + \frac{1}{10}$ **2.** $\frac{2}{3} + \frac{1}{6}$ **3.** $\frac{11}{12} - \frac{3}{4}$ **4.** $\frac{9}{14} - \frac{1}{2}$

5. $\frac{13}{20} - \frac{1}{4} + \frac{2}{5}$ **6.** $\frac{11}{15} - \frac{2}{5} + \frac{1}{3}$

For Exercises 7–16, find the indicated sum or difference. When possible, reduce your answer to lowest terms. See Examples 1 and 2.

7. $\frac{11}{3z} - \frac{2}{3z}$ **8.** $\frac{7}{2z} + \frac{3}{2z}$

9. $\frac{6}{2w+7} + \frac{7}{2w+7}$ **10.** $\frac{4}{2y+1} + \frac{5}{2y+1}$

11. $\frac{19a}{5a-3} - \frac{11a}{5a-3}$

12. $\frac{13b}{7b-4} - \frac{5b}{7b-4}$

13. $\frac{5c+9}{2c+3} - \frac{c+3}{2c+3}$

14. $\frac{7d+5}{3d+2} - \frac{d+1}{3d+2}$

15. $\frac{x+2}{x^2+8x+15} + \frac{x+8}{x^2+8x+15}$

16. $\frac{y+1}{y^2+7y+12} + \frac{y+5}{y^2+7y+12}$

For Exercises 17–60, find the indicated sum or difference. When possible, reduce your answer to lowest terms. See Examples 3–5.

17. $\frac{7}{3a} + \frac{2}{5a}$

18. $\frac{5}{7b} + \frac{3}{2b}$

19. $\frac{2}{3p} - \frac{1}{p}$

20. $\frac{1}{2q} - \frac{1}{q}$

21. $\frac{5}{2r} + \frac{1}{r^2}$

22. $\frac{3}{4s} + \frac{2}{s^2}$

23. $\frac{1}{x} - \frac{1}{y}$

24. $\frac{1}{w} + \frac{1}{z}$

25. $\frac{4}{15a^2b} + \frac{2}{25ab^2}$

26. $\frac{5}{12cd^2} + \frac{4}{15c^2d}$

27. $\frac{2}{x+5} - \frac{1}{x}$

28. $\frac{3}{y+7} - \frac{1}{y}$

29. $\frac{7}{w-2} + \frac{3}{2-w}$

30. $\frac{5}{z-1} + \frac{2}{1-z}$

31. $\frac{7}{b+3} + \frac{4}{b-3}$

32. $\frac{5}{a+1} + \frac{1}{a-1}$

33. $\frac{y}{y+3} + \frac{y-1}{y-3}$

34. $\frac{z}{z+4} + \frac{z+1}{z-4}$

35. $\frac{x}{x+1} - \frac{2x}{x+2}$

36. $\frac{y}{y+2} - \frac{3y}{y+6}$

37. $\frac{3w}{3w+1} - \frac{2w}{2w+1}$

38. $\frac{5z}{5z+1} - \frac{2z}{2z+1}$

39. $\frac{y+1}{y^2+5y+6} + \frac{y}{y^2-4}$

40. $\frac{x-2}{x^2-1} + \frac{x-1}{x^2+3x+2}$

41. $\frac{b+3}{b^2+8b+16} - \frac{b-3}{b^2-16}$

42. $\frac{a+1}{a^2-9} - \frac{a+1}{a^2+5a+6}$

43. $\frac{2c+3}{c^2-4c+3} + \frac{c+9}{c^2+2c-3}$

44. $\frac{2d+7}{d^2+4d-5} - \frac{d+3}{d^2+6d+5}$

45. $\frac{2q+7}{q^2+q-2} - \frac{q+3}{q^2+3q+2}$

46. $\frac{2p+3}{p^2+3p-18} + \frac{p+9}{p^2+9p+18}$

47. $\frac{3x+5}{2x^2+3x-14} + \frac{x+5}{2x^2+11x+14}$

48. $\frac{3y+10}{y^2+6y+8} + \frac{y+7}{y^2+5y+4}$

49. $\dfrac{5w+1}{w^2+w-20}-\dfrac{3w+2}{w^2-2w-8}$

50. $\dfrac{z-19}{z^2+6z+5}+\dfrac{z+6}{z^2+3z+2}$

51. $\dfrac{a+3b}{a^2-b^2}-\dfrac{a-b}{a^2+2ab+b^2}$

52. $\dfrac{a+b}{a^2-2ab+b^2}-\dfrac{a-2b}{a^2-b^2}$

53. $\dfrac{p-q}{p^2+2pq+q^2}-\dfrac{2p+q}{p^2-q^2}$

54. $\dfrac{2p-q}{p^2-q^2}-\dfrac{p+q}{p^2-2pq+q^2}$

55. $\dfrac{2}{7}-\dfrac{y}{y+3}+\dfrac{y}{y-3}$

56. $\dfrac{1}{5}+\dfrac{x}{x-1}-\dfrac{2x}{x+1}$

57. $\dfrac{4}{x+4}-\dfrac{4}{x-4}+\dfrac{2x}{x^2-16}$

58. $\dfrac{3}{x+3}-\dfrac{3}{x-3}+\dfrac{2x^2}{x^2-9}$

59. $\dfrac{3w^2-7w+2}{w^2+2w-15}-\dfrac{w}{w+5}-\dfrac{w-2}{w-3}$

60. $\dfrac{3z^2+5z}{z^2+z-12}-\dfrac{z}{z+4}-\dfrac{z+2}{z-3}$

Ratios and Proportions with Applications 4.5

In a local professional building there are five physicians and seven lawyers. One way of comparing the number of physicians to the number of lawyers is to say that the ratio of physicians to lawyers is 5 to 7. The ratio 5 to 7 is written as 5:7 or more commonly as the fraction $\dfrac{5}{7}$.

> The *ratio* of the number a to the number b ($b \neq 0$) is the quotient $\dfrac{a}{b}$.

Ratios are often encountered in our daily experiences. Thus we talk about the ratio of Democrats to Republicans in Congress or the ratio of female students to male students at a school. We usually reduce ratios to the lowest terms.

Example 1. There are 450 female students and 360 male students in the freshman class at a local state college.

(a) What is the ratio of female students to male students?

(b) What is the ratio of male students to the total number of students in the freshman class?

Solution

(a) $\dfrac{450}{360}$ or $\dfrac{5}{4}$, read "5 to 4."

(b) The total number of students in the freshman class is $450 + 360 = 810$. Thus the ratio of male students to this total is $\frac{360}{810} = \frac{4}{9}$, read "4 to 9."

Example 2. What is the ratio of states of the United States with a name that does not begin with the letter A to states with a name that does begin with A?

Solution: There are 50 states in the United States. Alabama, Alaska, Arizona, and Arkansas are the four states with a name that begins with A. Thus there are $50 - 4 = 46$ other states. The ratio we seek is $\frac{46}{4} = \frac{23}{2}$, read "23 to 2."

It is common practice to leave the 1 in the denominator of a fraction when it represents a ratio such as "5 to 1". Thus the ratio "5 to 1" is written as $\frac{5}{1}$.

Example 3. John is 4 feet 8 inches tall. His sister Cindy is 42 inches tall. What is the ratio of Cindy's height to John's height?

Solution: First we express both heights in the same unit of measure. If we convert John's height to inches, the result is 56. Therefore, the ratio we seek is $\frac{42}{56} = \frac{3}{4}$.

A ratio is used to compare two quantities. A statement of the *equality* of two ratios is called a *proportion*.

If $\frac{a}{b}$ and $\frac{c}{d}$ are two equal ratios, then $\frac{a}{b} = \frac{c}{d}$ is called a proportion.

Each of the four quantities a, b, c, and d in the proportion $\frac{a}{b} = \frac{c}{d}$ is called a *term of the proportion*. The quantities a and d are called the *extremes* of the proportion. The quantities b and c are called the *means* of the proportion. For example, in the proportion

$$\frac{4}{9} = \frac{12}{27}$$

the extremes are 4 and 27, and the means are 9 and 12.

If both sides of the proportion $\frac{a}{b} = \frac{c}{d}$ are multiplied by bd, the result is $ad = bc$. That is, in any proportion, the product of the means equals the product of the extremes.

The Fundamental Property of Proportions

> If a, b, c, and d are real numbers with $b \neq 0$ and $d \neq 0$ and $\frac{a}{b} = \frac{c}{d}$, then $ad = bc$.

The fundamental property of proportions is used to solve a proportion when one of its terms is unknown.

Example 4. Solve the proportion $\frac{3}{7} = \frac{x}{28}$ for x.

Solution: Applying the fundamental property of proportions, we may write

$$3 \cdot 28 = 7x.$$

Thus

$$7x = 84$$

or

$$x = 12.$$

Check that $x = 12$ is the correct answer.

$$\frac{12}{28} = \frac{3}{7}.$$

As the following examples illustrate, proportions can often be used to solve practical problems.

Example 5. A pilot is flying a small plane from Rock Hill to Hilton Head, South Carolina. On her map the two airports are 7.5 inches apart. She knows that on the map 1.25 inches represents 30 miles. What is the distance in miles between the two airports?

Solution: From the given information we may write the proportion

$$\frac{1.25}{30} = \frac{7.5}{x},$$

where x is the distance in miles between the two airports. Applying the fundamental property of proportions, we get

$$\begin{aligned}(1.25)x &= 30(7.5)\\ 1.25x &= 225\\ x &= 180.\end{aligned}$$

Therefore, the airports are 180 miles apart.

Example 6. The adult dosage for a liquid cold medicine is proportional to the weight of the patient. A 120-pound person is instructed to take 1.5 fluid ounces of the medicine every 6 hours. What is the dosage for a 200-pound person?

Solution: Let d be the dosage for a 200-pound person. From the information given we know that $\frac{120}{1.5} = \frac{200}{d}$ or

$$120d = (1.5)200.$$

Thus

$$120d = 300$$

or

$$d = 2.5.$$

The correct dosage for a 200-pound person is 2.5 fluid ounces every 6 hours.

EXERCISES 4.5

For Exercises 1–6, write the ratio as a fraction. Reduce the ratio to lowest terms. See Example 1.

1. 8 to 10 **2.** 9 to 12 **3.** 7 to 11 **4.** 9 to 13 **5.** 420 to 300 **6.** 400 to 450

For Exercises 7–12, refer to the following table to determine the appropriate ratio. The table gives the average depth in feet of one ocean and three seas. Reduce all answers to lowest terms. In the exercises, depth means average depth.

Ocean or Sea	Average Depth (Feet)
Arctic Ocean	4200
Mediterranean Sea	4500
South China Sea	5400
Caribbean Sea	8400

7. The depth of the Arctic Ocean to the depth of the Mediterranean Sea.
8. The depth of the Mediterranean Sea to the South China Sea.
9. The depth of the Arctic Ocean to the depth of the South China Sea.
10. The depth of the Arctic Ocean to the depth of the Caribbean Sea.
11. The depth of the South China Sea to the depth of the Caribbean Sea.
12. The depth of the Mediterranean Sea to the depth of the Caribbean Sea.

For Exercises 13–20, write the ratio indicated as a fraction reduced to lowest terms. See Examples 1–3.

13. The ratio of states of the United States with a name that begins with the letter M to the states with a name that does not begin with M.
14. The ratio of states of the United States with a name that begins with the letter N to states with a name that does not begin with N.
15. Samantha's height is 56 inches. Her brother James is 6 feet 5 inches tall. What is the ratio of Samantha's height to James' height?
16. Hector is 5 feet 4 inches tall. His brother Segundo is 56 inches tall. What is the ratio of Segundo's height to Hector's height?
17. What is the ratio of the months of the year with a total of at most 30 days to the months of the year with more than 30 days?
18. What is the ratio of the months of the year with more than 30 days to the months of the year with exactly 30 days?
19. In the word "Mississippi," what is the ratio of the number of occurrences of vowels to the number of occurrences of consonants?
20. In the word "mathematics," what is the ratio of the number of occurrences of consonants to the number of occurrences of vowels?

For Exercises 21–40, solve the proportion. See Example 4.

21. $\frac{x}{3} = \frac{7}{1}$
22. $\frac{y}{4} = \frac{3}{1}$
23. $\frac{w}{4} = \frac{42}{24}$
24. $\frac{z}{3} = \frac{21}{9}$
25. $\frac{a}{5} = \frac{6}{15}$
26. $\frac{b}{11} = \frac{16}{44}$
27. $\frac{48}{30} = \frac{y}{10}$
28. $\frac{16}{40} = \frac{x}{25}$
29. $\frac{5}{16} = \frac{x}{8}$
30. $\frac{7}{50} = \frac{r}{11}$
31. $\frac{9}{7} = \frac{81}{p}$
32. $\frac{8}{5} = \frac{40}{q}$
33. $\frac{28}{x} = \frac{7}{3}$
34. $\frac{39}{y} = \frac{13}{2}$
35. $\frac{a+5}{2} = \frac{6}{3}$
36. $\frac{b+9}{4} = \frac{6}{3}$
37. $\frac{c-4}{6} = \frac{5}{3}$
38. $\frac{d-5}{6} = \frac{5}{3}$
39. $\frac{r-1}{7} = \frac{r+1}{5}$
40. $\frac{s-4}{5} = \frac{s+2}{3}$

For Exercises 41–50, set up and solve an appropriate proportion. See Examples 5 and 6.

41. On a map, 2 inches represents 90 miles. If two towns are 5 inches apart on the map, how far apart are they in actual miles?

42. On a map, 3 inches represents 100 miles. If two towns are 7.5 inches apart on the map, how far apart are they in actual miles?

43. On a map, 3 inches represents 80 miles. If two towns are actually 360 miles apart, how many inches are between their locations on the map?

44. On a map, 2 inches represents 70 miles. If two towns are actually 315 miles apart, how many inches are between their locations on the map?

45. The instructions on a prescription for a 40-pound dog state that the animal is to receive 180 milliliters of the medicine each morning. If the ratio of the dosage to the weight of the animal is the same for all dogs, what is the correct dosage for a 70-pound dog?

46. A new diet suggests that you drink 8 ounces of pineapple juice for every 40 pounds of body weight. How many ounces of pineapple juice should a 210-pound person drink?

47. The ratio of male students to female students at a local college is 5 to 7. If there are 560 female students at the college, what is the total student enrollment?

48. A two-cycle engine requires an oil-to-gas ratio, in ounces, of 1 to 20. How many ounces of oil should be added to 5 gallons of gas? [HINT: One gallon equals 128 ounces.]

49. If 2 pounds of coffee will make 160 cups of coffee, how many pounds of coffee are needed to make 1000 cups of coffee?

50. If 50 pounds of lawn fertilizer will cover 2200 square feet, how many pounds are needed to cover 8000 square feet?

4.6 Equations Containing Rational Expressions

If an equation contains a rational expression, the usual method of solution is first to multiply each term of the equation by the least common denominator (LCD) of these rational expressions. However, if the LCD contains a variable, the resulting equation may have solutions that are not solutions to the original equation. Such solutions are called *extraneous solutions*. They can be easily identified because they will fail to satisfy the original equation. The process of multiplying each term of an equation by the LCD is called *clearing fractions*.

Example 1. Solve $\frac{5}{x-2} + \frac{2}{3} = \frac{1}{2}$.

Solution: We clear fractions by multiplying each term of the equation by the LCD, which is $3 \cdot 2(x - 2) = 6(x - 2)$. The result of this multiplication is

$$6(x-2)\left[\frac{5}{x-2} + \frac{2}{3}\right] = 6(x-2)\left(\frac{1}{2}\right)$$

$$6(x-2)\frac{5}{(x-2)} + \overset{2}{6}(x-2)\frac{2}{3} = 3(x-2)$$

$$30 + 4(x-2) = 3(x-2)$$

$$30 + 4x - 8 = 3x - 6$$

$$4x - 3x = -6 - 30 + 8$$

$$x = -28.$$

We check that -28 is the solution by letting $x = -28$ in the original equation. Thus

$$\frac{5}{-28-2} + \frac{2}{3} = \frac{1}{2}$$

$$\frac{5}{-30} + \frac{2}{3} = \frac{1}{2}$$

$$\frac{-5}{30} + \frac{20}{30} = \frac{1}{2}$$

$$\frac{15}{30} = \frac{1}{2}$$

$$\frac{1}{2} = \frac{1}{2}. \qquad \text{a true statement}$$

Therefore, -28 is the solution to the original equation.

The procedure of multiplying each side of an equation by the LCD is justified by the multiplication property of equality, which was discussed in Section 2.2.

Suppose that each term of an equation, which we shall call "equation 1," is multiplied by the same expression and we call the result "equation 2." It may be shown that any solution of equation 1 is also a solution of equation 2. As the next example illustrates, *extraneous solutions* are solutions to equation 2 that are not solutions to equation 1.

Example 2. Solve $\dfrac{x-2}{x-3} = 4 + \dfrac{1}{x-3}$.

Solution: We clear fractions by multiplying each term by the LCD, $x - 3$. This results in the following equation:

$$(x-3)\frac{x-2}{(x-3)} = (x-3)4 + (x-3)\frac{1}{(x-3)}$$

$$x - 2 = 4x - 12 + 1$$

$$x - 4x = -11 + 2$$

$$-3x = -9$$

$$x = 3.$$

If we substitute 3 for x in the original equation, the result is

$$\frac{1}{0} = 4 + \frac{1}{0}.$$

But division by zero is undefined. We conclude that 3 is *not* a solution to the original equation. This is an example of an extraneous solution. Since 3 is the only possible solution and it is extraneous, the original equation has no solution.

Example 2 indicates that if we multiply each term of an equation by an expression that contains a variable, we *must* check any proposed solutions in the original equation.

Example 3. Solve $\dfrac{6}{x-5} - \dfrac{2}{x+2} = \dfrac{5}{3x-15}$.

Solution: The LCD is $3(x-5)(x+2)$. Multiplying each term of the equation by this LCD results in the equation

$$3\cancel{(x-5)}(x+2)\frac{6}{\cancel{(x-5)}} - 3(x-5)\cancel{(x+2)}\frac{2}{\cancel{(x+2)}}$$

$$= \cancel{3(x-5)}(x+2)\frac{5}{\cancel{3(x-5)}}$$

$$\begin{aligned} 18(x+2) - 6(x-5) &= 5(x+2) \\ 18x + 36 - 6x + 30 &= 5x + 10 \\ 12x + 66 &= 5x + 10 \\ 12x - 5x &= 10 - 66 \\ 7x &= -56 \\ x &= -8. \end{aligned}$$

The reader should check that -8 is the solution to the original equation.

EXERCISES 4.6

For Exercises 1–20, solve the equation and check your answer. See Example 1.

1. $\dfrac{4}{w+4} + \dfrac{1}{5} = 1$ **2.** $\dfrac{5}{z+1} + \dfrac{3}{4} = 1$ **3.** $\dfrac{7}{x-1} + \dfrac{1}{2} = 1$

4. $\dfrac{4}{y-2} + \dfrac{1}{3} = 1$ **5.** $\dfrac{7}{2} = \dfrac{3}{x} + 1$ **6.** $\dfrac{5}{3} = \dfrac{2}{y} - 1$

7. $\dfrac{1}{p} + 5 = \dfrac{8}{3}$ **8.** $\dfrac{4}{q} - \dfrac{2}{5} = 2$ **9.** $\dfrac{7}{2w} - \dfrac{1}{9} = \dfrac{2}{w}$

10. $\dfrac{1}{2z} - \dfrac{3}{8} = -\dfrac{3}{z}$ **11.** $\dfrac{3}{r} - \dfrac{1}{2r} = \dfrac{1}{6}$ **12.** $\dfrac{5}{s} - \dfrac{3}{2s} = \dfrac{1}{6}$

13. $\frac{7}{3s} - \frac{1}{2s} = \frac{1}{6}$
14. $\frac{5}{2t} - \frac{2}{3t} = \frac{1}{6}$
15. $\frac{5}{w + 5} - \frac{1}{3} = \frac{1}{2}$

16. $\frac{8}{x + 3} - \frac{1}{5} = \frac{1}{3}$
17. $\frac{1}{u - 4} + \frac{3}{5} = \frac{1}{3}$
18. $\frac{5}{v - 5} + \frac{2}{3} = \frac{3}{4}$

19. $\frac{1}{y + 1} + \frac{1}{7} = \frac{1}{6}$
20. $\frac{1}{z + 3} + \frac{1}{5} = \frac{1}{4}$

For Exercises 21–30, solve the equation and check your answer. Identify any extraneous solutions. See Example 2.

21. $\frac{p + 2}{p - 5} + 1 = \frac{7}{p - 5}$
22. $\frac{q - 1}{q - 2} + 1 = \frac{1}{q - 2}$
23. $\frac{2r + 6}{r + 4} + 1 = \frac{-2}{r + 4}$

24. $\frac{5s + 3}{s + 2} + 2 = \frac{-7}{s + 2}$
25. $\frac{w - 5}{w + 5} + \frac{3}{5} = \frac{-10}{w + 5}$
26. $\frac{2x - 5}{x - 11} + 2 = \frac{17}{x - 11}$

27. $\frac{y + 5}{y + 6} = \frac{3}{4} - \frac{1}{y + 6}$
28. $\frac{7}{2} - \frac{3(z + 3)}{z + 2} = \frac{-8}{z + 2}$
29. $\frac{1}{3} + \frac{y - 1}{y + 2} = \frac{5}{y + 2}$

30. $2 - \frac{6}{w + 5} = \frac{w + 1}{w + 5}$

For Exercises 31–50, solve the equation. Identify any extraneous solutions. See Example 3.

31. $\frac{-1}{y + 2} + \frac{1}{(y + 2)(y + 3)} = \frac{1}{y + 3}$
32. $\frac{1}{x - 1} + \frac{1}{(x - 1)(x + 2)} = \frac{-1}{x + 2}$

33. $\frac{1}{z + 2} - \frac{9}{(z + 2)(z - 5)} = \frac{-1}{z - 5}$
34. $\frac{3}{w + 4} - \frac{9}{(w + 4)(w - 3)} = \frac{-2}{w - 3}$

35. $\frac{1}{p - 4} - \frac{2p}{p^2 - 16} = \frac{2}{p + 4}$
36. $\frac{q + 18}{q^2 - 36} + \frac{6}{q - 6} = \frac{-1}{q + 6}$

37. $\frac{2}{2z - 1} + \frac{1}{z + 1} = \frac{13}{2z^2 + z - 1}$
38. $\frac{2}{2x + 1} - \frac{5}{3x - 2} = \frac{-1}{6x^2 - x - 2}$

39. $\frac{5}{z + 2} - \frac{2z - 8}{z^2 - 4} = \frac{1}{z - 2}$
40. $\frac{7}{v + 1} + \frac{6v}{v^2 - 1} = \frac{3}{v - 1}$

41. $\frac{u}{u - 4} - \frac{u}{u + 3} = \frac{14}{u^2 - u - 12}$
42. $\frac{v}{v - 5} - \frac{v}{v + 3} = \frac{16}{v^2 - 2v - 15}$

43. $\frac{t}{t + 3} + \frac{2t}{t - 1} = \frac{3t^2 - 15}{t^2 + 2t - 3}$
44. $\frac{x}{x + 7} + \frac{2x}{x - 3} = \frac{3x^2 + 33}{x^2 + 4x - 21}$

45. $\frac{y - 1}{y + 2} - \frac{y + 1}{y - 1} = \frac{14}{y^2 + y - 2}$
46. $\frac{w - 2}{w + 3} - \frac{w + 1}{w - 2} = \frac{17}{w^2 + w - 6}$

47. $\dfrac{z+1}{z+2}+\dfrac{z+2}{z+1}=\dfrac{2z^2+5}{z^2+3z+2}$

48. $\dfrac{x+3}{x+1}+\dfrac{x+1}{x+3}=\dfrac{2x^2+10}{x^2+4x+3}$

49. $\dfrac{2y-1}{y+2}-\dfrac{y+3}{y-3}=\dfrac{y^2+21}{y^2-y-6}$

50. $\dfrac{2z-3}{z+4}-\dfrac{z+2}{z-2}=\dfrac{z^2-28}{z^2+2z-8}$

4.7 Applications and Word Problems

Many problems related to economics, engineering, or the sciences involve rational expressions. In this section we discuss some of these applications as well as some word problems that involve rational expressions. We begin with some simple word problems involving rational expressions.

Example 1. When the same number is subtracted from both the numerator and denominator of $\dfrac{19}{14}$ and the resulting fraction is reduced, the result is $\dfrac{3}{2}$. Find the number.

Solution: Let n be the number. From the information given we get

$$\frac{19-n}{14-n}=\frac{3}{2}.$$

We clear fractions by multiplying each side of the equation by $2(14-n)$.

$$\begin{aligned} \cancel{2(14-n)}\frac{19-n}{\cancel{(14-n)}} &= 2(14-n)\frac{3}{2} \\ 2(19-n) &= 3(14-n) \\ 38-2n &= 42-3n \\ n &= 4. \end{aligned}$$

CHECK: Let $n=4$ in $\dfrac{19-n}{14-n}$; then

$$\frac{19-4}{14-4}=\frac{15}{10}=\frac{3}{2}.$$

Thus 4 is the correct answer.

Example 2. When the reciprocal of a number is added to the reciprocal of three times the number, the result is $\dfrac{2}{9}$. Find the number.

Solution: Let x be the number. Thus we have

$$\frac{1}{x}+\frac{1}{3x}=\frac{2}{9}.$$

Multiply each side of the equation by $9x$ to clear fractions.

$$9x\left(\frac{1}{x}+\frac{1}{3x}\right)=9x\cdot\frac{2}{9}$$

$$9x\cdot\frac{1}{x}+9x\cdot\frac{1}{3x}=9x\cdot\frac{2}{9}$$

$$9+3=2x$$

$$12=2x$$

$$6=x.$$

Thus

$$x=6.$$

CHECK: Let $x=6$, then

$$\frac{1}{x}+\frac{1}{3x}=\frac{1}{6}+\frac{1}{18}=\frac{3}{18}+\frac{1}{18}=\frac{4}{18}=\frac{2}{9}.$$

Thus 6 is the correct answer.

Example 3. The Suwannee River in northern Florida has many long stretches of open water in which there is a steady current. On one of these open stretches of water, Fred can take his motorboat 2 miles upstream (against the current) in the same amount of time that he can travel 3 miles downstream. If Fred's boat can travel at 15 miles per hour in the still water of a lake, what is the speed of the river's current?

Solution: The correct formula to use in this problem is $d = rt$ (distance = rate · time). If we let x denote the speed of the current, the boat's rate of travel

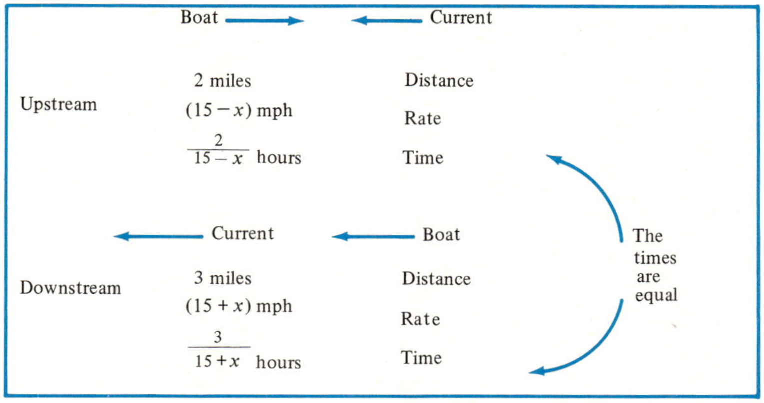

Figure 4.1

upstream is $(15 - x)$ miles per hour because it is going against the current. Similarly, the *downstream* rate of travel is $(15 + x)$ miles per hour because it is going with the current. Since the upstream travel time *equals* the downstream travel time, we will solve $d = rt$ for time t. The result is $t = \frac{d}{r}$ (time equals distance ÷ rate). Because Fred travels 2 miles upstream at a rate of $(15 - x)$ miles per hour, his *upstream* travel time is $\frac{2}{15 - x}$. Similarly, his *downstream* travel time is $\frac{3}{15 - x}$. Figure 4.1 summarizes this information. Since the times are equal, we may write

$$\frac{2}{15 - x} = \frac{3}{15 + x}.$$

If we use the LCD of $(15 - x)(15 + x)$ and clear fractions, the result is

$$\begin{aligned} 2(15 + x) &= 3(15 - x) \\ 30 + 2x &= 45 - 3x \\ 5x &= 15 \\ x &= 3. \end{aligned}$$

Thus the speed of the current is 3 miles per hour.

Example 4. An old photocopy machine can make 600 copies of a single page in 30 minutes. It takes a new machine 20 minutes to do the same job. If both machines are assigned to the same task simultaneously:
(a) How long will it take to produce a total of 600 copies?
(b) How many copies will each machine make?

Solution

(a) We will consider the 600 copies as the total job and determine what *part* of the job each machine does in 1 minute. Since the old machine makes 600 copies in 30 minutes, in 1 minute it makes $\frac{600}{30} = 20$ copies. Thus each minute it does $\frac{20}{600} = \frac{1}{30}$ of the job. Another way of viewing this is to note that if it takes 30 minutes to do a job, then $\frac{1}{30}$ of the job is done each minute. In a similar fashion we can determine that the new machine does $\frac{1}{20}$ of the job each minute. If we let x denote the number of minutes that it takes to do the job *using both* machines, then in 1 minute both machines can do $\frac{1}{x}$ of the job. Thus in *each minute* we know that

$$\frac{1}{30} + \frac{1}{20} = \frac{1}{x}.$$

$$\begin{pmatrix}\text{amount of the}\\ \text{job done by}\\ \text{the old machine}\end{pmatrix} + \begin{pmatrix}\text{amount of the}\\ \text{job done by}\\ \text{the new machine}\end{pmatrix} = \begin{pmatrix}\text{amount of the}\\ \text{job done by}\\ \text{both machines}\end{pmatrix}$$

If we clear fractions by multiplying each term by the LCD, $60x$, the result is

$$60x \cdot \frac{1}{30} + 60x \cdot \frac{1}{20} = 60x \cdot \frac{1}{x}$$
$$2x + 3x = 60$$
$$5x = 60$$
$$x = 12.$$

Thus it will take the two machines 12 minutes to do the job.

(b) The old machine makes 20 copies each minute. Thus in 12 minutes it will make $20(12) = 240$ copies. Similarly, the new machine makes $30(12) = 360$ copies.

If you had toast for breakfast or used an iron to touch up some clothing last night, you were aided by electrical *resistance*. Resistance is a measurable property of metallic devices that is exhibited when an electric current is passed through the device. Electrical resistance is measured in ohms. In toasters and irons, the result of resistance is heat. The symbol for a resistor is -/\/\/\-. Figure 4.2 shows two resistors that are connected in

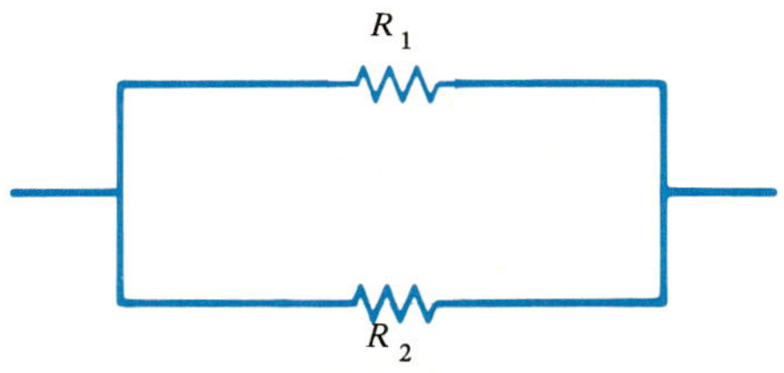

Figure 4.2

parallel in part of an electrical circuit. When two resistors are connected in parallel, the total resistance in that part of the circuit is computed using the formula

$$\frac{1}{R} = \frac{1}{R_1} + \frac{1}{R_2}.$$

In the formula, R is the total resistance in ohms and R_1 and R_2 are the resistances in ohms of the first and second resistors, respectively.

Example 5. A 10-ohm resistor is connected in parallel to a 15-ohm resistor. What is the total resistance in that part of the circuit?

Solution: Let $R_1 = 10$ and $R_2 = 15$; then

$$\frac{1}{R} = \frac{1}{R_1} + \frac{1}{R_2}$$
$$= \frac{1}{10} + \frac{1}{15}.$$

We clear fractions by multiplying each side of this equation by $30R$.

$$30R\left(\frac{1}{R}\right) = 30R\left(\frac{1}{10} + \frac{1}{15}\right)$$
$$30 = 3R + 2R$$
$$5R = 30$$
$$R = 6.$$

Thus a single resistor of 6 ohms is equal to two resistors of 10 ohms and 15 ohms when they are connected in parallel.

EXERCISES 4.7

Solve Exercises 1–20. See Examples 1 and 2.

1. When the same number is subtracted from both the numerator and denominator of $\frac{17}{23}$ and the resulting fraction is reduced, the result is $\frac{2}{3}$. Find the number.
2. When the same number is subtracted from both the numerator and denominator of $\frac{21}{29}$ and the resulting fraction is reduced, the result is $\frac{5}{9}$. Find the number.
3. When the same number is added to both the numerator and denominator of $\frac{7}{9}$ and the resulting fraction is reduced, the result is $\frac{8}{9}$. Find the number.
4. When the same number is added to both the numerator and denominator of $\frac{11}{15}$ and the resulting fraction is reduced, the result is $\frac{6}{7}$. Find the number.
5. A certain number is added to the numerator of $\frac{4}{23}$ and three times the number is subtracted from the denominator. If the result is reduced, the answer is 2. Find the number.
6. A certain number is added to the numerator of $\frac{5}{29}$ and three times the number is subtracted from the denominator. If the result is reduced, the answer is $\frac{2}{5}$. Find the number.
7. The numerator of a fraction is 5 less than the denominator. When 3 is added to both the numerator and denominator, the reduced result is $\frac{2}{3}$. Find the original fraction.
8. The numerator of a fraction is 8 less than the denominator. When 9 is added to both the numerator and denominator, the reduced result is $\frac{3}{5}$. Find the original fraction.
9. The denominator of a fraction is 6 more than the numerator. When 3 is added to the numerator and 8 is subtracted from the denominator, the reduced result is $\frac{4}{3}$. Find the original fraction.
10. The denominator of a fraction is 21 more than the numerator. When 13 is subtracted from the numerator and 14 is added to the denominator, the reduced result is $\frac{1}{4}$. Find the original fraction.

11. When the reciprocal of a number is added to the reciprocal of twice the number, the result is $\frac{1}{2}$. Find the number.

12. When the reciprocal of a number is added to the reciprocal of 5 times the number, the result is $\frac{2}{5}$. Find the number.

13. When twice the reciprocal of a number is added to the reciprocal of twice the number, the result is $\frac{1}{4}$. Find the number.

14. When three times the reciprocal of a number is added to the reciprocal of three times the number, the result is $\frac{1}{6}$. Find the number.

15. When four times the reciprocal of a number is added to the reciprocal of three times the number, the result is $\frac{1}{3}$. Find the number.

16. When five times the reciprocal of a number is added to the reciprocal of four times the number, the result is $\frac{3}{4}$. Find the number.

17. When the reciprocals of a number, twice the number and three times the number are summed, the sum of the three reciprocals is $\frac{11}{24}$. Find the number.

18. When the reciprocals of a number, twice the number and three times the number are summed, the sum of the three reciprocals is $\frac{11}{30}$. Find the number.

19. When the reciprocals of a number, three times the number and five times the number are summed, the sum of the three reciprocals is $\frac{23}{105}$. Find the number.

20. When the reciprocals of a number, three times the number and five times the number are summed, the sum of the three reciprocals is $\frac{23}{45}$. Find the number.

Solve Exercises 21–28. See Example 3.

21. John's motorboat cruises at the rate of 12 miles per hour in the still water of a lake. On a stretch of the Ohio River he can travel 3 miles upstream in the same amount of time that he travels 5 miles downstream. What is the speed of the river's current?

22. Yvonne owns a motorboat that can cruise at 14 miles per hour in the still water of a lake. On a stretch of open water of the middle Mississippi her boat can travel 5 miles upstream in the same amount of time that it can travel 9 miles downstream. What is the speed of the river's current?

23. Clare is using a motorboat on the Susquehanna River. The boat can go 4 miles upstream in the same time that it can go 9 miles downstream. If the boat's speed in still water is 13 miles per hour, what is the speed of the river's current?

24. Robert is using a motorboat on the Delaware River just north of Philadelphia. The boat can go 7 miles upstream in the same amount of time that it goes 11 miles downstream. If the speed of the boat in still water is 9 miles per hour, what is the speed of the river's current?

25. A small plane has a cruising speed of 140 miles per hour in still air. One day there is a steady wind blowing from the northwest. A pilot notices that it takes him the same amount of time to fly 15 miles directly against the wind as it does to fly 20 miles with the wind. What is the speed of the wind?

26. A small plane cruises at 135 miles per hour in still air. On a day when there is a steady wind from the southwest, a pilot notices that she can fly 20 miles directly against the wind in the same amount of time that she flies 25 miles with the wind. What is the speed of the wind?

27. A particular model of a commercial airplane has a cruising speed of 475 miles per hour in still air. One day there was a steady wind from the west. The pilot of an eastbound flight noticed that he went 50 miles in the same amount of time that a similar west-bound plane went 45 miles. What was the speed of the wind?

28. On a day when there was a steady wind of 20 miles per hour from the east, two planes with the same still-air cruising speed were flying in opposite directions. The crew of the west-bound flight recorded that their plane went 11 miles in the same amount of time that the east-bound flight went 10 miles. What was the cruising speed of the plane?

Solve Exercises 29–38. See Example 4.

29. An old model of an automated ice cream machine can fill 600 containers in 7 hours. A new model of the machine can fill 600 containers in 3 hours.
(a) How long will it take both machines to do the job of filling 600 containers?
(b) How many containers will the new machine fill when both machines do the job?

30. A small computer can complete a payroll in 6 hours. A larger and faster computer can complete the same payroll in 4 hours.
(a) If the two computers work on the same payroll, how long will the job take?
(b) What fraction of the payroll will the small computer process when both machines do the payroll?

31. A new photocopy machine can make 1000 copies in 30 minutes. An older machine takes 45 minutes to make 1000 copies.
(a) How long will it take both machines to make a total of 1000 copies?
(b) How many copies does the new machine make when both copiers do the job?

32. A school has two optical scanners that can be used to grade multiple-choice exams. Model A can grade 1200 papers in 60 minutes. Model B can grade 1200 papers in 40 minutes.
(a) How long will it take both machines to grade a total of 1200 papers?
(b) How many copies does Model A grade when both machines do the grading?

33. A swimming pool has two pipes that can be used to fill the pool. The large pipe can fill the pool in 10 hours, whereas the small pipe fills the pool in 15 hours.
(a) How long does it take to fill the pool if both pipes are used?
(b) When both pipes are used, what fraction of the pool is filled in 1 hour?
(c) What fraction of the pool is filled by the larger pipe when both pipes are used to fill the pool?

34. A large pipe can fill a swimming pool in 12 hours. A hose can fill the same pool in 20 hours.
(a) How long does it take to fill the pool if both the pipe and the hose are used?
(b) In 1 hour, how much of the pool is filled if both the pipe and the hose are used?
(c) What fraction of the pool is filled by the hose when both the pipe and the hose are used to fill the pool?

35. The spray tank on a tree surgeon's truck can be filled by a hose in 80 minutes. A larger hose can fill the tank in 48 minutes.
(a) How long will it take for both hoses to fill the tank?
(b) In 10 minutes, what fraction of the tank is filled if both hoses are used?
(c) What fraction of the tank is filled by the larger hose when both hoses are used to fill the tank?

36. The water tank on a forest service truck can be filled from a hose in 2 hours. It takes 3 hours to fill the same tank with a smaller hose.

(a) How long does it take to fill the tank if both hoses are used?
(b) In 1 hour, how much of the tank is filled if both hoses are used?
(c) What fraction of the tank's capacity is filled by the larger hose when both hoses are used to fill the tank?

37. The container for a chemical bath can be filled by a faucet in 2.5 minutes. The drain pipe can empty the full container in 3 minutes. If the drain is accidentally left open while the faucet is left on, how long will it take to fill the container?

38. It takes 5 minutes for the drain pipe to empty a full sink of water. A faucet can fill the empty sink in 4 minutes. If the drain is left open, how long does it take to fill an empty sink?

Solve Exercises 39–42. See Example 5.

39. Two resistances of 9 and 15 ohms, respectively, are connected in parallel. What single resistance is equal to these two resistances?

40. Two resistances of 10 and 12 ohms, respectively, are connected in parallel. What single resistance is equal to these two resistances?

41. A 10-ohm resistor is to be connected to another resistor in parallel. If the total resistance in the circuit is to be 2.5 ohms, at how many ohms should the second resistor be rated?

42. An 8-ohm resistor is to be connected in parallel to another resistor. If the total resistance in the circuit is to be 6 ohms, at how many ohms should the second resistor be rated?

Complex Fractional Expressions 4.8

A fractional expression with a numerator or denominator that contains one or more fractions is called a *complex fractional expression*. Thus

$$\frac{7}{\frac{2}{3}}, \quad \frac{\frac{2}{5}}{\frac{3}{4}}, \quad \frac{6+\frac{4}{3x}}{2x}, \quad \frac{\frac{3y}{4}}{y-\frac{1}{2}}, \quad \frac{\frac{1}{z}-\frac{1}{3}}{\frac{1}{6}+\frac{1}{2z}}$$

are all examples of complex fractional expressions. Such expressions are also called complex fractions. Complex fractions are often difficult to interpret and use until they have been rewritten as fractional expressions so that

1. There are no fractions in either the numerator or the denominator.
2. They have been reduced to lowest terms.

Any complex fraction that has been rewritten to satisfy conditions 1 and 2 is said to be *simplified*.

There are two methods that are usually used to simplify complex fractional expressions:

Method 1. Multiply the numerator and denominator by the LCD of all fractions in the numerator and denominator of the complex fractional expression.

Method 2. Treat the complex fractional expression as a division problem and multiply the numerator by the reciprocal of the denominator.

Method 1 can be easily applied to any complex fraction. It is particularly useful when either the numerator or denominator contains the sum or difference of fractions with different denominators. These two methods will be demonstrated in the following examples.

Example 1. Simplify $\dfrac{6 + \dfrac{4}{3x}}{2x}$.

Solution: Method 1 will be best for this problem. The LCD of the fractions in the numerator and denominator is $3x$. Multiplying the numerator and denominator by the LCD, we get

$$\frac{3x\left(6 + \dfrac{4}{3x}\right)}{3x(2x)} = \frac{18x + 4}{6x^2} \qquad \text{distributive property}$$

$$= \frac{\not{2}(9x + 2)}{\not{2}(3x^2)} \qquad \text{simplify}$$

$$= \frac{9x + 2}{3x^2}.$$

Example 2. Simplify $\dfrac{\dfrac{x + 5}{3x}}{\dfrac{x - 5}{x}}$.

Solution: Here we use Method 2. Multiplying the numerator $\dfrac{x + 5}{3x}$ by the reciprocal of the denominator $\dfrac{x - 5}{x}$, we get

$$\frac{x + 5}{3\not{x}_{1}} \cdot \frac{\overset{1}{\not{x}}}{x - 5} = \frac{x + 5}{3(x - 5)}.$$

Example 3. Simplify $\dfrac{1 - \dfrac{3}{2x}}{\dfrac{3}{4x} - \dfrac{1}{2}}$.

Solution: We will use Method 1. The LCD of all fractions in either the numerator or the denominator is $4x$. Thus we have

$$\frac{4x\left(1 - \dfrac{3}{2x}\right)}{4x\left(\dfrac{3}{4x} - \dfrac{1}{2}\right)} = \frac{4x - 6}{3 - 2x}$$

$$= \frac{-2\cancel{(3 - 2x)}}{\cancel{(3 - 2x)}}$$

$$= -2.$$

Methods 1 or 2 can also be used to simplify complex fractions that contain expressions with negative exponents. For example, consider $\dfrac{(xy)^{-1}}{x^{-2} - y^{-2}}$. Using the definition of negative exponents, we can rewrite this as

$$\frac{\dfrac{1}{xy}}{\dfrac{1}{x^2} - \dfrac{1}{y^2}}.$$

The LCD of all fractions in the complex fraction is x^2y^2. Thus, to simplify, we multiply the numerator and denominator by x^2y^2.

$$\frac{x^2y^2\left(\dfrac{1}{xy}\right)}{x^2y^2\left(\dfrac{1}{x^2} - \dfrac{1}{y^2}\right)} = \frac{xy}{y^2 - x^2}.$$

Example 4. Simplify $\dfrac{qp^{-1} - p}{p^2q^{-1} - 1}$.

Solution: First note that $\dfrac{qp^{-1} - p}{p^2q^{-1} - 1} = \dfrac{\dfrac{q}{p} - p}{\dfrac{p^2}{q} - 1}$.

Next apply Method 1 using the LCD, pq.

$$\frac{pq\left(\frac{q}{p} - p\right)}{pq\left(\frac{p^2}{q} - 1\right)} = \frac{q^2 - p^2q}{p^3 - pq}$$

$$= \frac{q(q - p^2)}{p(p^2 - q)}$$

$$= \frac{-q(p^2 - q)}{p(p^2 - q)}$$

$$= -\frac{q}{p}.$$

EXERCISES 4.8

For Exercises 1–18, simplify the complex fraction. Use either Method 1 or Method 2. See Examples 1 and 2.

1. $\dfrac{\frac{5}{9}}{\frac{2}{7}}$ 2. $\dfrac{\frac{3}{5}}{\frac{2}{9}}$ 3. $\dfrac{\frac{2}{9}}{\frac{4}{3}}$ 4. $\dfrac{\frac{5}{8}}{\frac{15}{4}}$ 5. $\dfrac{\frac{4x}{3}}{\frac{6x}{5}}$

6. $\dfrac{\frac{21}{4y}}{\frac{7}{2y}}$ 7. $\dfrac{\frac{-2x}{3y}}{\frac{4}{-y}}$ 8. $\dfrac{\frac{3w}{-5z}}{\frac{-w}{10}}$ 9. $\dfrac{\frac{a}{b}}{\frac{a^2}{b^3}}$ 10. $\dfrac{\frac{c}{d}}{\frac{c^3}{d^2}}$

11. $\dfrac{\frac{7p^4}{18q}}{\frac{14p}{9q^3}}$ 12. $\dfrac{\frac{5r^3}{s^2}}{\frac{10r}{s}}$ 13. $\dfrac{\frac{a+1}{2a}}{\frac{a-1}{a}}$ 14. $\dfrac{\frac{b-3}{6b}}{\frac{b+3}{2b}}$ 15. $\dfrac{\frac{2c+6d}{c^2d}}{\frac{c+3d}{cd^2}}$

16. $\dfrac{\frac{3r+12s}{rs^2}}{\frac{r+4s}{r^2s}}$ 17. $\dfrac{\frac{p^2-q^2}{pq}}{\frac{p+q}{p}}$ 18. $\dfrac{\frac{u+v}{u}}{\frac{u^2-v^2}{uv}}$

For Exercises 19–42, use Method 1 to simplify the complex fraction. See Example 3.

19. $\dfrac{\frac{3y}{4}}{y - \frac{1}{3}}$ **20.** $\dfrac{\frac{4z}{3}}{z - \frac{1}{4}}$ **21.** $\dfrac{\frac{6p}{5q}}{q - \frac{q}{2p}}$ **22.** $\dfrac{\frac{5r}{7s}}{s - \frac{s}{5r}}$

23. $\dfrac{p + \frac{1}{q}}{q + \frac{1}{p}}$ **24.** $\dfrac{u - \frac{1}{v}}{v - \frac{1}{u}}$ **25.** $\dfrac{1 - \frac{4}{5w}}{\frac{3}{2} - \frac{6}{5w}}$ **26.** $\dfrac{1 - \frac{2}{3z}}{\frac{5}{2} - \frac{5}{3z}}$

27. $\dfrac{\frac{2x^2}{y} - 4x}{\frac{3x}{5y} - \frac{6}{5}}$ **28.** $\dfrac{5y + \frac{20y^2}{x}}{\frac{4}{3} + \frac{16y}{3x}}$ **29.** $\dfrac{3b - \frac{1}{2b}}{b + \frac{1}{3b}}$ **30.** $\dfrac{2c + \frac{1}{3c}}{c - \frac{1}{2c}}$

31. $\dfrac{\frac{1}{u} - \frac{1}{v}}{\frac{1}{uv}}$ **32.** $\dfrac{\frac{1}{rs}}{\frac{1}{r} + \frac{1}{s}}$ **33.** $\dfrac{\frac{1}{x} - \frac{1}{y}}{\frac{1}{xy} - \frac{1}{x}}$ **34.** $\dfrac{\frac{1}{w} + \frac{1}{wz}}{\frac{1}{w} + \frac{1}{z}}$

35. $\dfrac{\frac{x}{3} - \frac{3}{y}}{\frac{2}{x} + \frac{y}{2}}$ **36.** $\dfrac{\frac{w}{4} + \frac{4}{z}}{\frac{z}{3} - \frac{3}{w}}$ **37.** $\dfrac{\frac{1}{x - 1} + \frac{x}{x + 1}}{\frac{1}{x + 1} - \frac{x}{x - 1}}$ **38.** $\dfrac{\frac{1}{y + 1} - \frac{1}{y - 1}}{\frac{1}{y + 1} + \frac{y}{y - 1}}$

39. $\dfrac{\frac{3}{w + 1} + \frac{2}{w}}{\frac{-3}{w + 1} + \frac{8}{w}}$ **40.** $\dfrac{\frac{4}{z + 2} - \frac{3}{z}}{\frac{1}{z + 2} + \frac{2}{z}}$ **41.** $\dfrac{\frac{1}{a + 3} - \frac{1}{a - 3}}{\frac{1}{a + 3} + \frac{2}{a - 3}}$ **42.** $\dfrac{\frac{2}{b - 5} - \frac{2}{b + 5}}{\frac{1}{b + 5} + \frac{4}{b - 5}}$

For Exercises 43–54, simplify the complex fractional expression. See Example 4.

43. $\dfrac{a^{-1} + b}{a + b^{-1}}$ **44.** $\dfrac{c - d^{-1}}{c^{-1} - d}$ **45.** $\dfrac{p^{-1} + q^{-1}}{p^{-1} - q^{-1}}$ **46.** $\dfrac{2u^{-1} - v^{-1}}{u^{-1} - 2v^{-1}}$

47. $\dfrac{w^{-2}}{w^{-2} - z^{-2}}$ **48.** $\dfrac{x^{-2} + y^{-2}}{y^{-2}}$ **49.** $\dfrac{t^{-1} - u^{-2}t}{t^{-2} - u^{-2}}$ **50.** $\dfrac{vw^{-2} - v^{-1}}{w^{-2} - v^{-2}}$

51. $\dfrac{(ab)^{-1}}{a^{-2} + b^{-2}}$ **52.** $\dfrac{(cd)^{-2}}{c^{-2} - d^{-2}}$ **53.** $(x^{-1} + y^{-1})^{-1}$ **54.** $(w^{-1} - z^{-1})^{-1}$

Key Words

Clearing fractions
Common denominator
Complex fractional expression
Extraneous solution
Extremes of a proportion
Least common denominator (LCD)
Means of a proportion
Proportion
Ratio
Rational expression
Reciprocal
Reduced form
Reduced to lowest terms
Remainder

CHAPTER 4 REVIEW

SECTION 4.1

Reduce each expression to lowest terms.

1. $\frac{26}{39}$ **2.** $\dfrac{20y^3}{12y}$ **3.** $\dfrac{22x^2w^5}{55x^4w}$ **4.** $\dfrac{q^2 - 3q}{q^2 - 2q - 3}$ **5.** $\dfrac{3z^2 - 12}{7z + 14}$ **6.** $\dfrac{3 - a}{a - 3}$

SECTION 4.2

Perform the divisions.

7. $(20v^4 - 15v^2 + 5v) \div 5v$ **8.** $(12x^3 + 28x^2 + 15x + 3) \div (2x + 3)$

SECTION 4.3

Compute the products and quotients and reduce your answers to lowest terms.

9. $\dfrac{9x}{y^2} \cdot \dfrac{y^3}{6x^4}$ **10.** $\dfrac{3cd^2}{4a^3b^2} \div \dfrac{9c^2d}{2a^2b^3}$ **11.** $\dfrac{w - 1}{w + 3} \cdot \dfrac{w^2 + 6w + 9}{w^2 - 1}$

12. $\dfrac{2x + 12}{x^2 - 4} \div \dfrac{3x + 18}{x - 2}$ **13.** $\dfrac{w^2 + 2w - 3}{w^2 + 4w + 3} \cdot \dfrac{w^2 + w - 6}{w^2 - 3w + 2}$ **14.** $\dfrac{2z - 6}{z^2 - 6z + 9} \div \dfrac{9 - 3z}{z^2 - 9}$

15. $\left(\dfrac{y^2 + y - 6}{y^2 - 25} \div \dfrac{y + 3}{y + 2}\right) \cdot \dfrac{y + 5}{y^2 - 4}$

SECTION 4.4

Find the following sums or differences and reduce your answers to lowest terms.

16. $\dfrac{3x + 7}{x + 3} - \dfrac{2x + 4}{x + 3}$ **17.** $\dfrac{3}{6z} + \dfrac{1}{z^2}$ **18.** $\dfrac{2w + 5}{w^2 + 5w + 6} + \dfrac{2w + 3}{w^2 + 5w + 6}$

19. $\dfrac{1}{a + 1} - \dfrac{2}{a + 2}$ **20.** $\dfrac{-4z - 19}{z^2 - z - 12} - \dfrac{1}{z + 3} + \dfrac{z + 1}{z - 4}$

SECTION 4.5

For Exercises 21 and 22, use the following information to write the requested ratio reduced to lowest terms.

A small canning company employs a total of 42 persons, of which 14 are males. Of these employees 6 are supervisors and 2 of these are females.

21. What is the ratio of the number of female employees to the number of male employees?

22. What is the ratio of the number of female supervisors to the number of male supervisors?

Solve for the unknown in the following proportions.

23. $\dfrac{x}{11} = \dfrac{15}{33}$ **24.** $\dfrac{v+1}{4} = \dfrac{v-2}{7}$

Solve the following word problem.

25. On a map, 3 inches represents 200 miles. If two towns are 5.4 inches apart on the map, how far apart are they in actual miles?

SECTION 4.6

Solve the equations. Identify any extraneous solutions.

26. $\dfrac{2}{3x} - \dfrac{5}{2x} = \dfrac{1}{6}$ **27.** $\dfrac{1}{w+3} - \dfrac{1}{4} = -\dfrac{1}{5}$ **28.** $\dfrac{1}{z+2} - \dfrac{5z+11}{z^2-4} = \dfrac{1}{z-2}$

29. $\dfrac{2x+1}{x-3} + \dfrac{x-2}{x+5} = \dfrac{3x^2+5x+13}{x^2+2x-15}$

SECTION 4.7

Solve each word problem.

30. When the same number is subtracted from both the numerator and denominator of $\frac{19}{25}$ and the resulting fraction is reduced, the result is $\frac{4}{7}$. Find the number.

31. Bill's motorboat cruises at the rate of 10 miles per hour in still water. On a stretch of river he can travel 4 miles upstream in the same amount of time that he travels 6 miles downstream. What is the speed of the river's current?

32. A small computer can complete a payroll in 7 hours. A larger and faster computer can complete the same payroll in 3 hours.
(a) If the two computers work on the same payroll, how long will the job take?
(b) What fraction of the payroll will the larger computer process when both do the job?

SECTION 4.8

Simplify the complex fractions.

33. $\dfrac{\frac{8x}{9}}{\frac{4x}{3}}$ **34.** $\dfrac{\frac{a^2-b^2}{ab}}{\frac{a-b}{a}}$ **35.** $\dfrac{\frac{x}{3}}{x-\frac{2}{x}}$ **36.** $\dfrac{\frac{1}{a+4}-\frac{1}{a-4}}{\frac{1}{a+4}+\frac{2}{a-4}}$

CHAPTER 4 TEST

For Problems 1–3, reduce the rational expression to lowest terms.

1. $\dfrac{27a^3}{12a}$ **2.** $\dfrac{80x^2y^3z}{60x^3y^2z^2}$ **3.** $\dfrac{u^2 - 9}{u^2 + 5u + 6}$

For Problems 4–7, find the product or quotient. Reduce all answers to lowest terms.

4. $\dfrac{9w^3}{12t^2} \cdot \dfrac{t^3}{6w^2}$ **5.** $\dfrac{v^2 - 4}{v + 1} \cdot \dfrac{v^2 - v - 2}{v^2 - 4v + 4}$ **6.** $\dfrac{3a + 15}{a - 2} \div \dfrac{a + 5}{6a - 12}$

7. $\dfrac{x^2 - 2x - 8}{x^2 + x - 12} \div \dfrac{x^2 - 3x - 10}{x^2 - 8x + 15}$

For Problems 8–13, find the sum or difference. When possible, reduce your answer to lowest terms.

8. $\dfrac{2x + 3}{x + 5} - \dfrac{x - 2}{x + 5}$ **9.** $\dfrac{5}{4y} + \dfrac{1}{y^2}$ **10.** $\dfrac{3}{x + 2} - \dfrac{1}{x}$

11. $\dfrac{w}{w + 2} + \dfrac{w + 1}{w - 2}$ **12.** $\dfrac{x + 5}{x^2 - 4} - \dfrac{1}{x^2 + 4x + 4}$ **13.** $\dfrac{p + 3}{p^2 - 1} - \dfrac{p + 2}{p^2 + 4p + 3}$

For Problems 14 and 15, use the following information to write the requested ratio reduced to lowest terms.

A local police force has 30 members of which 25 are males. Six members have the rank of detective and 10 members are rookies.

14. What is the ratio of the number of females to the number of males on the police force?

15. What is the ratio of the number of rookies to the number of detectives on the police force?

For Problems 16 and 17, solve the proportion.

16. $\dfrac{36}{75} = \dfrac{x}{600}$ **17.** $\dfrac{y - 3}{5} = \dfrac{y + 1}{3}$

For Problems 18 and 19, solve the equation and identify any extraneous solutions.

18. $\dfrac{3}{2s} - \dfrac{1}{4} = \dfrac{5}{s}$ **19.** $\dfrac{5}{z - 2} - \dfrac{z + 25}{z^2 + 5z - 14} = \dfrac{2}{z + 7}$

For Problems 20 and 21, simplify the complex fraction.

20. $\dfrac{a + \dfrac{1}{a}}{a - \dfrac{1}{a}}$ **21.** $\dfrac{\dfrac{3}{x + 1} - \dfrac{1}{x}}{\dfrac{1}{x + 1} + \dfrac{1}{x}}$

For Problems 22 and 23, perform the division.

22. $(24z^3 - 18z^2 + 6z) \div (3z)$

23. $(5x^4 - 2x^3 - 25x^2 + x + 1) \div (x^2 + 2x)$

Solve Problems 24 and 25.

24. The numerator of a certain fraction is 7 less than the denominator. When 10 is subtracted from the numerator and 3 is added to the denominator, the reduced result is $\frac{1}{3}$. Find the original fraction.

25. A soda manufacturer owns two filling machines. One machine can fill 800 bottles in 1 hour (60 minutes), and the other machine can fill 800 bottles in 40 minutes. If both machines are working at the same time, how long will it take to do the job of filling a total of 800 bottles?

CHAPTER 5

Roots, Radicals, and Rational Exponents

The average distance d of a planet from the Sun and the time T that it takes a planet to make a revolution around the Sun are related by the formula

$$T = d^{3/2}.$$

In the formula T is measured in years and d = 1 for the Earth. The fraction 3/2 appearing in the formula is called a rational exponent and it is related to roots and radicals. Since the planet Jupiter is a little more than 5 times the distance from the Sun than the Earth, using this formula we are able to compute that Jupiter takes about 12 years to make one revolution about the Sun.

National Aeronautics and Space Administration

In this chapter we begin with a discussion of square roots, cube roots, and higher roots of real numbers. The root of a number is represented by a radical. We will also present methods to evaluate and simplify expressions involving radicals.

The discussion of roots and radicals leads to a definition of rational or fractional exponents. Thus the concept of an exponent will have a meaning not only for integers, but for fractions as well.

Finally, in our discussion of the roots of a real number, we find that the square root of a negative real number cannot be a real number. In order for negative numbers to have square roots, we extend the set of real numbers by introducing the complex number system. As we shall see, the term "imaginary number" is associated with the complex numbers. Despite the use of the descriptive terms "imaginary" and "complex," this "new" system of numbers is just as mathematically sound as the system of real numbers. In fact, the complex numbers play an important role in the applied sciences, such as electrical engineering and aeronautical engineering.

5.1 Roots and Radicals

If x and b are real numbers such that

$$x^n = b,$$

where n is a positive integer ($n > 1$), then x is called an nth *root of b*. In particular, if $x^2 = b$, x is called a *square root of b;* and if $x^3 = b$, x is called a *cube root of b*.

Example 1. Find all real numbers such that the number is a
(a) Square root of 25. (b) Cube root of 8.
(c) Cube root of -8. (d) Fourth root of 81.

Solution

(a) 5 is a square root of 25 because $5^2 = 25$. Also, -5 is a square root of 25, since $(-5)^2 = 25$.
(b) 2 is a cube root of 8 because $2^3 = 8$.
(c) -2 is a cube root of -8 because $(-2)^3 = -8$.
(d) 3 is a fourth root of 81 because $3^4 = 81$; moreover, -3 is also a fourth root of 81, since $(-3)^4 = 81$.

We call 2nd, 4th, 6th roots, and so forth, *even* roots. The *odd* roots are 3rd, 5th, 7th roots, and so on. As Example 1 indicates, *positive* numbers have *two even* real roots, one positive and the other negative. An even root of 0 is 0. We note that *even* roots of negative numbers are *not* real numbers. For example, if n is an even integer, there is *no* real number x such that $x^n = -4$, since x^n is always nonnegative. In particular, for $n = 2$, $x^2 \neq -4$. Therefore, there is no real number that is a square root of -4. Negative numbers do have even roots but they belong to a set of numbers that will be discussed at the end of this chapter.

Example 1 also illustrates that *all* real numbers have exactly *one odd* real root. The odd real root of a positive number is positive. The odd real root of a negative number is negative.

The principal nth root of a real number b is represented by the *radical* $\sqrt[n]{b}$. The symbol $\sqrt[n]{}$ is called the *radical sign;* n is the *index* or *order* of the *radical;* and b is the *radicand.* For square roots the index 2 is omitted and we write $\sqrt{b}$ instead of $\sqrt[2]{b}$.

The *principal* nth root of a *positive* number, whether n is even or odd, is the *positive* root. For example, $\sqrt{9} = 3$, $\sqrt[3]{8} = 2$. The *principal* nth root of a *negative* number is *negative* provided that n is *odd*. Thus $\sqrt[3]{-8} = -2$. To represent the negative nth root of a real number $b (b > 0)$ when n is *even,* we write $-\sqrt[n]{b}$. Consequently, to indicate the negative square root of 9, we write $-\sqrt{9} = -3$.

By definition, if b is positive, $\sqrt{b^2} = b$, since the square of b is b^2. In general, for all real numbers b we write

(1) $$\sqrt{b^2} = |b|.$$

The absolute value is required to ensure that we have written the principal value, which is *positive*. Let us consider the radical $\sqrt{(-3)^2}$. From line (1) we have that

$$\sqrt{(-3)^2} = |-3| = 3.$$

The latter result is consistent with the fact that $\sqrt{(-3)^2} = \sqrt{9} = 3$. Note that $\sqrt{(-3)^2} \neq -3$.

In general, for any real number b:

If *n is an even positive integer,* then

$$\sqrt[n]{b^n} = |b|.$$

If *n is an odd positive integer,* then

$$\sqrt[n]{b^n} = b.$$

The next two examples provide concrete examples of the definitions discussed above.

Example 2. Find the root indicated. Assume that variables represent positive real numbers.

(a) $\sqrt{36x^2}$. (b) $\sqrt[3]{\dfrac{8}{27}}$. (c) $\sqrt[5]{-32}$. (d) $-\sqrt[4]{81z^8}$. (e) $\sqrt{(-8)^2}$.

Solution

(a) $\sqrt{36x^2} = 6x$ because $(6x)^2 = 36x^2$.

(b) $\sqrt[3]{\dfrac{8}{27}} = \dfrac{2}{3}$; note that $\left(\dfrac{2}{3}\right)^3 = \dfrac{8}{27}$.

(c) $\sqrt[5]{-32} = -2$ since $(-2)^5 = -32$.

(d) $-\sqrt[4]{81z^8} = -3z^2$ since $(-3z^2)^4 = 81z^8$.

(e) $\sqrt{(-8)^2} = |-8| = 8$.

Example 3. Which of the following expressions represent real numbers? Evaluate those that do represent real numbers.
(a) $\sqrt[4]{-16}$. (b) $-\sqrt[4]{16}$. (c) $\sqrt[6]{-64}$. (d) $\sqrt[5]{-243}$.

Solution

(a) $\sqrt[4]{-16}$ is not a real number. Even roots of negative numbers are not real numbers.
(b) $-\sqrt[4]{16} = -(2) = -2$. Note that $(-2)^4 = 16$.
(c) $\sqrt[6]{-64}$ is not a real number, for the same reason as that given in part (a).
(d) $\sqrt[5]{-243} = -3$, since $(-3)^5 = -243$.

In the following table we summarize in a convenient form the definition of the principal nth root for all real numbers.

Principal nth Root of b, $\sqrt[n]{b}$

Let n be an integer greater than 1, and b a real number. Then if

n is:	b is Positive	b is Negative	$b = 0$
Even	$\sqrt[n]{b}$ is positive	$\sqrt[n]{b}$ is not a real number	$\sqrt[n]{0} = 0$
Odd	$\sqrt[n]{b}$ is positive	$\sqrt[n]{b}$ is negative	$\sqrt[n]{0} = 0$

Note that, by definition, $(\sqrt[n]{b})^n = b$ provided that b is not negative when n is an even integer.

EXERCISES 5.1

For Exercises 1–40, find the root. Assume that all variables represent positive real numbers. If an expression does not represent a real number, so state. See Examples 1–3.

1. $\sqrt{81}$ **2.** $-\sqrt{49}$ **3.** $-\sqrt{144}$ **4.** $\sqrt{169}$ **5.** $\sqrt[3]{-1}$
6. $\sqrt[4]{-1}$ **7.** $\sqrt[3]{125}$ **8.** $\sqrt[3]{-27}$ **9.** $\sqrt[5]{32}$ **10.** $\sqrt[5]{-243}$
11. $-\sqrt{-1}$ **12.** $\sqrt{-100}$ **13.** $-\sqrt[6]{-1}$ **14.** $\sqrt[5]{-1}$ **15.** $\sqrt{(-9)^2}$
16. $\sqrt{(-10)^2}$ **17.** $\sqrt{25z^2}$ **18.** $\sqrt{100y^2}$ **19.** $-\sqrt{\frac{1}{4}}$ **20.** $\sqrt{\frac{25}{49}}$
21. $\sqrt[4]{81x^4}$ **22.** $\sqrt[4]{625y^4}$ **23.** $\sqrt[3]{\frac{27}{125}}$ **24.** $-\sqrt[4]{\frac{16}{81}}$ **25.** $\sqrt{x^2y^2}$
26. $\sqrt{\dfrac{x^2}{y^2}}$ **27.** $-\sqrt[3]{125a^3b^3}$ **28.** $\sqrt[3]{-216x^3y^3}$ **29.** $\sqrt[6]{(-5)^6}$ **30.** $\sqrt[8]{(-7)^8}$
31. $\sqrt{x^4}$ **32.** $\sqrt[3]{y^9}$ **33.** $\sqrt[3]{y^6}$ **34.** $\sqrt[4]{x^8}$ **35.** $\sqrt[3]{5^6}$
36. $\sqrt[4]{3^8}$ **37.** $\sqrt{-3^2}$ **38.** $\sqrt[4]{-3^4}$ **39.** $\sqrt[4]{(x+y)^4}$ **40.** $\sqrt[6]{-128x^6}$

Rational Exponents 5.2

The concept of the root of a real number can be used to define *rational* or *fractional exponents* of a number. We begin our discussion with rational exponents that have a numerator of 1. Consider an expression with a fractional exponent such as $9^{1/2}$. What meaning can we give to such a symbol? An answer to this question is provided by the fact that we want the properties of integer exponents, developed in Chapter 3, to remain valid for rational exponents. (Remember that integers are rational numbers.) Thus, if the rule for exponents, $(a^m)^n = a^{mn}$, is to be valid for rational exponents, we should have

(1) $$(9^{1/2})^2 = 9^{(1/2)\cdot 2} = 9^1 = 9.$$

Next recall that for real numbers x and b, where $b > 0$,

$$x^2 = b$$

means that x is a *square root* of b. To be consistent, we define $9^{1/2}$ to be a *square root* of 9, since $(9^{1/2})^2 = 9$. But 9 has two square roots, $+3$ and -3. We define $9^{1/2}$ to be the *principal square root,* that is, $9^{1/2} = \sqrt{9} = 3$.

Generalizing the discussion above, if n is a positive integer ($n > 1$) and x is a real number, then

$$(x^{1/n})^n = x^{(1/n)\cdot n} = x \qquad \text{provided that } x \text{ is not negative when } n \text{ is even.}$$

Thus we see that $x^{1/n}$ is the *principal* nth *root of* x provided that it is a real number. The previous discussion motivates the following definition.

> If x is a real number and n is a positive integer ($n > 1$), then
>
> $$x^{1/n} = \sqrt[n]{x} \qquad \text{provided that } x > 0, \text{ when } n \text{ is even.}$$

We use this definition in the next example.

Example 1. Evaluate where possible.

(a) $27^{1/3}$. (b) $-16^{1/2}$. (c) $(-16)^{1/2}$. (d) $(-64)^{1/3}$. (e) $\left(\dfrac{9}{25}\right)^{1/2}$.

Solution

(a) $27^{1/3} = \sqrt[3]{27} = 3$.

(b) $-16^{1/2} = -(\sqrt{16}) = -(4) = -4$.

(c) $(-16)^{1/2} = \sqrt{-16}$ is not a real number. Even roots of negative numbers are not real.

(d) $(-64)^{1/3} = \sqrt[3]{-64} = -4$.

(e) $\left(\dfrac{9}{25}\right)^{1/2} = \dfrac{3}{5}$.

As we gain experience with fractional exponents, the step involving radicals can be omitted, as in Example 1 part (e).

We are now in a position to define $x^{m/n}$ for x a real number and $\frac{m}{n}$ a rational number, where m and n are *positive* integers with $n > 1$. We observe again that if the property for exponents, $(a^m)^n = a^{mn}$, is to hold for rational numbers, we must have

$$x^{m/n} = x^{(1/n)\cdot m} = (x^{1/n})^m = (\sqrt[n]{x})^m$$

provided that x is not negative when n is even. With these restrictions we state the following *definition* for *positive rational exponents.*

Positive Rational Exponents

> If x is a real number, and m and n are positive integers with $n > 1$,
>
> $$x^{m/n} = (\sqrt[n]{x})^m \qquad \text{provided that } \sqrt[n]{x} \text{ is a real number.}$$

To *extend* the definition given above to powers with rational exponents that are *negative,* we observe that if m and n are *positive* integers with $n > 1$, then

$$\frac{m}{-n} = -\frac{m}{n} = \frac{-m}{n}$$

are negative rational numbers.

Again assuming $(a^m)^n = a^{mn}$ is valid for all rational numbers, we see that

$$x^{m/(-n)} = x^{-(m/n)} = x^{(-m)/n} = (x^{1/n})^{-m} = \left(\sqrt[n]{x}\right)^{-m} = \frac{1}{(\sqrt[n]{x})^m} = \frac{1}{x^{m/n}} \quad (x \neq 0).$$

Consequently, we *define* for x real $(x \neq 0)$, and m and n positive integers,

$$\mathbf{x^{-m/n} = \frac{1}{x^{m/n}}} \qquad \text{provided that } \sqrt[n]{x} \text{ is real } (x \neq 0).$$

Example 2. Evaluate where possible.
(a) $25^{3/2}$. (b) $16^{-1/2}$. (c) $(-27)^{4/3}$. (d) $(-4)^{3/2}$.

Solution

(a) $25^{3/2} = (\sqrt{25})^3 = (5)^3 = 125.$

(b) $16^{-1/2} = \dfrac{1}{16^{1/2}} = \dfrac{1}{\sqrt{16}} = \dfrac{1}{4}.$

(c) $(-27)^{4/3} = (\sqrt[3]{-27})^4 = (-3)^4 = 81.$

(d) $(-4)^{3/2} = (\sqrt{-4})^3$ is not a real number, since $\sqrt{-4}$ is not a real number.

The following theorem, which we state without proof, gives another interpretation of $x^{m/n}$.

Theorem 1

If x is a real number and $\frac{m}{n}$ is a rational number reduced to lowest terms with $n > 1$, then

$$x^{m/n} = \sqrt[n]{x^m}.$$

By the definition of $x^{m/n}$, Theorem 1, and provided that $\frac{m}{n}$ is a rational number reduced to lowest terms, we have that

$$x^{m/n} = (\sqrt[n]{x})^m = \sqrt[n]{x^m} \qquad \text{provided that } \sqrt[n]{x} \text{ is real.}$$

The preferred way of writing $x^{m/n}$, in terms of radicals as either $(\sqrt[n]{x})^m$ or $\sqrt[n]{x^m}$, often depends on whether we are evaluating numerical expressions or rewriting variables with rational exponents as radicals. There is no clear-cut rule; the choice depends on each particular situation. The following example should provide some guidelines for expressing $x^{m/n}$ in terms of radicals.

Example 3. Rewrite the following expressions in radical form. Evaluate where possible. Assume that all variables represent positive real numbers.
(a) $8^{2/3}$. (b) $y^{2/5}$. (c) $16^{3/4}$. (d) $(xy)^{5/6}$.

Solution

(a) We have two methods for evaluating the given expression.

Method 1. $8^{2/3} = (\sqrt[3]{8})^2 = (2)^2 = 4.$ — definition of $x^{m/n}$

Method 2. $8^{2/3} = \sqrt[3]{8^2} = \sqrt[3]{64} = 4.$ — Theorem 1

However, we see that the first form is preferable for evaluating $8^{2/3}$, since it is easier to evaluate $\sqrt[3]{8}$ rather than $\sqrt[3]{64}$.

(b) $y^{2/5} = \sqrt[5]{y^2}$ is usually preferred to $(\sqrt[5]{y})^2$.

(c) We choose to evaluate the expression as follows:

$$16^{3/4} = (\sqrt[4]{16})^3 = (2)^3 = 8.$$

We note that $16^{3/4} = \sqrt[4]{16^3} = \sqrt[4]{4096} = 8$ gives the same result but is a much more difficult way to proceed.

(d) $(xy)^{5/6} = \sqrt[6]{(xy)^5} = \sqrt[6]{x^5y^5}$ is usually preferred to $(\sqrt[6]{xy})^5$.

Before we extend to rational exponents the properties of exponents developed in Chapter 3 for integers, there is a point about rational exponents that should be discussed. An example will illustrate this point.

Note that $\frac{1}{2}$ and $\frac{2}{4}$ represent the same rational number, that is, $\frac{1}{2} = \frac{2}{4}$. However, we must not automatically assume that

$$x^{1/2} = x^{2/4}.$$

The expression $x^{1/2}$ represents the square root of x, whereas $x^{2/4}$ means the fourth root of x^2. It can be proven for $x \geq 0$ (x positive or zero) that they are the same. However, $x^{1/2}$ is not the same as $x^{2/4}$ if $x < 0$. For example, consider the particular expressions $(-4)^{1/2}$ and $(-4)^{2/4}$. We see that $(-4)^{1/2} = \sqrt{-4}$ is not a real number. But $(-4)^{2/4} = \sqrt[4]{(-4)^2} = \sqrt[4]{16} = 2$. Therefore, $(-4)^{1/2} \neq (-4)^{2/4}$.

The discussion above indicates that difficulties arise when dealing with *negative* bases and *rational* exponents. However, if we avoid even roots of negative numbers, the rules for integer exponents can be extended to rational or fractional exponents. The following properties of rational exponents can be proven by careful use of the properties for integer exponents developed in Chapter 3 and the definition of $a^{m/n}$.

Properties of Rational Exponents

If a and b are real numbers, s and t are rational numbers, and a and b are not negative when s and t involve even roots, then

(1) $a^s \cdot a^t = a^{s+t}$. (2) $a^0 = 1 \quad (a \neq 0)$.

(3) $a^{-s} = \dfrac{1}{a^s} \quad (a \neq 0)$. (4) $(a \cdot b)^s = a^s \cdot b^s$.

(5) $(a^s)^t = a^{st}$. (6) $\dfrac{a^s}{a^t} = a^{s-t} \quad (a \neq 0)$.

(7) $\left(\dfrac{a}{b}\right)^s = \dfrac{a^s}{b^s} \quad (b \neq 0)$.

Note that we included $a^0 = 1$ $(a \neq 0)$ in the theorem for completeness. The numbering of the properties of rational exponents is the same as the numbering of the corresponding properties for integer exponents given at the end of Chapter 3.

The following examples illustrate the application of the properties listed above. Remember again that an integer is also a rational number.

Example 4. Simplify.
(a) $(9^4)^{1/2}$. (b) $2^{1/2} \cdot 8^{1/2}$. (c) $(-27)^{-1/3}$. (d) $5^{1/3} \cdot 5^{2/3}$.

Solution: To simplify the expressions, we use the properties of rational exponents .

(a) $(9^4)^{1/2} = 9^{4(1/2)} = 9^2 = 81$.

(b) $2^{1/2} \cdot 8^{1/2} = (2 \cdot 8)^{1/2} = 16^{1/2} = 4.$

(c) $(-27)^{-1/3} = \dfrac{1}{(-27)^{1/3}} = \dfrac{1}{\sqrt[3]{-27}} = \dfrac{1}{-3}.$

(d) $5^{1/3} \cdot 5^{2/3} = 5^{1/3+2/3} = 5^1 = 5.$

Example 5. Simplify. Express all answers with positive exponents. Assume that all variables represent positive real numbers.

(a) $4x^{1/2} \cdot 5x^{1/3}$. (b) $(16x^2y^4)^{3/2}$. (c) $\left(\dfrac{a^{1/2}}{b^3}\right)^2$. (d) $\left(\dfrac{x^{1/3}}{y^{1/6}}\right)^{-3}$.

Solution: We apply the properties of rational exponents to simplify the expressions.

(a) Using Property 1, we see that

$$4x^{1/2} \cdot 5x^{1/3} = (4 \cdot 5)x^{1/2} \cdot x^{1/3} = 20x^{1/2+1/3} = 20x^{3/6+2/6} = 20x^{5/6}.$$

(b) $(16x^2y^4)^{3/2} = 16^{3/2}(x^2)^{3/2}(y^4)^{3/2}$ Property 4

$= (\sqrt{16})^3x^{2(3/2)}y^{4(3/2)}$ Property 5; and $a^{m/n} = (\sqrt[n]{a})^m$

$= 64x^3y^6.$

(c) $\left(\dfrac{a^{1/2}}{b^3}\right)^2 = \dfrac{(a^{1/2})^2}{(b^3)^2} = \dfrac{a}{b^6}.$ Properties 7 and 5

(d) $\left(\dfrac{x^{1/3}}{y^{1/6}}\right)^{-3} = \left(\dfrac{y^{1/6}}{x^{1/3}}\right)^3$ $\left(\dfrac{a}{b}\right)^{-n} = \left(\dfrac{b}{a}\right)^n$

$= \dfrac{(y^{1/6})^3}{(x^{1/3})^3} = \dfrac{y^{1/2}}{x}.$ Properties 7 and 5

EXERCISES 5.2

For Exercises 1–8, evaluate the root, when possible. See Example 1.

1. $36^{1/2}$ **2.** $100^{1/2}$ **3.** $(-125)^{1/3}$ **4.** $81^{1/4}$

5. $(-16)^{1/4}$ **6.** $(-32)^{1/5}$ **7.** $-64^{1/6}$ **8.** $(-64)^{1/6}$

For Exercises 9–24, evaluate the root, when possible. See Examples 2 and 3.

9. $16^{3/2}$ **10.** $9^{3/2}$ **11.** $8^{4/3}$ **12.** $(-27)^{2/3}$ **13.** $100^{-1/2}$

14. $64^{-1/2}$ **15.** $(-25)^{3/2}$ **16.** $(-16)^{3/4}$ **17.** $81^{3/2}$ **18.** $27^{2/3}$

19. $(\frac{16}{81})^{3/4}$ **20.** $(\frac{64}{125})^{2/3}$ **21.** $(-1)^{5/6}$ **22.** $(-1)^{3/5}$ **23.** $16^{-3/2}$

24. $9^{-5/2}$

For Exercises 25–32, rewrite the expression in radical form. Assume that all variables represent positive real numbers. See Example 3.

25. $x^{3/4}$ **26.** $y^{2/3}$ **27.** $(xy)^{2/3}$ **28.** $(ab)^{4/5}$ **29.** $(2y)^{2/3}$

30. $(3x)^{2/3}$ **31.** $(x^2 + y^2)^{1/2}$ **32.** $(x^2 - y^2)^{1/3}$

For Exercises 33–56, simplify the expression. Write the answer using positive exponents when possible. Assume that all variables represent positive numbers. See Examples 4 and 5.

33. $(3^{1/2})^6$ **34.** $(5^6)^{1/3}$ **35.** $2^{1/3} \cdot 4^{1/3}$ **36.** $2^{1/2} \cdot 50^{1/2}$

37. $7^{1/4} \cdot 7^{3/4}$ **38.** $3^{1/2} \cdot 3^{-1/2}$ **39.** $5^{4/3} \cdot 5^{-1/3}$ **40.** $\dfrac{9^{5/2}}{9^{3/2}}$

41. $2x^{1/4} \cdot 3x^{1/2}$ **42.** $3y^{1/2} \cdot 6y^{1/6}$ **43.** $a^{-1/2}a^{3/2}$ **44.** $b^{7/4}b^{-3/4}$

45. $\dfrac{y^{5/3}}{y^{2/3}}$ **46.** $\dfrac{z^{4/3}}{z^{2/3}}$ **47.** $(x^3y^6)^{2/3}$ **48.** $(a^8b^4)^{5/4}$

49. $\left(\dfrac{x^{1/3}}{y^{2/3}}\right)^6$ **50.** $\left(\dfrac{a^{3/4}}{b^{1/4}}\right)^8$ **51.** $\left(\dfrac{a^{1/2}}{b^{3/4}}\right)^{-2}$ **52.** $\left(\dfrac{x^{1/3}}{y^{1/2}}\right)^{-6}$

53. $(4x^4y^6)^{3/2}$ **54.** $(16a^8b^{12})^{3/4}$ **55.** $\left(\dfrac{8x^{-3}y^6}{z^9}\right)^{1/3}$ **56.** $\left(\dfrac{16z^8}{x^4y^{-12}}\right)^{1/4}$

5.3 Properties of Radicals

In this section we state and prove two properties involving products and quotients of radicals having the same index. These properties are useful when we are multiplying and dividing algebraic expressions containing radicals. Also, they are used to express radicals in simpler forms. Both properties follow immediately from the properties of rational exponents. We first state and prove the property involving products.

Theorem 2. Product Rule for Radicals

> If x and y are nonnegative real numbers, then
> $$\sqrt[n]{xy} = \sqrt[n]{x}\,\sqrt[n]{y}.$$

Proof

$$\sqrt[n]{xy} = (xy)^{1/n} = x^{1/n}y^{1/n} = \sqrt[n]{x}\,\sqrt[n]{y}.$$

The theorem states that the nth root of a *product* is the *product* of the nth roots of its *positive factors*. We illustrate the application of this property in the next example.

Example 1. Find each product. Assume that all variables represent positive real numbers.

(a) $\sqrt{2}\,\sqrt{18}$. (b) $\sqrt[3]{4x^2}\,\sqrt[3]{2x}$. (c) $\sqrt[4]{\frac{1}{8}}\,\sqrt[4]{\frac{1}{2}}$. (d) $\sqrt[4]{x^3y}\,\sqrt[4]{xy^3}$.

Solution

(a) $\sqrt{2}\,\sqrt{18} = \sqrt{2 \cdot 18} = \sqrt{36} = 6$.

(b) $\sqrt[3]{4x^2}\,\sqrt[3]{2x} = \sqrt[3]{(4x^2)(2x)} = \sqrt[3]{8x^3} = 2x$. note that $(2x)^3 = 8x^3$

(c) $\sqrt[4]{\frac{1}{8}}\,\sqrt[4]{\frac{1}{2}} = \sqrt[4]{\frac{1}{8} \cdot \frac{1}{2}} = \sqrt[4]{\frac{1}{16}} = \frac{1}{2}$. note that $\left(\frac{1}{2}\right)^4 = \frac{1}{16}$

(d) $\sqrt[4]{x^3y}\,\sqrt[4]{xy^3} = \sqrt[4]{(x^3y)(xy^3)} = \sqrt[4]{x^4y^4} = \sqrt[4]{(xy)^4} = xy$.

We now state the property that pertains to the quotient of two radicals with the same index or order.

Theorem 3. Quotient Rule for Radicals

> If x and y are positive real numbers, then
> $$\sqrt[n]{\frac{x}{y}} = \frac{\sqrt[n]{x}}{\sqrt[n]{y}}.$$

Proof

$$\sqrt[n]{\frac{x}{y}} = \left(\frac{x}{y}\right)^{1/n} = \frac{x^{1/n}}{y^{1/n}} = \frac{\sqrt[n]{x}}{\sqrt[n]{y}}.$$

We can describe this property informally in the following terms:

The *n*th root of a *fraction* is the *n*th root of the *numerator* divided by the *n*th root of the *denominator*.

We use this property in the next example.

Example 2. Simplify. Assume that all variables represent positive real numbers.

(a) $\frac{\sqrt{250}}{\sqrt{10}}$. (b) $\sqrt[3]{\frac{27}{125}}$. (c) $\frac{\sqrt{x^3}}{\sqrt{x}}$. (d) $\frac{\sqrt[3]{2y^5}}{\sqrt[3]{16y^2}}$.

Solution

(a) $\frac{\sqrt{250}}{\sqrt{10}} = \sqrt{\frac{250}{10}} = \sqrt{25} = 5$.

(b) $\sqrt[3]{\frac{27}{125}} = \frac{\sqrt[3]{27}}{\sqrt[3]{125}} = \frac{3}{5}$.

(c) $\dfrac{\sqrt{x^3}}{\sqrt{x}} = \sqrt{\dfrac{x^3}{x}} = \sqrt{x^2} = x.$

(d) $\dfrac{\sqrt[3]{2y^5}}{\sqrt[3]{16y^2}} = \sqrt[3]{\dfrac{2y^5}{16y^2}} = \sqrt[3]{\dfrac{y^3}{8}} = \dfrac{\sqrt[3]{y^3}}{\sqrt[3]{8}} = \dfrac{y}{2}$. Note also that

$$\sqrt[3]{\frac{y^3}{8}} = \sqrt[3]{\left(\frac{y}{2}\right)^3} = \frac{y}{2}.$$

A rational number is a *perfect nth power* if it is the nth power of some rational number. Examples of perfect nth powers are

$$49 = 7^2, \qquad \frac{25}{36} = \left(\frac{5}{6}\right)^2, \qquad 125 = 5^3, \qquad \frac{16}{81} = \left(\frac{2}{3}\right)^4.$$

A rational expression is called a *perfect nth power* if it is the nth power of some rational expression. For example, $27x^6y^{12}$ is a perfect third power because

$$27x^6y^{12} = (3x^2y^4)^3.$$

Perfect second and third powers are called *perfect squares* and *perfect cubes*, respectively.

We call an algebraic expression that contains radicals a *radical expression*. We now state the conditions for the simplified form of a radical expression.

Simplified Form of a Radical Expression

An expression containing radicals is in *simplified form* if the following conditions are satisfied:

1. The radicand of a radical does not contain a factor raised to a power equal to or greater than the index of the radical.
2. The denominator of a fraction does not contain a radical.
3. A radical does not have a fraction as a radicand.
4. The index of a radical and the power of the radicand have no factors in common.

The next example contains radicals violating one of conditions 1–4 for the simplified form of a radical expression. Using Theorems 2 and 3, we show how we can express these radicals in simplified form.

Example 3. Write the following radical expressions in simplified form.

(a) $\sqrt{50}$. (b) $\dfrac{1}{\sqrt{5}}$. (c) $\sqrt[3]{\dfrac{7}{4}}$. (d) $\sqrt[6]{2^3}$.

Solution

(a) The radicand can be factored as follows:

$$50 = 25 \cdot 2 = 5^2 \cdot 2.$$

Note that the index of the given radical is 2 and that 2 equals a power of a factor of the radicand, 5^2. This violates condition 1 for the simplified form of a radical. To express the given radical in simplified form, we use the product rule for radicals and write

$$\sqrt{50} = \sqrt{5^2 \cdot 2} = \sqrt{5^2} \cdot \sqrt{2} = 5\sqrt{2}.$$

Often, we simplify radicals such as $\sqrt{50}$ as follows:

$$\sqrt{50} = \sqrt{25 \cdot 2} = \sqrt{25}\,\sqrt{2} = 5\sqrt{2}.$$

(b) The expression $\dfrac{1}{\sqrt{5}}$ is a fraction with a radical in the denominator. This violates condition 2. We can eliminate the radical from the denominator by multiplying both numerator and denominator of the fraction by $\sqrt{5}$. (This does not change the value of the fraction, since this is equivalent to multiplying the fraction by 1.) As a result, we see that

$$\frac{1}{\sqrt{5}} = \frac{1}{\sqrt{5}} \cdot \frac{\sqrt{5}}{\sqrt{5}} = \frac{\sqrt{5}}{\sqrt{5^2}} = \frac{\sqrt{5}}{5}. \quad \longleftarrow \text{ simplified form of } \frac{1}{\sqrt{5}}$$

(c) The radicand of $\sqrt[3]{\dfrac{7}{4}}$ is a fraction. This violates condition 3. We need to express the given radical without a fraction under the radical sign. To do this, we rewrite $\dfrac{7}{4}$ as a fraction with a denominator that is a *perfect cube*. Then we use the quotient rule for radicals.

$$\sqrt[3]{\frac{7}{4}} = \sqrt[3]{\frac{7}{4} \cdot \frac{2}{2}} = \sqrt[3]{\frac{14}{8}} = \frac{\sqrt[3]{14}}{\sqrt[3]{8}}$$

$$= \frac{\sqrt[3]{14}}{2}. \quad \longleftarrow \text{ simplified form of } \sqrt[3]{\frac{7}{4}}$$

(d) The radical $\sqrt[6]{2^3}$ has index 6 and a radicand with power 3. The numbers 6 and 3 have a common factor 2. This violates condition 4. To simplify $\sqrt[6]{2^3}$, we express it in terms of rational exponents and proceed as follows:

$$\sqrt[6]{2^3} = 2^{3/6} = 2^{1/2} = \sqrt{2}. \quad \longleftarrow \text{ simplified form of } \sqrt[6]{2^3}$$

In Example 3(b) we simplified the radical expression $\dfrac{1}{\sqrt{5}}$ by eliminating the radical in the denominator and by expressing $\dfrac{1}{\sqrt{5}}$ as $\dfrac{\sqrt{5}}{5}$. The process of removing a radical from the denominator is called ***rationalizing the denominator.***

Example 4. Rationalize the denominator: $\dfrac{1}{\sqrt[3]{3x^2}}$ $(x \neq 0)$.

Solution: To rationalize the denominator of the given expression, we multiply the numerator and denominator by a radical so that the denominator is the cube root of a perfect cube. To do this, we observe that $3x^2(9x) = 27x^3 = (3x)^3$ is a perfect cube. As a result, we proceed as follows to rationalize the denominator of the given expression:

$$\frac{1}{\sqrt[3]{3x^2}} \cdot \frac{\sqrt[3]{9x}}{\sqrt[3]{9x}} = \frac{\sqrt[3]{9x}}{\sqrt[3]{(3x^2)(9x)}} = \frac{\sqrt[3]{9x}}{\sqrt[3]{(3x)^3}} = \frac{\sqrt[3]{9x}}{3x}.$$

The following example involves simplifying other radical expressions that contain variables.

Example 5. Simplify. Assume that all variables represent positive real numbers.

(a) $\sqrt[3]{32x^4y^7}$. (b) $\sqrt[4]{\dfrac{a^2}{2b}}$.

Solution

(a)
$$\begin{aligned}\sqrt[3]{32x^4y^7} &= \sqrt[3]{(8x^3y^6)(4xy)}\\ &= \sqrt[3]{(2xy^2)^3(4xy)}\\ &= \sqrt[3]{(2xy^2)^3}\,\sqrt[3]{4xy} = 2xy^2\sqrt[3]{4xy}.\end{aligned}$$

(b) We multiply numerator and denominator of the radical so that the denominator is a perfect fourth power.

$$\begin{aligned}\sqrt[4]{\frac{a^2}{2b}} = \sqrt[4]{\frac{a^2}{2b} \cdot \frac{8b^3}{8b^3}} &= \sqrt[4]{\frac{8a^2b^3}{16b^4}}\\ &= \frac{\sqrt[4]{8a^2b^3}}{\sqrt[4]{16b^4}} = \frac{\sqrt[4]{8a^2b^3}}{2b}.\end{aligned}$$

EXERCISES 5.3

For Exercises 1–20, simplify the expression. Assume that all variables represent positive real numbers. See Examples 1 and 2.

1. $\sqrt{3}\,\sqrt{27}$ **2.** $\sqrt{2}\,\sqrt{8}$ **3.** $\sqrt[3]{9y^2}\,\sqrt[3]{3y}$ **4.** $\sqrt[3]{2a}\,\sqrt[3]{4a^2}$ **5.** $\sqrt[4]{\tfrac{1}{3}}\,\sqrt[4]{\tfrac{1}{27}}$

6. $\sqrt[5]{\tfrac{1}{2}}\,\sqrt[5]{\tfrac{1}{16}}$ **7.** $\sqrt{\dfrac{1}{x}}\,\sqrt{x^3}$ **8.** $\sqrt{\dfrac{1}{y^2}}\,\sqrt{y^6}$ **9.** $\sqrt[3]{ab}\,\sqrt[3]{a^2b^2}$ **10.** $\sqrt[4]{x^2y^2}\,\sqrt[4]{x^2y^2}$

11. $\dfrac{\sqrt{75}}{\sqrt{3}}$ **12.** $\dfrac{\sqrt{128}}{\sqrt{2}}$ **13.** $\sqrt[3]{\tfrac{64}{27}}$ **14.** $\sqrt[4]{\tfrac{81}{625}}$ **15.** $\dfrac{\sqrt[3]{a^4}}{\sqrt[3]{a}}$

16. $\dfrac{\sqrt[5]{y^7}}{\sqrt[5]{y^2}}$ **17.** $\dfrac{\sqrt{a^3b}}{\sqrt{ab^3}}$ **18.** $\dfrac{\sqrt[3]{xy^4}}{\sqrt[3]{x^4y}}$ **19.** $\dfrac{\sqrt[4]{162a^3}}{\sqrt[4]{2a^7}}$ **20.** $\dfrac{\sqrt[5]{32b^4}}{\sqrt[5]{b^9}}$

For Exercises 21–52, simplify the expression. Assume that all variables represent positive real numbers. See Examples 3–5.

21. $\sqrt{98}$ **22.** $\sqrt{200}$ **23.** $\dfrac{1}{\sqrt{6}}$ **24.** $\dfrac{1}{\sqrt{7}}$ **25.** $\sqrt[3]{\frac{5}{9}}$

26. $\sqrt[3]{\frac{2}{25}}$ **27.** $\sqrt{5^4}$ **28.** $\sqrt[12]{3^4}$ **29.** $\dfrac{1}{\sqrt{2x}}$ **30.** $\dfrac{1}{\sqrt{3a}}$

31. $\sqrt{\dfrac{3}{a}}$ **32.** $\sqrt{\dfrac{4}{y}}$ **33.** $\sqrt{x^2y}$ **34.** $\sqrt{a^2b^3}$ **35.** $\sqrt[3]{54a^4}$

36. $\sqrt[3]{16b^5}$ **37.** $\sqrt[4]{a^5b^7}$ **38.** $\sqrt[5]{a^8b^2}$ **39.** $\sqrt[8]{(x^2)^4}$ **40.** $\sqrt[9]{(y^3)^3}$

41. $\sqrt{\dfrac{50x^3}{y^2}}$ **42.** $\sqrt{\dfrac{18a^3}{b^4}}$ **43.** $\sqrt[3]{\dfrac{1}{a}}$ **44.** $\sqrt[3]{\dfrac{1}{a^2}}$ **45.** $\sqrt{27x^3y^4}$

46. $\sqrt{48a^2b^3}$ **47.** $\sqrt{a^5b^7}$ **48.** $\sqrt{x^9y^4}$ **49.** $\sqrt[3]{16x^5y^7}$ **50.** $\sqrt[3]{54x^6y^8}$

51. $\sqrt[4]{\dfrac{a^4}{8b^2}}$ **52.** $\sqrt[4]{\dfrac{x^3}{4y^3}}$

For Exercises 53–58, rationalize the denominator. See Example 4.

53. $\dfrac{1}{\sqrt[3]{2x}}$ **54.** $\dfrac{1}{\sqrt[3]{4y^2}}$ **55.** $\dfrac{1}{\sqrt[3]{3y^4}}$ **56.** $\dfrac{1}{\sqrt[3]{16x^5}}$ **57.** $\dfrac{\sqrt[3]{5}}{\sqrt[3]{x^2}}$ **58.** $\dfrac{\sqrt[3]{7}}{\sqrt[3]{x^5}}$

For Exercises 59 and 60, verify each statement.

59. Use the numbers 3 and 4 to show that $\sqrt{a^2 + b^2} \neq \sqrt{a^2} + \sqrt{b^2}$.

60. Use the numbers 25 and 144 to show that $\sqrt{a + b} \neq \sqrt{a} + \sqrt{b}$.

Sums and Differences of Radicals 5.4

Radicals having the same index and radicand are called *like radicals*. Thus $\sqrt{5}$ and $\sqrt[3]{5}$ are unlike radicals because their indices are different. The cube roots $\sqrt[3]{10}$ and $\sqrt[3]{5}$ are unlike radicals because their radicands are different. Expressions containing terms with *like* or *identical* radicals may be added or subtracted by using the distributive property. Consider the expression

$$4\sqrt{5} + 3\sqrt{5}.$$

The terms of this expression, $4\sqrt{5}$ and $3\sqrt{5}$, are *like terms* because each term contains the same radical $\sqrt{5}$. We can combine these terms just as we combined like terms of polynomials by using the distributive property for real numbers. Thus

$$\begin{aligned} 4\sqrt{5} + 3\sqrt{5} &= (4 + 3)\sqrt{5} && \text{distributive property} \\ &= 7\sqrt{5}. && \text{addition} \end{aligned}$$

When we combine like terms of the simplified form of a radical expression and obtain an expression with fewer terms, we also say that we have *simplified* the original radical expression.

Example 1. Simplify the following expressions.
(a) $5\sqrt{2} + 7\sqrt{2} - 3\sqrt{2}$. (b) $7\sqrt[3]{ab} - 10\sqrt[3]{ab}$.

Solution
(a) We apply the extended distributive property to the expression.

$$\begin{aligned} 5\sqrt{2} + 7\sqrt{2} - 3\sqrt{2} &= (5 + 7 - 3)\sqrt{2} && \text{distributive property} \\ &= 9\sqrt{2}. \end{aligned}$$

(b) $$\begin{aligned} 7\sqrt[3]{ab} - 10\sqrt[3]{ab} &= (7 - 10)\sqrt[3]{ab} && \text{distributive property} \\ &= -3\sqrt[3]{ab}. \end{aligned}$$

We note that in a radical expression such as $3\sqrt{5}$, the 3 is called the coefficient of $\sqrt{5}$. Similarly, in an expression such as $3x\sqrt{y}$, $3x$ is the coefficient of the radical. Observe, from Example 1, that like radicals may be *combined* by merely adding or subtracting the *coefficients* of the *like radicals*.

Example 2. Simplify. Assume that all variables represent positive real numbers.
(a) $2\sqrt[3]{xy} + 5\sqrt[3]{xy} - \sqrt[3]{x^2y} - 3\sqrt[3]{xy}$. (b) $2x\sqrt{y} - 5x\sqrt{y}$.

Solution
(a) $$\begin{aligned} 2\sqrt[3]{xy} + 5\sqrt[3]{xy} - \sqrt[3]{x^2y} - 3\sqrt[3]{xy} &= (2 + 5 - 3)\sqrt[3]{xy} - \sqrt[3]{x^2y} \\ &= 4\sqrt[3]{xy} - \sqrt[3]{x^2y}. \end{aligned}$$

(b) $$\begin{aligned} 2x\sqrt{y} - 5x\sqrt{y} &= (2x - 5x)\sqrt{y} \\ &= -3x\sqrt{y}. \end{aligned}$$

More generally, it is sometimes possible to simplify a radical expression by first rewriting the terms containing *radicals* in simplified form and then combining like terms. We illustrate this in the following examples.

Example 3. Simplify.
(a) $7\sqrt{8} + \sqrt{50} - \sqrt{72}$. (b) $5\sqrt[3]{x^4y} - 2x\sqrt[3]{xy}$.

Solution

(a) $$\begin{aligned} 7\sqrt{8} + \sqrt{50} - \sqrt{72} &= 7\sqrt{4\cdot 2} + \sqrt{25\cdot 2} - \sqrt{36\cdot 2} \\ &= 7\sqrt{4}\,\sqrt{2} + \sqrt{25}\,\sqrt{2} - \sqrt{36}\,\sqrt{2} \\ &= 14\sqrt{2} + 5\sqrt{2} - 6\sqrt{2} \\ &= 13\sqrt{2}. \end{aligned}$$

(b) $$\begin{aligned} 5\sqrt[3]{x^4y} - 2x\sqrt[3]{xy} &= 5\sqrt[3]{x^3(xy)} - 2x\sqrt[3]{xy} \\ &= 5\sqrt[3]{x^3}\sqrt[3]{xy} - 2x\sqrt[3]{xy} \\ &= 5x\sqrt[3]{xy} - 2x\sqrt[3]{xy} = 3x\sqrt[3]{xy}. \end{aligned}$$

Example 4. Simplify $\dfrac{5}{4}\sqrt[3]{16} - \dfrac{1}{\sqrt[3]{4}}$.

Solution: First we simplify each term of the radical expression to see if it is possible to combine terms.

$$\begin{aligned} \frac{5}{4}\sqrt[3]{16} - \frac{1}{\sqrt[3]{4}} &= \frac{5}{4}\sqrt[3]{8\cdot 2} - \frac{1}{\sqrt[3]{4}}\,\frac{\sqrt[3]{2}}{\sqrt[3]{2}} \\ &= \frac{5}{4}\sqrt[3]{8}\,\sqrt[3]{2} - \frac{\sqrt[3]{2}}{\sqrt[3]{8}} \\ &= \frac{5}{4}(2)\sqrt[3]{2} - \frac{\sqrt[3]{2}}{2} \\ &= \frac{5}{2}\sqrt[3]{2} - \frac{\sqrt[3]{2}}{2} = \left(\frac{5}{2} - \frac{1}{2}\right)\sqrt[3]{2} = 2\sqrt[3]{2}. \end{aligned}$$

Thus $\dfrac{5}{4}\sqrt[3]{16} - \dfrac{1}{\sqrt[3]{4}}$ simplifies to $2\sqrt[3]{2}$.

In summary, when simplifying radical expressions, do the following:

1. Rewrite each term containing a radical in simplified form.
2. Combine terms with like radicals by adding or subtracting the coefficients of like radicals.

EXERCISES 5.4

For Exercises 1–20, simplify the expression. Assume that all variables represent positive real numbers. See Examples 1 and 2.

1. $7\sqrt{3} + 5\sqrt{3}$ **2.** $8\sqrt{7} + 2\sqrt{7}$ **3.** $10\sqrt{6} - 7\sqrt{6}$ **4.** $5\sqrt{11} - 7\sqrt{11}$

5. $5\sqrt{5} - 2\sqrt{5} + 8\sqrt{5}$
6. $9\sqrt{3} + 2\sqrt{3} - 7\sqrt{3}$
7. $8\sqrt[3]{4} + 2\sqrt[3]{4}$
8. $9\sqrt[3]{2} - 7\sqrt[3]{2}$
9. $10\sqrt{x} - \sqrt{x} - 3\sqrt{x}$
10. $7\sqrt[3]{a} - 2\sqrt[3]{a} - \sqrt[3]{a}$
11. $8\sqrt[3]{ab} - 4\sqrt[3]{ab} - 2\sqrt[3]{ab}$
12. $\sqrt[3]{xz} - 4\sqrt[3]{xz} + 2\sqrt[3]{xz}$
13. $5\sqrt{x} - 2\sqrt{x} + 3\sqrt{y} - \sqrt{x}$
14. $-7\sqrt{y} - \sqrt{y} - 3\sqrt{z} + 10\sqrt{y}$
15. $4\sqrt[3]{a} - 2\sqrt{a} + \sqrt[3]{a} + 3\sqrt{a}$
16. $3\sqrt{xy} - \sqrt[3]{xy} + 4\sqrt{xy} - \sqrt[3]{xy}$
17. $7a\sqrt{b} - 2a\sqrt{b} + \sqrt{b}$
18. $8x\sqrt[3]{y} - \sqrt[3]{y} + 2x\sqrt[3]{y}$
19. $8u\sqrt[3]{v} - 6u\sqrt[3]{v} - 2\sqrt[3]{v}$
20. $7x\sqrt[3]{z} + 9x\sqrt[3]{z} - 8\sqrt[3]{z}$

For Exercises 21–38, simplify the expression. See Example 3.

21. $3\sqrt{5} + \sqrt{20} - 2\sqrt{5}$
22. $4\sqrt{2} + 6\sqrt{2} - \sqrt{32}$
23. $\sqrt{18} - \sqrt{8} - \sqrt{50}$
24. $\sqrt{12} + \sqrt{27} + \sqrt{48}$
25. $\sqrt{45} + \sqrt{180} + \sqrt{20}$
26. $5\sqrt{48} + 2\sqrt{75} - \sqrt{12}$
27. $\sqrt{8y^3} + y\sqrt{72y}$
28. $n\sqrt{48n} - \sqrt{27n^3}$
29. $4s\sqrt{80s} + \sqrt{20s^3}$
30. $\sqrt{12x^3} - x\sqrt{27x}$
31. $\sqrt{200x^2y} + \sqrt{50x^2y}$
32. $\sqrt{25a^2b^3} + \sqrt{36a^2b^3}$
33. $7\sqrt[3]{a^4} - 2a\sqrt[3]{a}$
34. $3n\sqrt[3]{n^2} + 2\sqrt[3]{n^5}$
35. $2\sqrt[3]{x^7} + 3x^2\sqrt[3]{x}$
36. $3\sqrt[3]{a^4b} - a\sqrt[3]{ab}$
37. $8\sqrt[3]{x^7y^6} + x^2y^2\sqrt[3]{x}$
38. $2\sqrt[4]{x^5y} + 3x\sqrt[4]{xy}$

For Exercises 39–44, simplify the expression. See Example 4.

39. $\frac{2}{3}\sqrt{3} + \frac{1}{\sqrt{3}}$
40. $\frac{4}{5}\sqrt{5} + \frac{1}{\sqrt{5}}$
41. $\frac{7}{4}\sqrt{8} - \frac{1}{\sqrt{2}}$
42. $\frac{4}{6}\sqrt{12} - \frac{1}{\sqrt{3}}$
43. $\frac{2}{x^2}\sqrt{x^3} - \frac{1}{\sqrt{x}}$
44. $\frac{2}{x^2}\sqrt[3]{x^4} - \frac{1}{\sqrt[3]{x^2}}$

5.5 Products and Quotients Involving Radicals

In this section we discuss the product and quotient of two radical expressions. Generally, the radical expressions will contain only two terms such as

$$5 + \sqrt{7}, \qquad 3\sqrt{2} - \sqrt{6x}, \qquad \sqrt[3]{4x} + \sqrt[3]{2}.$$

We will call a radical expression with two terms a *binomial radical expression*.

We shall discuss multiplication first. Multiplication of two or more radicals is accomplished by using the product rule for radicals; $\sqrt[n]{a}\,\sqrt[n]{b} = \sqrt[n]{ab}$. It is important to note that when multiplying two radicals using the product rule, *both radicals must have the same index*.

Example 1. Multiply and then simplify if necessary.
(a) $\sqrt{2}(3\sqrt{12} + \sqrt{27})$. (b) $\sqrt[3]{2x}(\sqrt[3]{4x} - \sqrt[3]{32x})$.

Solution: We apply the distributive property to perform the multiplication.

(a)
$$\begin{aligned} \sqrt{2}(3\sqrt{12} + \sqrt{27}) &= 3\sqrt{2}\sqrt{12} + \sqrt{2}\sqrt{27} \\ &= 3\sqrt{24} + \sqrt{54} \\ &= 6\sqrt{6} + 3\sqrt{6} \qquad \sqrt{24} = \sqrt{4}\sqrt{6};\ \sqrt{54} = \sqrt{9}\sqrt{6} \\ &= 9\sqrt{6}. \end{aligned}$$

(b)
$$\begin{aligned} \sqrt[3]{2x}(\sqrt[3]{4x} - \sqrt[3]{32x}) &= \sqrt[3]{2x}\sqrt[3]{4x} - \sqrt[3]{2x}\sqrt[3]{32x} \\ &= \sqrt[3]{8x^2} - \sqrt[3]{64x^2} \\ &= 2\sqrt[3]{x^2} - 4\sqrt[3]{x^2} \\ &= -2\sqrt[3]{x^2}. \end{aligned}$$

In the next example we discuss multiplying two binomial radical expressions, such as $(\sqrt{3} + 2)(\sqrt{12} + 3)$. We can apply the distributive property twice to multiply such expressions. However, in Chapter 3 we saw that the *distributive property* was the basis for multiplying binomials using the FOIL "method." As a result, we can use the FOIL rule to multiply binomials whose terms involve radicals. We shall also make use of the special products formulas developed in Chapter 3, such as $(a + b)^2 = a^2 + 2ab + b^2$.

Example 2. Multiply the binomial radical expressions and simplify the answer if necessary.

(a) $(\sqrt{3} + 2)(\sqrt{12} + 3)$. (b) $(5 + \sqrt{a})^2$. (c) $(\sqrt{x} - \sqrt{3})^2$.

Solution

(a) We apply the FOIL method to the indicated binomial product.

$$\begin{aligned} & \quad\ \ \text{F} \quad + \quad \text{O} \quad + \quad \text{I} \quad + \quad \text{L} \\ & \quad\ \ \downarrow \qquad\quad \downarrow \qquad\quad \downarrow \qquad\quad \downarrow \\ (\sqrt{3} + 2)(\sqrt{12} + 3) &= \sqrt{3}\sqrt{12} + \sqrt{3}(3) + 2\sqrt{12} + 2(3) \\ &= \sqrt{36} + 3\sqrt{3} + 2(2\sqrt{3}) + 6 \qquad \sqrt{12} = 2\sqrt{3} \\ &= 6 + 3\sqrt{3} + 4\sqrt{3} + 6 \\ &= 12 + 7\sqrt{3}. \end{aligned}$$

(b)
$$\begin{aligned} (5 + \sqrt{a})^2 &= 5^2 + 2(5)\sqrt{a} + (\sqrt{a})^2 \qquad (a + b)^2 = a^2 + 2ab + b^2 \\ &= 25 + 10\sqrt{a} + a. \end{aligned}$$

(c)
$$\begin{aligned} (\sqrt{x} - \sqrt{3})^2 &= (\sqrt{x})^2 - 2(\sqrt{x})(\sqrt{3}) + (\sqrt{3})^2 \qquad (a - b)^2 = a^2 - 2ab + b^2 \\ &= x - 2\sqrt{3x} + 3. \end{aligned}$$

In the next example we multiply radical expressions of the form $(a + b)(a - b)$. To do this simply, we use the special products formula

$$(a + b)(a - b) = a^2 - b^2.$$

We note that factors $(a + b)$ and $(a - b)$ are called *conjugates* of each other.

Example 3. Multiply and then simplify if necessary.
(a) $(\sqrt{6} + 2)(\sqrt{6} - 2)$. (b) $(\sqrt{x} - \sqrt{7})(\sqrt{x} + \sqrt{7})$.

Solution

(a)
$$\begin{aligned}(\sqrt{6} + 2)(\sqrt{6} - 2) &= (\sqrt{6})^2 - (2)^2\\ &= 6 - 4\\ &= 2.\end{aligned}$$

(b)
$$\begin{aligned}(\sqrt{x} - \sqrt{7})(\sqrt{x} + \sqrt{7}) &= (\sqrt{x})^2 - (\sqrt{7})^2\\ &= x - 7.\end{aligned}$$

Notice in Example 3 that the product of conjugates involving square roots results in an expression *without radicals*. We use this property of conjugates to rationalize the denominator of a fraction with a binomial radical expression. We note that *rationalizing* the denominator of a fraction with a binomial radical expression means rewriting the fraction in a form without radicals in the denominator. We can also view this result as *division* by a binomial radical expression.

Example 4. Rationalize the denominators of the following expressions.

(a) $\dfrac{4}{3 - \sqrt{7}}$. (b) $\dfrac{12}{\sqrt{6} + \sqrt{2}}$.

Solution

(a) To rationalize the denominator of the fraction $\dfrac{4}{3 - \sqrt{7}}$, we multiply the numerator and denominator of the fraction by the conjugate of the denominator, $3 + \sqrt{7}$. As we saw in Example 3, the product of conjugates involving square roots is free of radicals.

$$\begin{aligned}\frac{4}{3 - \sqrt{7}} = \frac{4}{(3 - \sqrt{7})} \cdot \frac{(3 + \sqrt{7})}{(3 + \sqrt{7})} &= \frac{4(3 + \sqrt{7})}{3^2 - (\sqrt{7})^2}\\ &= \frac{4(3 + \sqrt{7})}{9 - 7}\\ &= \frac{4(3 + \sqrt{7})}{2}\\ &= 2(3 + \sqrt{7}).\end{aligned}$$

(b)
$$\begin{aligned}\frac{12}{\sqrt{6} + \sqrt{2}} = \frac{12}{(\sqrt{6} + \sqrt{2})} \cdot \frac{(\sqrt{6} - \sqrt{2})}{(\sqrt{6} - \sqrt{2})} &= \frac{12(\sqrt{6} - \sqrt{2})}{6 - 2}\\ &= \frac{12(\sqrt{6} - \sqrt{2})}{4}\\ &= 3(\sqrt{6} - \sqrt{2}).\end{aligned}$$

The next example illustrates that rationalizing a denominator may sometimes result in a simpler algebraic expression.

Example 5. Rationalize the denominator. Assume that all variables represent positive real numbers.

$$\frac{1}{\sqrt{x+1}-\sqrt{x}}.$$

Solution

$$\begin{aligned}\frac{1}{\sqrt{x+1}-\sqrt{x}} &= \frac{1}{(\sqrt{x+1}-\sqrt{x})}\cdot\frac{(\sqrt{x+1}+\sqrt{x})}{(\sqrt{x+1}+\sqrt{x})}\\ &= \frac{\sqrt{x+1}+\sqrt{x}}{(\sqrt{x+1})^2-(\sqrt{x})^2}\\ &= \frac{\sqrt{x+1}+\sqrt{x}}{x+1-x} = \frac{\sqrt{x+1}+\sqrt{x}}{1}\\ &= \sqrt{x+1}+\sqrt{x}.\end{aligned}$$

Thus $\dfrac{1}{\sqrt{x+1}-\sqrt{x}} = \sqrt{x+1}+\sqrt{x}$.

EXERCISES 5.5

For Exercises 1–36, multiply and simplify if necessary. Assume that all variables represent positive real numbers. See Examples 1–3.

1. $\sqrt{2}(2+\sqrt{2})$
2. $\sqrt{5}(3-\sqrt{5})$
3. $\sqrt{3}(\sqrt{12}+\sqrt{3})$
4. $\sqrt{2}(\sqrt{8}+\sqrt{32})$
5. $\sqrt{7}(2\sqrt{7}-3)$
6. $\sqrt{11}(3\sqrt{11}+2)$
7. $3\sqrt{6}(\sqrt{6}-\sqrt{24})$
8. $2\sqrt{5}(\sqrt{20}-\sqrt{80})$
9. $3\sqrt{2}(\sqrt{18}-\sqrt{50})$
10. $\sqrt{x}(\sqrt{x^3}-\sqrt{x})$
11. $\sqrt{a^3}(\sqrt{a}-\sqrt{a^3})$
12. $\sqrt{x^3}(\sqrt{x^3}+\sqrt{x^5})$
13. $\sqrt[3]{4}(\sqrt[3]{2}+\sqrt[3]{16})$
14. $\sqrt[3]{3}(\sqrt[3]{9}-\sqrt[3]{72})$
15. $\sqrt[4]{2x^2}(\sqrt[4]{8x}-\sqrt[4]{128x})$
16. $\sqrt[4]{8a^2}(\sqrt[4]{2a^2}-\sqrt[4]{32a^2})$
17. $(\sqrt{5}+1)(\sqrt{5}+2)$
18. $(\sqrt{7}+2)(\sqrt{7}-3)$
19. $(\sqrt{2}+3)(\sqrt{8}+1)$
20. $(\sqrt{2}-3)(\sqrt{18}+1)$
21. $(\sqrt{x}+2)^2$
22. $(\sqrt{y}+3)^2$
23. $(\sqrt{a}-\sqrt{b})^2$
24. $(\sqrt{x}+\sqrt{y})^2$
25. $(\sqrt{2}+\sqrt{8})^2$
26. $(\sqrt{2}+\sqrt{32})^2$
27. $(\sqrt{3}-\sqrt{27})^2$
28. $(\sqrt{125}-\sqrt{5})^2$
29. $(\sqrt{5}+1)(\sqrt{5}-1)$
30. $(\sqrt{10}+2)(\sqrt{10}-2)$
31. $(\sqrt{7}-3)(\sqrt{7}+3)$
32. $(\sqrt{11}-4)(\sqrt{11}+4)$
33. $(\sqrt{3}+2)(\sqrt{3}-2)$
34. $(\sqrt{13}-\sqrt{5})(\sqrt{13}+\sqrt{5})$
35. $(\sqrt{x}+\sqrt{2})(\sqrt{x}-\sqrt{2})$
36. $(\sqrt{y}+\sqrt{3})(\sqrt{y}-\sqrt{3})$

For Exercises 37–50, rationalize the denominator. Simplify when necessary. See Example 4.

37. $\dfrac{1}{2 + \sqrt{3}}$ **38.** $\dfrac{1}{3 - \sqrt{8}}$ **39.** $\dfrac{1}{\sqrt{10} - 3}$ **40.** $\dfrac{1}{\sqrt{5} + 2}$

41. $\dfrac{22}{4 + \sqrt{5}}$ **42.** $\dfrac{6}{1 + \sqrt{7}}$ **43.** $\dfrac{2}{\sqrt{6} + \sqrt{5}}$ **44.** $\dfrac{3}{\sqrt{11} + \sqrt{10}}$

45. $\dfrac{2}{\sqrt{7} - \sqrt{5}}$ **46.** $\dfrac{6}{\sqrt{13} - \sqrt{10}}$ **47.** $\dfrac{\sqrt{10} + 1}{\sqrt{5} - \sqrt{2}}$ **48.** $\dfrac{\sqrt{6} - 2}{\sqrt{3} + \sqrt{2}}$

49. $\dfrac{\sqrt{3} + \sqrt{2}}{\sqrt{3} - \sqrt{2}}$ **50.** $\dfrac{\sqrt{5} + 2}{\sqrt{5} - 2}$

For Exercises 51–62, rationalize the denominator. Simplify when necessary. Assume that all variables represent positive real numbers. See Example 5.

51. $\dfrac{6}{\sqrt{x + 3} - \sqrt{x}}$ **52.** $\dfrac{4}{\sqrt{y + 2} - \sqrt{y}}$ **53.** $\dfrac{x - 1}{\sqrt{x} + 1}$

54. $\dfrac{z - 9}{\sqrt{z} + 3}$ **55.** $\dfrac{x - y}{\sqrt{x} - \sqrt{y}}$ **56.** $\dfrac{r - s}{\sqrt{r} + \sqrt{s}}$

57. $\dfrac{x}{\sqrt{2x} - \sqrt{x}}$ **58.** $\dfrac{4y}{\sqrt{5y} + \sqrt{y}}$ **59.** $\dfrac{y^2 - z^2}{(y + z)(\sqrt{y} + \sqrt{z})}$

60. $\dfrac{x^2 - y^2}{(x + y)(\sqrt{x} - \sqrt{y})}$ **61.** $\dfrac{z(z^2 - 1)}{\sqrt{z^3} - \sqrt{z}}$ **62.** $\dfrac{y^3(y^2 - 1)}{\sqrt{y^5} + \sqrt{y^3}}$

5.6 Equations Involving Radicals

Equations such as

$$\sqrt{x + 3} = 2, \qquad \sqrt{5y + 1} - \sqrt{y} = 5, \qquad \sqrt[3]{z + 1} = 3$$

are called *radical equations*. Notice that in such equations the variable appears under the radical sign.

To solve radical equations, we attempt to eliminate the radicals appearing in the equation. Then the usual methods are used to solve the resulting equation. To eliminate radicals in an equation, we use the following theorem.

Theorem 4

If $a = b$ and n is a positive integer, then

$$a^n = b^n.$$

When we apply Theorem 4 to solve an equation, we may introduce extra "solutions," which are often referred to as *extraneous* solutions. For example, let us consider a simple equation that does not contain radicals such as

(1) $$x = 3.$$

If we square each side of equation (1), then by Theorem 4 we have

(2) $$x^2 = 3^2 \quad (n = 2) \quad \text{or} \quad x^2 = 9.$$

Equation (2) has two solutions, $+3$ and -3. However, *only* 3 is the solution to the original equation (1), $x = 3$. Thus, squaring both sides of the original equation $x = 3$ introduced the extraneous solution -3. Therefore, whenever we solve an equation by raising both sides of the equation to the same power, we should check all apparent solutions in the original equation.

In the following examples we show procedures for solving radical equations.

Example 1. Solve $\sqrt{2x + 6} = 4$.

Solution: To solve this equation, we square both sides of the equation to eliminate the radical.

$$\begin{aligned} (\sqrt{2x+6})^2 &= 4^2 && \text{Theorem 4} \\ 2x + 6 &= 16 \\ 2x &= 10 \\ x &= 5. \end{aligned}$$

The solution of the last equation is 5. We check to see if 5 is the solution to the original equation.

$$\begin{aligned} \sqrt{2(5) + 6} &= 4 && \text{replace } x \text{ with } 5 \\ \sqrt{16} &= 4 \\ 4 &= 4. \end{aligned}$$

Thus we see that 5 is the desired solution.

Example 2. Solve $\sqrt{3y + 1} + 5 = 10$.

Solution: If we square both sides of the equation as given, we will not eliminate the radical. To eliminate the radical, we must first isolate the radical on one side of the equation and then square both sides.

$$\begin{aligned} \sqrt{3y+1} + 5 &= 10 \\ \sqrt{3y+1} &= 5 && \text{add } -5 \text{ to both sides} \\ (\sqrt{3y+1})^2 &= 5^2 && \text{square both sides} \\ 3y + 1 &= 25 \\ 3y &= 24 \\ y &= 8. \end{aligned}$$

A check in the original equation would show that 8 satisfies the equation. Confirm that this is so.

Example 3. Solve $\sqrt{z + 1} = -2$.

Solution: The equation has no solution because the left side of the equation cannot be negative. Recall that a radical, such as $\sqrt{z + 1}$, with an *even index* (provided that it represents a real number) is nonnegative by definition. However, suppose that we did not recognize this and tried to solve the equation. What would the result be? Squaring both sides, we obtain

$$\begin{aligned} (\sqrt{z + 1})^2 &= (-2)^2 \\ z + 1 &= 4 \\ z &= 3. \end{aligned}$$

The number 3 appears to be the required solution. However, if we check $z = 3$ in the original equation, we see that

$$\begin{aligned} \sqrt{3 + 1} &= -2 \\ \sqrt{4} &= -2 \\ 2 &= -2. \qquad \textit{false statement} \end{aligned}$$

Therefore, 3 is an *extraneous* solution. Thus we have confirmed that the original equation has no solution.

The next example shows how to deal with radical equations that contain more than one radical.

Example 4. Solve $\sqrt{x + 16} - \sqrt{x} = 2$.

Solution: If we square both sides of the equation, we will not eliminate radicals. To eliminate radicals, we isolate the radicals on one side of the equation, one at a time, and then square both sides.

$$\begin{aligned} \sqrt{x + 16} - \sqrt{x} &= 2 \\ \sqrt{x + 16} &= \sqrt{x} + 2 && \text{add } \sqrt{x} \text{ to both sides} \\ (\sqrt{x + 16})^2 &= (\sqrt{x} + 2)^2 && \text{square both sides} \\ x + 16 &= x + 4\sqrt{x} + 4 && (a + b)^2 = a^2 + 2ab + b^2 \\ -4\sqrt{x} &= -12 && \text{combine like terms} \\ \sqrt{x} &= 3 && \text{divide both sides by } -4 \\ (\sqrt{x})^2 &= 3^2 && \text{square both sides} \\ x &= 9. \end{aligned}$$

Check that the required solution is 9.

Sometimes we can solve radical equations involving higher-order radicals by the method of eliminating radicals. We illustrate this in the next example, involving cube roots.

Example 5. Solve $\sqrt[3]{3x - 1} = 2$.

Solution

$$(\sqrt[3]{3x - 1})^3 = 2^3 \qquad \text{Theorem 4; cube both sides}$$
$$3x - 1 = 8$$
$$3x = 9$$
$$x = 3.$$

Raising both sides of an equation to an odd power does not result in extraneous solutions. However, we should check our answer to see if we made any errors in solving the equation. A check does show that 3 is the required solution.

EXERCISES 5.6

For Exercises 1–42, solve the equation. See Examples 1–5.

1. $\sqrt{x} = 4$
2. $\sqrt{y} = 7$
3. $\sqrt{z} + 1 = 7$
4. $\sqrt{x} + 2 = 8$
5. $\sqrt{z - 2} = 3$
6. $\sqrt{z + 3} = 5$
7. $\sqrt{3x - 6} = 3$
8. $\sqrt{5x + 6} = 4$
9. $\sqrt{7y + 8} = 6$
10. $\sqrt{8z + 1} = 5$
11. $\sqrt{5x + 11} + 7 = 13$
12. $\sqrt{6x + 7} - 3 = 4$
13. $\sqrt{4t + 1} - 4 = 1$
14. $\sqrt{10x + 11} + 5 = 14$
15. $\sqrt{3x + 1} = -5$
16. $\sqrt{5x + 1} = -6$
17. $\sqrt{x + 5} - \sqrt{x} = 1$
18. $\sqrt{x + 7} - \sqrt{x} = 1$
19. $\sqrt{x - 4} - 2 = \sqrt{x}$
20. $\sqrt{x - 9} + 3 = \sqrt{x}$
21. $\sqrt{4x + 9} = \sqrt{5x - 1}$
22. $\sqrt{3x + 7} = \sqrt{4x + 1}$
23. $\sqrt{9x + 10} = -\sqrt{10x + 4}$
24. $\sqrt{5x + 21} = -\sqrt{6x + 9}$
25. $\sqrt[3]{y} = 4$
26. $\sqrt[3]{z} = -4$
27. $\sqrt[3]{3x + 2} = 2$
28. $\sqrt[3]{2x + 6} = 2$
29. $\sqrt[3]{5x - 3} = 3$
30. $\sqrt[3]{7x + 1} = -3$
31. $\sqrt[3]{3x + 7} = -2$
32. $\sqrt[3]{5x + 25} = 5$
33. $\sqrt[3]{x^2 - 7} = -2$
34. $\sqrt[3]{x^2 - 26} = -3$
35. $\sqrt[4]{5x + 1} = 2$
36. $\sqrt[4]{5x + 21} = 3$
37. $\dfrac{1}{\sqrt{x}} = \dfrac{1}{2}$
38. $\dfrac{1}{\sqrt{y}} = \dfrac{1}{3}$
39. $\dfrac{1}{\sqrt{z - 1}} = \dfrac{1}{3}$
40. $\dfrac{1}{\sqrt{7x + 2}} = \dfrac{1}{4}$
41. $\dfrac{1}{\sqrt[3]{3x - 1}} = \dfrac{1}{2}$
42. $\dfrac{1}{\sqrt[3]{4x + 3}} = \dfrac{1}{3}$

Introduction to Complex Numbers 5.7

In Section 5.1 we observed that there is no real number x such that

$$x^2 = -5,$$

since x^2 is always positive or zero. Therefore, any square root of -5 cannot be a real number. In this section we extend the set of real numbers so that negative numbers, such as -5, have square roots defined for them. To accomplish this, we introduce a new set of numbers called the complex numbers. We designate the set of complex numbers with the letter $\mathbb{C}$.

Complex Numbers [$\mathbb{C}$]

> A *complex number* is a number that can be expressed in the form
> $$a + bi,$$
> where a and b are real numbers, and $i^2 = -1$.

Some examples of complex numbers are

$$3 - 2i, \qquad 0 + 4i, \qquad \sqrt{3} + \frac{1}{2}i, \qquad 0 + 0 \cdot i, \qquad 2 + 0 \cdot i.$$

Sometimes it is preferable to write $a + ib$ rather than $a + bi$. For example, we write $3 + i\sqrt{5}$ rather than $3 + \sqrt{5}i$ so that it is clear that i is not part of the radicand.

A complex number written in the form $a + bi$ or $a + ib$ is said to be in *standard form*. The complex number $\dfrac{1}{2 + 3i}$ is not in standard form.

We shall see that the complex numbers, $\mathbb{C}$, include the set of real numbers, $\mathbb{R}$. Thus, to be consistent, the *sum* and *product* of two complex numbers are *defined* below so that the axioms for the real numbers (listed in Section 1.4), such as closure, commutativity, associativity, and so forth, remain valid for complex numbers. We also *define* when two complex numbers are *equal*.

Basic Properties of Complex Numbers

> Equality: $a + bi = c + di$ if and only $a = c$ and $b = d$.
>
> Addition: $(a + bi) + (c + di) = (a + c) + (b + d)i$.
>
> Multiplication: $(a + bi)(c + di) = (ac - bd) + (ad + bc)i$.

Because the complex numbers conform to the ordinary rules of algebra for real numbers, it is not necessary to memorize the formal rules given above for addition and multiplication. We can add and multiply complex numbers as if they were ordinary binomials with the agreement that wherever i^2 occurs, we replace it with -1.

The following example illustrates the preceding remarks.

Example 1. Perform the operations indicated. Express the answer in standard form.
(a) $(3 + 5i) + (4 - 3i)$. (b) $(2 - 3i)(4 + i)$.

Solution

(a) $(3 + 5i) + (4 - 3i) = 3 + 5i + 4 - 3i$
$= 7 + 2i.$ combine like terms

(b) $(2 - 3i)(4 + i) = 2(4) + 2(i) + (-3i)(4) + (-3i)i$ FOIL method
$= 8 + 2i - 12i - 3i^2$
$= 8 - 10i - 3(-1)$ $i^2 = -1$
$= 11 - 10i.$

The real numbers a and b are called the *real part* and the *imaginary part* of $a + bi$, respectively. When either the real part or imaginary part of a complex number equals zero, we obtain particular subsets of complex numbers with the following associated names:

Real and Pure Imaginary Numbers

Real Numbers: $b = 0$	Pure Imaginaries: $a = 0,\ b \neq 0$
$a + 0i = a.$	$0 + bi = bi.$

Thus we see that the real numbers are a subset of the complex numbers.

The additive identity and multiplicative identity for complex numbers are $0 + 0i = 0$ and $1 + 0i = 1$, respectively. Thus the identity elements are the same for both the complex and real numbers. We note that the additive inverse of $a + bi$ is $(-a) + (-b)i$ because their sum is 0. As a result, we *define* subtraction for complex numbers as follows:

Subtraction of Complex Numbers

$$(a + bi) - (c + di) = (a - c) + (b - d)i.$$

Example 2. Find the difference.

$$(7 - 2i) - (2 - 6i).$$

Solution: Again we treat the complex numbers as "ordinary" binomials. As a result, we have

$(7 - 2i) - (2 - 6i) = 7 - 2i - 2 + 6i$
$= 5 + 4i.$ combine like terms

The complex numbers $a + bi$ and $a - bi$ are called *complex conjugates* of each other. For example, $2 + 3i$ and $2 - 3i$ are complex conjugates. We observe that $(2 + 3i)(2 - 3i) = 2^2 - (3i)^2 = 4 - 9i^2 = 4 - 9(-1) = 13$. Thus the product of

$(2 + 3i)$ and $(2 - 3i)$ is a real number. In general, the product of two complex conjugates

$$(a + bi)(a - bi) = a^2 + b^2.$$

is a *real number*. This fact proves useful when we divide complex numbers.

Example 3. Find the product of $4 - 3i$ and its conjugate.

Solution: The conjugate of $4 - 3i$ is $4 + 3i$. Their product is

$$\begin{aligned}(4 - 3i)(4 + 3i) &= 4^2 - (3i)^2 \\ &= 16 + 9 = 25.\end{aligned}$$

We now discuss *division* of complex numbers. To evaluate the quotient $\dfrac{a + bi}{c + di}$, we *need to express it in standard form, $A + Bi$*. The quotient $\dfrac{a + bi}{c + di}$ is defined as the product of $a + bi$ and the reciprocal $\dfrac{1}{c + di}$. This method of obtaining the quotient in standard form is somewhat difficult. An equivalent and efficient way of obtaining the quotient is to multiply the numerator and denominator of $\dfrac{a + bi}{c + di}$ by the conjugate of the denominator as shown in the following equation:

$$\frac{a + bi}{c + di} = \frac{a + bi}{c + di} \cdot \frac{(c - di)}{(c - di)}.$$

We illustrate this procedure for evaluating quotients of complex numbers in the following example.

Example 4. Perform the operation indicated and write the result in standard form.

(a) $\dfrac{5 - i}{3 + 2i}$. (b) $\dfrac{8 + 12i}{4i}$.

Solution

(a)
$$\begin{aligned}\frac{5 - i}{3 + 2i} &= \frac{(5 - i)}{(3 + 2i)} \cdot \frac{(3 - 2i)}{(3 - 2i)} \\ &= \frac{15 - 10i - 3i + 2i^2}{9 - 4i^2} \\ &= \frac{13 - 13i}{13} \\ &= \frac{13(1 - i)}{13} \\ &= 1 - i.\end{aligned}$$

Thus

$$\frac{5 - i}{3 + 2i} = 1 - i.$$

(b) The conjugate of the denominator $4i$ is $-4i$. Thus we proceed as follows:

$$\begin{aligned}\frac{8 + 12i}{4i} &= \frac{(8 + 12i)}{4i} \cdot \frac{(-4i)}{(-4i)} \\ &= \frac{-32i - 48i^2}{-16i^2} \\ &= \frac{-32i + 48}{16} = 3 - 2i.\end{aligned}$$

We see that $\dfrac{8 + 12i}{4i} = 3 - 2i.$

We introduced the set of complex numbers, $\mathbb{C}$, so that negative numbers such as -5 would have square roots defined for them. Recall that by definition a number x is a square root of -5 if

$$x^2 = -5.$$

As we have already mentioned, x cannot be a real number. Now, however, let us consider the pure imaginary (complex number) $i\sqrt{5}$. We see that

$$\begin{aligned}(i\sqrt{5})^2 &= i^2(\sqrt{5})^2 \\ &= (-1)(5) \\ &= -5.\end{aligned}$$

Thus, by definition, the complex number $i\sqrt{5}$ is a square root of -5. Similarly, we see that $-i\sqrt{5}$ is also a square root of -5 because $(-i\sqrt{5})^2 = (-i)^2(\sqrt{5})^2 = i^2(5) = -5$. Therefore, -5 has two complex roots. We denote the two square roots of -5 by $\sqrt{-5}$ and $-\sqrt{-5}$ and set in a natural way

$$\sqrt{-5} = i\sqrt{5} \qquad \text{and} \qquad -\sqrt{-5} = -i\sqrt{5}.$$

In general, we have the following *definition:*

The Square Roots of a Negative Number

If b is a real number and $b > 0$, then

$$\sqrt{-b} = i\sqrt{b}$$

and

$$-\sqrt{-b} = -i\sqrt{b}.$$

In particular,

$$\sqrt{-1} = i \qquad \text{and} \qquad -\sqrt{-1} = -i,$$

since

$$i^2 = -1 \qquad \text{and} \qquad (-i)^2 = i^2 = -1.$$

Example 5. Write the following in standard form, $a + bi$ or $a + ib$.
(a) $\sqrt{-10}$. (b) $-\sqrt{-81}$. (c) $5 + \sqrt{-9}$.

Solution

(a) $\sqrt{-10} = i\sqrt{10}$.
(b) $-\sqrt{-81} = -(i\sqrt{81}) = -(i9) = -9i$.
(c) $5 + \sqrt{-9} = 5 + i\sqrt{9} = 5 + i3 = 5 + 3i$.

When dealing with radicals involving *negative* radicands, we must be careful when using the product rule, $\sqrt{xy} = \sqrt{x}\,\sqrt{y}$, provided that $x \geq 0$, $y \geq 0$. We note that

$$\sqrt{-2}\,\sqrt{-18} = (i\sqrt{2})(i\sqrt{18}) = i^2\sqrt{2}\,\sqrt{18} = i^2\sqrt{36} = -6$$

gives the *correct* answer. However, if we write

$$\sqrt{-2}\,\sqrt{-18} = \sqrt{(-2)(-18)} = \sqrt{36} = 6,$$

we obtain an *incorrect* answer. Thus it is necessary when working with radicals involving negative radicands such as $\sqrt{-b}$ $(b > 0)$ to *write it first* in standard form as $i\sqrt{b}$.

Example 6. Simplify and write the following in standard form.
(a) $\sqrt{-40}$. (b) $\sqrt{-4}\,\sqrt{-25}$. (c) $\dfrac{\sqrt{-48}}{\sqrt{3}}$.

Solution

(a) $\sqrt{-40} = i\sqrt{40} = i\sqrt{4}\,\sqrt{10} = i2\sqrt{10}$.

(b) $\sqrt{-4}\,\sqrt{-25} = (i\sqrt{4})(i\sqrt{25}) = (i2)(i5) = (i^2)10 = -10$.

(c) $\dfrac{\sqrt{-48}}{\sqrt{3}} = \dfrac{i\sqrt{48}}{\sqrt{3}} = i\sqrt{\dfrac{48}{3}} = i\sqrt{16} = i4 = 4i$.

In concluding this section, we outline in Figure 5.1 the complex number system and its subsets.

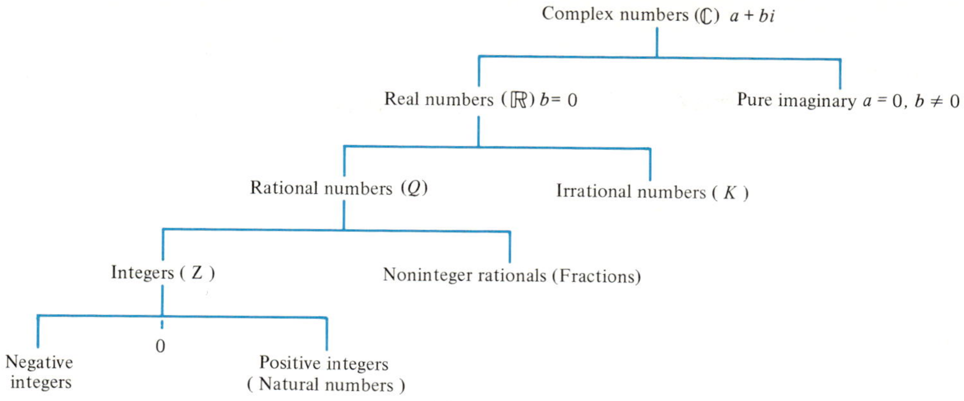

Figure 5.1

EXERCISES 5.7

For Exercises 1–20, perform the operation indicated. Write the answer in standard form. See Examples 1 and 2.

1. $(2 + i) + (3 + 2i)$
2. $(4 + 5i) + (3 + 4i)$
3. $(3 - 5i) + (2 + 4i)$
4. $(5 + 6i) + (3 - 5i)$
5. $(-3 + 7i) + (-2 + 3i)$
6. $(-1 + 2i) + (-4 - 2i)$
7. $(9 + 8i) - (5 + 3i)$
8. $(12 + 7i) - (10 + 5i)$
9. $(10 - 6i) - (7 + 6i)$
10. $(11 + 2i) - (4 - 2i)$
11. $(8 + 2i) - (8 - 2i)$
12. $(9 - i) - (9 + i)$
13. $(2 + 3i) + (5 + i) + 3i$
14. $(7 + 2i) + (2 + 3i) + 4i$
15. $(4 + 3i) + (2 + 2i) - 5i$
16. $(5 + 2i) + (3 + 7i) - 3i$
17. $(1 + i) + (1 - i)$
18. $(3 + 2i) + (3 - 2i)$
19. $(1 + i) - (1 - i)$
20. $(3 + 2i) - (3 - 2i)$

For Exercises 21–32, multiply the two complex numbers. Write the answer in standard form. See Example 1.

21. $3i(2 + 3i)$
22. $2i(5 + 4i)$
23. $5i(\frac{1}{5} - \frac{2}{5}i)$
24. $7i(\frac{3}{7} - \frac{4}{7}i)$
25. $(4 + 2i)(1 + 2i)$
26. $(10 + 2i)(1 + 5i)$
27. $(3 + 2i)(4 + 3i)$
28. $(1 - 2i)(1 - 3i)$
29. $(3 - i)(3 - i)$
30. $(2 + i)(2 + i)$
31. $(4 + 3i)(3 + 4i)$
32. $(5 + 2i)(2 + 5i)$

For Exercises 33–40, find (a) the conjugate of the complex number and (b) the product of the conjugates. See Example 3.

33. $1 + i$
34. $2 + 5i$
35. $2 - 4i$
36. $3 - i$
37. $3i$
38. $-4i$
39. $5 + 2i$
40. $4 + 2i$

For Exercises 41–52, find the quotient. Write the answer in standard form. See Example 4.

41. $\frac{4 - 8i}{2i}$ **42.** $\frac{18 + 9i}{-3i}$ **43.** $\frac{10}{1 + 3i}$ **44.** $\frac{13}{2 - 3i}$ **45.** $\frac{25i}{3 - 4i}$

46. $\frac{10i}{1 + 2i}$ **47.** $\frac{1 + i}{1 - i}$ **48.** $\frac{3 + 2i}{4 + 2i}$ **49.** $\frac{7 - i}{4 + 3i}$ **50.** $\frac{5 + 10i}{2 + i}$

51. $\frac{4 + 2i}{1 + i}$ **52.** $\frac{13 + 26i}{3 - 2i}$

For Exercises 53–78, write the answer in standard form, a + bi or a + ib. See Examples 5 and 6.

53. $\sqrt{-16}$ **54.** $\sqrt{-49}$ **55.** $\sqrt{-64}$ **56.** $\sqrt{-100}$ **57.** $-\sqrt{-4}$

58. $-\sqrt{-36}$ **59.** $\sqrt{-7}$ **60.** $\sqrt{-11}$ **61.** $6 - \sqrt{-1}$ **62.** $3 + \sqrt{-1}$

63. $2 + \sqrt{-100}$ **64.** $3 - \sqrt{-121}$ **65.** $\sqrt{-32}$ **66.** $\sqrt{-48}$ **67.** $-\sqrt{-18}$

68. $-\sqrt{-27}$ **69.** $\sqrt{-2}\,\sqrt{-8}$ **70.** $\sqrt{-1}\,\sqrt{-4}$ **71.** $\sqrt{-3}\,\sqrt{3}$ **72.** $\sqrt{7}\,\sqrt{-7}$

73. $\sqrt{-2}\,\sqrt{-32}$ **74.** $\sqrt{-18}\,\sqrt{-2}$ **75.** $\frac{\sqrt{-75}}{\sqrt{3}}$ **76.** $\frac{\sqrt{20}}{\sqrt{-5}}$ **77.** $\frac{\sqrt{128}}{\sqrt{-2}}$

78. $\frac{\sqrt{-54}}{\sqrt{6}}$

79. The number i is called the unit imaginary. Find the first eight powers of i. [HINT: $i^1 = i$, $i^2 = -1$, $i^3 = i^2 \cdot i = -i$.]

80. Based on your answers to Exercise 79, fill in the question marks.

(a) $i^1 = i^5 = i^9 = \cdots = \underline{\ ?\ }$ **(b)** $i^2 = i^6 = i^{10} = \cdots = \underline{\ ?\ }$

(c) $i^3 = i^7 = i^{11} = \cdots = \underline{\ ?\ }$ **(d)** $i^4 = i^8 = i^{12} = \cdots = \underline{\ ?\ }$

(e) $i^{36} = \underline{\ ?\ }$ **(f)** $i^{71} = \underline{\ ?\ }$

Key Words

Complex numbers
- Standard form
- Pure imaginary
- Real part
- Imaginary part
- Conjugate

Radical equations
- Extraneous solutions

Radical expressions
- Simplified form

Radicals
- Index
- Order
- Like

Radical sign

Radicand

Rational exponents

Rationalizing the denominator

Roots
- Square
- Cube
- *n*th
- Principal *n*th

CHAPTER 5 REVIEW

SECTION 5.1

Find each root.

1. $\sqrt{81}$ **2.** $\sqrt[3]{-27}$ **3.** $\sqrt[3]{x^6}$ **4.** $\sqrt[4]{16y^8}$

SECTION 5.2

Evaluate each root.

5. $81^{1/2}$ **6.** $(-8)^{1/3}$ **7.** $9^{3/2}$ **8.** $16^{-1/2}$

Simplify the expressions.

9. $b^{-1/4}b^{2/4}$ **10.** $\dfrac{a^{5/3}}{a^{2/3}}$ **11.** $(36x^2y^6)^{1/2}$ **12.** $\dfrac{64x^{3/2}y^{4/3}}{16x^{1/2}y}$ **13.** $\left(\dfrac{x^{3/4}}{y^{2/3}}\right)^{12}$ **14.** $\left(\dfrac{s^{3/4}}{t^{3/2}}\right)^{-1/3}$

SECTION 5.3

Simplify.

15. $\sqrt{8}\sqrt{2}$ **16.** $\sqrt[3]{4z^2}\sqrt[3]{2z}$ **17.** $\dfrac{\sqrt[4]{32x^5}}{\sqrt[4]{2x}}$ **18.** $\sqrt{300}$ **19.** $\dfrac{1}{\sqrt{7}}$ **20.** $\sqrt[3]{16y^4}$

SECTION 5.4

Simplify.

21. $5\sqrt{3} - 2\sqrt{3} + 4\sqrt{3}$ **22.** $2\sqrt{12} + 5\sqrt{27}$ **23.** $\sqrt[3]{y^4} + y\sqrt[3]{8y}$

SECTION 5.5

Simplify.

24. $(\sqrt{3} + 5)(\sqrt{3} - 2)$ **25.** $(\sqrt{3} + 2)(\sqrt{12} + 1)$ **26.** $\dfrac{7}{3 + \sqrt{2}}$ **27.** $\dfrac{\sqrt{8} + \sqrt{2}}{\sqrt{8} - \sqrt{2}}$

SECTION 5.6

Solve the equations.

28. $\sqrt{3y + 10} + 7 = 12$ **29.** $\sqrt[3]{5x + 12} = -2$ **30.** $\sqrt{x + 7} - \sqrt{x} = 1$

31. $\dfrac{1}{\sqrt{2x + 5}} = \dfrac{1}{3}$

SECTION 5.7

Perform the operations indicated.

32. $(3 + 5i) + (6 - 2i)$ **33.** $(9 - 7i) - (6 - 3i) - 5i$ **34.** $(2 + 5i)(3 + 4i)$

35. $(4 - i)(4 + i)$ **36.** $\dfrac{50}{3 - 4i}$ **37.** $\dfrac{5 - i}{1 + i}$

Simplify.

38. $\sqrt{-36}$ **39.** $-\sqrt{-4}$ **40.** $\sqrt{-8}\,\sqrt{-2}$

CHAPTER 5 TEST

In the following problems, assume that all variables represent positive real numbers.

For Problems 1–7, find the root.

1. $-\sqrt{100}$ **2.** $\sqrt[3]{27x^3}$ **3.** $\sqrt[4]{x^{12}}$ **4.** $\sqrt{(-4)^2}$ **5.** $64^{1/2}$ **6.** $(-125)^{2/3}$ **7.** $16^{2/4}$

For Problems 8–19, simplify the expression.

8. $x^{1/6} \cdot x^{3/6}$ **9.** $(8x^6y^3)^{1/3}$ **10.** $\dfrac{12x^{4/3}y^{3/2}}{3x^{2/3}y}$ **11.** $\left(\dfrac{x^{1/2}}{y^{1/5}}\right)^{10}$

12. $\sqrt{2x}\,\sqrt{8x}$ **13.** $\sqrt[3]{y^2}\,\sqrt[3]{y}$ **14.** $\dfrac{1}{\sqrt[4]{y^3}}$ **15.** $\dfrac{\sqrt[3]{x^4y}}{\sqrt[3]{xy^4}}$

16. $\sqrt{75x^3}$ **17.** $\sqrt[3]{24x^4y^5}$ **18.** $2\sqrt{45} + 5\sqrt{20}$ **19.** $\sqrt[3]{n^4} + n\sqrt[3]{27n}$

For Problems 20 and 21, multiply and then simplify if necessary.

20. $(\sqrt{3} + 2)(\sqrt{3} - 3)$ **21.** $(\sqrt{10} - 8)(\sqrt{10} + 8)$

For Problems 22 and 23, solve the equation.

22. $\sqrt{5y + 10} + 4 = 9$ **23.** $\sqrt[3]{4x - 7} = -3$

For Problems 24–27, perform the operation indicated.

24. $(3 + 7i) + (5 - 2i)$ **25.** $(8 - 6i) - (3 + 2i) + 5i$ **26.** $(2 + 5i)(3 - i)$ **27.** $\dfrac{3 - i}{2 + i}$

For Problems 28–30, simplify the expression.

28. $\sqrt{-64}$ **29.** $-\sqrt{-25}$ **30.** $\sqrt{-5}\,\sqrt{-20}$

CHAPTER 6

Solving Quadratic Equations

Quadratic equations are used in business. For example, a local manufacturer claims that the cost of producing a garden tool is $C = x^2 - 18x + 126$. If he sells the tools for \$7 each, then his revenue is given by the equation $R = 7x$. By solving the equation $x^2 - 25x + 126 = 0$, he can determine the two breakeven points for this product. Between these two points he makes a profit. Outside of the points he operates at a loss.

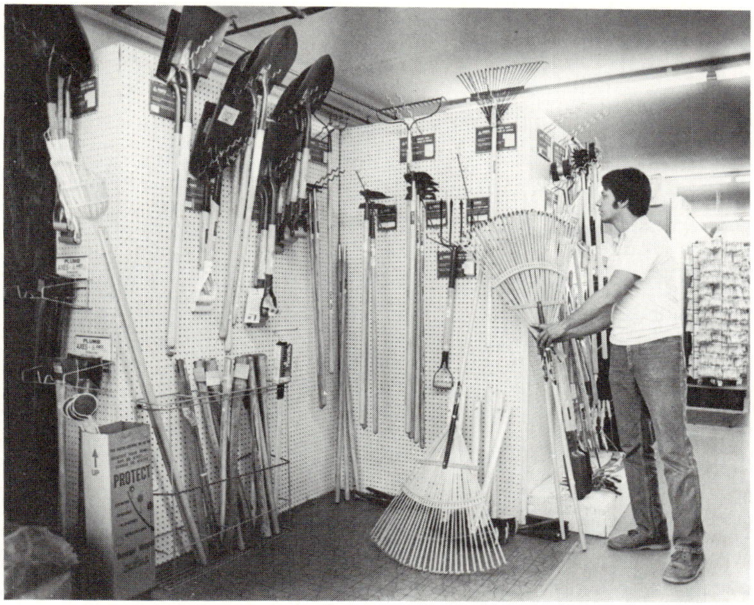

In this chapter we use our knowledge of solving first-degree equations, factoring, radicals, and complex numbers to develop methods for solving quadratic equations. We limit our discussion to quadratic equations that have real coefficients and one variable. Also discussed in this chapter are applications that involve quadratic equations and quadratic inequalities.

6.1 Solution of Quadratic Equations by Factoring

In this section we learn how to use factoring to solve *quadratic equations*. We now give the *definition* of a quadratic equation.

Quadratic Equation

> Any equation that can be put in the form $ax^2 + bx + c = 0$, where a, b, and c are real numbers, with $a \neq 0$, is called a *quadratic equation*. This is called the *standard form* of a quadratic equation.

The following equations are examples of quadratic equations:

$$x^2 - 8x + 15 = 0, \qquad 2x^2 - 8 = 0, \qquad x^2 - 2x = 0, \qquad 2x^2 - x = 3.$$

The key property of complex numbers that allows us to solve quadratic equations by factoring is the *factor property of zero,* which we now state.

Factor Property of Zero

> If a and b are complex numbers and $a \cdot b = 0$, then either $a = 0$ or $b = 0$.

We will now illustrate how the factor property of zero may be used to solve quadratic equations. The values of x that satisfy an equation such as

$$x^2 - 7x + 10 = 0$$

are called the *solutions* or *roots* of the equation. The set of solutions is called the *solution set*. When we consider $x^2 - 7x + 10 = 0$ in its present form, it is not apparent what values of x satisfy the equation. However, if we express $x^2 - 7x + 10$ as the product of two factors, we can then use the factor property of zero to find the solutions. Thus, factoring $x^2 - 7x + 10$, we see that

$$x^2 - 7x + 10 = (x - 2)(x - 5) = 0.$$

But the factor property of zero tells us that the product $(x - 2)(x - 5)$ can equal 0 only if at least one of the factors equals 0. Thus we see that either

$$\begin{aligned} x - 2 &= 0 \quad \text{or} \quad & x - 5 &= 0 \\ x &= 2 \quad \text{or} \quad & x &= 5. \end{aligned}$$

To check that 2 and 5 are solutions, we replace x in the original equation with these values.

Let $x = 2$:
$$\begin{aligned} x^2 - 7x + 10 &= 0 \\ 2^2 - 7(2) + 10 &= 0 \\ 4 - 14 + 10 &= 0 \\ 0 &= 0. \quad \text{true} \end{aligned}$$

Let $x = 5$:
$$\begin{aligned} x^2 - 7x + 10 &= 0 \\ 5^2 - 7(5) + 10 &= 0 \\ 25 - 35 + 10 &= 0 \\ 0 &= 0. \quad \text{true} \end{aligned}$$

We conclude that the solution set is $\{2, 5\}$.

The previous discussion illustrated how the factor property of zero can be applied when solving a quadratic equation that is in standard form. If we can factor a quadratic expression as the product of two first-degree factors, then by applying the factor property of zero, we see that at least one of the factors must be zero. Thus we can replace the quadratic equation with two first-degree equations, which we then solve separately.

Example 1. Solve the equation $3x^2 + 15x = 0$.

Solution: Since all the nonzero terms are already on the left, we factor

$$3x^2 + 15x$$

and rewrite the equation as

$$3x(x + 5) = 0.$$

Next we set each of the factors equal to 0 and solve the resulting equations.

$$\begin{aligned} 3x &= 0 \quad \text{or} \quad x + 5 = 0 \\ x &= 0 \quad \text{or} \quad x = -5. \end{aligned}$$

The reader should check that these two values satisfy the original equation and thereby verify that the solution set is $\{-5, 0\}$.

Example 2. Solve the equation $20 = 2w^2 + 3w$.

Solution: First we write the equation as $2w^2 + 3w - 20 = 0$. Next, we factor and rewrite the equation as

$$(2w - 5)(w + 4) = 0.$$

Solving the equations that result from setting each of the factors equal to zero, we get

$$\begin{aligned} 2w - 5 &= 0 \quad \text{or} \quad w + 4 = 0 \\ 2w &= 5 \quad \text{or} \quad w = -4 \\ w &= \frac{5}{2} \quad \text{or} \quad w = -4. \end{aligned}$$

Thus the solution set is $\left\{-4, \frac{5}{2}\right\}$.

In the next example we illustrate that the *factor property of zero* can be used to solve some equations other than quadratic equations.

Example 3. Solve the equation $x(x + 2)(x - 4) = 7x$.

Solution: Notice that $x(x + 2)(x - 4)$ does not equal zero. Therefore, before we can apply the factor property of zero, we must first perform the indicated multiplications and collect all nonzero terms on one side of the equation. That is,

$$\begin{aligned} x(x + 2)(x - 4) &= 7x \\ x(x^2 - 2x - 8) &= 7x \\ x^3 - 2x^2 - 8x &= 7x \\ x^3 - 2x^2 - 15x &= 0. \end{aligned}$$

If we remove the common factor x, we have $x(x^2 - 2x - 15) = 0$. Factoring the quadratic expression $x^2 - 2x - 15$ as $(x - 5)(x + 3)$, we see that the original equation may be rewritten as

$$x(x - 5)(x + 3) = 0.$$

Setting each factor equal to zero and solving the resulting equations, we get

$$\begin{array}{ccccc} x = 0 & \text{or} & x - 5 = 0 & \text{or} & x + 3 = 0 \\ x = 0 & \text{or} & x = 5 & \text{or} & x = -3. \end{array}$$

The reader should check each of these proposed solutions in the original equation and thereby verify that $\{-3, 0, 5\}$ is the solution set.

As Example 3 illustrated, the factor property of zero can be used to solve equations other than quadratic equations. The next two examples demonstrate how this property can be used to solve equations that contain rational expressions (see Example 4) and equations that contain radical expressions (see Example 5).

Example 4. Solve $\dfrac{x + 1}{x - 2} + \dfrac{1}{x - 1} = \dfrac{3}{x^2 - 3x + 2}$.

Solution: To solve this rational equation, we will clear fractions by multiplying each side of the equation by the LCD. Since the factors of $x^2 - 3x + 2$ are $(x - 2)$ and $(x - 1)$, the LCD is $(x - 2)(x - 1)$. Thus we obtain

$$\cancel{(x - 2)}(x - 1)\frac{x + 1}{\cancel{x - 2}} + (x - 2)\cancel{(x - 1)}\frac{1}{\cancel{x - 1}} = \cancel{(x - 2)(x - 1)}\frac{3}{\cancel{(x - 2)(x - 1)}}$$

$$\begin{aligned} (x - 1)(x + 1) + (x - 2) &= 3 \\ x^2 - 1 + x - 2 &= 3 \\ x^2 + x - 6 &= 0. \end{aligned}$$

The last equation is a quadratic equation that can be solved by factoring.

$$x^2 + x - 6 = 0$$
$$(x + 3)(x - 2) = 0.$$

Thus either

$$x + 3 = 0 \quad \text{or} \quad x - 2 = 0$$
$$x = -3 \quad \text{or} \quad x = 2.$$

We see that -3 and 2 are the possible solutions. A check would reveal that -3 is a solution to the original equation. We now check to see if 2 is also a solution.

Let $x = 2$: Notice that the denominator of the first term on the left side of the original equation,

$$\frac{x+1}{x-2} \text{ is 0 when } x = 2\text{; that is, } \frac{2+1}{2-2} = \frac{3}{0}.$$

Since a rational expression is undefined when its denominator is 0, 2 is *not* a solution to the original equation. Thus 2 is an extraneous solution. Therefore, -3 is the only solution to the original equation, and the solution set is $\{-3\}$.

Example 5. Solve the equation $\sqrt{5y - 5} + 1 = y$.

Solution: If we first isolate the term $\sqrt{5y - 5}$ on the left side and then square both sides, we will eliminate the radical. Thus

$$\sqrt{5y - 5} = y - 1 \qquad \text{subtract 1 from each side}$$
$$(\sqrt{5y - 5})^2 = (y - 1)^2 \qquad \text{square both sides}$$
$$5y - 5 = y^2 - 2y + 1.$$

Writing this quadratic equation in standard form and then factoring, we get

$$y^2 - 7y + 6 = 0$$
$$(y - 6)(y - 1) = 0.$$

Thus either

$$y - 6 = 0 \quad \text{or} \quad y - 1 = 0$$
$$y = 6 \quad \text{or} \quad y = 1.$$

Therefore, the two possible solutions are 6 and 1. We now check 6 and 1 in the original equation.

For $y = 1$ we have:

$$\sqrt{5(1) - 5} + 1 = 1$$
$$\sqrt{5 - 5} + 1 = 1$$
$$0 + 1 = 1$$
$$1 = 1. \qquad \text{true}$$

For $y = 6$ we have:

$$\sqrt{5(6) - 5} + 1 = 6$$
$$\sqrt{25} + 1 = 6$$
$$5 + 1 = 6$$
$$6 = 6. \qquad \text{true}$$

Thus the solution set is $\{1, 6\}$.

EXERCISES 6.1

For Exercises 1–10, use the factor property of zero to find the solution set.

1. $(x - 3)(x - 7) = 0$ **2.** $(y - 1)(y - 4) = 0$ **3.** $(x + 2)(x - 3) = 0$

4. $(z + 5)(z - 2) = 0$ **5.** $(p + 1)(p + 9) = 0$ **6.** $(q + 3)(q + 10) = 0$

7. $(3x - 10)(2x + 5) = 0$ **8.** $(2w - 1)(3w + 4) = 0$ **9.** $x(x - 1)(2x + 7) = 0$

10. $z(z - 4)(3z + 5) = 0$

For Exercises 11–50, solve the equation. See Examples 1–3.

11. $x^2 - x = 0$ **12.** $2y^2 - 6y = 0$ **13.** $4w^2 + 12w = 0$

14. $5z^2 + 45z = 0$ **15.** $7x^2 = 21x$ **16.** $3y^2 = 9y$

17. $w^2 + 6w + 8 = 0$ **18.** $z^2 + 5z + 6 = 0$ **19.** $28 = p^2 + 3p$

20. $q^2 = 12 + q$ **21.** $t^2 + t = 42$ **22.** $s^2 = 5s + 24$

23. $u^2 + 28 = -16u$ **24.** $v^2 + 50 = 15v$ **25.** $6x^2 - 5x + 1 = 0$

26. $3y^2 - 7y - 6 = 0$ **27.** $2w^2 + 7w - 15 = 0$ **28.** $10z^2 - 7z + 1 = 0$

29. $4r^2 - 25r = -6$ **30.** $2s^2 - 15 = s$ **31.** $12z^3 + 8z^2 - 15z = 0$

32. $8w^3 + 14w^2 - 15w = 0$ **33.** $9p^3 = 12p^2 - 4p$ **34.** $-10q^2 + 8q = 25q^3$

35. $x^3 - 16x = 0$ **36.** $y^3 - 9y = 0$ **37.** $w^3 - 144w = 0$

38. $z^3 - 100z = 0$ **39.** $(y - 4)(y + 2) = 7$ **40.** $(w + 1)(w - 5) = -8$

41. $5t(2t + 5) = 15$ **42.** $4r(4r + 2) = 15$ **43.** $2p(6p - 5) = 2$

44. $4y(4y + 9) = -18$ **45.** $x(x - 3)^2 = 4x$ **46.** $y(y - 5)^2 = 9y$

47. $p(p + 3) - 12 = 4p$ **48.** $q(q + 2) - 8 = 4q$ **49.** $2r(r + 1) = 5r + 2$

50. $3s(s + 23) = 19s + 10$

For Exercises 51–62, solve by first clearing fractions and then solving the resulting quadratic equation. Check your answer. See Example 4.

51. $2 + \dfrac{1}{y} = \dfrac{1}{y^2}$ **52.** $3 - \dfrac{7}{w} = \dfrac{-2}{w^2}$

53. $\dfrac{10}{x^2} + 1 = \dfrac{-7}{x}$ **54.** $\dfrac{5}{y^2} + \dfrac{4}{y} = 1$

55. $\dfrac{6}{p^2 - 36} + \dfrac{p}{p - 6} = \dfrac{-1}{p + 6}$ **56.** $\dfrac{q}{q - 4} + \dfrac{4}{q^2 - 16} = \dfrac{-1}{q + 4}$

57. $\dfrac{w}{w + 5} + \dfrac{w}{w - 5} = \dfrac{50}{w^2 - 25}$ **58.** $\dfrac{1}{x - 5} - \dfrac{13}{x^2 - 25} = \dfrac{x}{x + 5}$

59. $\dfrac{z}{z-3} - \dfrac{30}{z^2-9} = \dfrac{5}{z+3}$

60. $\dfrac{y}{y-2} - \dfrac{12}{y^2-4} = \dfrac{3}{y+2}$

61. $\dfrac{y}{y-5} + \dfrac{2y}{y+5} = \dfrac{2}{y^2-25}$

62. $\dfrac{w}{w+3} + \dfrac{2w}{w-3} = \dfrac{-2w+2}{w^2-9}$

For Exercises 63–72, solve the equation by first eliminating the radical. See Example 5.

63. $\sqrt{x^2+9} = 5$

64. $\sqrt{y^2+16} = 5$

65. $\sqrt{z^2+144} = 13$

66. $\sqrt{w^2+225} = 17$

67. $\sqrt{4p+12} = p+3$

68. $\sqrt{5q-1} = q+1$

69. $\sqrt{4r+1} - r = 1$

70. $\sqrt{4s-5} + 2 = 2s$

71. $2x = 1 - \sqrt{5x-1}$

72. $y = 7 + \sqrt{2y+1}$

Exercises 73–78 require much more factoring than Exercises 51–62. However, they are solved in the same general manner. That is, factor each denominator, clear fractions, and then solve the resulting quadratic equation. See Example 4.

73. $\dfrac{2x}{x^2-4x-21} + \dfrac{-x}{x^2-10x+21} = \dfrac{-8}{x^2-9}$

74. $\dfrac{y-3}{y^2+3y-10} + \dfrac{2}{y^2+7y+10} = \dfrac{-2}{y^2-4}$

75. $\dfrac{3w}{w^2+5w+6} = \dfrac{4}{w^2-4} - \dfrac{w}{w^2+w-6}$

76. $\dfrac{5z}{z^2+2z-3} = \dfrac{2}{z^2+z-2} + \dfrac{3z}{z^2+5z+6}$

77. $\dfrac{3r-6}{r^2+r-2} + \dfrac{r+4}{r^2+2r-3} = \dfrac{3r+18}{r^2+5r+6}$

78. $\dfrac{3s}{s^2+9s+20} + \dfrac{s+2}{s^2+7s+12} = \dfrac{6s}{s^2+8s+15}$

Solution of Quadratic Equations by Completing the Square 6.2

To solve the quadratic equation $x^2 = 25$, we note that the equation can be read as "what number x, when squared, will equal 25?" Thus, from our work in Chapter 5, we know that the solutions are

$$x = \sqrt{25} \qquad \text{or} \qquad x = -\sqrt{25}$$
$$= 5 \qquad\qquad\qquad = -5.$$

The following theorem generalizes the preceding discussion.

Square Root Theorem

If b is a real number and $x^2 = b$, then

$$x = \sqrt{b} \quad \text{or} \quad x = -\sqrt{b}.$$

Example 1. Solve the following equations.
(a) $y^2 = 11$. (b) $z^2 = 12$. (c) $w^2 = -49$.

Solution

(a) Direct application of the square root theorem yields

$$y = -\sqrt{11} \quad \text{or} \quad y = \sqrt{11}.$$

Therefore, the solution set is $\{-\sqrt{11}, \sqrt{11}\}$.

(b) In this case we get

$$z = -\sqrt{12} = -2\sqrt{3} \quad \text{or} \quad z = \sqrt{12} = 2\sqrt{3}.$$

Thus the solution set is $\{-2\sqrt{3}, 2\sqrt{3}\}$.

(c) When we apply the square root theorem this time, we must use complex numbers to represent the answer. That is, $\sqrt{-49} = 7i$ and $-\sqrt{-49} = -7i$. Thus the solutions are $-7i$ and $7i$.

When applying the square root theorem to an equation such as $x^2 = a$, we usually write the solutions compactly as $\pm\sqrt{a}$ and, if possible, simplify. The expression $\pm\sqrt{a}$ is read "plus or minus the square root of a."

Example 2. Solve the following equations.
(a) $(z - 2)^2 = 25$. (b) $(2x - 3)^2 = 9$.

Solution: The left side of each of these equations is the square of a binomial. Thus by the square root theorem we may write for the given equations.

(a)
$$z - 2 = \pm 5.$$

That is,

$$\begin{aligned} z - 2 &= 5 \quad \text{or} \quad z - 2 = -5 \\ z &= 7 \quad \text{or} \quad z = -3. \end{aligned}$$

Thus the solution set is $\{-3, 7\}$.

(b)
$$2x - 3 = \pm 3.$$

That is,

$$\begin{aligned} 2x - 3 &= 3 \quad \text{or} \quad 2x - 3 = -3 \\ 2x &= 6 \quad \text{or} \quad 2x = 0 \\ x &= 3 \quad \text{or} \quad x = 0. \end{aligned}$$

Therefore, the solution set is $\{0, 3\}$.

Thus far in the examples of this section, we have been able to apply the square root theorem because the side of each equation that involved the unknown was a perfect square. In the remainder of the section we show how to solve any quadratic equation by writing it in the form

$$\underset{\text{square of a binomial}}{(x + a)^2} = \underset{\text{constant}}{b}$$

and then applying the square root theorem. This method is called *solving a quadratic equation by completing the square.*

To write one side of a quadratic equation as the square of a binomial, it must be a perfect square trinomial. Thus, as discussed in Section 3.8, the first-degree term of the trinomial must equal twice the product of the two terms of the binomial. But if the leading coefficient of a quadratic trinomial is 1, it will be a perfect square trinomial if the constant term is the square of one-half the coefficient of the first degree term. That is, in

$$(x + a)^2 = x^2 + 2ax + a^2$$

note that

$$a^2 = \left[\frac{1}{2}(2a)\right]^2$$

We will demonstrate the method of completing the square by using it to find the roots of

$$x^2 - 12x - 9 = 0.$$

Note that $x^2 - 12x - 9$ is not factorable using integer coefficients, and therefore the method of factoring cannot easily be used. The first step is to isolate the terms with variables on one side of the equality sign (usually the left side). Thus we have

(1) $$x^2 - 12x = 9.$$

For the left side of this equation to be a perfect square trinomial (square of a binomial), it must have a constant term that equals the square of one-half the coefficient of the first-degree term. That is,

$$\left[\frac{1}{2}(-12)\right]^2 = (-6)^2 = 36.$$

Adding 36 to each side of equation (1), we obtain

(2) $$x^2 - 12x + 36 = 9 + 36.$$

Now the left side of the equation is a perfect square trinomial. Factoring the left side of (2), we have

(3) $$(x - 6)^2 = 45.$$

Since the left side of equation (3) is a perfect square, we can now apply the square root theorem, which tells us that the solutions are given by

$$x - 6 = \pm\sqrt{45} = \pm 3\sqrt{5}$$

or

$$x = 6 \pm 3\sqrt{5}.$$

We will check the proposed solution $6 - 3\sqrt{5}$ in the original equation. The reader should check $6 + 3\sqrt{5}$.

CHECK: Replacing x with $6 - 3\sqrt{5}$ in $x^2 - 12x - 9 = 0$, we get

$$\begin{aligned} (6 - 3\sqrt{5})^2 - 12(6 - 3\sqrt{5}) - 9 &= 0 \\ (36 - 36\sqrt{5} + 45) - (72 - 36\sqrt{5}) - 9 &= 0 \\ 81 - 36\sqrt{5} - 72 + 36\sqrt{5} - 9 &= 0 \\ 0 &= 0. \quad \text{true} \end{aligned}$$

Thus the solution set is $\{6 - 3\sqrt{5}, 6 + 3\sqrt{5}\}$.

Example 3. Solve $y^2 = 50 - 20y$ by completing the square and check your answer.

Solution: First, we rewrite the equation with all the variable terms on the left and the constant term on the right of the equality symbol. Thus we have

$$y^2 + 20y = 50.$$

Next, we add the square of one-half the coefficient of the first-degree term, 20, to each side of the equation. Thus we add $\left[\frac{1}{2}(20)\right]^2 = (10)^2 = 100$ to each side. The result is

$$y^2 + 20y + 100 = 50 + 100$$

or

$$(y + 10)^2 = 150.$$

The square root theorem tells us that the solutions to the last equation are given by

$$y + 10 = \pm\sqrt{150}$$

or

$$y + 10 = \pm 5\sqrt{6}.$$

That is,

$$y = -10 + 5\sqrt{6} \quad \text{or} \quad y = -10 - 5\sqrt{6}.$$

The reader should verify that these are solutions to the original equation. We conclude that the solution set is $\{-10 \pm 5\sqrt{6}\}$.

The procedures that we have been applying assume that the leading coefficient is 1. If the leading coefficient of a quadratic equation is not 1, *divide each term of the equation by this coefficient* before completing the square.

Example 4. Solve $3y^2 + 2y - 3 = 0$ using the method of completing the square.

Solution: To begin, we add 3 to each side of the equation. Thus

$$3y^2 + 2y - 3 + 3 = 0 + 3$$

or

$$3y^2 + 2y = 3.$$

Dividing each term by the leading coefficient 3, we get

$$y^2 + \frac{2}{3}y = 1.$$

Next we complete the square by adding

$$\left[\frac{1}{2}\left(\frac{2}{3}\right)\right]^2 = \left(\frac{1}{3}\right)^2 = \frac{1}{9}$$

to each side of the equation. Thus we obtain

$$y^2 + \frac{2}{3}y + \frac{1}{9} = 1 + \frac{1}{9}$$

or

$$\left(y + \frac{1}{3}\right)^2 = \frac{10}{9}.$$

Applying the square root theorem to the last equation, we get

$$y + \frac{1}{3} = \sqrt{\frac{10}{9}} \quad \text{or} \quad y + \frac{1}{3} = -\sqrt{\frac{10}{9}}$$

$$= \frac{\sqrt{10}}{3} \qquad\qquad = \frac{-\sqrt{10}}{3}.$$

Thus we have

$$y = -\frac{1}{3} + \frac{\sqrt{10}}{3} \quad \text{or} \quad y = -\frac{1}{3} - \frac{\sqrt{10}}{3}$$

$$= \frac{-1 + \sqrt{10}}{3} \qquad\qquad = \frac{-1 - \sqrt{10}}{3}.$$

Therefore, the solution set is $\left\{\frac{-1 \pm \sqrt{10}}{3}\right\}$.

Example 5. Solve $2z^2 - 2z + 1 = 0$ by completing the square.

Solution: First we subtract 1 from each side of the equation and then we divide each term by the leading coefficient of 2. Thus we have

$$2z^2 - 2z = -1$$

$$z^2 - z = -\frac{1}{2}.$$

We complete the square by adding $\left[\frac{1}{2}(-1)\right]^2 = \frac{1}{4}$ to each side of the equation.

$$z^2 - z + \frac{1}{4} = -\frac{1}{2} + \frac{1}{4}$$

$$\left(z - \frac{1}{2}\right)^2 = -\frac{1}{4}.$$

Applying the square root theorem, we get

$$z - \frac{1}{2} = \sqrt{\frac{-1}{4}} \quad \text{or} \quad z - \frac{1}{2} = -\sqrt{\frac{-1}{4}}.$$

Thus

$$z - \frac{1}{2} = \frac{-i}{2} \quad \text{or} \quad z - \frac{1}{2} = \frac{i}{2} \quad \left(\sqrt{\frac{-1}{4}} = i\sqrt{\frac{1}{4}} = i \cdot \frac{1}{2} = \frac{i}{2}\right).$$

That is,

$$z = \frac{1}{2} - \frac{i}{2} \quad \text{or} \quad z = \frac{1}{2} + \frac{i}{2}.$$

We conclude that the solution set is $\left\{\frac{1 \pm i}{2}\right\}$.

EXERCISES 6.2

For Exercises 1–30, solve the equation by applying the square root theorem. Simplify all radicals and check your answers. See Examples 1 and 2.

1. $x^2 = 64$ **2.** $y^2 = 81$ **3.** $w^2 = \frac{9}{4}$ **4.** $z^2 = \frac{16}{25}$

5. $x^2 = .04$ **6.** $y^2 = .09$ **7.** $p^2 = 75$ **8.** $q^2 = 32$

9. $x^2 - 13 = 0$ **10.** $y^2 - 7 = 0$ **11.** $t^2 - 300 = 0$ **12.** $u^2 - 200 = 0$

13. $w^2 = -9$ **14.** $z^2 = -16$ **15.** $x^2 + 50 = 0$ **16.** $y^2 + 18 = 0$

17. $(x - 4)^2 = 49$
18. $(y - 2)^2 = 36$
19. $(x + 3)^2 = 81$
20. $(y + 2)^2 = 25$
21. $(2t - 6)^2 = 100$
22. $(2x - 4)^2 = 64$
23. $(4u - 2)^2 = 36$
24. $(4v - 2)^2 = 100$
25. $(5v + 10)^2 = 625$
26. $(5w + 10)^2 = 400$
27. $(z - 3)^2 = -4$
28. $(w - 5)^2 = -9$
29. $(2x + 3)^2 = -25$
30. $(2y + 5)^2 = -49$

For Exercises 31–44, solve the equation by completing the square. See Example 3.

31. $x^2 + 10x = 3$
32. $y^2 + 6y = 1$
33. $w^2 - 8w = 4$
34. $z^2 - 4z = 1$
35. $y^2 + 20y = 25$
36. $x^2 - 16x = 4$
37. $u^2 - 1 = 14u$
38. $v^2 - 9 = -18v$
39. $p^2 + 18p + 6 = 0$
40. $q^2 - 14q + 1 = 0$
41. $z^2 + 1 = 8z$
42. $w^2 - 11 = -6w$
43. $x^2 = 24x + 6$
44. $y^2 = 2y + 5$

For Exercises 45–56, solve the equation by completing the square. See Example 4.

45. $4x^2 + 12x - 4 = 0$
46. $2u^2 + u - 4 = 0$
47. $8y^2 + 4y - 4 = 0$
48. $4w^2 - 2w - 1 = 0$
49. $2u^2 + 7u + 3 = 0$
50. $2v^2 + 9v - 5 = 0$
51. $3x^2 - x - 4 = 0$
52. $3y^2 + 4y - 15 = 0$
53. $2v^2 - 6 = -3v$
54. $3u^2 - 4 = 2u$
55. $5z^2 - 4 = 3z$
56. $5w^2 - 2 = -w$

For Exercises 57–64, solve the equation by completing the square. See Example 5.

57. $x^2 + 2x + 26 = 0$
58. $y^2 + 2y + 17 = 0$
59. $w^2 - 4w + 31 = 0$
60. $z^2 - 6z + 59 = 0$
61. $u^2 - 4u + 13 = 0$
62. $v^2 - 6v + 34 = 0$
63. $3z^2 + 10z + 23 = 0$
64. $9w^2 + 24w + 91 = 0$

For Exercises 65–70, multiply each term by the least common denominator to clear fractions, then solve the resulting quadratic equation using the method of completing the square.

65. $1 - \frac{6}{x} = \frac{3}{x^2}$
66. $1 - \frac{4}{x} = \frac{8}{x^2}$
67. $1 + \frac{4}{z} = \frac{2}{z^2}$
68. $1 + \frac{6}{y} = \frac{2}{y^2}$
69. $\frac{5}{u - 2} + \frac{2}{u + 3} = 1$
70. $\frac{7}{v + 5} + \frac{2}{v - 2} = 1$

The Quadratic Formula 6.3

As you have discovered, the method of completing the square is often tedious to apply. Since this method can be used to solve any quadratic equation in one variable, a logical next step is to apply it to the general quadratic equation, $ax^2 + bx + c = 0$. In doing this, we derive one of the best-known formulas of mathematics, the *quadratic formula*. This formula is so important and so widely used that it should be committed to memory.

The Quadratic Formula

If a, b, and c are real constants with $a \neq 0$, the solutions to $ax^2 + bx + c = 0$ are given by

$$x = \frac{-b + \sqrt{b^2 - 4ac}}{2a} \quad \text{or} \quad x = \frac{-b - \sqrt{b^2 - 4ac}}{2a}.$$

Proof. To solve $ax^2 + bx + c = 0$ using the method of completing the square, we will subtract c from both sides of the equation and then divide each term by a. That is,

$$ax^2 + bx = -c$$

$$x^2 + \frac{b}{a}x = \frac{-c}{a}.$$

Next, we add the square of one-half the coefficient of the first-degree term to each side. That is, adding

$$\left[\frac{1}{2}\left(\frac{b}{a}\right)\right]^2 = \frac{b^2}{4a^2}$$

to each side of the equation, we get

$$x^2 + \frac{b}{a}x + \frac{b^2}{4a^2} = \frac{-c}{a} + \frac{b^2}{4a^2}.$$

The left side of this equation is now a perfect square trinomial; therefore, it can be written as the square of a binomial. Thus we have

$$\left(x + \frac{b}{2a}\right)^2 = \frac{-c}{a} + \frac{b^2}{4a^2}.$$

The next step is to add the fractions on the right side of the equation by rewriting $\frac{-c}{a}$ with a denominator of $4a^2$. That is,

$$\left(x + \frac{b}{2a}\right)^2 = \frac{-c}{a} \cdot \frac{4a}{4a} + \frac{b^2}{4a^2}$$

$$= \frac{-4ac}{4a^2} + \frac{b^2}{4a^2}.$$

Thus

$$\left(x + \frac{b}{2a}\right)^2 = \frac{b^2 - 4ac}{4a^2}.$$

We now apply the square root theorem and get

$$x + \frac{b}{2a} = \sqrt{\frac{b^2 - 4ac}{4a^2}} \quad \text{or} \quad x + \frac{b}{2a} = -\sqrt{\frac{b^2 - 4ac}{4a^2}}.$$

Thus

$$x + \frac{b}{2a} = \pm\sqrt{\frac{b^2 - 4ac}{4a^2}}$$

$$= \pm\frac{\sqrt{b^2 - 4ac}}{2a}.$$

Therefore,

$$x = \frac{-b}{2a} \pm \frac{\sqrt{b^2 - 4ac}}{2a}$$

or

$$x = \frac{-b \pm \sqrt{b^2 - 4ac}}{2a}.$$

Recall that any equation of the form $ax^2 + bx + c = 0$, where a, b, and c are real constants with $a \neq 0$, is called a *quadratic equation in standard form*. Consider each of the following quadratic equations:

$$x^2 - 7x + 10 = 0, \qquad 3x^2 = 48, \qquad 20 = 2x^2 + 3x.$$

The first equation is in standard form with $a = 1$, $b = -7$, and $c = 10$. In the second equation, if we subtract 48 from each side of the equation, the result is $3x^2 - 48 = 0$, which is in standard form with $a = 3$, $b = 0$, and $c = -48$. The reader should verify that the third equation can be put in standard form with $a = 2$, $b = 3$, and $c = -20$.

To solve a specific quadratic equation, we put it in standard form and compare it to $ax^2 + bx + c = 0$ to determine the values of a, b, and c. For example, to solve $2x^2 = 5x - 2$, we first write it in standard form as

$$2x^2 - 5x + 2 = 0.$$

Comparing this to

$$ax^2 + bx + c = 0,$$

we see that $a = 2$, $b = -5$, and $c = 2$. We now substitute these values in the quadratic formula. Thus

$$x = \frac{-b \pm \sqrt{b^2 - 4ac}}{2a}$$

$$= \frac{-(-5) \pm \sqrt{(-5)^2 - 4(2)(2)}}{2(2)}$$

$$= \frac{5 \pm \sqrt{25 - 16}}{4}$$

$$= \frac{5 \pm \sqrt{9}}{4} = \frac{5 \pm 3}{4}.$$

As a result we have

$$x = \frac{5 + 3}{4} = \frac{8}{4} = 2 \quad \text{or} \quad x = \frac{5 - 3}{4} = \frac{2}{4} = \frac{1}{2}.$$

When each of these numbers is substituted in the original equation, the result is a true statement. Thus the solution set is $\left\{\frac{1}{2}, 2\right\}$.

Example 1. Use the quadratic formula to solve $3x^2 - 6x - 1 = 0$.

Solution: Comparing $3x^2 - 6x - 1 = 0$ to $ax^2 + bx + c = 0$, we see that $a = 3$, $b = -6$, and $c = -1$. Using these values in the quadratic formula, we get

$$\begin{aligned} x &= \frac{-(-6) \pm \sqrt{(-6)^2 - 4(3)(-1)}}{2(3)} \\ &= \frac{6 \pm \sqrt{48}}{6} \\ &= \frac{6 \pm 4\sqrt{3}}{6} = \frac{3 \pm 2\sqrt{3}}{3}. \end{aligned}$$

Thus the solution set is $\left\{\frac{3 - 2\sqrt{3}}{3}, \frac{3 + 2\sqrt{3}}{3}\right\}$.

Example 2. Solve $x^2 - 2x + 3 = 0$ using the quadratic formula and check the answer.

Solution: Here we see that $a = 1$, $b = -2$, and $c = 3$. Thus

$$\begin{aligned} x &= \frac{-(-2) \pm \sqrt{(-2)^2 - 4(1)(3)}}{2(1)} \\ &= \frac{2 \pm \sqrt{-8}}{2} = \frac{2 \pm 2i\sqrt{2}}{2} = 1 \pm i\sqrt{2}. \end{aligned}$$

CHECK

Let $x = 1 - i\sqrt{2}$:

$$\begin{aligned} x^2 - 2x + 3 &= 0 \\ (1 - i\sqrt{2})^2 - 2(1 - i\sqrt{2}) + 3 &= 0 \\ 1 - 2i\sqrt{2} - 2 - 2 + 2i\sqrt{2} + 3 &= 0 \\ 0 &= 0. \quad \text{true} \end{aligned}$$

A similar check would show that $1 + i\sqrt{2}$ also satisfies the equation. We conclude that the solution set of the equation $x^2 - 2x + 3 = 0$ is $\{1 - i\sqrt{2}, 1 + i\sqrt{2}\}$.

Example 3. Solve $(x - 4)(x - 3) = 7$.

Solution: To use the quadratic formula, we must first put the equation in standard form. If we perform the indicated multiplication and collect like terms, we get

$$\begin{aligned}(x - 4)(x - 3) &= 7\\ x^2 - 7x + 12 &= 7\\ x^2 - 7x + 5 &= 0.\end{aligned}$$

It is now clear that $a = 1$, $b = -7$, and $c = 5$. Thus the solutions are given by

$$\begin{aligned}x &= \frac{-(-7) \pm \sqrt{(-7)^2 - 4(1)(5)}}{2(1)}\\ &= \frac{7 \pm \sqrt{29}}{2}.\end{aligned}$$

Therefore, the solution set is $\left\{\dfrac{7 - \sqrt{29}}{2}, \dfrac{7 + \sqrt{29}}{2}\right\}$.

Example 4. Solve $\dfrac{y}{y - 3} + y = 2$ by first clearing fractions.

Solution: We can clear fractions by multiplying each term by $(y - 3)$. Thus we get

$$\begin{aligned}\cancel{(y-3)}\frac{y}{\cancel{y-3}} + (y - 3)y &= (y - 3)2\\ y + y^2 - 3y &= 2y - 6\\ y^2 - 4y + 6 &= 0.\end{aligned}$$

Solving this last equation by the quadratic formula results in

$$\begin{aligned}y &= \frac{-(-4) \pm \sqrt{(-4)^2 - 4(1)(6)}}{2}\\ &= \frac{4 \pm \sqrt{16 - 24}}{2}\\ &= \frac{4 \pm \sqrt{-8}}{2}\\ &= \frac{4 \pm 2i\sqrt{2}}{2} = 2 \pm i\sqrt{2}.\end{aligned}$$

From this it is clear that the solution set of the original equation is $\{2 + i\sqrt{2}, 2 - i\sqrt{2}\}$.

We note that in the quadratic formula,

$$x = \frac{-b \pm \sqrt{b^2 - 4ac}}{2a},$$

the radicand

$$b^2 - 4ac$$

is called the **discriminant.** If we are interested only in the nature of the solutions of a quadratic equation with rational coefficients, the discriminant is very helpful. For example, if the discriminant is negative, we are taking the square root of a negative number and both solutions will be *complex numbers*. In fact, the solutions will be *complex conjugates* when $b^2 - 4ac < 0$. Similarly, if $b^2 - 4ac$ is positive and a perfect square, both solutions are rational numbers and the equation could be solved by factoring. The following table lists the possible relationships between the discriminant and the nature of the solutions of a quadratic equation $ax^2 + bx + c = 0$, where a, b, and c are rational constants with $a \neq 0$.

Discriminant ($b^2 - 4ac$)	Solutions
Negative	Two complex numbers that are complex conjugates
Positive and a perfect square	Two distinct rational numbers
Positive and not a perfect square	Two distinct irrational numbers
Zero	One rational number, sometimes called a double, or multiple, solution

Example 5. For each of the following equations, use the discriminant to describe the number and nature of the solutions.

(a) $4x^2 - 2x + 3 = 0$. (b) $\frac{1}{2}y^2 + y - 2 = 0$. (c) $3z^2 - 2z - 5 = 0$.

Solution: For each equation we will evaluate the discriminant, $b^2 - 4ac$, and refer to the preceding table.

(a) Here $a = 4$, $b = -2$, and $c = 3$; thus $b^2 - 4ac = (-2)^2 - 4(4)(3) = -44 < 0$. Thus there are two distinct solutions and they are complex conjugates.

(b) Since $a = \frac{1}{2}$, $b = 1$, and $c = -2$, we have $b^2 - 4ac = 1^2 - 4\left(\frac{1}{2}\right)(-2) = 5$. Since 5 is positive but not a perfect square, there are two distinct irrational solutions.

(c) In this case $a = 3$, $b = -2$, and $c = -5$. Thus $b^2 - 4ac = (-2)^2 - 4(3)(-5) = 64$. Since 64 is a positive perfect square, there are two distinct rational solutions.

EXERCISES 6.3

For Exercises 1–12, use the quadratic formula to solve the equation. See Examples 1 and 2.

1. $4x^2 - 3x - 1 = 0$
2. $2y^2 - 3y - 5 = 0$
3. $4z^2 - 8z - 5 = 0$
4. $w^2 + 3w + 2 = 0$
5. $u^2 + 4u + 1 = 0$
6. $v^2 - 2v - 5 = 0$
7. $p^2 - 2p - 3 = 0$
8. $q^2 + 3q + 1 = 0$
9. $2t^2 - 3t - 2 = 0$
10. $6s^2 - s - 2 = 0$
11. $y^2 - 2y + 5 = 0$
12. $x^2 + 4x + 13 = 0$
13. $x^2 + 2x - 5 = 0$
14. $2y^2 + 5y + 2 = 0$
15. $z^2 + 10z + 125 = 0$
16. $w^2 - 6w + 58 = 0$
17. $5w^2 - 3w - 1 = 0$
18. $z^2 - 10z + 12 = 0$
19. $u^2 + 2u + 6 = 0$
20. $v^2 - v + 1 = 0$
21. $p^2 + 8p + 6 = 0$
22. $q^2 - 4q - 2 = 0$
23. $x^2 + x + 4 = 0$
24. $y^2 + 2y + 5 = 0$
25. $t^2 - 3t - 3 = 0$
26. $3r^2 - r - 1 = 0$
27. $3w^2 + w + 2 = 0$
28. $2z^2 + z + 1 = 0$
29. $5v^2 - 8v + 1 = 0$
30. $7u^2 - 2u - 1 = 0$

For Exercises 31–40, first write the equation in standard form and then use the quadratic formula to find the solution set. See Example 3.

31. $2x^2 = 6x - 3$
32. $y^2 = -5y - 2$
33. $(z - 4)(z - 2) = 12$
34. $(w + 5)(w + 1) = 10$
35. $(u - 4)(u - 1) = -1$
36. $(2v + 1)(v - 1) = 1$
37. $w(w + 8) = -65$
38. $z(z + 14) = -74$
39. $z^2 - 6z = -25$
40. $x^2 - 10x = -34$

For Exercises 41–50, first clear fractions and write the resulting equation in standard form. Then solve the equation by using the quadratic formula. See Example 4.

41. $\dfrac{2w}{w + 2} + w = 2$
42. $\dfrac{2y}{y + 4} + y = 4$
43. $\dfrac{3p}{p + 1} + 2p = 1$
44. $\dfrac{2q}{q + 3} + q = 2$
45. $\dfrac{6x}{x + 2} + x = 1$
46. $\dfrac{4y}{y + 2} + y = 1$
47. $\dfrac{3w}{w + 2} - w = 3$
48. $\dfrac{4z}{z + 3} - z = 3$
49. $\dfrac{x}{x - 2} + x = 2$
50. $\dfrac{y}{y - 4} + y = 2$

For Exercises 51–58, use the discriminant to determine the number and nature of the solutions of the equation. See Example 5.

51. $x^2 + 3x - 5 = 0$
52. $y^2 + 5y - 3 = 0$
53. $2w^2 - w + 1 = 0$
54. $3z^2 - 2z + 5 = 0$
55. $3p^2 + 5p - 2 = 0$
56. $4q^2 - 17q - 15 = 0$
57. $4y^2 - 20y + 25 = 0$
58. $9x^2 - 30x + 25 = 0$

6.4 Applications Involving Quadratic Equations

When many practical problems are expressed with mathematical symbolism, the result is a quadratic equation. Although any quadratic equation can be solved by using the quadratic formula, it is sometimes easier to use the method of factoring. We begin this section by considering problems that involve the use of the *Pythagorean theorem,* which relates the squares of the lengths of the three sides of a right triangle. A right triangle is a triangle that has an angle with a measure of 90° (ninety degrees). The Pythagorean theorem is stated here without proof.

Pythagorean Theorem

In a right triangle (see Figure 6.1), if the length of the *hypotenuse* (the side opposite the right angle) is denoted by c and the lengths of the other two sides (the *legs*) are denoted by a and b, then

$$a^2 + b^2 = c^2.$$

That is, in a right triangle, the sum of the squares of the lengths of the two legs is equal to the square of the length of the hypotenuse.

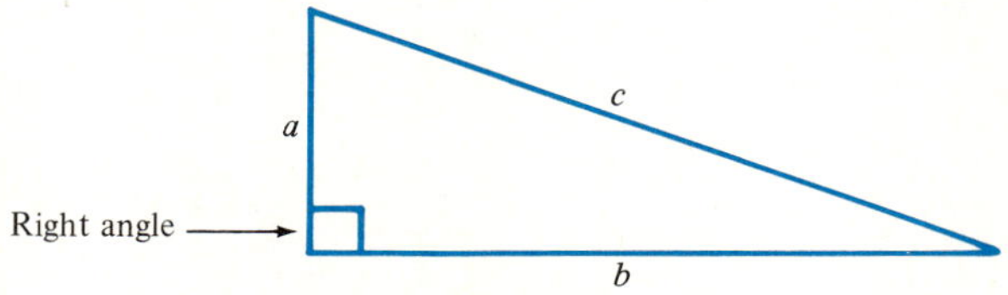

Figure 6.1

Example 1. In a right triangle, the lengths of the two legs are shorter than the length of the hypotenuse by 4 inches and 8 inches, respectively. What is the length of the hypotenuse?

Solution: If we let x denote the length of the hypotenuse, as shown in Figure 6.2, then the lengths of the two legs are $x - 4$ and $x - 8$, respectively.

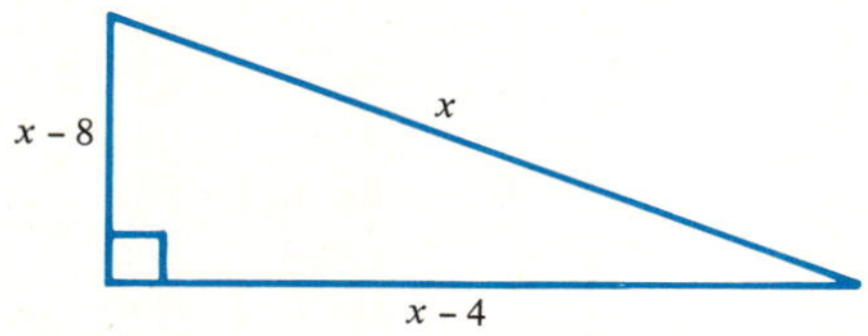

Figure 6.2

From the Pythagorean theorem we have

$$(x - 8)^2 + (x - 4)^2 = x^2$$

or

$$x^2 - 16x + 64 + x^2 - 8x + 16 = x^2.$$

Thus

$$x^2 - 24x + 80 = 0.$$

Since $80 = 20 \cdot 4$ and $20 + 4 = 24$, we can solve this equation by factoring. That is, we rewrite the equation as

$$(x - 4)(x - 20) = 0.$$

From this we see that either

$$x = 4 \qquad \text{or} \qquad x = 20.$$

However, if we use $x = 4$, the legs would have "lengths" $x - 4 = 0$ and $x - 8 = -4$, respectively. Since the lengths of the sides of a triangle are positive numbers, $x = 4$ is not a solution to the original problem. A similar check with $x = 20$ yields positive numbers for the lengths of the sides of the triangle. Thus the length of the hypotenuse is 20 inches.

Example 2. When the squares of three consecutive integers are summed, the result is 194. Find the integers.

Solution: If we let x denote the first integer, then $x + 1$ and $x + 2$ will denote the other two integers. Since the sum of the squares of these three integers is 194, we can write

$$x^2 + (x + 1)^2 + (x + 2)^2 = 194$$

or

$$x^2 + x^2 + 2x + 1 + x^2 + 4x + 4 = 194.$$

Thus

$$3x^2 + 6x - 189 = 0$$

or simplifying the equation by dividing *both* sides of the equation by 3, we find that

$$x^2 + 2x - 63 = 0.$$

Factoring the last equation, we obtain

$$(x + 9)(x - 7) = 0.$$

Thus

$$x + 9 = 0 \qquad \text{or} \qquad x - 7 = 0.$$

Therefore,

$$x = -9 \qquad \text{or} \qquad x = 7.$$

If $x = -9$, then $x + 1 = -8$ and $x + 2 = -7$. If $x = 7$, then $x + 1 = 8$ and $x + 2 = 9$. Hence there are two sets of three consecutive integers with the property that the sum of their squares is 194. They are

$$\{-9, -8, -7\} \quad \text{and} \quad \{7, 8, 9\}.$$

Figure 6.3 illustrates the situation after an object has been thrown directly upward from a platform that is h feet above the ground. If the object has been thrown with an initial velocity of v feet per second, then its *height*, y, above the ground after t seconds is given by $y = -16t^2 + vt + h$.

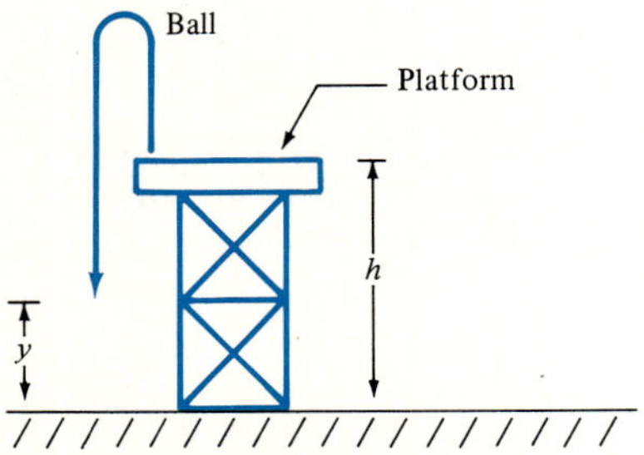

Figure 6.3

Example 3. A ball is thrown directly upward from a position that is 74 feet above the ground. If the initial velocity of the ball is 40 feet per second, determine how long it takes before the ball:
(a) Is exactly 50 feet above the ground.
(b) Hits the ground.

Solution: For this problem $h = 74$ and $v = 40$. Substituting these values in the equation

$$y = -16t^2 + vt + h,$$

we get

$$y = -16t^2 + 40t + 74.$$

(a) To determine how long it takes before the ball is exactly $y = 50$ feet above the ground, we need to solve the equation

$$50 = -16t^2 + 40t + 74$$

or

$$16t^2 - 40t - 24 = 0$$

or dividing both sides of the equation by 8, we get

$$2t^2 - 5t - 3 = 0.$$

Factoring the left side of the last equation, we get

$$(2t + 1)(t - 3) = 0.$$

Thus either

$$2t + 1 = 0 \quad \text{or} \quad t - 3 = 0$$
$$2t = -1 \qquad\qquad t = 3.$$
$$t = -\frac{1}{2}$$

In this problem time is measured with positive numbers. Thus we will consider only the positive solution $t = 3$. Thus, in 3 seconds, the ball will be 50 feet above the ground.

(b) Observe that when the ball hits the ground, $y = 0$. Thus we need to solve the equation

$$0 = -16t^2 + 40t + 74$$

or

$$8t^2 - 20t - 37 = 0.$$

We solve this equation using the quadratic formula with $a = 8$, $b = -20$, and $c = -37$. Thus

$$\begin{aligned} t &= \frac{-b \pm \sqrt{b^2 - 4ac}}{2a} \\ &= \frac{-(-20) \pm \sqrt{(-20)^2 - 4(8)(-37)}}{2(8)} \\ &= \frac{20 \pm \sqrt{1584}}{16} \qquad \left(\sqrt{1584} = \sqrt{144 \cdot 11} = 12\sqrt{11}\right) \\ &= \frac{5 \pm 3\sqrt{11}}{4}. \end{aligned}$$

Again, since time is measured with positive numbers, we will only consider the solution $\frac{5 + 3\sqrt{11}}{4}$. Using the square root table or a *calculator,* we see that the ball will hit the ground in approximately 3.74 seconds.

Recall from Section 2.3 that the formula $d = rt$ is used to solve distance–rate–time problems. As the next example illustrates, these problems sometimes result in a quadratic equation.

Example 4. On a river that has a steady current of 3.5 miles per hour, it takes two men a total of 9 hours to paddle 8 miles upstream and return to their starting point. At what rate can the men paddle in still water?

Solution: If we let x denote the rate at which the men can paddle in still water, their rate against the current (upstream) is $x - 3.5$ miles per hour and their rate downstream is $x + 3.5$ miles per hour. Note that they travel 8 miles upstream and 8 miles downstream. We summarize this information below using the formula $d = rt$. In the summary we let t_1 (read "t sub-one") denote the time it takes to paddle 8 miles upstream. Similarly, t_2 (read "t sub-two") denotes the time it takes to paddle 8 miles downstream.

$$d = r \cdot t$$

$$\text{upstream:} \quad 8 = (x - 3.5)t_1 \quad \text{or} \quad \frac{8}{x - 3.5} = t_1$$

$$\text{downstream:} \quad 8 = (x + 3.5)t_2 \quad \text{or} \quad \frac{8}{x + 3.5} = t_2.$$

But we also know that $t_1 + t_2 = 9$. That is,

$$\frac{8}{x - 3.5} + \frac{8}{x + 3.5} = 9.$$

Clearing fractions, we get

$$8(x + 3.5) + 8(x - 3.5) = 9(x - 3.5)(x + 3.5).$$

After simplifying the equation above, we get

$$9x^2 - 16x - 110.25 = 0.$$

Applying the quadratic formula results in

$$x = \frac{-(-16) \pm \sqrt{256 - 4(9)(-110.25)}}{2(9)}$$

$$= \frac{16 \pm \sqrt{4225}}{18}$$

$$= \frac{16 \pm 65}{18}.$$

Thus the possible solutions are $x = \frac{81}{18} = 4.5$ or $x = -\frac{49}{18}$. Since the men's paddling rate is a positive number, we find that the men's paddling rate in still water is 4.5 miles per hour.

EXERCISES 6.4

Solve Exercises 1–8. See Example 1.

1. In a right triangle the two legs are shorter than the hypotenuse by 1 and 8 inches, respectively. What is the length of the hypotenuse?

2. In a right triangle the two legs are shorter than the hypotenuse by 2 and 4 inches, respectively. What is the length of the hypotenuse?

3. In a right triangle one leg is 3 feet shorter than the hypotenuse. If the other leg is 6 feet shorter than the hypotenuse, how long is the hypotenuse?

4. One leg of a right triangle is 1 foot shorter than the hypotenuse. If the other leg is 18 feet shorter than the hypotenuse, how long is the hypotenuse?

5. One leg of a right triangle is twice as long as the other leg. If the hypotenuse is 10 inches long, what is the length of the shorter leg?

6. One leg of a right triangle is three times as long as the other leg. If the hypotenuse is 20 inches long, what is the length of the shorter leg?

7. The hypotenuse of a right triangle is 13 feet long. If the sum of the lengths of the two legs is 15 feet, what is the length of the shorter leg?

8. The hypotenuse of a right triangle is 20 feet long. If the sum of the lengths of the two legs is 24 feet, how long is the shorter leg?

Solve Exercises 9–16. See Example 2.

9. The sum of the squares of two consecutive integers is 145. What are the integers?

10. The sum of the squares of two consecutive integers is 265. What are the integers?

11. The sum of the squares of what three consecutive integers is 110?

12. The sum of the squares of what three consecutive integers is 149?

13. The product of two consecutive integers is 210. What are the integers?

14. The product of two consecutive integers is 462. What are the integers?

15. Find two consecutive odd integers that have a product of 195.

16. Find two consecutive even integers that have a product of 168.

Solve Exercises 17–24. See Example 3.

17. A ball is thrown directly upward from a platform that is 96 feet above the ground. If the initial velocity of the ball is 80 feet per second, when will the ball hit the ground?

18. A ball is thrown directly upward from a platform that is 36 feet above the ground. If the initial velocity of the ball is 64 feet per second, when will the ball hit the ground?

19. A bullet is fired directly upward with an initial velocity of 800 feet per second. When will the bullet return to the ground?

20. A bullet is fired directly upward from the roof of a building that is 816 feet above the ground. If the initial velocity of the bullet is 800 feet per second, when will the bullet hit the ground?

21. An object is thrown directly upward from a platform that is 50 feet above the ground. If the object's initial velocity is 40 feet per second, when will it be 30 feet above the ground?

22. An object is thrown directly upward from the roof of a building that is 90 feet above the ground. If the object's initial velocity is 64 feet per second, when will it be 50 feet above the ground?

23. An arrow is shot directly upward from the roof of a building that is 75 feet above the ground. If the initial velocity of the arrow is 120 feet per second, when will it be 50 feet above the ground?

24. An arrow is shot directly upward from a platform that is 30 feet above the ground. If the initial velocity of the arrow is 200 feet per second, when will it be 15 feet above the ground?

Solve Exercises 25–30. See Example 4.

25. On a river that has a steady current of 2 miles per hour, two canoeists can travel 8 miles upstream and return to their starting point in 3 hours. What is their rate of travel while paddling in still water?

26. Two canoeists can paddle 12 miles upstream and return to their starting point in 5 hours. If the river has a steady current of 1 mile per hour, what is their rate of travel while paddling in still water?

27. On a river that has a steady current of 3 miles per hour, a small outboard motorboat can travel 10 miles upstream and return in $3\frac{1}{2}$ hours. At what rate would the outboard motorboat travel in still water?

28. An outboard motorboat can travel 9 miles upstream and return to where it started in 4 hours. If the river's current is a steady 3 miles per hour, at what rate would the boat travel in still water?

29. A large motorboat can travel 8 miles upstream and return in 54 minutes. If the river's current is 2 miles per hour, what is the boat's rate of travel in still water? [HINT: Change 54 minutes to hours.]

30. A cabin cruiser can travel 6 miles upstream and return in 50 minutes. If the river's current is 3 miles per hour, what is the boat's rate of travel in still water? [HINT: Change 50 minutes to hours.]

6.5 Solving Quadratic Inequalities

In Sections 2.5 and 2.6 we studied first-degree inequalities. In this section we discuss how to solve quadratic (second-degree) inequalities in one variable. A *quadratic inequality* has the following form:

$$ax^2 + bx + c \leq 0,$$

where a, b, and c are real numbers with $a \neq 0$. Other quadratic inequalities can be formed by replacing $\leq$ with $\geq$, $>$, or $<$.

The efficient solution of the quadratic inequalities that we discuss in this section requires skill at factoring and a good understanding of the rules that govern the multiplication of signed numbers, discussed in Chapter 1. In particular, it is necessary to remember that

1. The product of two positive numbers is a positive number.
2. The product of two negative numbers is a positive number.
3. The product of a positive number and a negative number is a negative number.

For example, in order to solve $x^2 + x - 20 > 0$, we first factor the polynomial $x^2 + x - 20$ as $(x + 5)(x - 4)$. Thus we have $(x + 5)(x - 4) > 0$. But the product $(x + 5)(x - 4)$ will be greater than 0, that is, positive, only when *both* factors are *positive* or when *both* factors are *negative*. Note that

$x + 5 > 0$	when $x > -5$		$x - 4 > 0$	when $x > 4$
$x + 5 < 0$	when $x < -5$	and	$x - 4 < 0$	when $x < 4$
$x + 5 = 0$	when $x = -5$		$x - 4 = 0$	when $x = 4$.

We now show how this information can be used to conveniently determine where the two factors have the same sign. We use the number line and place + signs where a factor is positive, − signs where a factor is negative, and a vertical line where a factor is 0. Thus for the factors $(x + 5)$ and $(x - 4)$, we have the diagram shown in Figure 6.4.

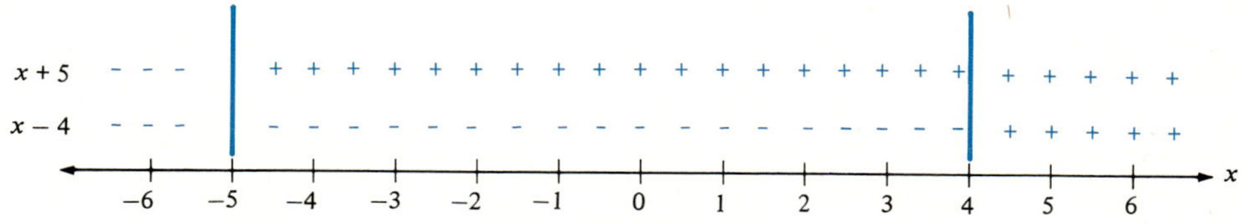

Figure 6.4

From the diagram it is easy to see that *both* factors are *negative* when $x < -5$. Also, *both* factors are *positive* when $x > 4$. Thus the solution set of $x^2 + x - 20 > 0$ is $\{x \mid x < -5 \text{ or } x > 4\}$. The graph of this solution set is shown in Figure 6.5.

Figure 6.5

Example 1. Solve $x^2 - 2x - 15 \geq 0$ and sketch the graph of the solution set.

Solution: Since $x^2 - 2x - 15 = (x + 3)(x - 5)$, we can rewrite the original inequality as $(x + 3)(x + 5) \geq 0$. The correct diagram for this problem is shown in Figure 6.6. In this problem we want all values of x that will cause *both* factors to

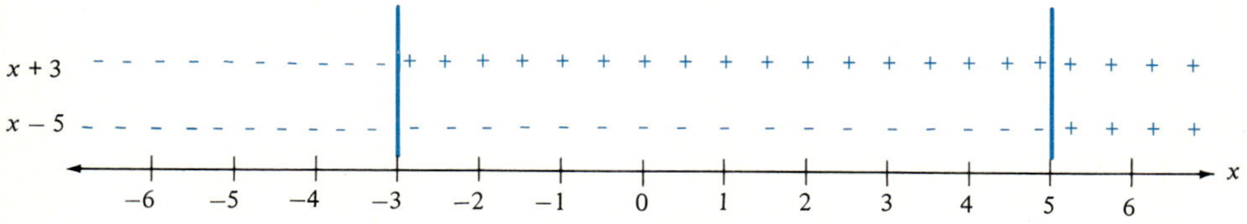

Figure 6.6

have the same sign or that will cause *one* of the factors to be *zero*. Thus the desired solution set is $\{x \mid x \leq -3 \text{ or } x \geq 5\}$. The graph of this solution set is shown in Figure 6.7.

Figure 6.7

Example 2. Solve $x^2 - x - 2 \le 0$ and graph the solution set.

Solution: We can factor the left side and rewrite the inequality as $(x + 1)(x - 2) \le 0$. Using the factors $(x + 1)$ and $(x - 2)$, we construct the diagram shown in Figure 6.8. In order for $x^2 - x - 2 = (x + 1)(x - 2)$ to be nega-

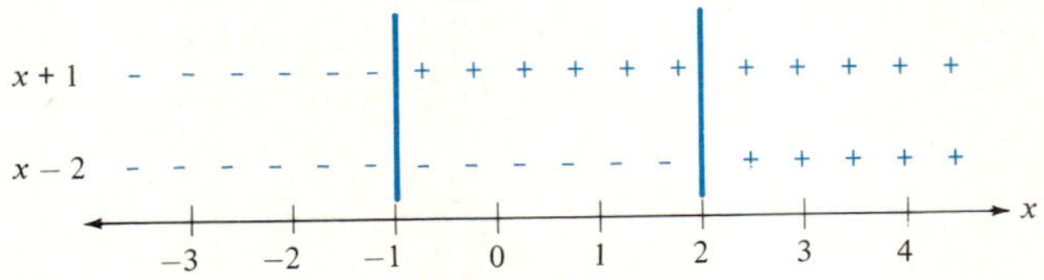

Figure 6.8

tive or zero, the factors $(x + 1)$ and $(x - 2)$ must have *opposite* signs or *one* of them must be 0. Thus the solution set is $\{x \mid -1 \le x \le 2\}$ (see Figure 6.9).

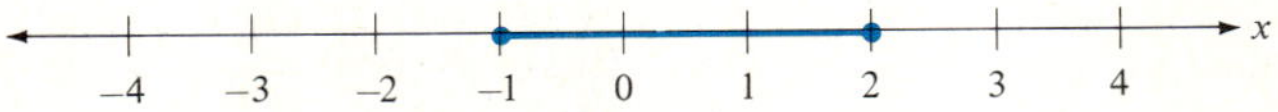

Figure 6.9

Example 3. Solve the inequalities: (a) $x^2 + 8x + 16 \ge 0$. (b) $x^2 - 6x + 9 < 0$.

Solution

(a) If we factor the left side, we can rewrite the inequality as $(x + 4)^2 \ge 0$. But this is true for all values of x, so the solution set is the set of *all* real numbers $\mathbb{R}$.

(b) Since $x^2 - 6x + 9 = (x - 3)^2$, we can rewrite the inequality as $(x - 3)^2 < 0$. But $(x - 3)^2$ is *never* negative and therefore the solution set of $x^2 - 6x + 9 < 0$ is the empty set $\varnothing$.

The following steps may be used to solve a quadratic inequality where the solution set consists of real numbers.

Steps for Solving a Quadratic Inequality

Step 1. Write the inequality so that the right side is 0. Such an inequality may be of the form $ax^2 + bx + c > 0$. Other forms can be obtained by replacing $>$ with $<$, $\ge$, or $\le$.

Step 2. Find values that make the quadratic expression 0. This is done by factoring or by using the quadratic formula.

Step 3. Use the number line and make a diagram that indicates where each factor is positive, negative, or 0.

Step 4. From the diagram in Step 3, determine where the product of the factors is positive, negative, or 0.

As the next example illustrates, third-degree inequalities can sometimes be solved in a manner that is very similar to that used to solve quadratic inequalities.

Example 4. Solve $(x + 2)(x - 3)(2x - 9) < 0$.

Solution: Note that the product of three factors is negative when either all three factors are negative or exactly one of the factors is negative. The diagram in Figure 6.10 will help us to determine the solution set. From the diagram it is easy to see

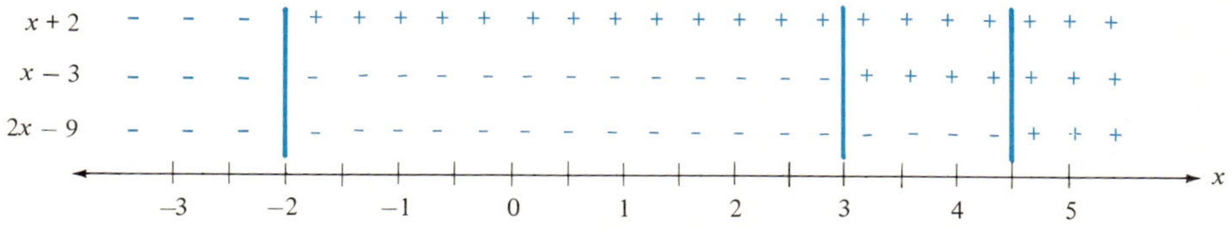

Figure 6.10

that the desired solution set is $\{x \mid x < -2 \text{ or } 3 < x < 4.5\}$. The graph of this set is shown in Figure 6.11.

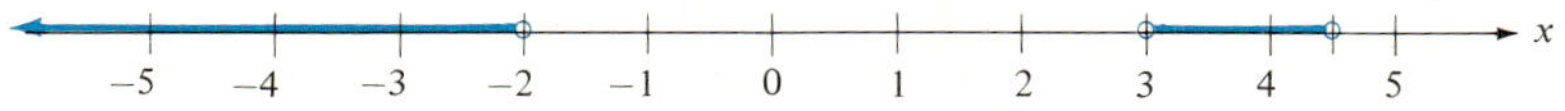

Figure 6.11

The next two examples illustrate how some inequalities that involve rational expressions may be solved.

Example 5. Solve $\dfrac{6}{x - 4} > -1$.

Solution: First we rewrite the inequality so that the right side is 0. Thus we have

$$\frac{6}{x - 4} + 1 > 0.$$

At this stage of the solution we might consider clearing fractions by multiplying by $(x - 4)$. However, we would then have to keep track of where $(x - 4)$ is positive or negative in order to reverse the direction of the inequality when $(x - 4)$ is negative. An alternative and simpler method is to get a common denominator and add.

$$\frac{6}{x - 4} + 1 > 0$$

$$\frac{6}{x - 4} + \frac{x - 4}{x - 4} > 0$$

$$\frac{x + 2}{x - 4} > 0.$$

But $\dfrac{x+2}{x-4}$ is positive when $(x + 2)$ and $(x - 4)$ have the same sign. The diagram in Figure 6.12 will help us determine where $(x + 2)$ and $(x - 4)$ have the same sign. From the diagram we see that the solution set is $\{x \mid x < -2 \text{ or } x > 4\}$.

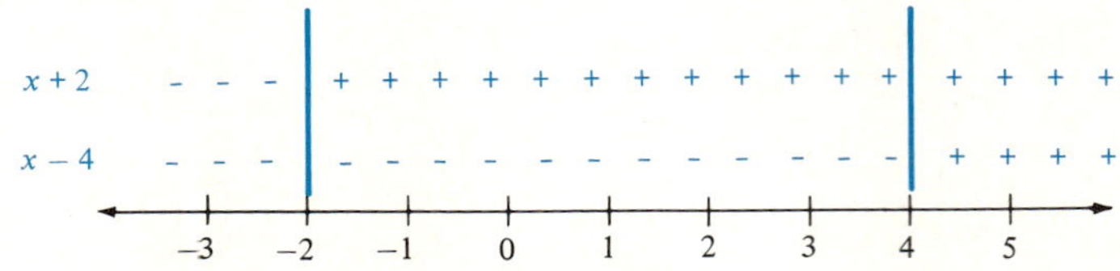

Figure 6.12

Example 6. Solve $\dfrac{4}{x+1} > \dfrac{2}{x-2}$.

Solution: First, we subtract $\dfrac{2}{x-2}$ from each side. Next, we combine the two rational expressions by writing them with a common denominator.

$$\frac{4}{x+1} - \frac{2}{x-2} > 0$$

$$\frac{4}{x+1}\cdot\frac{x-2}{x-2} - \frac{2}{x-2}\cdot\frac{x+1}{x+1} > 0$$

$$\frac{4x - 8 - 2x - 2}{(x+1)(x-2)} > 0$$

$$\frac{2(x-5)}{(x+1)(x-2)} > 0.$$

Figure 6.13 shows the signs of the three expressions $(x - 5)$, $(x + 1)$, and $(x - 2)$.

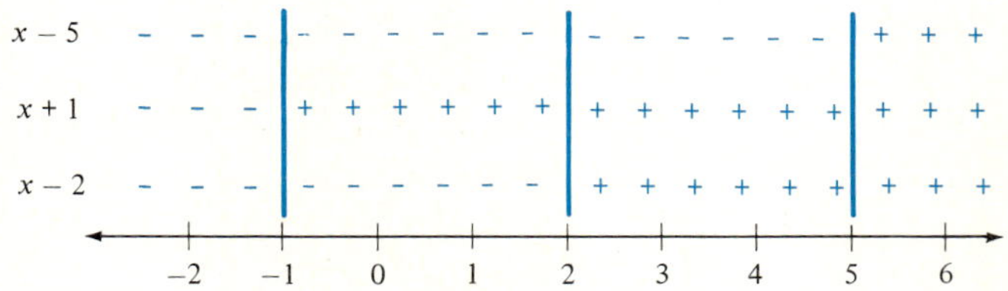

Figure 6.13

Since we want the quotient $\dfrac{2(x-5)}{(x+1)(x-2)}$ to be positive, either all of the expressions are positive or exactly two of them are negative. Referring to Figure 6.13, we see that the solution set is $\{x \mid -1 < x < 2 \text{ or } x > 5\}$.

EXERCISES 6.5

For Exercises 1–16, solve the inequality and graph the solution set. See Examples 1 and 2.

1. $x^2 - 2x - 8 \geq 0$ **2.** $x^2 - x - 12 \geq 0$ **3.** $x^2 - 1 > 0$ **4.** $x^2 - 4 > 0$

5. $x^2 - 3x - 4 \leq 0$ **6.** $x^2 - x - 6 \leq 0$ **7.** $x^2 + 4x + 3 < 0$ **8.** $x^2 + 6x + 8 < 0$

9. $2x^2 + 7x - 4 < 0$ **10.** $2x^2 - x - 3 > 0$ **11.** $3x^2 + 4x - 4 \geq 0$ **12.** $3x^2 + 8x - 3 \leq 0$

13. $x^2 + 5x > 0$ **14.** $x^2 + 2x < 0$ **15.** $x^2 - 4x \leq 0$ **16.** $x^2 - 3x \geq 0$

For Exercises 17–24, solve the inequality. See Examples 3 and 4.

17. $x^2 + 10x + 25 < 0$ **18.** $x^2 - 12x + 36 < 0$

19. $x^2 - 8x + 16 \geq 0$ **20.** $x^2 + 2x + 1 \geq 0$

21. $(x - 3)(x + 1)(x + 4) > 0$ **22.** $(x - 2)(x - 4)(x + 5) > 0$

23. $(x + 2)(x - 1)(x + 1) < 0$ **24.** $(x + 3)(x + 1)(x - 4) < 0$

For Exercises 25–40, solve the inequality. See Examples 5 and 6.

25. $\frac{5}{x - 2} > 1$ **26.** $\frac{3}{x - 1} > -1$ **27.** $\frac{4}{x + 3} < -1$ **28.** $\frac{5}{x + 1} < -1$

29. $\frac{1}{x + 2} > 3$ **30.** $\frac{1}{x + 4} > 2$ **31.** $\frac{2}{x + 1} < 4$ **32.** $\frac{3}{x + 2} < 2$

33. $\frac{3}{x + 1} > \frac{1}{x - 3}$ **34.** $\frac{3}{x + 3} > \frac{2}{x - 1}$ **35.** $\frac{5}{x + 1} < \frac{4}{x + 2}$ **36.** $\frac{4}{x + 2} < \frac{3}{x + 1}$

37. $\frac{5}{x + 5} > \frac{3}{x - 1}$ **38.** $\frac{5}{x + 3} > \frac{3}{x - 3}$ **39.** $\frac{4}{x + 2} < \frac{2}{x - 2}$ **40.** $\frac{4}{x + 3} < \frac{2}{x - 3}$

Key Words

Completing the square
Discriminant
Factor property of zero
Pythagorean theorem
Quadratic equation
Quadratic formula
Quadratic inequality
Root
Square root theorem
Standard form

CHAPTER 6 REVIEW

SECTION 6.1

Use the factor property of zero to solve the equations.

1. $(x - 4)(x + 2) = 0$ **2.** $7x^2 + 21x = 0$ **3.** $w^2 - 28 = 3w$

4. $(x + 2)(x + 6) = -3$ **5.** $1 + \dfrac{18}{z^2} = \dfrac{11}{z}$ **6.** $\dfrac{x}{x + 5} + \dfrac{2x - 25}{(x - 2)(x + 5)} = \dfrac{2}{x - 2}$

7. $\sqrt{x^2 - 11} = 5$ **8.** $\sqrt{3v + 1} - v = 1$

SECTION 6.2

Use the square root theorem to solve the equations.

9. $x^2 = 49$ **10.** $y^2 = 72$ **11.** $(x + 2)^2 = 25$ **12.** $(2u + 1)^2 = 49$

Solve the equations by completing the square.

13. $x^2 + 10x = 25$ **14.** $w^2 + 8w = 4$

SECTION 6.3

Use the quadratic formula to solve the equations.

15. $x^2 + 4x + 1 = 0$ **16.** $x^2 - x - 3 = 0$ **17.** $3w^2 - w - 2 = 0$ **18.** $x^2 + x + 2 = 0$

First clear fractions and then use the quadratic formula to solve each equation.

19. $\dfrac{2x}{x + 1} - x = 5$ **20.** $\dfrac{x}{x - 2} + 3 = \dfrac{1}{x}$

SECTION 6.4

Solve each word problem.

21. One leg of a right triangle is three times as long as the other leg. If the hypotenuse is 30 inches long, what is the length of the shorter leg?

22. The sum of the squares of two consecutive integers is 290. What are the integers?

23. The product of two consecutive odd integers is 423. What are the integers?

24. A ball is thrown directly upward from a platform that is 48 feet above the ground. If the initial velocity of the ball is 64 feet per second, when will the ball hit the ground?

25. Two canoeists can paddle 12 miles upstream and return to their starting point in 6 hours. If the river has a steady current of $1\frac{1}{2}$ miles per hour, what is their rate of travel while paddling in still water?

SECTION 6.5

Solve the inequalities and graph their solution sets.

26. $x^2 - 4x - 5 \geq 0$ **27.** $x^2 + 11x + 30 < 0$ **28.** $2x^2 - 9x - 5 > 0$

29. $x^2 - 8x + 16 < 0$ **30.** $(x - 1)(x + 2)(x + 3) \leq 0$ **31.** $\dfrac{3}{x + 1} > 2$

32. $\dfrac{2}{x + 4} < \dfrac{1}{x - 3}$

CHAPTER 6 TEST

For Problems 1–6, solve the equation.

1. $(x + 2)(x - 3) = 0$
2. $y^2 + 5y + 6 = 0$
3. $z^3 = 9z$
4. $w(w - 4) = 12$
5. $\frac{4}{w^2} + \frac{3}{w} - 1 = 0$
6. $\frac{x + 1}{x - 3} + \frac{2}{x + 2} = \frac{2x}{x^2 - x - 6}$

For Problems 7–10, use the square root theorem to solve the equation. Simplify all radicals.

7. $x^2 = 72$
8. $(y - 1)^2 = -4$
9. $x^2 + 4x = 1$
10. $2y^2 - 5 = -6y$

For Problems 11–13, use the quadratic formula to solve the equation.

11. $x^2 - 3x + 1 = 0$
12. $3y^2 + 5y + 1 = 0$
13. $w^2 + w = -1$

For Problem 14, first eliminate the radicals and then solve the resulting equation.

14. $x - \sqrt{2x - 3} = 1$

Solve Problems 15 and 16.

15. One side of a right triangle is 2 inches longer than the other side. If the hypotenuse is 5 inches longer than the shortest side, what is the length of the hypotenuse?
16. On a river that has a steady current of 2.5 miles per hour, it takes a small motorboat a total of 4.5 hours to go 7 miles downstream and return to the starting point. What is the speed of the boat in still water?

For Problems 17–20, solve the inequality.

17. $x^2 - 5x + 6 > 0$
18. $x^2 - 2x - 8 \leq 0$
19. $(x + 2)(x - 1)(x - 3) < 0$
20. $\frac{5}{x + 3} > 1$

CHAPTER 7

Equations in Two Variables and Graphing

Generally, a weatherman gives the temperature for the day using two scales Celsius (C) and Fahrenheit (F). The two scales are related by the following equation in two variables

$$F = \frac{9}{5}C + 32.$$

Thus, if the temperature is 20° Celsius, we can determine from this equation that the corresponding Fahrenheit temperature is 68°.

National Oceanic and Atmospheric Administration

In this chapter we study first-degree equations and inequalities in two variables. To prepare for a discussion on the graphs of equations in two variables, we begin the chapter with a discussion of the rectangular or Cartesian coordinate system in the geometric plane. As a result, we find that first-degree equations in two variables are associated with straight lines in the plane. We conclude the chapter with a discussion on graphing simple linear inequalities in two variables.

7.1 The Rectangular Coordinate System

In Chapter 1 we discussed the real number line. Recall that each point on the number line has a unique real number associated with it, called its *coordinate*. Conversely, each real number is associated with a point on the number line called its *graph*. In this manner we identified the set of points on a line with the set of real numbers. In a similar way, it proves useful in algebra and plane geometry to associate with each *point* in the plane, an *ordered pair* of real numbers. To accomplish this, we draw two real number lines, one horizontal and the other one vertical, intersecting at right angles at their origins, as shown in Figure 7.1.

We call the horizontal line the *x-axis* and the vertical line the *y-axis*. The positive side of the x-axis extends to the right. The y-axis has its positive side in the upward direction. We use arrows to indicate the positive direction of each axis, as shown in Figure 7.1.

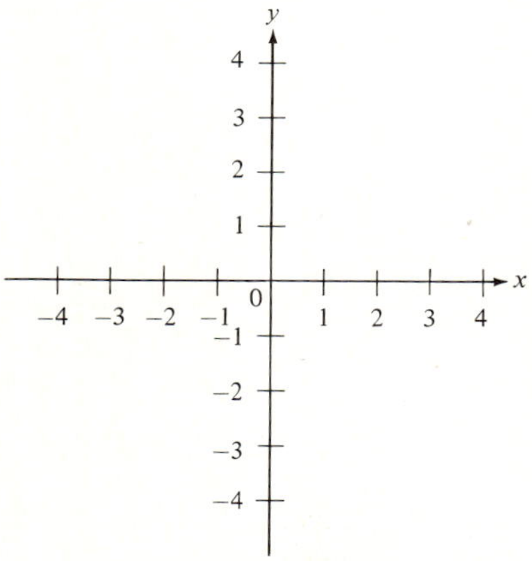

Figure 7.1

The two axes form a reference system for associating the points in the plane with *ordered pairs of real numbers* (a, b). Note that a is the first member of the ordered pair and b is the second member of the ordered pair.

To locate the point associated with a given ordered pair of real numbers (a, b), we proceed in the following manner. At a on the x-axis we draw a line parallel to the

y-axis. Then at b on the y-axis we draw a line parallel to the x-axis. The intersection of these two lines determines point P, as shown in Figure 7.2. The notation $P(a, b)$ is often used to designate the correspondence between point P and the associated ordered pair (a, b).

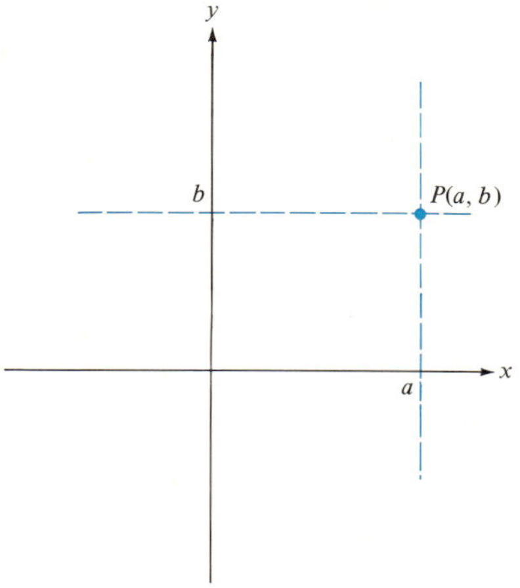

Figure 7.2

Conversely, suppose that we are given a point P in the plane as shown in Figure 7.2 and wish to find the ordered pair of numbers associated with P. To do this, we pass horizontal and vertical lines through point P. We see that these lines meet the x-axis and y-axis at a and b, respectively. We then associate the ordered pair (a, b) with the given point P. Notice that the value associated with the x-axis appears *first* in the parentheses, (,).

This method of associating ordered pairs of real numbers (a, b) with points in the plane is called a *rectangular* or *Cartesian coordinate system*.

In general, given a rectangular coordinate system, an ordered pair of numbers (a, b) determines a unique point P in the plane. Conversely, for any point P there is associated a unique ordered pair of real numbers, (a, b). The first number of the ordered pair, a, is called the *x-coordinate* or *abscissa* of the point P. The second member, b, is called the *y-coordinate* or *ordinate* of the point P.

The *point* corresponding to an ordered pair (a, b) is called the *graph* of (a, b). Locating a point with specific coordinates such as (2, 3) in a rectangular coordinate system is called *plotting* or *graphing* the point. We frequently identify the point P with its pair of coordinates (a, b) and speak of "the point (a, b)."

The coordinate system divides the plane into four regions called *quadrants,* labeled counterclockwise I, II, III, and IV, as shown in Figure 7.3. By convention, points on the x-axis and y-axis do not belong to any of the quadrants. We note, however, that points lying on the x-axis have coordinates of the form $(a, 0)$ and those on the y-axis have coordinates of the form $(0, b)$. The intersection of the coordinate axes is called

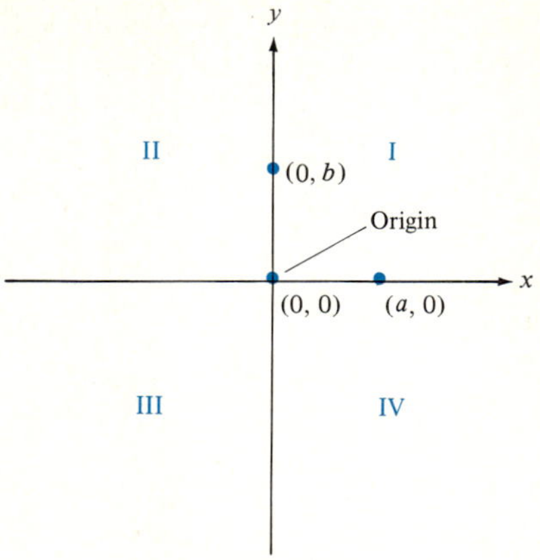

Figure 7.3

the *origin* and has coordinates (0, 0) as shown in Figure 7.3. In Figure 7.4 we have graphed four ordered pairs of real numbers, one in each quadrant.

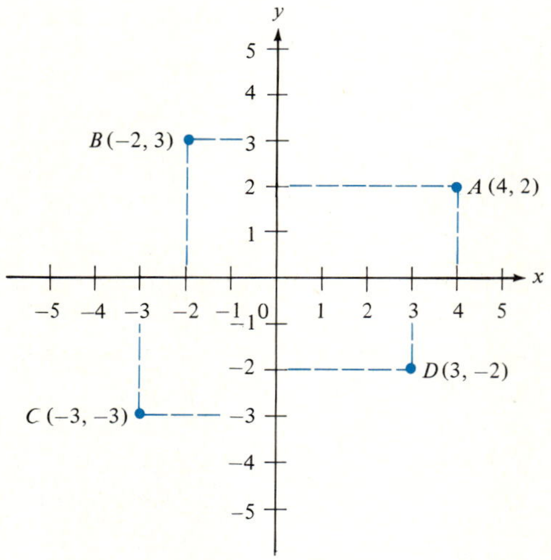

Figure 7.4

We now outline an equivalent way of graphing an ordered pair of real numbers, (a, b), without drawing horizontal or vertical lines. In this alternative method, to *locate* the point associated with (a, b) we start at the origin and, provided $a \neq 0$, we move *horizontally* along the x-axis. We move to the *right* if a is *positive* or to the *left* if a is *negative* until we reach a on the x-axis. Then if $b \neq 0$, we move in a *vertical* direction b units *up* if b is *positive*, or b units *down* if b is *negative*. We illustrate this technique in the following example.

Example 1. Graph the following ordered pairs on a rectangular coordinate system:

$$(-3, -4), \quad (3, 4), \quad (5, 0), \quad (-2, 3), \quad (4, -3).$$

Solution: To graph $(-3, -4)$, we start at the origin and move to -3 on the *x-axis*. Then we go down 4 units to arrive at the point corresponding to $(-3, -4)$. We use a similar procedure to graph the remaining ordered pairs. The graphs of all five ordered pairs are shown in Figure 7.5.

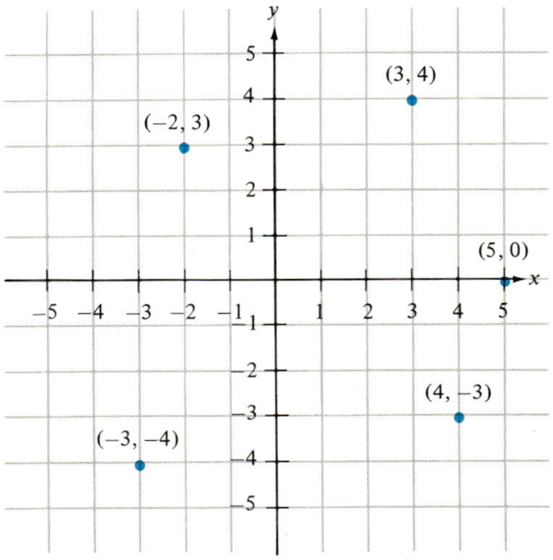

Figure 7.5

A similar procedure to that outlined in Example 1 and the discussion preceding Example 1 can be used to find the coordinates of a given point relative to a rectangular coordinate system. The next example shows how to do this.

Example 2. Find the coordinates of the points P, Q, R, S, and T shown in the rectangular coordinate system of Figure 7.6.

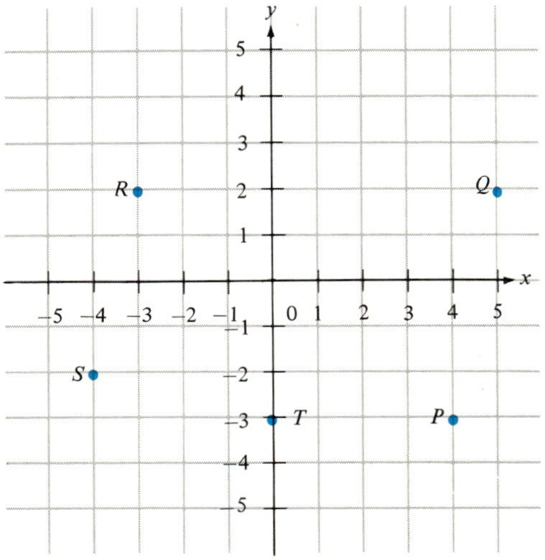

Figure 7.6

Solution: Observe that P can be located by starting at the origin and moving 4 units to the right (positive) on the x-axis and then down (negative) 3 units. Therefore, the coordinates of P are $(4, -3)$. We indicate this correspondence between point P and its coordinates by writing $P(4, -3)$. Proceeding in a similar manner, we see that we have $Q(5, 2)$, $R(-3, 2)$, $S(-4, -2)$, and $T(0, -3)$.

EXERCISES 7.1

For Exercises 1–20, plot the ordered pairs of numbers on the same rectangular coordinate system. See Example 1.

1. $(2, 3)$	**2.** $(4, 3)$	**3.** $(-4, 2)$	**4.** $(-6, 4)$	**5.** $(-3, -5)$
6. $(-4, -6)$	**7.** $(4, -4)$	**8.** $(5, -5)$	**9.** $(4, 0)$	**10.** $(6, 0)$
11. $(0, 6)$	**12.** $(0, 5)$	**13.** $(-5, -2)$	**14.** $(-3, 3)$	**15.** $(0, -6)$
16. $(-2, 0)$	**17.** $(0, 0)$	**18.** $(0, -2)$	**19.** $(-3, 0)$	**20.** $(-6, -2)$

For Exercises 21–30, find the coordinates of each labeled point. See Example 2.

21. *A*	**22.** *B*	**23.** *C*	**24.** *D*	**25.** *E*
26. *F*	**27.** *G*	**28.** *H*	**29.** *I*	**30.** *J*

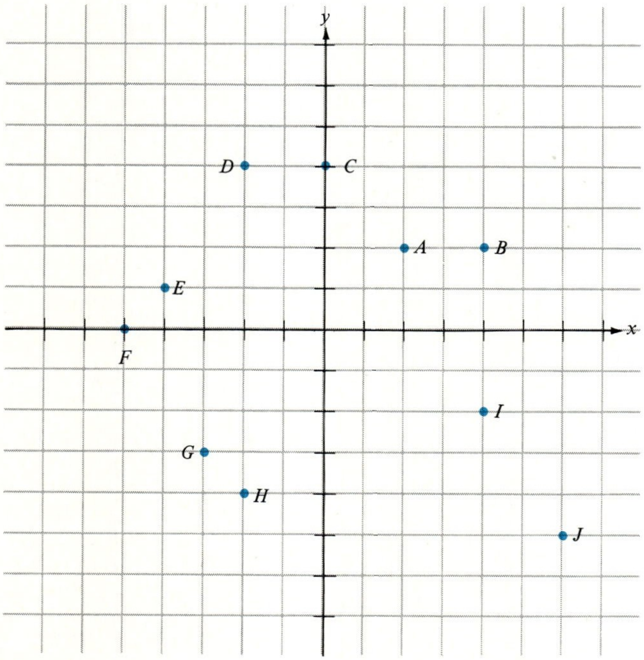

Solve Exercises 31–35.

31. In which quadrants do the coordinates of a point have
(a) Both signs negative? **(b)** Both signs positive? **(c)** The same sign?

32. In which quadrants does a point have
(a) negative abscissa, positive ordinate?
(b) positive abscissa, negative ordinate?
(c) coordinates with opposite signs?

33. **(a)** What is the x-coordinate of any point that is located on the y-axis?
(b) What is the y-coordinate of any point that is located on the x-axis?

34. The figure shown is a 3 by 3 square with sides parallel to the coordinate axes. The coordinates of the vertex A are given. Find the coordinates of the vertices labeled B, C, and D.

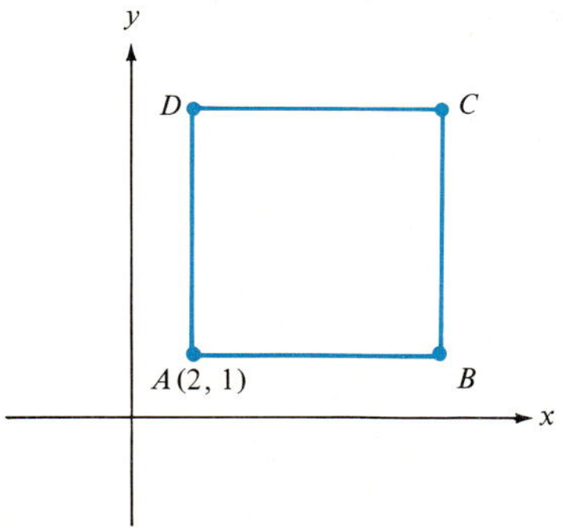

35. The figure shown is a 4 by 3 rectangle with sides parallel to the coordinate axes. The coordinates of vertex A are given. Find the coordinates of the vertices labeled B, C, and D.

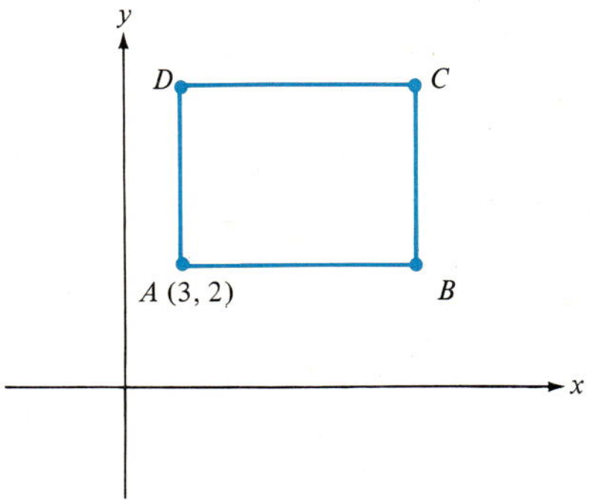

Linear Equations in Two Variables and Their Graphs 7.2

A *linear equation* in two variables, x and y, is an equation that can be put into the form

$$Ax + By = C,$$

where A, B, and C are real numbers (A and B are not both zero).

Examples of linear equations in two variables, x and y, are

$$3x + 2y = 5, \qquad 4x - y = 2, \qquad y = 3x + 2, \qquad x + y - 1 = 0.$$

We note that $y = 3x + 2$ and $x + y - 1 = 0$ are linear equations because using the rules of algebra, we can rewrite these equations in the form $Ax + By = C$.

A *solution* of a linear equation $Ax + By = C$ is an ordered pair of real numbers, (a, b), such that when x is replaced by a, and y is replaced by b, the result is a true statement. We will assume, unless otherwise stated, that the domain or replacement set for x and y is the set of real numbers.

In general, when an ordered pair of real numbers, (a, b), is a solution to an equation in two variables, we say that (a, b) *satisfies* the equation. The set of all ordered pairs, (a, b), that satisfy the equation is called the *solution set* of the equation.

To obtain a clear understanding of the terms defined above, let us apply them to a particular linear equation, such as $3x + y = 6$. We see that the ordered pair $(1, 3)$ is a *solution* to $3x + y = 6$ because

$$\begin{aligned} 3(1) + 3 &= 6 \\ 6 &= 6. \qquad \textit{a true statement} \end{aligned}$$

We can easily check that some other solutions to the equation, $3x + y = 6$, are $(0, 6)$, $(3, -3)$, and $\left(-\frac{2}{3}, 8\right)$.

A natural question to ask is: Are there other ordered pairs that satisfy the equation $3x + y = 6$ in addition to those given above? Also, if there are more solutions, what is the procedure for finding them?

We answer the questions raised above by noting first that the equation $3x + y = 6$ states a fixed relationship between x and y. Thus, if we assign a value to x or y, the value of the other variable is determined. For example, if we assign the value 2 to x in the equation $3x + y = 6$, the corresponding y-value is the solution of the equation

$$3(2) + y = 6.$$

Solving for y, we find that

$$\begin{aligned} 6 + y &= 6 \\ y &= 0. \end{aligned}$$

Thus the ordered pair $(2, 0)$ is also a solution to the equation $3x + y = 6$. Similarly, we may assign an arbitrary value to y in the equation $3x + y = 6$ and determine the corresponding x-value. For example, letting $y = 9$, we obtain the equation

$$3x + 9 = 6.$$

Solving the last equation for x, we obtain

$$\begin{aligned} 3x &= 6 - 9 \\ 3x &= -3 \\ x &= -1. \end{aligned}$$

Thus $(-1, 9)$ is another solution to $3x + y = 6$.

The procedure just discussed for finding solutions to the linear equation $3x + y = 6$ suggests that the equation has infinitely many solutions. In fact, a linear equation $Ax + By = C$, in two variables x and y, does have an *infinite number of solutions*.

Example 1. Find ordered pairs, (x, y), that are solutions to

$$2x + 3y = 8$$

for the given values of x or y.
(a) $x = 1$. (b) $y = 3$. (c) $x = -2$.

Solution

(a) If $x = 1$, we obtain

$$\begin{aligned} 2(1) + 3y &= 8 \\ 2 + 3y &= 8 \\ y &= 2. \end{aligned}$$

The ordered pair $(1, 2)$ is a solution.

(b) Let $y = 3$ in the equation $2x + 3y = 8$. We see that

$$\begin{aligned} 2x + 3(3) &= 8 \\ 2x + 9 &= 8 \\ 2x &= -1 \\ x &= -\frac{1}{2}. \end{aligned}$$

Thus $\left(-\frac{1}{2}, 3\right)$ satisfies the equation.

(c) For $x = -2$ we have

$$\begin{aligned} 2(-2) + 3y &= 8 \\ -4 + 3y &= 8 \\ 3y &= 12 \\ y &= 4. \end{aligned}$$

Therefore, $(-2, 4)$ is also a solution.

We have observed that the solution set of a linear equation, say $2x + y = 4$, contains an infinite number of ordered pairs of numbers (a, b). Therefore, we cannot list or write all the solutions of a linear equation such as $2x + y = 4$. However, since the solutions are ordered pairs of real numbers, they can be graphed or plotted on a rectangular coordinate system. By plotting just a few solutions of $2x + y = 4$, we will be able to obtain a geometric figure or "picture" of the entire solution set of $2x + y = 4$. To do this, we proceed as follows.

First we find some specific ordered pairs that satisfy the equation. It can easily be verified that

$$(-1, 6), \quad (0, 4), \quad (2, 0), \quad (3, -2), \quad (4, -4)$$

are all solutions to the equation $2x + y = 4$. The graphs of these ordered pairs are shown in Figure 7.7.

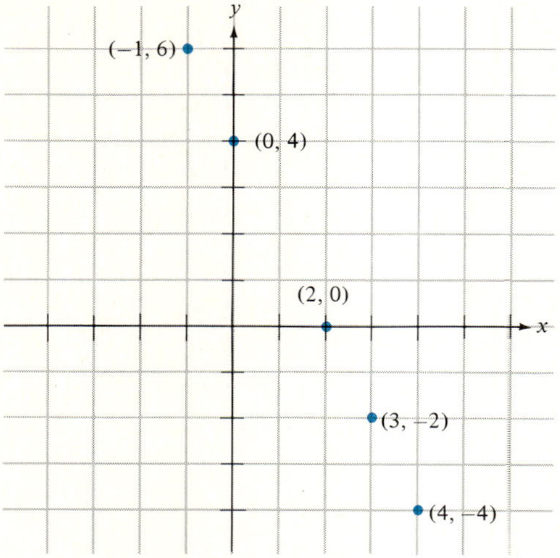

Figure 7.7

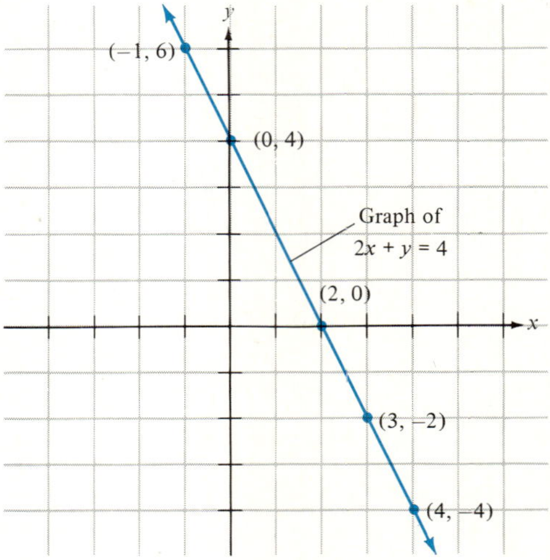

Figure 7.8

Observe that the points in Figure 7.7 appear to lie on a straight line. In fact, the graphs of all the solutions to the equation $2x + y = 4$ lie on a straight line passing through these points, as shown in Figure 7.8. Furthermore, *any point* on this line has coordinates that satisfy the equation $2x + y = 4$. Consequently, we say that the *straight line* is the *graph of* $2x + y = 4$. Thus the straight line shown in Figure 7.8 is the geometric figure representing all the solutions to the equation $2x + y = 4$. In general, we have the following theorem, which we state without proof.

Theorem 1

The *graph of a linear equation* in two variables x and y,

$$Ax + By = C,$$

where A, B, and C are real number constants (A and B not both 0),

is a straight line.

With respect to Theorem 1, we note that

1. Any point on the straight line has coordinates (a, b) that satisfy the equation $Ax + By = C$.
2. Any solution (a, b) to $Ax + By = C$ will plot onto the straight line.

We know from plane geometry that two distinct points determine a line. Thus, to obtain the graph of a linear equation in two variables in a rectangular coordinate system, we usually follow the steps listed next.

Steps for Graphing a Linear Equation

Step 1. Find two different ordered pairs that are solutions to the equation.

Step 2. Plot or graph the two ordered pairs on a rectangular coordinate system.

Step 3. Draw a straight line through the two points obtained in Step 2.

In addition to the steps outlined above, we usually find a *third solution* to the equation as a *checkpoint*. The graph of the third solution should be on the straight line that we have drawn. Sometimes we graph all three solutions at once because if the line that we have drawn does not pass through all three points, an error has been made in finding the solutions.

Example 2. Graph the equation $3x - y = 6$.

Solution: First we find two solutions to the given equation. We usually choose values of x or y to substitute into the equation so that the computations involved are relatively simple. Thus we

$$\text{Let } x = 0: \quad \begin{aligned} 3(0) - y &= 6 \\ 0 - y &= 6 \\ y &= -6. \end{aligned} \qquad \text{Let } y = 0: \quad \begin{aligned} 3x - 0 &= 6 \\ 3x &= 6 \\ x &= 2. \end{aligned}$$

Therefore, $(0, -6)$ and $(2, 0)$ are two solutions.

We now find a third solution as a checkpoint.

$$\text{Let } x = 1: \quad \begin{aligned} 3(1) - y &= 6 \\ -y &= 3 \\ y &= -3. \end{aligned}$$

We see that $(1, -3)$ is another solution.

Next, we plot the solutions $(0, -6)$, $(2, 0)$, and $(1, -3)$ on a rectangular coordinate system and draw the line passing through the corresponding points. The graph of $3x - y = 6$ is shown in Figure 7.9.

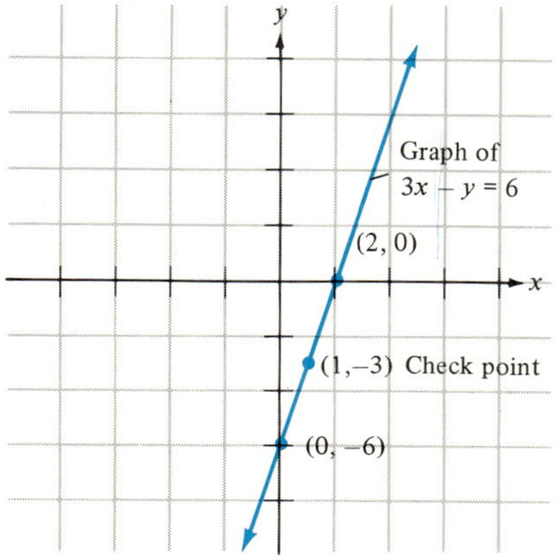

Figure 7.9

The x-coordinate of the point at which a graph intersects the x-axis is called the *x-intercept*. Similarly, the y-coordinate of the point where a graph intersects the y-axis is called the *y-intercept*. Thus we see that in Example 2, the x and y intercepts are 2 and -6, respectively.

As Example 2 illustrated, it is usually easier to plot the graph of a linear equation by finding the intercepts *provided* that it has two intercepts. In general, for a given linear equation, $Ax + By = C$:

1. To find the y-intercept, let $x = 0$, and solve for y.
2. To find the x-intercept, let $y = 0$, and solve for x.

In some special cases a linear equation has only one intercept, as discussed in the next example.

Example 3. Graph the following equations on a rectangular coordinate system.

$$x = 4 \qquad \text{and} \qquad y = -3.$$

Solution: The equation $x = 4$, considered as an equation in two variables, x and y, is equivalent to $x + 0 \cdot y = 4$. Note that for all values of y, $0 \cdot y = 0$. Thus the corresponding value of x is always 4. Some solutions to $x + 0 \cdot y = 4$ are (4, 0), (4, 1), and (4, 2). The graph of $x = 4$ is the *vertical line* shown in Figure 7.10. Note that there is *only* an x-intercept and that it equals 4.

Similarly, $y = -3$ is equivalent to the equation $0 \cdot x + y = -3$. When any number is substituted for x in the equation, the corresponding y value is -3. Some solutions to $0 \cdot x + y = -3$ are $(-2, -3)$, $(0, -3)$, and $(2, -3)$. The graph of $y = -3$ is the *horizontal line* shown in Figure 7.10. Notice also that the graph of this type of equation only has one intercept: the y-intercept, which is -3.

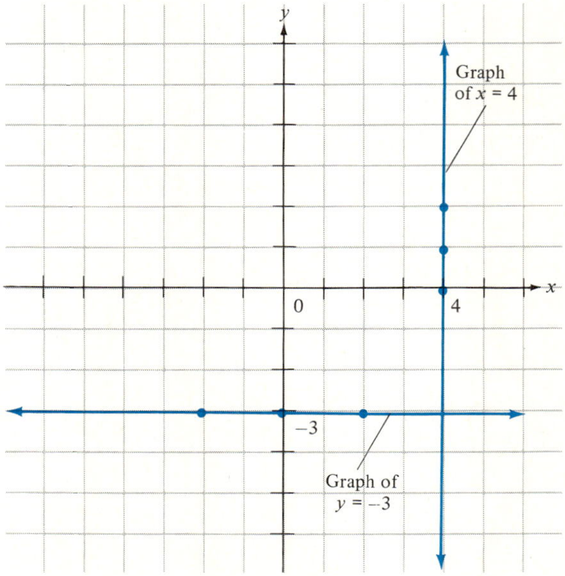

Figure 7.10

In the next example we graph a linear equation written in the form

$$y = mx + b \qquad m \text{ and } b \text{ are constants.}$$

We note that this form occurs quite frequently in mathematics, and we shall have more to say about it in the following sections of this chapter.

Example 4. Graph $y = 2x - 4$.

Solution: Observe that y is explicitly equal to an expression involving only x. Therefore, to find three solutions we simply assign values to x. Thus we could let $x = 0, 1, 2$ and then find the corresponding y values. The result is shown in the following table of values.

x	0	1	2
y	-4	-2	0

Plotting the ordered pairs $(0, -4)$, $(1, -2)$, and $(2, 0)$, we obtain the graph of $y = 2x - 4$ shown in Figure 7.11.

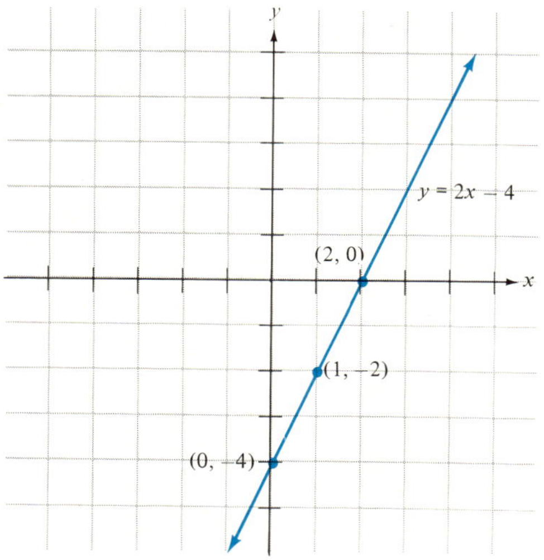

Figure 7.11

EXERCISES 7.2

For Exercises 1–14, find ordered pairs of numbers, (x, y), that satisfy the equation for the specified values of x or y. See Example 1.

1. $x + y = 3$ $\quad x = 1;\ y = 1;\ x = -1$

2. $x - y = 5$ $\quad x = -1;\ y = -3;\ x = 3$

3. $x - 3y = 6$ $\quad x = 0;\ y = 0;\ y = 1$

4. $4x - y = 8$ $\quad x = 0;\ y = 0;\ y = 4$

5. $3x + 2y = 5$ $\quad x = \frac{1}{3};\ y = -\frac{1}{2};\ y = -2$

6. $4x - 5y = 1$ $\quad x = \frac{1}{4};\ y = -\frac{1}{5};\ x = -1$

7. $y = 3x + 1$ $x = 0; x = 1; x = -1$

8. $y = 2x - 5$ $x = 1; x = 2; x = 3$

9. $x + 0 \cdot y = 3$ $y = 2; y = 4; y = -3$

10. $0 \cdot x + y = 7$ $x = -2; x = 2; x = 5$

11. $6x - 4y = 12$ $x = 0; y = 0; y = 3$

12. $7x + 2y = 14$ $x = 0; y = 0; y = -7$

13. $y = \frac{1}{2}x + 3$ $x = 0; x = 2; x = -2$

14. $y = \frac{2}{3}x - 1$ $x = 0; x = 3; x = 6$

For Exercises 15–34, graph the equation on a rectangular coordinate system. See Examples 2–4.

15. $x + y = 4$ **16.** $x - y = 2$ **17.** $2x - y = 4$ **18.** $2x + y = 6$

19. $x + 3y = 6$ **20.** $x - 3y = -6$ **21.** $x = 3$ **22.** $y = 2$

23. $y = -4$ **24.** $x = -2$ **25.** $y = 2x + 1$ **26.** $y = -2x + 1$

27. $y = 3x$ **28.** $y = 4x$ **29.** $y = \frac{1}{2}x$ **30.** $y = -\frac{1}{3}x$

31. $y - x = 0$ **32.** $y + x = 0$ **33.** $x = 0$ **34.** $y = 0$

For Exercises 35–42, write a linear equation expressing the relationship between the variables x and y. [HINT: *The word "is" translates symbolically into "=." Also review the special English phrases and their translation into mathematical expressions given in Section 2.7.*]

35. y is four times x.

36. y is twice x.

37. y is 3 more than twice x.

38. y is 5 less than twice x.

39. The sum of x and y is 4.

40. The difference of x and y is 7.

41. Three times y is 2 more than four times x.

42. Twice y is 5 less than twice x.

7.3 The Slope of a Straight Line

In the preceding section we saw that the graph of a linear equation is a straight line. Conversely, for any given straight line on a rectangular coordinate system there is a corresponding linear equation in two variables. In the next section we will discuss how to find the equation of a line given certain information about the line. However, to do this, we need to introduce the concept of the slope of a line. The *slope* of a line is a number that measures its steepness relative to the x-axis.

To motivate the definition of the slope of a line, we consider the line shown in Figure 7.12. The points $P(x_1, y_1)$ and $Q(x_2, y_2)$ represent any two points on the line. $P(x_1, y_1)$ represents a point with coordinates x_1 and y_1 (read "x sub-one and y sub-one"). Similarly, $Q(x_2, y_2)$ represents a second point Q with coordinates x_2 and y_2.

Notice in Figure 7.12 that as we move on the line from point P to point Q, the y-variable changes from y_1 to y_2. The change is given by the quantity $(y_2 - y_1)$. The corresponding change in the x-variable is the quantity $(x_2 - x_1)$. These changes in variables may be *positive,* representing an *increase; negative,* representing a *decrease;* or *zero,* representing neither an increase nor a decrease. Now if we compare the change

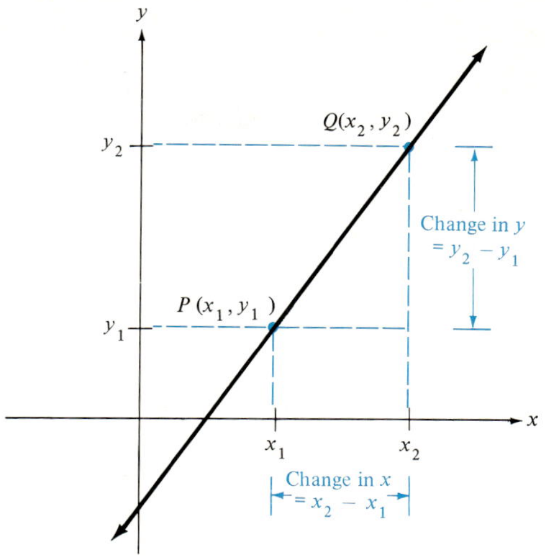

Figure 7.12

in y to the change in x, we can obtain an idea of the steepness of the line. We illustrate these observations graphically by considering two specific lines, such as those shown in Figures 7.13 and 7.14. Notice that both lines pass through the same point, $P(2, 1)$.

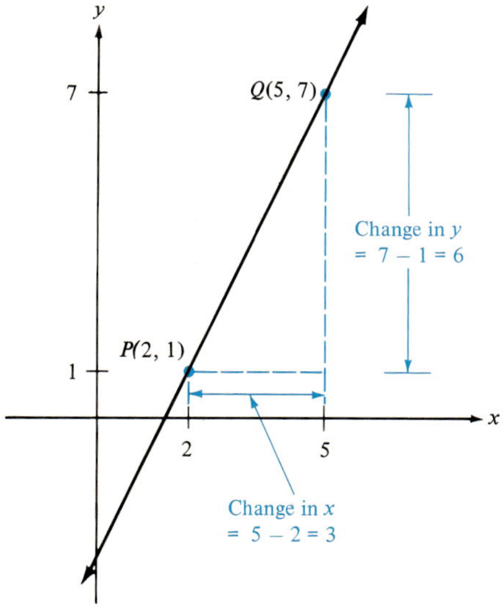

Figure 7.13

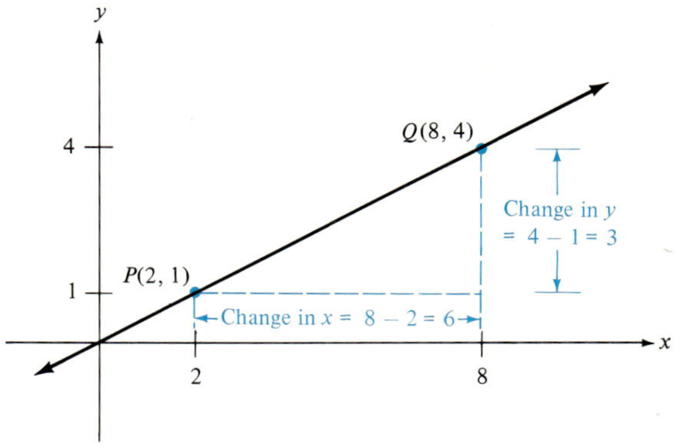

Figure 7.14

Now if we form the *ratio* of the change in y to the change in x for the "steeper" line in Figure 7.13, we obtain the following number:

$$\frac{\text{change in } y \text{ from } P \text{ to } Q}{\text{change in } x \text{ from } P \text{ to } Q} = \frac{7 - 1}{5 - 2} = \frac{6}{3} = 2.$$

Forming the same ratio for the line in Figure 7.14 that rises "slowly" from left to right, we obtain

$$\frac{\text{change in } y \text{ from } P \text{ to } Q}{\text{change in } x \text{ from } P \text{ to } Q} = \frac{4-1}{8-2} = \frac{3}{6} = \frac{1}{2}.$$

Thus we see that the ratio

$$\frac{\text{change in } y}{\text{change in } x} = \frac{6}{3} = \frac{2}{1} = 2$$

for the "steeper" line is a *larger* number than the corresponding ratio of $\frac{1}{2}$ for the "slowly rising" line. We note in connection with the previous discussion that the ratios that we obtained for each line would be the same no matter what two points we choose on the line. In general, if P and Q are two points on a line, the ratio

$$\frac{\text{change in the } y\text{-coordinate from } P \text{ to } Q}{\text{change in the } x\text{-coordinate from } P \text{ to } Q}$$

is *a number that enables us to measure* the "steepness" or "slope" of a line. It is customary to use the letter m to denote the slope of a line.

The previous discussion motivates the following definition for measuring the slope of a line:

Slope of a Line

If $P(x_1, y_1)$ and $Q(x_2, y_2)$ are any two points on a line, then the slope m of the line is

$$m = \frac{y_2 - y_1}{x_2 - x_1} = \frac{\text{change in } y}{\text{change in } x}; \qquad x_1 \neq x_2.$$

The formula for the slope may also be written in the form

$$m = \frac{y_1 - y_2}{x_1 - x_2} \qquad \text{provided that } x_1 \neq x_2.$$

The last formula can be obtained from the original formula, $m = \frac{y_2 - y_1}{x_2 - x_1}$, by multiplying the numerator and denominator by -1.

In the following examples we show how to compute the slope of a line given the coordinates of two points on the line. Furthermore, we shall see that the slope of a line may be positive, zero, negative, or not defined. To compute the slope, it is not necessary to graph the line; all that we need is the formula for the slope m. However, we show the graphs in each example in order to associate the sign of the slope with the way the line "rises" or "falls" as we move from left to right along the x-axis.

Example 1. Find the slope of the line passing through the points (3, 1) and (5, 4).

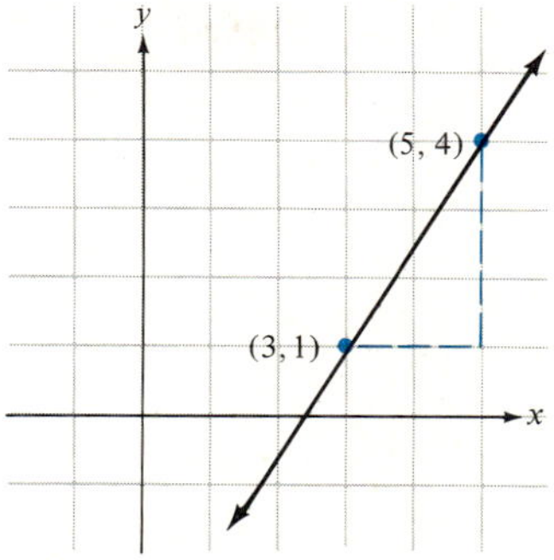

Figure 7.15

Solution: The line is shown in Figure 7.15. It does not matter which coordinates we label as (x_2, y_2) and which are labeled as (x_1, y_1), as long as we subtract consistently. Thus if we set

$$(x_2, y_2) = (5, 4) \qquad \text{and} \qquad (x_1, y_1) = (3, 1),$$

then

$$x_2 = 5,\ y_2 = 4 \qquad \text{and} \qquad x_1 = 3,\ y_1 = 1.$$

Substituting these values in the formula for the slope, we see that

$$m = \frac{y_2 - y_1}{x_2 - x_1} = \frac{4 - 1}{5 - 3} = \frac{3}{2}.$$

Thus the slope of the line is $\frac{3}{2}$. This ratio means that an *increase* of 2 units in x, as we move from one point on the line to another point, is accompanied by an *increase* of 3 units in y.

Example 1 illustrates that lines that rise from *left to right* have *positive slopes*. Conversely, lines with positive slopes are lines that rise from left to right.

Example 2. Find the slope of line passing through the points $(-3, 4)$ and $(2, 1)$.

Solution: Let $(x_2, y_2) = (2, 1)$ and $(x_1, y_1) = (-3, 4)$. Substituting in the slope formula, we obtain

$$m = \frac{1 - 4}{2 - (-3)} = \frac{-3}{5}.$$

Thus, for each *increase* of 5 units $(+5)$ in x, as we move from one point on the line to another point on the line, there is a corresponding *decrease* of 3 units (-3) in y, as illustrated in Figure 7.16.

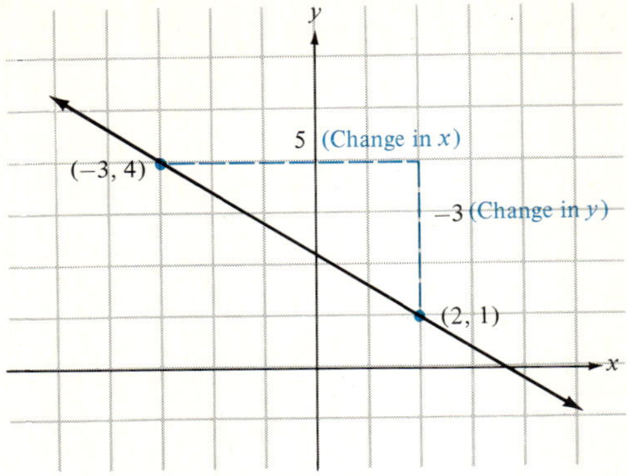

Figure 7.16

In Example 2, as we move from left to right the line is "falling" and has a negative slope. In general, lines that fall from *left to right* have *negative slopes,* and vice versa.

Example 3. Find the slope of
(a) The horizontal line passing through the points (−2, 4) and (3, 4).
(b) The vertical line passing through the points (3, 4) and (3, −2).

Solution: The two lines are shown in Figure 7.17.

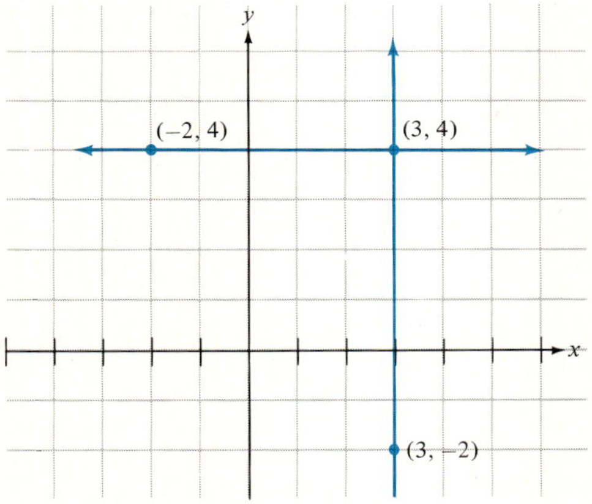

Figure 7.17

(a) Set $(x_2, y_2) = (-2, 4)$ and $(x_1, y_1) = (3, 4)$. Then

$$m = \frac{y_2 - y_1}{x_2 - x_1} = \frac{4 - 4}{-2 - 3} = \frac{0}{-5} = 0.$$

The slope of the given horizontal line is 0.
(b) If $(x_2, y_2) = (3, 4)$ and $(x_1, y_1) = (3, -2)$, then

$$m = \frac{y_2 - y_1}{x_2 - x_1} = \frac{4 - (-2)}{3 - 3} = \frac{6}{0} \quad \textit{is not defined.}$$

Remember that division by 0 is not defined. Thus we see that the *slope* of the given *vertical* line *is not defined*.

The results obtained in Example 3 generalize to all vertical and horizontal lines. As a result, we have the following statements.

> All horizontal lines have slope $m = 0$.
>
> The slope of a vertical line is not defined.

We emphasize that horizontal lines do have a slope defined for them: it is 0, a perfectly good number. We might also mention that the larger the number—exclusive of sign—associated with the slope, the more nearly vertical the line.

A summary of the previous discussion relative to slopes is shown in Figure 7.18.

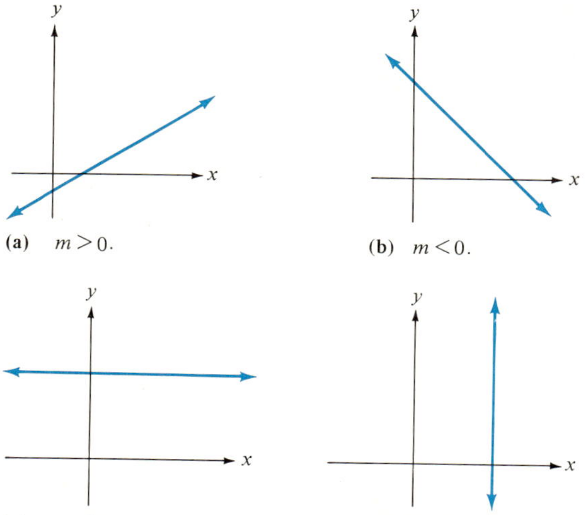

(a) $m > 0$. (b) $m < 0$. (c) $m = 0$. (d) m is undefined.

Figure 7.18

We conclude this section with a discussion of the slopes of parallel and perpendicular lines. In plane geometry two lines, designated L_1 and L_2, are

1. *Parallel* ($L_1 \parallel L_2$) if and only if the lines L_1 and L_2 *never intersect*.
2. *Perpendicular* ($L_1 \perp L_2$) if and only if the lines L_1 and L_2 *intersect at right angles*.

It can be shown that the following statements are true for any two lines L_1 and L_2 *having* slopes m_1 and m_2, respectively.

L_1 is parallel to L_2 ($L_1 \parallel L_2$) if and only if $m_1 = m_2$.
L_1 is perpendicular to L_2 ($L_1 \perp L_2$) if and only if $m_1 \cdot m_2 = -1$.

Figure 7.19 illustrates these statements.

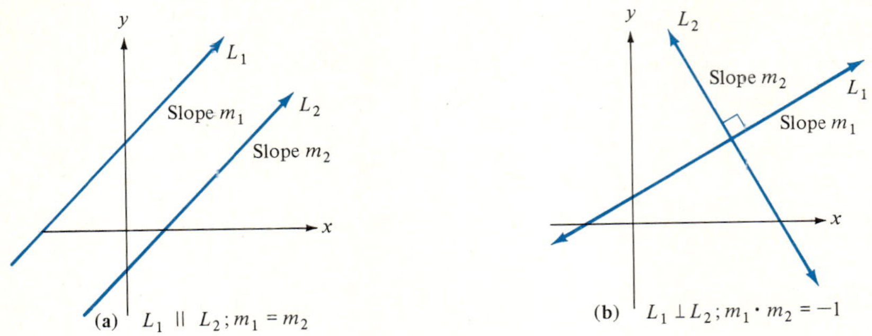

Figure 7.19

Example 4. Line L passes through the points (2, 4) and (3, 8).
(a) Find the slope of a line M that is parallel to line L.
(b) Find the slope of a line N that is perpendicular to line L.

Solution: First we determine the slope of line L:

$$\text{slope of line } L = m = \frac{8-4}{3-2} = \frac{4}{1} = 4.$$

(a) Since line M is parallel to line L, it has the same slope, 4.
(b) Let us designate the slope of line N, m_1. Since $N \perp L$, we have

$$m_1 \cdot 4 = -1 \qquad \text{or} \qquad m_1 = \frac{-1}{4} = -\frac{1}{4}.$$

We see from Example 4(b) that, alternatively, we may describe the relationship between the slopes of perpendicular lines by saying that they are *negative reciprocals*.

EXERCISES 7.3

For Exercises 1–14, when possible compute the slope of the line passing through the pair of points. Identify any lines that do not have a slope. See Examples 1–3.

1. (1, 2), (4, 5) **2.** (3, 7), (4, 9) **3.** (6, 1), (8, 4)

4. (2, 6), (5, 10) **5.** $(\frac{1}{2}, \frac{5}{4}), (\frac{3}{2}, \frac{2}{4})$ **6.** $(\frac{5}{3}, \frac{1}{5}), (\frac{8}{3}, \frac{3}{5})$

7. (−3, 5), (3, −1) **8.** (−4, 2), (1, −8) **9.** (9, 2), (6, 2)

10. (−10, 5), (−10, 7) **11.** (−5, 1), (−5, 3) **12.** (7, −4), (10, −4)

13. (0, 2), (5, 0) **14.** (−8, 0), (0, 4)

For Exercises 15–20, for the line passing through the specified points, determine the slope of any line (a) parallel to the given line; (b) perpendicular to the given line. See Example 4.

15. (1, 3), (2, 6) **16.** (−2, −3), (0, 7) **17.** (5, −2), (8, −8)

18. (1, −1), (3, −9) **19.** (−5, 0), (0, 3) **20.** (−2, 0), (0, −3)

For Exercises 21–24, answer the same questions given for Exercises 15–20. However, be careful, since the slope may not always be defined.

21. (1, 3), (2, 3) **22.** (6, 1), (6, 3) **23.** (7, 3), (7, 6) **24.** (−4, 2), (5, 2)

For Exercises 25–30, find the slope of the line shown in the figure.

25.

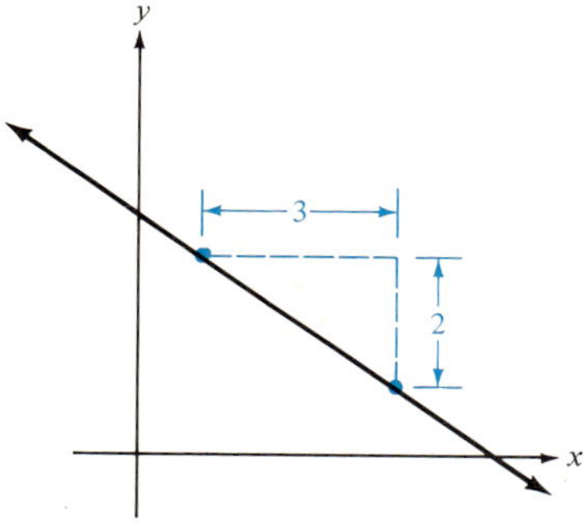

26.

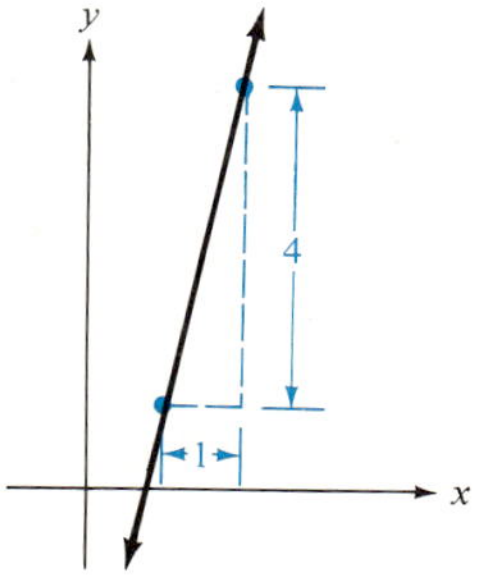

27.

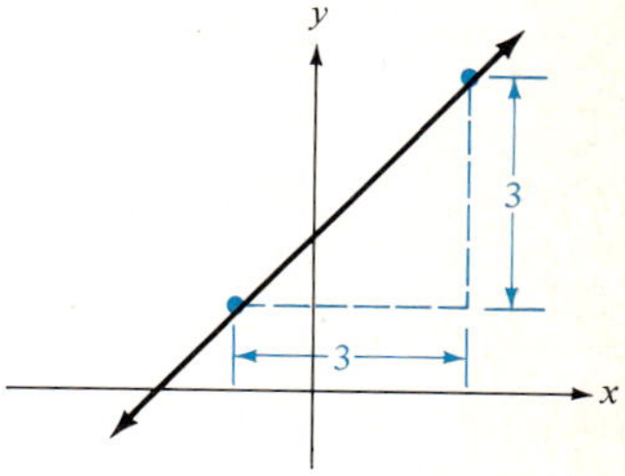

28.

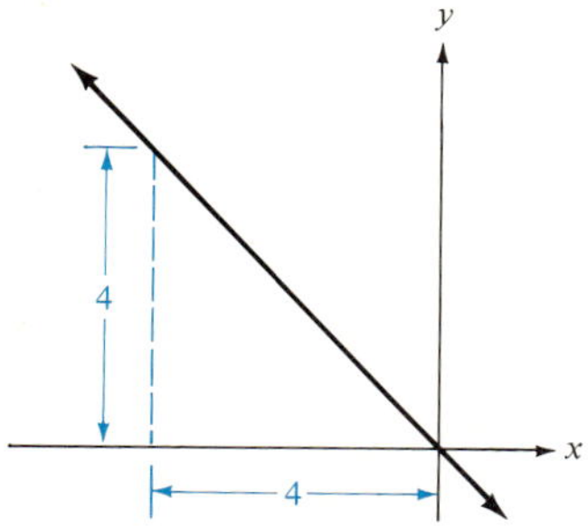

29.

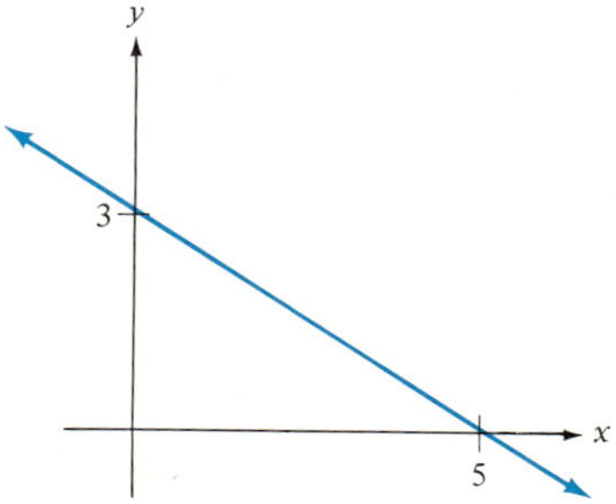

30.

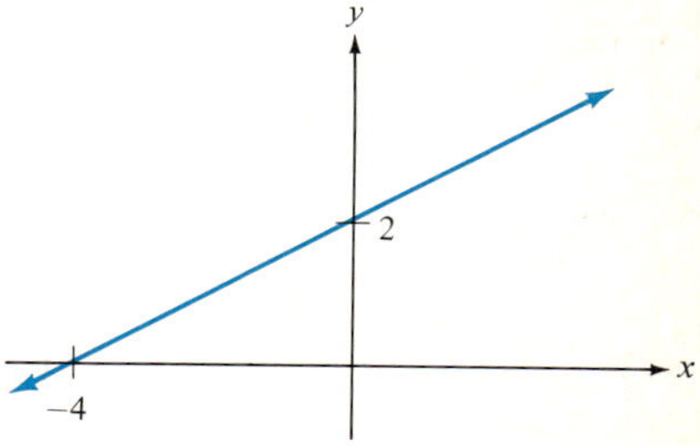

For Exercises 31–38, find two solutions of the linear equation and then determine the slope of the corresponding graph. Do not graph the equation.

31. $y = 2x$ **32.** $y = -3x$ **33.** $y = -4x + 1$ **34.** $y = 5x - 3$

35. $y = \frac{3}{2}x$ **36.** $y = -\frac{1}{2}x$ **37.** $y = \frac{2}{3}x - 1$ **38.** $y = -\frac{5}{2}x + 3$

Formulas for the Equation of a Line 7.4

In this section we discuss how to find the equation of a line given certain information about the line. What sort of information do we need about a line in order to write its equation? To answer this question, let us consider the line L in Figure 7.20.

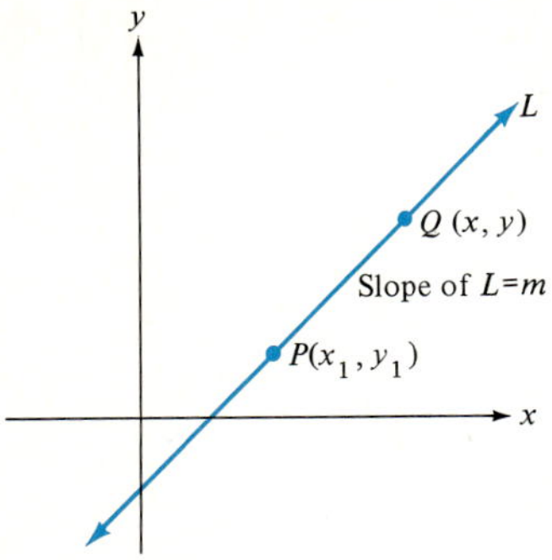

Figure 7.20

Figure 7.20 indicates that we know the following information about line L:

1. It passes through point P with known coordinates (x_1, y_1).
2. The slope of the line is m.

The point $Q(x, y)$ in the figure represents any point on the line L with general coordinates (x, y). To write the equation of line L means to write an equation relating the coordinates x and y of *any* point on the line. To determine this equation, we use the *known* point (x_1, y_1) and slope m together with the slope formula. Remember that (x_1, y_1) and (x, y) are coordinates of points *on* the line L. Thus they satisfy the formula for the slope of the line as follows:

$$\frac{y - y_1}{x - x_1} = m \qquad x \neq x_1.$$

Multiplying both sides of the equation above by $(x - x_1)$, we obtain the equation of the line L:

$$y - y_1 = m(x - x_1).$$

Note that in the last equation we removed the restriction $x \neq x_1$, since (x_1, y_1) satisfies the equation.

Summarizing, if we know

1. the coordinates (x_1, y_1) of a point on a line, and
2. the slope, m of the line,

then the equation of the line is given by the following formula, called the *point-slope form* of the equation of a straight line.

Point-Slope Form

$$y - y_1 = m(x - x_1).$$

We note that many times in order to be brief we identify the equation of a line with the line itself. Thus we often use the phrase "the line $2x + 3y = 6$" to mean "the line represented by the equation $2x + 3y = 6$."

Example 1. Find the equation of the line passing through the point (2, 4) with a slope of 3.

Solution: We use the point-slope form of the equation of a straight line. First we set

$$x_1 = 2, \qquad y_1 = 4, \qquad m = 3.$$

Then substituting in the formula

$$y - y_1 = m(x - x_1),$$

we obtain

(1) $$y - 4 = 3(x - 2).$$

We write equation (1) in simpler form as follows:

$$y - 4 = 3x - 6 \qquad \text{distributive property}$$

(2) $$y = 3x - 2. \qquad \text{add 4 to both sides}$$

Either equation (1) or (2) is the desired equation. However, equation (2) is usually the preferred form.

Suppose that we are given a line with slope m and point $(0, b)$ where the line intersects the y-axis, as shown in Figure 7.21. Recall that b is the y-intercept of the line. Given point $(0, b)$ and slope m, we can use the point-slope form of a line to obtain

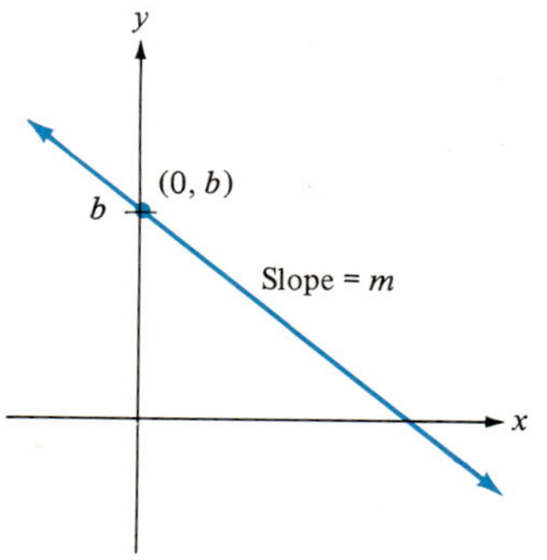

Figure 7.21

a very simple and useful form of the equation of a straight line. To this end we let

$$(x_1, y_1) = (0, b) \qquad \text{or} \qquad x_1 = 0 \quad \text{and} \quad y_1 = b.$$

Remember that we know that the slope of the line is m. Substituting in the point-slope form

$$y - y_1 = m(x - x_1),$$

we obtain

$$y - b = m(x - 0) \qquad \text{since} \qquad x_1 = 0; \quad y_1 = b.$$

Solving for y, we obtain

$$y - b = mx$$
$$y = mx + b. \qquad \text{add } b \text{ to both sides}$$

Conversely, it can be shown that if the equation of a straight line is expressed in the form $y = mx + b$, then the slope of the line is m and the y-intercept is b. Thus $y = mx + b$ is called the *slope-intercept form* of the equation of a straight line.

To summarize:

If a line has slope m and y-intercept b, its equation can be written as

Slope-Intercept Form

$$y = mx + b.$$

Example 2. Find the equation of the line with slope $-\frac{3}{2}$ and y-intercept 5.

Solution: We use the slope-intercept form

$$y = mx + b$$

with $m = -\frac{3}{2}$ and $b = 5$. Substituting for m and b in the formula above, we obtain the desired equation:

$$y = -\frac{3}{2}x + 5.$$

We observe that most of the time we express the equation of a line in the slope-intercept form, $y = mx + b$. Note that the *coefficient of* x in this form *is the slope of the line,* and the *constant term is the y-intercept*. Thus the slope-intercept form is important because it enables us to determine the slope of a line from its equation. For example, suppose that we are given the equation of a line in the form $ax + by = c$, provided that $b \neq 0$, and wish to determine the slope of this line. We can do this by rewriting the equation in slope-intercept form, as shown in the next example.

Example 3. Find the slope of the line represented by the equation

$$6x - 2y = 4.$$

Solution: We rewrite the equation in slope-intercept form, $y = mx + b$. To do this, we may solve for y as follows:

$$-2y = -6x + 4 \qquad \text{add } -6x \text{ to both sides}$$
$$y = 3x - 2. \qquad \text{divide both sides by } -2$$

The last equation is the slope-intercept form, $y = mx + b$, of the original equation.

In this form we see that the slope, m, of the given line is 3. Notice also that the y-intercept is -2.

The following example illustrates how to find the equation of a line given two distinct points on the line.

Example 4. Find the equation of the line passing through the points (2, 1) and (6, 4).

Solution: First we use the two points to compute the slope of the line. Then using the slope, together with either one of the two given points, we utilize the *point-slope form* to find the equation of the line. Following this procedure, we let

$$(x_1, y_1) = (2, 1) \qquad \text{and} \qquad (x_2, y_2) = (6, 4).$$

Thus

$$m = \frac{y_2 - y_1}{x_2 - x_1} = \frac{4 - 1}{6 - 2} = \frac{3}{4}.$$

Next, using the point-slope form

$$y - y_1 = m(x - x_1)$$

with $(x_1, y_1) = (2, 1)$ and $m = \frac{3}{4}$, we obtain

$$y - 1 = \frac{3}{4}(x - 2).$$

We express the last equation in a simpler form by rewriting it in slope-intercept form. As a result, we have

$$y = \frac{3}{4}x - \frac{1}{2}.$$

If we had substituted (6, 4) in the point-slope form rather than (2, 1), we would get

$$y - 4 = \frac{3}{4}(x - 6).$$

This equation also simplifies to $y = \frac{3}{4}x - \frac{1}{2}$. This result illustrates that *no matter which point we select,* we get the same slope-intercept form of the equation.

Now we consider the equations of horizontal and vertical lines. Consider Figure 7.22 with a horizontal line and a vertical line passing through an arbitrary point (a, b). Recall that all horizontal lines have slope 0. If we were to use the point-slope form for the horizontal line passing through (a, b) with slope $m = 0$, we would obtain the following general rule.

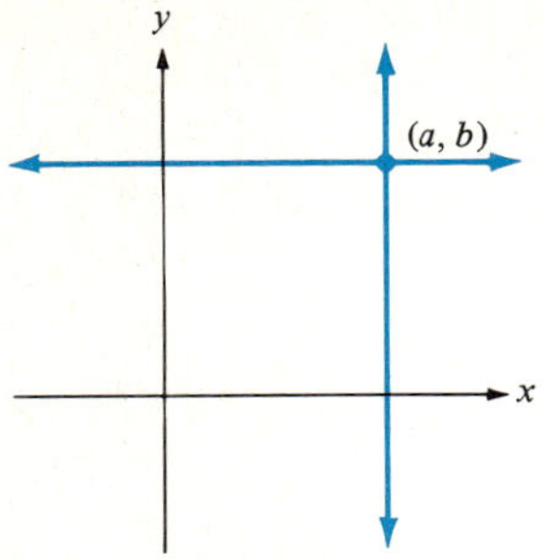

Figure 7.22

An equation of the *horizontal line* through (a, b) is

$$y = b.$$

Recall that vertical lines *do not* have slopes defined for them. However, vertical lines *do* have equations. In general,

An equation of the *vertical line* through (a, b) is

$$x = a.$$

The following example illustrates the application of the last two rules to particular horizontal and vertical lines.

Example 5. Find the equations of the lines passing through the points.
(a) $(-4, 7)$, $(5, 7)$. (b) $(3, 2)$, $(3, -4)$.

Solution

(a) Notice that distinct points with the *same y-coordinates* determine a *horizontal line*. Thus the two points determine a horizontal line. The equation of the horizontal line passing through $(-4, 7)$ and $(5, 7)$ is

$$y = 7.$$

(b) Two distinct points with the *same x-coordinates* determine a *vertical line*. Thus the line passing through the points $(3, 2)$ and $(3, -4)$ is a vertical line. Therefore, the equation of the line is

$$x = 3.$$

Example 6. The equation of a line L is $6x - 3y = 12$. Find the equation of the line
(a) Parallel to line L and passing through the point $(-3, 5)$.
(b) Perpendicular to line L and passing through the point $(4, 3)$.

Solution: First, we find the slope of line L by putting its equation into slope-intercept form. As a result, we obtain $y = 2x - 4$. Thus we see that the slope of line L is 2.

(a) We know that parallel lines have the same slope. Therefore, a line parallel to L must also have slope 2. To find the required equation, we set $m = 2$ and $(x_1, y_1) = (-3, 5)$ in the point-slope form

$$y - y_1 = m(x - x_1).$$

As a result, we obtain the required equation,

$$y - 5 = 2[x - (-3)] \quad \text{or} \quad y - 5 = 2(x + 3).$$

The last equation can be expressed in simpler form as

$$y = 2x + 11.$$

(b) Nonvertical perpendicular lines have slopes that are negative reciprocals of each other. Therefore, the line passing through $(4, 3)$ perpendicular to the given line L with slope 2 must have slope $-\frac{1}{2}$. Using the point-slope form, we find that the equation of this line is

$$y - 3 = -\frac{1}{2}(x - 4).$$

The last equation written in slope-intercept form is

$$y = -\frac{1}{2}x + 5.$$

EXERCISES 7.4

For Exercises 1–12, find the equation of the line passing the point with the specified slope m. See Example 1.

1. $(1, 2)$; $m = 5$
2. $(2, 7)$; $m = 4$
3. $(-1, 5)$; $m = -3$
4. $(-1, 5)$; $m = -2$
5. $(-3, 0)$; $m = -1$
6. $(-1, -3)$; $m = 8$
7. $(6, -1)$; $m = -\frac{1}{3}$
8. $(4, 2)$; $m = -\frac{1}{4}$
9. $(5, -3)$; $m = -\frac{2}{3}$
10. $(-9, -2)$; $m = \frac{1}{5}$
11. $(\frac{1}{5}, 1)$; $m = 3$
12. $(\frac{1}{7}, \frac{5}{3})$; $m = 7$

For Exercises 13–24, find the equation of the line with the given slope, m, and y-intercept, b. See Example 2.

13. $m = 2, b = 3$
14. $m = 8, b = 1$
15. $m = 1, b = -3$
16. $m = -9, b = 1$

17. $m = 0, b = 8$ **18.** $m = \frac{1}{2}, b = 0$ **19.** $m = -\frac{3}{5}, b = -2$ **20.** $m = 6, b = -4$

21. $m = .25, b = .3$ **22.** $m = -.4, b = -.15$ **23.** $m = -.72, b = .8$ **24.** $m = .7, b = -.8$

For Exercises 25–36, find the slope of the line represented by the equation, when possible. See Example 3.

25. $4x + y = 3$ **26.** $9x + y = -1$ **27.** $10x - 5y = 20$ **28.** $9x - 3y = 12$

29. $6x + 2y = -6$ **30.** $4x + 4y = -16$ **31.** $5x - 2y = 6$ **32.** $7x - 3y = 15$

33. $y = 3$ **34.** $x = 2$ **35.** $x + y = 8$ **36.** $5x + 3y = 15$

For Exercises 37–50, find the equation of the line passing through the two points. See Examples 4 and 5.

37. (1, 2), (2, 5) **38.** (1, 1), (3, −3) **39.** (2, −4), (−1, 11)

40. (3, 7), (2, 3) **41.** (0, −3), (−2, −5) **42.** (3, 5), (6, 7)

43. (5, 0), (0, 3) **44.** (0, 2), (2, 0) **45.** (5, −3), (5, 2)

46. (7, 3), (9, 3) **47.** (6, −8), (7, −8) **48.** (−7, 2), (−7, 3)

49. (3, 0), (−4, 0) **50.** (0, 4), (0, 5)

For Exercises 51–56, find the equation of the line satisfying the stated conditions. See Example 6.

51. The line passing through (1, 5) and parallel to $y = 3x - 1$.

52. The line passing through (−1, 4), perpendicular to $y = \frac{1}{2}x + 4$.

53. The line passing through (0, 3), perpendicular to a line with slope $-\frac{5}{2}$.

54. The line passing through (0, 5) parallel to a line with slope −6.

55. The line parallel to $4x - 2y = 10$ containing the point (3, 2).

56. The line perpendicular to $x + 5y = 15$ containing the point (2, 3).

For Exercises 57–64, find the x and y intercepts of the equation, then graph the equation. The form of the equation given is called the intercept form.

57. $\frac{x}{2} + \frac{y}{3} = 1$ **58.** $\frac{x}{3} + \frac{y}{5} = 1$ **59.** $\frac{x}{-4} + \frac{y}{3} = 1$ **60.** $\frac{x}{5} + \frac{y}{-2} = 1$

61. $\frac{x}{-3} + \frac{y}{-4} = 1$ **62.** $\frac{x}{-2} + \frac{y}{-3} = 1$ **63.** $\frac{x}{4} + \frac{y}{4} = 1$ **64.** $\frac{x}{-5} + \frac{y}{-5} = 1$

Solve Exercises 65 and 66 using the definition of the slope of a line. You should be able to solve them without any computation.

65. The formula $F = \frac{9}{5}C + 32$ relates the Celsius and Fahrenheit temperature scales. If there is a temperature change of 5° on a Celsius thermometer, what is the corresponding change on a Fahrenheit thermometer?

66. The total cost C of producing x units of a certain type battery is given by $C = \frac{3}{4}x + 800$. The number $\frac{3}{4}$ represents the *production cost* in dollars *per battery,* and 800 represents fixed costs in dollars. For every four batteries produced, what is the corresponding *production cost?*

Graphing Linear Inequalities in Two Variables 7.5

In this section we discuss the graphs of the solution set of first-degree inequalities *in two variables*. A few examples of *first-degree* or *linear inequalities in two variables* are as follows:

$$x + y \leq 4, \qquad y < 2x - 3, \qquad y \geq x + 1.$$

Because there are two variables involved, we will graph the solution sets on a rectangular coordinate system.

In general, a linear inequality in two variables x and y is an equality that can be put in the form

$$Ax + By \leq C,$$

where A, B, and C are real numbers (A and B not both 0). In the definition above, the inequality sign $\leq$ may be replaced by any one of the inequality symbols $<$, $\geq$, $>$.

A *solution to an inequality* in two variables, x and y, is an ordered pair of numbers (a, b) such that when x and y are replaced by a and b, respectively, the resulting statement is true. The ordered pair (a, b) is said to *satisfy* the inequality.

The *graph of an inequality in two variables* is the set of all points on a rectangular coordinate system whose coordinates (a, b) satisfy the inequality.

The following example illustrates a technique used to graph linear inequalities in two variables.

Example 1. Graph the inequality $-2x + y \leq 4$.

Solution: To graph the inequality, we begin by graphing the equation $-2x + y = 4$. The graph is a line in the rectangular coordinate plane. Note in Figure 7.23 that the line divides the plane into two *regions* or *half-planes*. One half-plane lies above the line, the other below the line. The line is called the *boundary* of each of the half-planes. It is *not* part of either half-plane. This is illustrated in Figure 7.23.

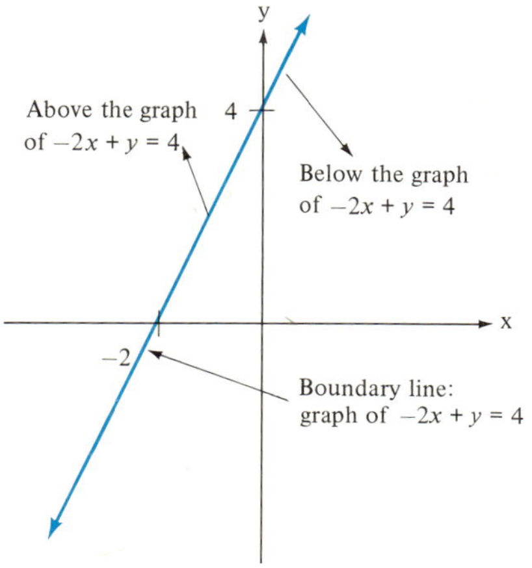

Figure 7.23

One of the half-planes represents the graph of the restricted inequality $-2x + y < 4$. The other half-plane contains points whose coordinates do *not* satisfy the inequality. All points on the line have coordinates that satisfy $-2x + y = 4$. Thus the *boundary line with one of the half-planes is the graph* of $-2x + y \leq 4$. To determine which half-plane to choose, we need select only a *test point* in one of the half-planes. If the coordinates of the test point satisfy the inequality, we have chosen the correct half-plane. If the coordinates of the test point do not satisfy the inequality, we select the other half-plane. Thus, for $-2x + y \leq 4$, we choose a convenient point such as the origin, (0, 0). Usually, the origin is a good point to choose as a test point. Replacing both x and y with 0 in $-2x + y \leq 4$, we obtain

$$-2 \cdot 0 + 0 \leq 4 \quad \text{or} \quad 0 \leq 4. \qquad \text{a true statement}$$

Thus the graph of $-2x + y \leq 4$ consists of the half-plane below the line (containing the origin), together with the boundary line. We indicate this by shading the region below the line, as shown in Figure 7.24.

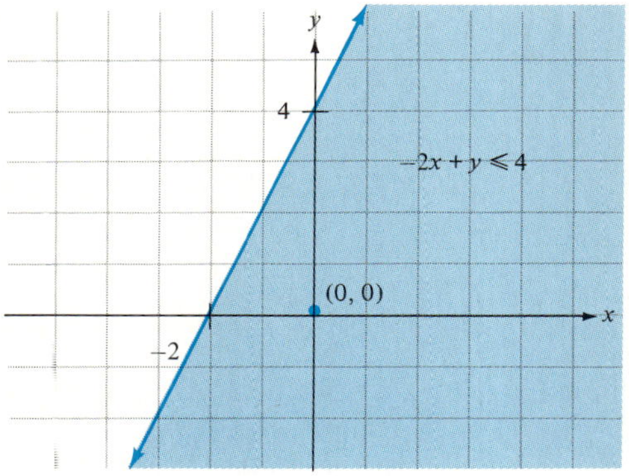

Figure 7.24

Note in Example 1 that the shaded region in Figure 7.24 extends indefinitely and is not confined to the page.

We apply the techniques developed in Example 1 in the following example.

Example 2. Graph $y > -3x$.

Solution: First we graph $y = -3x$, which is the boundary line. We *cannot* choose (0, 0) as a test point, since it is on the boundary line and therefore does not lie in either one of the half-planes. Therefore, we choose another test point. Since the inequality is $y > -3x$, we choose a point above the boundary line, such as (0, 2). Substituting these coordinates in the inequality, we obtain

$$2 > -3(0) \quad \text{or} \quad 2 > 0. \qquad \textit{a true statement}$$

Note that the original inequality, $y > -3x$, does not include the symbol of equality.

Thus points on the boundary line are *not* part of the graph of $y > -3x$. We indicate this in the graph by using *dashed* lines for the boundary, as shown in Figure 7.25.

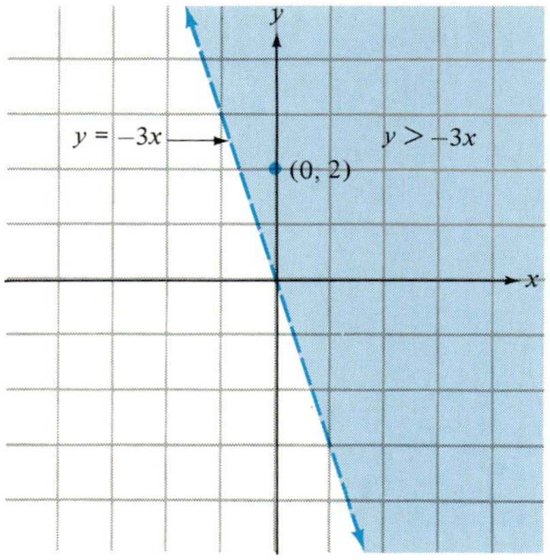

Figure 7.25

Example 3. Graph the inequality $x < 4$ on a rectangular coordinate system.

Solution: If we write the $x < 4$ in the form $x + 0 \cdot y < 4$, we see that it represents an inequality in two variables. We graph the corresponding equation $x + 0 \cdot y = 4$ or $x = 4$, which is a vertical line. Because the inequality does not include $x = 4$, we draw a dashed line. In this case the half-plane lies either to the left or to the right of the boundary line, $x = 4$. Notice that all points (x, y) with $x < 4$ lie to the left of the line. Using (0, 0) as a test point confirms that this is so. The graph of $x < 4$ is shown in Figure 7.26.

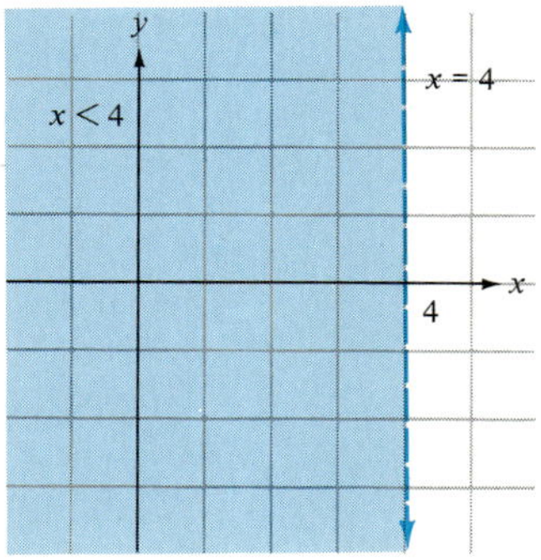

Figure 7.26

In general, to *graph the solution set of a linear equality in two variables, such as,*

$$Ax + By \leq C,$$

do the following.

1. Replace the symbol of inequality ($\leq$, $<$, $\geq$, $>$) with the equality symbol, $=$. Graph the resulting equation, which is a straight line. This is the boundary line between the two half-planes. Draw a solid line if the inequality symbol is $\leq$ or $\geq$. If the symbol is $<$ or $>$, draw a dashed line.
2. Choose a test point in one of the two half-planes. Do not choose as a test point a point on the boundary line.
3. If the coordinates of the test point satisfy the given inequality, "shade in" the half-plane that the point lies in. If the coordinates do not satisfy the inequality, shade in the other half-plane.

EXERCISES 7.5

For Exercises 1–40, graph the linear inequality. See Examples 1–3.

1. $x + y \leq 4$	**2.** $x + y > 3$	**3.** $x - y > 2$	**4.** $x - y \leq 5$
5. $y \geq x$	**6.** $y < 2x$	**7.** $-3x + y < 6$	**8.** $2x + y \geq 4$
9. $x < 5$	**10.** $x \geq -4$	**11.** $y \geq 3$	**12.** $y < -2$
13. $x + 2y > 4$	**14.** $x - 3y < 3$	**15.** $y < 3x + 1$	**16.** $y \geq 4x + 2$
17. $3x - 2y \geq 6$	**18.** $2x - 4y < 8$	**19.** $x \geq 0$	**20.** $y \leq 0$
21. $y \geq 0$	**22.** $x \leq 0$	**23.** $-3x + 5y \leq 15$	**24.** $4x + 3y \geq 12$
25. $2x + 3y < 6$	**26.** $7x - 2y > 14$	**27.** $3x - 7y > 21$	**28.** $5x - 3y < 15$
29. $3x - 8y \geq 24$	**30.** $11x + 2y \leq 22$	**31.** $2x - 5y \leq 10$	**32.** $4x - 7y \geq 28$
33. $5x + 6y \geq 30$	**34.** $3x - 4y \leq 24$	**35.** $5x - 8y < 40$	**36.** $9x - 4y > 36$
37. $3x - 2y < 18$	**38.** $6x - 5y > 30$	**39.** $3x + 6y \geq 24$	**40.** $5x + 7y \leq 35$

Key Words

Graph of a linear equation
Horizontal line
Linear equation
- Point-slope form
- Slope-intercept form

Linear inequality
- Boundary
- Half-plane

Parallel lines
Perpendicular lines
Rectangular (Cartesian) coordinate system
- Abscissa
- Ordered pair
- Ordinate
- Origin
- Quadrant
- *x*-axis; *y*-axis
- *x*-coordinate; *y*-coordinate

Slope
Vertical line
x-intercept
y-intercept

CHAPTER 7 REVIEW

SECTION 7.1

Plot the following ordered pairs of numbers on the same rectangular coordinate system.

1. (3, 4) **2.** (−2, 5) **3.** (−3, −3) **4.** (4, −2) **5.** (−2, 1)

Find the coordinates of each of the labeled points shown in the figure.

6. *A* **7.** *B* **8.** *C* **9.** *D* **10.** *E*

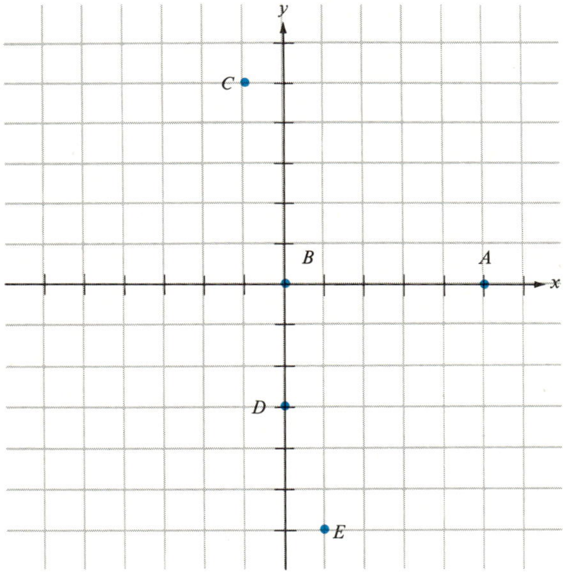

11. What are the coordinates of vertices *A* and *C* of the square shown in the figure?

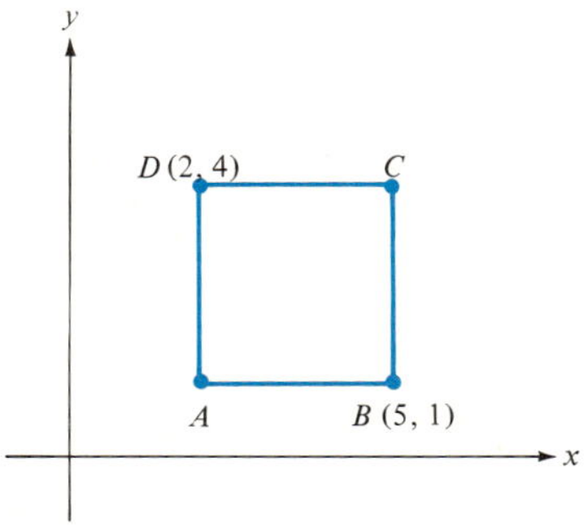

12. What are the coordinates of vertices A and C of the rectangle shown in the figure?

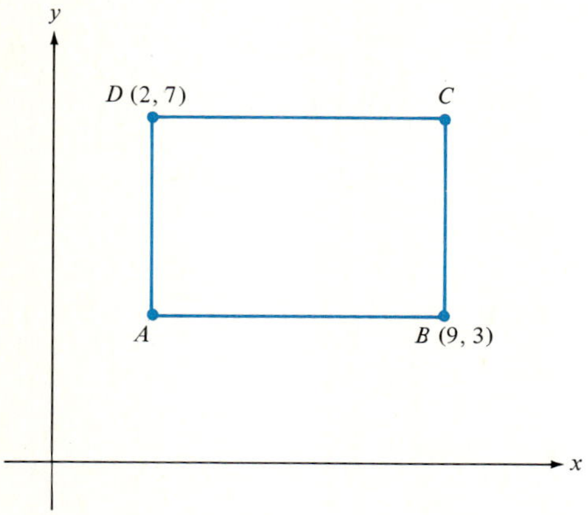

SECTION 7.2

Find ordered pairs of numbers, (x, y), that satisfy the given equation for the specified value of x or y.

13. $2x - y = 6$ $\quad x = 0;\ \ y = 1;\ \ y = 2$

14. $3x + 4y = -24$ $\quad x = 4;\ \ x = -4;\ \ y = 3$

Graph the equations.

15. $x + 2y = 4$ **16.** $3x + 2y = 6$ **17.** $-5x + 3y = 15$

Determine the following equation.

18. Write an equation that indicates that "the sum of three times x and twice y is 14."

SECTION 7.3

Compute the slope of the line that passes through the given pair of points.

19. $(5, 3), (7, 6)$ **20.** $(0, 7), (4, 0)$ **21.** $(\frac{2}{3}, \frac{1}{4}), (\frac{1}{3}, \frac{1}{2})$

22. What is the x-coordinate of the point $(x, 5)$ if it is on a line with no slope defined for it that also contains the point $(-2, -1)$?

23. What is the y-coordinate of the point $(4, y)$ if it is on a line with zero slope that also contains the point $(-2, -1)$?

24. Given that a line L contains the points $(5, 3)$ and $(7, 2)$ determine the slope of any line that is **(a)** Parallel to L. **(b)** Perpendicular to L.

Find the slopes of the lines shown in the figures.

25.

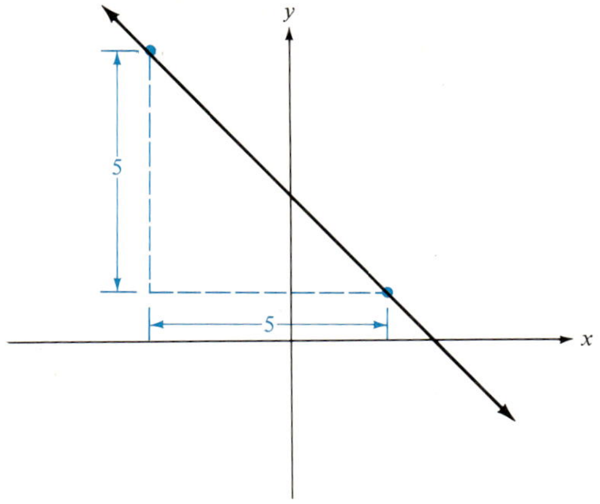

26.

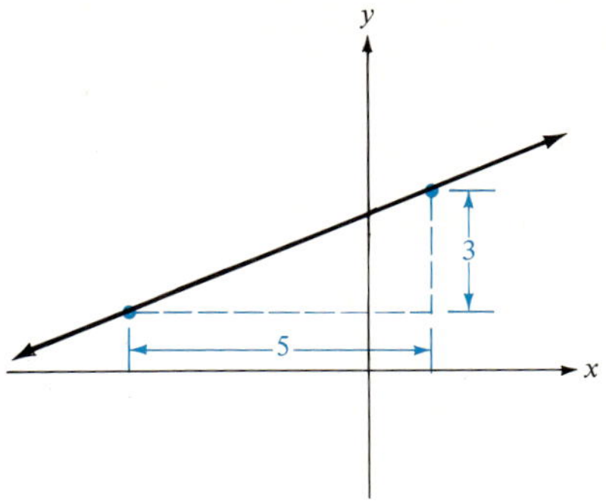

27.

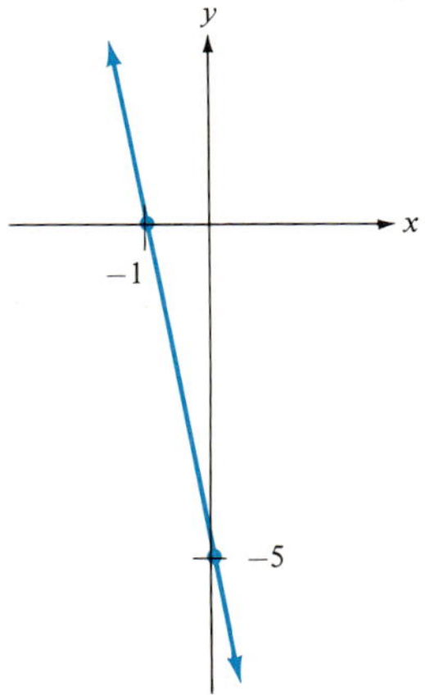

28.

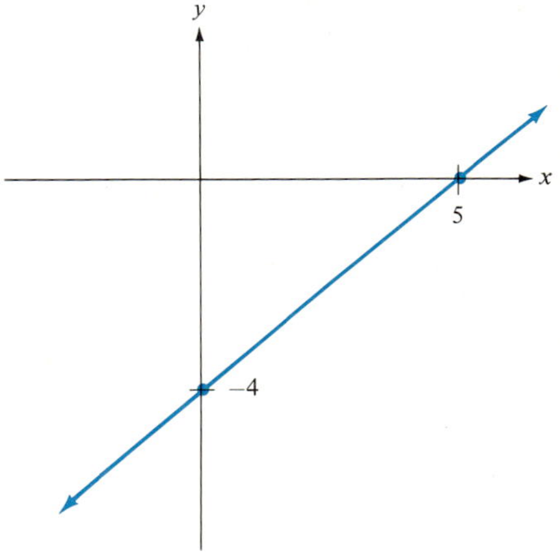

Solve by finding two solutions to the equation.

29. Find the slope of the line whose equation is $3x - y = -2$.

SECTION 7.4

Determine the equation of each line.

30. The line with slope $m = 2$ that passes through the point (3, 1).

31. The line with slope $m = \frac{4}{3}$ that passes through the point (−3, 2).

32. The line with slope $m = -4$ and y-intercept $b = 6$.

33. The line with slope $m = .65$ and y-intercept $b = .31$.

34. The line passing through the point (0, 1) and parallel to the graph of $y = 5x - 4$.

35. The line perpendicular to the graph of $y = \frac{3}{4}x - 5$ and passing through (6, 1).

Find the intercepts.

36. What are the x and y intercepts of the line that has the equation $5x - 2y = -10$?

37. What are the x and y intercepts of the line with the equation $\frac{x}{5} + \frac{y}{4} = 1$?

Solve using the definition of the slope of a line.

38. The formula $V = \frac{7}{4}T + 410$ relates the volume, V, of a certain gas in cubic centimeters to the temperature, T, in degrees Celsius. If there is a temperature increase of 8°C, what is the corresponding change in the volume?

39. The length L, of a particular spring is related to the weight, W, that is suspended from the spring by the formula $L = \frac{5}{28}W + 15$, where L is measured in inches and W is measured in ounces. If there is an increase of 28 ounces in W, what is the corresponding change in L?

SECTION 7.5

Sketch the graphs of the inequalities.

40. $8x + 3y \leq 24$ **41.** $-4x + 9y > 36$

CHAPTER 7 TEST

1. On the same rectangular coordinate system, plot the following points: $(-3, -2)$, $(0, 1)$, $(-2, 0)$, $(6, -1)$, and $(-5, 2)$.

2. The rectangle shown in the figure has its sides parallel to the x and y axes and has a length of 6 and a width of 5. What are the coordinates of vertices A, C, and D?

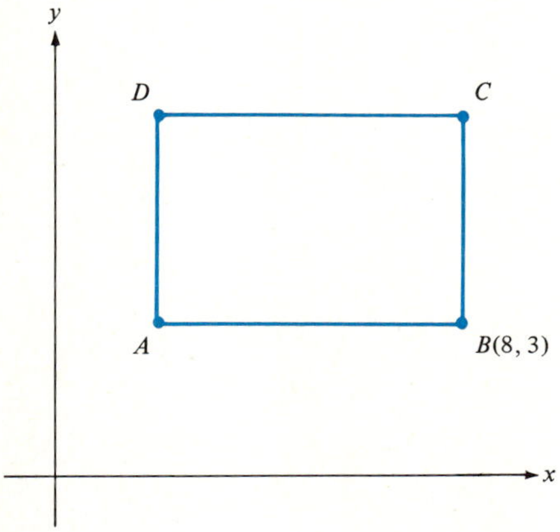

For Problems 3 and 4, complete the ordered pairs so that they satisfy the given equation.

3. $5x - y = 10$; $(3, \ )$, $(\ , 10)$, $(2, \ )$, $(\ , -10)$

4. $2x + 6y = 20$; $(0, \ )$, $(\ , 0)$, $(-4, \ )$, $(\ , 3)$

For Problems 5 and 6, sketch the graph of the equation.

5. $3x + 4y = 12$ **6.** $-7x + 3y = -21$

For Problems 7–9, determine the slope of the line passing through the points.

7. (3, 4), (7, −4) **8.** (2, 5), (−3, 5) **9.** (6, 5), (3, 4)

10. Write an equation that corresponds to the statement: "five times x plus twice y equals 17."

For Problems 11–18, find the equation of the line.

11. Through (3, 2) with slope $m = -4$.

12. With slope $m = -5$ and y-intercept 2.

13. With a slope of $\frac{3}{4}$ and a y-intercept of $\frac{7}{5}$.

14. Passing through the points (7, 6) and (7, 11).

15. Passing through the points (3, 1) and (1, 5).

16. Through (0, 4) and parallel to $y = -3x + 1$.

17. Parallel to $y = \frac{4}{7}x - 5$ and with y-intercept 3.

18. Through (0, 5) and perpendicular to $y = -\frac{1}{4}x + 1$.

For Problems 19 and 20, find the coordinates.

19. What is the y-coordinate of the point $(-2, y)$ if it is on a line with zero slope that also contains the point (5, 3)?

20. What is the x-coordinate of the point $(x, 2)$ if it is on a line with no slope defined for it that also contains the point (6, −4)?

For Problems 21 and 22, find the intercepts.

21. Determine the x and y intercepts of $\dfrac{x}{4} + \dfrac{y}{-5} = 1$.

22. What are the x and y intercepts of the line that has the equation $-2x + 7y = 14$?

For Problems 23 and 24 find the change in each variable.

23. Given the equation $y = \frac{7}{12}x + 10$, what change in y will correspond to an increase of 12 in x?

24. The equation $S = \frac{2}{3}R + 42$ is used to predict the score, S, that a person will obtain on an aptitude test given that the person obtained a score of R on a different test. What change in S corresponds to an increase of 3 in R?

For Problems 25 and 26, sketch the graph of the inequality.

25. $3x - 4y < 24$ **26.** $5x + 2y \geq 10$

CHAPTER 8

Second-Degree Equations in Two Variables and Conic Sections

Conic sections have been studied since ancient times and are also used for many practical applications in the modern world. One of the conic sections with many applications is the parabola. For example, the headlights of automobiles and locomotives have parabolic reflectors behind the source of light. Because the reflectors are parabolic, they reflect intense beams of light.

In Chapter 7 we discussed the relationship between a first-degree equation in two variables, $Ax + By = C$, and a straight line. In this chapter we discuss the equations of certain curves involving second-degree equations in two variables. Some examples of second-degree equations in two variables are as follows:

$$y = x^2, \qquad x^2 + y^2 = 4, \qquad 5x^2 + 10y^2 = 20, \qquad x^2 - y^2 = 1.$$

Aside from some special cases, the curves associated with second-degree equations in two variables are called *conic sections* or *conics*. The main conics that we shall study are the *circle, parabola, ellipse,* and *hyperbola*. The name "conic section" or "conics" is derived from the fact that these curves can be obtained by cutting a cone with a plane as shown in Figure 8.1.

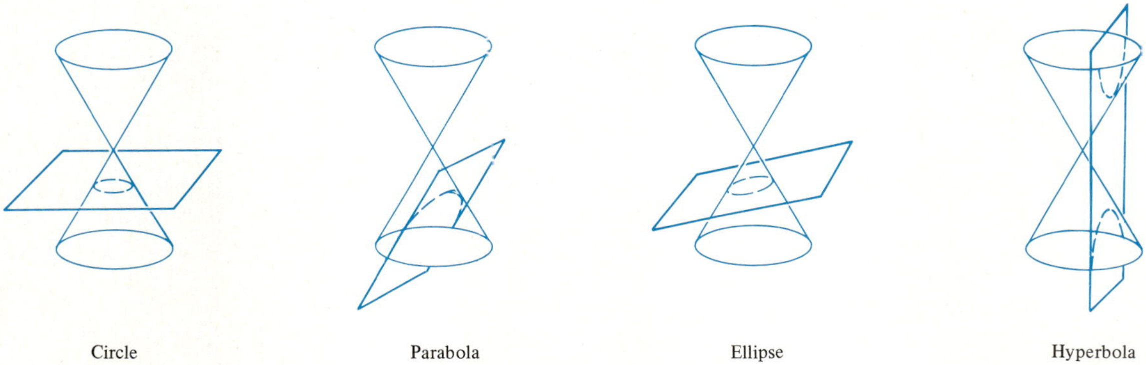

Figure 8.1

8.1 Distance Formula; The Circle

In this section we find the equation of a circle relative to the Cartesian coordinate system, given certain information about the circle. Before we do so, we need to develop a formula for determining the distance between two points in a rectangular (Cartesian) coordinate system.

Let us first define the distance between two points

$$P(x_1, y_1) \qquad \text{and} \qquad R(x_2, y_1)$$

on a horizontal line. At the same time, let us define the distance between two points

$$R(x_2, y_1) \qquad \text{and} \qquad Q(x_2, y_2)$$

on a vertical line. These special cases are shown in Figures 8.2 and 8.3.

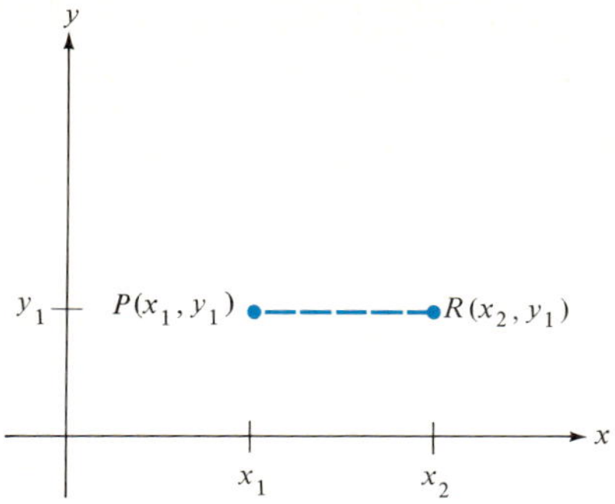

Figure 8.2

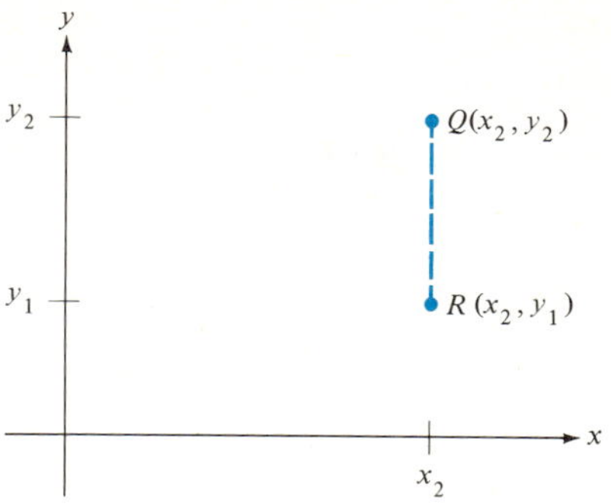

Figure 8.3

We define the distance between $P(x_1, y_1)$ and $R(x_2, y_1)$, symbolized by $d(P, R)$, as follows:

$$d(P, R) = |x_2 - x_1|.$$

Similarly, we define the distance between points $R(x_2, y_1)$ and $Q(x_2, y_2)$ on the vertical line shown in Figure 8.3 by

$$d(R, Q) = |y_2 - y_1|.$$

Observe that $d(P, R) = |x_2 - x_1| = |x_1 - x_2| = d(R, P)$.

We now raise the following question: How do we find the distance between *any* two points $P(x_1, y_1)$ and $Q(x_2, y_2)$ that do not necessarily lie on a horizontal or vertical line, as shown in Figure 8.4? To provide an answer to this question, we draw a horizontal line segment through P and a vertical line segment through Q meeting at point

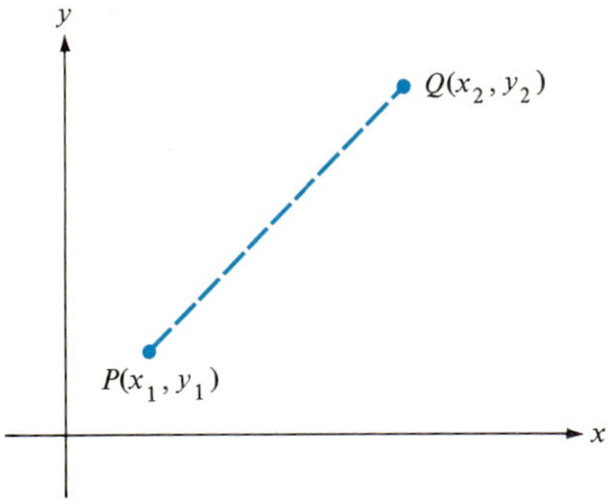

Figure 8.4

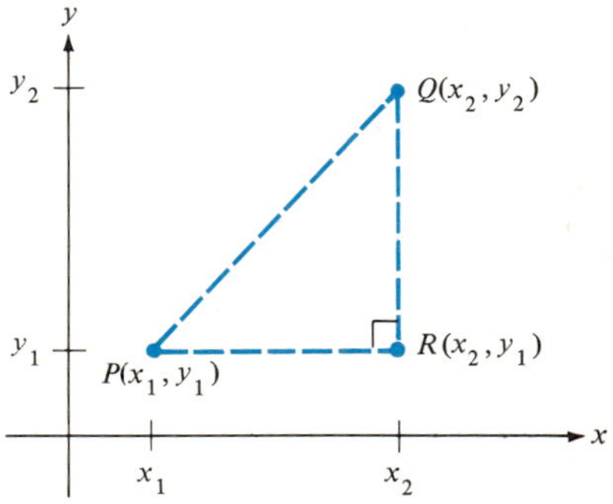

Figure 8.5

$R(x_2, y_1)$. Thus we form a right triangle $\triangle PRQ$, as shown in Figure 8.5. Next we use the Pythagorean theorem, which states that for a right triangle such as $\triangle PRQ$ in Figure 8.5, we have the relationship

(1) $$[d(P, Q)]^2 = [d(P, R)]^2 + [d(R, Q)]^2.$$

By definition we know that $d(P, R) = |x_2 - x_1|$ and $d(R, Q) = |y_2 - y_1|$. Thus, substituting these quantities for $d(P, R)$ and $d(R, Q)$ in line (1), we get

(2) $$[d(P, Q)]^2 = |x_2 - x_1|^2 + |y_2 - y_1|^2.$$

Since it can be shown that $|x|^2 = x^2$; we can rewrite the equation on line (2) as follows:

$$[d(P, Q)]^2 = (x_2 - x_1)^2 + (y_2 - y_1)^2.$$

Taking positive roots of the last equation, we obtain the *distance formula,*

$$d(P, Q) = \sqrt{(x_2 - x_1)^2 + (y_2 - y_1)^2}.$$

Although we have used Figure 8.5 to derive the distance formula, the formula is independent of the positions of P and Q in the coordinate plane. Also, it is convenient to use the letter d, instead of $d(P, Q)$, to designate the distance between any two points P and Q in a coordinate plane. The previous discussion is summarized in the following statement.

Distance Formula

The distance, d, between any two points (x_1, y_1) and (x_2, y_2) is given by

$$d = \sqrt{(x_2 - x_1)^2 + (y_2 - y_1)^2}.$$

Note that the formula above includes our special cases. For if the points P and Q are on a *horizontal* line (therefore, $y_1 = y_2$), then

$$d = \sqrt{(x_2 - x_1)^2 + 0^2} = \sqrt{(x_2 - x_1)^2} = |x_2 - x_1|.$$

Similarly, if P and Q are on a *vertical* line (thus $x_1 = x_2$), then

$$d = \sqrt{0^2 + (y_2 - y_1)^2} = \sqrt{(y_2 - y_1)^2} = |y_2 - y_1|.$$

Example 1. Find the distance between the points $(-2, 3)$ and $(1, 7)$.

Solution: We use the distance formula

$$d = \sqrt{(x_2 - x_1)^2 + (y_2 - y_1)^2}.$$

It does not matter which of the points we designate (x_2, y_2) or (x_1, y_1). We choose to set $(x_1, y_1) = (-2, 3)$ and $(x_2, y_2) = (1, 7)$. Then, substituting in the distance formula, we obtain

$$\begin{aligned} d &= \sqrt{(1 - (-2))^2 + (7 - 3)^2} \\ &= \sqrt{3^2 + 4^2} \\ &= \sqrt{25} = 5. \end{aligned}$$

We now use the distance formula to find the equation of a circle. First, we state the definition of a circle as a set of points in a plane.

Definition of a Circle

A *circle* is the set of all points that are a fixed distance from a fixed point. The fixed distance is called the *radius* of the circle; the fixed point is called the *center* of the circle.

Figure 8.6 shows a circle relative to a rectangular coordinate system. The *center* is the point $C(h, k)$ and $P(x, y)$ represents *any* point on the circle. The letter r represents

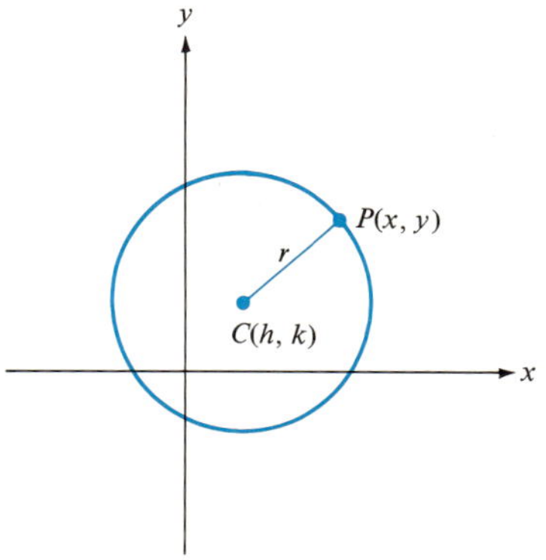

Figure 8.6

the radius of the circle. By definition the *distance* from $P(x, y)$ to $C(h, k)$ is *always equal* to r. Thus, if we let $(x_2, y_2) = (x, y)$, $(x_1, y_1) = (h, k)$, and $d = r$ in the distance formula

$$d = \sqrt{(x_2 - x_1)^2 + (y_2 - y_1)^2}$$

we obtain

$$r = \sqrt{(x - h)^2 + (y - k)^2}.$$

Squaring both sides of the last equation, we get

$$r^2 = (x - h)^2 + (y - k)^2$$

or

$$(x - h)^2 + (y - k)^2 = r^2.$$

Summarizing, we have

Equation of a Circle

The equation of a circle with center at (h, k) and radius r is given by

$$(x - h)^2 + (y - k)^2 = r^2.$$

We shall call the equation above the *standard form of the equation of a circle*.

If the center of the circle is at the origin, $h = 0$ and $k = 0$. The equation of the circle then has the simple form

$$x^2 + y^2 = r^2.$$

Example 2. Find the equation of the circle with center at $(-2, 5)$ with radius 4.

Solution: The equation of the circle is obtained by substituting $h = -2$, $k = 5$, and $r = 4$ in the formula for the circle:

$$(x - h)^2 + (y - k)^2 = r^2.$$

Thus the equation of the circle is

$$[x - (-2)]^2 + (y - 5)^2 = 4^2$$

or

$$(x + 2)^2 + (y - 5)^2 = 16.$$

Example 3. Find the equation of the circle with center at the origin and radius equal to 3. Sketch the graph of the equation on a rectangular coordinate system.

Solution: The equation of the circle with radius $r = 3$ with center at the origin is given by

$$x^2 + y^2 = 3^2$$

or

$$x^2 + y^2 = 9.$$

Since the center is at (0, 0) and the radius $r = 3$, we see that the circle must cut the x-axis at 3 and -3. The same holds for the y-axis. The graph is sketched in Figure 8.7.

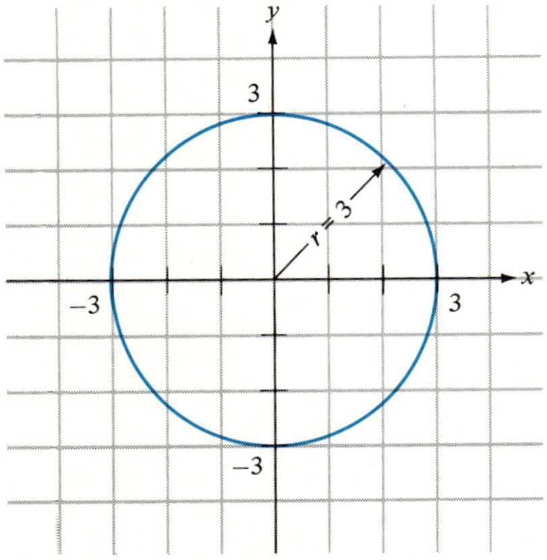

Figure 8.7

If a second-degree equation in two variables, such as x and y, represents a circle, then the numerical coefficients of the x^2 and y^2 terms *must be* the same. In the next example we illustrate how to determine the center and radius of a circle from an equation that represents a circle. Before reading Example 4 it may prove useful to review the method of completing the square in Section 6.2.

Example 4. Find the center and radius of the circle represented by

$$x^2 + y^2 + 4x - 6y - 12 = 0.$$

Solution: To find the coordinates of the center of the circle and its radius, we rewrite the given equation in the form

$$(x - h)^2 + (y - k)^2 = r^2. \qquad \text{center at } (h, k)\text{; radius equal to } r$$

To accomplish this, we first group together those terms containing x and then those terms containing y on one side of the equation. Constant terms are isolated on the other side of the equation. As a result, for the given equation we have

$$x^2 + 4x + y^2 - 6y = 12.$$

Next, we add appropriate constants to both sides of the equation to complete the square with respect to the x-variable and also with respect to the y-variable.

$$x^2 + 4x + 4 + y^2 - 6y + 9 = 12 + 4 + 9$$

Factoring the resulting perfect square trinomials in x and y, we obtain

$$(x + 2)^2 + (y - 3)^2 = 25$$

or

$$[x - (-2)]^2 + (y - 3)^2 = 5^2.$$

We see from the last equation that the center of the circle is located at

$$(h, k) = (-2, 3)$$

and has a radius $r = 5$.

EXERCISES 8.1

For Exercises 1–12, find the distance between the points. See Example 1.

1. (2, 3), (8, 11) **2.** (3, 4), (5, 7) **3.** (−2, 6), (−2, 9) **4.** (4, −5), (6, −6)

5. (0, −3), (−4, 0) **6.** (12, 0), (0, −5) **7.** (3, −4), (7, −4) **8.** (5, 3), (5, 10)

9. (0, 0), (−6, 8) **10.** (−7, 0), (3, 0) **11.** (12, 3), (0, −2) **12.** (5, 7), (6, 8)

For Exercises 13–24, determine the standard form of the equation of the circle with center at the given point (h, k) and radius r. See Example 2.

13. $(3, 4)$, $r = 3$ **14.** $(2, 5)$, $r = 2$ **15.** $(-5, 1)$, $r = \sqrt{3}$ **16.** $(7, -2)$, $r = \sqrt{5}$

17. $(0, 0)$, $r = 5$ **18.** $(-1, 0)$, $r = 6$ **19.** $(0, 3)$, $r = \sqrt{13}$ **20.** $(-2, -6)$, $r = 1$

21. $(-4, -4)$, $r = 7$ **22.** $(0, 0)$, $r = \sqrt{6}$ **23.** $(\sqrt{2}, 1)$, $r = 4$ **24.** $(3, -\sqrt{5})$, $r = \sqrt{3}$

For Exercises 25–32, determine the equation of a circle with center at the origin and with given radius r. Sketch the graph of the equation on a rectangular coordinate system. See Example 3.

25. $r = 4$ **26.** $r = 5$ **27.** $r = 2$ **28.** $r = 1$

29. $r = \sqrt{2}$ **30.** $r = \sqrt{5}$ **31.** $r = 6$ **32.** $r = 7$

For Exercises 33–50, find the center and radius of the circle represented by the equation. See Example 4.

33. $(x - 2)^2 + (y - 1)^2 = 25$ **34.** $(x - 4)^2 + (y - 3)^2 = 9$

35. $x^2 + y^2 = 10$ **36.** $x^2 + y^2 = 7$

37. $(x + 3)^2 + (y - 1)^2 = 5$ **38.** $(x + 4)^2 + (y + 5)^2 = 3$

39. $x^2 + y^2 - 2x - 4y = 4$ **40.** $x^2 + y^2 - 4x - 6y = 12$

41. $x^2 + y^2 - 8x + 2y - 8 = 0$ **42.** $x^2 + y^2 + 10x - 10y - 14 = 0$

43. $2x^2 + 2y^2 - 8 = 0$ **44.** $5x^2 + 5y^2 - 45 = 0$

45. $x^2 + y^2 - 12x = 0$ **46.** $x^2 + y^2 - 14y = 0$

47. $x^2 + y^2 - 3x = 0$ **48.** $x^2 + y^2 - y = 0$

49. $x^2 + y^2 - x - y = \frac{1}{2}$ **50.** $x^2 + y^2 - 5x - 3y - \frac{2}{4} = 0$

Solve Exercises 51–54.

51. Find the equation of the circle with center at the origin passing through the point $(8, -6)$.

52. Find the equation of the circle with center at the origin passing through the point $(-4, -3)$.

53. Find the equation of the circle with center at $(2, 1)$ and passing through the point $(5, 5)$.

54. Find the equation of the circle with center at $(4, 5)$ and passing through the point $(10, 13)$.

8.2 Equations of Parabolas and Their Graphs

The parabola is a conic section, as mentioned at the beginning of this chapter. Parabolic curves are used in the design of automobile headlights, searchlights, telescopes, and radar antennas, to mention a few applications. In physical nature we find that such diverse phenomena as the paths of some comets and the path of projectiles on earth are nearly parabolic curves.

The definition of a parabola viewed as a curve lying in a plane follows.

Definition of a Parabola

> *A parabola* is the set of all points, *P*, whose distance from a fixed point, *F*, is always equal to its distance from a fixed line, *L*. The fixed point is called the *focus*, and the fixed line is called the *directrix*.

Figure 8.8 shows a parabola *opening upward* relative to a rectangular coordinate system. In general, the line through the focus at right angles to the directrix is called the *axis of the parabola*. We call the axis of the parabola the *axis of symmetry* because it divides the parabola into two parts that are mirror images of each other. In Figure 8.8 the axis or axis of symmetry is the y-axis. The point on the parabola that is equidistant

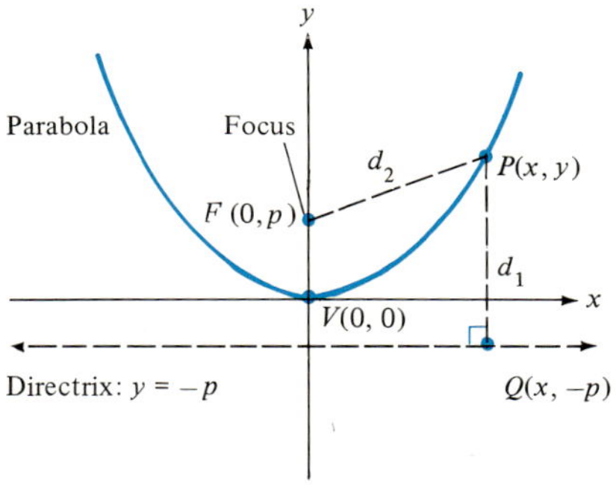

Figure 8.8

from the focus and the directrix is called the *vertex of the parabola*. The parabola shown in Figure 8.8 has its vertex, $V(0, 0)$, located at the origin of the coordinate system; the focus, F, is located at $(0, p)$ (p is positive); and the equation of the directrix is $y = -p$.

We shall now derive an equation for the parabola shown in Figure 8.8. By definition a point P with coordinates (x, y) is on the parabola if and only if

(1) $$d_1 = d_2. \qquad \text{see Figure 8.8}$$

But $d_1 = |y - (-p)| = |y + p|$ and $d_2 = \sqrt{(x - 0)^2 + (y - p)^2}$. Therefore, it follows that

$$|y + p| = \sqrt{x^2 + (y - p)^2}.$$

Squaring both sides of the equation above, we obtain

$$(y + p)^2 = x^2 + (y - p)^2 \qquad \text{note that } |y + p|^2 = (y + p)^2$$
$$y^2 + 2py + p^2 = x^2 + y^2 - 2py + p^2,$$

which simplifies to

$$4py = x^2.$$

As a result of the previous discussion, we have the following statement.

Equation of a Parabola

The equation of a parabola that *opens upward* with *vertex at the origin,* $V(0, 0)$, focus at $(p, 0)$ and with the *y-axis as the axis of symmetry,* is given by

$$4py = x^2 \qquad (p > 0).$$

The equation above is called the *standard form of the equation of a parabola.*

When sketching a parabola on a coordinate system, we do so by plotting the vertex and points, (x, y), that are on opposite sides of, but at the same distance (equidistant) from, the axis of symmetry. We say that these points are *symmetric* with respect to the axis of symmetry. Note that this method for sketching a parabola does not require that we locate the focus or directrix of the parabola. We illustrate this technique for sketching a parabola in the following example.

Example 1. Sketch the graph of $\frac{1}{2}y = x^2$.

Solution: The equation is the standard form of a parabola that opens upward, since the coefficient of y in the equation is $\frac{1}{2} = 4p > 0$. The vertex is located at $(0, 0)$. The axis of symmetry *is* the y-axis. Thus, to sketch the parabola, we plot points, (x, y), on opposite sides but equidistant from the y-axis. To do this, we select values of x on the x-axis that are on opposite sides of, but equidistant from, the origin. For example, we let

$$x = 1, -1.$$

Next, we substitute these values into the equation

$$\frac{1}{2}y = x^2 \qquad \text{or} \qquad y = 2x^2$$

to find the corresponding y-values. The resulting ordered pairs

$$(1, 2), \qquad (-1, 2)$$

satisfy the equation and are symmetric with respect to the y-axis. We plot these ordered pairs and the vertex $(0, 0)$, and sketch in the parabola as shown in Figure 8.9.

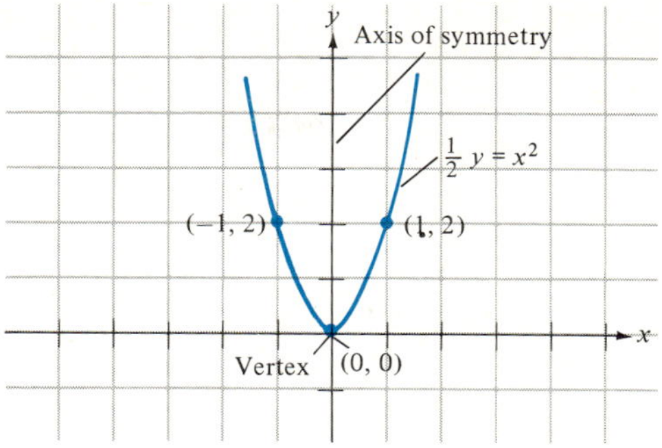

Figure 8.9

Parabolas situated in a rectangular coordinate system with the vertex at the origin and one of the coordinate axes (x- or y-axis) as an axis of symmetry may also open *down,* to the *right,* or to the *left*. Figure 8.10 shows such parabolas with their corresponding equations written in standard form.

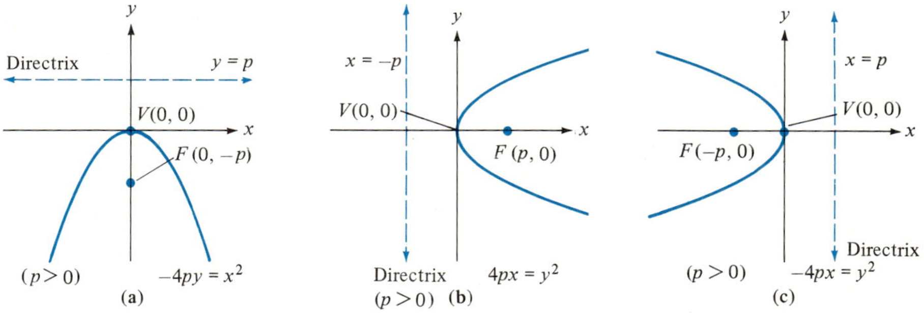

Figure 8.10

Example 2. Graph $-3y = x^2$.

Solution: The equation $-3y = x^2$ is the standard form of a parabola with the y-axis as the axis of symmetry, vertex at the origin, and *opening down*. See Figure 8.10(a). The equation can be rewritten as $y = -\frac{1}{3}x^2$. As a result, we see that it is convenient for computational purposes to let $x = 3, -3$ in the given equation. Thus we obtain the following ordered pairs that satisfy the given equation: $(3, -3)$, $(-3, -3)$. We use these points together with the vertex point $(0, 0)$ to sketch the graph of $-3y = x^2$ shown in Figure 8.11.

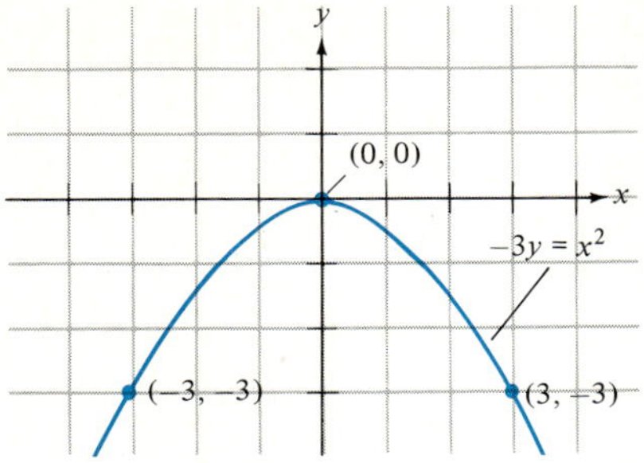

Figure 8.11

Example 3. Sketch the graphs of the following equations.

(a) $-\frac{1}{3}x = y^2$. (b) $2x = y^2$.

Solution: The axis of symmetry for both equations is the *x-axis*. See Figure 8.10, parts (c) and (b). Thus we seek to plot points on the graphs of the given equations that are symmetric relative to the x-axis. To accomplish this, we select convenient values of y to substitute in each equation and then compute the corresponding values of x. For the equation given in part (a), we let

$$y = 1, \; -1.$$

For the equation in part (b), we let

$$y = 2, \; -2.$$

Plotting the resulting ordered pairs and the vertex point (0, 0), we sketch the two parabolas as shown in Figure 8.12.

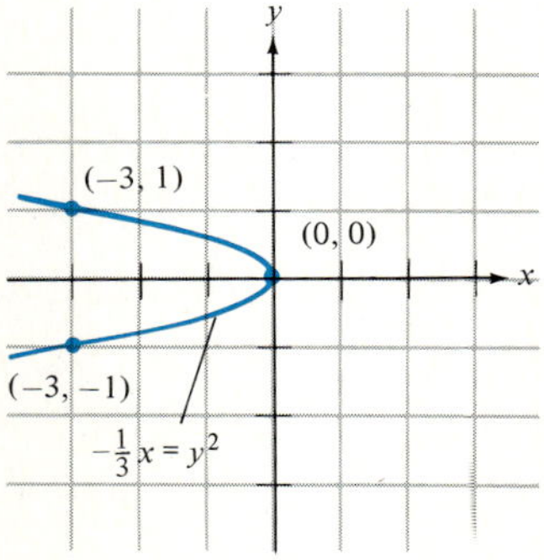

(a)

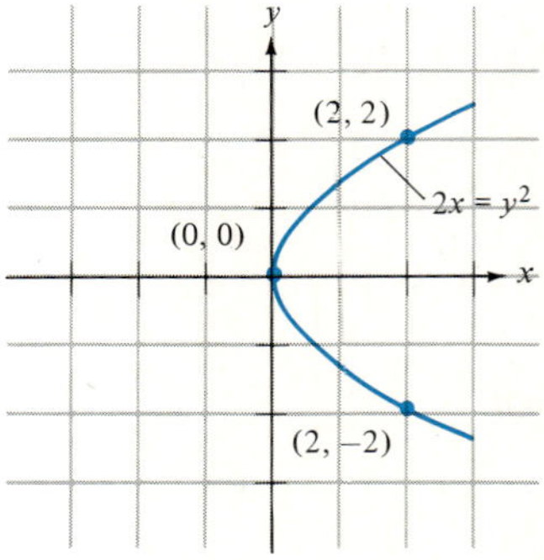

(b)

Figure 8.12

We now consider the equation of a parabola with the axis of symmetry *parallel* to a coordinate axis. We shall confine our discussion to parabolas with axis of symmetry parallel to the y-axis and vertex V at (h, k), as shown in Figure 8.13.

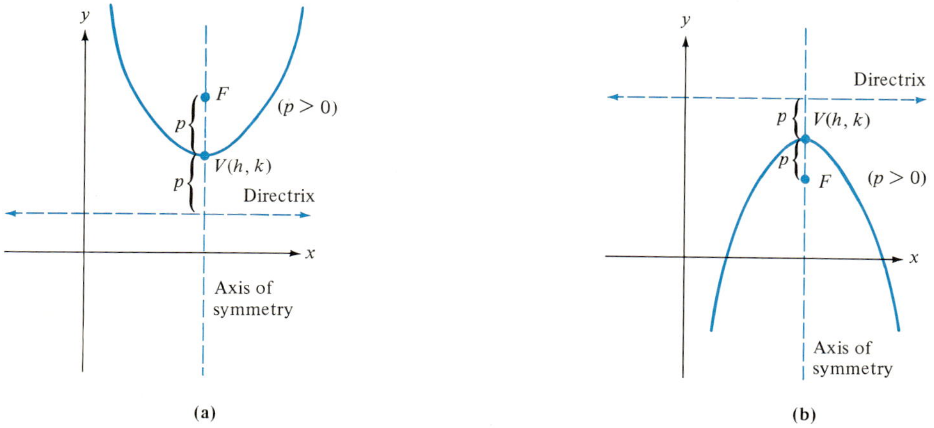

Figure 8.13

By using the definition of the parabola and the distance formula, it can be shown that equations for the parabolas shown in Figure 8.13 are as follows:

Equations of a Parabola with Vertex (h, k) and Axis Parallel to the y-Axis

Parabola Opens Upward	*Parabola Opens Downward*
$4p(y - k) = (x - h)^2 \quad (p > 0)$	$-4p(y - k) = (x - h)^2 \quad (p > 0)$

We note that the *equation of the axis of symmetry* for the parabolas given above is $x - h = 0$ or $x = h$. We will make use of the axis of symmetry when we sketch the graph of a parabola.

Sometimes the term *parabola* is used interchangeably with the phrase "the equation of the parabola." The same is true of "axis of symmetry" and "equation of axis of symmetry." We use this convention in the next two examples.

Example 4. Given the parabola $\frac{1}{2}(y - 2) = (x - 3)^2$, determine (a) the vertex and (b) the axis of symmetry. Sketch the parabola.

Solution: The equation is in the standard form of a parabola with its axis parallel to the y-axis: $4p(y - k) = (x - h)^2$. We see by inspection that for the equation $\frac{1}{2}(y - 2) = (x - 3)^2$, we have:

(a) $h = 3$ and $k = 2$. Thus the vertex is at (3, 2).
(b) The equation of the axis of symmetry is $x - 3 = 0$ or $x = 3$.

To sketch the parabola, we plot the vertex and pairs of points that are symmetric to the axis $x = 3$. To do this, we substitute into the equation values of x that are on opposite sides of, but equidistant from, the axis of symmetry $x = 3$. Thus we substitute values of x such as $x = 2$ and $x = 4$ into the equation. The resulting ordered pairs are (2, 4) and (4, 4). We use these points together with the vertex (3, 2) to sketch the parabola as shown in Figure 8.14.

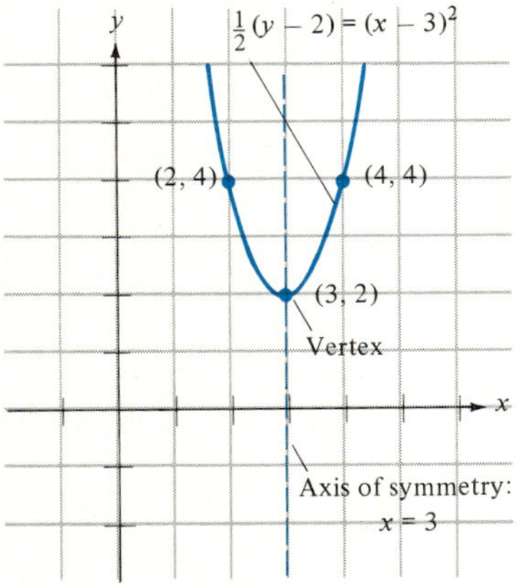

Figure 8.14

If we simplify the equations

(2) $$4p(y - k) = (x - h)^2 \quad \text{or} \quad -4p(y - k) = (x - h)^2$$

of parabolas with axes parallel to the y-axis, by squaring the right side of each equation, we obtain a second-degree equation in two variables of the form

(3) $$y = ax^2 + bx + c,$$

where a, b, and c are constants ($a \neq 0$). On the other hand, if we are given equation (3), by completing the square in x, we can arrive at an equation in standard form as given in (2). We shall illustrate this technique in Example 5. Therefore, the equation $y = ax^2 + bx + c$ represents

a parabola with axis of symmetry parallel to the y-axis.

We note that if

1. $a > 0$, the parabola *opens upward.*
2. $a < 0$, the parabola *opens downward.*

Example 5. Express the equation of the parabola $y = -3x^2 + 12x - 13$ in standard form. Find (a) the vertex and (b) the axis of symmetry. Sketch the parabola.

Solution: We begin by isolating all terms not containing x on one side of the equation.

$$y + 13 = -3x^2 + 12x.$$

Next, we factor the expression in x so that the coefficient of x^2 is 1.

$$y + 13 = -3(x^2 - 4x).$$

We complete the square for the expression $x^2 - 4x$ by adding 4 within the parentheses. Since each term in the parentheses is multiplied by -3, to compensate for this we add -12 to the left side of the equation. Thus we have that

$$y + 13 - 12 = -3(x^2 - 4x + 4)$$
$$y + 1 = -3(x - 2)^2$$
$$-\frac{1}{3}(y + 1) = (x - 2)^2.$$

The last equation is the desired equation. We see that (a) the vertex is at $(2, -1)$ and (b) the axis of symmetry is $x - 2 = 0$ or $x = 2$. Since the coefficient of x^2 is -3, and therefore negative, the parabola *opens downward*. To sketch the parabola, we proceed as we did in Example 4 by using the equation

$$-\frac{1}{3}(y + 1) = (x - 2)^2$$

to plot points symmetric about the axis of symmetry, $x = 2$. The resulting graph is shown in Figure 8.15.

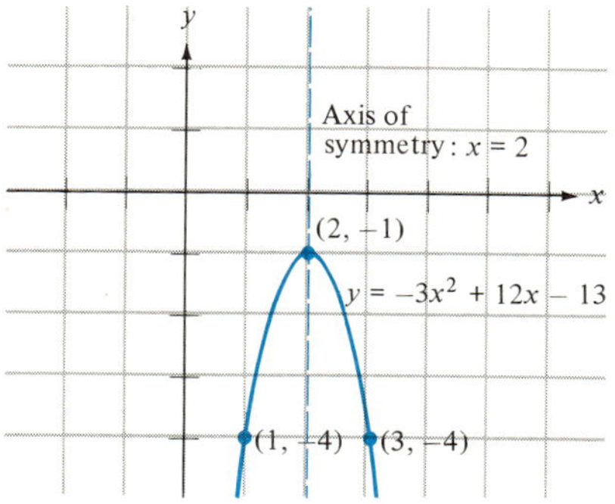

Figure 8.15

We shall further discuss the equation $y = ax^2 + bx + c$ and its graph later in the text in Chapter 10. Furthermore, we shall also discuss alternative methods for sketching its graph.

EXERCISES 8.2

For Exercises 1–20, sketch the graph of the equation. See Examples 1–3.

1. $\frac{1}{4}y = x^2$ **2.** $\frac{1}{3}y = x^2$ **3.** $-\frac{1}{3}y = x^2$ **4.** $-\frac{1}{2}y = x^2$ **5.** $2y = x^2$

6. $3y = x^2$ **7.** $y = x^2$ **8.** $y = -x^2$ **9.** $-\frac{1}{4}x = y^2$ **10.** $-\frac{1}{2}x = y^2$

11. $\frac{1}{3}x = y^2$ **12.** $\frac{1}{2}x = y^2$ **13.** $4x = y^2$ **14.** $3x = y^2$ **15.** $x = -y^2$

16. $x = y^2$ **17.** $y = -4x^2$ **18.** $y = 5x^2$ **19.** $x = 5y^2$ **20.** $x = -4y^2$

For Exercises 21–32, determine (a) the vertex of the parabola and (b) the axis of symmetry. Draw the axis of symmetry and sketch the parabola. See Example 4.

21. $\frac{1}{3}(y - 1) = (x - 3)^2$ **22.** $\frac{1}{2}(y - 3) = (x - 2)^2$ **23.** $-\frac{1}{2}(y - 2) = (x - 4)^2$

24. $-\frac{1}{4}(y - 4) = (x - 3)^2$ **25.** $(y - 3) = (x + 2)^2$ **26.** $\frac{1}{4}(y + 4) = (x - 3)^2$

27. $2(y - 1) = x^2$ **28.** $-3(y + 2) = x^2$ **29.** $-2y = (x + 3)^2$

30. $y = (x + 4)^2$ **31.** $\frac{1}{2}(y + 2) = (x + 3)^2$ **32.** $\frac{1}{4}(y + 4) = (x + 2)^2$

For Exercises 33–41, express the equation of the parabola in standard form. Determine (a) the vertex and (b) the axis of symmetry. Sketch the parabola. See Example 5.

33. $y = x^2 - 2x + 3$ **34.** $y = x^2 - 4x + 5$ **35.** $y = -x^2 - 2x$

36. $y = -x^2 - 6x - 11$ **37.** $y = x^2 + 2$ **38.** $y = -x^2 - 1$

39. $y = 2x^2 + 12x + 20$ **40.** $y = 4x^2 + 16x + 12$ **41.** $y = -3x^2 + 18x - 29$

8.3 Graphing Ellipses and Hyperbolas

The last two conic sections that we discuss are the *ellipse* and the *hyperbola*. We shall treat these conics less formally than we treated the circle and parabola. Furthermore, we will restrict our discussion to ellipses and hyperbolas having simple equations. We discuss the ellipse first.

We note that the orbits of the planets about the sun and other physical phenomena have nearly elliptical paths. We now define the ellipse as a set of points in the plane.

Definition of an Ellipse

> An *ellipse* is the set of all points, P, such that the sum of distances from P to two fixed points, F_1 and F_2, is constant. The two fixed points are called the *foci* (plural of focus).

In Figures 8.16 and 8.17 we show ellipses relative to a rectangular coordinate system. In Figure 8.16 the foci F_1 and F_2 of the ellipse are located on the x-axis equidistant from the origin. For the ellipse shown in Figure 8.17, the foci are on the y-axis equidistant from the origin. Notice in both Figures 8.16 and 8.17 that the x-intercepts are labeled a and $-a$ $(a > 0)$, with the y-intercepts labeled b and $-b$ $(b > 0)$. Both ellipses are said to be *centered at* the origin. Our treatment of ellipses will be confined to ellipses with center at the origin and whose foci both lie on the same coordinate axis.

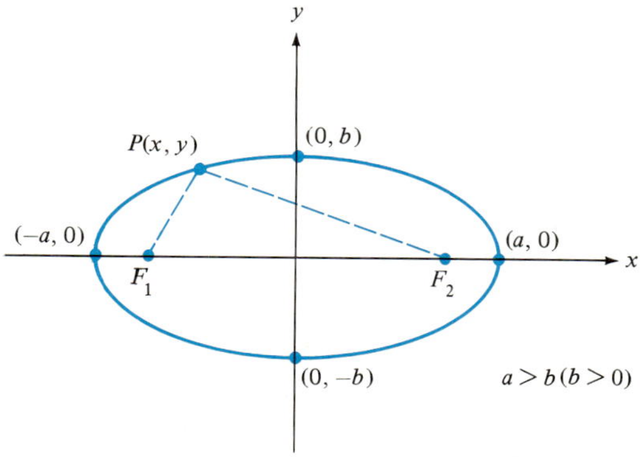

Figure 8.16

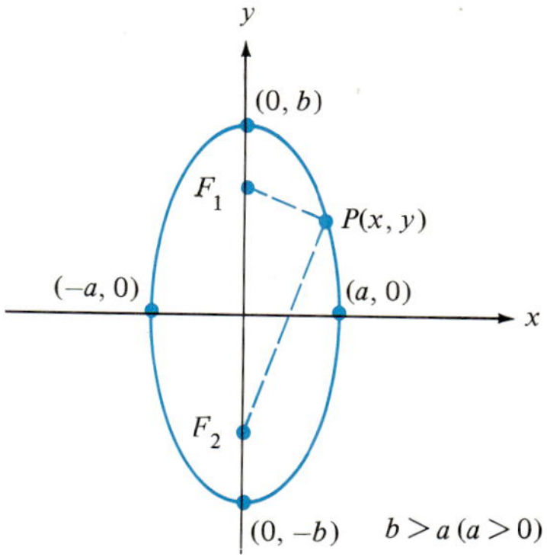

Figure 8.17

Using the definition of an ellipse and the distance formula, it can be shown that the equation of an ellipse as shown in either Figure 8.16 or Figure 8.17 is given by

Equation of an Ellipse

$$\frac{x^2}{a^2} + \frac{y^2}{b^2} = 1 \qquad (a > 0,\ b > 0).$$

Center at (0, 0); foci on the same axis.

The equation above is called the *standard form of the equation of an ellipse*.

A sketch of an ellipse can be made by using the intercepts and drawing a smooth oval curve through them. We illustrate this technique in the following example.

Example 1. Sketch the graph of $\frac{x^2}{25} + \frac{y^2}{9} = 1$.

Solution: The equation can be written in the standard form of an ellipse as

$$\frac{x^2}{5^2} + \frac{y^2}{3^2} = 1.$$

The x-intercepts are -5 and 5. The y-intercepts are -3 and 3. A sketch of the graph is shown in Figure 8.18.

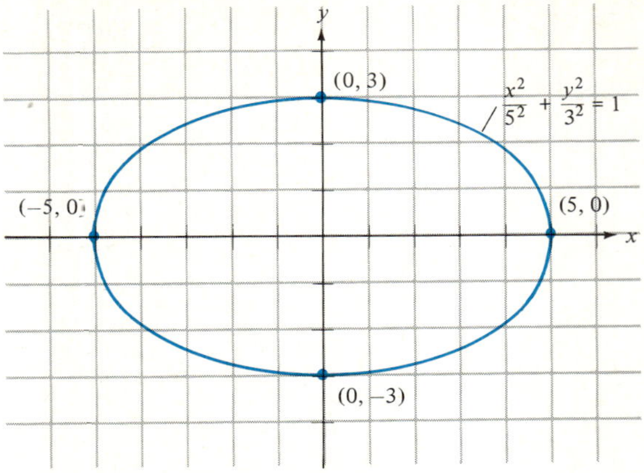

Figure 8.18

Sometimes the equation of an ellipse is not given in standard form. We establish that a given equation is that of an ellipse by rewriting it in the standard form of an ellipse. The following example illustrates this point.

Example 2. Sketch the graph of $8x^2 + 2y^2 = 32$.

Solution: We divide both sides of the equation by 32 and proceed as follows:

$$\frac{8x^2}{32} + \frac{2y^2}{32} = 1$$

$$\frac{x^2}{4} + \frac{y^2}{16} = 1$$

$$\frac{x^2}{2^2} + \frac{y^2}{4^2} = 1.$$

The last equation is the standard form of the equation of an ellipse with x-intercepts -2 and 2 and y-intercepts -4 and 4. The sketch of the ellipse is shown in Figure 8.19.

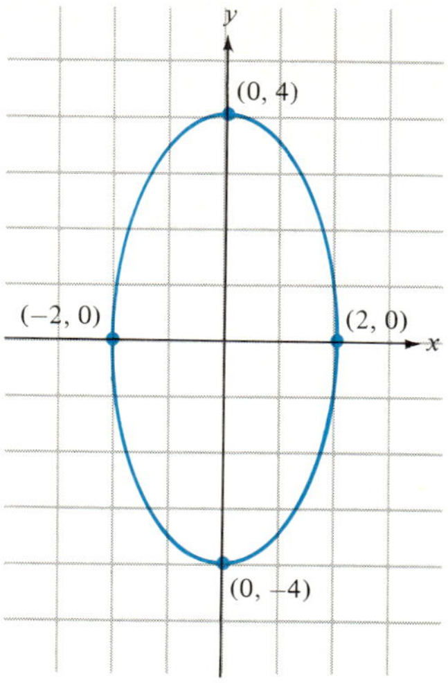

Figure 8.19

A hyperbola is the set of points in the plane that is defined as follows.

Definition of a Hyperbola

> A *hyperbola* is the set of all points, P, such that the difference of the distances from P to two fixed points, F_1 and F_2, is a positive constant. The two fixed points are called the *foci*.

Figure 8.20 shows a hyperbola with two foci F_1 and F_2 on the x-axis that are equidistant from the origin. The two curves that make up the hyperbola are called the *branches of the hyperbola*. The two branches of the hyperbola always lie between two straight lines called the *asymptotes* of the *hyperbola*. The asymptotes are not part of the hyperbola, but we say that the hyperbola "approaches" the asymptotes. The *center of the hyperbola* as situated in Figure 8.20 is the origin, (0, 0).

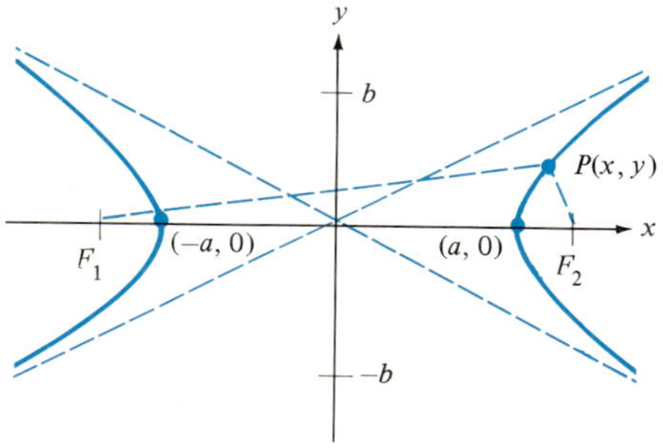

Figure 8.20

It can be shown that the equation of a hyperbola with center at the origin as shown in Figure 8.20 is as follows.

Equation of a Hyperbola (Type 1)

$$\frac{x^2}{a^2} - \frac{y^2}{b^2} = 1 \qquad (a > 0,\ b > 0).$$

Foci on x-axis; Center at (0, 0).

The equation above is called the *standard form of the equation of a hyperbola*. The equations of the two asymptotes are given by

$$y = \frac{b}{a}x \qquad \text{and} \qquad y = -\frac{b}{a}x \qquad (a > 0,\ b > 0).$$

Note that the hyperbola shown in Figure 8.20 has x-intercepts $-a$ and a. There are no y-intercepts because if $x = 0$, the resulting equation in y is

$$\frac{0^2}{a^2} - \frac{y^2}{b^2} = 1 \qquad \text{or} \qquad y^2 = -b^2.$$

But $-b^2$ is negative and thus the equation $y^2 = -b^2$ has no real solutions.

A sketch of a hyperbola can be made using its intercepts and asymptotes. We illustrate this technique in the following example.

Example 3. Sketch the graph of $\frac{x^2}{9} - \frac{y^2}{4} = 1$.

Solution: The equation can be expressed in the standard form of a hyperbola as

$$\frac{x^2}{3^2} - \frac{y^2}{2^2} = 1.$$

The equations of the asymptotes are $y = \pm\frac{2}{3}x$. We now show an easy way to graph the asymptotes. *Draw a rectangle* with sides parallel to the coordinate axes having x-intercepts -3 and 3 and y-intercepts -2 and 2. The asymptotes are obtained by drawing lines through opposite corners of the rectangle. Using the x-intercepts of the hyperbola -3 and 3 and the asymptotes, we sketch the two branches of the hyperbola as shown in Figure 8.21.

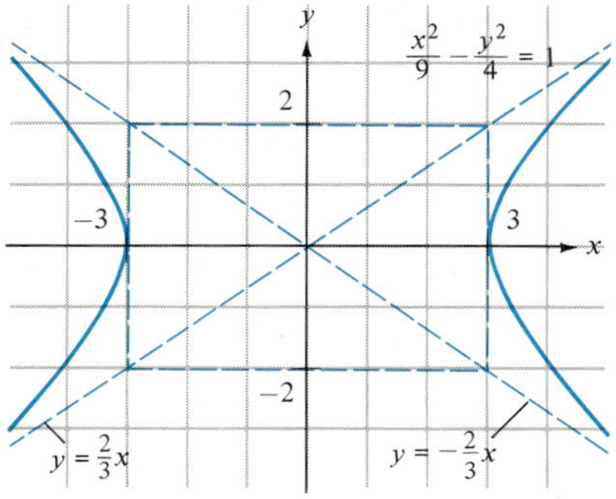

Figure 8.21

In Figure 8.22 the hyperbola is located with its foci on the y-axis equidistant from the origin. The y-intercepts are labeled $-b$ and b $(b > 0)$. There are no x-intercepts.

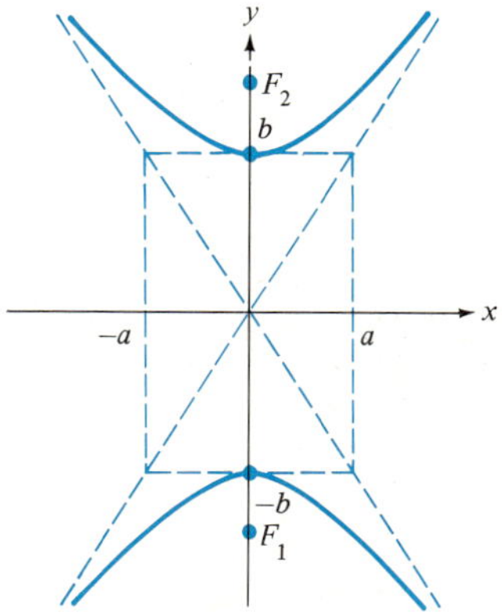

Figure 8.22

The *standard form* of the equation of the hyperbola shown in Figure 8.22 is given below.

Equation of a Hyperbola (Type 2)

$$\frac{y^2}{b^2} - \frac{x^2}{a^2} = 1 \qquad (a > 0,\ b > 0).$$

Foci on the y-axis; Center at (0, 0)

The asymptote equations are $y = \pm\frac{b}{a}x$.

We use the same procedures as in Example 3 to sketch the graph of a hyperbola that intersects the y-axis.

Example 4. Sketch the graph of $16y^2 - 9x^2 = 144$.

Solution: If we divide both sides of the given equation by 144, we obtain

$$\frac{16y^2}{144} - \frac{9x^2}{144} = 1 \qquad \text{not in standard form}$$

$$\frac{y^2}{9} - \frac{x^2}{16} = 1$$

$$\frac{y^2}{3^2} - \frac{x^2}{4^2} = 1. \qquad \text{standard form}$$

The last equation is the equation of a hyperbola with y-intercepts -3 and 3. The sketch of the hyperbola is shown in Figure 8.23.

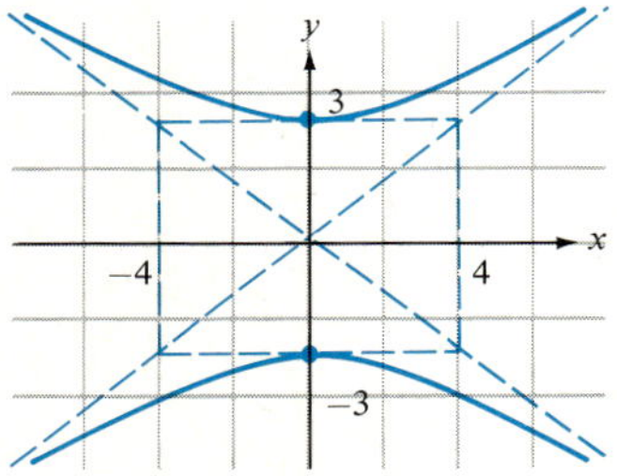

Figure 8.23

When sketching a hyperbola, care must be exercised to make sure that we have the correct intercepts.

EXERCISES 8.3

For Exercises 1–12, sketch the graph of the equation. Indicate the intercepts. See Examples 1 and 2.

1. $\frac{x^2}{16} + \frac{y^2}{4} = 1$ **2.** $\frac{x^2}{36} + \frac{y^2}{9} = 1$ **3.** $\frac{x^2}{9} + \frac{y^2}{49} = 1$ **4.** $\frac{x^2}{4} + \frac{y^2}{25} = 1$

5. $x^2 + 9y^2 = 36$ **6.** $x^2 + 4y^2 = 4$ **7.** $16x^2 + y^2 = 64$ **8.** $4x^2 + y^2 = 144$

9. $4x^2 + 25y^2 = 100$ **10.** $4x^2 + 49y^2 = 196$ **11.** $\frac{x^2}{36} + \frac{y^2}{8} = 1$ **12.** $\frac{x^2}{64} + \frac{y^2}{17} = 1$

For Exercises 13–26, sketch the graph of the equation. Show the asymptotes. See Examples 3 and 4.

13. $\frac{x^2}{16} - \frac{y^2}{4} = 1$ **14.** $\frac{x^2}{36} - \frac{y^2}{9} = 1$ **15.** $\frac{y^2}{25} - \frac{x^2}{4} = 1$ **16.** $\frac{y^2}{49} - \frac{x^2}{9} = 1$

17. $x^2 - 4y^2 = 100$ **18.** $x^2 - 16y^2 = 64$ **19.** $x^2 - 25y^2 = 25$ **20.** $y^2 - 36x^2 = 36$

21. $x^2 - y^2 = 1$ **22.** $y^2 - x^2 = 1$ **23.** $4x^2 - 36y^2 = 144$ **24.** $16y^2 - 4x^2 = 64$

25. $\dfrac{x^2}{64} - \dfrac{y^2}{10} = 1$ **26.** $\dfrac{y^2}{100} - \dfrac{x^2}{50} = 1$

For Exercises 27–38, state whether the equation represents a circle, a parabola, an ellipse, or a hyperbola. [HINT: *Review the standard forms of the conic sections.*]

27. $x^2 + y^2 = 9$ **28.** $x^2 - y^2 = 16$ **29.** $9x^2 + 4y^2 = 36$

30. $9x^2 + 16y^2 - 144 = 0$ **31.** $y^2 - x^2 - 4 = 0$ **32.** $3x^2 + 3y^2 = 27$

33. $9x^2 + 4y^2 = 36$ **34.** $4x^2 + y - 1 = 0$ **35.** $y^2 - x^2 - 1 = 0$

36. $y = 5x^2 - 1$ **37.** $2x^2 + 2y^2 = 50$ **38.** $y = x^2 - 2x + 2$

39. If $a = b$ in the standard form of the ellipse $x^2/a^2 + y^2/b^2 = 1$, we obtain the equation of a circle. Determine

(a) The standard form of the equation of a circle.

(b) The radius of the circle.

Graphing Second-Degree Inequalities 8.4

In this section we conclude this chapter with a discussion on graphing *second-degree inequalities in two variables*. Examples of such inequalities are

$$x^2 + y^2 < 9, \qquad y \geq 3x^2 - 1, \qquad x^2 - y^2 \leq 1.$$

The method for graphing the solution set of a second-degree inequality is similar to that used for graphing linear inequalities. That is, we first graph the boundary curve (either solid or dashed). Next, we use an appropriate *test point* to "determine" the region in the plane that represents the graph of the inequality. We illustrate this procedure in the following examples.

Example 1. Graph $x^2 + y^2 < 25$.

Solution: First we graph the boundary curve $x^2 + y^2 = 25$. The boundary is a circle with center at (0, 0) and radius 5. It separates the plane in two regions, one inside the circle, the other outside the circle as shown in Figure 8.24. We draw the circle *dashed* because it is not part of the graph of $x^2 + y^2 < 25$. We choose as a test point (0, 0). Substituting for x and y in $x^2 + y^2 < 25$, we obtain

$$0^2 + 0^2 < 25 \qquad \text{or} \qquad 0 < 25. \qquad \textit{true statement}$$

Thus the graph of $x^2 + y^2 < 25$ is the region *inside the circle*. The graph is shown in Figure 8.24.

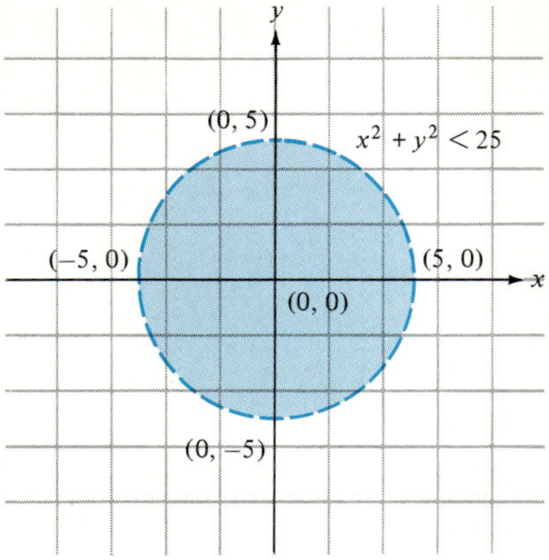

Figure 8.24

Example 2. Graph $y \leq x^2 - 1$.

Solution: The boundary is the parabola $y = x^2 - 1$. If we write the equation $y = x^2 - 1$ in standard form, we obtain $y + 1 = x^2$. As a result, we see that the parabola has vertex at $(0, -1)$; the axis of symmetry is the y-axis; and it opens upward. The test point (0, 0) does *not* satisfy the inequality $y \leq x^2 - 1$ since $0 < 0^2 - 1$ or $0 < -1$ is a *false statement*. Thus we shade in the region *not containing* the test point, as shown in Figure 8.25. Observe that we draw the parabola as a solid curve, since it is part of the graph.

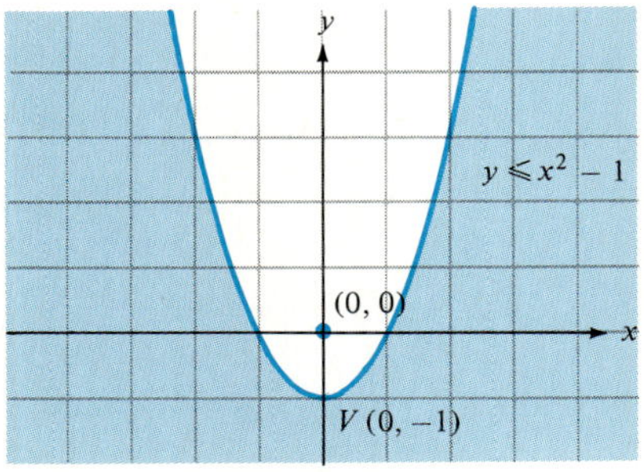

Figure 8.25

Example 3. Graph $\dfrac{x^2}{16} - \dfrac{y^2}{25} \leq 1$.

Solution: The boundary for the graph of the inequality is a hyperbola with vertices on the x-axis at $(-4, 0)$ and $(4, 0)$. The graph of the hyperbola is shown in Figure 8.26. We choose as a test point (0, 0) and find that it satisfies the inequality. Thus the graph is the region between the two branches of the hyperbola, as shown in

Figure 8.27. Note that we draw the hyperbola curve solid because it is part of the graph of the inequality.

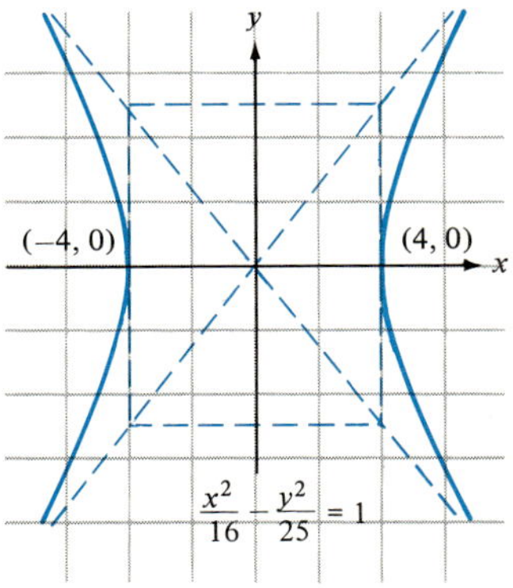

Figure 8.26

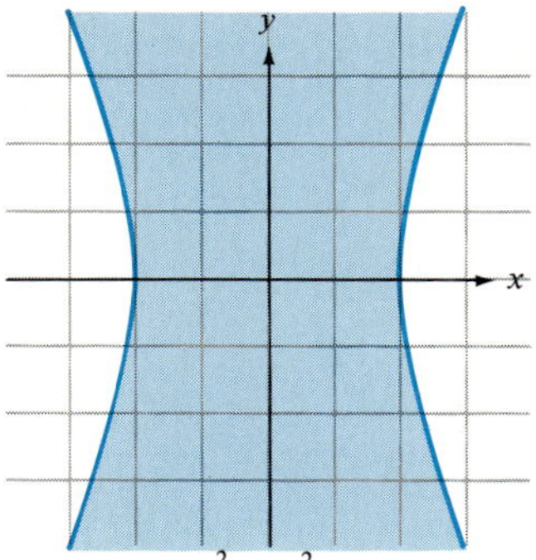

Figure 8.27

EXERCISES 8.4

For Exercises 1–30, graph the second-degree inequality. See Examples 1–3.

1. $x^2 + y^2 < 16$

2. $x^2 + y^2 \leq 9$

3. $x^2 + y^2 \geq 4$

4. $x^2 + y^2 > 25$

5. $y > x^2 + 2$

6. $y \leq -x^2 + 3$

7. $y < x^2$

8. $y \geq x^2$

9. $\frac{x^2}{4} + \frac{y^2}{9} \leq 1$

10. $\frac{x^2}{9} + \frac{y^2}{4} > 1$

11. $4x^2 + y^2 > 16$

12. $9x^2 + y^2 < 36$

13. $4x^2 + 9y^2 < 36$

14. $25x^2 + 4y^2 \geq 100$

15. $\frac{x^2}{9} - \frac{y^2}{4} \leq 1$

16. $\frac{x^2}{4} - \frac{y^2}{9} < 1$

17. $x^2 - y^2 < 9$

18. $x^2 - y^2 \leq 4$

19. $25y^2 - 16x^2 \geq 400$

20. $4y^2 - 16x^2 > 64$

21. $(x - 2)^2 + (y - 2)^2 < 4$

22. $(x - 3)^2 + (y - 2)^2 > 9$

23. $(x + 3)^2 + (y - 3)^2 > 25$

24. $(x + 1)^2 + (y - 2)^2 < 16$

25. $y^2 - x^2 \geq -1$

26. $y^2 - x^2 \leq -1$

27. $-y \leq -x^2 + 2$

28. $-y < x^2 - 3$

29. $(y - 1) \geq (x - 1)^2$

30. $(y + 1) \leq -(x - 2)^2$

Key Words

Circle
- Center
- Radius

Conic section
Distance formula
Ellipse
Focus
Hyperbola
- Asymptotes

Parabola
- Vertex
- Axis of symmetry

Second-degree inequality

CHAPTER 8 REVIEW

SECTION 8.1

Find the distance between the pairs of points.

1. (3, 4) and (0, 0) **2.** (4, −2) and (8, 1)
3. (6, −7) and (12, 1) **4.** (−5, 2) and (7, −3)

Determine the standard form of the equation of the circle with center at the given point and with radius r.

5. (5, 2), $r = 2$ **6.** (−3, 4), $r = 4$ **7.** (−8, −2), $r = 5$

Determine the equation of the circle with center at the origin and with the given radius r. Sketch the graph of the circle on a rectangular coordinate system.

8. $r = 3$ **9.** $r = 4$ **10.** $r = \sqrt{3}$

Find the center and radius of the circles represented by the equations.

11. $(x - 3)^2 + (x - 1)^2 = 16$ **12.** $(x + 4)^2 + (y - 2)^2 = 9$ **13.** $(x + 6)^2 + (y - 3)^2 = 4$
14. $x^2 + y^2 = 5$ **15.** $x^2 + y^2 - 4x - 6y = -9$ **16.** $x^2 + y^2 + 2x - 8y + 8 = 0$
17. $x^2 + y^2 - 10y + 9 = 0$ **18.** $6x^2 + 6y^2 - 24 = 0$

Solve the following.

19. Find the equation of the circle with center at the origin passing through the point (−3, 4).

20. Find the equation of the circle with center at (5, −2) and passing through the point (9, 1).

SECTION 8.2

Sketch the graph of the parabolas.

21. $\frac{1}{2}y = x^2$ **22.** $-\frac{1}{3}y = x^2$ **23.** $-\frac{1}{4}x = y^2$ **24.** $3x = y^2$

For each parabola, determine (a) the vertex of the parabola and (b) the axis of symmetry. (c) Sketch the parabola.

25. $\frac{1}{2}(y + 2) = (x - 2)^2$ **26.** $-\frac{1}{4}(y - 4) = (x + 3)^2$

27. $y = (x + 3)^2$ **28.** $y = -(x - 2)^2$

For each parabola, determine (a) the vertex and (b) the axis of symmetry. (c) Sketch the parabola.

29. $y = x^2 - 6x + 11$ **30.** $y = -x^2 - 2x - 5$

SECTION 8.3

Sketch the graph of each equation.

31. $x^2 + 4y^2 = 16$ **32.** $4x^2 + 9y^2 = 36$ **33.** $4x^2 - 25y^2 = 100$ **34.** $y^2 - x^2 = 9$

For each equation, state whether the equation represents a circle, parabola, ellipse, or hyperbola.

35. $3x^2 + 3y^2 = 12$ **36.** $y^2 - x^2 = 4$ **37.** $4x^2 + 10y^2 = 100$

38. $(x - 5)^2 + (y + 2)^2 = 9$ **39.** $5x^2 - 2y^2 = 50$ **40.** $9x^2 + 25y^2 - 225 = 0$

SECTION 8.4

Graph each inequality.

41. $x^2 + y^2 < 16$ **42.** $4x^2 + 9y^2 \leq 36$ **43.** $y + 2 > (x - 3)^2$

44. $4x^2 - 16y^2 > 64$ **45.** $x^2 + y^2 > 25$ **46.** $y \leq -x^2$

CHAPTER 8 TEST

1. Find the distance between the points $(-3, 4)$ and $(1, 7)$.

2. Write the standard form of the equation of the circle with center at $(3, -5)$ and radius $r = 4$.

3. Find the equation of the circle with center at the origin and radius $r = 8$.

4. For the circle with the equation $x^2 + y^2 + 4x - 2y = -1$, determine **(a)** the coordinates of the center and **(b)** the radius.

5. Find the equation of the circle with center at $(1, 2)$ and passing through the point $(4, 6)$.

For Problems 6 and 7, sketch the graph of the parabola.

6. $y = -x^2$ **7.** $y = (x - 2)^2$

8. For the parabola $\frac{1}{4}(y - 1) = (x - 3)^2$, determine **(a)** the coordinates of the vertex; **(b)** the equation of the axis of symmetry. **(c)** Sketch the parabola.

9. For the parabola $y = x^2 + 4x + 3$, determine **(a)** the coordinates of the vertex and **(b)** the equation of the axis of symmetry. **(c)** Sketch the parabola.

For Problems 10 and 11, sketch the graph of the equation.

10. $x^2 + 4y^2 = 36$ **11.** $4y^2 - 9x^2 = 36$

For Problems 12–20, identify the corresponding conic section.

12. $4y = x^2$ **13.** $x^2 - y^2 = 25$ **14.** $x^2 + y^2 = 49$

15. $y - x^2 = 9$ **16.** $4x^2 + y^2 = 36$ **17.** $4x^2 + 4y^2 = 12$

18. $y^2 - x^2 - 16 = 0$ **19.** $4x^2 + 3y^2 = 12$ **20.** $x^2 - y = 0$

For Problems 21–25, sketch the graph of the inequality.

21. $x^2 + y^2 < 4$ **22.** $y \geq -x^2 + 2$ **23.** $9x^2 + 4y^2 \leq 36$

24. $y^2 - x^2 \geq 1$ **25.** $x^2 - y^2 \leq 4$

CHAPTER 9

Systems of Equations and Inequalities

Large logging companies must often build forest roads in order to provide a way of getting the harvested lumber out of the forest. The better a road is built, the more it costs to construct. However, a better road reduces transportation costs. The solution to a system of equations will help the company determine what type of road to build.

U.S. Department of Agriculture

In Chapters 7 and 8 we studied equations with two variables. In Chapter 7 we considered linear equations and their graphs. We begin this chapter with a discussion of several methods that are used to solve simultaneously two linear equations in two variables, such as

$$\begin{aligned} 5x + 6y &= 60 \\ -x + 2y &= 4. \end{aligned}$$

This set of two equations is an example of a *system of two linear equations in two variables*. In this chapter we also discuss how to solve *systems of three linear equations in three variables,* systems of linear inequalities, and systems of equations that involve second-degree equations. Systems of equations are often used to solve problems that are encountered in business, chemistry, engineering, and statistics.

9.1 Systems of Two Linear Equations in Two Variables

A system of two equations in two variables has the form

$$\begin{aligned} ax + by &= h \qquad && a \text{ and } b \text{ are not both } 0 \\ cx + dy &= k. && c \text{ and } d \text{ are not both } 0 \end{aligned}$$

To solve such a system means to find all ordered pairs (x, y) that *simultaneously* satisfy both equations. For this reason, such systems are also called *simultaneous systems of linear equations*. For example, consider the system

(1)
$$\begin{aligned} 5x + 6y &= 60 \\ -x + 2y &= 4. \end{aligned}$$

Note that (10, 7) is a solution to the equation $-x + 2y = 4$ because $-10 + 2(7) = -10 + 14 = 4$. But (10, 7) does not satisfy the equation $5x + 6y = 60$ because $5(10) + 6(7) = 50 + 42 = 92 \neq 60$. Thus (10, 7) is not a solution to the system (1). We shall now discuss a method that will give us a solution to system (1).

One method of solving the system given by (1) is called the *graphical method*. The following discussion illustrates the method in terms of system (1). The first step is to graph each equation as shown in Figure 9.1. Since we want all ordered pairs (x, y) that satisfy both equations, we want the coordinates of all points common to the two lines. Reading from the graph, we see that there is only one point common to the two lines, (6, 5). To check that (6, 5) is the solution, we substitute 6 for x and 5 for y in each equation in (1). That is,

$$5(6) + 6(5) = 30 + 30 = 60 \qquad \textit{true}$$

and

$$-6 + 2(5) = -6 + 10 = 4. \qquad \textit{true}$$

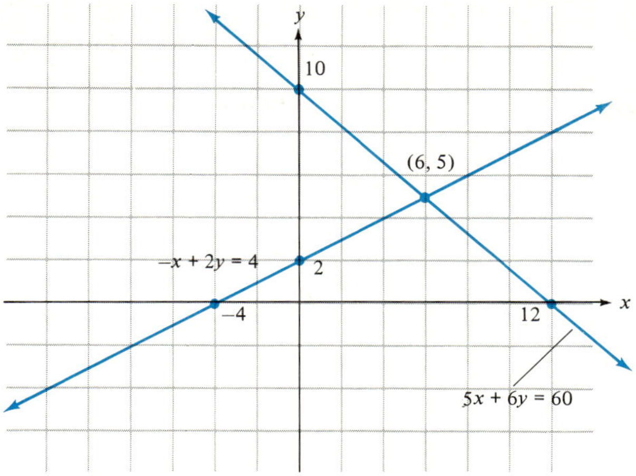

Figure 9.1

Since the substitutions resulted in a true statement for *each* equation, we conclude that the solution set of the system is {(6, 5)}.

Example 1. Solve the following system by graphing:

$$\begin{aligned} x + y &= 6 \\ x - 2y &= 12. \end{aligned}$$

Solution: The graph of each equation is shown in Figure 9.2. From the graphs,

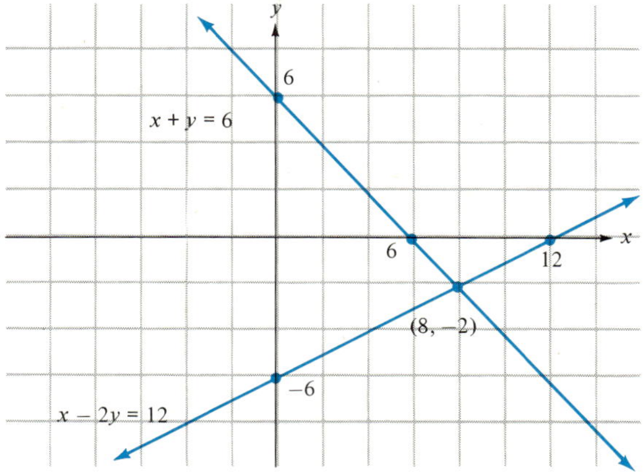

Figure 9.2

it is clear that (8, −2) is their point of intersection. It is easy to check that (8, −2) satisfies each equation. Thus (8, −2) is the solution to the system.

Figure 9.3 illustrates the three different ways that the graphs of two linear equations can be related when they are sketched on the same Cartesian coordinate system. As a result we see that three possibilities can occur when solving the associated system of equations. In Figure 9.3(a), the lines intersect in exactly one point. Thus the corresponding system of equations has exactly one solution. In this case the system is said to be *independent;* it is also said to be *consistent*. In Figure 9.3(b) the lines are parallel. Thus the lines do not have a point in common and therefore there is no solution to the system. In this case the system of equations is said to be *inconsistent;* it is also said to be *independent*. In Figure 9.3(c) the two lines are the same. Thus they have an infinite number of points in common. Therefore, the corresponding system of equations has an infinite number of solutions. In this case the system is said to be *dependent* and *consistent*.

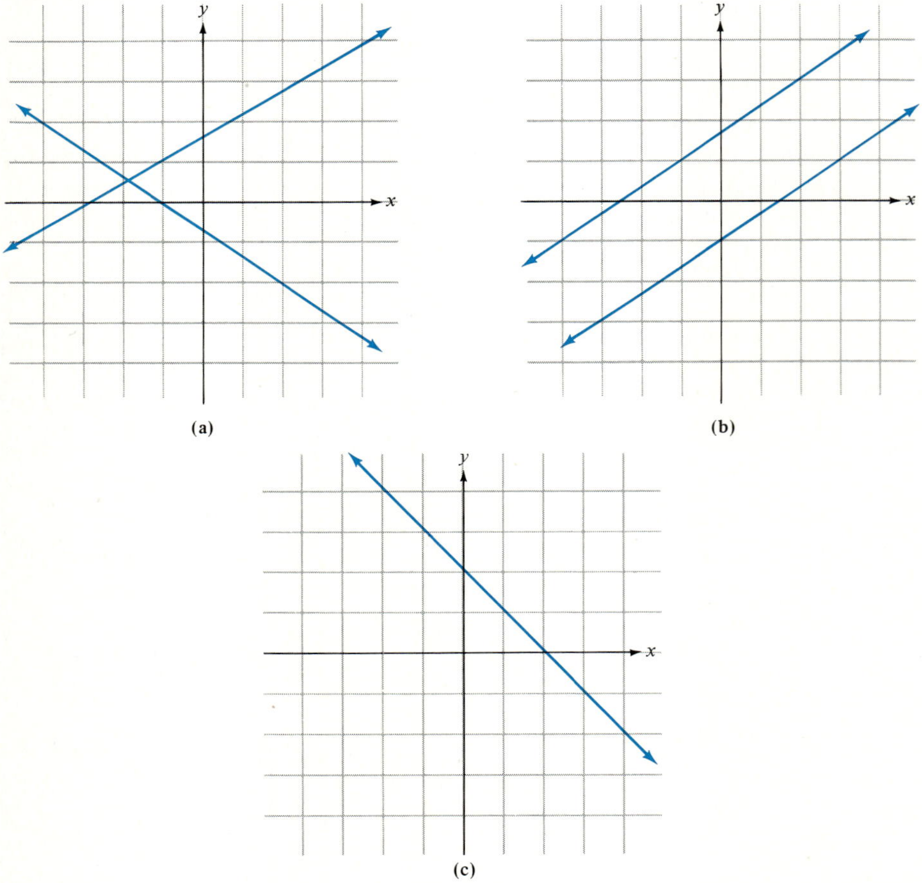

Figure 9.3

Example 2. Solve each system by graphing. Identify any system that is inconsistent or dependent.

(a) $x + y = 1$ $\quad x - y = 7.$ (b) $x + 2y = 3$ $\quad 3x + 6y = 9.$ (c) $2x + y = 6$ $\quad 2x + y = 4.$

Solution: The solutions are given with the aid of Figure 9.4.

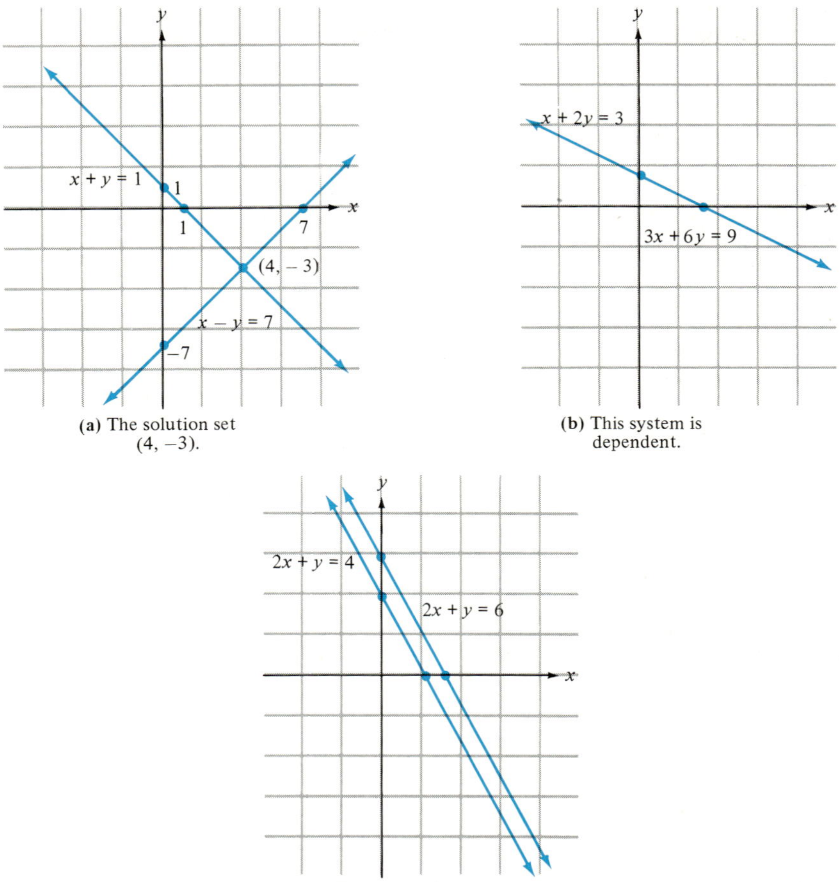

(a) The solution set (4, −3).

(b) This system is dependent.

(c) This system is inconsistent.

Figure 9.4

We note that any point on the line in part (b) is a solution to the corresponding system. If we solve either equation for x, the result is $x = 3 - 2y$. Thus the solution set may be written as $\{(x, y) \mid x = 3 - 2y\}$. In part (c), since the two lines are parallel, we see that there are no solutions to the system.

Although the graphical method may be used to solve systems of equations, it is often difficult to accurately read a solution from the graph. There are two algebraic methods that avoid the problem of accuracy associated with the graphical method. These methods are the *addition method* and the *substitution method*. These methods are also known as methods of *elimination*. These methods involve replacing one system of equations with a simpler system that has the same solution set. Systems of equations that have the *same* solution set are called *equivalent systems*. The process is repeated until a system with an obvious solution is obtained.

To solve a system of equations using the addition method, we use the addition property of equality (Theorem 1 of Chapter 2) and the following theorem.

Theorem 1

> When any of the following operations are applied to a system of equations, the result is an equivalent system of equations.
>
> 1. Interchange any two equations.
> 2. Replace any equation with a nonzero multiple of itself. By this we mean, multiply *both* sides of the equation by the same nonzero number.
> 3. Replace any equation with the sum of that equation and any multiple of a different equation from the system.
> 4. Solve one equation for one variable in terms of the other variables and substitute this result in the other equations.

The next example illustrates the application of Theorem 1 using the *addition method* for solving a system of equations.

Example 3. Solve using the *addition method*.

(1) $$3x + 4y = 5$$
(2) $$2x - y = 7.$$

Solution: If we multiply equation (2) by 4 and then add the result to equation (1), we will eliminate the variable y. Thus we first replace the original system with the equivalent system

(3) $3x + 4y = 5$ original equation (1)
(4) $8x - 4y = 28.$ 4 *times* equation (2)

If we add equations (3) and (4), the result is

$$11x = 33.$$

Thus $x = 3$. To find the value of y we substitute 3 for x in either equation (1) or equation (2) and solve for y. Because the coefficient of y in equation (2) is 1, we use equation (2). The result is

$$\begin{aligned} 2x - y &= 7 \\ 2(3) - y &= 7 \\ 6 - y &= 7 \\ -y &= 1 \\ y &= -1. \end{aligned}$$

Thus the solution set of the system is $\{(3, -1)\}$.

Example 4. Solve by the *addition method*.

(1) $$x - 3y = -2$$
(2) $$-2x + 6y = 7.$$

Solution: Replace this system with the equivalent system

(3) $2x - 6y = -4$ two *times* equation (1)
(4) $-2x + 6y = 7.$ original equation (2)

The result of adding equations (3) and (4) is

$$0 = 3. \qquad \textit{false}$$

Whenever the addition method leads us to a false statement, the original system does not have a solution and therefore is *inconsistent*. The graphs of the equations of such a system are a pair of parallel lines. Thus there are no ordered pairs that satisfy both equations. Therefore, the solution set is ∅.

The next two examples illustrate the *substitution* method.

Example 5. Use the *substitution method* to solve

(1) $$7x + 2y = 8$$
(2) $$x - y = -1.$$

Solution: In the substitution method we first solve one of the equations for one of the variables in terms of the other variable. Thus the result of solving (2) for x in terms of y is

$$x = y - 1.$$

Substituting $y - 1$ for x in (1), we get

(1)
$$\begin{aligned} 7x + 2y &= 8 \\ 7(y - 1) + 2y &= 8 \\ 7y - 7 + 2y &= 8 \\ 9y &= 15 \\ y &= \frac{15}{9} = \frac{5}{3}. \end{aligned}$$

Using $y = \frac{5}{3}$ in (2), we now solve for x. That is,

(2)
$$\begin{aligned} x - y &= -1 \\ x - \frac{5}{3} &= -1 \\ x &= \frac{2}{3}. \end{aligned}$$

Thus the solution set is $\left\{\left(\frac{2}{3}, \frac{5}{3}\right)\right\}$.

Example 6. Solve by the *substitution method.*

$$\begin{aligned} (1)\quad -3x + y &= 2 \\ (2)\quad 6x - 2y &= -4. \end{aligned}$$

Solution: Solving (1) for y in terms of x results in

$$y = 3x + 2.$$

Replacing y with $3x + 2$ in (2), we get

$$\begin{aligned} (2)\quad 6x - 2y &= -4 \\ 6x - 2(3x + 2) &= -4 \\ 6x - 6x - 4 &= -4 \\ -4 &= -4. \quad \text{true} \end{aligned}$$

Because both variables have been eliminated and the resulting expression is a true statement the original equations are *dependent*. That is, the graphs of the two equations are the same line. Thus the coordinates of all the points on the line satisfy both equations. Therefore, the solution set consists of all ordered pairs associated with points on the line. One way to represent the solution set is $\{(x, y) \mid y = 3x + 2\}$.

EXERCISES 9.1

For Exercises 1–10, use the graphical method to solve the system of equations. Identify any system that is dependent or inconsistent. See Examples 1 and 2.

1. $x - y = -3$, $2x + y = 6$

2. $x - y = 2$, $2x + y = 7$

3. $5x + 2y = 10$, $x - 2y = 2$

4. $2x - 3y = 6$, $x - 3y = 6$

5. $3x - 4y = -12$, $x - 2y = -8$

6. $5x - 3y = 15$, $x - 2y = -4$

7. $3x - 5y = 15$, $6x - 10y = 30$

8. $4x - 3y = 12$, $12x - 9y = 36$

9. $x + 3y = 24$, $2x + 6y = 12$

10. $2x + y = 4$, $4x + 2y = 6$

For Exercises 11–30, use the addition method to solve the system of equations. Identify any system that is inconsistent or dependent. See Examples 3 and 4.

11. $x + y = 5$, $x - y = 1$

12. $x + y = 9$, $x - y = 3$

13. $2x + y = 2$, $3x - y = 8$

14. $2x + y = -1$, $x - y = 10$

15. $4x + y = -5$, $3x - 4y = -18$

16. $3x - 4y = 30$, $5x + y = 4$

17. $x - 5y = 0$, $5x + 3y = 14$

18. $x - 3y = 0$, $3x + 4y = 13$

19. $4x - 5y = 6$, $3x - 4y = 4$

20. $3x - 4y = -1$, $2x - 3y = -3$

21. $6x - 4y = 1$, $-4x + 3y = 1$

22. $3x + 4y = 5$, $2x + 5y = -6$

23. $2x - 5y = 19$, $3x + 4y = -6$

24. $2x + 3y = 11$, $7x + 5y = -11$

25. $7x - 5y = 3$, $4x - 3y = -4$

26. $5x - 2y = -20$, $2x + 3y = 11$

27. $\begin{aligned} 3x + 4y &= 1 \\ -6x - 8y &= 3 \end{aligned}$ **28.** $\begin{aligned} 7x - 2y &= 1 \\ -14x + 4y &= 3 \end{aligned}$ **29.** $\begin{aligned} 3x - 4y &= 7 \\ -12x + 16y &= -28 \end{aligned}$ **30.** $\begin{aligned} 6x - 5y &= -4 \\ -18x + 15y &= 12 \end{aligned}$

For Exercises 31–40, use the substitution method to solve the system of equations. Identify any system that is dependent or inconsistent. See Examples 5 and 6.

31. $\begin{aligned} x - 2y &= 3 \\ 2x + y &= 16 \end{aligned}$ **32.** $\begin{aligned} 2x - 5y &= 1 \\ x - 4y &= -1 \end{aligned}$ **33.** $\begin{aligned} x - y &= 0 \\ 9x - 2y &= 28 \end{aligned}$ **34.** $\begin{aligned} 4x + 7y &= 45 \\ x - 2y &= 0 \end{aligned}$

35. $\begin{aligned} 3x - 7y &= -7 \\ 12x - 7y &= -7 \end{aligned}$ **36.** $\begin{aligned} 5x + 4y &= 5 \\ 3x + 2y &= 1 \end{aligned}$ **37.** $\begin{aligned} 3x + 2y &= 5 \\ 12x + 8y &= 10 \end{aligned}$ **38.** $\begin{aligned} 3x + y &= 4 \\ 9x + 3y &= 8 \end{aligned}$

39. $\begin{aligned} 2x - 3y &= 7 \\ 4x - 6y &= 14 \end{aligned}$ **40.** $\begin{aligned} 6x - 3y &= 1 \\ 12x - 6y &= 2 \end{aligned}$

For Exercises 41–50, use any method discussed in this section to solve the system of equations.

41. $\begin{aligned} 5x - 2y &= 20 \\ 5x + y &= 5 \end{aligned}$ **42.** $\begin{aligned} 4x + 3y &= 24 \\ x - y &= 6 \end{aligned}$ **43.** $\begin{aligned} 7x + 3y &= 21 \\ x - y &= 3 \end{aligned}$ **44.** $\begin{aligned} 4x - 3y &= 12 \\ x + y &= 3 \end{aligned}$

45. $\begin{aligned} 5x - 2y &= 6 \\ 2x + 3y &= 10 \end{aligned}$ **46.** $\begin{aligned} 7x - 3y &= 44 \\ 4x - 5y &= 12 \end{aligned}$ **47.** $\begin{aligned} 3x - y &= 18 \\ -6x + 2y &= -24 \end{aligned}$ **48.** $\begin{aligned} 5x + 2y &= 20 \\ 10x + 4y &= 40 \end{aligned}$

49. $\begin{aligned} 7x + 2y &= 4 \\ 10x + 3y &= 5 \end{aligned}$ **50.** $\begin{aligned} 5x - 8y &= 7 \\ 3x - 5y &= 3 \end{aligned}$

Systems of Three Linear Equations in Three Variables 9.2

In this section we discuss the *elimination method* for solving a system of three linear equations in three variables, such as

$$\begin{aligned} &(1) \qquad & 2x + y - 3z &= -4 \\ &(2) \qquad & x + 2y + z &= 5 \\ &(3) \qquad & 3x + y + 4z &= 1. \end{aligned}$$

The solution set of a system of three equations in three variables is the set of all *ordered triplets* (x, y, z) that simultaneously satisfy all three equations. We illustrate the method using the system of equations described by (1), (2), and (3). But first we note that subtraction is defined in terms of addition. Thus we may replace any equation in the system with the difference of that equation and any multiple of a different equation from the system. We make use of this fact in solving the given system.

The first step is to use two of the equations and eliminate one of the variables. If we subtract equation (1) from equation (3), the resulting equation will not contain the variable y. That is,

$$\begin{aligned} &(3) \qquad & 3x + y + 4z &= 1 \\ &(1) \qquad & 2x + y - 3z &= -4 \\ \hline &(4) \qquad & x \qquad + 7z &= 5. \end{aligned}$$

Next, we select a different pair of equations and eliminate the *same* variable. For example, if we subtract equation (2) from two times equation (1), the result is

$$\begin{array}{lrl} (5) & 4x + 2y - 6z = -8 & \text{two } \textit{times} \text{ equation (1)} \\ (2) & \underline{x + 2y + z = 5} & \\ (6) & 3x - 7z = -13. & \end{array}$$

We now consider equations (4) and (6) as a system of two equations in two variables. We solve for the variables x and z using one of the methods of Section 9.1. If we use the addition method, the result is

$$\begin{array}{lr} (4) & x + 7z = 5 \\ (6) & \underline{3x - 7z = -13} \\ & 4x = -8. \end{array}$$

Solving the last equation, we obtain $x = -2$. Substituting -2 for x in equation (4), we get

$$\begin{aligned} x + 7z &= 5 \\ -2 + 7z &= 5 \\ 7z &= 7 \\ z &= 1. \end{aligned}$$

Finally, we substitute -2 for x and 1 for z in one of the three original equations and solve for the remaining variable y. If we use equation (1), the result is

$$\begin{aligned} 2x + y - 3z &= -4 \\ 2(-2) + y - 3(1) &= -4 \\ -4 + y - 3 &= -4 \\ y &= 3. \end{aligned}$$

The reader should verify that the ordered triplet $(-2, 3, 1)$ satisfies each of the original three equations. It would then follow that $\{(-2, 3, 1)\}$ is the solution set of the original system.

The preceding discussion suggests the following *procedure for solving* a system of three equations in three variables.

Step 1. Select any two of the equations and eliminate one of the three variables by the addition or substitution method. The result is one equation in the other two variables.

Step 2. Select a different pair of equations from those used in Step 1. Use these two equations to eliminate the *same* variable that was eliminated in Step 1. The result is another equation in the other two variables.

Step 3. Use the two equations that were the result of Steps 1 and 2 and eliminate one of the remaining two variables. If possible, solve the resulting equation for this variable. If an identity such as $0 = 0$ results, the system is *dependent*. If a false statement such as $0 = 3$ results, the system is *inconsistent*.

Step 4. If a value is obtained for one of the variables, substitute it in either of the equations with two variables and solve for the other variable.

Step 5. Substitute the values of the two known variables in any of the original equations and solve for the remaining variable.

Step 6. Check your solution in all three of the original equations.

Example 1. Solve the system

(1) $x + y + z = 3$
(2) $x - y - z = 1$
(3) $-x + y - z = 2.$

Solution: The results of adding equation (1) to equation (2) and equation (2) to equation (3) are

$$\begin{array}{r} x + y + z = 3 \\ \underline{x - y - z = 1} \\ 2x = 4 \end{array} \quad \text{and} \quad \begin{array}{r} x - y - z = 1 \\ \underline{-x + y - z = 2} \\ -2z = 3. \end{array}$$

Thus

$$x = 2 \quad \text{and} \quad z = -\frac{3}{2}.$$

Replacing x with 2 and z with $-\frac{3}{2}$ in equation (1) results in

$$\begin{aligned} x + y + z &= 3 \\ 2 + y - \frac{3}{2} &= 3 \\ y &= \frac{5}{2}. \end{aligned}$$

Therefore, the solution set is $\left\{\left(2, \frac{5}{2}, -\frac{3}{2}\right)\right\}$.

Example 2. Solve the system

(1) $x + 2y + 3z = 6$
(2) $x + 3y + 6z = 6$
(3) $2x + 4y + 6z = 9.$

Solution: The result of subtracting equation (1) from equation (2) is

(2) $x + 3y + 6z = 6$
(1) $\underline{x + 2y + 3z = 6}$
(4) $y + 3z = 0.$

Next, multiply equation (2) by 2 and obtain

(5) $2x + 6y + 12z = 12.$

Now subtract equation (3) from equation (5).

$$\begin{array}{lrcl} (5) & 2x + 6y + 12z &=& 12 \\ (3) & 2x + 4y + 6z &=& 9 \\ \hline (6) & 2y + 6z &=& 3. \end{array}$$

Next, we subtract two times equation (4) from equation (6).

$$\begin{array}{rl} 2y + 6z = 3 & \text{equation (6)} \\ 2y + 6z = 0 & 2 \textit{ times} \text{ equation (4)} \\ \hline 0 = 3. & \textit{false} \end{array}$$

When we obtain a false statement such as $0 = 3$, we conclude that the system is inconsistent. Thus there is no solution and the solution set is the empty set, $\varnothing$.

Example 3. Solve the system

$$\begin{array}{lrcl} (1) & 2x + 3y - 4z &=& 12 \\ (2) & \frac{2}{3}x + y - \frac{4}{3}z &=& 4 \\ (3) & -4x - 6y + 8z &=& -24. \end{array}$$

Solution: If we multiply equation (2) by 3, the result is equation (1). Similarly, if we multiply equation (3) by $-\frac{1}{2}$, we also get equation (1). Thus all three equations are the same and the system is dependent. Since all three equations are equivalent, any ordered triplet that satisfies one of the equations will also satisfy the other two equations. We write the solution set as $\{(x, y, z) \mid 2x + 3y - 4z = 12\}$.

It can be shown that the graph of a linear equation in three variables is a plane in three-dimensional space. Thus to solve a system of three linear equations in three variables by graphing, we must graph three planes. For this reason the graphical method is not useful for solving linear systems with more than two variables.

Recall that the solution set of a system of three linear equations in three variables is the set of all ordered triplets (x, y, z) that simultaneously satisfy all three equations. But the graph of each equation is a plane. Thus, geometrically, the solution set is all points in three-dimensional space that belong to all three planes. For the purpose of illustrating some of the different ways that three planes can be related, we include Figures 9.5 through 9.7. In each figure we assume that the three planes are the graphs of three equations. The following comments about each figure refer to the system of equations represented by the planes.

In Figure 9.5, the three planes intersect in a single point P, which is the solution to the system.

In Figure 9.6, there is no point common to all three planes. Another way that three planes could have no common point occurs when all three planes are parallel. In both situations the associated system has no solution and is inconsistent.

In Figure 9.7, there are an infinite number of points common to all three planes because the three planes intersect in a line, as shown. Another way we can have an infinite number of solutions occurs when all three planes are the same.

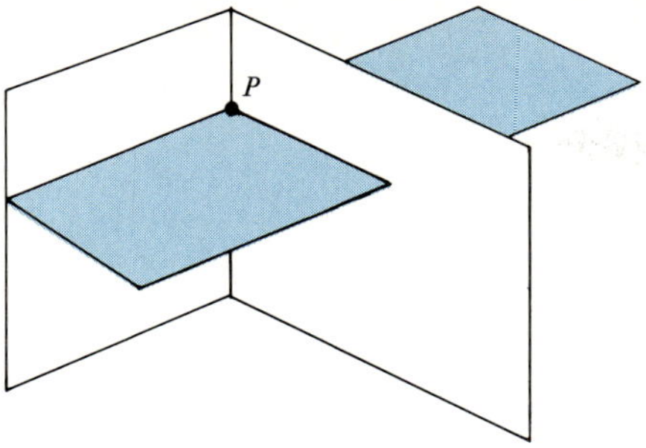

Figure 9.5

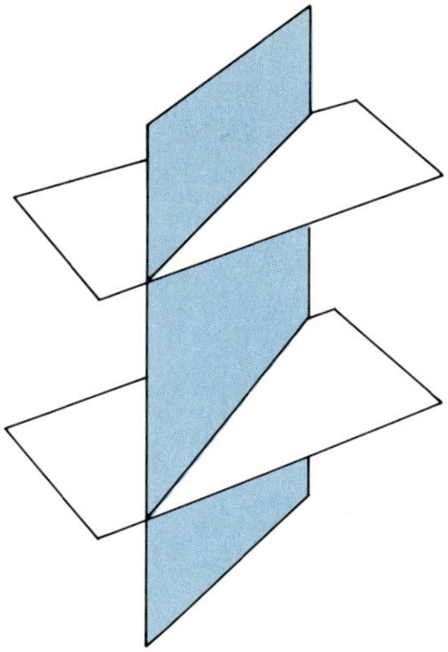

Figure 9.6

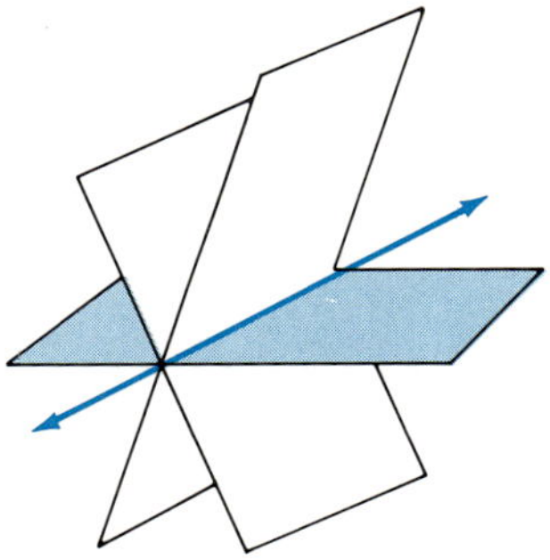

Figure 9.7

EXERCISES 9.2

For Exercises 1–30, solve the system of equations. Identify any system that is inconsistent. See Examples 1–3.

1. $\begin{aligned} x + y - 3z &= 1 \\ x + 2y + 3z &= 1 \\ x + 2y + 6z &= 4 \end{aligned}$

2. $\begin{aligned} x - y - z &= -9 \\ 2x + y - z &= -10 \\ x + 2y + z &= 4 \end{aligned}$

3. $\begin{aligned} x + 2y + 3z &= 13 \\ 2x - y + 3z &= 12 \\ x - 3y + z &= 2 \end{aligned}$

4. $\begin{aligned} x - 2y + z &= -2 \\ 3x + y + z &= 10 \\ 2x + 2y + 3z &= 7 \end{aligned}$

5. $\begin{aligned} x + 3y - 2z &= -8 \\ 3x + y + z &= -1 \\ 2x - y + 3z &= 5 \end{aligned}$

6. $\begin{aligned} x + 4y - z &= -6 \\ 2x + 2y + z &= 3 \\ 3x - 2y + 2z &= 15 \end{aligned}$

7. $\begin{aligned} 3x + y - 2z &= 3 \\ 2x - y &= -2 \\ x + z &= 3 \end{aligned}$

8. $\begin{aligned} 4x - y + 3z &= 4 \\ 2x + z &= 3 \\ 3y + 2z &= 11 \end{aligned}$

9. $\begin{aligned} 5x + 2z &= -11 \\ 3x + y - 2z &= 5 \\ 2x + y + 3z &= -9 \end{aligned}$

10. $\begin{aligned} 5x - y &= -11 \\ 2x + y + z &= -4 \\ 3x + 2y - z &= -3 \end{aligned}$

11. $\begin{aligned} 2x + y &= 2 \\ 3x + z &= 2 \\ y - 2z &= 0 \end{aligned}$

12. $\begin{aligned} 3x + 2y &= 4 \\ 2x - z &= 1 \\ 2y + z &= 2 \end{aligned}$

13. $\begin{aligned} 3x + y - z &= 3 \\ 3x - 2y &= -1 \\ 2y + 3z &= -1 \end{aligned}$

14. $\begin{aligned} 3x - 2y + z &= 4 \\ 6x + y - z &= 0 \\ 2y + 3z &= 1 \end{aligned}$

15. $\begin{aligned} 2x + 3y + 2z &= 3 \\ 3x + y - 4z &= -13 \\ 2x - 2y + z &= -4 \end{aligned}$

16. $\begin{aligned} 3x - 2y + 3z &= 2 \\ 2x - y + 2z &= 1 \\ 3x + 2y + z &= 2 \end{aligned}$

17. $\begin{aligned} 4x + 2y - z &= 8 \\ 2x + y - \tfrac{1}{2}z &= 4 \\ -4x - 2y + z &= -8 \end{aligned}$

18. $\begin{aligned} 6x - 3y + z &= 12 \\ 2x - y + \tfrac{1}{3}z &= 4 \\ 12x - 6y + 2z &= 24 \end{aligned}$

19. $\begin{aligned} \tfrac{1}{2}x + y - \tfrac{1}{2}z &= 3 \\ x + 2y - z &= 6 \\ 2x + 4y - 2z &= 12 \end{aligned}$

20. $\begin{aligned} \tfrac{1}{3}x - \tfrac{2}{3}y + z &= 1 \\ x - 2y + 3z &= 3 \\ -3x + 6y - 9z &= -9 \end{aligned}$

21. $\begin{aligned} x - 2y + z &= 8 \\ x + 3y + 2z &= 8 \\ 2x - 4y + 2z &= 4 \end{aligned}$

22. $\begin{aligned} 2x + y - z &= 4 \\ x - y + 2z &= 2 \\ 2x - 2y + 4z &= 10 \end{aligned}$

23. $\begin{aligned} 2x - 3y + z &= 10 \\ x + 3y - 2z &= 2 \\ 3x - z &= 4 \end{aligned}$

24. $\begin{aligned} 3x + 2y - z &= 7 \\ x - 2y + 2z &= 3 \\ 7x + 2y &= 10 \end{aligned}$

25. $\begin{aligned} 5x + 2y + 3z &= 16 \\ 3x - 4y + 2z &= -1 \\ 2x - 3y + 5z &= -5 \end{aligned}$

26. $\begin{aligned} 3x - 5y + 2z &= -15 \\ 4x + 2y - 3z &= 20 \\ 5x - 3y + 2z &= -9 \end{aligned}$

27. $\begin{aligned} 6x + 2y + 3z &= 2 \\ 2x + 5y + 2z &= 9 \\ 3x - 2y + 4z &= 8 \end{aligned}$

28. $\begin{aligned} 4x + 6y + 3z &= -12 \\ 2x + 3y + 2z &= -7 \\ 5x + 2y + 4z &= -21 \end{aligned}$

29. $\begin{aligned} 6x + 4y + z &= 1 \\ 3x + 6y - z &= -3 \\ 9x - 2y + 4z &= 6 \end{aligned}$

30. $\begin{aligned} 6x - 3y - z &= 3 \\ 4x + 6y + 3z &= 3 \\ 2x + 9y + 2z &= 0 \end{aligned}$

9.3 Applied Problems

A large variety of practical problems involves more than one unknown quantity. Many of these problems can be modeled by a system of linear equations. Thus, by solving the associated system of equations, we solve the practical problem. Often the most difficult aspect of solving practical problems or word problems is translating the given information into mathematical statements. As with most activities, becoming skilled at solving word problems takes practice. Application of the following steps will help you to become skilled at solving word problems. Some steps may have to be repeated before the problem has been solved.

Step 1. Carefully read the problem and determine the unknown quantities.

Step 2. Select letters such as x, y, or z to represent the unknown quantities.

Step 3. Reread the problem and use the information to write the necessary equations. If there are only two unknowns, it is usually possible to write two equations.

Step 4. Solve the system of equations that results from Step 3.

Step 5. Check your answers against the information in the problem, not the equations from Step 3.

The following examples illustrate the use of these steps when solving word problems.

Example 1. The sum of two numbers is 56. The larger number is 5 more than twice the smaller number. Find the two numbers.

Solution: If we let x be the smaller number and y the larger number, then from the given information we get the following two equations:

(1) $x + y = 56$ the sum of the numbers is 56
(2) $y = 2x + 5.$ the larger is 5 more than twice the smaller

Substituting $2x + 5$ for y in equation (1) results in

$$\begin{aligned} x + (2x + 5) &= 56 \\ 3x + 5 &= 56 \\ 3x &= 51 \\ x &= 17. \end{aligned}$$

But

$$y = 2x + 5.$$

Thus

$$\begin{aligned} y &= 2(17) + 5 \\ &= 39. \end{aligned}$$

The two numbers are 17 and 39.

CHECK: Note that the sum of 17 and 39 is 56, and 39 is 5 more than twice 17.

Example 2. Kim Lih took his two children to a family restaurant that serves fast food and also has an electronic game arcade. The food cost \$13.65 and Kim paid for it with a \$20 bill. He received his change in \$.40 tokens and nickels. If the number of tokens is one more than twice the number of nickels, how many tokens and how many nickels did Kim receive?

Solution: Let x denote the number of nickels and y denote the number of tokens. The total value of the tokens is $\$.40(y)$. The total value of the nickels is $\$.05(x)$. Since the food cost \$13.65, the change Kim received from a \$20 bill was \$6.35. Thus

$$.05x + .40y = 6.35$$

or equivalently, multiplying each side of the equation by 100, we get

(1) $$5x + 40y = 635.$$

Since each coefficient in equation (1) is divisible by 5, we can simplify that equation and write

(2) $$x + 8y = 127.$$

But the number of tokens is one more than twice the number of nickels. That is,

(3) $$y = 2x + 1.$$

Thus, to solve the given problem, we have to solve the following system of equations.

(2) $$x + 8y = 127$$
(3) $$y = 2x + 1$$

Substituting $2x + 1$ for y in equation (2), we get

$$\begin{aligned} x + 8(2x + 1) &= 127 \\ x + 16x + 8 &= 127 \\ 17x &= 119 \\ x &= 7 \text{ nickels.} \end{aligned}$$

Since $y = 2x + 1$, we get

$$\begin{aligned} y &= 2(7) + 1 \\ &= 15. \end{aligned}$$

Thus Kim received 7 nickels and 15 tokens. Note that 15 is one more than twice 7. Also, the total value of the nickels is 7(\$.05) = \$.35 and the total value of the tokens is 15(\$.40) = \$6.00, and \$.35 + \$6.00 = \$6.35.

Example 3. A chemist has two bottles that contain solutions of hydrogen chloride. One bottle contains a 5% solution (5% hydrogen chloride); the other contains a 25% solution. How many ounces of each solution must be mixed in order to obtain 20 ounces of a 15% solution?

Solution: Let x denote the number of ounces of the 5% solution and y denote the number of ounces of the 25% solution needed. Hence we see that

(1) $$x + y = 20.$$

Now 5% of x ounces and 25% of y ounces is pure hydrogen chloride. Thus we can write the following equation:

(2) $$.05x \quad + \quad .25y \quad = \quad .15(20).$$

$$\begin{pmatrix}\text{ounces of pure} \\ \text{hydrogen chloride} \\ \text{in } x \text{ ounces of} \\ \text{the 5\% solution}\end{pmatrix} + \begin{pmatrix}\text{ounces of pure} \\ \text{hydrogen chloride} \\ \text{in } y \text{ ounces of} \\ \text{the 25\% solution}\end{pmatrix} = \begin{pmatrix}\text{ounces of pure} \\ \text{hydrogen chloride} \\ \text{in 20 ounces of} \\ \text{the 15\% solution}\end{pmatrix}$$

Since $.15(20) = 3$, we can rewrite equation (2) as

$$.05x + .25y = 3.$$

If we multiply each side of the last equation by 100 in order to clear decimals, the result is

$$5x + 25y = 300.$$

Since 5 is a factor of each term of the equation, we divide each term by 5 and get

(3) $$x + 5y = 60.$$

Next we solve the system

$$(1) \qquad x + y = 20$$
$$(3) \qquad x + 5y = 60.$$

Subtracting equation (1) from equation (3), we get

$$4y = 40$$
$$y = 10.$$

Substituting 10 for y in equation (1) results in

$$x + 10 = 20$$
$$x = 10.$$

thus we should use 10 ounces of each of the solutions in order to obtain 20 ounces of a 15% solution.

CHECK: Note that $10 + 10 = 20$. Also note that 5% of 10 is .5 and 25% of 10 is 2.5. Thus the total amount of pure hydrogen chloride is $2.5 + .5 = 3$ ounces. Since 3 is 15% of 20, our answer is correct.

Example 4. In order to maintain the present distribution of sales of radios, tape recorders, and television sets, a manufacturer uses the following guidelines for production. The number of television sets is equal to the number of radios plus twice the number of tape recorders. The number of radios is five times the number of tape recorders. If the total number of radios, tape recorders, and television sets is 1300, how many of each item should be manufactured?

Solution: Let x, y, and z denote the number of radios, tape recorders, and television sets, respectively. Since the total number of items to be manufactured is 1300, we may write

$$(1) \qquad x + y + z = 1300.$$

The statement "the number of televisions is equal to the number of radios plus twice the number of tape recorders" becomes

$$z = x + 2y$$

or

$$(2) \qquad x + 2y - z = 0.$$

Finally, since there are five times as many radios as tape recorders, we write

$$x = 5y$$

or

$$(3) \qquad x - 5y = 0.$$

Thus we need to solve the following system.

$$\begin{aligned} &(1) \quad & x + y + z &= 1300 \\ &(2) \quad & x + 2y - z &= 0 \\ &(3) \quad & x - 5y &= 0. \end{aligned}$$

Because equation (3) involves only two variables, we can use it to substitute for one of the variables in equations (1) and (2). From equation (3) we see that $x = 5y$. The result of substituting $5y$ for x in each of equations (1) and (2) is the following system

$$\begin{aligned} &(4) \quad & 6y + z &= 1300 \\ &(5) \quad & 7y - z &= 0. \end{aligned}$$

Now if we add these last two equations, we obtain

$$\begin{aligned} 13y &= 1300 \\ y &= 100. \end{aligned}$$

Substituting 100 for y in either equation (4) or (5) and solving for z results in

$$z = 700.$$

Finally, substituting 100 for y in equation (3) results in

$$x = 500.$$

Thus the manufacturer should produce 500 radios, 100 tape recorders, and 700 television sets.

Example 5. A boat can travel 9 miles upstream in 3 hours. The same boat can travel 22 miles downstream in 2 hours. Find the speed of the boat in still water and the speed of the current.

Solution: Let x denote the speed of the boat in still water and let y denote the speed of the current. The speed of the boat *going upstream* is

$$\underset{\begin{pmatrix}\text{speed of the}\\ \text{boat in still water}\end{pmatrix}}{x} \quad - \quad \underset{\begin{pmatrix}\text{speed of}\\ \text{the current}\end{pmatrix}}{y}.$$

Similarly, the speed of the boat *going downstream* is $x + y$. Recall that distance equals rate times time. That is,

$$d = r \cdot t.$$

From the information given we may write

$$9 = (x - y)3 \qquad \text{and} \qquad 22 = (x + y)2.$$

Simplifying, we get

$$\begin{aligned} x - y &= 3 \\ x + y &= 11. \end{aligned}$$

Solving this system by the addition method results in

$$2x = 14 \qquad \text{or} \qquad x = 7.$$

It follows that $y = 4$. Thus the speed of the boat in still water is 7 miles per hour and the speed of the current is 4 miles per hour.

EXERCISES 9.3

Solve Exercises 1–10. See Example 1.

1. The sum of two numbers is 42. The larger number is 3 more than twice the smaller number. Find the numbers.

2. The sum of two numbers is 74. The larger number is 5 more than twice the smaller number. Find the numbers.

3. The sum of two numbers is 74. The larger number is 6 more than three times the smaller number. Find the numbers.

4. The sum of two numbers is 64. The larger number is 12 more than three times the smaller number. Find the numbers.

5. Find two numbers with a sum of 328 and a difference of 80.

6. Find two numbers with a sum of 309 and a difference of 63.

7. Find three numbers with a sum of 54 if the third number is 5 less than twice the first number and the second number is 3 more than the first number.

8. Find three numbers with a sum of 90 if the difference of the first two is 5 and the difference of the last two is 8.

9. Find three numbers with a sum of 108 where the second is 2 less than four times the first and the first is $\frac{1}{2}$ the difference of the second and third.

10. Find three numbers with a sum of 97 where the third is 1 less than five times the first and the first is $\frac{1}{2}$ the difference of the third and second.

Solve Exercises 11–18. See Example 2.

11. Jim has a collection of dimes and nickels with a total face value of \$4.15. There are 60 coins in the collection. How many of each coin are in the collection?

12. Lori has a collection of 62 coins. If her collection consists of nickels and dimes and their total face value is \$4.05, how many of each coin are in the collection?

13. Vivian has a collection of nickels and dimes with a total face value of \$4.50. There are 63 coins in all. How many of each coin are in the collection?

14. Scott has a collection of 50 coins consisting of dimes and quarters. How many of each type of coin are in the collection if the total face value of the coins is \$7.25?

15. Marie collects stamps. In her collection she has a total of 70 stamps with either a \$.05 or a \$.40 face value. If the total face value of the stamps is \$10.50, how many of each stamp are in the collection?

16. Isabel has a stamp collection in which there is a total of 70 stamps with either a \$.15 or a \$.05 face value. If the total face value of these stamps is \$7.50, how many of each stamp are in the collection?

17. A small carnival sells two types of tokens that can be used for activities connected with the carnival. One type of token costs \$.75 and the other type costs \$.25. If you spend \$12.00 and buy 20 tokens, how many of each type did you buy?

18. A traveling carnival sells two types of tokens that can be used for rides at the carnival. The two types of tokens are worth \$.60 and \$.90, respectively. If you spend \$15.00 and receive 22 tokens, how many of each token did you receive?

Solve Exercises 19–24. See Example 3.

19. A chemist has two solutions of sodium chloride; one is a 10% sodium chloride solution and the other is a 40% solution. If she wants 40 ounces of a 25% solution, how much of the 10% solution and how much of the 25% solution should be used?

20. Nidia has two solutions of alcohol; one is a 30% alcohol solution and the other is an 80% solution. How many ounces of each solution are needed to make 20 ounces of a 50% solution?

21. A jewelry manufacturer has available a 65% platinum alloy and a 15% platinum alloy. How much of each alloy should be mixed to obtain 60 grams of a 35% platinum alloy?

22. A jeweler has available two silver alloys. One is 25% silver and the other is 65% silver. How much of each alloy should be mixed to obtain 80 ounces of an alloy that is 40% silver?

23. On a 20-question true-or-false test the score is based on 7 points for each correct answer minus 2 points for each incorrect answer. If a student answers all the questions and gets a score of 77, how many correct answers did the student have?

24. On a 30-question multiple-choice test each correct answer is worth 4 points and each incorrect or unanswered question results in minus 1 point. If a student earns a score of 85, how many questions were answered correctly?

Solve Exercises 25–30. See Example 4.

25. A fish tank contains a total of 34 fish. Furthermore, there are three species in the tank: yellowfins, red spots, and catfish. There are two more yellowfins than the total of the other two species. Also, there are six times as many yellowfins as there are catfish. How many of each species are in a tank?

26. A farmer has 110 acres for planting sweet corn, potatoes, and onions. The number of acres planted with sweet corn is 10 less than the number of acres planted with the other two crops. His profit per acre of onions was \$170, per acre of sweet corn it was \$280, and per acre of potatoes it was \$120. If his total profit was \$22,150, how many acres of each crop did he plant?

27. A store that specializes in coffee mixes 20-pound lots of a house blend using Arabian, Colombian, and Tanzanian coffees. The store sells the special blend for \$4.00 a pound. How many pounds of each coffee are in the blend if the store sells Arabian coffee for \$4.60 per pound, Colombian and Tanzanian coffees for \$3.80 per pound, and the amount of Colombian coffee used is four times the amount of Tanzanian coffee used?

28. A gourmet food store makes a special blend of coffee from Arabian, Hawaiian, and Indonesian coffees. They mix the blend in 30-pound lots and sell it for $4.39 per pound. How many pounds of each coffee are used if the store sells Arabian coffee for $4.10 per pound, Hawaiian coffee for $4.20 per pound, Indonesian coffee for $5.10 per pound, and the amount of Indonesian coffee used is 1 pound more than the amount of Hawaiian coffee used?

29. What are the values of a, b, and c given that the points $(-1, 4)$, $(1, 0)$, and $(2, 7)$ are on the graph of $y = ax^2 + bx + c$? HINT: Substitute each ordered pair in the equation to obtain a system of equations in a, b, and c.

30. What are the values of a, b, and c given that the points $(-1, 13)$, $(1, 3)$, and $(2, 5)$ belong to the graph of $y = ax^2 + bx + c$? HINT: Substitute each ordered pair in the equation to obtain a system of equations in a, b, and c.

Solve Exercises 31–36. See Example 5.

31. A boat can travel 18 miles upstream in 3 hours. The same boat can travel 18 miles downstream in $1\frac{1}{2}$ hours. Find the speed of the boat in still water and the speed of the current.

32. Brendan can travel by boat 12 miles upstream to visit Diane in $1\frac{1}{3}$ hours. He makes the return trip downstream in $\frac{4}{5}$ hour. What is the speed of his boat in still water, and what is the speed of the current?

33. A plane travels 225 miles in the same time that a train travels 35 miles. Find the speed of the train and the speed of the plane if the plane's speed is 30 miles per hour faster than six times the train's speed.

34. A boat can travel 30 miles in the same time that a small plane can travel 350 miles. If the speed of the plane is 8 miles per hour more than 11 times the speed of the boat, find the speed of both the boat and the plane.

35. Two trains start toward each other from cities that are 200 miles apart. One train travels 10 miles per hour faster than the other train. How fast is each train traveling if they pass in 2 hours?

36. Two cars travel toward each other from towns that are 190 miles apart. One car travels 5 miles per hour faster than the other car. How fast is each car traveling if the cars pass in 2 hours?

Systems of Linear Inequalities in Two Variables 9.4

In Section 7.5 we studied the graph of a single linear inequality in two variables. In this section we illustrate how to determine the graph of the solution set to a system of linear inequalities in two variables.

A *system of linear inequalities* in two variables consists of two or more inequalities such as

$$\begin{aligned} ax + by &\le p \\ cx + dy &\le q, \end{aligned}$$

where a, b, p, c, d, and q are real numbers. Other systems of inequalities result if either of the $\leq$ symbols is replaced with $<$, $>$, or $\geq$. We also assume that in each inequality at least one of the variable terms has a nonzero coefficient.

The *solution set* of a system of linear inequalities is the set of all ordered pairs (x, y) that satisfy *all* of the inequalities of the system. Recall from Section 7.5 that the graph of any linear inequality in two variables is a half-plane or a half-plane with its boundary line. We use the following steps to find the graph of the solution set of a system of linear inequalities in two variables.

Step 1. Graph each linear inequality on the same rectangular coordinate system.

Step 2. Determine the intersection of all the graphs in Step 1. This is the graph of the solution set.

Step 3. Sketch a graph that shows only the solution set.

Example 1. Graph the following system of linear inequalities.

$$\begin{aligned} x + 2y &\geq 6 \\ x - y &> -3. \end{aligned}$$

Solution: As shown in Section 7.5, first sketch the boundary line of each inequality. If the inequality is $\leq$ or $\geq$, draw a *solid* line. If the inequality is $<$ or $>$, draw a *dashed* line. Next, for each inequality select a convenient test point and determine the half-plane that represents the solution set. Finally, determine the intersection of the graphs of the solution sets. The end result is shown in Figure 9.8.

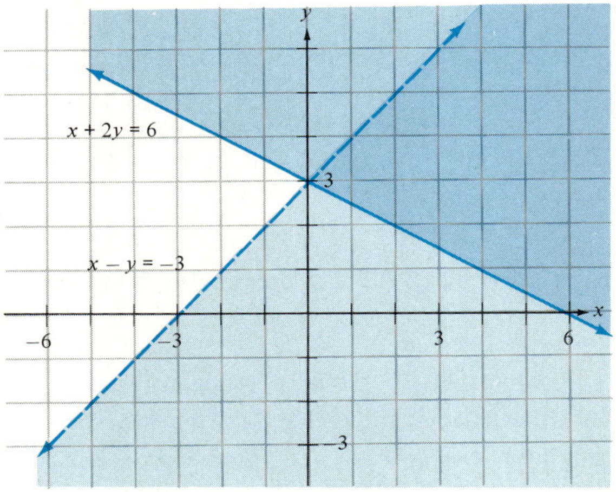

Figure 9.8

The graph of the solution set of the original system of linear inequalities is the intersection or overlap of the two graphs in Figure 9.8. For reasons of clarity, a graph of the solution set only is shown in Figure 9.9.

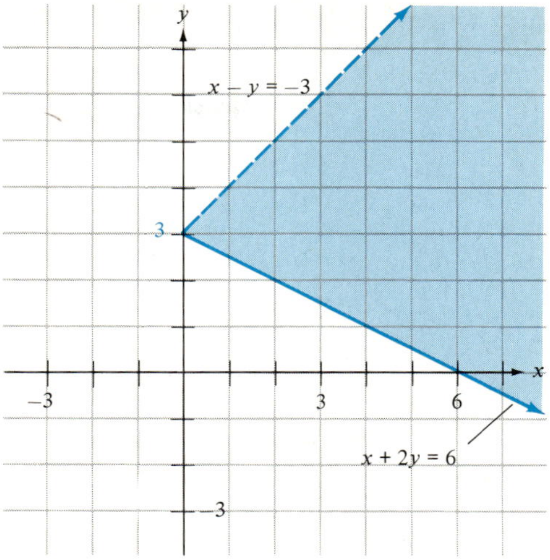

Figure 9.9

Many applications in business and economics involve linear inequalities in which the variables are nonnegative. This forces the graph of the solution set to be restricted to the first quadrant together with the positive x and y axes and the origin.

Example 2. Graph the solution set to the following system of linear inequalities.

$$\begin{aligned} x &\geq 0 \\ y &\geq 0 \\ 5x + 2y &\leq 10. \end{aligned}$$

Solution: First we sketch the graph of $5x + 2y \leq 10$ as shown in Figure 9.10.

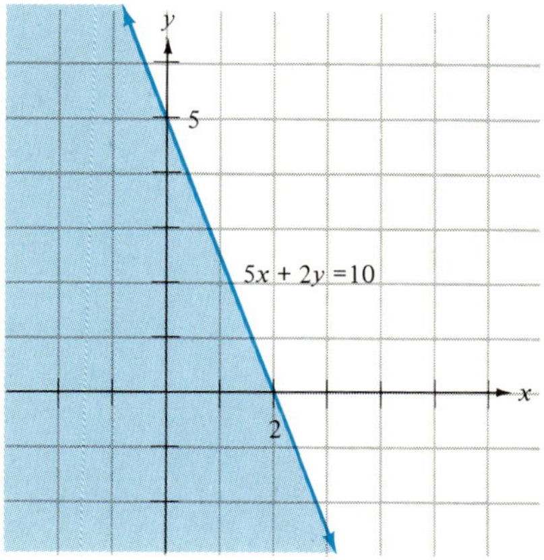

Figure 9.10

But the other two inequalities in the system,

$$x \geq 0 \qquad \text{and} \qquad y \geq 0,$$

tell us the graph of the solution set is restricted to the first quadrant and the nonnegative parts of the x and y axes, as shown in Figure 9.11.

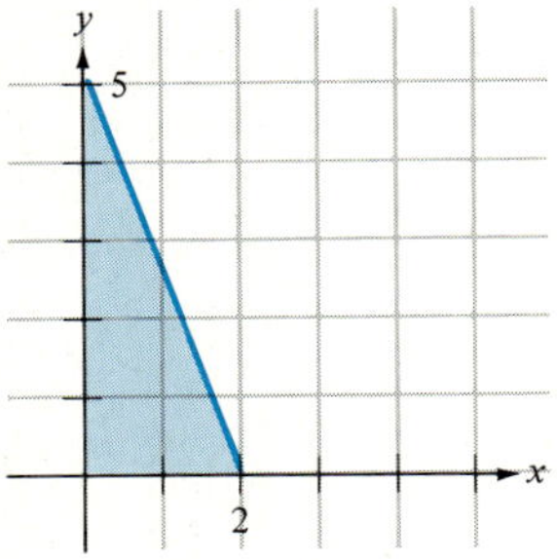

Figure 9.11

Example 3. Graph the solution set of

$$\begin{aligned} x &\geq 0 \\ y &\geq 0 \\ x + y &\leq 120 \\ 2x + y &\leq 200. \end{aligned}$$

Solution: Combining the methods illustrated by Examples 1 and 2 results in the graph shown by the shaded region in Figure 9.12.

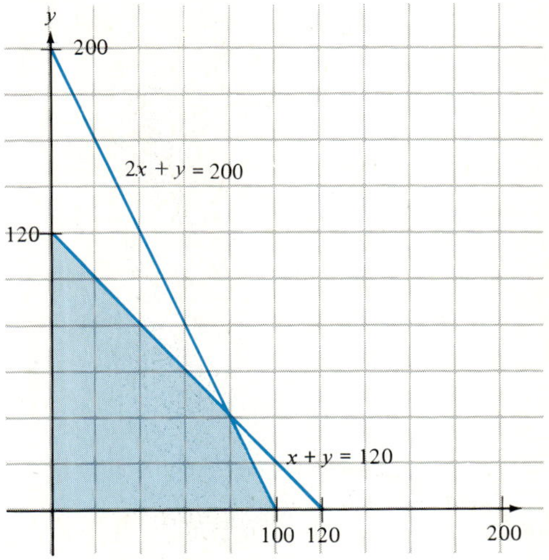

Figure 9.12

EXERCISES 9.4

For Exercises 1–12, graph the solution set of the system of linear inequalities. See Example 1.

1. $x + y \le 8$, $x + 2y \le 12$

2. $x + y \le 6$, $2x + y \le 10$

3. $2x + y \ge 8$, $x + y \le 6$

4. $2x + y \le 8$, $x + y \ge 4$

5. $2x + y \le -4$, $x - 2y \le 3$

6. $x + y \le 10$, $2x + y \le 12$

7. $3x + 4y \le 12$, $-x + y \le 3$

8. $2x + 3y < 12$, $2x + y \le 12$

9. $x + y \le 7$, $3x - 2y < -4$

10. $x + y < 3$, $2x - 3y \le 6$

11. $5x + 4y > 40$, $-x + y \ge 1$

12. $3x + 5y < 15$, $3x - 2y \ge -6$

For Exercises 13–30, graph the solution set of the system of linear inequalities. See Examples 2 and 3.

13. $x \ge 0,\ y \ge 0$, $3x + 4y \le 12$

14. $x \ge 0,\ y \ge 0$, $5x + 2y \le 10$

15. $x \ge 0,\ y \ge 0$, $2x + 4y < 10$

16. $x \ge 0,\ y \ge 0$, $x + 3y < 9$

17. $x \ge 0,\ y \ge 0$, $x + y \le 10$, $x + 3y \le 12$

18. $x \ge 0,\ y \ge 0$, $x + y \le 4$, $3x + y \le 6$

19. $x \ge 0,\ y \ge 0$, $3x + 4y \le 12$, $4x + 3y \le 12$

20. $x \ge 0,\ y \ge 0$, $2x + 3y \le 6$, $3x + 2y \le 6$

21. $x \ge 0,\ y \ge 0$, $7x + 3y < 21$, $3x + 5y < 15$

22. $x \ge 0,\ y \ge 0$, $x + 2y < 16$, $2x + y < 11$

23. $x \ge 0,\ y \ge 0$, $5x + 6y \le 30$, $3x - 2y > 4$

24. $x \ge 0,\ y \ge 0$, $6x + 5y < 30$, $3x - y \ge 6$

25. $x \ge 0,\ y \ge 0$, $x + 2y < 14$, $x - y > 1$

26. $x \ge 0,\ y \ge 0$, $5x + 2y < 14$, $2x - y < 1$

27. $x \ge 0,\ y \ge 0$, $x + 3y \le 9$, $x + 3y \ge 3$

28. $x \ge 0,\ y \le 0$, $2x + y \le 6$, $2x + y \ge 2$

29. $x \ge 0,\ y \ge 0$, $7x + 3y \le 42$, $7x + 3y > 21$

30. $x \ge 0,\ y \ge 0$, $2x + 3y \le 12$, $2x + 3y > 6$

Systems of Equations Involving Second-Degree Equations 9.5

Thus far in this chapter we have been studying systems of linear equations or linear inequalities. We now consider systems of two equations in two variables in which at least one of the equations is a second-degree equation. It is often possible to solve such systems using the *addition* or *substitution* method. To solve some systems, a combination of both methods is used. It is also very helpful to sketch the graph of each equation in order to determine the number of solutions.

Example 1. Solve the system

(1) $$x^2 - y = -1$$

(2) $$-x + y = 3.$$

Solution: Rewriting equations (1) and (2) as

(3) $$y = x^2 + 1$$
(4) $$y = x + 3,$$

we recognize that the graph of equation (3) is a parabola and the graph of equation (4) is a straight line. The sketch of the graphs of these equations is shown in Figure 9.13. Notice that the sketch indicates that there are two solutions. When one equation in a system is linear and the other is not linear, the substitution method is often the best method of solution. Thus, if we replace the variable y in equation (3) with $x + 3$, the result is

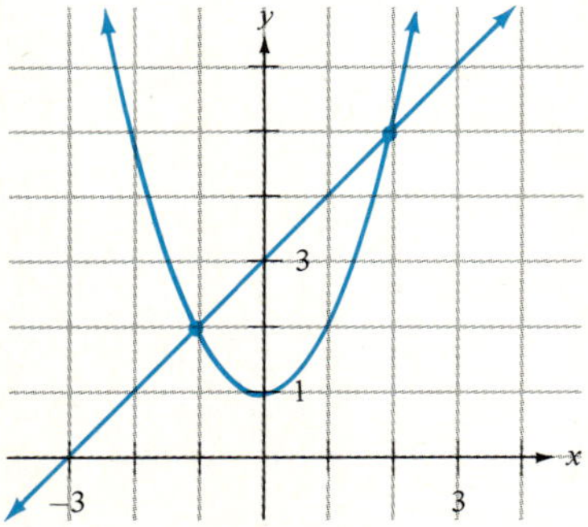

Figure 9.13

$$x + 3 = x^2 + 1$$
$$x^2 - x - 2 = 0$$
$$(x - 2)(x + 1) = 0.$$

From the last equation we see that

$$x = 2 \quad \text{or} \quad x = -1.$$

Now using $y = x + 3$, we see that when $x = 2$, then $y = 5$; and when $x = -1$, then $y = 2$. Thus the solution set for the given system is $\{(2, 5), (-1, 2)\}$.

Example 2. Solve the system

(1) $$x^2 + y^2 = 25$$
(2) $$7x - y = 25.$$

Solution: Recognizing that the graph of equation (1) is a circle and the graph of equation (2) is a straight line, we sketch their graphs in Figure 9.14. The sketch indicates that the system has two solutions. Solving equation (2) for y results in

$$y = 7x - 25.$$

Thus

$$y^2 = (7x - 25)^2$$
$$= 49x^2 - 350x + 625.$$

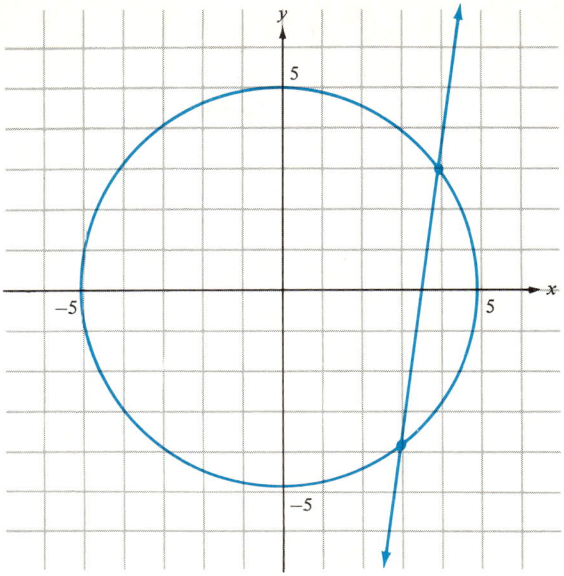

Figure 9.14

Substituting this last result for y^2 in equation (1) results in

$$x^2 + 49x^2 - 350x + 625 = 25$$

or

$$50x^2 - 350x + 600 = 0.$$

Dividing both sides by 50, we get

$$x^2 - 7x + 12 = 0,$$

which can be solved by factoring. That is,

$$(x - 3)(x - 4) = 0.$$

Thus

$$x = 3 \qquad \text{or} \qquad x = 4.$$

Finally, if we substitute 3 and 4 for x in equation (2), we get the corresponding y-values. Thus when $x = 3$, $y = -4$; and when $x = 4$, $y = 3$. We conclude that the solution set to the given system is $\{(3, -4), (4, 3)\}$.

Example 3. Solve the system

(1) $$x^2 + y^2 = 12$$

(2) $$x^2 - y^2 = 2.$$

Solution: In this case the addition method results in the single equation

$$2x^2 = 14$$

or

$$x^2 = 7$$

Thus $x = \pm\sqrt{7}$.

Substituting $x = \pm\sqrt{7}$ into either equation (1) or (2) will produce the corresponding y-values. If we use equation (1), we get:

When $x = \sqrt{7}$

(1) $x^2 + y^2 = 12$

$$(\sqrt{7})^2 + y^2 = 12$$
$$7 + y^2 = 12$$
$$y^2 = 5.$$

Thus

$$y = \pm\sqrt{5}.$$

When $x = -\sqrt{7}$

(1) $x^2 + y^2 = 12$

$$(-\sqrt{7})^2 + y^2 = 12$$
$$7 + y^2 = 12$$
$$y^2 = 5.$$

Thus

$$y = \pm\sqrt{5}.$$

Therefore, we have found four solutions to the original system and the solution set is

$$\{(\sqrt{7}, \sqrt{5}), (\sqrt{7}, -\sqrt{5}), (-\sqrt{7}, \sqrt{5}), (-\sqrt{7}, -\sqrt{5})\}.$$

The sketch of the two graphs shown in Figure 9.15 confirms that there are exactly four solutions.

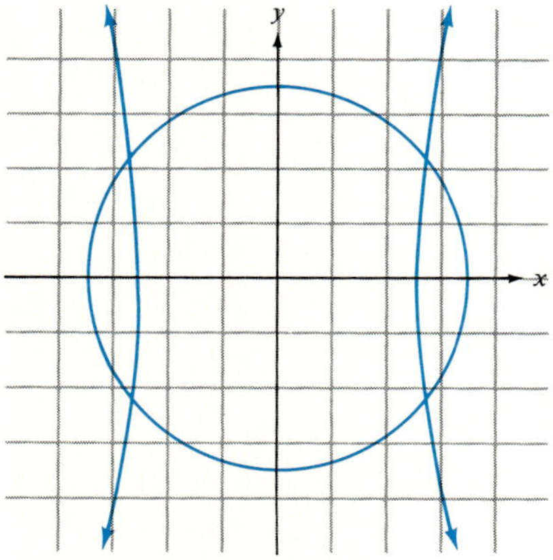

Figure 9.15

Example 4. The sum of the squares of two numbers is 34. The product of the same two numbers is 15. Find the numbers.

Solution: Let x and y denote the two numbers. From the given information we can write the following system

(1) $$x^2 + y^2 = 34$$
(2) $$xy = 15.$$

A sketch of the graphs of these two equations is shown in Figure 9.16. (The graph of $xy = 15$ is called a rectangular hyperbola.) Notice that the sketch indicates there

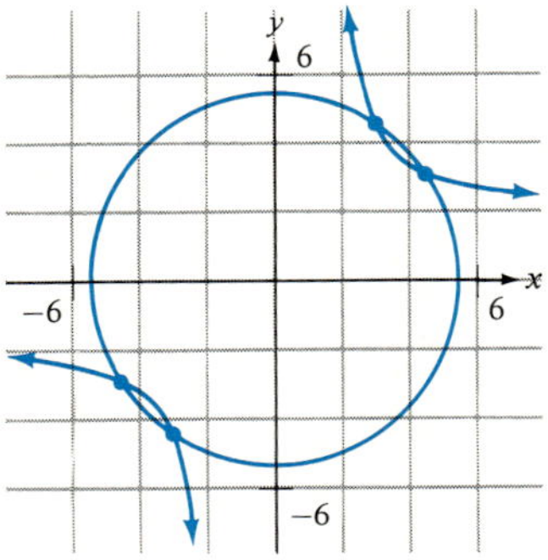

Figure 9.16

are exactly four solutions. If we solve equation (2) for x, we get $x = \frac{15}{y}$. Substituting $\frac{15}{y}$ for x in (1) results in

$$\left(\frac{15}{y}\right)^2 + y^2 = 34$$

or

$$\frac{225}{y^2} + y^2 = 34.$$

Clearing fractions and simplifying, we get

$$225 + y^4 = 34y^2$$

or

$$y^4 - 34y^2 + 225 = 0.$$

But this last equation can be solved by factoring as follows:

$$(y^2 - 25)(y^2 - 9) = 0$$
$$(y - 5)(y + 5)(y - 3)(y + 3) = 0.$$

Thus the possible y values are $y = 5$, $y = -5$, $y = 3$, or $y = -3$. Substituting these values in (2) and solving for x results in the following solution set:

$$\{(3, 5), (-3, -5), (5, 3), (-5, -3)\}.$$

It is easy to check that all four of these solutions satisfy the original problem.

EXERCISES 9.5

For Exercises 1–24, solve the system of equations using the addition or the substitution method. See Examples 1–3.

1. $x^2 - y = 1$
$x - y = -1$

2. $x^2 - y = -3$
$3x - y = -1$

3. $x^2 - y = 5$
$3x + y = -1$

4. $x^2 - y = 6$
$2x + y = 2$

5. $x^2 - y = -4$
$2x + y = 3$

6. $x^2 - y = -9$
$4x + y = 5$

7. $y = x^2 + 5x$
$y = 2x$

8. $y = x^2 + 4x$
$y = x$

9. $y = x^2 + 4x + 4$
$y - x = 2$

10. $y = x^2 - 2x + 1$
$y - x = -1$

11. $x^2 + y^2 = 1$
$x - y = 1$

12. $x^2 + y^2 = 1$
$-x + y = 1$

13. $x^2 - y^2 = 3$
$x + y = 3$

14. $x^2 - y^2 = 8$
$x + y = 4$

15. $y = x^2 + 3x$
$y = 2x^2 + 5x - 8$

16. $y = x^2 + 4x$
$y = 2x^2 + 5x - 2$

17. $x^2 + y^2 = 20$
$x^2 - y^2 = 2$

18. $x^2 + y^2 = 10$
$x^2 - y^2 = 2$

19. $3x^2 + y^2 = 12$
$2x^2 + y^2 = 10$

20. $4x^2 + 3y^2 = 12$
$x^2 + 3y^2 = 9$

21. $4x^2 + 9y^2 = 36$
$9x^2 + 4y^2 = 36$

22. $5x^2 + 4y^2 = 20$
$4x^2 + 5y^2 = 20$

23. $x^2 + 2y^2 = 8$
$2x^2 + y^2 = 10$

24. $x^2 + 3y^2 = 8$
$2x^2 + y^2 = 6$

For Exercises 25–30, write and solve a system of equations. See Example 4.

25. The sum of the squares of two numbers is 52. The product of the same two numbers is 24. Find the numbers.

26. The sum of the squares of two numbers is 53. The product of the same two numbers is 14. Find the numbers.

27. The difference of the squares of two numbers is 33. The sum of the squares of the same two numbers is 65. Find the numbers.

28. The difference of the squares of two numbers is 9. The sum of the squares of the same two numbers is 41. Find the numbers.

29. One number is two more than the square of another number. The sum of the two numbers is 14. Find the numbers.

30. One number is one more than the square of another number. The sum of the two numbers is 21. Find the numbers.

9.6 Solution of Systems of Linear Equations by Determinants

In Sections 9.1 and 9.2 we learned how to use the methods of addition and substitution to solve systems of linear equations. In this section we show how to use *determinants* to solve systems of two equations in two variables. We begin by introducing the idea of a *matrix*.

A *matrix* is a rectangular array of numbers such as

$$\begin{bmatrix} 1 & 3 \\ 4 & 7 \\ 6 & 2 \end{bmatrix} \quad \text{or} \quad \begin{bmatrix} -1 & 3 & 5 \\ 0 & -10 & 12 \end{bmatrix}.$$

Each number in the matrix is called an element of the matrix. The first matrix has three rows and two columns, the second matrix has two rows and three columns (rows are horizontal, columns are vertical). The size of a matrix is described in terms of its rows and columns. For example, the following matrix with three rows and four columns,

$$\begin{bmatrix} 1 & 3 & 4 & 2 \\ 0 & -7 & 11 & -5 \\ 6 & 8 & 23 & 9 \end{bmatrix},$$

is a 3×4 (read "three by four") matrix. A matrix is a *square* matrix if it has the same number of rows as columns.

Associated with each square matrix is a unique *number* called the *determinant* of the matrix. For the general 2×2 matrix

$$\begin{bmatrix} a & b \\ c & d \end{bmatrix}$$

the determinant of the matrix is denoted by

$$\begin{vmatrix} a & b \\ c & d \end{vmatrix}$$

and the value of this determinant is defined to be

$$\begin{vmatrix} a & b \\ c & d \end{vmatrix} = ad - bc.$$

Example 1. Evaluate the determinant of each of the following "matrices" (plural of "matrix").

(a) $\begin{bmatrix} 7 & 4 \\ 5 & 3 \end{bmatrix}$. (b) $\begin{bmatrix} 3 & 10 \\ 1 & -5 \end{bmatrix}$. (c) $\begin{bmatrix} 2 & 4 \\ 3 & 6 \end{bmatrix}$.

Solution: Remember that $\begin{vmatrix} a & b \\ c & d \end{vmatrix} = ad - bc$.

(a) For this matrix $a = 7$, $b = 4$, $c = 5$, and $d = 3$. Thus the corresponding determinant is

$$\begin{vmatrix} 7 & 4 \\ 5 & 3 \end{vmatrix} = 7 \cdot 3 - 4 \cdot 5 = 21 - 20 = 1.$$

(b) $\begin{vmatrix} 3 & 10 \\ 1 & -5 \end{vmatrix} = 3(-5) - 10 \cdot 1 = -15 - 10 = -25.$

(c) $\begin{vmatrix} 2 & 4 \\ 3 & 6 \end{vmatrix} = 2 \cdot 6 - 4 \cdot 3 = 12 - 12 = 0.$

In order to show how determinants can be used to solve a linear system of equations, we now apply the addition method to a general system of two linear equations in two variables. Consider the system in the form

(1) $$a_1x + b_1y = c_1$$
(2) $$a_2x + b_2y = c_2.$$

If we multiply equation (1) by b_2 and equation (2) by $-b_1$, we will be able to eliminate the y variable. Thus we have

(3) $$a_1b_2x + b_1b_2y = c_1b_2$$
(4) $$-a_2b_1x - b_1b_2y = -c_2b_1.$$

Adding (3) and (4) results in

(5) $$(a_1b_2 - a_2b_1)x = c_1b_2 - c_2b_1.$$

Now if $a_1b_2 - a_2b_1 \neq 0$, we can divide both sides of equation (5) by $a_1b_2 - a_2b_1$ and solve for x. The result is

$$x = \frac{c_1b_2 - c_2b_1}{a_1b_2 - a_2b_1}.$$

To solve for y, we return to the original system and multiply equation (1) by $-a_2$ and equation (2) by a_1, which results in

(6) $$-a_1a_2x - a_2b_1y = -a_2c_1$$
(7) $$a_1a_2x + a_1b_2y = a_1c_2.$$

Solving for y by adding equations (6) and (7) results in

$$(a_1b_2 - a_2b_1)y = a_1c_2 - a_2c_1$$

or

$$y = \frac{a_1c_2 - a_2c_1}{a_1b_2 - a_2b_1}$$

provided that $a_1b_2 - a_2b_1 \neq 0$. In summary, the solution to the general system given by (1) and (2) is

$$x = \frac{c_1b_2 - c_2b_1}{a_1b_2 - a_2b_1} \quad \text{and} \quad y = \frac{a_1c_2 - a_2c_1}{a_1b_2 - a_2b_1}.$$

These solutions can be conveniently represented using determinants as

$$x = \frac{c_1b_2 - c_2b_1}{a_1b_2 - a_2b_1} = \frac{\begin{vmatrix} c_1 & b_1 \\ c_2 & b_2 \end{vmatrix}}{\begin{vmatrix} a_1 & b_1 \\ a_2 & b_2 \end{vmatrix}}$$

and

$$y = \frac{a_1c_2 - a_2c_1}{a_1b_2 - a_2b_1} = \frac{\begin{vmatrix} a_1 & c_1 \\ a_2 & c_2 \end{vmatrix}}{\begin{vmatrix} a_1 & b_1 \\ a_2 & b_2 \end{vmatrix}}$$

provided that

$$\begin{vmatrix} a_1 & b_1 \\ a_2 & b_2 \end{vmatrix} = a_1b_2 - a_2b_1 \neq 0.$$

The three determinants in the solution are conveniently denoted by

$$\begin{vmatrix} a_1 & b_1 \\ a_2 & b_2 \end{vmatrix} = D, \qquad \begin{vmatrix} c_1 & b_1 \\ c_2 & b_2 \end{vmatrix} = D_x, \qquad \begin{vmatrix} a_1 & c_1 \\ a_2 & c_2 \end{vmatrix} = D_y.$$

The elements of D are the *coefficients* of the variables from the original system of equations. To get the elements of D_x we start with D and replace the coefficients of x (the first column of D) with the *constant terms* c_1 and c_2 from the original system. Similarly, to get the elements of D_y, we replace the coefficients of y in D with the constant terms c_1 and c_2. We summarize these results by stating the following theorem.

Cramer's Rule (2 × 2 System)

The system of equations

$$\begin{aligned} a_1x + b_1y &= c_1 \\ a_2x + b_2y &= c_2 \end{aligned}$$

has solutions given by

$$x = \frac{D_x}{D}, \qquad y = \frac{D_y}{D} \qquad \text{provided that } D \neq 0,$$

where

$$D = \begin{vmatrix} a_1 & b_1 \\ a_2 & b_2 \end{vmatrix}, \qquad D_x = \begin{vmatrix} c_1 & b_1 \\ c_2 & b_2 \end{vmatrix}, \qquad D_y = \begin{vmatrix} a_1 & c_1 \\ a_2 & c_2 \end{vmatrix}.$$

When $D \neq 0$ Cramer's rule will produce the *unique solution* to a system of equations. When $D = 0$ the system is either *inconsistent* or *dependent* and Cramer's rule does not apply.

Example 2. Use Cramer's rule to solve the system

$$\begin{aligned} 3x + 4y &= 5 \\ 2x + 7y &= -1. \end{aligned}$$

Solution: For this system the determinants are

$$D = \begin{vmatrix} 3 & 4 \\ 2 & 7 \end{vmatrix} = 3 \cdot 7 - 4 \cdot 2 = 21 - 8 = 13$$

$$D_x = \begin{vmatrix} 5 & 4 \\ -1 & 7 \end{vmatrix} = 5 \cdot 7 - 4(-1) = 35 + 4 = 39$$

$$D_y = \begin{vmatrix} 3 & 5 \\ 2 & -1 \end{vmatrix} = 3(-1) - 5 \cdot 2 = -3 - 10 = -13.$$

Thus the solution is given by

$$x = \frac{D_x}{D} = \frac{39}{13} = 3, \qquad y = \frac{D_y}{D} = \frac{-13}{13} = -1.$$

Therefore, the solution set is $\{(3, -1)\}$.

Example 3. Use Cramer's rule to solve the system

$$\begin{aligned} 5x - 2y &= 4 \\ 3x + y &= 1. \end{aligned}$$

Solution: For this system we have

$$D = \begin{vmatrix} 5 & -2 \\ 3 & 1 \end{vmatrix} = 5 - (-6) = 11$$

$$D_x = \begin{vmatrix} 4 & -2 \\ 1 & 1 \end{vmatrix} = 4 - (-2) = 6$$

$$D_y = \begin{vmatrix} 5 & 4 \\ 3 & 1 \end{vmatrix} = 5 - 12 = -7.$$

Thus

$$x = \frac{D_x}{D} = \frac{6}{11}, \qquad y = \frac{D_y}{D} = \frac{-7}{11}.$$

Therefore, the solution set is $\left\{\left(\frac{6}{11}, -\frac{7}{11}\right)\right\}$.

EXERCISES 9.6

For Exercises 1–10, evaluate the determinant. See Example 1.

1. $\begin{vmatrix} 5 & 4 \\ 7 & 6 \end{vmatrix}$ **2.** $\begin{vmatrix} 5 & 7 \\ 3 & 5 \end{vmatrix}$ **3.** $\begin{vmatrix} 8 & 5 \\ 7 & 4 \end{vmatrix}$ **4.** $\begin{vmatrix} 9 & 8 \\ 5 & 4 \end{vmatrix}$ **5.** $\begin{vmatrix} 6 & 9 \\ 8 & 12 \end{vmatrix}$

6. $\begin{vmatrix} 3 & 8 \\ 6 & 16 \end{vmatrix}$ **7.** $\begin{vmatrix} \frac{3}{4} & \frac{1}{2} \\ 1 & 2 \end{vmatrix}$ **8.** $\begin{vmatrix} \frac{5}{8} & \frac{3}{4} \\ 1 & 6 \end{vmatrix}$ **9.** $\begin{vmatrix} 1.5 & 2 \\ 2.5 & 3 \end{vmatrix}$ **10.** $\begin{vmatrix} 2.5 & 2 \\ 3.5 & 3 \end{vmatrix}$

For Exercises 11–30, use Cramer's rule to solve the system. See Examples 2 and 3.

11. $2x + y = 1$, $3x + 2y = 4$

12. $4x + y = 1$, $7x + y = 2$

13. $3x + 4y = 4$, $2x + 3y = 2$

14. $5x + 3y = 2$, $3x + 2y = 4$

15. $5x + 2y = 3$, $3x + y = 1$

16. $4x + 3y = 5$, $3x + 2y = 2$

17. $5x + 3y = 8$, $7x + 4y = 2$

18. $6x + 5y = 10$, $5x + 4y = 6$

19. $3x - 2y = 4$, $5x + y = 1$

20. $2x - 5y = 4$, $3x + 2y = 1$

21. $x - y = 5$, $2x - 3y = 6$

22. $x - y = 4$, $3x - 5y = 10$

23. $7x + 10y = 15$, $3x + 2y = 3$

24. $6x + 7y = 21$, $4x + 2y = 7$

25. $9x + 4y = 10$, $5x + 3y = 15$

26. $3x + 8y = 10$, $2x + 3y = 4$

27. $7x - 9y = 11$, $2x + 5y = 20$

28. $5x - 8y = 12$, $3x + 7y = 10$

29. $21x + 4y = 30$, $26x + 5y = 37$

30. $4x + 15y = 48$, $5x + 19y = 61$

Determinants of the Third Order (Optional) 9.7

In this section we show how to find the value of the determinant of a 3×3 matrix. We shall refer to such a determinant as a "3×3 determinant." We also use determinants to solve systems of three linear equations in three unknowns. A 3×3 determinant is called a determinant of the third order. The value of a 3×3 determinant is given by

$$(1) \quad \begin{vmatrix} a_1 & b_1 & c_1 \\ a_2 & b_2 & c_2 \\ a_3 & b_3 & c_3 \end{vmatrix} = a_1b_2c_3 + a_2b_3c_1 + a_3b_1c_2 - a_1b_3c_2 - a_2b_1c_3 - a_3b_2c_1 .$$

Because this formula is difficult to remember, alternative methods for evaluating determinants are used. The method we now describe is called *the expansion of a determinant by minors*.

Associated with each element of a 3×3 matrix is a 2×2 *determinant* called the *minor* of that element. This minor is obtained by deleting the row and column that contain the element. For example, in the matrix

$$\begin{bmatrix} a_1 & b_1 & c_1 \\ a_2 & b_2 & c_2 \\ a_3 & b_3 & c_3 \end{bmatrix}$$

the minor of c_3 is obtained by deleting row 3 and column 3,

$$\begin{bmatrix} a_1 & b_1 & \cancel{c_1} \\ a_2 & b_2 & \cancel{c_2} \\ \cancel{a_3} & \cancel{b_3} & \cancel{c_3} \end{bmatrix}$$

and forming the 2×2 determinant

$$\begin{vmatrix} a_1 & b_1 \\ a_2 & b_2 \end{vmatrix} = a_1b_2 - a_2b_1 .$$

Similarly, to get the minor of b_2 we delete row 2 and column 2,

$$\begin{bmatrix} a_1 & b_1 & c_1 \\ a_2 & b_2 & c_2 \\ a_3 & b_3 & c_3 \end{bmatrix}$$

and write the 2 × 2 determinant

$$\begin{vmatrix} a_1 & c_1 \\ a_3 & c_3 \end{vmatrix} = a_1c_3 - a_3c_1.$$

In the preceding discussion, because there was no ambiguity, we referred to the minors of the specific elements c_3 and b_2. It is also customary to identify a minor in terms of a particular row and column. For example, we might say "the minor of the element in row 2 and column 3."

Example 1. In the matrix $\begin{bmatrix} 1 & 2 & 3 \\ 4 & 5 & 6 \\ 7 & 8 & 9 \end{bmatrix}$ find and evaluate the minor of (a) 2. (b) 7.

Solution: The minor of an element is the determinant of the 2 × 2 matrix that results after deleting the row and column in the original matrix that contains that element. Thus

(a) The minor of 2 is $\begin{vmatrix} 4 & 6 \\ 7 & 9 \end{vmatrix} = 36 - 42 = -6.$

(b) The minor of 7 is $\begin{vmatrix} 2 & 3 \\ 5 & 6 \end{vmatrix} = 12 - 15 = -3.$

We now show how minors can be used to evaluate a determinant. Note that by rearranging the terms on the right-hand side of

(1) $$\begin{vmatrix} a_1 & b_1 & c_1 \\ a_2 & b_2 & c_2 \\ a_3 & b_3 & c_3 \end{vmatrix} = a_1b_2c_3 + a_2b_3c_1 + a_3b_1c_2 - a_1b_3c_2 - a_2b_1c_3 - a_3b_2c_1$$

and factoring, we can write

(2) $$\begin{vmatrix} a_1 & b_1 & c_1 \\ a_2 & b_2 & c_2 \\ a_3 & b_3 & c_3 \end{vmatrix} = a_1(b_2c_3 - b_3c_2) - a_2(b_1c_3 - b_3c_1) + a_3(b_1c_2 - b_2c_1).$$

Thus

(3) $$\begin{vmatrix} a_1 & b_1 & c_1 \\ a_2 & b_2 & c_2 \\ a_3 & b_3 & c_3 \end{vmatrix} = a_1\begin{vmatrix} b_2 & b_3 \\ c_2 & c_3 \end{vmatrix} - a_2\begin{vmatrix} b_1 & b_3 \\ c_1 & c_3 \end{vmatrix} + a_3\begin{vmatrix} b_1 & b_2 \\ c_1 & c_2 \end{vmatrix}.$$

Note that the value of the determinant of a 3×3 matrix can be found by (a) multiplying each element in column 1 by its minor and then (b) combining these products as shown in (3).

This way of evaluating the determinant of a matrix is called *expansion by minors of the first column of the matrix.*

Example 2. By expanding by minors of the first column, evaluate the determinant

$$\begin{vmatrix} 1 & 2 & 3 \\ 4 & 5 & 6 \\ 7 & 8 & 9 \end{vmatrix}.$$

Solution: Referring to (3), we may write

$$\begin{aligned} \begin{vmatrix} 1 & 2 & 3 \\ 4 & 5 & 6 \\ 7 & 8 & 9 \end{vmatrix} &= 1\begin{vmatrix} 5 & 6 \\ 8 & 9 \end{vmatrix} - 4\begin{vmatrix} 2 & 3 \\ 8 & 9 \end{vmatrix} + 7\begin{vmatrix} 2 & 3 \\ 5 & 6 \end{vmatrix} \\ &= 1(45 - 48) - 4(18 - 24) + 7(12 - 15) \\ &= 1(-3) - 4(-6) + 7(-3) \\ &= -3 + 24 - 21 \\ &= 0. \end{aligned}$$

It is possible to rearrange the terms of (1) and factor out the elements of any one row or any one column. Thus *it is possible to evaluate a* 3×3 *determinant using the expansion by minors of any particular row or column.* In (3), which is the expansion of the determinant of a 3×3 matrix by minors of the first column, notice that we changed the sign of the element in row 2, column 1. When we use a different row or column, we must be able to determine the sign associated with each minor. One way to do this is to use the following pattern of signs:

$$\begin{matrix} + & - & + \\ - & + & - \\ + & - & + \end{matrix}.$$

Beginning with a + in the first row, first column position, the signs alternate as we go across any row or down any column. The following example illustrates how this pattern of signs enters into the evaluation of a determinant by the expansion by minors.

Example 3. Evaluate $\begin{vmatrix} -4 & 3 & -1 \\ 2 & 5 & -3 \\ 7 & 1 & -2 \end{vmatrix}.$

Solution: If we expand by minors of the *second row* and use the sign pattern for that row, we obtain

$$\begin{vmatrix} -4 & 3 & -1 \\ 2 & 5 & -3 \\ 7 & 1 & -2 \end{vmatrix} = -(2)\begin{vmatrix} 3 & -1 \\ 1 & -2 \end{vmatrix} + 5\begin{vmatrix} -4 & -1 \\ 7 & -2 \end{vmatrix} - (-3)\begin{vmatrix} -4 & 3 \\ 7 & 1 \end{vmatrix}$$

$$= -2(-6 + 1) + 5(8 + 7) + 3(-4 - 21)$$
$$= -2(-5) + 5(15) + 3(-25)$$
$$= 10.$$

Example 4. Evaluate $\begin{vmatrix} 4 & 0 & -2 \\ 10 & 3 & 11 \\ 7 & 0 & 1 \end{vmatrix}$.

Solution: When a matrix contains one or more zeros, it is easier to evaluate the determinant by expanding along a row or column that has the most zeros. Thus, for this determinant, we will expand by the second column:

$$\begin{vmatrix} 4 & 0 & -2 \\ 10 & 3 & 11 \\ 7 & 0 & 1 \end{vmatrix} = -0\begin{vmatrix} 10 & 11 \\ 7 & 1 \end{vmatrix} + 3\begin{vmatrix} 4 & -2 \\ 7 & 1 \end{vmatrix} - 0\begin{vmatrix} 4 & -2 \\ 10 & 11 \end{vmatrix}$$
$$= 3\begin{vmatrix} 4 & -2 \\ 7 & 1 \end{vmatrix} = 3(4 + 14) = 54.$$

The preceding discussion on 3×3 determinants enables us to state Cramer's rule for a system of three linear equations in three unknowns.

Cramer's Rule (3×3 System)

The system of linear equations

$$a_1x + b_1y + c_1z = d_1$$
$$a_2x + b_2y + c_2z = d_2$$
$$a_3x + b_3y + c_3z = d_3$$

has the solution

$$x = \frac{D_x}{D}, \qquad y = \frac{D_y}{D}, \qquad z = \frac{D_z}{D},$$

where

$$D = \begin{vmatrix} a_1 & b_1 & c_1 \\ a_2 & b_2 & c_2 \\ a_3 & b_3 & c_3 \end{vmatrix}, \qquad D_x = \begin{vmatrix} d_1 & b_1 & c_1 \\ d_2 & b_2 & c_2 \\ d_3 & b_3 & c_3 \end{vmatrix},$$

$$D_y = \begin{vmatrix} a_1 & d_1 & c_1 \\ a_2 & d_2 & c_2 \\ a_3 & d_3 & c_3 \end{vmatrix}, \qquad D_z = \begin{vmatrix} a_1 & b_1 & d_1 \\ a_2 & b_2 & d_2 \\ a_3 & b_3 & d_3 \end{vmatrix},$$

provided that $D \neq 0$.

If $D = 0$, Cramer's rule does not apply and the system is either *inconsistent* or *dependent*.

In Cramer's rule (3 × 3), notice that the elements of D are the coefficients of the variables in the original system. To get the elements of D_x we start with D and replace the coefficients of x (the first column of D) with the constant terms d_1, d_2, and d_3 from the original system. To get the elements of D_y and D_z, we start with D and replace the coefficients of y and z, respectively, with the constant terms d_1, d_2, d_3.

Example 5. Use Cramer's rule to solve

$$\begin{aligned} 2x - y - z &= 4 \\ x + y + 3z &= 5 \\ 5x - 3y + z &= 1. \end{aligned}$$

Solution: We begin by evaluating D. If we expand by row 1, we obtain

$$D = \begin{vmatrix} 2 & -1 & -1 \\ 1 & 1 & 3 \\ 5 & -3 & 1 \end{vmatrix} = 2\begin{vmatrix} 1 & 3 \\ -3 & 1 \end{vmatrix} - (-1)\begin{vmatrix} 1 & 3 \\ 5 & 1 \end{vmatrix} + (-1)\begin{vmatrix} 1 & 1 \\ 5 & -3 \end{vmatrix}$$

$$= 2(10) + 1(-14) - 1(-8)$$
$$= 14.$$

Similarly, we can evaluate D_x by expanding by column 3. Thus

$$D_x = \begin{vmatrix} 4 & -1 & -1 \\ 5 & 1 & 3 \\ 1 & -3 & 1 \end{vmatrix} = -1\begin{vmatrix} 5 & 1 \\ 1 & -3 \end{vmatrix} - 3\begin{vmatrix} 4 & -1 \\ 1 & -3 \end{vmatrix} + 1\begin{vmatrix} 4 & -1 \\ 5 & 1 \end{vmatrix}$$

$$= -1(-16) - 3(-11) + 1(9)$$
$$= 16 + 33 + 9$$
$$= 58.$$

Similarly,

$$D_y = 84 \quad \text{and} \quad D_z = -24.$$

Applying Cramer's rule, we get

$$x = \frac{D_x}{D} = \frac{58}{14} = \frac{29}{7}, \qquad y = \frac{D_y}{D} = \frac{84}{14} = 6, \qquad z = \frac{D_z}{D} = \frac{-24}{14} = \frac{-12}{7}.$$

Thus the solution set is $\left\{\left(\frac{29}{7}, 6, -\frac{12}{7}\right)\right\}$.

Although Cramer's rule can be used to solve systems of four equations in four unknowns, the method of eliminating variables is usually more efficient.

EXERCISES 9.7

For Exercises 1–6, refer to $\begin{bmatrix} 5 & 4 & -2 \\ 3 & 2 & 20 \\ 1 & 10 & 7 \end{bmatrix}$ *and find the value of the minor of the indicated element. See Example 1.*

1. 5 **2.** 10 **3.** −2 **4.** 1 **5.** 7 **6.** 20

For Exercises 7–12, evaluate the determinant. Expand by a row or column of your choice. See Examples 2–4.

7. $\begin{vmatrix} 3 & 0 & 2 \\ -1 & 6 & 7 \\ 4 & 2 & 1 \end{vmatrix}$

8. $\begin{vmatrix} 4 & -2 & 0 \\ 3 & 5 & 0 \\ 10 & 0 & 11 \end{vmatrix}$

9. $\begin{vmatrix} 0 & 1 & 8 \\ 2 & 0 & -5 \\ 0 & 6 & 3 \end{vmatrix}$

10. $\begin{vmatrix} 12 & -5 & 3 \\ 1 & 0 & 2 \\ -2 & 1 & -1 \end{vmatrix}$

11. $\begin{vmatrix} 3 & 2 & -1 \\ 5 & 7 & 8 \\ 4 & -5 & 6 \end{vmatrix}$

12. $\begin{vmatrix} 5 & 1 & -3 \\ 6 & 2 & 7 \\ 4 & 3 & -2 \end{vmatrix}$

For Exercises 13–22, use Cramer's rule to solve the system of equations. See Example 5.

13. $\begin{aligned} x + 2y + z &= 3 \\ 2x + 3z &= 1 \\ y - 2z &= 2 \end{aligned}$

14. $\begin{aligned} x + 3y + 2z &= 4 \\ 2x - z &= 1 \\ 2y + z &= 3 \end{aligned}$

15. $\begin{aligned} 3x + z &= 10 \\ 2y - 3z &= 5 \\ 5x + 2z &= 10 \end{aligned}$

16. $\begin{aligned} 2x + z &= 8 \\ 3y + 2z &= 5 \\ 3x - z &= 2 \end{aligned}$

17. $\begin{aligned} x + y - 2z &= 3 \\ 2x - y + z &= 2 \\ x + 2y - 4z &= 4 \end{aligned}$

18. $\begin{aligned} x - y + 3z &= 5 \\ 3x + y - z &= 13 \\ x + 2y - 2z &= 6 \end{aligned}$

19. $\begin{aligned} 2x + 3y + z &= 2 \\ 3x + y + 2z &= 1 \\ 5x - 4y + z &= -11 \end{aligned}$

20. $\begin{aligned} 3x + 2y - z &= 3 \\ 4x - y + 3z &= -1 \\ 2x + 3y + 2z &= -5 \end{aligned}$

21. $\begin{aligned} 4x - 5y + 2z &= 10 \\ 3x + 2y + 4z &= 5 \\ 5x - 3y + 3z &= 10 \end{aligned}$

22. $\begin{aligned} 3x + 5y - 2z &= 12 \\ 4x - 2y + 3z &= 5 \\ 3x + 4y - 5z &= 10 \end{aligned}$

9.8 Solving Linear Systems Using an Augmented Matrix (Optional)

Earlier in this chapter we discussed how the method of addition (elimination) could be used to solve a system of linear equations. In this section we discuss the addition method in terms of the *augmented matrix* that is associated with a system of linear equations. For example, if the system of equations is

$$\begin{aligned} 7x + 5y &= 1 \\ 4x + 7y &= 2, \end{aligned}$$

then the associated augmented matrix is

$$\left[\begin{array}{cc|c} 7 & 5 & 1 \\ 4 & 7 & 2 \end{array}\right].$$

Notice that the coefficients of the variable terms are placed to the left of the vertical bar and the constant terms are placed to the right of the bar.

Example 1. For each system of equations, write the associated augmented matrix.

(a) $\begin{aligned} 4x - 3y &= 11 \\ 5x + 2y &= 8. \end{aligned}$ (b) $\begin{aligned} 3x + y + z &= 4 \\ x - y + 2z &= 5 \\ 4x + 3y + 7z &= 11. \end{aligned}$

Solution

(a) $\left[\begin{array}{cc|c} 4 & -3 & 11 \\ 5 & 2 & 8 \end{array}\right].$ (b) $\left[\begin{array}{ccc|c} 3 & 1 & 1 & 4 \\ 1 & -1 & 2 & 5 \\ 4 & 3 & 7 & 11 \end{array}\right].$

Example 2. Use the variables x and y and write the system of equations that is associated with each augmented matrix.

(a) $\left[\begin{array}{cc|c} 11 & -3 & 6 \\ 2 & 1 & 7 \end{array}\right].$ (b) $\left[\begin{array}{cc|c} 1 & 5 & 3 \\ 0 & 1 & 2 \end{array}\right].$

Solution

(a) $\begin{aligned} 11x - 3y &= 6 \\ 2x + y &= 7. \end{aligned}$ (b) $\begin{aligned} x + 5y &= 3 \\ y &= 2. \end{aligned}$

Consider the system of equations

(1) $x + 2y = 3$
(2) $3x + 5y = 7.$

Now we solve this system using the addition method. On the left we write the systems of equations, on the right the corresponding augmented matrix. Note that for each operation performed on the system there is a corresponding operation on the rows of the augmented matrix. We begin with the original system:

(1) $x + 2y = 3$
(2) $3x + 5y = 7$ $\quad\left[\begin{array}{cc|c} 1 & 2 & 3 \\ 3 & 5 & 7 \end{array}\right].$

Next, we write the system that results from subtracting three times equation (1) from equation (2):

(1) $x + 2y = 3$
(3) $-y = -2$ $\quad\left[\begin{array}{cc|c} 1 & 2 & 3 \\ 0 & -1 & -2 \end{array}\right].$

In the last system if we replace equation (3) with -1 times equation (3), the result is

(1) $x + 2y = 3$
(4) $y = 2$ $\quad\left[\begin{array}{cc|c} 1 & 2 & 3 \\ 0 & 1 & 2 \end{array}\right].$

This last system is easy to solve because from equation (4) it is clear that $y = 2$. Replacing y with 2 in equation (1) results in

$$\begin{aligned} x + 2(2) &= 3 \\ x &= -1. \end{aligned}$$

Thus the solution set is $\{(-1, 2)\}$.

In the remainder of this section we discuss systems of equations that have unique solutions. Our goal is to start with an arbitrary system and produce an equivalent system with specific characteristics. Also, instead of working with the actual systems of equations we will work with the associated augmented matrices.

Consider a system of three equations in three unknowns that has an augmented matrix with the following *characteristics*.

1. The element in row 1, column 1 is 1 and the rest of column 1 consists of 0's.
2. The element in row 2, column 2 is 1 and the element in row 3, column 2 is 0.
3. The element in row 3, column 3 is 1.

Below on the left is such an augmented matrix. On the right is the associated system of equations.

$$\left[\begin{array}{ccc|c} 1 & a & b & d \\ 0 & 1 & c & e \\ 0 & 0 & 1 & f \end{array}\right] \qquad \begin{array}{lr} (1) & x + ay + bz = d \\ (2) & y + cz = e \\ (3) & z = f. \end{array}$$

It is easy to solve systems such as that described by equations (1)–(3). This is because equation (3) gives the value of z that can be used in equation (2) to find the value of y. Finally, once z and y are known, equation (1) will give the value of x.

Theorem 1 (Section 9.1) listed operations that when applied to a system of equations produced an equivalent system of equations. Recall that equivalent systems have the same solution set. When we use augmented matrices, instead of operating on equations, we operate with the rows of the matrix. We now list the *row operations* that may be applied to an augmented matrix. They will produce another augmented matrix that corresponds to an equivalent system of equations.

Elementary Row Operations

1. Interchange any two rows.
2. Replace any row with a nonzero multiple of itself.
3. Replace any row with the sum of that row and any multiple of a different row.

It is instructive to compare these elementary row operations with the operations described in Theorem 1. The following examples illustrate how to use the augmented matrix and elementary row operations to solve a system of linear equations.

Example 3. Use elementary row operations to solve the system

$$\begin{aligned} x - 5y &= -3 \\ 3x - 4y &= 2. \end{aligned}$$

Solution: The augmented matrix for this system is

$$\left[\begin{array}{cc|c} 1 & -5 & -3 \\ 3 & -4 & 2 \end{array}\right]$$

We now apply elementary row operations to this matrix until we obtain a matrix that has characteristics (1) and (2), restricted to a system of two equations in two variables, that were described on page 370. That is, we want to get a matrix that

1. Has a 1 in row 1, column 1 and 0 in row 2, column 1.
2. Has a 1 in row 2, column 2.

$$\left[\begin{array}{cc|c} 1 & -5 & -3 \\ 3 & -4 & 2 \end{array}\right]$$ the original augmented matrix

$$\left[\begin{array}{cc|c} 1 & -5 & -3 \\ 0 & 11 & 11 \end{array}\right]$$ add (-3) times row 1 to row 2 to obtain the new row 2

$$\left[\begin{array}{cc|c} 1 & -5 & -3 \\ 0 & 1 & 1 \end{array}\right]$$ row 2 has been replaced by $\frac{1}{11}$ times row 2

This augmented matrix corresponds to the system

$$\begin{aligned} x - 5y &= -3 \\ y &= 1. \end{aligned}$$

From the second equation it is clear that $y = 1$. Substituting 1 for y in the first equation results in

$$\begin{aligned} x - 5(1) &= -3 \\ x - 5 &= -3 \\ x &= 2. \end{aligned}$$

Thus the solution set is $\{(2, 1)\}$.

Example 4. Use elementary row operations to solve the system

$$\begin{aligned} 3x - y + 5z &= 6 \\ x - 4y - 2z &= -20 \\ 2x - y + 6z &= 5. \end{aligned}$$

Solution

$$\left[\begin{array}{ccc|c} 3 & -1 & 5 & 6 \\ 1 & -4 & -2 & -20 \\ 2 & -1 & 6 & 5 \end{array}\right]$$ the original augmented matrix

$$\left[\begin{array}{ccc|c} 1 & -4 & -2 & -20 \\ 3 & -1 & 5 & 6 \\ 2 & -1 & 6 & 5 \end{array}\right]$$ rows 1 and 2 were interchanged in order to get a 1 in row 1, column 1

$\begin{bmatrix} 1 & -4 & -2 & | & -20 \\ 0 & 11 & 11 & | & 66 \\ 2 & -1 & 6 & | & 5 \end{bmatrix}$ new row 2 is the result of adding (−3) times row 1 to old row 2

$\begin{bmatrix} 1 & -4 & -2 & | & -20 \\ 0 & 11 & 11 & | & 66 \\ 0 & 7 & 10 & | & 45 \end{bmatrix}$ new row 3 is the result of adding (−2) times row 1 to old row 3

$\begin{bmatrix} 1 & -4 & -2 & | & -20 \\ 0 & 1 & 1 & | & 6 \\ 0 & 7 & 10 & | & 45 \end{bmatrix}$ to make the first nonzero element in row 2 equal to 1, we have divided each element of row 2 by 11

$\begin{bmatrix} 1 & -4 & -2 & | & -20 \\ 0 & 1 & 1 & | & 6 \\ 0 & 0 & 3 & | & 3 \end{bmatrix}$ new row 3 is the result of adding (−7) times row 2 to old row 3

$\begin{bmatrix} 1 & -4 & -2 & | & -20 \\ 0 & 1 & 1 & | & 6 \\ 0 & 0 & 1 & | & 1 \end{bmatrix}$ row 3 has been replaced by $\frac{1}{3}$ times row 3

Notice that we now have an augmented matrix that corresponds to the system

$$\begin{aligned} x - 4y - 2z &= -20 \\ y + z &= 6 \\ z &= 1. \end{aligned}$$

Thus $z = 1$. Substituting 1 for z in the second equation, we get

$$\begin{aligned} y + 1 &= 6 \\ y &= 5. \end{aligned}$$

Finally, substitute 5 for y and 1 for z in the first equation. Thus

$$\begin{aligned} x - 4(5) - 2(1) &= -20 \\ x - 20 - 2 &= -20 \\ x &= 2. \end{aligned}$$

Therefore, the solution set is $\{(2, 5, 1)\}$.

EXERCISES 9.8

For Exercises 1–6, write the augmented matrix that corresponds to the system of equations. See Example 1.

1. $\begin{aligned} 4x + 5y &= 3 \\ 7x - 8y &= 15 \end{aligned}$

2. $\begin{aligned} 11x + 2y &= 20 \\ 5x - 3y &= 4 \end{aligned}$

3. $\begin{aligned} 3x - y &= -2 \\ -x + 2y &= 1 \end{aligned}$

4. $\begin{aligned} 5x - y &= -1 \\ -x + 3y &= 0 \end{aligned}$

5. $\begin{aligned} x + 2y - 3z &= 1 \\ 2x - 3y + 4z &= 5 \\ 3x + 5y - z &= 7 \end{aligned}$

6. $\begin{aligned} x + 7y - z &= 0 \\ 4x - y + 3z &= 10 \\ 2x + y + 5z &= 7 \end{aligned}$

For Exercises 7–10, write the system of equations that corresponds to the augmented matrix. See Example 2.

7. $\left[\begin{array}{cc|c} 3 & 4 & 10 \\ 2 & -1 & 20 \end{array}\right]$ **8.** $\left[\begin{array}{cc|c} 2 & -5 & 6 \\ 0 & 1 & -1 \end{array}\right]$ **9.** $\left[\begin{array}{ccc|c} 3 & 5 & 7 & 6 \\ 2 & 0 & 1 & 4 \\ -2 & 3 & 6 & 4 \end{array}\right]$ **10.** $\left[\begin{array}{ccc|c} 2 & 1 & 0 & 3 \\ 6 & -2 & 1 & 0 \\ 4 & 3 & 5 & 2 \end{array}\right]$

For Exercises 11–18, use an augmented matrix and elementary row operations to solve the system of equations. See Example 3.

11. $\begin{aligned} x + 2y &= 13 \\ 3x - y &= -3 \end{aligned}$ **12.** $\begin{aligned} x - 3y &= 5 \\ 7x + 5y &= 9 \end{aligned}$ **13.** $\begin{aligned} x + 4y &= 3 \\ -2x + 3y &= 16 \end{aligned}$ **14.** $\begin{aligned} x + 6y &= 11 \\ 2x + 5y &= 1 \end{aligned}$

15. $\begin{aligned} 3x + 4y &= -8 \\ x + y &= -3 \end{aligned}$ **16.** $\begin{aligned} 3x + 2y &= 11 \\ x + y &= 2 \end{aligned}$ **17.** $\begin{aligned} 3x + 2y &= 2 \\ 6x - 8y &= -2 \end{aligned}$ **18.** $\begin{aligned} 11x + y &= 2 \\ 3x - 3y &= 3 \end{aligned}$

For Exercises 19–24, use an augmented matrix and elementary row operations to solve the system of equations. See Example 4.

19. $\begin{aligned} x + 2y + z &= -1 \\ 2x - 3y - 5z &= 5 \\ 2y + 3z &= 6 \end{aligned}$ **20.** $\begin{aligned} x + 3y + 2z &= -7 \\ 3x - 2y &= 11 \\ 4y + 5z &= -6 \end{aligned}$ **21.** $\begin{aligned} x + y + z &= 2 \\ x + 3y + 2z &= 1 \\ 3x + y - z &= 4 \end{aligned}$

22. $\begin{aligned} x + y + z &= 2 \\ x + 4y + 3z &= -2 \\ 2x + y - 4z &= 0 \end{aligned}$ **23.** $\begin{aligned} 4x + 2y + 3z &= 1 \\ 4x + y + z &= 0 \\ x + z &= 0 \end{aligned}$ **24.** $\begin{aligned} 2x + 3y - z &= 3 \\ 2x + y - 3z &= 2 \\ x + z &= 0 \end{aligned}$

Key Words

Addition method
Augmented matrix
Consistent system
Cramer's rule
Dependent system
Determinant
Elimination method
Equivalent systems
Expansion by minors
Graphical method
Inconsistent system
Independent system
Linear system
Matrix
Minor
Ordered pair
Ordered triplet
Row operations
Simultaneous equations
Substitution method
Systems of equations
Systems of inequalities

CHAPTER 9 REVIEW

SECTION 9.1

Use the graphical method to solve each system of equations.

1. $\begin{aligned} -x + y &= 1 \\ x + y &= 5 \end{aligned}$ **2.** $\begin{aligned} x + 6y &= 18 \\ x - y &= 4 \end{aligned}$

Solve each system of equations using either the addition method or the substitution method. If a system is dependent or inconsistent, label it accordingly.

3. $x + y = 7$, $x - y = 3$

4. $2x + y = 9$, $5x + 2y = 19$

5. $3x - 2y = 5$, $-6x + 4y = 10$

6. $7x + 3y = 9$, $3x - 5y = 7$

7. $9x - 6y = 15$, $-6x + 4y = -10$

8. $6x - 4y = 3$, $3x + 8y = 4$

SECTION 9.2

Solve each system of equations. Identify any systems that are dependent or inconsistent.

9. $x - y + z = 0$, $x + y - z = 0$, $x - 2y + z = 4$

10. $x - 2y + 3z = 11$, $2x + 2y + z = -1$, $3x + y + 4z = 7$

11. $2x + y + 3z = 7$, $3x + 2y + z = 5$, $3x + 2y + 2z = 7$

12. $2x + 3y - z = -4$, $-4x - 6y + 2z = 8$, $6x + 9y - 3z = -12$

13. $3x + 2y + z = 3$, $9x - 2y - z = 1$, $3x + 2y - 3z = 1$

14. $3x + y - 2z = -3$, $-6x - 2y + 4z = 5$, $3x + y + 4z = 7$

SECTION 9.3

Solve each word problem.

15. Find two numbers with a sum of 80 and a difference of 6.

16. Find three numbers with a sum of 83 if the third number is 3 more than the sum of the first and the second, and the second is 4 less than three times the first.

17. A chemist has two solutions of alcohol; one is a 40% alcohol solution and the other is a 90% solution. If she wants 20 ounces of a 50% solution, how many ounces of each of the two solutions should be used?

18. A jar contains 160 marbles. The total weight of the marbles is 324 ounces. Some of the marbles weigh 2.5 ounces each and the rest weigh 1.5 ounces each. How many of each type of marble are there in the jar?

19. A boat can travel 5 miles upstream in 30 minutes and 5 miles downstream in 20 minutes. What is the speed of the current, and what is the speed of the boat in still water?

20. What are the values of a, b, and c given that the points $(-1, 0)$, $(1, 6)$, and $(2, 3)$ are on the graph of $y = ax^2 + bx + c$?

SECTION 9.4

Graph the solution set of each system of linear inequalities.

21. $x + y \le 7$, $2x + 3y \le 18$

22. $x + y > 5$, $2x + y < 8$

23. $x \ge 0, y \ge 0$, $3x + y \le 9$, $x + 2y \le 6$

SECTION 9.5

Solve each system of equations.

24. $x^2 - y = 2$, $2x - y = -1$

25. $4x^2 + 9y^2 = 36$, $9x^2 + 4y^2 = 36$

Solve the following word problem.

26. The sum of two numbers is 25. The difference of the squares of the same two numbers is 75. Find the numbers.

SECTION 9.6

Evaluate each determinant.

27. $\begin{vmatrix} 8 & 9 \\ 5 & 6 \end{vmatrix}$ **28.** $\begin{vmatrix} 7 & 3 \\ 5 & 2 \end{vmatrix}$ **29.** $\begin{vmatrix} -2 & 3 \\ 5 & 1 \end{vmatrix}$ **30.** $\begin{vmatrix} \frac{2}{3} & 6 \\ \frac{1}{3} & 12 \end{vmatrix}$

Use Cramer's rule to solve each system of equations.

31. $\begin{aligned} 3x + y &= 1 \\ 5x + 2y &= 1 \end{aligned}$ **32.** $\begin{aligned} 2x + 3y &= 5 \\ 3x + 4y &= 6 \end{aligned}$ **33.** $\begin{aligned} 4x - 5y &= 11 \\ 6x + 2y &= 7 \end{aligned}$ **34.** $\begin{aligned} 5x + 7y &= 10 \\ 6x + 11y &= 20 \end{aligned}$

SECTION 9.7

Refer to the matrix

$$\begin{bmatrix} 3 & -1 & 7 \\ 5 & 10 & 2 \\ 1 & 4 & 0 \end{bmatrix}$$

and find the minor of the indicated element.

35. 4 **36.** 7

Evaluate each determinant.

37. $\begin{vmatrix} 3 & 0 & 2 \\ 5 & 1 & 7 \\ 1 & 2 & -1 \end{vmatrix}$ **38.** $\begin{vmatrix} 5 & 6 & 4 \\ -3 & 7 & -2 \\ 2 & 2 & 3 \end{vmatrix}$

Use Cramer's rule to solve each system of equations.

39. $\begin{aligned} x \quad\quad + 3z &= 4 \\ y - 2z &= 0 \\ x + 2y + z &= 2 \end{aligned}$ **40.** $\begin{aligned} 3x + 2y - 3z &= -4 \\ 4x + 6y + 3z &= 1 \\ 5x - 2y - 3z &= -8 \end{aligned}$

SECTION 9.8

Write the augmented matrix and then use elementary row operations to solve each system of equations.

41. $\begin{aligned} x + 3y &= 2 \\ 4x + 13y &= 9 \end{aligned}$ **42.** $\begin{aligned} 2x + 5y &= 0 \\ 6x + 5y &= 20 \end{aligned}$

43. $\begin{aligned} x + 3y + z &= 5 \\ y + 4z &= 6 \\ 2x + 7y + 5z &= 12 \end{aligned}$ **44.** $\begin{aligned} 3x + y + 20z &= 17 \\ x \quad\quad + 6z &= 5 \\ 2x + 2y + 15z &= 12 \end{aligned}$

CHAPTER 9 TEST

For Problems 1–6, solve the system of equations using either the addition method or the substitution method. Identify any systems that are inconsistent or dependent.

1. $\begin{aligned} x + y &= 6 \\ x - y &= -2 \end{aligned}$

2. $\begin{aligned} x + 2y &= 9 \\ 3x - y &= 13 \end{aligned}$

3. $\begin{aligned} 2x + 3y &= 11 \\ 5x - 2y &= -20 \end{aligned}$

4. $\begin{aligned} 4x + 3y &= 6 \\ 8x + 6y &= 10 \end{aligned}$

5. $\begin{aligned} x - 2y + 3z &= 4 \\ -x + y + 6z &= 0 \\ 3x + 2y &= 1 \end{aligned}$

6. $\begin{aligned} 3x + 2y - 4z &= 12 \\ -6x - 4y + 8z &= -24 \\ \tfrac{3}{2}x + y - 2z &= 6 \end{aligned}$

For Problems 7 and 8, write and solve a system of equations.

7. The sum of two numbers is 62. The larger number is equal to 16 more than the smaller number. Find the numbers.

8. Maria has a small collection of 36 stamps. The stamps have either a face value of \$.25 or \$.35. The total face value of her collection is \$10.40. How many of each type stamp are in her collection?

9. Graph the solution set of

$$\begin{aligned} x \geq 0,\ y &\geq 0 \\ x + 2y &\leq 8 \\ 2x + 3y &\leq 12 \end{aligned}$$

10. Solve the system

$$\begin{aligned} y &= x^2 + 3x \\ y &= 2x^2 - 4 \end{aligned}$$

11. Evaluate each determinant.

(a) $\begin{vmatrix} 3 & 6 \\ 1 & 8 \end{vmatrix}$ **(b)** $\begin{vmatrix} 2 & 4 & 1 \\ 3 & -1 & 0 \\ -2 & 5 & 6 \end{vmatrix}$

12. Use Cramer's rule to solve each system of equations.

(a) $\begin{aligned} 4x + 2y &= 3 \\ 3x + 3y &= 1 \end{aligned}$ **(b)** $\begin{aligned} x \quad\quad + z &= 2 \\ 2x + y \quad\quad &= 1 \\ x - y + z &= 7 \end{aligned}$

13. Solve the system of equations using an augmented matrix and elementary row operations.

$$\begin{aligned} x + 2y - z &= -1 \\ 2x + 5y + z &= 16 \\ x + 2y + 2z &= 14 \end{aligned}$$

CHAPTER 10
Functions

For a person in an airplane the approximate distance, D, to the horizon is a function of, or depends on, the height, h, of the airplane above the ground. The function is specified by the equation $D = 1.25\sqrt{h}$. For a person in an airplane 4900 feet above the ground, the horizon is about 87.5 miles distant.

Kerwin B. Roche—Freelance Photographer's Guild

CHAPTER 10
CHAPTER 10
CHAPTER 10
CHAPTER 10
CHAPTER 10
CHAPTER 10
CHAPTER 10
CHAPTER 10
CHAPTER 10
CHAPTER 10
CHAPTER 10
CHAPTER 10
CHAPTER 10
CHAPTER 10
CHAPTER 10
CHAPTER 10
CHAPTER 10
CHAPTER 10
CHAPTER 10
CHAPTER 10
CHAPTER 10
CHAPTER 10
CHAPTER 10
CHAPTER 10
CHAPTER 10
CHAPTER 10
CHAPTER 10
CHAPTER 10
CHAPTER 10
CHAPTER 10
CHAPTER 10
CHAPTER 10
CHAPTER 10
CHAPTER 10

The mathematical concept of a function is of basic importance in most of modern mathematics. Functions have a wide variety of application in the physical and social sciences. We introduce the definition of a function as a special type of relation between two variables. In addition, we will familiarize ourselves with the notation associated with functions. We shall also become acquainted with some special types of functions. At the close of the chapter we discuss variation, which is a general name for terminology usually used in the sciences to describe certain relationships among variables that often involve functions.

10.1 Relations

By a relation we mean something like marriage (between men and women), being the mother of (between people), belonging to (between elements and sets), and being equal (between numbers). Our discussion of relations will be confined to relations involving numbers. We will see that the language of sets proves very useful for formulating the concept of a relation in mathematics. We illustrate this in the following discussion.

Consider the following sets A and B:

$$A = \{3, 5, 6\} \qquad \text{and} \qquad B = \{2, 4, 7\}.$$

Let the variables x and y represent any element of sets A and B, respectively. Now let us consider the following *relationship* between the elements or members of A and B;

$$x \text{ is less than } y \qquad (x < y).$$

The relationship "*less than*" between members of A and B may be thought of as the set of **all** *ordered pairs* (x, y) for which the statement "x is less than y" is *true*. Thus the relation "*less than*" between the members of A and B is equivalent to the set of ordered pairs

$$R = \{(x, y) \mid x \in A,\ y \in B, \text{ and } x < y\}$$

read "the set of (x, y) such that x belongs to A, y belongs to B, and x is less than y." Since the number 3 in set A is *less than* the number 4 in set B, we have that the ordered pair (3, 4) *belongs to* set R. That is, $(3, 4) \in R$. Can you determine other ordered pairs that belong to R? The set of ordered pairs

$$R = \{(3, 4), (3, 7), (5, 7), (6, 7)\}$$

defines the relation "*less than*" between the members of the sets

$$A = \{3, 5, 6\} \qquad \text{and} \qquad B = \{2, 4, 7\}.$$

Thus since (5, 7) belongs to set R, we know that $5 < 7$. It is sometimes convenient to say that *R is a relation from A to B*.

The previous discussion suggests a close connection between relations and sets and motivates the following definition of a relation.

Definition of a Relation

> A relation is a set of ordered pairs.

The *domain* of a relation is the set of first components of the ordered pairs making up the relation and the *range* of the relation is the set of second components of the ordered pairs of the relation. Thus, for the relation

$$R = \{(3, 4), (3, 7), (5, 7), (6, 7)\},$$

we have that

$$\text{domain of } R = \{3, 5, 6\}; \qquad \text{range of } R = \{4, 7\}.$$

The *graph* of a relation is the set of points in the coordinate plane associated with the ordered pairs of the relation.

As we shall see shortly, there are several ways to define a relation. When the relation consists of a small number of ordered pairs, we list them, as shown in the following example.

Example 1. Determine the domain and range of the relation

$$R = \{(-1, -1), (1, 2), (3, 2), (3, 4)\}.$$

Show its graph.

Solution: The domain of R is $\{-1, 1, 3\}$ and the range is $\{-1, 2, 4\}$. The graph of R is shown in Figure 10.1.

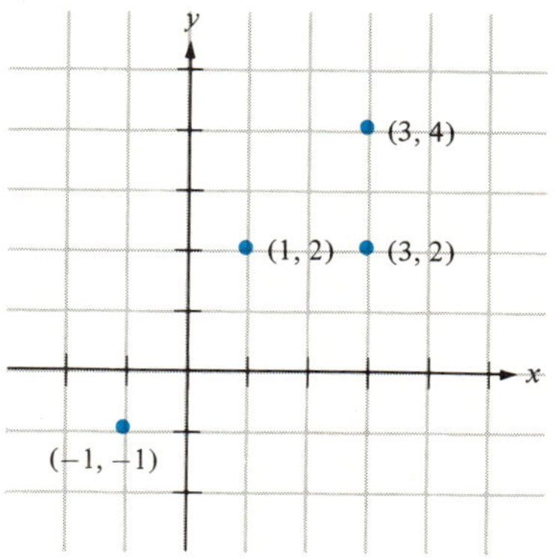

Figure 10.1

Often, a relation is defined as consisting of all those ordered pairs of real numbers satisfying a certain equation or inequality. Examples of such relations are

$$\{(x, y) \mid x^2 + y^2 = 9\}$$

and

$$\{(x, y) \mid y < 3x + 1\}.$$

For such relations we cannot list all the ordered pairs of the relation because there are an *infinite* number of them. However, when a relation is defined by *an equation,* we can obtain the graph of the relation by plotting a few ordered pairs whose components satisfy the equation. Then we join the points by a smooth curve that represents the ordered pairs in the relation. Note that the graph of a relation is the same as the graph of the equation defining the relation. To plot the graph of a relation defined by an *inequality,* we use the procedures developed in Sections 7.5 and 8.4 for graphing inequalities. The following examples illustrate the preceding remarks.

Example 2. Graph the relation

$$R = \{(x, y) \mid 4x^2 + 16y^2 = 64\}.$$

Determine the domain and range of the relation from its graph.

Solution: The equation defining the relation $4x^2 + 16y^2 = 64$ is the equation of an ellipse with its center at the origin. The standard form of the given equation is

$$\frac{x^2}{16} + \frac{y^2}{4} = 1 \quad \text{or} \quad \frac{x^2}{4^2} + \frac{y^2}{2^2} = 1.$$

Thus we see the ellipse has x-intercepts -4 and 4. The y-intercepts are -2 and 2. The graph is shown in Figure 10.2, where $P(x, y)$ represents any point on the ellipse. If we inspect the graph, we observe that as $P(x, y)$ moves around the ellipse, the x-coordinate always lies between the x-intercepts -4 and 4. Similarly, the y-coordinate ranges between the y-intercepts -2 and 2. Thus, for the given relation R, we have that

$$\text{domain of } R = \{x \mid -4 \le x \le 4\} \quad \text{and} \quad \text{range of } R = \{y \mid -2 \le y \le 2\}.$$

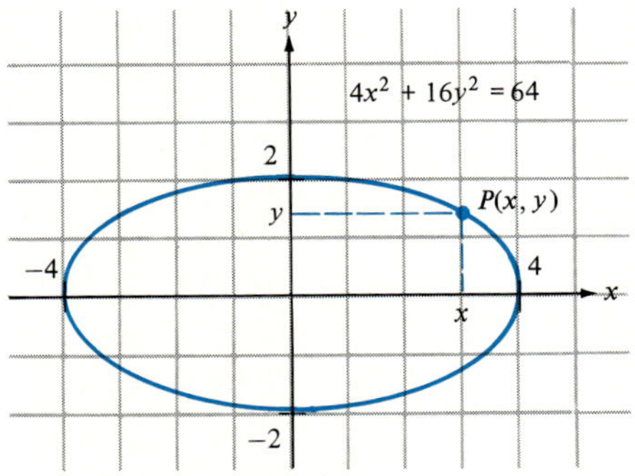

Figure 10.2

Example 3. Graph the relation

$$S = \{(x, y) \mid y < x + 1\}.$$

Determine the domain and range of the relation from its graph.

Solution: The graph of the relation is shown in Figure 10.3. Since there is no restriction on the values of x and y, we see that the

$$\text{domain of } S = \{x \mid x \text{ is a real number}\}$$
$$\text{range of } S = \{y \mid y \text{ is a real number}\}.$$

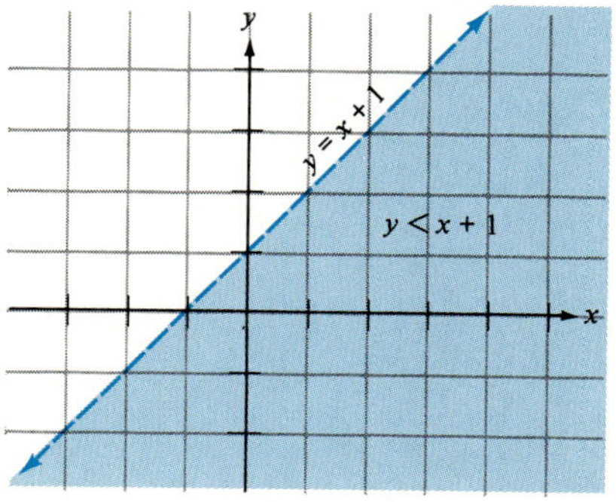

Figure 10.3

Example 4. Graph the relation

$$T = \{(x, y) \mid y = -x + 2\}.$$

Determine its domain and range.

Solution: The equation defining the relation is

$$y = -x + 2.$$

The graph of the relation is the graph of the linear equation $y = -x + 2$ shown in Figure 10.4.

$$\text{domain of } T = \{x \mid x \text{ is a real number}\}$$
$$\text{range of } T = \{y \mid y \text{ is a real number}\}.$$

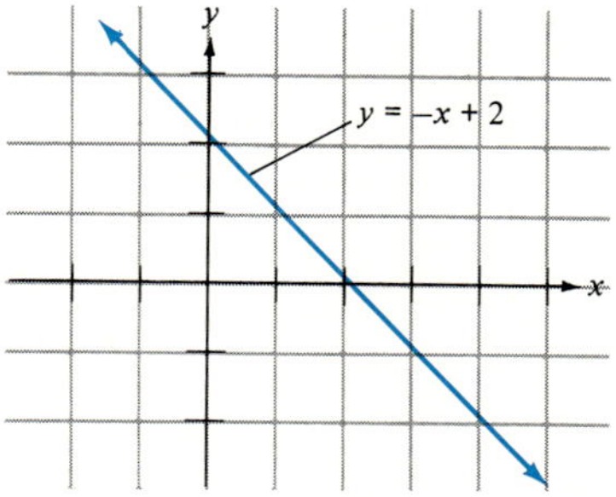

Figure 10.4

EXERCISES 10.1

For Exercises 1–10, determine the domain (D) and the range (R) of the relation. See Example 1.

1. $\{(-2, -4), (-1, -2), (1, 2)\}$

2. $\{(-2, -1), (2, 1), (4, 2)\}$

3. $\{(-3, -4), (-3, 4), (3, 4), (3, -4)\}$

4. $\{(4, -2), (4, 2), (9, -3), (9, 3)\}$

5. $\{(0, 0), (2, 3), (4, 3)\}$

6. $\{(-2, 5), (2, 5), (3, 5)\}$

7. $\{(-2, 4), (2, 4), (3, 9)\}$

8. $\{(7, 2), (7, 3), (5, 4)\}$

9. $\{(0, 8), (0, -8), (10, 6)\}$

10. $\{(6, 5), (6, 7), (6, 9)\}$

For Exercises 11–24, (a) sketch the graph of the relation, (b) find the domain, and (c) find the range. See Examples 2–4.

11. $R = \{(x, y) \mid x^2 + y^2 = 4\}$

12. $R = \{(x, y) \mid x^2 + y^2 = 9\}$

13. $S = \{(x, y) \mid x^2/4 + y^2/9 = 1\}$

14. $S = \{(x, y) \mid x^2/9 + y^2/25 = 1\}$

15. $T = \{(x, y) \mid y = x\}$

16. $T = \{(x, y) \mid y > x + 3\}$

17. $U = \{(x, y) \mid y < 2x + 4\}$

18. $U = \{(x, y) \mid y = -3x + 6\}$

19. $V = \{(x, y) \mid y^2 = x\}$

20. $V = \{(x, y) \mid y = x^2\}$

21. $W = \{(x, y) \mid y = -x^2\}$

22. $W = \{(x, y) \mid y^2 = -x\}$

23. $X = \{(x, y) \mid (x - 1)^2 + (y - 1)^2 = 1\}$

24. $X = \{(x, y) \mid (x - 2)^2 + (y - 2)^2 = 4\}$

Exercises 25–30 specify relations R from set A to B, where $A = \{2, 4, 9\}$ and $B = \{2, 4, 9\}$. For each exercise, use one of the phrases "equal to," "not equal to," "less than," "greater than," "square of," and "square root of" to describe the relation. [HINT: Each relation is described by exactly one of the previous phrases.]

25. $R = \{(2, 2), (4, 4), (9, 9)\}$

26. $R = \{(4, 2), (9, 2), (9, 4)\}$

27. $R = \{(2, 4), (2, 9), (4, 9)\}$

28. $R = \{(2, 4)\}$

29. $R = \{(4, 2)\}$

30. $R = \{(2, 4), (2, 9), (4, 2), (4, 9), (9, 2), (9, 4)\}$

10.2 Functions

In mathematics, science, and other subjects, such as economics, we are interested in how variables are related. For example, we might want to know one of the following:

1. How the distance, d, an object travels is related to its speed, s.
2. How the cost, C, of a certain item is related to the number, x, of items produced.
3. How the temperature, T, is related to altitude (height), h.
4. How the population, P, is related to or depends on time, t.

There is a special kind of *relation* between two variables that is called a *function*. The main idea behind a function is that as soon as we assign a value to the first variable, the corresponding value of the second variable is *fixed*. Thus a function *pairs together* the values of two variables such that for *each* value of the first variable, there is *one and only one* value of the second variable. A function may be formally defined as follows:

Definition of a Function

> A *function* is a relation consisting of ordered pairs (x, y) such that to each value of the first variable, x, there corresponds a unique (one and only one) value of the second variable y.

It is customary to state the definition above in the following equivalent form:

> A *function* is a relation in which no two different ordered pairs (x, y) have the same first component.

Sometimes it is more convenient to use this second definition of a function, but it is the first definition given above that states the way a function is usually thought of in actual practice.

Since a function is a relation, the *domain* of the function is the set of x-values. The *range* of the function is the corresponding set of y-values.

Example 1. Show that the relation

$$\{(x, y) \mid y = x^2\}$$

with domain $\{-2, -1, 0, 1, 2\}$ is a function and identify the range.

Solution: If we let x assume the values $-2, -1, 0, 1$, and 2 in the equation $y = x^2$, the corresponding values of y are 0, 1, and 4. The relation as defined is the set of ordered pairs

$$\{(-2, 4), (-1, 1), (0, 0), (1, 1), (2, 4)\}.$$

The relation *is* a function, since no two different ordered pairs have the same first component. The range of the function is $\{0, 1, 4\}$.

Note, in Example 1, that a function may have two or more ordered pairs with the same second components such as $(-2, 4)$ and $(2, 4)$. This is permitted by the definition of a function. What *is not permitted* is the occurrence of two different ordered pairs with the same first components, such as $(4, -2)$ and $(4, 2)$. There are equations defining a relation that produce more than one y-value for each x-value. Therefore, such equations do not define a function. We illustrate such a case in the following example.

Example 2. Show that the relation

$$\{(x, y) \mid y^2 = x\}$$

with domain $\{0, 4\}$ is not a function.

Solution: We see that when we let $x = 0$ in the equation $y^2 = x$ defining the relation, we obtain $y^2 = 0$ or $y = 0$. Thus to $x = 0$ there corresponds one y value, 0. However, when $x = 4$ we have

$$y^2 = 4 \qquad \text{or} \qquad y = \pm 2.$$

Thus the ordered pairs of the relation are

$$\{(0, 0), (4, -2), (4, 2)\}.$$

Since two ordered pairs $(4, -2)$ and $(4, 2)$ have the same first component but different second components, the relation is *not* a function.

The letters f, g, h (or F, G, H) are usually used to designate functions. For example, consider a relation that is also a function such as

$$f = \{(x, y) \mid y = 2x + 3\}.$$

The unique y such that $(x, y) \in f$ is designated by $f(x)$. The notation $f(x)$ is read "the *value of the function f* at x" or simply "f of x" and refers to the value of y in the range of the function that *corresponds* to the domain value x. Therefore, we write $f(x) = y$ or $y = f(x)$. Thus, for $y = 2x + 3$, we write

$$f(x) = 2x + 3.$$

Accordingly, we speak of "the function f whose values are $f(x) = 2x + 3$" or "whose values are $y = 2x + 3$." It is customary to abbreviate this and speak of "the function $f(x) = 2x + 3$" or "the function $y = 2x + 3$." Furthermore, we usually drop the set-builder notation when specifying a function, such as

$$f = \{(x, y) \mid y = 2x + 3\}$$

and simply write

$$f(x) = 2x + 3 \text{ or } y = 2x + 3$$

to designate the function f. These abbreviations are acceptable as long as we do not confuse the concept of the function f, which is a set of ordered pairs (x, y), with its set of range values y. For example, the symbol $f(2)$ represents the value of the function f at 2. Usually, we define a function of x using an algebraic expression involving x. Thus, to find $f(2)$, we replace x with 2 wherever we see x in the expression defining the function. With this understanding, for the function $f(x) = 2x + 3$, we have

$$f(2) = 7. \qquad \text{because} \qquad f(2) = 2(2) + 3 = 7$$

Also,

$$f(3) = 9 \qquad \text{because} \qquad f(3) = 2(3) + 3 = 9$$

and so on.

The notation $f(x)$ is called *functional notation*. Note that it *does not mean f* times a number x. When we write

$$y = f(x),$$

we say that *y is a function of x*. The variable x that generates the ordered pairs (x, y) of a function f is called the *independent* variable. The second variable y is called the *dependent* variable.

Example 3. Given the function $g(x) = x^2 + 2$, find (a) $g(0)$. (b) $g(3)$. (c) $g(-2)$.

Solution: Remember that the number in parentheses replaces x in the formula for the function g. Consequently,

(a) $g(0) = 0^2 + 2 = 2$; therefore, $g(0) = 2$.

(b) $g(3) = 3^2 + 2 = 11$; therefore, $g(3) = 11$.

(c) $g(-2) = (-2)^2 + 2 = 6$; therefore, $g(-2) = 6$.

Example 4. Given $f(x) = x^2$, find

(a) $f(x + h)$. (b) $f(x + h) - f(x)$. (c) $\dfrac{f(x + h) - f(x)}{h}$.

Solution

(a) $f(x + h) = (x + h)^2 = x^2 + 2hx + h^2$.

(b) $f(x + h) - f(x) = x^2 + 2hx + h^2 - x^2 = 2hx + h^2$.

(c) Since we have $f(x + h) - f(x)$ in part (b), we divide by h as follows:

$$\frac{f(x + h) - f(x)}{h} = \frac{2hx + h^2}{h} = \frac{2hx}{h} + \frac{h^2}{h} = 2x + h.$$

In this book, if a rule or equation defining y as a function of x is given without specifically stating what values x can assume, then the domain of the function is assumed to be *all real numbers x* for which there is *exactly one* corresponding *real number y*. Note that when a function is specified by an algebraic expression, we exclude from the domain any numbers that would result in an imaginary number or division by 0. The following examples illustrate how to determine the domain of a function of x when the domain is not specified.

Example 5. Specify the domain of the function $y = \dfrac{1}{x - 6}$.

Solution: Note that if we replace x by 6, we obtain 0 as a denominator. But division by 0 is not defined. Thus the domain of the function is

$$\text{domain} = \{x \mid x \text{ is a real number and } x \neq 6\}.$$

Example 6. Specify the domain of the function $y = \sqrt{x - 4}$.

Solution: The square root of a negative number is not a real number; it is an imaginary number. For example, if $x = 0$, we have $\sqrt{0 - 4} = \sqrt{-4} = 2i$. Thus we wish to assign values to x that make the expression under the radical sign positive or 0. As a result, we must have $x - 4 \geq 0$ or $x \geq 4$. The domain of the function is

$$\text{domain} = \{x \mid x \text{ is real and } x \geq 4\}.$$

Since a function is a relation, the definition of the *graph* of a function is the same as that for a relation. If we are given the graph of a relation, it is possible to determine from the graph whether or not it is the graph of a function. If a relation is a function, then no two different ordered pairs (x, y) have the same first component. Graphically, this means that a vertical line can intersect the graph of a function in at most one point. Thus we are led to the following test.

Vertical Line Test

> If a vertical line intersects a graph of a relation in two or more points, the graph is *not* the graph of y as a function of x.

The following example illustrates the application of this rule.

Example 7. Determine if the graphs of the relations given in Figure 10.5 represent functions of x.

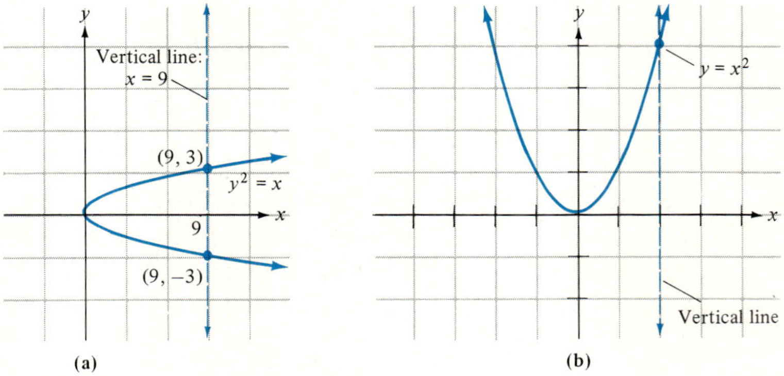

Figure 10.5

Solution

(a) We see that a vertical line such as $x = 9$ will intersect the graph of the relation $y^2 = x$ in two points $(9, 3)$ and $(9, -3)$. Therefore, by the vertical line test, the graph does not represent y as a function of x. Notice that because the vertical line intersects the graph in two points, this means that the relation contains two ordered pairs with the same first component, such as (9, 3) and (9, −3). In that case by definition, the relation is not a function.

(b) The graph in Figure 10.5(b) does represent the graph of y as a function of x, since *any* vertical line intersects the graph in exactly one point.

EXERCISES 10.2

For Exercises 1–16, determine if the equation defines a function of x with the given domain. Find the range of all functions. See Examples 1 and 2.

1. $y = 4x;\ \{0, 1, 2, 3\}$

2. $y = -2x;\ \{-1, 0, 1, 2\}$

3. $y = 3x - 1;\ \{-2, 0, 2\}$

4. $y = -5x + 1;\ \{-1, 0, 1\}$

5. $y = x^2 + 1;\ \{-3, 0, 3\}$

6. $y = x^2 - 8;\ \{-4, 0, 4\}$

7. $|y| = x;\ \{1, 2, 3\}$

8. $y^2 = x - 1;\ \{1, 5, 10\}$

9. $y = |x|;\ \{-5, -3, 0, 2\}$

10. $y = x^3;\ \{-2, -1, 0, 1, 2\}$

11. $y = 4;\ \{x \mid x \text{ is a real number}\}$

12. $y^2 = 25 - x^2;\ \{-4, -3, 0\}$

13. $y^3 = x;\ \{-8, -1, 0, 1, 8\}$

14. $y = \sqrt{x};\ \{0, 4, 9, 16\}$

15. $y^2 = x^2;\ \{0, 1, 2, 3\}$

16. $|y - 1| = x;\ \{9, 2, 3\}$

For Exercises 17–22, find the function values. See Example 3.

17. $f(x) = 3x - 1$ **(a)** $f(0)$ **(b)** $f(-1)$ **(c)** $f(2)$

18. $f(x) = 4x + 2$ **(a)** $f(0)$ **(b)** $f(-2)$ **(c)** $f(3)$

19. $g(x) = 3x^2$ **(a)** $g(-1)$ **(b)** $g(0)$ **(c)** $g(-2)$

20. $g(x) = \sqrt{x}$ **(a)** $g(4)$ **(b)** $g(16)$ **(c)** $g(25)$

21. $h(x) = \sqrt{25 - x^2}$ **(a)** $h(0)$ **(b)** $h(-3)$ **(c)** $h(-4)$

22. $h(x) = \dfrac{x^2 - 1}{x + 1}$ **(a)** $h(-4)$ **(b)** $h(-5)$ **(c)** $h(0)$

For Exercises 23–28, find (a) $f(x + h)$*; (b)* $f(x + h) - f(x)$*; and (c)* $\dfrac{f(x + h) - f(x)}{h}$. *See Example 4.*

23. $f(x) = 3x + 1$

24. $f(x) = 6x - 2$

25. $f(x) = 3x^2$

26. $f(x) = 5x^2$

27. $f(x) = x^2 + x$

28. $f(x) = x^2 - x$

For Exercises 29–42, determine the domain of the function of x. See Examples 5 and 6.

29. $y = \dfrac{5}{x}$

30. $y = \dfrac{7}{x - 2}$

31. $y = \sqrt{x + 3}$

32. $y = \sqrt{x - 5}$

33. $y = \dfrac{1}{(x + 1)(x - 5)}$

34. $y = \dfrac{1}{(x - 6)(x - 2)}$

35. $y = x^2 + 5$

36. $y = x^3 - 1$

37. $y = \dfrac{1}{x^2 - 4}$

38. $y = \dfrac{1}{x^2 - x}$

39. $y = \sqrt{3x - 1}$

40. $y = \dfrac{1}{\sqrt{2x - 1}}$

41. $y = \dfrac{1}{|x| - 2}$

42. $y = \dfrac{1}{x^2 - 2x + 1}$

For Exercises 43–51, use the vertical line test to determine if the graph is the graph of a function. See Example 7.

43.

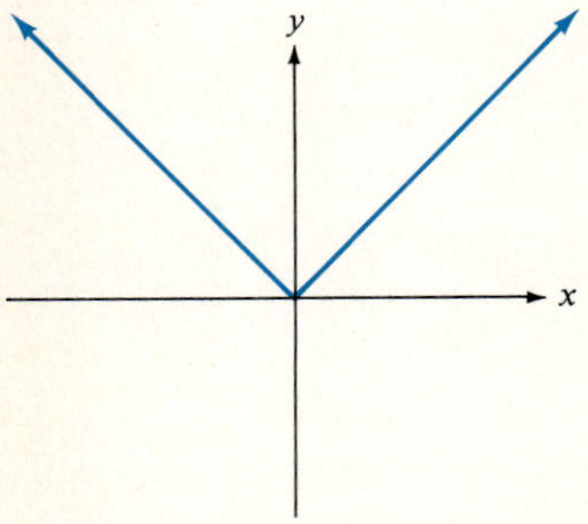

44.

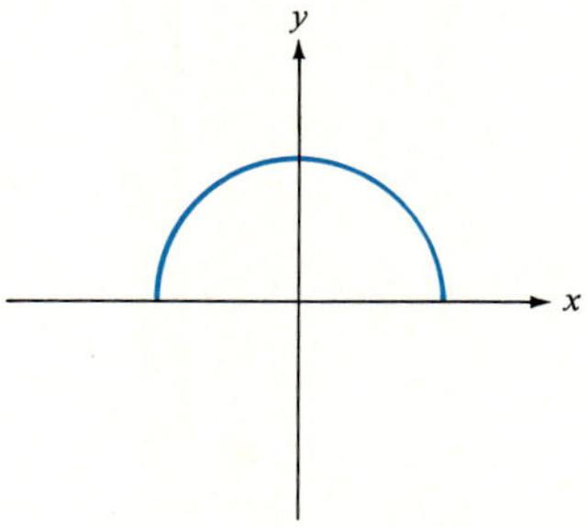

45.

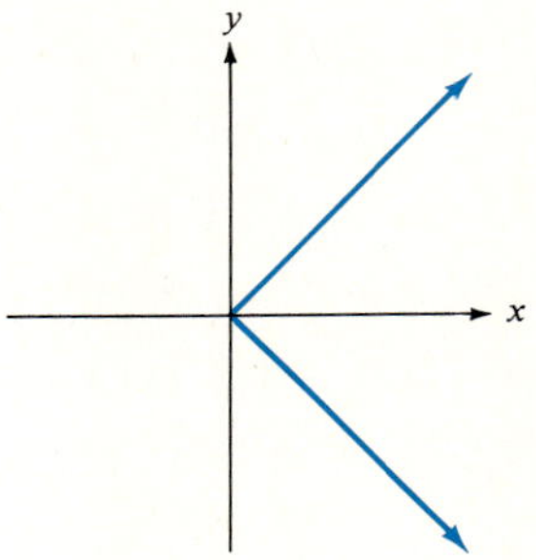

46.

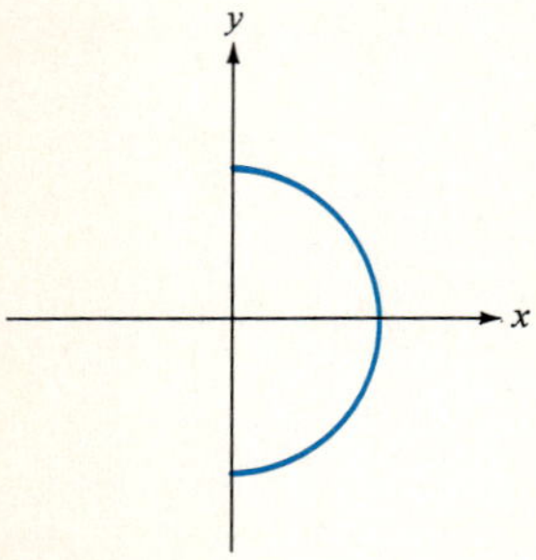

47.

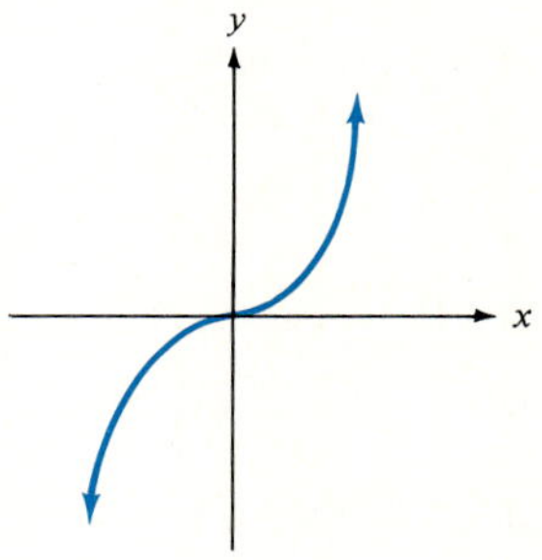

48.

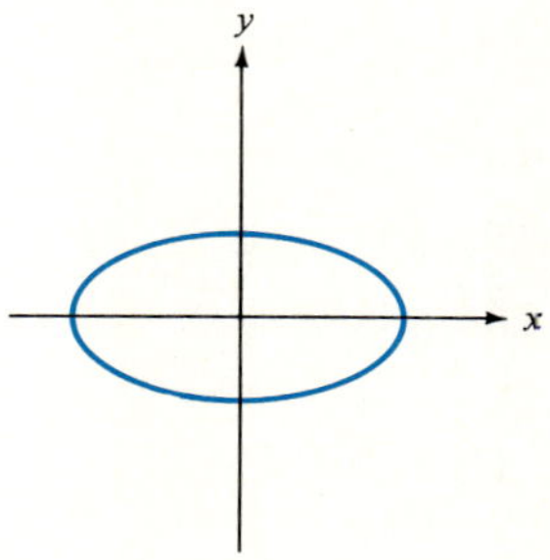

49.

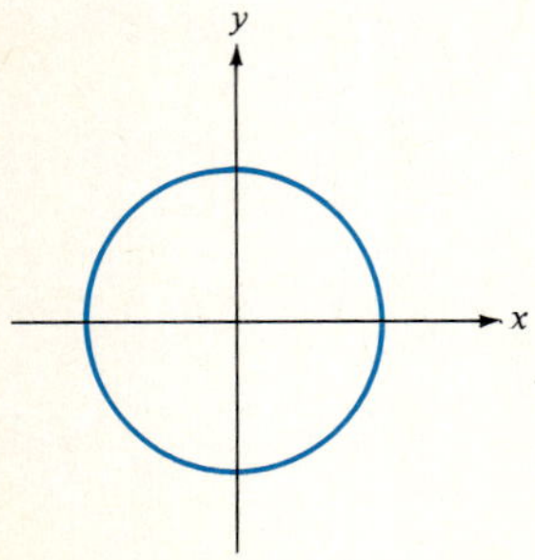

50.

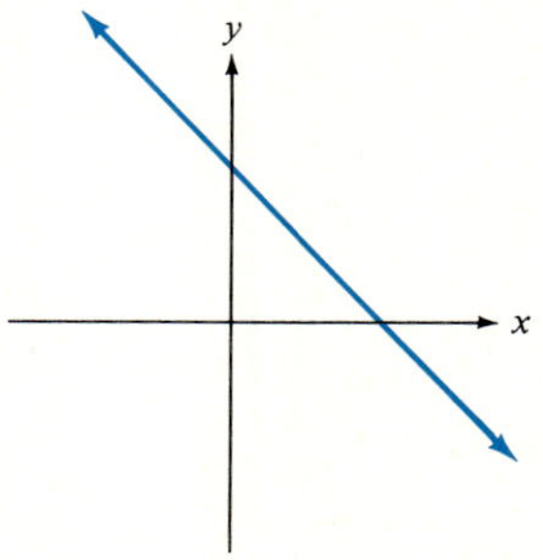

51.

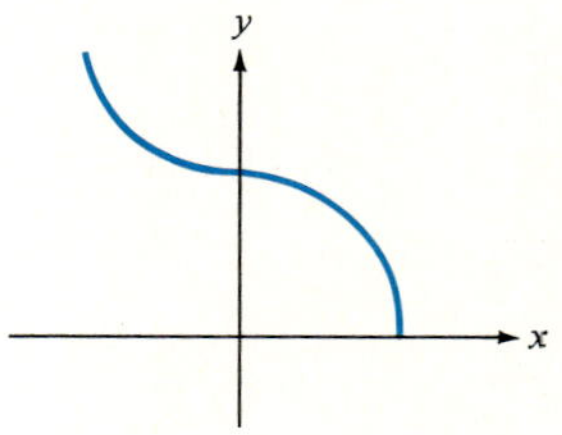

Quadratic Functions 10.3

A *quadratic function* is any function $y = f(x)$ defined by an equation of the form

$$f(x) = ax^2 + bx + c \qquad \text{or} \qquad y = ax^2 + bx + c \qquad (a \neq 0),$$

where a, b, and c are constants.

In Section 8.2 we noted that the graph of the equation $y = ax^2 + bx + c$ $(a \neq 0)$ was a parabola with axis of symmetry parallel to the y-axis. Typical examples of such parabolas with the vertex V at (h, k) are shown in Figure 10.6. Note that the parabola *opens upward* if $a > 0$ and *opens downward* if $a < 0$.

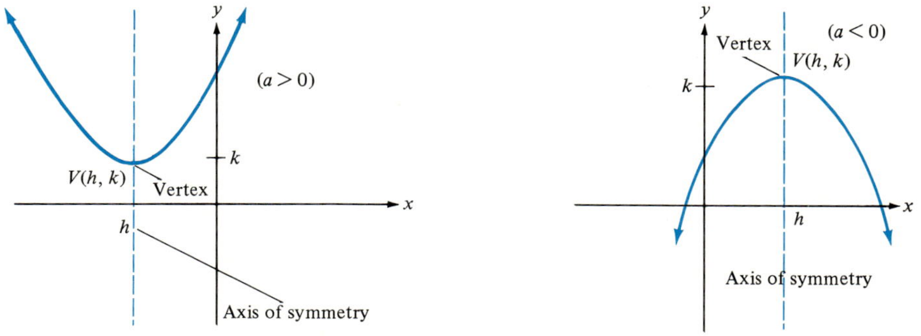

(a) (b)

Figure 10.6

From the preceding remarks, we see that the *graph* of a quadratic function f, defined by the equation $y = f(x) = ax^2 + bx + c$, is a *parabola*. Notice, in Figure 10.6, that the vertex of a parabola is either the *lowest* or *highest* point on the parabola. This suggests that the *minimum* (smallest) or *maximum* (largest) *value* of a quadratic function $f(x) = ax^2 + bx + c$ is the y-value, k, of the vertex of the corresponding parabola, and that *this value occurs* at the x-value of the vertex, h. In Section 8.2 (see Example 5) we used the technique of completing the square to determine the vertex (h, k) of a parabola with equation $y = ax^2 + bx + c$. In this section we develop a formula that will enable us to easily determine the vertex (h, k) of the parabola $y = ax^2 + bx + c$ *in terms of the constants a, b, and c.*

If we were to apply the completing square process to the general quadratic equation $y = ax^2 + bx + c$ $(a \neq 0)$, we would obtain

(1) $$y = a\left(x + \frac{b}{2a}\right)^2 + \frac{4ac - b^2}{4a}.$$

For simplicity, we let

(2) $$h = -\frac{b}{2a} \qquad \text{and} \qquad k = \frac{4ac - b^2}{4a}$$

and rewrite equation (1) as

(3) $$y = a(x - h)^2 + k$$

or equivalently,

(4) $$y - k = a(x - h)^2.$$

In Section 8.2 we saw that equation (4) is equivalent to the standard form of a parabola with *vertex* at (h, k) and *axis of symmetry* $x = h$. Thus, from (2), we have that the vertex of the parabola is

$$V(h, k) = V\left(-\frac{b}{2a}, \frac{4ac - b^2}{4a}\right); \qquad \text{with axis of symmetry } x = -\frac{b}{2a}.$$

We now summarize the previous discussion on quadratic functions.

Quadratic Function

> The graph of a quadratic function $f(x) = ax^2 + bx + c$ is a parabola with vertex at $\left(-\frac{b}{2a}, \frac{4ac - b^2}{4a}\right)$; axis of symmetry $x = -\frac{b}{2a}$; the function has a minimum value if $a > 0$, a maximum value if $a < 0$. The minimum or maximum value of the function is $\frac{4ac - b^2}{4a}$ and occurs when $x = -\frac{b}{2a}$.

The y-coordinate of the vertex of the parabola $y = ax^2 + bx + c$ can be obtained by using the formula $\frac{4ac - b^2}{4a}$, or simply by substituting $x = -\frac{b}{2a}$ into the equation $y = ax^2 + bx + c$ and solving for y. Note that the value of the y-coordinate obtained either way is also the maximum or minimum value of the function $f(x) = ax^2 + bx + c$.

Example 1. For $f(x) = 4x^2 - 12x + 2$, find

(a) The vertex of the parabola.

(b) The minimum or maximum value of $f(x)$.

Solution

(a) For this function $a = 4$, $b = -12$, and $c = 2$. The x-coordinate of the vertex is $-\frac{b}{2a} = -\frac{-12}{2(4)} = \frac{3}{2}$. To find the y-coordinate, we substitute $x = \frac{3}{2}$ into the equation:

$$y = f\left(\frac{3}{2}\right) = 4\left(\frac{3}{2}\right)^2 - 12\left(\frac{3}{2}\right) + 2 = -7.$$

The vertex is $\left(\frac{3}{2}, -7\right)$.

(b) Since $a = 4 > 0$, $f(x)$ has a *minimum* value of -7 when $x = \frac{3}{2}$.

In Section 8.2, when sketching the parabola $y = ax^2 + bx + c$, we used the technique of completing the square. In Example 2 we show how to *sketch quickly* the graph of a quadratic function $f(x) = ax^2 + bx + c$ without using the completing the square process. To accomplish this, we make use of the equation $x = -\dfrac{b}{2a}$ for finding the x-coordinate of the vertex of the parabola and also use the x or y intercepts of the graph. We note that, as we shall shortly see, some parabolas of this type do not have x-intercepts.

Example 2

(a) Sketch the graph of $f(x) = -x^2 + 6x - 5$.
(b) Find the maximum or minimum value of $f(x)$.

Solution

(a) For the given equation, $a = -1$, $b = 6$, and $c = -5$. The x-coordinate of the vertex of the parabola is $-\dfrac{b}{2a} = -\dfrac{6}{2(-1)} = 3$. For $x = 3$,

$$y = f(3) = -(3)^2 + 6(3) - 5 = 4.$$

The *vertex* is at (3, 4). The axis of symmetry is $x = -\dfrac{b}{2a} = 3$. As an aid to sketching the graph, we find the x and y intercepts of the graph. The graph will intersect the y-axis when $x = 0$. Thus, to find the y-intercept, we substitute $x = 0$ into the given equation.

$$y = f(0) = -(0)^2 - 6(0) - 5. \qquad -5 \textit{ is the y-intercept}$$

The graph will intersect the x-axis only when $y = 0$. Therefore, we set $y = 0$ in the equation $y = f(x) = -x^2 + 6x - 5$ to obtain the following quadratic equation in one variable:

$$0 = -x^2 + 6x - 5$$

or

$$0 = x^2 - 6x + 5.$$

Solving the equation by either factoring or using the quadratic formula, we find that the solutions to the equation are $x = 1$ and $x = 5$, which are the *x-intercepts* of the graph. Because $a = -1 < 0$, we know that the parabola *opens downward*. Since the vertex is at (3, 4), the x-intercepts at 1 and 5, and the y-intercept at -5, we can therefore quickly sketch the graph of

$$y = -x^2 + 6x - 5,$$

as shown in Figure 10.7. This is the desired graph of $f(x) = -x^2 + 6x - 5$.

(b) Since $a = -1 < 0$, the function f has *maximum* value 4 when $x = 3$.

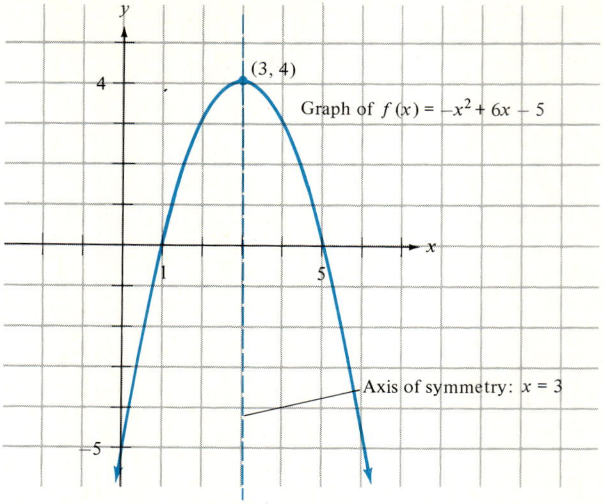

Figure 10.7

Example 2 illustrates that when we substitute $x = 0$ in the equation

$$y = ax^2 + bx + c \tag{5}$$

to find the y-intercept, the result will be the constant term c. To find the x-intercepts, we substitute $y = 0$ in equation (5) and solve the corresponding quadratic equation in one variable:

$$ax^2 + bx + c = 0. \tag{6}$$

However, if the solutions to equation (6) are complex numbers, the parabola does not have any x-intercepts, since all points on the x-axis have real numbers as coordinates. In that case, we may sketch the parabola by plotting points with x-values equidistant from the axis of symmetry as was done in Section 8.2. These points will be symmetric with respect to the axis of symmetry. The following example gives a concrete example of the previous remarks.

Example 3

(a) Graph $f(x) = 2x^2 + 8x + 9$.

(b) Find the minimum or maximum value of $f(x)$.

Solution

(a) Here $a = 2$, $b = 8$, and $c = 9$. Using the procedures outlined in Example 2, we find that the vertex is at $(-2, 1)$; with axis of symmetry $x = -2$. The y-intercept of the graph is 9. To find the x-intercepts, we solve the equation

$$2x^2 + 8x + 9 = 0.$$

Using the quadratic formula $x = \dfrac{-b \pm \sqrt{b^2 - 4ac}}{2a}$ yields the complex number solutions $(-4 \pm i\sqrt{2})/2$. Thus the graph *does not* have x-intercepts. Notice, however, that 0 and -4 on the x-axis are equidistant from the axis of symmetry, $x = -2$. If we substitute $x = -4$ into the equation $y = f(x) = 2x^2 + 8x + 9$, the corresponding y-value is 9. Therefore, $(-4, 9)$ is on the parabola representing the graph of $f(x)$. We already know that the y-intercept 9 corresponds to $x = 0$. Plotting the vertex $(-2, 1)$, the y-intercept 9, the point $(-4, 9)$, and using the fact that the parabola *opens upwards* (since $a = 2 > 0$) enables us to sketch the parabola shown in Figure 10.8.

(b) Because $a = 2 > 0$, the given function has a minimum value 1 at $x = -2$. That is, $f(-2) = 1$ is the smallest value f assumes.

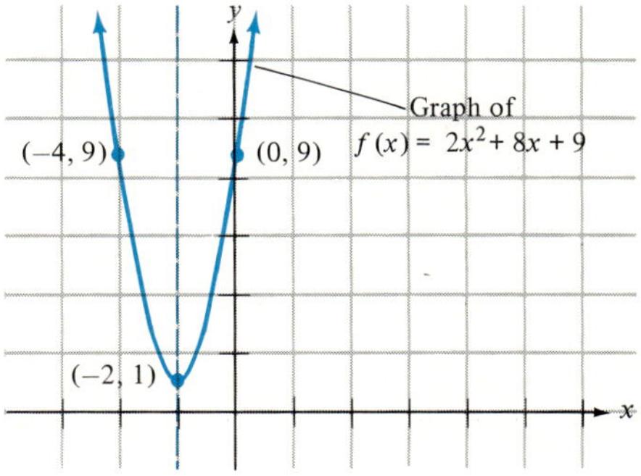

Figure 10.8

Problems involving the maximum and minimum value of a quadratic function can be easily solved using the techniques developed in this section.

Example 4. What is the area of the largest rectangular field that can be enclosed by 400 feet of fencing?

Solution: Let x and y represent the width and length, respectively, of the rectangle, as shown in Figure 10.9. We see that the *perimeter* $= 2x + 2y = 400$ or

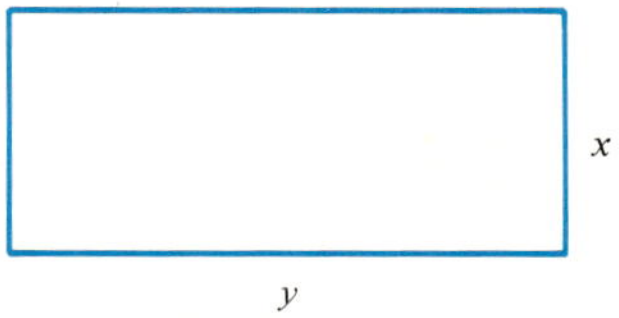

Figure 10.9

$x + y = 200$. Let A denote the *area* of the field; then $A = xy$. Since $x + y = 200$ or $y = 200 - x$, we substitute for y in $A = xy$ to obtain the quadratic function

$$A = x(200 - x) = -x^2 + 200x.$$

Now we need to find the x-value that will make A, the area, a maximum. For this quadratic function $a = -1$, $b = 200$, and $c = 0$. Since $a = -1 < 0$, the function A has a *maximum* value that occurs at

$$x = -\frac{b}{2a} = -\frac{200}{2(-1)} = 100.$$

To find the maximum area, A, we substitute $x = 100$ into the equation

$$A = -x^2 + 200x$$

to obtain

$$A = 200(100) - 100^2 = 10{,}000 = 100^2.$$

Thus the maximum area $A = 100^2$ when $x = 100$, as shown in Figure 10.10. Note that since $x + y = 200$, $y = 100$ when $x = 100$. So we see that the maximum area is enclosed by a *square* field.

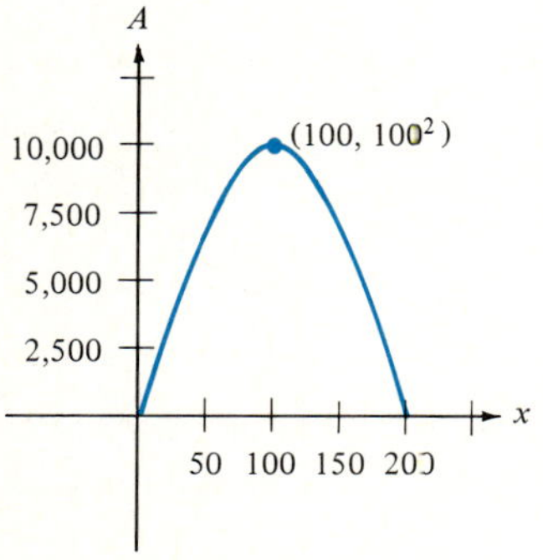

Figure 10.10

EXERCISES 10.3

For Exercises 1–12, find (a) the vertex of the graph of f and (b) the minimum or maximum value of f. See Example 1.

1. $f(x) = 3x^2 - 6x + 5$ **2.** $f(x) = 2x^2 - 8x + 3$ **3.** $f(x) = -4x^2 - 24x - 30$

4. $f(x) = -5x^2 - 10x + 2$ **5.** $f(x) = 5x^2 + \frac{7}{2}$ **6.** $f(x) = -3x^2 + 12$

7. $f(x) = -13x^2$ **8.** $f(x) = 17x^2$ **9.** $f(x) = 7x^2 + 28x$

10. $f(x) = -x^2 - 2x$ **11.** $f(x) = 2x^2 - .4x + .52$ **12.** $f(x) = -3x^2 - 7.2x + 4$

For Exercises 13–24, (a) graph the function f and (b) find the minimum or maximum value of f. See Examples 2 and 3.

13. $f(x) = x^2 - 4x$

14. $f(x) = -x^2 - 6x$

15. $f(x) = -2x^2 - 6x$

16. $f(x) = 4x^2 - 8x$

17. $f(x) = x^2 - 6x + 8$

18. $f(x) = -x^2 - 2x + 3$

19. $f(x) = -\frac{1}{4}x^2 - x + 3$

20. $f(x) = \frac{1}{2}x^2 - x - 4$

21. $f(x) = 2x^2 - 4x + 5$

22. $f(x) = -2x^2 - 4x - 3$

23. $f(x) = 4x^2 - 12x + 8$

24. $f(x) = -2x^2 + 2x + 4$

Solve Exercises 25–30. See Example 4.

25. Find two positive numbers whose sum is 20 and whose product, P, is a maximum. [HINT: Let x and $20 - x$ represent the numbers.]

26. Find two numbers whose difference is 30 and whose product, P, is a minimum. [HINT: Let x and $30 + x$ represent the numbers.]

27. A maintenance man has 100 feet of fencing. He wants to use the fencing to enclose a rectangular region against a warehouse wall. Find the largest area, A, that he can enclose.

28. A baseball pitcher throws a ball vertically upward into the air with an initial velocity of 96 feet per second. The height after t seconds is given by $h = -16t^2 + 96t$. Find
(a) The number of seconds it takes to reach the maximum height.
(b) The maximum height that the ball will reach.

29. A manufacturer determines that the cost, C, in dollars of producing x pocket calculators per day is given by the equation $C = 2x^2 - 200x + 5150$.
(a) How many calculators should be produced daily so that the cost, C, is a minimum?
(b) What is the minimum cost?

30. A company determines that the profit, P, in dollars of selling x home computers per day is given by $P = -\frac{2}{3}x^2 + 80x - 600$.
(a) How many computers should be sold daily to make the maximum profit?
(b) What is the maximum profit?

Different Types of Functions 10.4

An important part of mathematics is devoted to the study of functions. To facilitate this study, it proves useful to classify functions by certain categories or types. In this section we discuss some of the types of functions that are frequently encountered in mathematics.

Polynomial functions are functions that are defined using polynomial expressions, such as $3x^4 - 5x^2 + 1$ and $5x^3 - x + 2$. Examples of polynomial functions are

$$f(x) = 3x^4 - 5x^2 + 1 \quad \text{and} \quad g(x) = 5x^3 - x + 3.$$

The simplest polynomial functions are the *constant, linear,* and *quadratic* functions. We begin our discussion with the constant function. A *constant* function f is any function of the form

$$f(x) = c, \qquad \text{where } c \text{ is a real number constant.}$$

We note that a constant c, when $c \neq 0$, is a polynomial of degree 0. When $c = 0$, we have the *zero polynomial*. It is customary not to assign a degree to the zero polynomial.

Recall that we use the notation $y = f(x)$ to represent the values of a function f at a domain value x. Thus the *graph* of a constant function $f(x) = c$ is always a horizontal line whose equation is $y = c$. The domain of the constant function $f(x) = c$ is the set of all real numbers x and the range is the constant c. In Example 1 we illustrate the preceding remarks by discussing a particular constant function.

Example 1

(a) Graph the function $f(x) = 3$.

(b) Evaluate $f(2)$, $f(-3)$, and $f(0)$.

Solution

(a) Since $y = f(x)$, to graph the function f defined by $f(x) = 3$ we graph the equation $y = 3$. The graph of $y = 3$ is the horizontal line shown in Figure 10.11.

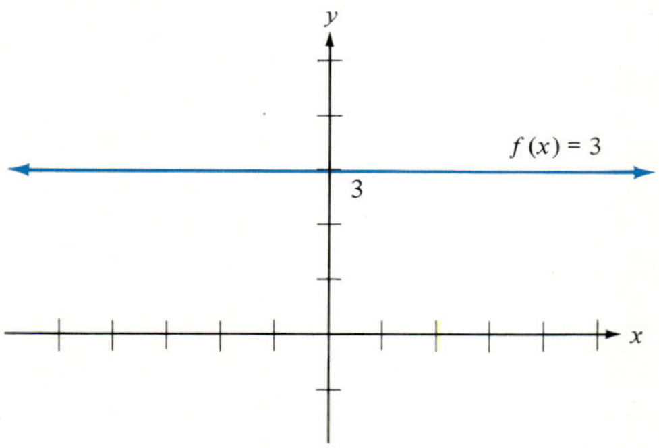

Figure 10.11

(b) Since $f(x) = 3$ for all real x, we have that

$$f(2) = 3, \qquad f(-3) = 3, \qquad f(0) = 3.$$

The range of f is the set $\{3\}$.

A function f is called a *linear function* if

$$f(x) = ax + b, \qquad \text{where } a \text{ and } b \text{ are constants, } a \neq 0.$$

The function is described by the word ''linear'' because its *graph* is a nonvertical straight line. Recall that in Chapter 7 we discussed in detail the graphs of equations of the form $y = mx + b$, which were straight lines with slope m. Thus the graph of a function defined by the equation $y = f(x) = ax + b$ is a straight line with slope $m = a$. Notice that the expression $ax + b$ $(a \neq 0)$ is a polynomial of the *first degree*. Both the domain and range of a linear function are the set of real numbers.

Example 2

(a) Graph the function $f(x) = \frac{1}{2}x - 1$.

(b) Evaluate $f(4)$ and $f(-2)$.

Solution

(a) As we did in Example 1, we let $y = f(x)$. Thus, to graph the given function, we graph the equation $y = \frac{1}{2}x - 1$. The graph of the function is shown in Figure 10.12.

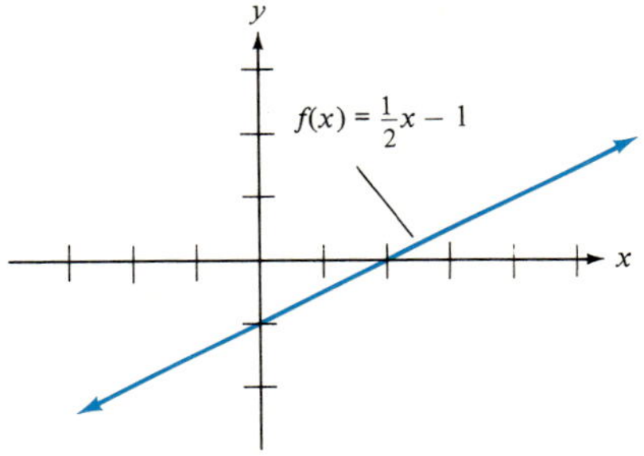

Figure 10.12

(b) Since $f(x) = \frac{1}{2}x - 1$, we have that

$$f(4) = \frac{1}{2}(4) - 1 = 1 \qquad \text{and} \qquad f(-2) = \frac{1}{2}(-2) - 1 = -2.$$

The *quadratic function* $f(x) = ax^2 + bx + c$ $(a \neq 0)$ is a polynomial function. Notice that the quadratic function is defined by a polynomial of *degree* 2. We shall not discuss the quadratic function in this section, since we discussed it in detail in the preceding section.

Polynomial functions of degree greater than 2 usually do not have special names. Instead, they are generally referred to as *polynomial functions*. The graphing of such functions is a topic studied in more advanced courses such as calculus. However, we should become acquainted with finding the range values of a polynomial function for specific values of x in its domain. In the next example we illustrate how to do this.

Example 3. For the polynomial function

$$g(x) = 2x^4 - 7x^2 - 3x + 2$$

find

(a) $g(1)$. (b) $g(-2)$. (c) $g(-1)$.

Solution

(a) $g(1) = -6$ because $g(1) = 2(1)^4 - 7(1)^2 - 3(1) + 2 = -6$.
(b) $g(-2) = 12$ because $g(-2) = 2(-2)^4 - 7(-2)^2 - 3(-2) + 2 = 12$.
(c) $g(-1) = 0$ because $g(-1) = 2(-1)^4 - 7(-1)^2 - 3(-1) + 2 = 0$.

The next two functions that we discuss are frequently encountered in mathematics, but they are not polynomial functions. These functions are

the *absolute value* function $f(x) = |x|$

and

the *square root* function $f(x) = \sqrt{x} \qquad x \geq 0$.

We discuss these functions in the following examples.

Example 4. If $f(x) = |x|$,
(a) Evaluate $f(0)$, $f(2)$, and $f(-2)$.
(b) Graph the function f.

Solution

(a) We see that $f(0) = |0| = 0$, $f(2) = |2| = 2$, and that $f(-2) = |-2| = 2$. Thus $f(0) = 0$, $f(2) = 2$, and $f(-2) = 2$.
(b) Graphing the function $f(x) = |x|$ is equivalent to graphing the equation $y = |x|$. To graph $y = |x|$, it is helpful to rewrite the equation without the absolute value bars. By definition, $|x| = x$ when x is positive or 0. Thus

$$y = |x| = x \quad \text{or} \quad y = x \qquad \textit{provided} \text{ that } x \geq 0.$$

The graph of $y = x$ when $x \geq 0$ is the solid straight line shown in Figure 10.13(a).

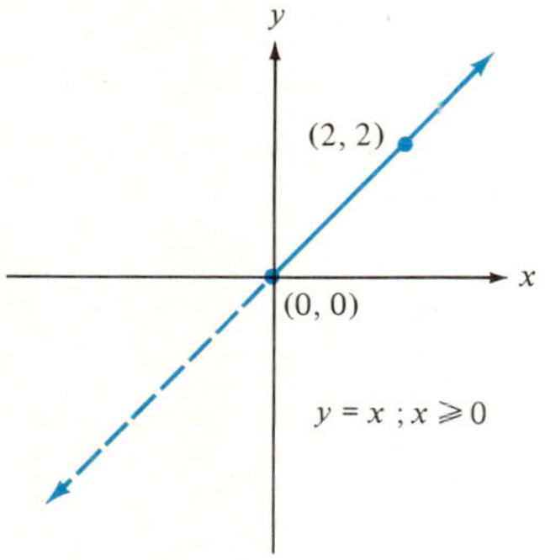

Figure 10.13 (a)

If x is negative, then $|x| = -x$. Therefore, we have

$$y = |x| = -x \quad \text{or} \quad y = -x \qquad \textit{provided} \text{ that } x < 0.$$

This graph is the solid line shown in Figure 10.13(b). Thus the complete graph of $y = |x|$ is the V-shaped graph shown in Figure 10.13(c).

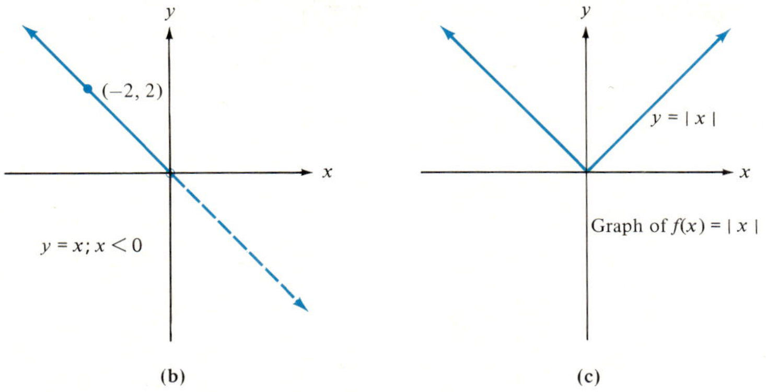

Figure 10.13 (b) and (c)

Example 5. If $f(x) = \sqrt{x}$; $x \geq 0$.
(a) Find $f(0)$, $f(1)$, and $f(4)$. (b) Graph the function f.

Solution

(a) We find that $f(0) = \sqrt{0} = 0$, $f(1) = \sqrt{1} = 1$, and $f(4) = \sqrt{4} = 2$. Thus $f(0) = 0$, $f(1) = 1$, and $f(4) = 2$.

(b) From the results of part (a) we observe that the graph of $y = f(x) = \sqrt{x}$ contains the points (0, 0), (1, 1), and (4, 2). If we were to plot these points on a rectangular coordinate system and draw a smooth curve through them, we would obtain what appears to be half of a parabola lying above the x-axis as shown in Figure 10.14(b). We now show that this is indeed so. We set $y = \sqrt{x}$. For $x \geq 0$, $\sqrt{x}$ is positive or zero. Therefore, y is *positive* or *zero*, since $y = \sqrt{x}$. To graph $y = \sqrt{x}$ we square both sides of the equation to obtain the equation of the parabola $y^2 = x$. The graph of the parabola is shown in Figure 10.14(a). To obtain the graph of $y = \sqrt{x}$, we *eliminate* all those points on the graph of the parabola for which y is *negative* [see Figure 10.14(b)].

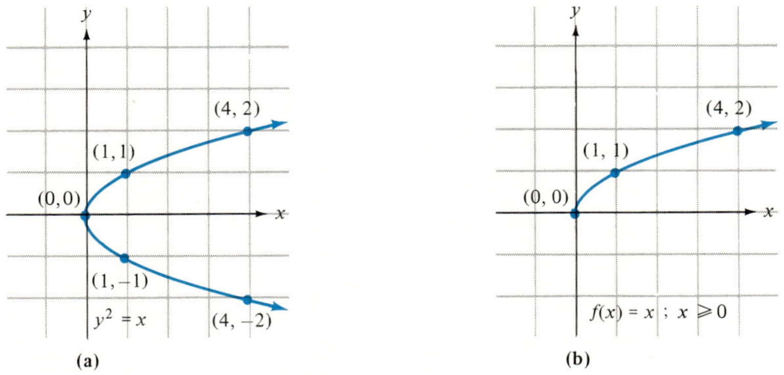

Figure 10.14

Not all functions are defined by equations involving only relatively simple equations. Sometimes a function is defined using several expressions, as shown in the next example.

In the next example we illustrate how to *graph* such functions.

Example 6. Graph the function defined as follows:

$$f(x) = \begin{cases} x - 2 & \text{when } x \geq 0 \\ -2 & \text{when } x < 0. \end{cases}$$

Solution: For $x \geq 0$ the values of the function f are given by $f(x) = x - 2$. Thus part of the graph of the function is obtained by graphing $y = x - 2$ provided that we restrict x to positive numbers or 0. For $x < 0$, $f(x) = -2$. Thus, for negative values of x, the graph of the function is obtained by graphing $y = -2$. The graph of the function is shown in Figure 10.15.

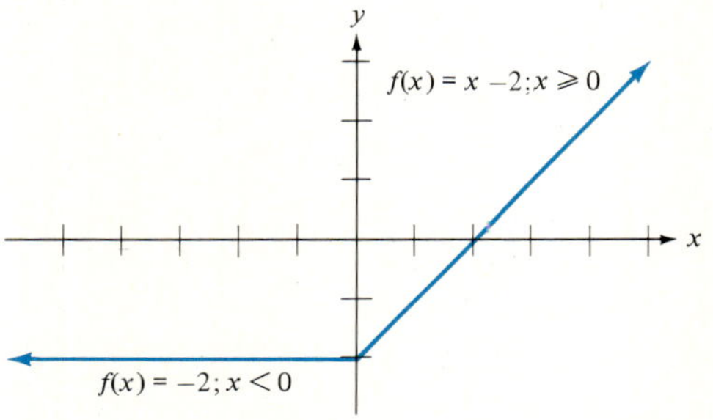

Figure 10.15

In the following example we show how to *evaluate* a function defined by two algebraic expressions for different values in its domain.

Example 7. Given the function

$$f(x) = \begin{cases} x^2 + 1 & \text{when } x \geq 0 \\ x - 1 & \text{when } x < 0 \end{cases}$$

find $f(3)$, $f(0)$, and $f(-4)$.

Solution: To evaluate the function at nonnegative values such as 3 and 0, we use the formula $f(x) = x^2 + 1$, which defines the function for $x \geq 0$ (x nonnegative). Thus

$$f(3) = 3^2 + 1 = 10 \qquad \text{and} \qquad f(0) = 0^2 + 1 = 1.$$

To evaluate $f(-4)$, we use the formula $f(x) = x - 1$, which defines the function for $x < 0$ (negative values of x). Therefore,

$$f(-4) = -4 - 1 = -5.$$

EXERCISES 10.4

For Exercises 1–12, classify the function as constant, linear, quadratic, or polynomial. Use the term that gives the most precise description. See Examples 1–3 and the discussion preceding these examples.

1. $f(x) = 2x + 3$ **2.** $f(x) = x^2 + x - 1$ **3.** $g(x) = 5$

4. $g(x) = 6x$ **5.** $g(x) = 2x^3 - x^2$ **6.** $h(x) = 5 - x$

7. $h(x) = 0$ **8.** $f(x) = 1 - x^2$ **9.** $f(x) = x^3 - 1$

10. $g(x) = 6x^4 - x^2 - 3$ **11.** $g(x) = x$ **12.** $h(x) = 1$

For Exercises 13–24, given the functions

$$f(x) = 7, \qquad g(x) = \tfrac{3}{4}x - 1, \qquad h(x) = x^3 - 2x^2 + 5,$$

find the indicated function values. See Examples 1–3.

13. $f(3)$ **14.** $f(-5)$ **15.** $g(-4)$ **16.** $g(8)$

17. $h(2)$ **18.** $h(0)$ **19.** $f(0)$ **20.** $g(-12)$

21. $h(-3)$ **22.** $h(-1)$ **23.** $f(10)$ **24.** $h(3)$

For Exercises 25–32, for the specified function f, (a) find f(0), f(2), and f(−2), and (b) graph the function. See Example 4.

25. $f(x) = |2x|$ **26.** $f(x) = |3x|$ **27.** $f(x) = |x| + 2$ **28.** $f(x) = |x| - 3$

29. $f(x) = -2|x|$ **30.** $f(x) = -4|x|$ **31.** $f(x) = |x - 2|$ **32.** $f(x) = |\tfrac{1}{2}x - 2|$

For Examples 33–38, for the specified function g, find (a) g(2), g(8), and g(18), and (b) graph the function. See Example 5.

33. $g(x) = \sqrt{2x}$ **34.** $g(x) = \sqrt{\tfrac{1}{2}x}$ **35.** $g(x) = -\sqrt{2x}$

36. $g(x) = -\sqrt{\tfrac{1}{2}x}$ **37.** $g(x) = \sqrt{x - 2}$ **38.** $g(x) = -\sqrt{x - 2}$

For Exercises 39–42, graph the function. See Example 6.

39. $f(x) = \begin{cases} x + 1 & x \geq 0 \\ 1 & x < 0 \end{cases}$ **40.** $g(x) = \begin{cases} 2x - 1 & x \geq 0 \\ -1 & x < 0 \end{cases}$

41. $h(x) = \begin{cases} -x + 2 & x \geq 0 \\ 2 & x < 0 \end{cases}$ **42.** $f(x) = \begin{cases} -2x + 3 & x \geq 0 \\ 3 & x < 0 \end{cases}$

For Exercises 43–46, for the function given, find f(1), f(0), and f(−2). See Example 7.

43. $f(x) = \begin{cases} x^3 + 2 & x \geq 0 \\ 1 - x^2 & x < 0 \end{cases}$ **44.** $f(x) = \begin{cases} x^4 + 1 & x \geq 0 \\ x^3 & x < 0 \end{cases}$

45. $f(x) = \begin{cases} 5x + 4 & x > 0 \\ 3 - x & x \leq 0 \end{cases}$ **46.** $f(x) = \begin{cases} x^2 + 3 & x > 0 \\ 2 - x^2 & x \leq 0 \end{cases}$

10.5 Combining Functions

We can combine numbers by adding, subtracting, multiplying, and dividing to obtain other numbers. Similarly, we may combine functions by adding, subtracting, multiplying, and dividing functions to obtain other functions. We will discuss the addition of two particular functions first and then the remaining operations (subtraction, multiplication, division) that can be performed on two functions will follow naturally.

Consider the functions f and g defined by $f(x) = x^2$ and $g(x) = \sqrt{x}$. The *sum* of the functions f and g is a new function denoted by $f + g$ and is defined by

$$\begin{aligned}(f + g)(x) &= f(x) + g(x)\\ &= x^2 + \sqrt{x}.\end{aligned}$$

To compute $(f + g)(4)$, we proceed as follows:

$$(f + g)(4) = 4^2 + \sqrt{4} = 16 + 2 = 18.$$

Thus $(f + g)(4) = 18$. We note that the domain of $f + g$ is the intersection of the domains of f and g (the set of numbers common to both domains). Since $f(x) = x^2$ is defined for all real numbers x, and $g(x) = \sqrt{x}$ for only nonnegative real numbers (positive or zero), the domain of $f + g$ is the set of nonnegative numbers. Sometimes it is convenient to denote the function $f + g$ by another letter, such as h, to emphasize that $h = f + g$ is a distinct function.

In general, given functions f and g where set D is the intersection of their domains, we define their

Sum $f + g$ by

$$(f + g)(x) = f(x) + g(x) \qquad x \in D.$$

Difference $f - g$ by

$$(f - g)(x) = f(x) - g(x) \qquad x \in D.$$

Product $f \cdot g$ by

$$(f \cdot g)(x) = f(x)g(x) \qquad x \in D.$$

Quotient $\dfrac{f}{g}$ by

$$\left(\frac{f}{g}\right)(x) = \frac{f(x)}{g(x)} \qquad \text{provided that } g(x) \neq 0 \text{ for } x \in D.$$

Example 1. Given the functions f and g defined by $f(x) = 3x + 2$ and $g(x) = \sqrt[3]{x}$, find the expressions defining the sum, difference, and product of f and g, and the quotient of f by g.

Solution

$$(f + g)(x) = 3x + 2 + \sqrt[3]{x}$$
$$(f - g)(x) = 3x + 2 - \sqrt[3]{x}$$
$$(fg)(x) = (3x + 2)\sqrt[3]{x}$$
$$\left(\frac{f}{g}\right)(x) = \frac{3x + 2}{\sqrt[3]{x}} \qquad \text{provided that } x \neq 0.$$

Example 2. Given the functions f and g defined by

$$f(x) = \sqrt{x} + 1 \qquad \text{and} \qquad g(x) = x^2 + 2,$$

find

(a) $(f + g)(4)$. (b) $(f - g)(4)$. (c) $(f \cdot g)(4)$. (d) $\left(\frac{f}{g}\right)(4)$.

Solution

(a) $(f + g)(x) = \sqrt{x} + 1 + x^2 + 2 = x^2 + \sqrt{x} + 3$. Thus

$$(f + g)(4) = 4^2 + \sqrt{4} + 3 = 16 + 2 + 3 = 21.$$

Consequently, $(f + g)(4) = 21$.

(b) $(f - g)(x) = (\sqrt{x} + 1) - (x^2 + 2) = -x^2 + \sqrt{x} - 1$. Letting $x = 4$, we see that

$$(f - g)(4) = -4^2 + \sqrt{4} - 1 = -16 + 2 - 1 = -15.$$

Thus $(f - g)(4) = -15$.

(c) $(f \cdot g)(x) = (\sqrt{x} + 1)(x^2 + 2)$. Thus

$$(f \cdot g)(4) = (\sqrt{4} + 1)(4^2 + 2) = (2 + 1)(16 + 2) = (3)(18) = 54.$$

Hence $(f \cdot g)(4) = 54$.

(d) $\left(\frac{f}{g}\right)(x) = \frac{\sqrt{x} + 1}{x^2 + 2}$. Therefore,

$$\left(\frac{f}{g}\right)(4) = \frac{\sqrt{4} + 1}{4^2 + 2} = \frac{3}{18} = \frac{1}{6}.$$

We conclude this section by discussing an important operation using two functions f and g to obtain another function, called the composition of f with g. Given two functions f and g, the *composition of f with g*, symbolized by $f \circ g$, is the function defined by

$$(f \circ g)(x) = f[g(x)]$$

provided that *the range of g is contained in the domain of f*.

Note that the composition of two functions $f \circ g$ should not be confused with the product of two functions $(f \cdot g)$. We illustrate how to compose a new function from two given functions f and g in the following example.

Example 3. Given $f(x) = x^2$ and $g(x) = x + 3$, find
(a) $(f \circ g)(x)$. (b) $(f \circ g)(2)$.

Solution

(a) By definition, $(f \circ g)(x) = f[g(x)]$. Consequently,

$$(f \circ g)(x) = f[g(x)] = f(x + 3) \qquad \text{since } g(x) = x + 3$$
$$= (x + 3)^2. \qquad f \text{ is the squaring function}$$

Thus $(f \circ g)(x) = (x + 3)^2$.

(b) Using the expression that we obtained in part (a) for $(f \circ g)(x)$, we have that

$$(f \circ g)(2) = (2 + 3)^2 = (5)^2 = 25.$$

Note, in Example 3, that the function $(f \circ g)$ has the same effect as applying the functions g and f successively. That is, $g(2) = 5$ and $f(5) = 25$.

We note that for two given functions f and g, the composition of f with g, $f \circ g$ may not necessarily be the same as the composition of g with f, $g \circ f$. We illustrate this fact in the next example.

Example 4. Given $f(x) = x + 5$ and $g(x) = \sqrt{x}$, find
(a) $(f \circ g)(x)$ and $(g \circ f)(x)$. (b) $(f \circ g)(4)$ and $(g \circ f)(4)$.

Solution

(a) $(f \circ g)(x) = f[g(x)] = f(\sqrt{x}) = \sqrt{x} + 5.$
$(g \circ f)(x) = g[f(x)] = g(x + 5) = \sqrt{x + 5}.$

(b) $(f \circ g)(4) = \sqrt{4} + 5 = 2 + 5 = 7.$
$(g \circ f)(4) = \sqrt{4 + 5} = \sqrt{9} = 3.$

EXERCISES 10.5

For Exercises 1–10, for the pair of functions f and g find the sum, difference, *and* product *of f and g, and* quotient *of f by g. See Example 1.*

1. $f(x) = x + 1$, $g(x) = x - 1$

2. $f(x) = 2x$, $g(x) = 3x + 1$

3. $f(x) = x^2$, $g(x) = x + 5$

4. $f(x) = 3x + 1$, $g(x) = x^3$

5. $f(x) = \dfrac{1}{x}$, $g(x) = x$

6. $f(x) = \dfrac{x}{x^2 + 2}$, $g(x) = \dfrac{1}{x}$

7. $f(x) = \sqrt{x}$, $g(x) = |x|$

8. $f(x) = 3|x|$, $g(x) = -\sqrt[3]{x}$

9. $f(x) = x^2 - 1$, $g(x) = x^2 + 1$

10. $f(x) = x^3$, $g(x) = x^2$

For Exercises 11–22, given $f(x) = 2x - 1$, $g(x) = |x| + 1$, *evaluate each expression. See Example 2.*

11. $(f + g)(3)$

12. $(f + g)(1)$

13. $(f - g)(-2)$

14. $(f - g)(-4)$

15. $(f \cdot g)(4)$

16. $(f \cdot g)(-1)$

17. $\left(\frac{f}{g}\right)(0)$

18. $\left(\frac{f}{g}\right)(1)$

19. $(g - f)(2)$

20. $(g - f)(-4)$

21. $(f \cdot f)(4)$

22. $(g \cdot g)(-5)$

In Exercises 23–30, for the pair of functions f and g, find (a) $(f \circ g)(x)$ *and (b)* $(f \circ g)(4)$. *See Example 3.*

23. $f(x) = 3x + 2$, $g(x) = x + 1$

24. $f(x) = -2x + 1$, $g(x) = 1 - x$

25. $f(x) = x^2$, $g(x) = x + 4$

26. $f(x) = x^3$, $g(x) = x^2$

27. $f(x) = \sqrt{x}$, $g(x) = 2x + 1$

28. $f(x) = |x|$, $g(x) = 3x - 2$

29. $f(x) = |x|$, $g(x) = -3$

30. $f(x) = 6$, $g(x) = x^2 + 1$

For Exercises 31–38, for the pair of functions f and g, find (a) $f \circ g$ *and (b)* $g \circ f$. *See Example 4(a).*

31. $f(x) = x^2$, $g(x) = x + 2$

32. $f(x) = x^3$, $g(x) = x + 1$

33. $f(x) = \sqrt{x}$, $g(x) = x^2$

34. $f(x) = 2x + 6$, $g(x) = \frac{1}{2}x - 3$

35. $f(x) = \frac{1}{3}x - 3$, $g(x) = 3x + 9$

36. $f(x) = x^3$, $g(x) = \sqrt[3]{x}$

37. $f(x) = x^2 + 3$, $g(x) = |x|$

38. $f(x) = \frac{1}{x^2}$, $g(x) = x^2$

Inverse of a Function 10.6

In this section we discuss the notion of the inverse of a function (provided that it has one). In order to discuss inverse functions, we must first become acquainted with the definition of a one-to-one function.

Remember that a function f is a set of ordered pairs $(x, y) \in f$ such that no two different ordered pairs have the same *first* component, x. A function can have several ordered pairs with the same second component y. However, if a function has the additional property that no two different ordered pairs have the same *second* component y, then it is called a *one-to-one function*.

Example 1. Determine whether each of the following functions is a one-to-one function.
(a) $f(x) = 3x + 1$; domain $= \{-1, 1, 2\}$.
(b) $f(x) = x^2 + 2$; domain $= \{-1, 1, 2\}$.

Solution
(a) We note that

$$f(-1) = 3(-1) + 1 = -3 + 1 = -2.$$

Thus $(-1, -2)$ belongs to the function f. Similarly, we find that $f(1) = 4$ and $f(2) = 7$. Thus the function f consists of the set of ordered pairs

$$f = \{(-1, -2), (1, 4), (2, 7)\}.$$

Since the function f has no two different ordered pairs with the same second component, it is by definition a one-to-one function.
(b) Following the procedure outlined in part (a), we see that the function consists of the set of ordered pairs

$$f = \{(-1, 3), (1, 3), (2, 6)\}.$$

The function is *not* one-to-one because it contains two ordered pairs $(-1, 3)$ and $(1, 3)$ with the *same second* component.

A one-to-one function f has an *inverse function,* denoted by f^{-1}, consisting of the set of ordered pairs obtained from f by *interchanging* the first and second components of each ordered pair in f.

Note that the symbol f^{-1}, read "f inverse," does *not* mean $\dfrac{1}{f}$. In the following example we illustrate how to obtain the inverse of a one-to-one function.

Example 2. If f is the function

$$f = \{(-1, -2), (0, 1), (1, 4), (2, 7)\},$$

find the inverse function f^{-1} of f.

Solution: If we interchange the first and second components of the ordered pairs in f, we obtain f^{-1}.

$$f^{-1} = \{(-2, -1), (1, 0), (4, 1), (7, 2)\}.$$

Notice in Example 2 that

1. The domain of f^{-1} is $\{-2, 1, 4, 7\}$, which is the range of f.
2. The range of f^{-1} is $\{-1, 0, 1, 2\}$, which is the domain of f.

In general, the *domain* of a function f^{-1} is the *range* of f and the *range* of f^{-1} is the *domain* of f.

We pause to note that if a function is not a one-to-one function, it cannot have an inverse. For example, consider the function f defined by

$$f = \{(-1, 3), (1, 3), (2, 6)\},$$

which we saw in Example 1(b) was not a one-to-one function. If we follow the procedure for obtaining an inverse function, by interchanging first and second components of the ordered pairs in f, we obtain the relation R, where

$$R = \{(3, -1), (3, 1), (6, 2)\}.$$

We see that R is not a function because, by definition, a function *cannot* have two different ordered pairs such as $(3, -1)$ and $(3, 1)$ with the same *first* component.

We now consider functions that are defined by an equation, such as $f(x) = x^3$ for an infinite set of real numbers x. Such functions consist of an infinite set of ordered pairs. Thus it would not be possible to list all the ordered pairs of the function in order to determine whether or not it is a one-to-one function. As a result, we could not determine if it had an inverse function. However, it may be possible "to see that" for each x in the domain of f, the equation defining f leads to *one* and *only one* y value. If it does, the function cannot have two different ordered pairs with the same second component and therefore it would be a one-to-one function. The graph of a function can be of aid in determining if it is a one-to-one function by applying the following test.

Horizontal Line Test

> If any horizontal line intersects the graph of a function in more than one point, the function is *not* a one-to-one function.

We illustrate the application of the rule above and the remarks preceding it in the following example.

Example 3. Determine if each of the following is a one-to-one function.
(a) $f(x) = x^2$, x any real number.
(b) $f(x) = x^3$, x any real number.

Solution

(a) We observe that $f(x) = x^2$ or $y = x^2$ leads to the same y-value for different values of x in the domain of f. For example, $f(-2) = (-2)^2 = 4$ and $f(2) = 2^2 = 4$. Thus the function contains two ordered pairs $(-2, 4)$ and $(2, 4)$ with the same second component. Therefore, the given function is not a one-to-one function. The graph of $f(x) = x^2$ or $y = x^2$ is the parabola shown in Figure 10.16. Notice that any horizontal line above the x-axis intersects the graph in two points.

(b) We note that the formula $f(x) = x^3$ will not lead to the same y-value for different x-values. To confirm this, we graph $f(x) = x^3$ or $y = x^3$ as shown in Figure 10.17. Since no horizontal line will intersect the graph of the given function in more than one point, $f(x) = x^3$ is a one-to-one function.

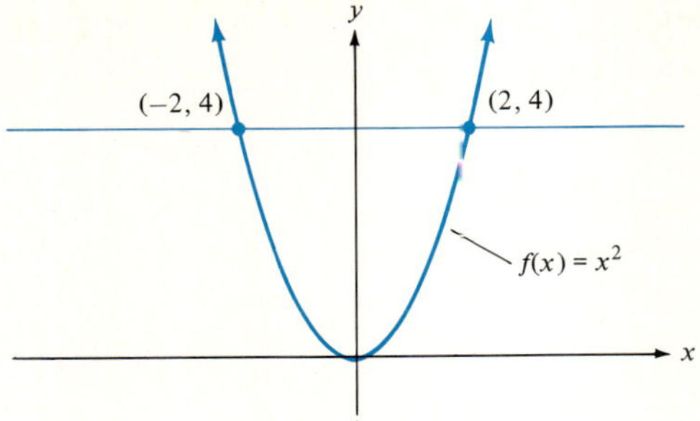

Figure 10.16

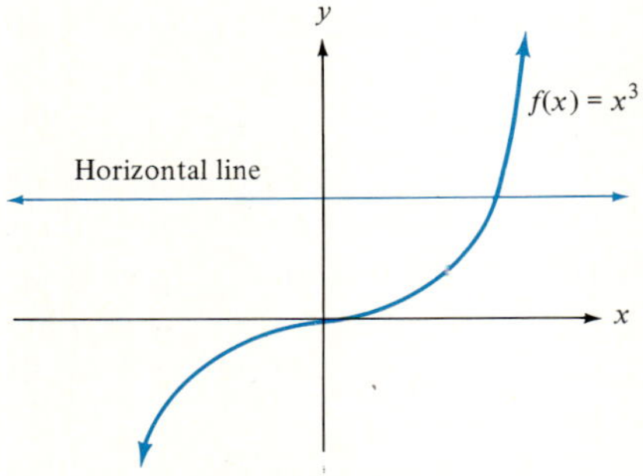

Figure 10.17

We know that the inverse of a one-to-one function f is obtained by interchanging the first component x and the second component y of the ordered pairs (x, y) that make up the function f. However, in the case of a function consisting of an infinite set of ordered pairs defined by a rule, such as $f(x) = 2x + 4$, this is not possible. But we can *sometimes determine the rule that defines the inverse* function by the following procedure.

Steps for Finding the Inverse of a One-to-One Function $y = f(x)$

Step 1. Solve $y = f(x)$ for x as a function of y if possible.

Step 2. The result is $x = f^{-1}(y)$, where y represents the domain of inverse function f^{-1}.

Step 3. Switch the variables to obtain $y = f^{-1}(x)$. Now x represents an element in the domain of f^{-1}, as is customary.

We illustrate the application of the procedure above in the following example.

Example 4. Given $f(x) = 2x + 4$, find the equation for the inverse function $f^{-1}(x)$.

Solution: We set $y = 2x + 4$.

Solve for x:

$$x = \frac{y - 4}{2} \qquad \text{or} \qquad x = \frac{1}{2}y - 2.$$

Switch variables:

$$y = \frac{1}{2}x - 2.$$

Thus

$$f^{-1}(x) = \frac{1}{2}x - 2.$$

In Example 4 we saw that for $f(x) = 2x + 4$, the inverse function is given by $f^{-1}(x) = \frac{1}{2}x - 2$. Note that $f(2) = 2(2) + 4 = 8$. However, on the other hand, $f^{-1}(8) = \frac{1}{2}(8) - 2 = 2$. Thus $f(2) = 8$ and $f^{-1}(8) = 2$. Consequently, we see that the effect of the inverse function "reverses" that of the original function. In general, if f is a one-to-one function with inverse function f^{-1}, the following statement holds:

$$f(a) = b \qquad \text{if and only if} \qquad f^{-1}(b) = a.$$

As a result of the previous discussion, we note that we can also regard f as the inverse function of f^{-1}, since it also "reverses" the effect of f^{-1}.

We now discuss how to obtain the graph of $y = f^{-1}(x)$ from the graph of $y = f(x)$. We know that

$$(a, b) \in f \qquad \text{if and only if} \qquad (b, a) \in f^{-1}.$$

It can be shown that the graphs of (a, b) and (b, a) are points that are "mirror images" of each other with respect to the line $y = x$ as shown in Figure 10.18.

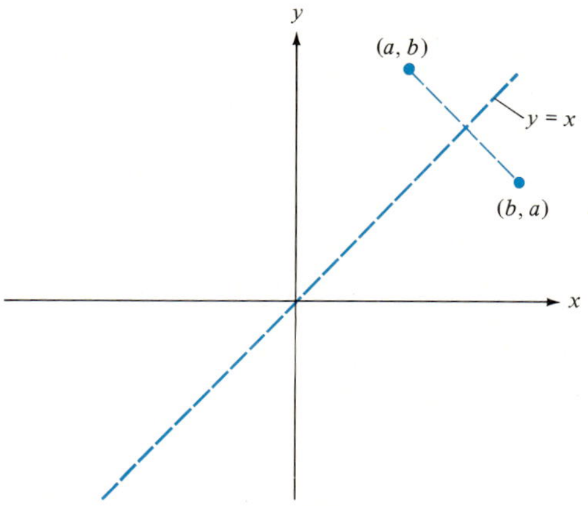

Figure 10.18

Formally, we say that (a, b) and (b, a) are *symmetric with respect to the line* $y = x$. We also say that the graphs of (a, b) and (b, a) are *reflections* of each other in the line $y = x$. Thus the points on the graph of f^{-1} may be obtained from the graph of f by reflecting the points on the graph of f in the line $y = x$. As a simple illustration of the preceding remarks, let us consider the graph of the inverse function of

$$f = \{(-3, 1), (-1, 2), (1, 3), (3, 5)\}.$$

In Figure 10.19, the graph of f consists of the points in black. Their corresponding reflections in the line $y = x$ are shown in blue and represent the graph of f^{-1}.

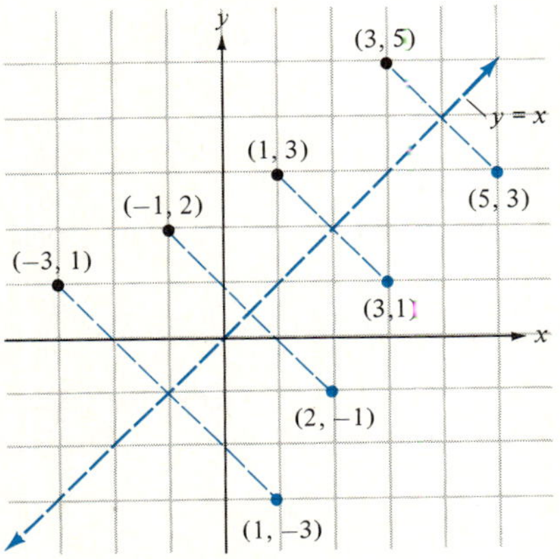

Figure 10.19

In Example 5 we illustrate how to find the graph of the inverse function of a function, whose graph is a smooth curve, using the technique of reflecting the graph of the original function in the line $y = x$.

Example 5. For the function $f(x) = x^2, x \geq 0$, find the inverse function $f^{-1}(x)$ and graph both f and f^{-1}.

Solution: Solve for x: Solving $y = x^2$ or $x^2 = y$ for x, we obtain $x = \pm\sqrt{y}$. Thus we have two solutions for x: $x = \sqrt{y}$ or $x = -\sqrt{y}$. Since we are given that $x \geq 0$ (x is positive or zero), we choose the positive solution

$$x = \sqrt{y}.$$

Switch variables:

$$y = \sqrt{x}.$$

Thus

$$f^{-1}(x) = \sqrt{x}.$$

The graph of $f(x) = x^2, x \geq 0$, is shown in Figure 10.20 in black. We reflect this graph in the line $y = x$ to obtain the graph of $f^{-1}(x) = \sqrt{x}$ shown in Figure 10.20

in blue. Notice that since (2, 4) is on the graph of $f(x) = x^2$, then (4, 2) must be on the graph of $f^{-1}(x) = \sqrt{x}$.

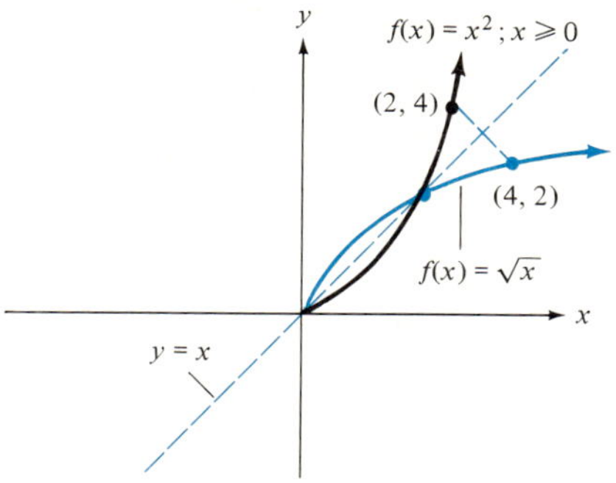

Figure 10.20

EXERCISES 10.6

For Exercises 1–12, given that each function has domain D = {−2, 1, 2, 3}, determine which functions are one-to-one functions. See Example 1.

1. $f(x) = 4x - 1$ **2.** $f(x) = -2x + 1$ **3.** $f(x) = x^4$ **4.** $f(x) = x^2$
5. $f(x) = x^3$ **6.** $f(x) = x^5$ **7.** $f(x) = |x|$ **8.** $f(x) = |x - 2|$
9. $f(x) = x^2 - 2x$ **10.** $f(x) = x - x^2$ **11.** $f(x) = 1 - x^2$ **12.** $f(x) = 3 - |x|$

For Exercises 13–20, for the function f, find the inverse function f^{-1}. See Example 2.

13. $f = \{(-2, 1), (0, 3), (1, 4)\}$ **14.** $f = \{(0, -3), (1, 2), (2, 7)\}$
15. $f = \{(1, 1), (2, 4), (3, 9)\}$ **16.** $f = \{(-2, -8), (0, 0), (2, 8)\}$
17. $f = \{(4, 2), (9, 3), (25, 5)\}$ **18.** $f = \{(1, -1), (4, -2), (9, -3)\}$
19. $f = \{(-2, 2), (-3, 3), (-5, 5)\}$ **20.** $f = \{(2, 0), (3, 1), (4, 2)\}$

For Exercises 21–28, given that each function has domain D = {all real numbers}, determine which functions are one-to-one functions. See Example 3.

21. $f(x) = 2x + 1$ **22.** $f(x) = 4x - 1$ **23.** $f(x) = |x|$ **24.** $f(x) = |x - 1|$
25. $f(x) = x^2 + 1$ **26.** $f(x) = -x^2 + 2$ **27.** $f(x) = x^4$ **28.** $f(x) = x^5$

For Exercises 29–42, for the function f, find f^{-1} and graph both f and f^{-1}. See Examples 4 and 5.

29. $f(x) = x + 3$ **30.** $f(x) = x - 4$ **31.** $f(x) = 3x - 6$
32. $f(x) = \frac{1}{2}x + 2$ **33.** $f(x) = x^2 + 1, x \geq 0$ **34.** $f(x) = x^2 + 2, x \geq 0$
35. $f(x) = |2x|, x \leq 0$ **36.** $f(x) = |3x|, x \geq 0$ **37.** $f(x) = x^3$

38. $f(x) = x^3 + 1$ **39.** $f(x) = \sqrt{x}, x \geq 0$ **40.** $f(x) = \sqrt[3]{x}$

41. $f(x) = \sqrt{9 - x^2}, 0 \leq x \leq 3$ **42.** $f(x) = \sqrt{4 - x^2}, -2 \leq x \leq 0$.

Solve Exercises 43–45. See the discussion immediately following Example 4.

43. If a function f has an inverse and $f(-4) = 6$, find $f^{-1}(6)$.

44. If $f^{-1}(10) = 5$, find $f(5)$.

45. Find $(f^{-1} \circ f)(5)$. **46.** Find $(f^{-1} \circ f)(8)$.

10.7 Variation

In the physical and social sciences and technology we often find that the term *function* is not used when describing certain special types of functions. Instead, descriptive phrases are used to specify how one variable changes or *varies* with respect to another variable (or variables). The word *variation* is often used as an alternative way of describing these special relationships between variables. In this section we define three different types of variation: *direct, indirect,* and *joint variation.*

If two variables, such as x and y, are related by an equation of the form

(1) $$y = kx \qquad k \text{ is a real constant} \quad (k \neq 0),$$

y is said to *vary directly* with x or y *varies as* x. We also say that y is *directly proportional to* x or simply y *is proportional to* x. The concept of direct variation can be extended to include powers of the variable x in equation (1). Thus, for $y = k\sqrt{x}$, we say that y *varies directly as the square root of* x.

Generalizing, we say that y *varies directly as* (or y *is proportional to*) the qth power of x if

$$y = kx^q \qquad (k \neq 0),$$

where q is a positive rational number and k is a real constant called the *constant of variation* (or *constant of proportionality*).

To illustrate the previous definition, we consider two formulas that are examples of direct variation.

1. The circumference of a circle varies directly as the length of its diameter D. The formula relating C and D is

$$C = \pi D,$$

where π is the constant of variation.

2. The vibrating frequency f of a vibrating string varies directly as the square root of the tension t of the string. Thus

$$f = k\sqrt{t},$$

where k is the constant of variation. This k depends on the nature of the string.

In variation problems we are usually given the type of variation and a set of corresponding values of the variables involved. From this information we can usually determine the constant of variation and the equation that relates the variables. We illustrate this procedure in the following examples. Notice that in direct variation problems, an *increase* in the absolute value of one variable results in an *increase* in the absolute value of the other variable.

Example 1. If y varies directly as x and $y = 20$ when $x = 4$, find the equation relating y and x. Also, find the value of y when $x = 6$.

Solution: Since y varies directly as x, we know that the function relating y and x is of the form

$$y = kx$$

for some constant k. Next, we need to determine the value of the constant k. But $y = 20$ when $x = 4$; therefore,

$$20 = k(4).$$

Solving this equation for k, we obtain $k = 5$. Thus the equation relating y and x is

$$y = 5x.$$

Substituting 6 for x in the last equation, we find that

$$y = 5(6) = 30.$$

Thus $y = 30$ when $x = 6$.

Example 2. For an observer in a small airplane the approximate distance, D (in miles), to the horizon varies directly as the square root of the height, h (in feet), of the airplane above the ground. At a height of 100 feet the horizon is at a distance of 12.5 miles. What is the distance from the horizon when the height of the airplane is 4900 feet?

Solution: Since D the distance to the horizon varies directly as the square root of the height, h, of the airplane above the ground, we know that

$$D = k\sqrt{h}$$

for some constant k. We also know that $D = 12.5$ when $h = 100$. Thus, substituting these values for D and h in the equation above, we have

$$12.5 = k\sqrt{100}$$
$$12.5 = k(10) \quad \text{or} \quad k = 1.25.$$

It follows that

$$D = 1.25\sqrt{h}.$$

When $h = 4900$, we see that

$$D = 1.25\sqrt{4900}$$
$$D = 1.25(70) = 87.5.$$

Therefore, we find that the horizon is 87.5 miles distant from an observer in an airplane that is 4900 feet above the ground.

Another common type of variation is *inverse variation*. The variables x and y *vary inversely* if

(2) $$xy = k \qquad k \text{ is a constant} \quad (k \neq 0).$$

If we write equation (2) as

$$y = \frac{k}{x},$$

we say that y *varies inversely as* x, or y *is inversely proportional to* x. As in direct variation the concept of inverse variation can be extended to include powers of a variable. Thus, if $y = \dfrac{k}{x^2}$, we say that y *varies inversely as the square of* x.

In general, y *varies inversely as* (or *is inversely proportional to*) the qth power of x if

$$y = \frac{k}{x^q} \qquad (k \neq 0),$$

where q is a positive rational number and k is a real number called the *constant of variation* (or *constant of proportionality*).

The procedures used to deal with inverse variation problems are similar to those that we used for dealing with direct variation problems. We illustrate this in the following examples. Note, however, that in *indirect* variation, an *increase* in the absolute value of one variable is accompanied by a *decrease* in the absolute value of the other variable.

Example 3. If y varies inversely as the cube of x and $y = \dfrac{1}{8}$ when $x = 4$, find the equation relating x and y. Also, find the value of y when $x = 2$.

Solution: We are given that y varies inversely as the cube of x. Thus, to find the equation relating x and y, we need to determine k in the equation

$$y = \frac{k}{x^3}.$$

Since $y = \dfrac{1}{8}$ when $x = 4$, we substitute these values in the equation above and solve the following equation for k:

$$\frac{1}{8} = \frac{k}{4^3}.$$

From this equation it follows that $k = 8$. Therefore, the equation relating x and y is

$$y = \frac{8}{x^3}.$$

Next, we substitute $x = 2$ in the last equation to obtain

$$\begin{aligned} y &= \frac{8}{2^3} \\ &= \frac{8}{8} = 1. \end{aligned}$$

Thus, when $x = 2$, $y = 1$.

The next example is an illustration of an "inverse square law" that governs many physical phenomena. This particular example is concerned with the inverse square law of illumination.

Example 4. The intensity of illumination I (measured in foot-candles) of a source of light on a wall varies inversely as the square of the distance d (in feet) from the source of light. If $I = 12$ foot-candles when $d = 5$ feet, what is I when $d = 10$ feet?

Solution: From the given information we know that

$$I = \frac{k}{d^2}.$$

To solve this problem, we need first to determine k, the constant of variation. Since $I = 12$ when $d = 5$, we have

$$12 = \frac{k}{5^2}$$

from which it follows that

$$k = 5^2(12) = 300.$$

Therefore,

$$I = \frac{300}{d^2}.$$

We can now find I when $d = 10$ by substituting $d = 10$ in the last equation. As a result, we see that

$$I = \frac{300}{10^2}$$

$$= \frac{300}{100} = 3.$$

Thus $I = 3$ foot-candles when $d = 10$ feet.

The third type of variation we discuss is that of *joint variation*. A variable w *varies jointly as x and y* if there exists a constant k ($k \neq 0$) (constant of variation) such that

$$w = kxy.$$

The concept of joint variation can be extended to two or more variables with positive rational exponents. In the next example we give an illustration of this extended concept of joint variation.

Example 5. Suppose that w varies jointly as x and $\sqrt{y}$. If $w = 18$ when $x = 2$ and $y = 9$, find w when $x = 5$ and $y = 16$.

Solution: From the information given we know that

(1) $$w = kx\sqrt{y}.$$ note that $\sqrt{y} = y^{1/2}$

First we find the value of k, by substituting $w = 18$, $x = 2$, and $y = 9$ in equation (1). As a result, we see that

$$18 = k(2)\sqrt{9}$$

from which it follows that $k = 3$. Thus

$$w = 3x\sqrt{y}.$$

When $x = 5$ and $y = 16$, we see that

$$w = 3(5)\sqrt{16}$$
$$= 3(5)(4) = 60.$$

Many variation problems involve combinations of direct, inverse, and joint variation. The following example involves both direct and inverse variation.

Example 6. The rate at which heat, H (calories per second), develops in a coil of electrical wire varies directly as the square of the voltage V (volts) supplied across its ends and inversely as the resistance R (ohms) of the coil of wire. If $H = 800$ calories per second when $V = 20$ volts and $R = .12$ ohm, find H when $V = 30$ volts and $R = .24$ ohm.

Solution

$$H = k\frac{V^2}{R},$$

where k is a constant to be determined. Since $H = 800$ when $V = 20$ and $R = .12$, we have that

$$800 = k \cdot \frac{(20)^2}{.12}$$
$$(.12)800 = k(400)$$
$$.24 = k.$$

Thus

$$H = .24\frac{V^2}{R}.$$

To find H when $V = 30$ and $R = .24$, we substitute in the last equation to obtain

$$H = .24\frac{(30)^2}{.24}.$$

It follows that $H = 900$ calories per second.

EXERCISES 10.7

Exercises 1–8 involve direction variation. Find the equation relating the given variables, and solve for the value of the variable specified. See Example 1 and the discussion preceding it.

1. y varies directly as x and $y = 15$ when $x = 5$. Find y when $x = 9$.

2. y varies directly as x and $y = 21$ when $x = 3$. Find y when $x = 8$.

3. y varies directly as the square of t and $y = 8$ when $t = 2$. Find y when $t = 7$.

4. y varies directly as the square of t and $y = 27$ when $x = 3$. Find y when $t = 5$.

5. s varies directly as the cube of t and $s = 24$ when $t = 2$. Find s when $t = 3$.

6. u varies directly as the cube of v and $u = 4$ when $v = 2$. Find u when $v = 4$.

7. w varies directly as the square root z and $w = 25$ when $z = 25$. Find w when $z = 64$.

8. r varies directly as the square root of t and $r = 16$ when $t = 4$. Find r when $t = 25$.

Exercises 9–18 involve inverse and joint variation. Find the equation relating the given variables and solve for the variable specified. See Examples 3 and 5.

9. y varies inversely as x and $y = 8$ when $x = 2$. Find y when $x = 8$.

10. y varies inversely as x and $y = 9$ when $x = 3$. Find y when $x = 9$.

11. s varies inversely as the square root of t and $s = 16$ when $t = 4$. Find s when $t = 16$.

12. p varies inversely as the square root of q and $p = 6$ when $q = 25$. Find p when $q = 36$.

13. d varies inversely as the square of s and $d = 25$ when $s = 2$. Find d when $s = 10$.

14. v varies inversely as the square of t and $v = 2$ when $t = 10$. Find v when $t = 5$.

15. w varies jointly as x and y and $w = 60$ when $x = 3$ and $y = 5$. Find w when $x = 7$ and $y = 2$.

16. w varies jointly as x and y and $w = 48$ when $x = 4$ and $y = 6$. Find w when $x = 5$ and $y = 7$.

17. z varies jointly as x^2 and y and $z = 60$ when $x = 2$ and $y = 3$. Find z when $x = 3$ and $y = 2$.

18. z varies jointly as $\sqrt{x}$ and y^2 and $z = 180$ when $x = 4$ and $y = 3$. Find z when $x = 25$ and $y = 2$.

Solve Exercises 19–28. See Examples 1–6.

19. The total daily revenue (R) that a small company receives varies directly as the number of units (x) of a product that the company sells. If $R = \$1100$ when $x = 200$, find R when $x = 300$.

20. The pressure (P) on an object submerged in a liquid varies directly with the depth (d) of the object below the surface of the liquid. If $P = 300$ pounds per square foot when $d = 6$ feet, what is the pressure P when $d = 15$ feet?

21. The distance (d) that an object falls from rest varies directly with the square of the time (t) during which it falls. If an object falls 144 feet in 3 seconds, how far will it fall in 5 seconds?

22. The vibrating frequency (f) of a vibrating string varies directly with the square root of the tension (t) of the string. If $f = 72$ vibrations per second when $t = 16$ pounds, find f when $t = 36$ pounds.

23. The velocity (v) of outflow of a liquid from an opening varies directly as the square root of the height (h) of the liquid above the opening. Write an equation expressing v in terms of h if $v = 40$ feet per second when $h = 25$ feet.

24. The volume (V) of a sphere is directly proportional to the cube of its radius (r). If $V = 36\pi$ cubic feet when $r = 3$ feet, find the formula for V in terms of r.

25. The weight (w) of a body above the earth varies inversely as the square of its distance (d) from the center of the earth. A body weighs 200 pounds on the earth's surface, which is approximately 4000 miles from the center of the earth. How much will it weigh 1000 miles above the earth's surface?

26. If the temperature of an ideal gas is kept constant, its volume (V) varies inversely as the pressure (P) to which the gas is subjected (Boyle's law). Suppose for a certain gas that $V = 4$ cubic feet when $P = 10$ pounds per square inch (psi). Find the volume of the gas V when $P = 8$ psi.

27. The power P (watts) in an electrical system varies directly as the square of the voltage V (volts) and inversely as the resistance R (ohms). Suppose in the system that $P = 500$ watts when $V = 100$ volts and $R = 20$ ohms. Find the power P when $V = 110$ volts and $R = 20$ ohms.

28. If a body moves in a circle, the centrifugal force (F) exerted on the body radially from the center of the circle varies directly as the square of the linear speed (v) of the body and inversely as the radius (r) of the circle. A stone tied to the end of a cord is whirled in a circle of radius 4 feet with a linear speed of 24π feet per second and with centrifugal force $F = 9\pi^2$ pounds. What is the centrifugal force if the stone is whirled in a circle of radius $r = 6$ with $v = 36\pi$ feet per second?

Key Words

Combination of functions
  Sum
  Difference
  Product
  Quotient
Composition of functions
Domain
Function
  One-to-one function
Horizontal line test
Inverse function
Proportional to
Quadratic function
  Maximum value
  Minimum value
Range
Reflection
Symmetric
Value of a function
Variation
  Direct
  Inverse
  Joint
Vertical line test

CHAPTER 10 REVIEW

SECTION 10.1

For each relation, determine the domain and the range.

1. $\{(-1, -3), (0, 0), (2, 6)\}$
2. $\{(-3, 18), (-1, 2), (3, 18)\}$
3. $\{(x, y) \mid x^2 + y^2 = 16\}$
4. $\{(x, y) \mid x^2/4 + y^2/9 = 1\}$
5. $\{(x, y) \mid y = 2x\}$
6. $\{(x, y) \mid y^2 = x\}$
7. $\{(x, y) \mid y = \frac{1}{2}x^2\}$
8. $\{(x, y) \mid y < x\}$

SECTION 10.2

For each equation, determine if the equation defines a function of x with the given domain. Find the range of all functions.

9. $y = 5x$; $\{0, 1, 2, 3\}$
10. $y = x^2 - 1$; $\{-2, 1, 2, 4\}$
11. $y = x^3$; $\{-1, 0, 1, 2\}$
12. $y^2 = 4x$; $\{0, 1, 4, 9\}$
13. $y = 3$; $\{-2, -1, 0, 1\}$
14. $|y| = |x|$; $\{-3, 0, 1, 3\}$

For each function, find the indicated function values.

15. $f(x) = 5x + 3$ **(a)** $f(0)$ **(b)** $f(-2)$ **(c)** $f(1)$

16. $g(x) = \dfrac{x + 1}{x - 1}$ **(a)** $g(0)$ **(b)** $g(2)$ **(c)** $g(-1)$

17. $h(x) = x^4 + 2$ **(a)** $h(-1)$ **(b)** $h(2)$ **(c)** $h(-2)$

For each function, determine the largest possible real number domain.

18. $y = \dfrac{1}{3 - x}$
19. $y = \sqrt{x + 2}$
20. $y = \dfrac{1}{(x - 5)(x - 7)}$

21. $y = x^2 + 10$ **22.** $y = \dfrac{1}{|x| - 1}$ **23.** $y = \dfrac{1}{x^2 - 3x + 2}$

For each graph, use the vertical line test to determine if they represent graphs of functions of x.

24.

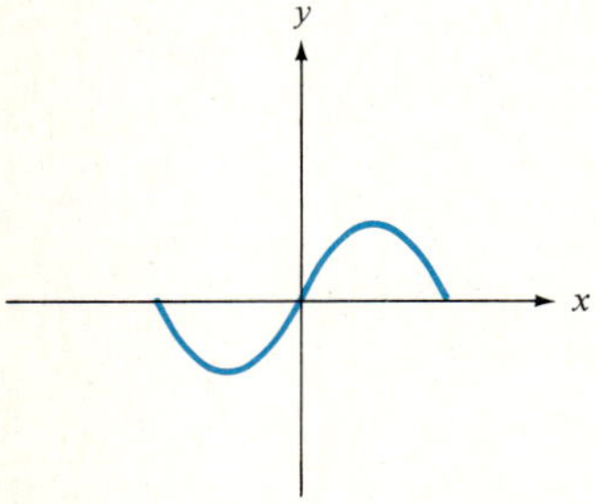

25.

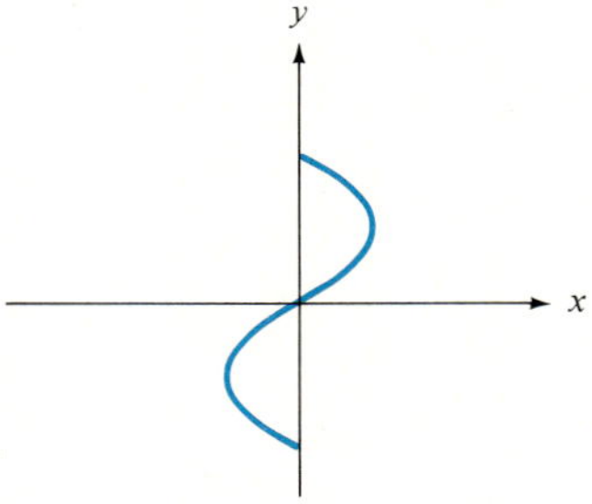

SECTION 10.3

For each quadratic function f, find (a) *the vertex of the graph of f and* (b) *the minimum or maximum value of f.*

26. $f(x) = -x^2 + 2x - 5$ **27.** $f(x) = x^2 - 6x + 10$

Solve each word problem.

28. Find two positive numbers whose sum is 30 and whose product is a maximum.

29. A steel ball is shot vertically up into the air with an initial velocity 160 feet per second. The height after t seconds is given by $h(t) = -16t^2 + 160t + 220$. Find
(a) The time it takes to reach the maximum height.
(b) The maximum height that the ball will reach.

SECTION 10.4

Graph each function.

30. $f(x) = -|x|$ **31.** $f(x) = \begin{cases} x + 2 & x \geq 0 \\ 2 & x < 0 \end{cases}$

SECTION 10.5

Find the values of each function if $f(x) = 3x + 2$ *and* $g(x) = \sqrt{x}$.

32. $(f + g)(1)$ **33.** $(f - g)(4)$ **34.** $(f \cdot g)(9)$ **35.** $\left(\dfrac{f}{g}\right)(16)$

For each pair of functions f and g, find (a) $(f \circ g)(x)$*; and (b)* $(g \circ f)(x)$.

36. $f(x) = x^2 - 1$, $g(x) = x + 2$ **37.** $f(x) = x^2$, $g(x) = x^3 + 1$

SECTION 10.6

For each function f, find the set of ordered pairs defining the inverse function, f^{-1}.

38. $f = \{(-1, -10), (0, -3), (1, 4)\}$ **39.** $f = \{(2, 4), (3, 9), (4, 16)\}$

For each function f, find the equation defining f^{-1} and graph both f and f^{-1}.

40. $f(x) = 3x - 9$ **41.** $f(x) = \frac{1}{4}x^2;\ x \geq 0$

SECTION 10.7

For each exercise, find the equation relating the given variables, and solve for the value of the variable specified.

42. y varies directly as x and $y = 10$ when $x = 2$. Find y when $x = 7$.

43. y varies inversely as the square of x and $y = 4$ when $x = 3$. Find y when $x = 6$.

44. w varies jointly as x and y and $w = 24$ when $x = 2$ and $y = 4$. Find w when $x = 5$ and $y = 3$.

45. The current I (amperes) in an electric circuit varies directly as the voltage V (volts) and inversely as the resistance R (ohms). The current $I = 5$ amperes when $V = 110$ volts and $R = 22$ ohms. Find the current I when $V = 120$ and $R = 30$.

46. The time T (seconds) that it takes a simple pendulum to make one swing back and forth varies directly as the square root of the length of the pendulum L (meters). For a pendulum with length $L = 1$ meter, the period T is approximately 2 seconds. Find the period T for a pendulum with length $L = 4$ meters.

CHAPTER 10 TEST

1. For the relation $R = \{(-5, 2), (3, 7), (5, 8), (6, 8)\}$, determine its domain and range.

2. Determine which of the following equations defines a function of x with domain $D = \{-1, 2, 7\}$. Find the range of any function.
(a) $y^2 = x + 2$. **(b)** $y = x^2 + 2$.

3. For the function $f(x) = 3x + 2$, find **(a)** $f(3)$, **(b)** $f(0)$, and **(c)** $f(-2)$.

4. For $f(x) = x^2$, find $f(x + h)$.

5. Determine the largest possible real number domains of the following functions of x.

(a) $y = \dfrac{1}{(x + 4)(x - 2)}$. **(b)** $y = \dfrac{1}{\sqrt{x - 4}}$.

6. Determine which of the following graphs are graphs of functions.

(a)

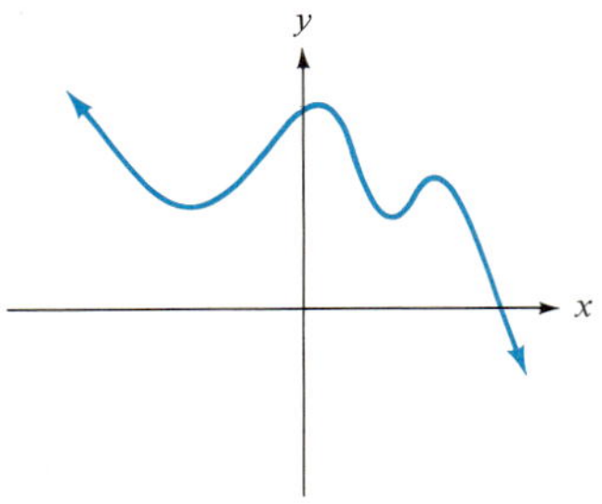

(b)

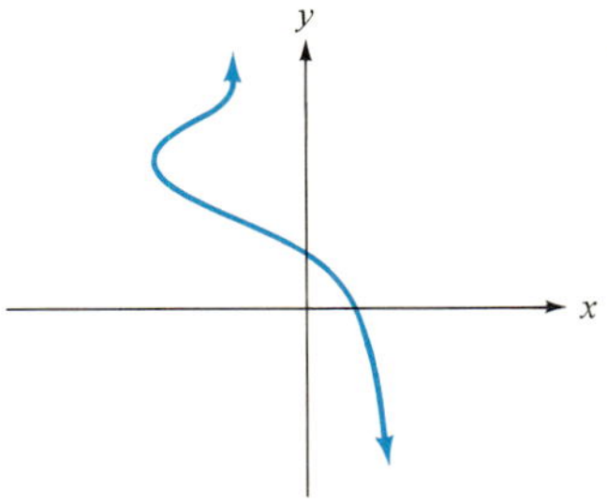

7. For $f(x) = -3x^2 + 6x + 2$, determine
(a) Whether f has a minimum or maximum value.
(b) The maximum or minimum value of f.

8. Find two positive numbers whose sum is 40 and whose product is a maximum.

9. For $f(x) = \begin{cases} x^2 + 1 & x \geq 0 \\ x + 5 & x < 0 \end{cases}$ find **(a)** $f(-1)$, **(b)** $f(0)$, and **(c)** $f(2)$.

10. Given $f(x) = 5x + 2$ and $g(x) = x^2$, find **(a)** $(f + g)(2)$, **(b)** $(f - g)(2)$, **(c)** $(f \cdot g)(2)$, and **(d)** $\left(\frac{f}{g}\right)(2)$.

11. For $f(x) = x^2$ and $g(x) = x + 1$, find **(a)** $(f \circ g)(3)$ and **(b)** $(g \circ f)(3)$.

12. If $f(x) = 3x + 6$, find the equation defining $f^{-1}(x)$.

13. z varies directly as x^2 and inversely as $\sqrt{y}$ and $z = 36$ when $x = 6$ and $y = 4$. Find z when $x = 8$ and $y = 16$.

14. The lift force (L) on an airplane wing by the atmosphere varies jointly as the product of the area (A) of the wing and the square of the velocity (v) of the airplane. $L = 1000$ pounds when $A = 100$ square feet and $v = 50$ miles per hour.
(a) Find the constant of proportionality.
(b) Write the formula for L in terms of the variables A and v.

CHAPTER 11

Exponential and Logarithmic Functions

Heavy rains have caused run-off from a local industrial site to pollute a reservoir that supplies some of the water to an urban area. The level of pollution is 15 times the accepted level. Engineers estimate that the natural cleansing action of the reservoir will reduce the pollution by 50% every 7 days. By solving the exponential equation $(0.5)^{x/7} = \frac{1}{15}$, the engineers are able to determine that it will be approximately 28 days before the water is safe to use again.

© 1981 Tom Kelly

In Chapter 5 we discussed expressions involving exponents in which the exponent was any rational number. Thus we are already familiar with expressions of the form $b^{m/n}$, where $b > 0$ and $n \neq 0$. In this chapter we consider expressions of the form b^x, where b is a positive constant and x is any real number. In addition, we discuss an alternative representation of exponents known as logarithms. We also consider some applied problems and equations that involve exponential and logarithmic expressions.

11.1 Exponential Functions

Consider the function $f(x) = 2^x$. Because the exponent involves a variable, the function is called an *exponential function*. From our work in Chapter 5 we know how to evaluate this function for rational values of x such as 3, 0, and $\frac{-5}{2}$, as shown by the following calculations: if $f(x) = 2^x$, then $f(3) = 2^3 = 8$, $f(0) = 2^0 = 1$, and $f\left(\frac{-5}{2}\right) = 2^{-5/2} = \frac{1}{2^{5/2}} = \frac{1}{(2^5)^{1/2}} = \frac{1}{\sqrt{32}} = \frac{1}{\sqrt{32}} \cdot \frac{\sqrt{2}}{\sqrt{2}} = \frac{\sqrt{2}}{8}$. Using a calculator or the table at the back of the book, we can get an approximate value for $\frac{\sqrt{2}}{8}$ such as $\frac{1.414}{8} = .17675$. Because this is an approximation we use the symbol "$\simeq$" and write

$$f\left(\frac{-5}{2}\right) = \frac{\sqrt{2}}{8} \simeq \frac{1.414}{8} = .17675.$$

The symbol "$\simeq$" is read as "is approximately equal to." Thus, using the definition for rational exponents, we can construct the following table of values for the function $f(x) = 2^x$.

x	-4	-3	-2	-1	0	1	2	3	4
$y = 2^x$	$\frac{1}{16}$	$\frac{1}{8}$	$\frac{1}{4}$	$\frac{1}{2}$	1	2	4	8	16

If we plot these ordered pairs, (x, y), the result is shown in Figure 11.1.

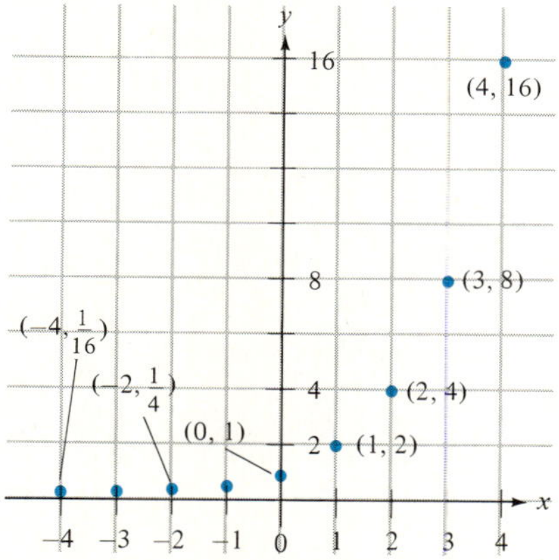

Figure 11.1

In Figure 11.1 notice that the scales on the two axes are different. This is because as x gets larger, the corresponding change in $y = 2^x$ is much greater, as illustrated by the following table.

	Change in x	x	y	Change in y
$y = 2^x$		3	8	
	$4 - 3 = 1$			$8 = 16 - 8$
		4	16	
	$5 - 4 = 1$			$16 = 32 - 16$
		5	32	
	$6 - 5 = 1$			$32 = 64 - 32$
		6	64	

Many calculators have keys that can be used to evaluate exponential expressions such as 2^x. If you have access to a calculator with a $\boxed{y^x}$ key, you could verify that $2^{5.21} \simeq 37.014$, $10^{1.432} \simeq 27.04$, and $7^{-.241} \simeq .626$. Notice that all of these exponents are rational numbers. If we want to extend the meaning of an exponent to all real numbers, we must give meaning to expressions involving irrational numbers as exponents such as $2^{\sqrt{2}}$, 2^{π}, and $5^{\sqrt[3]{-10}}$.

A precise definition of exponential functions that includes irrational exponents as well as rational exponents is discussed in advanced mathematics courses. As a result, it can be shown that the *exponential function*

$$f(x) = b^x, \qquad b > 0, \quad b \neq 1, \qquad b \text{ a constant}$$

is defined for all real numbers x. That is, the *domain* of the exponential function is the set of real numbers $\mathbb{R}$. However, the *range* is the set of positive real numbers. The following four properties characterize the exponential function for all real values of x.

Properties of the Exponential Function

$$f(x) = b^x, \qquad b > 0, \qquad b \neq 1.$$

E1. For each x, b^x is a unique real number.

E2. If $b^x = b^y$, then $x = y$.

E3. If $b > 1$ and $x < y < z$, then $b^x < b^y < b^z$.

E4. If $0 < b < 1$ and $x < y < z$, then $b^x > b^y > b^z$.

Property E1 tells us that the expression b^x defines a function. Property E2 tells us that the function is one-to-one. Property E3 tells us that if $b > 1$, then as x *increases,* the value of b^x increases. Property E4 tells us that when $0 < b < 1$, then as x *increases,* the value of b^x *decreases.*

It can be shown using Properties E1–E4 that the general shape of the graph of the exponential function $f(x) = b^x$ must look like one of the curves in Figure 11.2. For a particular value of b, Properties E1–E4 enable us to use rational values of x to graph an exponential function. Thus we locate some representative points using rational values of x and then draw a smooth curve through these points. The resulting curve will also

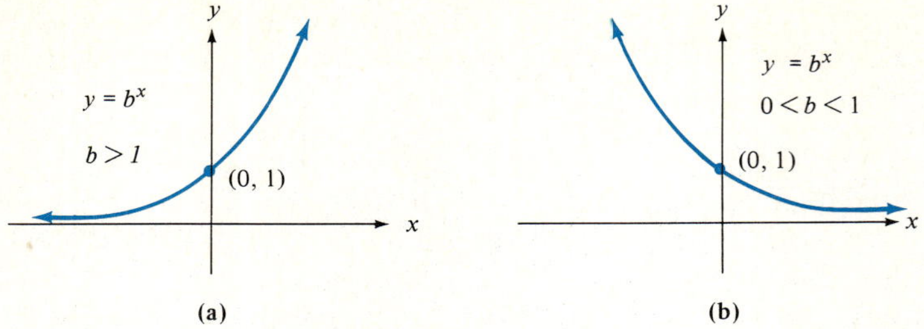

Figure 11.2

pass through points whose coordinates may be irrational numbers. The following examples illustrate the previous discussion.

Example 1. Sketch the graph of $f(x) = 2^x$.

Solution: Let $y = f(x)$; then using the points that were plotted in Figure 11.1, we sketch the graph as shown in Figure 11.3.

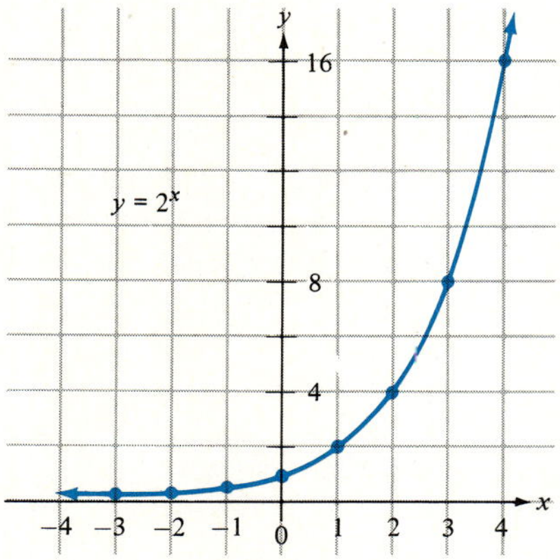

Figure 11.3

Example 2. Sketch the graph of $y = \left(\frac{1}{3}\right)^x$.

Solution: The following table contains some ordered pairs (x, y) satisfying the given equation that we will use to sketch a smooth curve as shown in Figure 11.4.

x	-2	-1	0	1	2
y	9	3	1	$\frac{1}{3}$	$\frac{1}{9}$

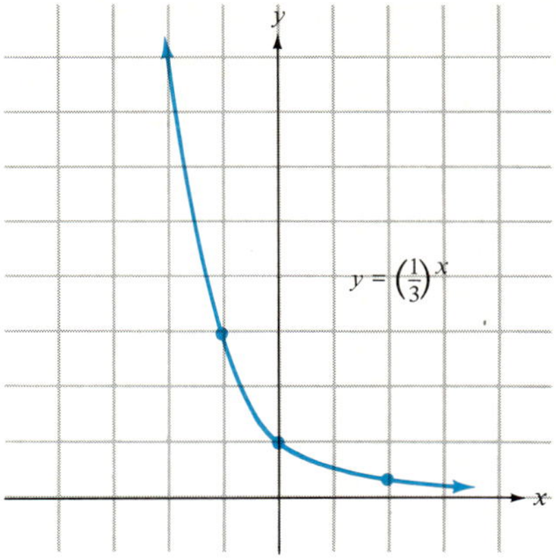

Figure 11.4

Example 3. Sketch the graph of $y = 3^{-x}$.

Solution: Since $3^{-x} = (3^{-1})^x = \left(\dfrac{1}{3}\right)^x$, this is the same graph as in Example 2.

Example 4. Sketch the graph of $y = 2^{3x-2}$.

Solution: First construct a table using convenient values of x in order to compute the corresponding y values of the given function. Then plot some of the ordered pairs (x, y) and draw a smooth curve through them as shown in Figure 11.5.

x	$\frac{-1}{3}$	0	$\frac{2}{3}$	1	$\frac{4}{3}$	2
y	$\frac{1}{8}$	$\frac{1}{4}$	1	2	4	16

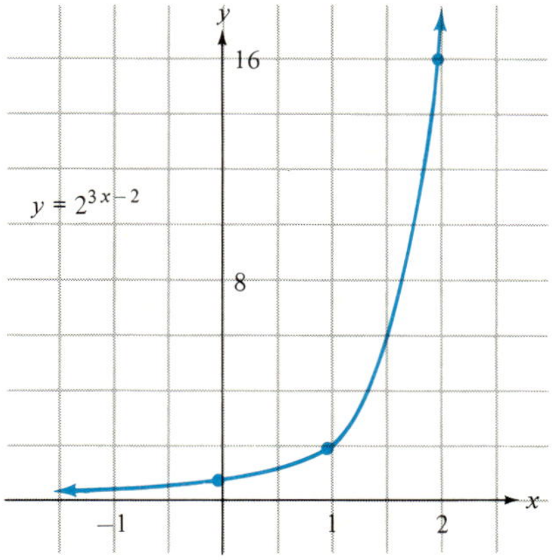

Figure 11.5

We now turn our attention to solving equations involving exponential expressions. Some equations that involve exponential expressions can be solved by applying Property E2, as shown in the next example. We shall also assume that the properties of rational exponents stated in Section 5.2 are valid for all real numbers. That is, they hold for rational and irrational exponents.

Example 5. Solve the following equations for x:

(a) $3^x = 81$. (b) $8^x = 4$. (c) $5^{-3x+2} = \frac{1}{5}$.

Solution

(a) Since $81 = 3^4$, the given equation may be rewritten as $3^x = 3^4$. But Property E2 states that the exponents must be equal. Thus $x = 4$.

(b) We rewrite 8 as 2^3 and express 4 as 2^2. Thus

$$8^x = 4$$

becomes

$$(2^3)^x = 2^2 \qquad \text{or} \qquad 2^{3x} = 2^2.$$

Thus, by Property E2, we have

$$3x = 2 \qquad \text{or} \qquad x = \frac{2}{3}.$$

(c) Note that $\frac{1}{5} = 5^{-1}$. Thus

$$5^{-3x+2} = \frac{1}{5}$$

becomes

$$5^{-3x+2} = 5^{-1}$$

Thus we have

$$-3x + 2 = -1$$

or

$$-3x = -3$$
$$x = 1.$$

Note that not all equations that involve exponentials can be solved in the manner illustrated in Example 5. For example, since 7 is not a rational power of 2, the equation $2^x = 7$ cannot be solved for x in this manner. Later in this chapter we will show how to solve equations such as $2^x = 7$.

EXERCISES 11.1

For Exercises 1–8, sketch the graph of the equation. See Examples 1–3.

1. $y = 3^x$ **2.** $y = 4^x$ **3.** $y = 6^x$ **4.** $y = 5^x$

5. $y = (\frac{1}{2})^x$ **6.** $y = (\frac{1}{5})^x$ **7.** $y = 4^{-x}$ **8.** $y = 6^{-x}$

For Exercises 9–14, sketch the graph of the equation. See Example 4.

9. $y = 2^{x+1}$ **10.** $y = 2^{x+3}$ **11.** $y = 3^{x-1}$

12. $y = 3^{x-2}$ **13.** $y = 3^{2x-1}$ **14.** $y = 2^{4x+1}$

For Exercises 15–20, use a calculator and sketch the graph of the equation. See Examples 1–4.

15. $y = 3.14^x$ **16.** $y = 2.72^x$ **17.** $y = 2.718^{x-3}$

18. $y = 3.14^{x+2}$ **19.** $y = 3.14^{2x+1}$ **20.** $y = 2.718^{3x-2}$

For Exercises 21–30, solve the equation. See Example 5.

21. $6^x = 36$ **22.** $2^x = 16$ **23.** $9^x = 27$ **24.** $25^x = 125$ **25.** $4^x = \frac{1}{32}$

26. $9^x = \frac{1}{27}$ **27.** $2^{4x+1} = \frac{1}{8}$ **28.** $3^{5x+3} = \frac{1}{9}$ **29.** $49^{x+1} = \frac{1}{7}$ **30.** $64^{x-1} = \frac{1}{8}$

Logarithmic Functions 11.2

The exponential function

$$f(x) = b^x \qquad b > 0, \quad b \neq 1$$

is a one-to-one function with a domain that consists of all real numbers $\mathbb{R}$. Thus the exponential function has an inverse. To find the inverse, we usually solve for x in terms of y and then switch variables. To find this inverse, we let $y = f(x)$, and first interchange x and y, and then solve for y. This is an equivalent method for finding the inverse of a function. As a result, we see that

$$\begin{aligned} f(x) &= b^x \\ y &= b^x && y = f(x) \\ x &= b^y. && \text{interchange } x \text{ and } y \end{aligned}$$

But we do not have a way to solve $x = b^y$ for y. Therefore, we introduce a new notation that enables us to define the inverse function of $y = b^x$ called the *logarithmic function*.

Logarithmic Function

If $b > 0$, $b \neq 1$, then $y = \log_b x$ is the same as $x = b^y$.

The expression $y = \log_b x$ is read "y equals the logarithm of x to the base b." It means that y is the exponent on b that makes b^y equal to x. Note that a *logarithm is an exponent*. In particular, if the exponent is $y = 1$, we have $x = b^1 = b$. Therefore,

$$\log_b b = 1.$$

Furthermore, since $b^0 = 1$, we also have

$$\log_b b^1 = 0.$$

Example 1. Rewrite each exponential form in logarithmic form.

(a) $5^2 = 25$. (b) $\left(\frac{3}{4}\right)^3 = \frac{27}{64}$.

Solution

(a) $\log_5 25 = 2$. (b) $\log_{3/4} \frac{27}{64} = 3$.

Example 2. Rewrite the following logarithmic forms in exponential form.

(a) $\log_2 32 = 5$. (b) $\log_4 \frac{1}{16} = -2$.

Solution

(a) $2^5 = 32$. (b) $4^{-2} = \frac{1}{16}$.

For $b > 1$, Figure 11.6 shows the graphs of $y = \log_b x$ and $y = b^x$, which are inverse functions. Note that the *domain* of the logarithmic function is the set of all

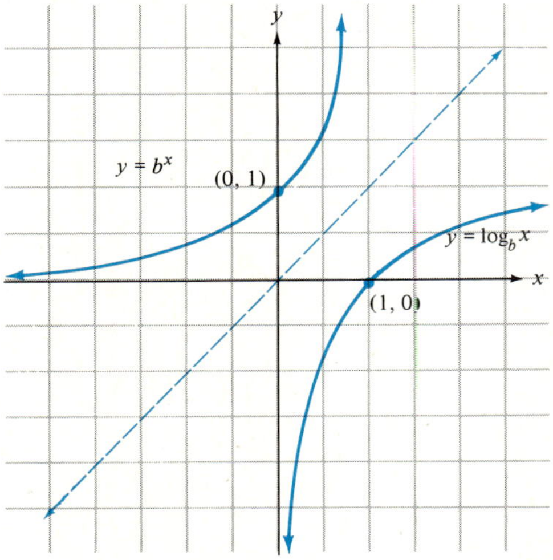

Figure 11.6

positive real numbers. Also note that the *range* of the logarithmic function is the set of real numbers $\mathbb{R}$. Since $f(x) = b^x$ is a one-to-one function, we note that $f(x) = \log_b x$ is also a one-to-one function.

Observe that in Figure 11.6 the graph of the inverse function $y = f(x) = \log_b x$ is the reflection of the graph of $y = b^x$ in the line $y = x$. In the next example we show how to obtain the graph of a logarithmic function using its inverse.

Example 3. Sketch the graph of $y = \log_3 x$.

Solution: We will sketch the function $y = 3^x$ and then reflect this in the line $y = x$ in order to get the graph of the inverse function $y = \log_3 x$ as shown in Figure 11.7. In the figure we have labeled some points on the graph of the function whose coordinates are easy to compute.

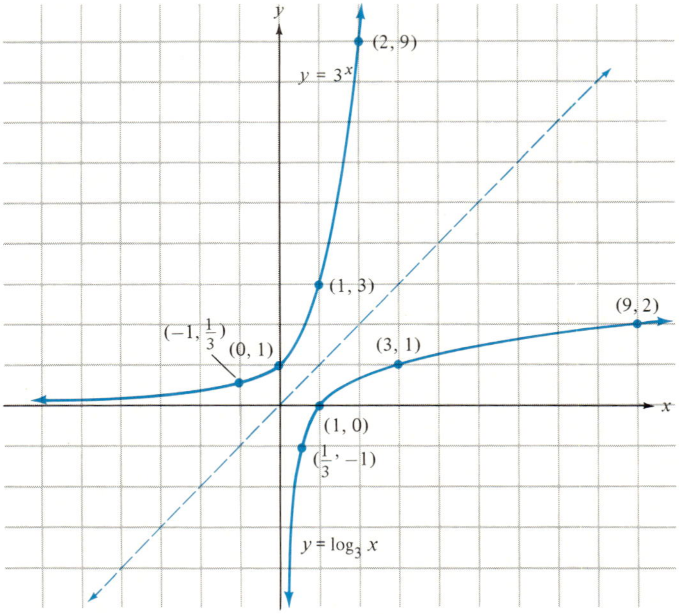

Figure 11.7

Example 4. Sketch the graph of $y = \log_{1/3} x$.

Solution: Figure 11.8 shows a sketch of the graph of $y = \left(\dfrac{1}{3}\right)^x$ and its inverse $y = \log_{1/3} x$. Notice that we used values of x that enabled us to easily compute the corresponding y values of $y = \left(\dfrac{1}{3}\right)^x$.

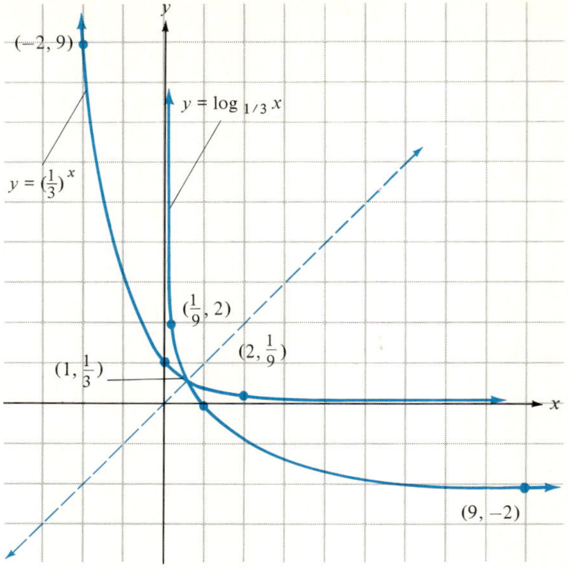

Figure 11.8

The shape of the graph of $y = \log_{1/3} x$ as shown in Figure 11.8 is representative of the shape of the graph of $y = \log_b x$ for $0 < b < 1$. Similarly, the shape of $y = \log_3 x$ as shown in Figure 11.7 is representative of the shape of $y = \log_b x$ for $b > 1$. Most applications involve a base $b > 1$.

The following examples illustrate how to solve some special equations that involve logarithms by first rewriting the equations in exponential form.

Example 5. Solve the equations for y.

(a) $y = \log_2 8$. (b) $y = \log_{10} 100{,}000$. (c) $y = \log_5 \left(\frac{1}{125}\right)$.

Solution

(a) First write $y = \log_2 8$ in exponential form as

$$2^y = 8.$$

Next write both sides of the equation as exponentials with a base of 2. Thus we have

$$2^y = 2^3. \qquad \text{because } 8 = 2^3$$

By Property E2 of Section 11.1, we know that the solution is

$$y = 3.$$

(b) Using the same steps that were used in part (a), we get

$$10^y = 100{,}000 \qquad \text{or} \qquad 10^y = 10^5.$$

Thus

$$y = 5.$$

(c) Note that $y = \log_5\left(\frac{1}{125}\right)$ is the same as

$$5^y = \frac{1}{125}.$$

Thus

$$5^y = 5^{-3}.$$

Therefore,

$$y = -3.$$

Example 6. Solve the equations for y.
(a) $y = \log_3 \sqrt{3}$. (b) $y = \log_2 \sqrt[3]{2^5}$.

Solution

(a) Note that $y = \log_3 \sqrt{3}$ is the same as $\sqrt{3} = 3^y$. But $\sqrt{3} = 3^{1/2}$. Thus we have $3^{1/2} = 3^y$ and it follows that $y = \frac{1}{2}$.

(b) The original equation may be rewritten as

$$2^y = \sqrt[3]{2^5} \qquad \text{or} \qquad 2^y = 2^{5/3}.$$

Thus

$$y = \frac{5}{3}.$$

The procedures illustrated by Examples 5 and 6 provide us with one method of evaluating a logarithmic expression, as shown in the next example.

Example 7. Evaluate the following expressions.
(a) $\log_9 81$. (b) $\log_4 32$.

Solution

(a) Let $y = \log_9 81$ and rewrite this equation in exponential form as $9^y = 81$. Next write each side of the exponential equation with base 9. Thus $9^y = 9^2$. It follows from Property E2 that $y = 2$. Therefore, we see that $\log_9 81 = 2$.

(b) Let $y = \log_4 32$. Thus $4^y = 32$. Now we write 32 as a power of 4. Note that $32 = 16 \cdot 2$. But $16 = 4^2$ and $2 = 4^{1/2}$. Thus $32 = 4^2 \cdot 4^{1/2} = 4^{5/2}$. Therefore, $4^y = 4^{5/2}$ and thus $y = \frac{5}{2}$.

An alternate way of solving the equation $4^y = 32$ would be to rewrite the equation as $(2^2)^y = 2^5$. From this we see that $2^{2y} = 2^5$. Thus $2y = 5$ and $y = \frac{5}{2}$.

Using either method, we conclude that $\log_4 32 = \frac{5}{2}$.

Later in this chapter we will discuss the use of tables and calculators to evaluate logarithms.

Example 8. Solve the equation for x.
(a) $\log_3 x = 4$. (b) $\log_5 x = -1$.

Solution

(a) Since $\log_3 x = 4$ is the same as $3^4 = x$, it follows that $x = 81$.

(b) Note that $\log_5 x = -1$ is the same as $5^{-1} = x$. Thus $x = \frac{1}{5}$.

EXERCISES 11.2

For Exercises 1–12, rewrite the expression in logarithmic form. See Example 1.

1. $2^4 = 16$ **2.** $3^3 = 27$ **3.** $10^2 = 100$ **4.** $10^4 = 10000$ **5.** $5^4 = 625$ **6.** $7^2 = 49$

7. $4^{-2} = \frac{1}{16}$ **8.** $2^{-3} = \frac{1}{8}$ **9.** $(\frac{2}{5})^3 = \frac{8}{125}$ **10.** $(\frac{3}{10})^2 = \frac{9}{100}$ **11.** $(\frac{3}{4})^{-2} = \frac{16}{9}$ **12.** $(\frac{5}{7})^{-2} = \frac{49}{25}$

For Exercises 13–24, rewrite the expression in exponential form. See Example 2.

13. $\log_3 9 = 2$ **14.** $\log_8 64 = 2$ **15.** $\log_{10} 100 = 2$ **16.** $\log_{10} 1000 = 3$

17. $\log_5 125 = 3$ **18.** $\log_6 216 = 3$ **19.** $\log_2 (\frac{1}{4}) = -2$ **20.** $\log_3 \frac{1}{9} = -2$

21. $\log_{1/2} \frac{1}{8} = 3$ **22.** $\log_{1/3} \frac{1}{27} = 3$ **23.** $\log_{1/2} 16 = -4$ **24.** $\log_{1/5} 25 = -2$

For Exercises 25–30, sketch the graph of the equation. See Examples 3 and 4.

25. $y = \log_2 x$ **26.** $y = \log_4 x$ **27.** $y = \log_{10} x$

28. $y = \log_5 x$ **29.** $y = \log_{1/2} x$ **30.** $y = \log_{1/4} x$

For Exercises 31–36, solve the equation by first rewriting it in exponential form. See Examples 5 and 6.

31. $y = \log_{10} 1000$ **32.** $y = \log_{10} 10{,}000$ **33.** $y = \log_2 16$

34. $y = \log_6 36$ **35.** $y = \log_{10} \frac{1}{100}$ **36.** $y = \log_5 \frac{1}{25}$

For Exercises 37–48, evaluate the logarithm. See Example 7.

37. $\log_{11} 121$ **38.** $\log_{12} 144$ **39.** $\log_5 625$ **40.** $\log_3 81$

41. $\log_{49} 7$ 42. $\log_{25} 5$ 43. $\log_5 \sqrt{5^7}$ 44. $\log_7 \sqrt{7^3}$
45. $\log_3 \sqrt{3^3}$ 46. $\log_9 \sqrt{9^5}$ 47. $\log_9 27$ 48. $\log_{25} 125$

For Exercises 49–60, solve the equation. See Example 8.

49. $\log_2 x = 5$ 50. $\log_2 x = 3$ 51. $\log_3 x = 2$ 52. $\log_3 x = 3$
53. $\log_{10} x = -1$ 54. $\log_{10} x = -2$ 55. $\log_4 x = -3$ 56. $\log_3 x = -2$
57. $\log_4 x = \dfrac{5}{2}$ 58. $\log_9 x = \dfrac{3}{2}$ 59. $\log_{16} x = \dfrac{3}{2}$ 60. $\log_{25} x = \dfrac{3}{2}$

Logarithmic Properties 11.3

In this section we discuss several important properties of logarithms. As will be shown, these properties permit us to change multiplication problems to associated addition problems, division problems to associated subtraction problems, and problems involving powers to multiplication problems. Keep in mind that a logarithm is an exponent. Because of this, you will see a similarity between the following properties of logarithms and the properties of exponents that we have previously discussed.

We state these properties in the following theorems. For all of these theorems we assume that x, y, and b are positive numbers with $b \neq 1$. The first theorem states that *the logarithm of the product xy equals the sum of the logarithms of x and y.*

Theorem 1

$$\log_b xy = \log_b x + \log_b y.$$

Proof. Let $p = \log_b x$ and $q = \log_b y$. Using the definition of a logarithm, we may write $b^p = x$ and $b^q = y$. Thus $xy = b^p \cdot b^q = b^{p+q}$. Again using the definition of a logarithm, we see that because

$$xy = b^{p+q},$$

we may write $\log_b xy = p + q$. Next we replace p with $\log_b x$ and q with $\log_b y$ to get $\log_b xy = \log_b x + \log_b y$, as claimed.

Example 1. Use Theorem 1 to rewrite each of the following logarithms as the sum of two or more logarithms.
(a) $\log_3 7 \cdot 5$. (b) $\log_{10} 19xy$. (c) $\log_5 75$.

Solution

(a) $\log_3 7 \cdot 5 = \log_3 7 + \log_3 5$.
(b) $\log_{10} 19xy = \log_{10} 19 + \log_{10} x + \log_{10} y$.
(c) $\log_5 75 = \log_5 3 \cdot 25 = \log_5 3 + \log_5 25$.

But we can proceed further, since $25 = 5^2$; thus $\log_5 25 = 2$. It follows that

$$\log_5 75 = \log_5 3 + \log_5 25 = \log_5 3 + 2.$$

The next theorem, which we state without proof, states that *the logarithm of the quotient* $\frac{x}{y}$ *equals the logarithm of the numerator,* x, *minus the logarithm of the denominator,* y.

Theorem 2

$$\log_b \frac{x}{y} = \log_b x - \log_b y.$$

Example 2. Use Theorem 2 to rewrite the following logarithms as the difference of two logarithms.

(a) $\log_8 \frac{17}{3}$. (b) $\log_{10} \frac{100}{z}$.

Solution

(a) $\log_8 \frac{17}{3} = \log_8 17 - \log_8 3$.

(b) $\log_{10} \frac{100}{z} = \log_{10} 100 - \log_{10} z$

$= 2 - \log_{10} z$. $\quad \log_{10} 100 = 2$ because $10^2 = 100$

The next theorem states a rule for evaluating *the logarithm of a power*.

Theorem 3

$$\log_b x^p = p \cdot \log_b x.$$

Proof. Let $t = \log_b x$; thus $b^t = x$, by the definition of a logarithm. Next we raise both sides of the equation, $x = b^t$, to the pth power and get

$$x^p = (b^t)^p = b^{pt}.$$

But the statement $x^p = b^{pt}$ is equivalent to

$$\log_b x^p = pt = p \cdot \log_b x. \quad (t = \log_b x)$$

Example 3. Use Theorem 3 to rewrite the following logarithms.
(a) $\log_5 7^3$. (b) $\log_{10} \sqrt{3}$. (c) $\log_3 x^{2/5}$.

Solution

(a) $\log_5 7^3 = 3 \log_5 7$.

(b) $\log_{10} \sqrt{3} = \log_{10} 3^{1/2} = \dfrac{1}{2} \log_{10} 3$.

(c) $\log_3 x^{2/5} = \dfrac{2}{5} \log_3 x$.

Example 4. Use Theorems 1–3 to rewrite the following logarithms as sums and differences of logarithms as far as possible.
(a) $\log_{10} \dfrac{5x}{11}$. (b) $\log_5 \dfrac{125z}{\sqrt{7}}$.

Solution

(a) $\log_{10} \dfrac{5x}{11} = \log_{10} 5x - \log_{10} 11$ $\qquad \log_b \dfrac{x}{y} = \log_b x - \log_b y$

$= \log_{10} 5 + \log_{10} x - \log_{10} 11.$ $\qquad \log_b xy = \log_b x + \log_b y$

(b) $\log_5 \dfrac{125z}{\sqrt{7}} = \log_5 125z - \log \sqrt{7}$ $\qquad \log_b \dfrac{x}{y} = \log_b x - \log_b y$

$= \log_5 125 + \log_5 z - \log_5 \sqrt{7}$ $\qquad \log_b xy = \log_b x + \log_b y$

$= \log_5 5^3 + \log_5 z - \log_5 7^{1/2}$

$= 3 + \log_5 z - \dfrac{1}{2} \log_5 7.$ $\qquad \log_b x^p = p \log_b x$

As the next examples illustrate, we can often use Theorems 1–3 to rewrite the sum and difference of logarithmic expressions as a single logarithm. In fact, this is one method that can be used to solve some equations that involve logarithms.

Example 5. Use Theorems 1–3 to rewrite $\log_7 11 + \dfrac{1}{2} \log_7 z - 2 \log_7 3$ as one logarithm.

Solution

$\log_7 11 + \dfrac{1}{2} \log_7 z - 2 \log_7 3$

$= \log_7 11 + \log_7 z^{1/2} - \log_7 3^2$ $\qquad \log_b x^p = p \log_b x$

$= \log_7 11 + \log_7 \sqrt{z} - \log_7 9$

$= \log_7 11\sqrt{z} - \log_7 9$ $\qquad \log_b xy = \log_b x + \log_b y$

$= \log_7 \dfrac{11\sqrt{z}}{9}.$ $\qquad \log_b \dfrac{x}{y} = \log_b x - \log_b y$

Example 6. Solve for x if

$$\log_5 x + \log_5 4 = 2.$$

Solution

$$\log_5 x + \log_5 4 = 2$$
$$\log_5 4x = 2. \qquad \log_b xy = \log_b x + \log_b y$$

Thus

$$4x = 5^2 \qquad \text{definition of a logarithm}$$
$$4x = 25$$
$$x = \frac{25}{4}.$$

The reader should check that $\frac{25}{4}$ satisfies the original equation.

Example 7. Find all values of x for which $\log_6 (x + 5) + \log_6 x = 2$.

Solution

$$\log_6 (x + 5) + \log_6 x = 2$$
$$\log_6 (x + 5)x = 2 \qquad \log_b xy = \log_b x + \log_b y$$
$$(x + 5)x = 6^2 \qquad \text{definition of a logarithm}$$
$$x^2 + 5x = 36$$
$$x^2 + 5x - 36 = 0$$
$$(x + 9)(x - 4) = 0.$$

Thus

$$x + 9 = 0 \quad \text{or} \quad x - 4 = 0$$
$$x = -9 \quad \text{or} \quad x = 4.$$

Recall that $\log x$ is defined only for $x > 0$. Thus $x = -9$ is an not a solution. It is easy to check that $x = 4$ satisfies the given equation.

EXERCISES 11.3

For Exercises 1–28, use Theorems 1–3 to rewrite each logarithm as the sum or difference of logarithms. See Examples 1–4.

1. $\log_5 11 \cdot 3$ **2.** $\log_2 13 \cdot 5$ **3.** $\log_{10} 17 \cdot 2$ **4.** $\log_7 17 \cdot 23$

5. $\log_6 5x$ **6.** $\log_{10} 3w$ **7.** $\log_3 18\sqrt{3}$ **8.** $\log_{10} 300\sqrt{10}$

9. $\log_2 24\sqrt[3]{2}$ **10.** $\log_5 250\sqrt[4]{5}$ **11.** $\log_{10} \frac{3}{7}$ **12.** $\log_7 \frac{11}{13}$

13. $\log_2 \frac{5}{x}$ **14.** $\log_{10} \frac{y}{7}$ **15.** $\log_7 \frac{49}{w}$ **16.** $\log_3 \frac{27}{z}$

17. $\log_{10} 3^4$ **18.** $\log_2 5^3$ **19.** $\log_5 7^2$ **20.** $\log_7 11^5$

21. $\log_5 \sqrt{2}$ **22.** $\log_{10} \sqrt[5]{7}$ **23.** $\log_7 w^{3/4}$ **24.** $\log_2 z^{5/3}$

25. $\log_{10} \frac{7w}{13}$ **26.** $\log_4 \frac{11x}{7}$ **27.** $\log_3 \frac{81y}{\sqrt{2}}$ **28.** $\log_4 \frac{64w}{\sqrt[3]{5}}$

For Exercises 29–40, rewrite the expression as one logarithm. See Example 5.

29. $\log_2 3 + \log_2 x$ **30.** $\log_3 y + \log_3 7$

31. $\log_5 11 - \log_5 w$ **32.** $\log_{10} z - \log_{10} 2$

33. $2 \log_5 4 + \log_5 w$ **34.** $3 \log_5 2 + \log_5 z$

35. $\log_{10} 12 - \frac{1}{2} \log_{10} 3$ **36.** $\log_{10} 6 - \frac{1}{2} \log_{10} 2$

37. $\log_{10} 4 + \frac{1}{2} \log_{10} w - 3 \log_{10} 2$ **38.** $\log_2 5 + \frac{1}{3} \log_2 x - 4 \log_2 3$

39. $\log_{10} 14 - \frac{1}{2} \log_{10} 7 + 2 \log_{10} x$ **40.** $\log_8 15 - \frac{1}{2} \log_8 5 + 3 \log_8 w$

For Exercises 41–50, solve the equation. See Examples 6 and 7.

41. $\log_4 x + \log_4 3 = 2$ **42.** $\log_3 2 + \log_3 x = 2$

43. $\log_{10} x + \log_{10} 5 = 0$ **44.** $\log_{10} 4 + \log_{10} x = 0$

45. $\log_5 x - \log_5 6 = 1$ **46.** $\log_2 x - \log_2 10 = 1$

47. $\log_{10} x + \log_{10} (x - 21) = 2$ **48.** $\log_6 x + \log_6 (x - 5) = 2$

49. $\log_2 x + \log_2 (x + 6) = 4$ **50.** $\log_6 x + \log_6 (x - 9) = 2$

Computations with Common Logarithms 11.4

Historically, logarithms were developed as an aid to computations with very large numbers such as those associated with astronomy and navigation. Today, because of the availability of hand calculators, logarithms are seldom used as a computational aid. However, it is still important to be able to work with logarithms to solve equations that involve exponentials or logarithms. Such equations are often encountered in applied problems. For example, in chemistry the pH (the acidity) of a solution is expressed using a logarithmic function. In this section we discuss base 10 logarithms, which are called *common logarithms*. When using a base of 10, we write $\log x$ for $\log_{10} x$.

Since the statement $\log_{10} x = r$ is equivalent to $x = 10^r$, it is easy to evaluate the common logarithm of powers of 10.

$$\log 1000 = 3 \quad \text{because} \quad 1000 = 10^3$$
$$\log 100 = 2 \quad \text{because} \quad 100 = 10^2$$

$$\begin{array}{llll} \log 10 &= 1 & \text{because} & 10 = 10^1 \\ \log 1 &= 0 & \text{because} & 1 = 10^0 \\ \log 0.1 &= -1 & \text{because} & .1 = 10^{-1} \\ \log 0.01 &= -2 & \text{because} & .01 = 10^{-2} \\ \log 0.001 &= -3 & \text{because} & .001 = 10^{-3} \end{array}$$

In order to find the common logarithm of numbers that are not powers of 10, we need a calculator with a [log] key or a table of common logarithms. In this section our primary focus is on the use of the Table of Common Logarithms given in Appendix C at the back of the book. If you are using a calculator, you should refer to the instruction manual for the appropriate information. By reading the examples and doing the exercises, you will gain a better understanding of the properties of the logarithmic function. Whether we use a calculator or a table of logarithms, *most* of the values we obtain are approximations. Nevertheless, it is common practice to use the equality symbol, =, instead of the approximation symbol, $\simeq$.

The Table of Common Logarithms in Appendix C gives the common logarithms of numbers *between* 1.00 and 9.99 in increments of .01. A portion of the table is given here. We now discuss how to read this table.

x	0	1	2	3	4	5	6	7	8	9
5.0	.6990	.6998	.7007	.7016	.7024	.7033	.7042	.7050	.7059	.7067
5.1	.7076	.7084	.7093	.7101	.7110	.7118	.7126	.7135	.7143	.7152
5.2	.7160	.7168	.7177	.7185	.7193	.7202	.7210	.7218	.7226	.7235
5.3	.7243	.7251	.7259	.7267	.7275	.7284	.7292	.7300	.7308	.7316
5.4	.7324	.7332	.7340	.7348	.7356	.7364	.7372	.7380	.7388	.7396
5.5	.7404	.7412	.7419	.7427	.7435	.7443	.7451	.7459	.7466	.7474
5.6	.7482	.7490	.7497	.7505	.7513	.7520	.7528	.7536	.7543	.7551
5.7	.7559	.7566	.7574	.7582	.7589	.7597	.7604	.7612	.7619	.7627
5.8	.7634	.7642	.7649	.7657	.7664	.7672	.7679	.7686	.7694	.7701
5.9	.7709	.7716	.7723	.7731	.7738	.7745	.7752	.7760	.7767	.7774
x	0	1	2	3	4	5	6	7	8	9

To find the common logarithm of a number such as 5.47, we read down the left-hand column until we reach the row labeled 5.4. Next we move across this row to the column labeled by 7 and get the value .7380. Thus log 5.47 = .7380. Refer to the portion of the Table of Common Logarithms given above and verify that

$$\log 5.12 = .7093$$
$$\log 5.40 = .7324$$

In order to find the common logarithm of a number that is not between 1.00 and 9.99, we first write the number in scientific notation as explained in Section 3.3. For example, to find the common logarithm of 352, we proceed as follows:

$$\begin{aligned}\log 352 &= \log (3.52 \times 10^2)\\ &= \log 3.52 + \log 10^2. \qquad \text{(Theorem 1)}\end{aligned}$$

First note that $\log 10^2 = 2$. From the Table of Common Logarithms in Appendix C we see that $\log 3.52 = .5465$. Thus

$$\begin{aligned}\log 352 &= .5465 + 2\\ &= 2.5465.\end{aligned}$$

The part of a common logarithm that is a decimal fraction between 0 and 1 is called the *mantissa*. We note that the mantissa is *always* a positive decimal fraction between 0 and 1. When a common logarithm is written with a mantissa, the integer part is called the *characteristic*. Its primary use is to determine the location of the decimal point in a number. Thus in

$$\log 352 = 2.5465$$

the characteristic is 2 and the mantissa is .5465. The mantissa is the value found in the Table of Common Logarithms in Appendix C. In general, if x is a positive number, then x can be written in scientific notation as

$$x = y \cdot 10^k \qquad \text{with } 1 \leq y < 10 \text{ and } k \text{ an integer.}$$

Thus

$$\begin{aligned}\log x &= \log (y \cdot 10^k)\\ &= \log y + \log 10^k\\ &= \underbrace{\log y}_{\text{mantissa}} + \underbrace{k}_{\text{characteristic}}.\end{aligned}$$

Example 1. Use the Table of Common Logarithms to find log 83,700.

Solution

$$\begin{aligned}\log 83{,}700 &= \log (8.37 \times 10^4)\\ &= \log 8.37 + \log 10^4\\ &= \underbrace{.9227}_{\text{mantissa}} + \underbrace{4}_{\text{characteristic}}\\ &= 4.9227\end{aligned}$$

We now consider the logarithm of a number between 0 and 1 such as .00192. Note that

$$\begin{aligned}\log .00192 &= \log (1.92 \times 10^{-3})\\ &= \log 1.92 + \log 10^{-3}\\ &= .2833 - 3.\end{aligned}$$

If we add .2833 and -3, we get -2.7117 and indeed you may use a calculator to verify that $10^{-2.7117} \simeq .00192$. However, if we use -2.7117 $(-2 - .7117)$ for log .00192 in conjunction with the Table of Common Logarithms, the decimal fraction (−.7117) does not correspond to log 1.92 because mantissas are positive. For computational reasons we often avoid writing a logarithm with a negative decimal fraction. It is customary to rewrite logarithms with negative characteristics by adding and subtracting 10 to the logarithm. For example, log .00192 = .2883 − 3 would be rewritten as follows.

$$\begin{aligned} \log .00192 &= .2833 - 3 \\ &= .2883 + 10 - 3 - 10 \\ &= .2833 + 7 + (-10) \\ &= 7.2833 - 10. \end{aligned}$$

We could also have written log .00192 as

$$5.2833 - 8 \qquad \text{characteristic is } 5 - 8 = -3$$

or

$$38.2833 - 41. \qquad \text{characteristic is } 38 - 41 = -3$$

Thus there are many ways to write the characteristic.

Example 2. Find log .078.

Solution

$$\begin{aligned} \log .078 &= \log (7.80 \times 10^{-2}) \\ &= \log 7.80 + \log 10^{-2} \\ &= .8921 - 2 \\ &= .8921 + 8 + (-10) \\ &= 8.8921 - 10. \end{aligned}$$

After you try a few problems, it will probably not be necessary to write down all the steps that were shown in Examples 1 and 2.

Example 3. Find log .451.

Solution: Notice that $.451 = 4.51 \times 10^{-1}$. Thus the characteristic is −1. From the table in Appendix C we see that log 4.51 = .6452. Thus log .451 = 9.6452 − 10.

It is also possible to reverse the process and find a number x if we are given log x. For example, if log x = .8222, then from the Table of Common Logarithms we see that the number x that has a common logarithm of .8222 is x = 6.64. The number 6.64 is called the **antilogarithm** or **antilog** of .8222 because log 6.64 = .8222.

Example 4. Find the following antilogarithms.
(a) antilog 5.8222 (b) antilog (6.8222 − 10).

Solution

(a) The mantissa is .8222. The number in the table that has this mantissa is 6.64. Since the characteristic is 5, the antilog of 5.8222 is $6.64 \times 10^5 = 664{,}000$.
(b) Proceeding as in part (a), but recognizing that the characteristic is −4, we see that

$$\begin{aligned}\text{antilog } (6.8222 - 10) &= 6.64 \times 10^{-4}\\ &= .000664\end{aligned}$$

We previously stated that the primary function of the characteristic is to locate the position of the decimal point. To clarify this claim, we use Example 4 to indicate a schematic method for finding antilogarithms. To find the

$$\text{antilog } 5.8222,$$

we first use the Table of Common Logarithms to find

$$\text{antilog } .8222 = 6.64$$

But since the characteristic is *positive,* that is 5, we *write*

$$\text{antilog } 5.8222 = 6.64000 = 664{,}000$$

where the arrow indicates the moving of the decimal point five positions to the *right*. Similarly, to find antilog (6.8222 − 10), we write

$$0006.64 = .00064$$

Notice that in this case, since the characteristic is $6 - 10 = -4$ is *negative,* the arrow indicates moving the decimal four positions to the *left*.

The following examples illustrate how the previous discussion and the properties of logarithms can be used to assist with involved computations. If you have access to a calculator, it is appropriate to use it to check your answers.

Example 5. Use logarithms to compute
(a) (32,400)(7500). (b) (8750)(.0000197).

Solution

(a) Let $x = (32{,}400)(7500)$; then

$$\begin{aligned}\log x &= \log 32{,}400 + \log 7500\\ &= 4.5105 + 3.8751\\ &= 8.3856\end{aligned}$$

We now seek to find x which is the *antilog* of 8.3856. Looking into the Table of Common Logarithms, we see that the number that corresponds to the mantissa

.3856 is 2.43. Since the characteristic is 8, it follows that

$$\begin{aligned} x &= 2.43 \times 10^8 \\ &= 243{,}000{,}000. \end{aligned}$$

(b) Let $x = (8750)(.0000197)$; then

$$\begin{aligned} \log x &= \log 8750 + \log .0000197 \\ &= 3.9420 + 5.2945 - 10 \\ &= 9.2365 - 10. \end{aligned}$$

To find x, we need the antilog of $9.2365 - 10$. Since the mantissa .2365 does not appear in the table, we choose the number in the table with the mantissa that best approximates .2365, which is 1.72. Since the characteristic is -1, it follows that $x = 1.72 \times 10^{-1} = .172$

As Example 5 illustrates, when the mantissa is not in the table, we use the mantissa nearest in value to it. Thus the resulting answer $x = .172$ is really an approximate answer. However, in this book we will continue to use = instead of $\simeq$ as in Example 5.

Example 6. Use logarithms to find (.00423)(.0297).

Solution: Let $x = (.00423)(.0297)$; thus

$$\begin{aligned} \log x &= \log .00423 + \log .0297 \\ &= 7.6263 - 10 + 8.4728 - 10. \end{aligned}$$

We now add the positive parts of each logarithm and the negative parts to get

$$\log x = 16.0991 - 20.$$

First note that the characteristic is $16 - 20 = -4$. The number in the table with mantissa that best approximates .0991 is 1.26. Thus

$$\begin{aligned} \text{antilog } (.0991 - 4) &= 0001.26 \\ &= .000126 \end{aligned}$$

Example 7. Use logarithms to find
(a) $(41.2)^{2.3}$. (b) $\sqrt[4]{.728}$.

Solution

(a) Let $x = (41.2)^{2.3}$; then

$$\begin{aligned} \log x &= \log (41.2)^{2.3} \\ &= 2.3 \log 41.2 \\ &= 2.3(1.6149) = 3.7143 \end{aligned}$$

Taking the antilog of 3.7143, we see that

$$x = 5.18 \times 10^3 = 5180.$$

(b) Let $x = \sqrt[4]{.728} = (.728)^{1/4}$. Thus

$$\log x = \frac{1}{4} \log .728$$

$$= \frac{1}{4}(9.8621 - 10).$$

Recall that we want the characteristic of log x to be an integer. Since 10 is not evenly divisible by 4, we rewrite log x as

$$\log x = \frac{1}{4}(7.8621 - 8)$$
$$= 1.\underbrace{9655}_{\text{mantissa}} - 2.$$

Since $1 - 2 = -1$, the characteristic is -1. The number in the table with the mantissa closest to .9655 is 9.24. Thus $x = 9.24 \times 10^{-1} = .924$

Example 8. Use logarithms to compute $\dfrac{727}{.302}$.

Solution: Let $x = \dfrac{727}{.302}$. Thus

$$\begin{aligned}\log x &= \log 727 - \log .302\\ &= 2.8615 - (9.4800 - 10)\\ &= 2.8615 - 9.4800 + 10\\ &= 3.3815\end{aligned}$$

The characteristic is 3 and the mantissa is .3815. Thus $x = 2.41 \times 10^3 = 2410$.

Example 9. Use logarithms to find $\dfrac{(1.68)^5\sqrt[3]{5410}}{(9.11)^2}$.

Solution: Let $x = \dfrac{(1.68)^5\sqrt[3]{5410}}{(9.11)^2}$; then

$$\begin{aligned}\log x &= 5 \log 1.68 + \frac{1}{3} \log 5410 - 2 \log 9.11\\ &= 5(.2253) + \frac{1}{3}(3.7332) - 2(.9595)\\ &= 1.1265 + 1.2444 - 1.9190\\ &= .4519\end{aligned}$$

Thus, since antilog $.4519 = 2.83$, we see that $x = 2.83$

EXERCISES 11.4

For Exercises 1–12, use the Table of Common Logarithms in Appendix C to find each logarithm. See Examples 1–3.

1. log 4.91 **2.** log 8.27 **3.** log 806 **4.** log 203
5. log 91,100 **6.** log 24,500 **7.** log .00178 **8.** log .0861
9. log .710 **10.** log .0409 **11.** log 87,600,000 **12.** log 1,490,000

For Exercises 13–24, find the antilogarithm of the number. See Example 4.

13. .5391 **14.** .6304 **15.** 1.8716 **16.** 1.9274
17. 4.8000 **18.** 5.9191 **19.** 9.1903 − 10 **20.** 8.5403 − 10
21. 6.8182 − 10 **22.** 4.9009 − 10 **23.** 4.0792 − 5 **24.** 1.0212 − 2

For Exercises 25–54, use logarithms to evaluate the expression. See Examples 5–9. Use a calculator to check your answer.

25. (70,500)(2130) **26.** (4110)(89,100) **27.** (319)(24.8)
28. (4.07)(76100) **29.** (.000217)(23.5) **30.** (.000876)(47.2)
31. (.00391)(.0402) **32.** (.00501)(.0917) **33.** (.00014)(.0053)
34. (.00036)(.0078) **35.** (.0000471)(.000111) **36.** (.000912)(.0000217)
37. $\sqrt{31.8}$ **38.** $\sqrt{90.9}$ **39.** $\sqrt[3]{92.9}$
40. $\sqrt[3]{2.43}$ **41.** $\sqrt[3]{.0817}$ **42.** $\sqrt{.317}$
43. $(4.07)^{3.4}$ **44.** $(.981)^{7.6}$ **45.** $(.0147)^{5.2}$
46. $(5.09)^{1.3}$ **47.** $\dfrac{8450}{1.43}$ **48.** $\dfrac{392{,}000}{94.7}$
49. $\dfrac{.163}{34.7}$ **50.** $\dfrac{.0214}{76.1}$ **51.** $\dfrac{(481)\sqrt{522}}{(11.1)^2}$
52. $\dfrac{(9.14)\sqrt[3]{50.5}}{(7.04)^3}$ **53.** $\dfrac{(52.6)^2\sqrt[3]{761}}{(4.13)^3}$ **54.** $\dfrac{(11.9)\sqrt[3]{5.90}}{(5.91)^2}$

11.5 Natural Logarithms and Exponential Equations

Many practical problems such as those that involve growth or decay are most easily solved using *natural logarithms,* which have a base of e. The letter e represents an irrational number. It can be shown that the value of e is approximately 2.71828. It is common practice to write $\ln x$ in place of $\log_e x$.

In this section we show how logarithms to bases other than 10 are related to common logarithms. In particular, we discuss the relationship between natural logarithms and common logarithms. Finally, we show how logarithms can be used to solve equations that involve exponential expressions. For a particular positive number x, the following rule provides us with a formula that enables us to change from $\log_b x$ to $\log_a x$, where a and b are different bases.

Change of Base Rule 1

If a, b, and x are positive numbers with $a \neq 1$ and $b \neq 1$, then

$$\log_a x = \frac{\log_b x}{\log_b a}.$$

Proof. Let $p = \log_a x$; then by definition, $a^p = x$. If we take the logarithm to the base b of each side of the equation $a^p = x$, we get

$$\log_b a^p = \log_b x$$

Thus

$$p \log_b a = \log_b x$$
$$(\log_a x)(\log_b a) = \log_b x. \qquad p = \log_a x$$

Therefore,

$$\log_a x = \frac{\log_b x}{\log_b a}.$$

If in the Change of Base Rule 1 we let $b = 10$, we obtain the following result, which enables us to convert the log of a number from base 10 to some other base a.

Change of Base Rule 2

If a and x are positive numbers and $a \neq 1$, then

$$\log_a x = \frac{\log x}{\log a}.$$

Example 1. Use Change of Base Rule 2 to find $\log_8 10$.

Solution: In the formula $\log_a x = \dfrac{\log x}{\log a}$, let $x = 10$ and $a = 8$. Thus

$$\log_8 10 = \frac{\log 10}{\log 8} = \frac{1}{.9031} = 1.1073$$

In Example 1 we could use logarithms for the computation $\frac{1}{.9031} = 1.1073$. However, it is usually accomplished by long division or with the aid of a calculator.

Example 2. Use Change of Base Rule 2 to find $\log_5 42$.

Solution: In the formula $\log_a x = \frac{\log x}{\log a}$, let $x = 42$ and $a = 5$. The result is

$$\begin{aligned}\log_5 42 &= \frac{\log 42}{\log 5}\\ &= \frac{1.6232}{.6990}\\ &= 2.3222\end{aligned}$$

We now state the following formula for converting from *common* logarithms to *natural* logarithms.

$$\mathbf{ln}\ \boldsymbol{x} = \mathbf{(2.3026)\ log}\ \boldsymbol{x}$$

The formula above is obtained by using the Change of Base Rule 2 with $a = e$ and the approximation $\log e = .4343$ as follows.

Approximating $\frac{1}{.4343}$ with (2.3026), we see that

$$\ln x = \frac{\log x}{\log e} = \frac{\log x}{.4343} = (2.3026) \log x.$$

Example 3. Evaluate the following natural logarithms.
(a) ln 5.62. (b) ln 971.

Solution

(a) In the formula $\ln x = (2.3026) \log x$, let $x = 5.62$. Thus

$$\begin{aligned}\ln 5.62 &= (2.3026) \log 5.62\\ &= (2.3026)(.7497)\\ &= 1.7263\end{aligned}$$

(b) For ln 971 we have
$$\begin{aligned}\ln 971 &= (2.3026) \log 971\\ &= (2.3026)(2.9872)\\ &= 6.8783\end{aligned}$$

The results in Example 3 can also be found using a calculator with a [ln] key or with a table of natural logarithms. If you have either of these, you should use it to check the answers to Example 3.

Many equations in science and business involve a variable in the exponent. Such equations are called *exponential equations*. When the exponent contains only one variable, logarithms can often be used to solve the equation. We illustrate this method of solution in the following examples.

Example 4. Solve $2^x = 91$. Express your answer to two decimal places.

Solution: If we take the log of both sides of this equation, we get

$$\begin{aligned} \log(2^x) &= \log 91 \\ x \cdot \log 2 &= \log 91 \\ x &= \frac{\log 91}{\log 2} \\ &= \frac{1.9590}{.3010} \simeq 6.51 \end{aligned}$$

Example 5. Solve $5^{2x+3} = 3.4$. Express your answer to two decimal places.

Solution: Taking the log of each side of the equation, we get

$$\begin{aligned} \log 5^{2x+3} &= \log 3.4 \\ (2x + 3)\log 5 &= \log 3.4 \\ 2x + 3 &= \frac{\log 3.4}{\log 5} \\ 2x &= \frac{\log 3.4}{\log 5} - 3 \\ x &= \frac{1}{2}\left(\frac{\log 3.4}{\log 5} - 3\right) \\ &= \frac{1}{2}\left(\frac{.5315}{.6990} - 3\right) \\ &\simeq \frac{1}{2}(.7604 - 3) \simeq -1.12 \end{aligned}$$

Thus

$$x \simeq -1.12$$

Example 6. Solve $e^{.23x} = 100$. Express your answer to two decimal places.

Solution: Because the exponential equation involves e, we will make use of natural logarithms

$$\begin{aligned} \ln e^{.23x} &= \ln 100 \\ (.23x)\ln e &= \ln 100 \\ .23x &= \ln 100 \\ &= (2.3026)\log 100 \\ &= (2.3026)(2) \\ &= 4.6052 \end{aligned}$$

$\ln x = (2.306)\log x$

Thus, dividing by .23, we obtain

$$x = \frac{4.6052}{.23} \simeq 20.02$$

EXERCISES 11.5

For Exercises 1–12, use the Change of Base Rule 2 and the Table of Common Logarithms in Appendix C to obtain an approximation for the logarithm. See Examples 1 and 2.

1. $\log_5 10$ **2.** $\log_5 12$ **3.** $\log_2 9$ **4.** $\log_2 7$

5. $\log_{12} 430$ **6.** $\log_{12} 715$ **7.** $\log_3 63$ **8.** $\log_3 401$

9. $\log_4 21$ **10.** $\log_7 37$ **11.** $\log_{25} 30{,}700$ **12.** $\log_{50} 7230$

For Exercises 13–20, use the conversion formula and a table of common logarithms to get an approximation for the natural logarithm. See Example 3.

13. $\ln 6$ **14.** $\ln 4$ **15.** $\ln 10$ **16.** $\ln 16$

17. $\ln 406$ **18.** $\ln 92$ **19.** $\ln 14.7$ **20.** $\ln 8.21$

For Exercises 21–32, use logarithms to solve the equation. Express your answer to two decimal places. See Examples 4–6.

21. $3^x = 10$ **22.** $5^x = 20$ **23.** $2^x = 19$ **24.** $7^x = 25$

25. $4^{x+1} = 7.2$ **26.** $6^{x-3} = 15$ **27.** $12^{3x+7} = 517$ **28.** $20^{5x-1} = 631$

29. $e^{.42x} = 200$ **30.** $e^{.57x} = 50$ **31.** $e^{2x-1} = 12$ **32.** $e^{3x+2} = 20$

11.6 Applied Problems

We conclude this chapter by considering applications that involve exponential equations. The first application we discuss is that of carbon dating.

Scientists know that when a living organism dies, the carbon 14 in the organism begins to decay and change to nitrogen 14. This decay occurs at a very slow rate. In fact, it takes approximately 5600 years for half of carbon 14 to change to nitrogen 14.

This is why we say that carbon 14 has a *half-life* of approximately 5600 years. The equation that relates the original amount (A_0) of carbon 14 to the amount (A) present t years later is

$$A = A_0 \cdot 2^{-t/5600}.$$

Example 1. The fossilized remains of a sea creature contain 4 milligrams of carbon 14. How much carbon 14 will be present in 300 years?

Solution: The original amount, A_0, of carbon 14 is 4 milligrams and $t = 300$ years. Substituting these values in the equation above results in

$$\begin{aligned} A &= 4 \cdot 2^{-300/5600} \\ &= 4 \cdot 2^{-3/56}. \end{aligned}$$

To find the value of A, we first take the common logarithm of each side of the last equation. Thus

$$\begin{aligned} \log A &= \log (4 \cdot 2^{-3/56}) \\ &= \log 4 - 3/56 \log 2 \\ &= .6021 - 3/56(.3010) \\ &= .5860 \end{aligned}$$

To find A, we need the antilog of .5860. The number in the Table of Common Logarithms with mantissa closest to .5860 is 3.85. Thus there will be $A \simeq 3.85$ milligrams remaining in 300 years.

Carbon 14 is not the only element that decays and changes into a different element. The half-life of such elements is the time it takes for half of the element to decay to a different element. Radium is a radioactive element that has a *half-life* of approximately 1600 years. For radium the equation that relates the original amount (A_0) to the amount (A) present after t years is

$$A = A_0 \cdot 2^{-t/1600}.$$

Example 2. How many milligrams of radium will be present after 500 years if the original amount is 50 milligrams?

Solution: Using $A_0 = 50$ and $t = 500$ in the equation above, we get

$$\begin{aligned} A &= 50 \cdot 2^{-500/1600} \\ &= 50 \cdot 2^{-5/16}. \end{aligned}$$

Thus

$$\begin{aligned} \log A &= \log 50 - 5/16 \log 2 \\ &= 1.6990 - 5/16(.3010) \\ &= 1.6049 \end{aligned}$$

Since antilog 1.6049 = 40.3 we have that the amount, A, of radium left after 500 years will be approximately 40.3 milligrams.

Example 3. A laboratory technician observes that in a particular culture of bacteria the number, N, of bacteria present after t minutes is given by the equation

$$N = N_0 \cdot 5^{.015t},$$

where t is time measured in minutes, N_0 is the initial number of bacteria and N is the number of bacteria present after t minutes. If $N_0 = 1000$, how long will it take for the bacteria to reach 5000 in number?

Solution: Replacing N with 5000 and N_0 with 1000, we get

$$5000 = 1000 \cdot 5^{.015t}.$$

Thus

$$5^{.015t} = 5.$$

If we take the common logarithm of each side of the last equation, we get

$$.015t \log 5 = \log 5.$$

Therefore,

$$.015t = 1$$

Thus

$$t = 66.7 \text{ minutes.}$$

When an amount of P dollars is invested in an account that pays interest at a rate of r percent compounded annually, the total amount of money, A, in the account after t years is given by the equation

$$A = P(1 + r)^t.$$

Example 4. After 20 years, how much money will be in an account if the initial investment is $2000 and the account pays 9% interest per year, compounded annually?

Solution: For this example, $P = 2000$, $t = 20$, and $r = .09$ (recall that 9% means $r = .09$). Therefore,

$$A = 2000(1 + .09)^{20}$$

or

$$A = 2000(1.09)^{20}.$$

Thus

$$\begin{aligned}\log A &= \log 2000 + 20 \log 1.09\\ &= 3.3010 + .7480\\ &= 4.0490.\end{aligned}$$

Taking the antilog of 4.0490, we see that

$$A \simeq \$11{,}200.$$

Thus after 20 years there will be approximately $11,200 in the account.

Example 5. How long will it take $3000 to double if it is invested in an account that pays 8% interest per year, compounded annually?

Solution: In this case we let $P = 3000$, $A = 6000$, and $r = .08$. Thus, using

$$A = P(1 + r)^t$$

we get

$$6000 = 3000(1 + .08)^t$$

or

$$2 = (1.08)^t.$$

Therefore,

$$\begin{aligned}\log 2 &= t \log (1.08)\\ .3010 &= (.0334)t.\end{aligned}$$

Thus

$$t \simeq 9.01$$

It takes a little more than 9 years for $3000 to double if the interest rate is 8% per year, compounded annually.

EXERCISES 11.6

Solve Exercises 1–6 using the equation $A = A_0 \cdot 2^{-t/5600}$. *See Example 1.*

1. A fossil contains 6 milligrams of carbon 14. How much carbon 14 will be left after 2000 years?
2. A fossil contains 3 milligrams of carbon 14. After 3000 years, how much carbon 14 will remain?
3. If a living organism contains 8 milligrams of carbon 14, how much carbon 14 will its fossilized remains contain 9000 years after it dies?
4. Suppose that a living organism contained 2 grams of carbon 14. If its fossilized remains are found 15,000 years after it dies, how much carbon 14 will they contain?

5. The fossilized remains of a small sea creature contain 12 milligrams of carbon 14. In how many years will the fossil contain 10 milligrams of carbon 14?

6. The fossilized remains of a small animal contain 15 milligrams of carbon 14. In how many years will the fossil contain 5 milligrams of carbon 14?

Solve Exercises 7–10 using the equation $A = A_0 \cdot 2^{-t/1600}$. *See Example 2.*

7. How many milligrams of radium will be present after 1000 years if the original amount is 30 milligrams?

8. How many milligrams of radium will be present after 800 years if the original amount is 80 milligrams?

9. How many years will it take 25 milligrams of radium to decay to 15 milligrams?

10. How many years will it take 12 milligrams of radium to decay to 10 milligrams?

Solve Exercises 11–14 using the equation $N = N_0 \cdot 5^{.015t}$ *to measure the number,* N, *of bacteria that are present after* t *minutes if the initial number of bacteria is* N_0. *See Example 3.*

11. If the original number of bacteria is 1000, when will there be 3000 bacteria?

12. If the original number of bacteria is 2000, when will there be 5000 bacteria?

13. If the original number of bacteria is 500, after 60 minutes, how many bacteria will be present?

14. How many bacteria will be present after 150 minutes if the original number of bacteria is 1200?

Solve Exercises 15–22 using the equation $A = P(1 + r)^t$. *See Examples 4 and 5.*

15. After 10 years, how much money will be in an account that pays 10% interest, compounded annually, if the initial investment was $3000?

16. How much money will be in an account after 12 years if the initial investment was $1000 and the account pays 7%, compounded annually?

17. How much will an initial investment of $2000 amount to in 15 years if interest is paid at the rate of 9%, compounded annually?

18. After 8 years, how much money will be in an account that pays 8%, compounded annually, if the initial investment was $500?

19. How long will it take $2000 to double if it is invested in an account that pays 9% interest per year, compounded annually?

20. How long will it take $1000 to double if it is invested in an account that pays 8% interest per year, compounded annually?

21. If you deposit $1500 in an account that pays 8% interest per year, compounded annually, how long will it take before the account has $4000 in it?

22. If you deposit $2500 in an account that pays 9% interest per year, compounded annually, how long will it take before the account has $6000 in it?

Key Words

Antilogarithm
Base
Characteristic
Exponential equation
Exponential function
Logarithm
  Common logarithm
  Natural logarithm
Logarithmic function
Mantissa

CHAPTER 11 REVIEW

SECTION 11.1

Sketch the graphs of the equations.

1. $y = 3^x$ **2.** $y = (\frac{1}{2})^x$ **3.** $y = 2^{3x+1}$

Solve the equations.

4. $7^x = 49$ **5.** $3^{2x-5} = 27$ **6.** $5^{4x+3} = \frac{1}{125}$

SECTION 11.2

Rewrite each expression in logarithmic form.

7. $5^4 = 625$ **8.** $4^{-2} = \frac{1}{16}$ **9.** $10^3 = 1000$

Rewrite each expression in exponential form.

10. $\log_9 81 = 2$ **11.** $\log_5 \frac{1}{25} = -2$ **12.** $\log_{1/3} \frac{1}{9} = 2$

Sketch the graphs of the following equations.

13. $y = \log_3 x$ **14.** $y = \log_5 x$

Evaluate the expressions.

15. $\log_5 \sqrt{5}$ **16.** $\log_{36} 6$ **17.** $\log_{25} 125$

Solve the equations.

18. $\log_4 x = 2$ **19.** $\log_3 x = 4$ **20.** $\log_{10} x = -2$

SECTION 11.3

Rewrite each expression as the sum or difference of logarithms.

21. $\log_4 7 \cdot 11$ **22.** $\log 3 \cdot 13 \cdot 19$ **23.** $\log \dfrac{3x}{5}$ **24.** $\log \dfrac{29}{7x}$

Solve the equations.

25. $\log_5 x + \log_5 9 = 2$ **26.** $\log_6 x + \log_6 (x - 5) = 2$

SECTION 11.4

Use the Table of Common Logarithms in Appendix C to find the value of each expression.

27. log 5.17 **28.** log 1210 **29.** log .0031

30. antilog 2.8675 **31.** antilog (7.9201 − 10)

Use logarithms to evaluate the following.

32. (3120)(70200) **33.** $\dfrac{.0871}{.000213}$ **34.** $(7.32)^{5.1}$ **35.** $\sqrt{.0681}$

SECTION 11.5

Find the value of the logarithms.

36. $\log_8 100$ **37.** $\log_2 10$ **38.** ln 20 **39.** ln 25

Use logarithms to solve the equations. Express your answer to two decimal places.

40. $2^x = 10$ **41.** $6^{2x+1} = 50$

SECTION 11.6

Solve each word problem.

42. Use the equation $A = A_0 \cdot 2^{-t/5600}$ to determine how many milligrams of carbon 14 will be in a fossil in 700 years if it currently contains 3 milligrams of carbon 14.

43. Use the equation $N = N_0 \cdot 5^{.015t}$ to determine how many bacteria will be present after 90 minutes if the original number of bacteria is 1000.

44. How long will it take \$3000 to triple in value if it is invested in an account that pays 9% per year, compounded annually?

CHAPTER 11 TEST

For Problems 1 and 2, sketch the graph of the equation.

1. $y = 2^x$ **2.** $y = \log_2 x$

For Problems 3–8, use the definition of a logarithm to find the value of the expression.

3. $\log_7 49$ **4.** $\log_2 16$ **5.** $\log_5 \frac{1}{25}$ **6.** $\log_{10} (.0001)$

7. $\log_5 \sqrt{5}$ **8.** $\log_6 \sqrt{\frac{1}{36}}$

For Problems 9–12, solve for x.

9. $\log_3 x = 4$ **10.** $\log_{64} x = -\frac{2}{3}$ **11.** $\log_3 x + \log_3 5 = 2$ **12.** $\log_8 5 + \frac{1}{2}\log_8 x = 3\log_8 5$

For Problems 13–16, use the Table of Common Logarithms in Appendix C to find the logarithm. For Problem 16, use $\ln x = (2.3026)\log x$.

13. $\log 3.71$ **14.** $\log .00284$ **15.** $\log 472000$ **16.** $\ln 52$

For Problems 17–20, use logarithms to find an approximate value for the expression.

17. $(.00473)(89{,}100)$ **18.** $\dfrac{4.13}{.627}$ **19.** $\sqrt[3]{42.1}$ **20.** $\dfrac{(7620)(1.09)^{1.4}}{(31.6)^2}$

For Problems 21–23, use logarithms to solve the equation.

21. $6^x = 50$ **22.** $5^{2x-3} = 120$ **23.** $e^{.38x} = 100$

Solve Problems 24 and 25.

24. Use the equation $A = A_0 \cdot 2^{-t/5600}$ to determine how many milligrams of carbon 14 will be in a fossil in 1000 years if it currently contains 3 milligrams.

25. Use the equation $A = P(1 + r)^t$ to determine how long it takes \$1000 to double if it is invested in an account that pays 9% interest per year, compounded annually.

CHAPTER 12
Sequences and Series; Counting Principles

A couple decides to put \$1 in a jar at the beginning of the month and then double the amount they put in for each succeeding month. The geometric sequence with the general term

$$\textit{amount} = 2^{n-1},$$

where n is the number of months, tells us that they may have to modify their savings plan because, for the eleventh month, they will have to put \$1024 into the jar.

In this chapter we present several independent topics that are frequently encountered in algebra.

First we present a discussion of arithmetic and geometric sequences and series of real numbers.

Next we discuss permutations and combinations. These topics are of importance in such mathematical disciplines as probability and statistics.

Finally, we conclude the chapter with a discussion of the binomial formula, which is used quite extensively in many branches of mathematics.

12.1 Sequences and Series

An *infinite sequence* is a function f whose domain is the set $\{1, 2, 3, \ldots\}$ of *all* positive integers. For example, consider the sequence

$$f(n) = \frac{1}{n}$$

with domain $\{1, 2, 3, \ldots\}$. To each positive integer n there corresponds a real number value $f(n)$. We list these range values as follows:

(1) $$f(1) = 1, \quad f(2) = \frac{1}{2}, \quad f(3) = \frac{1}{3}, \quad \ldots, \quad f(n) = \frac{1}{n}, \quad \ldots$$

where the dots at the end (ellipses) indicate that the sequence does not terminate. For infinite sequences it is customary to represent these function values using a letter such as a, b, or c in the following manner.

(2) $$a_1 = f(1), \quad a_2 = f(2), \quad a_3 = f(3), \quad \ldots, \quad a_n = f(n), \quad \ldots$$

Substituting the function values $f(1) = 1, f(2) = \frac{1}{2}, f(3) = \frac{1}{3}$, and so forth, from (1) into the equalities appearing in (2), we get

(3) $$a_1 = 1, \quad a_2 = \frac{1}{2}, \quad a_3 = \frac{1}{3}, \quad \ldots, \quad a_n = \frac{1}{n}, \quad \ldots$$

As a result, we obtain a collection of numbers which are *ordered* in the sense that

a_1 is called the *first term* of the sequence
a_2 is called the *second term* of the sequence
a_3 is called the *third term* of the sequence
⋮
a_n is called the nth *term* of the sequence
⋮

In general, we specify a sequence by writing a formula for the nth *term* or *general term*, a_n, of the sequence. The formula for a_n gives us the rule that will generate the terms of the sequence. Thus the sequence shown in (3) may be specified by writing the following formula for the nth term:

$$a_n = \frac{1}{n}.$$

The sequence can then be listed by assigning to n the values 1, 2, 3, . . . in the formula for a_n. As a result, we see that the sequence consists of the following terms taken in order

$$1, \quad \frac{1}{2}, \quad \frac{1}{3}, \quad \frac{1}{4}, \quad \ldots$$

Sometimes, a sequence is specified by writing the first few terms and then the general or nth term as follows:

$$1, \quad \frac{1}{2}, \quad \frac{1}{3}, \quad \ldots, \quad \frac{1}{n}, \quad \ldots$$

Example 1. List the first five terms of the infinite sequence whose general term is $a_n = 2^n$.

Solution: We assign to n in the formula the consecutive integer values from 1 to 5. Thus we obtain the first five terms of the sequence:

$$2^1, \quad 2^2, \quad 2^3, \quad 2^4, \quad 2^5$$

or equivalently,

$$2, \quad 4, \quad 8, \quad 16, \quad 32.$$

A *finite sequence* is a sequence whose domain is the set $\{1, 2, 3, \ldots, n\}$ consisting of the first positive n integers.

Example 2. Write the terms of the finite sequence with nth term $b_n = \frac{1}{n^2}$ and domain $\{1, 2, 3, 4, 5\}$.

Solution: To obtain the terms of the sequence, we use the formula $b_n = \frac{1}{n^2}$ and assign to n the consecutive positive integers in the given domain. Thus we have

$$\frac{1}{1}, \quad \frac{1}{2^2}, \quad \frac{1}{3^2}, \quad \frac{1}{4^2}, \quad \frac{1}{5^2}$$

or

$$1, \quad \frac{1}{4}, \quad \frac{1}{9}, \quad \frac{1}{16}, \quad \frac{1}{25}.$$

Associated with each sequence (finite or infinite) there is a *series*. A *finite series* is the indicated sum of the terms of a finite sequence. For example, consider the finite sequence with nth or general term $c_n = n^2$ and domain $\{1, 2, 3, 4, 5\}$. The terms of the sequence are $1^2, 2^2, 3^2, 4^2, 5^2$. The series corresponding to this sequence is

$$1^2 + 2^2 + 3^2 + 4^2 + 5^2. \tag{4}$$

To write a series such as (4) in compact form, we use the capital Greek letter Σ (called *sigma*) in the following way:

$$\sum_{k=1}^{5} k^2.$$

The symbol $\sum_{k=1}^{5} k^2$ directs us to assign the consecutive integers starting at 1 and ending with 5 to the expression k^2 and to write the indicated sum of the terms obtained. Thus we have

$$\sum_{k=1}^{5} k^2 = 1^2 + 2^2 + 3^2 + 4^2 + 5^2.$$

The letter k is called the *index of summation* or simply the *index*, and expressions such as $\sum_{k=1}^{5} k^2$ are said to be in *summation notation*.

Usually, the letters i, j, k, and also m, n are used as indices (plural of index) of summation. However, we may choose as an index in summation notation any letter that we find suitable to symbolize a given sum. Thus we may write equivalently

$$\sum_{i=1}^{5} i^2 = \sum_{j=1}^{5} j^2 = \sum_{k=1}^{5} k^2 = 1^2 + 2^2 - 3^2 + 4^2 + 5^2.$$

Note that when the letter i is used as an index, it does not represent the imaginary number i.

A *finite series* can always be evaluated by adding the terms of the series algebraically. We illustrate this in the following example.

Example 3. Write out the indicated sum

$$\sum_{k=1}^{4} (2k + 1)$$

and evaluate.

Solution

$$\sum_{k=1}^{4} (2k + 1) = [2(1) + 1] + [2(2) + 1] + [2(3) + 1] + [2(4) + 1]$$

$$= (2 + 1) + (4 + 1) + (6 + 1) + (8 + 1)$$
$$= 3 + 5 + 7 + 9$$
$$= 24.$$

Sometimes, for a finite series, the last term is not specified. For example, we may write the indicated sum of the first n positive even integers as follows:

(5) $$2 + 4 + 6 + 8 + \cdots + 2n.$$

Using summation notation, we may write the sum appearing in (5) as follows:

$$\sum_{i=1}^{n} 2i.$$

We note that we cannot evaluate a finite series such as the one given in (5) unless we assign a specific number to n.

Example 4. Write out the terms of the sum

$$\sum_{k=2}^{n} 3^k.$$

Solution: Notice that in the given summation notation, the index does not begin with the number 1. Thus we have

$$\sum_{k=2}^{n} 3^k = 3^2 + 3^3 + 3^4 + \cdots + 3^n.$$

To each *infinite* sequence $a_1, a_2, a_3, \ldots, a_n, \ldots$ there corresponds an *infinite* series denoted by

(6) $$a_1 + a_2 + a_3 + \cdots + a_n + \cdots .$$

To represent an infinite series such as (6) using summation notation, we introduce a special symbol as follows. Thus

$$\sum_{n=1}^{\infty} a_n = a_1 + a_2 + a_3 + \cdots ,$$

where the symbol ∞ tells us that this is an infinite series.

The topic of infinite series and defining the sum of an infinite series is discussed in more detail in advanced courses of mathematics. However, we shall discuss the sum of a special type of infinite series in Section 12.3.

Example 5. Write out the first four terms of the infinite series

$$\sum_{n=0}^{\infty} \frac{1}{2^n}.$$

Solution

$$\sum_{n=0}^{\infty} \frac{1}{2^n} = \frac{1}{2^0} + \frac{1}{2^1} + \frac{1}{2^2} + \frac{1}{2^3} + \cdots$$

$$= 1 + \frac{1}{2} + \frac{1}{4} + \frac{1}{8} + \cdots.$$

We note that the index of summation in summation notation for infinite series may sometimes start with 0 as in Example 5, or with any other value. For example, consider the series

$$\sum_{n=3}^{\infty} \frac{1}{2^n}.$$

EXERCISES 12.1

For Exercises 1–16, write out the first four terms of the given sequence with the specified n*th term. See Examples 1 and 2.*

1. $a_n = n^3$ **2.** $b_n = \frac{1}{n^2}$ **3.** $c_n = 2n$ **4.** $a_n = 2n + 1$

5. $a_n = \frac{1}{3^n}$ **6.** $a_n = \frac{3}{n}$ **7.** $b_n = (-1)^n$ **8.** $c_n = (-1)^{n+1}$

9. $a_n = \frac{1}{n+1}$ **10.** $b_n = \frac{1}{2n+1}$ **11.** $a_n = 5$ **12.** $b_n = 5n + 3$

13. $c_n = \frac{1}{\sqrt{n}}$ **14.** $b_n = (-1)^n + 1$ **15.** $a_n = (.1)^n$ **16.** $a_n = e^n$

For Exercises 17–28, evaluate each sum. See Example 3.

17. $\sum_{i=1}^{3} 2i$ **18.** $\sum_{i=1}^{4} 3i$ **19.** $\sum_{k=2}^{5} (3k - 1)$

20. $\sum_{k=3}^{6} (2k + 5)$ **21.** $\sum_{j=1}^{4} (-1)^j$ **22.** $\sum_{j=1}^{5} (-1)^{j+1}$

23. $\sum_{k=1}^{4} (-1)^k(2k)$ **24.** $\sum_{i=3}^{7} (-1)^i(2i - 5)$ **25.** $\sum_{k=1}^{3} 2^{-k}$

26. $\sum_{k=2}^{4} 3^{-k}$ **27.** $\sum_{k=1}^{4} (\sqrt{2})^{2k}$ **28.** $\sum_{i=1}^{4} (\sqrt{3})^{2i}$

For Exercises 29–36, write without using summation notation. For infinite series write down the first four terms of the series. See Examples 4 and 5.

29. $\sum_{k=1}^{n} k$ **30.** $\sum_{k=1}^{n} (2k + 1)$ **31.** $\sum_{k=1}^{n} \frac{1}{k(k + 1)}$

32. $\sum_{k=1}^{n} \left(1 + \frac{1}{k}\right)$ **33.** $\sum_{k=1}^{\infty} \frac{1}{k^3}$ **34.** $\sum_{k=1}^{\infty} \left(1 - \frac{1}{k}\right)$

35. $\sum_{n=0}^{\infty} x^n$ **36.** $\sum_{n=0}^{\infty} (-1)^n x^{2n}$ **37.** $\sum_{n=2}^{\infty} x^n$

Arithmetic Sequences and Series 12.2

Consider the finite sequence consisting of the following five terms:

$$2, 5, 8, 11, 14.$$

Notice that any two consecutive terms of this sequence differ by a constant number 3: for example, $5 - 2 = 3$, $8 - 5 = 3$, and so forth. A sequence with this property is called an arithmetic sequence.

In general, an *arithmetic sequence*

$$a_1, \quad a_2, \quad a_3, \quad \ldots, \quad a_n, \quad \ldots$$

is a sequence whose consecutive terms differ by the *same* constant number, which is denoted by d. The constant number d is called the *common difference* of the sequence. Arithmetic sequences are also called *arithmetic progressions*.

We note that in an arithmetic sequence, each term of the sequence can be obtained by adding the common difference d to the term preceding it in the sequence. We illustrate these remarks in Example 1. We note first that for the remainder of this chapter we will denote the terms of any particular sequence by a_1, a_2, a_3, and so forth.

Example 1. Given the infinite arithmetic sequence

$$-7, \quad -3, \quad 1, \quad 5, \quad 9, \quad \ldots$$

find

(a) The common difference.
(b) The eighth term, a_8, of the sequence.

Solution
(a) Since $-3 - (-7) = 4$, $1 - (-3) = 4$, $5 - 1 = 4$, and so forth, we see that the common difference is $d = 4$.
(b) To obtain the term a_8, we find first a_6, a_7, and finally a_8. To accomplish this, we add $d = 4$ to the fifth term of the sequence, $a_5 = 9$, and to each succeeding term until we arrive at a_8. As a result, we see that the first eight terms of the sequence are

$$-7, -3, 1, 5, 9, 13, 17, 21.$$

Thus the eighth term, a_8, of the given sequence is 21.

Example 1 suggests that it would prove very useful to have a general formula that would enable us to compute any term of an arithmetic sequence. We now develop such a formula.

Suppose that

$$a_1, \quad a_2, \quad a_3, \quad \ldots, \quad a_n, \quad \ldots$$

is an arithmetic sequence with common difference d. By the definition of an arithmetic sequence we see that

$$\begin{aligned} a_1 &= a_1 \\ a_2 &= a_1 + d \\ a_3 &= a_2 + d = a_1 + 2d && \text{since } a_2 = a_1 + d \\ a_4 &= a_3 + d = a_1 + 3d && \text{since } a_3 = a_1 + 2d \\ &\vdots \end{aligned}$$

Note in the list above that the coefficient of d is *one less* than the number of the term. This yields the following formula for the nth term, a_n, of an arithmetic sequence:

$$\mathbf{a_n = a_1 + (n - 1)d \qquad (n \geq 1),}$$

where d is the common difference and $n = 1, 2, 3, \ldots$.

Example 2. Find the term a_{20} of the arithmetic sequence

$$4, 7, 10, 13, \ldots.$$

Solution: The first term of the sequence is $a_1 = 4$. The common difference $d = a_2 - a_1 = 7 - 4 = 3$. That is, $d = 3$. Substituting $a_1 = 4$ and $d = 3$ into the formula for the nth term of an arithmetic sequence

$$a_n = a_1 + (n - 1)d,$$

we obtain as a formula for the *nth* term of the given sequence

$$a_n = 4 + (n - 1)3.$$

To find the term a_{20}, we substitute $n = 20$ in the last formula above. Thus we have

$$\begin{aligned} a_{20} = 4 + (20 - 1)3 &= 4 + (19)3 \\ &= 4 + 57 \\ &= 61. \end{aligned}$$

Therefore, $a_{20} = 61$.

Given a finite arithmetic sequence such as

$$5, 7, 9, 11, 13, 15,$$

if we wish to sum the corresponding series

$$5 + 7 + 9 + 11 + 13 + 15,$$

we would simply add the terms algebraically and find that

$$5 + 7 + 9 + 11 + 13 + 15 = 60.$$

However, a useful formula can be derived that will enable us to determine the *sum* of first n terms of a *finite* arithmetic sequence

$$a_1, a_2, a_3, \ldots, a_n.$$

We let S_n denote this sum as follows:

$$S_n = a_1 + a_2 + a_3 + \cdots + a_n.$$

But

(1) $$S_n = a_1 + (a_1 + d) + (a_1 + 2d) + \cdots + [a_1 + (n - 1)d].$$

We can also write the sum S_n in an equivalent way in reverse order by writing the *last* term first and subtracting d successively until we arrive at the *first* term a_1 as follows:

(2) $$S_n = a_n + (a_n - d) + (a_n - 2d) + \cdots + \underbrace{[a_n - (n - 1)d]}_{a_1}.$$

As a result of adding (1) and (2), the corresponding multiples of d add to 0 and we obtain

$$2S_n = (a_1 + a_n) + (a_1 + a_n) + (a_1 + a_n) + \cdots + (a_1 + a_n) \longleftarrow n \text{ terms}$$

or

$$2S_n = n(a_1 + a_n)$$

$$S_n = \frac{n}{2}(a_1 + a_n).$$

The last equation yields the desired formula.
The previous discussion gives the following result.

Sum of an Arithmetic Sequence; Formula I

If $a_1, a_2, a_3, \ldots, a_n$ is a finite arithmetic sequence and

$$S_n = a_1 + a_2 + a_3 + \cdots + a_n,$$

then

$$S_n = \frac{n}{2}(a_1 + a_n).$$

Example 3. Find the sum of the first 10 terms of an arithmetic sequence whose first term a_1 is 2 and whose last term, a_{10}, is 38.

Solution: From the information given we have that $a_1 = 2$, $a_{10} = 38$, and $n = 10$. Substituting these values in Formula I stated above,

$$S_n = \frac{n}{2}(a_1 + a_n),$$

we obtain

$$\begin{aligned} S_{10} &= \frac{10}{2}(2 + 38) \\ &= 5(40) = 200. \end{aligned}$$

Thus the sum of the first 10 terms of the given arithmetic sequence is 200.

Often we want to sum the first 8 or possibly the first 20 terms of an arithmetic sequence but do not know the value of the last term. For example, suppose that we wanted to sum the first 10 terms of the arithmetic sequence

$$3, 6, 9, 12, \ldots.$$

We could do this by computing a_{10} and then using the formula given in Formula I. However, there is an alternative form for finding the sum, S_n, of the first n terms of a given arithmetic sequence that does not depend on knowing the value of the last term a_n. We now derive this alternative form.

We know that

(3) $$a_n = a_1 + (n - 1)d$$

and also that

(4) $$S_n = \frac{n}{2}(a_1 + a_n).$$

If we substitute the formula for a_n appearing in (3) into the formula appearing in (4), we obtain the formula

$$S_n = \frac{n}{2}[a_1 + \underbrace{a_1 + (n-1)d}_{a_n}] = \frac{n}{2}[2a_1 + (n-1)d].$$

Thus we have established the following formula.

Sum of an Arithmetic Sequence; Formula II

If $a_1, a_2, \ldots, a_n$ is a finite arithmetic sequence and

$$S_n = a_1 + a_2 + a_3 + \cdots + a_n,$$

then

$$S_n = \frac{n}{2}[2a_1 + (n-1)d].$$

We apply Formula II in the following examples.

Example 4. Find the sum of the first 15 terms of the arithmetic sequence $-8, -5, -2, 1, \ldots$.

Solution: Here $a_1 = -8$, $d = 3$, and $n = 15$. Using Formula II,

$$S_n = \frac{n}{2}[2a_1 + (n-1)d],$$

we see that

$$\begin{aligned} S_{15} &= \frac{15}{2}[2(-8) + (15-1)3] \\ &= \frac{15}{2}[-16 + (14)3] \\ &= \frac{15}{2} \cdot (-16 + 42) \\ &= \frac{15}{2}(26) = 195. \end{aligned}$$

Thus the required sum is 195.

Example 5. A woman accepted a *part-time* office position at the beginning of a certain year at a starting salary of \$8000. In addition, it was agreed that she would receive an annual raise of \$500. What will her total earnings be at the end of 7 years?

Solution: The sequence of salaries that the woman earns is an arithmetic sequence with $a_1 = \$8000$, $a_2 = \$8500$, $a_3 = \$9000$, and so forth. The common difference $d = \$500$, and the number of terms $n = 7$. Since we know that $a_1 = 8000$, $d = 500$, and $n = 7$, we use Formula II to compute S_7.

$$S_7 = \frac{7}{2}[2(8000) + (7 - 1)500]$$

$$= \frac{7}{2}(16{,}000 + 3000)$$

$$= \frac{7}{2}(19{,}000) = 66{,}500.$$

Thus she will have earned a total of \$66,500 at the end of 7 years.

EXERCISES 12.2

For Exercises 1–8, determine if the given finite sequence is an arithmetic sequence. For any arithmetic sequence, find the common difference d and the next two terms. See Example 1.

1. 2, 4, 6, 8, 10, . . .
2. 5, 10, 15, 20, 25, . . .
3. 11, 8, 5, 2, −1, . . .
4. 12, 8, 4, 0, −4, . . .
5. 1, 8, 15, 20, 27, . . .
6. −17, −7, 3, 13, 23, . . .
7. −18, −12, −6, 0, 6, . . .
8. 3, 11, 19, 27, 30, . . .

For Exercises 9–18, find the indicated term of the arithmetic sequence with first term a_1 and common difference d. See Example 2.

9. $a_1 = 4$, $d = 7$; $a_{11} = ?$
10. $a_1 = 6$, $d = 5$; $a_{16} = ?$
11. $a_1 = -3$, $d = 4$; $a_{10} = ?$
12. $a_1 = -9$, $d = 7$; $a_{13} = ?$
13. $a_1 = 8$, $d = -5$; $a_{21} = ?$
14. $a_1 = 11$, $d = -3$; $a_{31} = ?$
15. $a_1 = -4$, $d = -10$; $a_{19} = ?$
16. $a_1 = -20$, $d = -2$; $a_{26} = ?$
17. $a_1 = 1$, $d = \frac{1}{3}$; $a_{31} = ?$
18. $a_1 = 2$, $d = \frac{1}{4}$; $a_{81} = ?$

For Exercises 19–24, find the indicated sum S_n of the arithmetic sequence with first term a_1 and last term a_n. See Example 3.

19. $a_1 = 2$, $a_{11} = 42$; S_{11}
20. $a_1 = 5$, $a_{13} = 41$; S_{13}
21. $a_1 = 1$, $a_{20} = 20$; S_{20}
22. $a_1 = 2$, $a_{10} = 20$; S_{10}
23. $a_1 = -3$, $a_{17} = 77$; S_{17}
24. $a_1 = -5$, $a_{22} = -89$; S_{22}

For Exercises 25–32, find the indicated sum S_n of the arithmetic sequence with first term a_1 and common difference d. See Example 4.

25. $a_1 = 6$, $d = 8$; S_{10}
26. $a_1 = 5$, $d = 3$; S_7
27. $a_1 = 15$, $d = 5$; S_5

28. $a_1 = 2, d = 10; S_8$ **29.** $a_1 = -4, d = -3; S_9$ **30.** $a_1 = -1, d = -1; S_{10}$

31. $a_1 = 1, d = 1; S_{15}$ **32.** $a_1 = 5, d = 4; S_{13}$

Solve Exercises 33–36. See Example 5.

33. The first year that a youngster is employed he saves \$200. In each succeeding year he saves \$125 more than he saved the preceding year. How much will he have saved in 10 years?

34. The first n positive integers 1, 2, 3, . . . , n form an arithmetic sequence.
(a) Find a formula for their sum $S_n = 1 + 2 + 3 \cdots + n$.
(b) Use the formula in part (a) to find the sum of the first 100 positive integers.

35. A freely falling body near the surface of the earth falls 16 feet during the first second, 48 feet during the second second, 80 feet during the third second, and so forth, in arithmetic sequence. Find
(a) How far it falls during the eighth second.
(b) The distance it fell by the end of the eighth second.

36. A piece of machinery is purchased by a factory at the beginning of a certain year at a cost of \$20,000. The machinery depreciates in value \$300 the first year, \$400 the second year, \$500 the third year, and continues in this way for 10 years. If the sum of the depreciations are applied to the original cost, what is the value of the machinery in 10 years?

Geometric Sequences and Series 12.3

The following sequence is an example of a geometric sequence

$$1, \frac{1}{2}, \frac{1}{2^2}, \frac{1}{2^3}, \frac{1}{2^4}, \ldots .$$

Notice that each term after the first is obtained by *multiplying* the preceding term by $\frac{1}{2}$.

In general, a *geometric sequence* is a sequence in which each term after the first is obtained by multiplying the preceding term by the same fixed (constant) number, denoted by r. The constant number r is called the *common ratio*.

Geometric sequences are also called *geometric progressions*. We observe that the common ratio of a geometric sequence can be determined by dividing a term (except the first) of the sequence by the term immediately preceding it in the sequence. We illustrate this procedure for finding the common ratio r of a geometric sequence in the following example.

Example 1. Find the common ratio of the geometric sequence

$$\frac{2}{3}, \frac{2}{9}, \frac{2}{27}, \frac{2}{81}, \ldots .$$

Also find a_5 and a_6.

Solution: If we divide any given term (except the first term) by the term immediately preceding it in the given sequence, we can determine the common ratio r. Thus, if we choose to divide the second term a_2 by the first term a_1, we see that

$$r = \frac{2}{9} \div \frac{2}{3} = \frac{2}{9} \cdot \frac{3}{2} = \frac{1}{3}.$$

Thus the common ratio $r = \frac{1}{3}$. Therefore,

$$a_5 = a_4 \cdot \frac{1}{3} = \frac{2}{81} \cdot \frac{1}{3} = \frac{2}{243} \qquad \text{and similarly} \qquad a_6 = \frac{2}{729}.$$

We now develop a formula for finding the nth term of a geometric sequence. The first n terms of a geometric sequence with first term a and common ratio r ($r \neq 0$) has the form

(1) $$a, \quad ar, \quad ar^2, \quad \ldots, \quad \underbrace{ar^{n-1}}_{n\text{th term } a_n}.$$

Thus we see that the nth term of geometric sequence (1) is given by the formula

$$\boldsymbol{a_n = ar^{n-1}, \qquad \text{where } n \geq 1,\ r \neq 0.}$$

Example 2. Find a_5 for the geometric sequence with first term $a = 3$ and common ratio $r = \frac{1}{2}$.

Solution: Here $n = 5$. We are given that $a = 3$ and $r = \frac{1}{2}$. Substituting $n = 5$ in the formula $a_n = ar^{n-1}$, we obtain

$$a_5 = 3\left(\frac{1}{2}\right)^{5-1} = 3\left(\frac{1}{2}\right)^4 = \frac{3}{16}.$$

Thus $a_5 = \frac{3}{16}$.

Example 3. Find the value of a_{10} in the geometric sequence 6, 18, 54, ...

Solution: The common ratio is

$$r = \frac{18}{6} = 3.$$

We know $a = 6$. Thus

$$\begin{aligned} a_{10} &= 6(3)^{10-1} \\ &= 6(3)^9. \end{aligned}$$

If we use logarithms or a calculator, we find that $a_{10} = 118{,}098$.

We now derive a formula for finding the *sum* S_n of the first n terms of a geometric sequence,

$$a, \quad ar, \quad ar^2, \quad ar^3, \quad \ldots, \quad ar^{n-1}.$$

The indicated sum of the first n terms is

(2) $$S_n = a + ar + ar^2 + ar^3 + \cdots + ar^{n-1}.$$

Multiplying both sides of equation (2) by the common ratio r, we obtain

(3) $$rS_n = ar + ar^2 + ar^3 + ar^4 + \cdots + ar^{n-1} + ar^n.$$

If we subtract the left and right sides of equation (3) from the corresponding sides of equation (2), we get

$$S_n - rS_n = a - ar^n$$

or equivalently,

(4) $$(1 - r)S_n = a(1 - r^n).$$

Dividing both sides of equation (4) by $(1 - r)$ provided $r \neq 1$ yields

$$S_n = \frac{a(1 - r^n)}{1 - r} \qquad (r \neq 1).$$

The preceding discussion establishes the following formula.

Sum of a Finite Geometric Sequence

If $a, ar, ar^2, \ldots, ar^{n-1}$ is a geometric sequence and

$$S_n = a + ar + ar^2 + \cdots + ar^{n-1} \qquad (r \neq 1),$$

then

$$S_n = a\frac{1 - r^n}{1 - r} \qquad (r \neq 1).$$

Example 4. Find the sum of the terms of the geometric sequence

$$\frac{1}{2}, \quad \frac{1}{4}, \quad \frac{1}{8}, \quad \frac{1}{16}, \quad \frac{1}{32}, \quad \frac{1}{64}.$$

Solution: We see that there are $n = 6$ terms and that $a = \frac{1}{2}$. The common ratio $r = \frac{1}{4} \div \frac{1}{2} = \frac{1}{4} \cdot \frac{2}{1} = \frac{1}{2}$. Thus, using the formula for finding the sum of a finite geometric sequence, we have

$$\begin{aligned} S_6 &= \frac{1}{2} \frac{1 - (\frac{1}{2})^6}{1 - \frac{1}{2}} \\ &= \frac{1}{2} \frac{1 - \frac{1}{64}}{\frac{1}{2}} \\ &= 1 - \frac{1}{64} = \frac{63}{64}. \end{aligned}$$

The desired sum is $\frac{63}{64}$.

Example 5. Find the sum of the first 10 terms of the geometric sequence with first term $a = \frac{1}{3}$ and common ratio $r = \frac{1}{3}$.

Solution

$$\begin{aligned} S_{10} &= \frac{1}{3} \frac{1 - (\frac{1}{3})^{10}}{1 - \frac{1}{3}} \\ &= \frac{1}{3} \frac{1 - (\frac{1}{3})^{10}}{\frac{2}{3}} \\ &= \frac{1 - (\frac{1}{3})^{10}}{2}. \end{aligned}$$

We may leave the value of S_{10} in exponent form as shown in the last line, because computing $\left(\frac{1}{3}\right)^{10}$ without the aid of logarithms or a calculator is cumbersome. However, even without a calculator we can give an approximate value for S_{10}. If we were to compute $\left(\frac{1}{3}\right)^{10}$, we would find that $\left(\frac{1}{3}\right)^{10} \simeq .0000169$, which is a small number, as we would expect. Thus, by neglecting the term $\left(\frac{1}{3}\right)^{10}$, we can approximate the sum S_{10} as follows:

$$S_{10} = \frac{1 - (\frac{1}{3})^{10}}{2} \simeq \frac{1}{2}.$$

We note that if we computed S_{10} with $\left(\frac{1}{3}\right)^{10}$, the actual result is

$$S_{10} = \frac{29{,}524}{59{,}049} \simeq .4999915.$$

The discussion in Example 5 suggests defining a formula for summing an *infinite geometric series,*

(5) $\displaystyle\sum_{n=1}^{\infty} ar^{n-1} = a + ar + ar^2 + ar^3 + \cdots + \cdots$ provided that $-1 < r < 1$.

We cannot find the sum of the geometric series (5) by adding the terms algebraically because there are an infinite number of terms involved.

However, the sum of the first n terms,

$$S_n = a + ar + ar^2 + \cdots + ar^{n-1},$$

where n is some finite positive integer, *can be* summed algebraically. In fact, we have the following formula for finding this *partial sum* S_n of an infinite geometric series:

$$S_n = a\frac{1 - r^n}{1 - r} \qquad (r \neq 1).$$

Now if r is a real number such that $-1 < r < 1$, then r is some fraction such as $\frac{1}{2}$. We shall observe that $\left(\frac{1}{2}\right)^n$ approaches 0 as the number n gets very large. For example,

$$\left(\frac{1}{2}\right)^{10} \simeq .0009766, \qquad \left(\frac{1}{2}\right)^{20} \simeq .000001, \qquad \left(\frac{1}{2}\right)^{30} \simeq .0000000009.$$

Thus we see that $\left(\frac{1}{2}\right)^n$ *approaches* 0 as n gets very large. Therefore, we can argue that for r such that $-1 < r < 1$, r^n *approaches* 0 for n very large and therefore

$$S_n = a\frac{1 - r^n}{1 - r} \qquad \textit{approaches} \qquad S = a\frac{1}{1 - r}.$$

We note that if $|r| \geq 1$, then r^n does *not* approach 0 for large values of n. Based on the preceding remarks, we are led to the following definition:

The *infinite geometric* series

(6) $$a + ar + ar^2 + ar^3 + \cdots + ar^{n-1} + \cdots$$

is defined to have a *sum* S where

(7) $$S = a \cdot \frac{1}{1 - r} \quad \text{provided that } -1 < r < 1.$$

We note that the infinite series (6) does not have a sum defined for it if $|r| \geq 1$ that is, when $r \leq -1$ or $r \geq 1$.

Example 6. Find the sum of the infinite geometric series

$$\sum_{n=0}^{\infty} \left(\frac{1}{2}\right)^n = 1 + \frac{1}{2} + \frac{1}{2^2} + \frac{1}{2^3} + \cdots .$$

Solution: For the given series $a = 1$ and $r = \dfrac{1}{2}$. Substituting in formula (7), we have

$$S = 1\frac{1}{1 - \frac{1}{2}} = \frac{2}{2 - 1} = 2.$$

Therefore,

$$\sum_{n=0}^{\infty} \left(\frac{1}{2}\right)^n = 2.$$

Example 7. A ball is dropped from a height of 81 feet and rebounds one-third of the distance it falls. If it continues to fall and rebound in this way, find its height when it reaches the top of the sixth bounce.

Solution: When solving verbal problems involving geometric sequences (and also arithmetic sequences), it is often very helpful to write down the first few terms of the sequence in order to determine the first term a and the common ratio r.

	$81,$	$81\left(\frac{1}{3}\right),$	$81\left(\frac{1}{3}\right)^2,$	$81\left(\frac{1}{3}\right)^3, \ldots$	$\longleftarrow$ geometric sequence
	↑	↑	↑	↑	
height:	initial height	top of first bounce	top of second bounce	top of third bounce	

Thus we see that the first term $a = 81$ and $r = \dfrac{1}{3}$. Notice the height of the ball at the top of the first bounce is the second term in the sequence of heights, the height of the ball at the top of the second bounce is the third term, and so forth. Thus, to find the height of the ball at the top of the sixth bounce, we need to find the seventh term of the sequence, a_7. Thus, for $n = 7$, we see that

$$a_7 = 81\left(\frac{1}{3}\right)^{7-1} = 81\left(\frac{1}{3}\right)^6 = \frac{1}{9}.$$

Therefore, the height of the ball at the top of the sixth bounce is $\dfrac{1}{9}$ of a foot.

Example 8. Express the rational number $.727\overline{72}$ as a fraction.

Solution: Note that $.727\overline{72} = .72 + .0072 + .000072 + \cdots$. Thus $.727\overline{72}$ represents an *infinite* geometric series with common ratio $r = \dfrac{.0072}{.72} = .01$ and first term $a = .72$. The sum of the series is

$$S = .72\frac{1}{1 - .01} = \frac{.72}{.99} = \frac{72}{99}.$$

As a result, we see that $.727\overline{72} = \dfrac{72}{99}$.

EXERCISES 12.3

For Exercises 1–8, find the common ratio of the geometric sequence and the term a_5. See Example 1.

1. 1, 4, 16, 64, . . .
2. 1, 5, 25, 125, . . .
3. $1, \frac{1}{4}, \frac{1}{16}, \frac{1}{64}, \ldots$
4. $1, \frac{1}{5}, \frac{1}{25}, \frac{1}{125}, \ldots$
5. .1, .01, .001, .0001, . . .
6. .2, .02, .002, .0002
7. 2, −4, 8, −16, . . .
8. 3, −9, 27, −81, . . .

For Exercises 9–20, for the geometric sequence with first term a and common ratio r find the indicated term. See Examples 2 and 3.

9. $a = 9, r = \frac{1}{3}; a_6$
10. $a = 4, r = \frac{1}{2}; a_7$
11. $a = 1, r = -1; a_{10}$
12. $a = 216, r = \frac{1}{6}; a_6$
13. $a = 3, r = \frac{2}{3}; a_6$
14. $a = -1, r = -\frac{1}{2}; a_7$
15. $a = \frac{7}{9}, r = 3; a_5$
16. $a = \frac{2}{25}, r = 5; a_6$
17. $a = \frac{1}{16}, r = 4; a_9$
18. $a = 3, r = 2; a_{11}$
19. $a = 5, r = 3; a_7$
20. $a = \frac{1}{25}, r = 5; a_7$

For Exercises 21–24, find the sum of the finite geometric sequence. See Example 4.

21. $1, \frac{1}{2}, \frac{1}{4}, \frac{1}{8}, \frac{1}{16}$
22. $1, \frac{1}{5}, \frac{1}{25}, \frac{1}{125}, \frac{1}{625}$
23. $1, \frac{1}{3}, \frac{1}{9}, \frac{1}{27}, \frac{1}{81}$
24. $1, \frac{1}{4}, \frac{1}{16}, \frac{1}{64}, \frac{1}{256}$

For Exercises 25–28, give an approximate value for the indicated sum S_n of the geometric sequence with first term a and common ratio r. See Example 5.

25. $a = \frac{1}{4}, r = \frac{1}{4}; S_{12}$
26. $a = \frac{1}{20}, r = \frac{1}{10}; S_{20}$
27. $a = 1, r = \frac{1}{5}; S_{10}$
28. $a = 1, r = \frac{1}{6}; S_{15}$

For Exercises 29–34, evaluate the sum of the infinite geometric series. See Example 6.

29. $\sum_{n=0}^{\infty} (\frac{3}{4})^n$ **30.** $\sum_{n=0}^{\infty} (\frac{2}{3})^n$ **31.** $\sum_{n=0}^{\infty} (\frac{2}{5})^n$

32. $\sum_{n=0}^{\infty} (\frac{5}{6})^n$ **33.** $\sum_{n=0}^{\infty} (-\frac{1}{2})^n$ **34.** $\sum_{n=0}^{\infty} (-\frac{1}{3})^n$

Solve Exercises 35 and 36. See Example 7.

35. A ball is dropped from a height of 64 feet and always rebounds $\frac{1}{2}$ the height it falls. Find the height it reaches at the top of the eighth bounce.

36. A piece of equipment purchased by a factory cost \$6250 at the beginning of a certain year. At the end of each year its value is $\frac{4}{5}$ of the value it had at the beginning of the year. What is its value at the end of 5 years?

For Exercises 37–40, use infinite series to express the rational number given in decimal form as a fraction. See Example 8.

37. $.22\overline{2}$ **38.** $.55\overline{5}$ **39.** $.6464\overline{64}$ **40.** $.5454\overline{54}$

12.4 Permutations and Combinations

When five people, such as Anne, Blaise, Carol, Jake, and Michael, run a race, the order in which the first two people finish the race is important. This is because Anne first, Jake second is different from Jake first, Anne second. When the same five people select two of their members to form a relay team, the order in which the two are selected is not important. This is because the relay team composed of Anne and Jake is the same as the relay team composed of Jake and Anne. In this section we are concerned with counting the number of different ways an event, or a sequence of events, can occur. Sometimes the order in which the events occur will be important; other times their order of occurrence will not be important. We begin with the following basic principle of counting.

Multiplication Principle

> If event A can occur in m ways and after it has occurred event B can occur in n ways, then the number of different ways that event A followed by event B can occur is $m \cdot n$.

As an application of the *multiplication principle,* suppose that a college cafeteria offers *three* main entrées: roast chicken, baked fish, and a vegetarian platter. In addition, it offers a choice of *five* desserts; fruit, ice cream, Jell-O, pie, and cake. The total number of different meals composed of an entrée followed by a dessert is determined as follows:

$$\underbrace{3}_{\substack{\text{number of choices}\\ \text{for the main entrée}}} \cdot \underbrace{5}_{\substack{\text{number of choices}\\ \text{for the dessert}}} = \underbrace{15}_{\substack{\text{total number of}\\ \text{different meals}}}$$

The 15 different meals are indicated in the ''tree diagram'' shown in Figure 12.1. A different meal is associated with each path of the tree diagram. For example, the paths that correspond to the two meals roast chicken/Jell-O and vegetarian platter/fruit are highlighted in Figure 12.1.

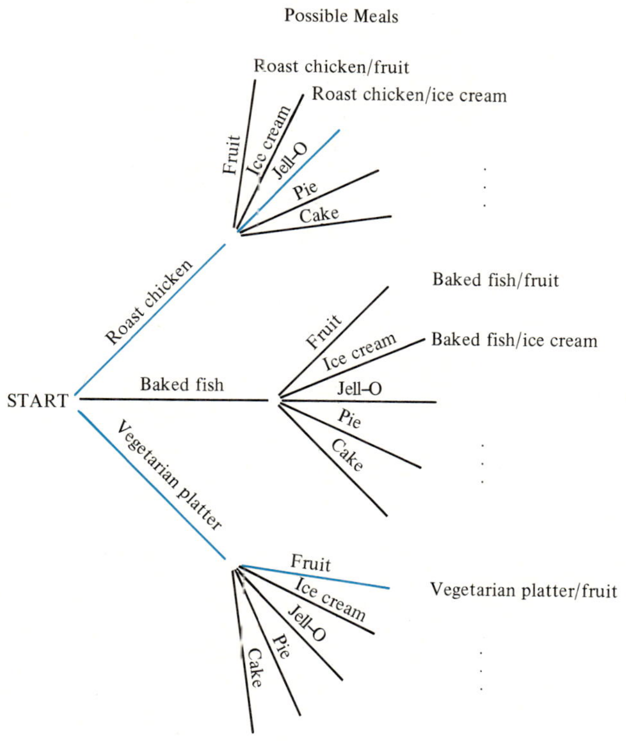

Figure 12.1

Although tree diagrams help us to understand the multiplication principle, they become impractical when dealing with events that can occur in a large number of different ways. The direct application of the multiplication principle is usually more efficient than listing all the possible outcomes with a tree diagram.

Example 1. A college dean is going to select and review one mathematics text and one chemistry text for a professional journal. He has 15 mathematics texts and 12 chemistry texts. In how many ways can he select one of each type of text?

Solution: Applying the multiplication principle, we get

$$\underbrace{15}_{\text{number of mathematics texts}} \cdot \underbrace{12}_{\text{number of chemistry texts}} = \underbrace{180}_{\text{total}}.$$

The next example illustrates how the multiplication principle can be extended to more than two events.

Example 2. A series of license plates is to be produced using two different letters of the alphabet followed by two different digits. How many such plates can be produced?

Solution: Since there are 26 letters in the alphabet, there are 26 choices for the first letter and 25 choices for the second letter. Similarly there are 10 choices for the first digit and 9 choices for the second digit. Extending the multiplication principle, we have the following result:

$$\underbrace{26}_{\text{first letter}} \cdot \underbrace{25}_{\text{second letter}} \cdot \underbrace{10}_{\text{first digit}} \cdot \underbrace{9}_{\text{second digit}} = 58{,}500.$$

We now consider situations where we want to count all the different ways an arrangement of k elements can be selected from n elements, $n \geq k$.

For example, there are six ordered arrangements consisting of two elements each that can be selected from the letters A, B, C. The six arrangements are

$$AB, \quad AC, \quad BA, \quad BC, \quad CA, \quad \text{and} \quad CB.$$

Note that the arrangement AB is different from the arrangement BA.

Each *different arrangement* of k elements selected from n elements is called a *permutation* of the n elements taken k at a time. The total number of permutations of n elements taken k at a time is denoted by ${}_nP_k$. Other commonly used symbols for ${}_nP_k$ are P_k^n and $P_{(n,k)}$. Following is a general formula for ${}_nP_k$.

Permutation Formula

The number of permutations of n elements taken k at a time is given by the formula

$${}_nP_k = n \cdot (n - 1) \cdot (n - 2) \cdot \cdots \cdot (n - k + 1).$$

Proof. The first element can be chosen in n ways. Thus there are $(n - 1)$ choices remaining for the second element. Similarly, there are $(n - 2)$ choices for the third element, and so on. Thus, if we apply the multiplication principle, we get

$$\text{the number of ways of choosing the } = \underbrace{n}_{\text{first element}} \cdot \underbrace{(n-1)}_{\text{second element}} \cdot \underbrace{(n-2)}_{\text{third element}} \cdot \cdots \cdot \underbrace{[n-(k-1)]}_{k\text{th element}}.$$

But $[n - (k - 1)] = n - k + 1$ and thus

$$_nP_k = n \cdot (n-1)(n-2) \cdot \cdots \cdot (n-k+1)$$

as claimed.

Example 3. Use the formula for $_nP_k$ to evaluate
(a) $_7P_3$. (b) $_{20}P_4$. (c) $_{12}P_7$.

Solution: Direct application of the Permutation Formula results in the following:
(a) $_7P_3 = 7 \cdot 6 \cdot 5 = 210$.
(b) $_{20}P_4 = 20 \cdot 19 \cdot 18 \cdot 17 = 116{,}280$.
(c) $_{12}P_7 = 12 \cdot 11 \cdot 10 \cdot 9 \cdot 8 \cdot 7 \cdot 6 = 3{,}991{,}680$.

Example 4. Using the letters of the word MATRIX, how many different arrangements of distinct letters can be formed if each arrangement contains
(a) Three letters? (b) Four letters? (c) Five letters?

Solution: First consider the letters M, A, T, R, I, X. Since we want different *arrangements* of letters, this is a permutation problem. Applying the Permutation Formula, we get
(a) $_6P_3 = 6 \cdot 5 \cdot 4 = 120$.
(b) $_6P_4 = 6 \cdot 5 \cdot 4 \cdot 3 = 360$.
(c) $_6P_5 = 6 \cdot 5 \cdot 4 \cdot 3 \cdot 2 = 720$.

Example 5. In preparation for a campaign, a politician must choose and give a priority ranking to four issues chosen from a list of nine issues. In how many ways can this be done?

Solution: Because the issues are to be ranked, order is involved. Thus this is a permutation problem. The answer is

$$_9P_4 = 9 \cdot 8 \cdot 7 \cdot 6 = 3024.$$

For any positive integer n, the symbol $n!$ is read as "*n factorial*" and denotes the product

$$n! = n \cdot (n-1) \cdot (n-2) \cdot \cdots \cdot 3 \cdot 2 \cdot 1.$$

For example,

$$\begin{aligned} 5! &= 5 \cdot 4 \cdot 3 \cdot 2 \cdot 1 = 120 \\ 4! &= 4 \cdot 3 \cdot 2 \cdot 1 \quad = 24 \\ 3! &= 3 \cdot 2 \cdot 1 \quad = 6 \\ 2! &= 2 \cdot 1 \quad = 2 \\ 1! &= 1 \quad = 1. \end{aligned}$$

The formula for ${}_nP_k$ can be expressed using factorial notation as follows:

(1) $$\boldsymbol{{}_nP_k = \frac{n!}{(n-k)!}}.$$

We can prove (1) by noting that

$$\frac{n!}{(n-k)!} = \frac{n(n-1)\cdots(n-k+1)\cancel{(n-k)}\cancel{(n-k-1)}\cdots\cancel{3}\cdot\cancel{2}\cdot\cancel{1}}{\cancel{(n-k)}\cancel{(n-k-1)}\cdots\cancel{3}\cdot\cancel{2}\cdot\cancel{1}}$$
$$= n(n-1)\cdots(n-k+1)$$

But the last expression is the original Permutation Formula given for ${}_nP_k$.

Thus we have shown that the formula for ${}_nP_k$ given in (1) is valid. This alternative way of writing the formula for ${}_nP_k$ is particularly convenient if you have a calculator with an $\boxed{n!}$ key. On many calculators with an $\boxed{n!}$ key, the answer to $n!$ for

$$14 \leq n \leq 69$$

is given in scientific notation. For $n > 69$, the use of the $\boxed{n!}$ key will result in an error indication. This is because $n!$ gets very large very quickly and for $n > 69$, $n!$ is too large for most calculators. On many calculators, to evaluate

$${}_{21}P_6 = \frac{21!}{(21-6)!} = \frac{21!}{15!},$$

you would enter 21, then press $\boxed{n!}$, then press $\boxed{\div}$, enter 15, press $\boxed{n!}$ and finally press $\boxed{=}$. The result should be ${}_{21}P_6 = 39{,}070{,}080$.

If we use the Permutation Formula given on page 480 to evaluate ${}_nP_n$, the result is

$${}_nP_n = n(n-1)(n-2)\cdots 3\cdot 2\cdot 1 = n!.$$

However, if we use the formula given in (1) above, since $n = k$ we get

$${}_nP_n = \frac{n!}{0!}.$$

Equating these two results, we see that $n! = \dfrac{n!}{0!}$. To maintain consistency, we define $0! = 1$.

The last topic we discuss in this section is how to count the number of subsets of k elements that can be selected *without regard to order* from a set of n elements, $n \geq k$. Each subset of k elements is called a *combination* of n elements taken k at a time. The total number of combinations of n elements taken k at a time is denoted by ${}_nC_k$. Other commonly used symbols for ${}_nC_k$ are C^n_k, $C_{(n,k)}$, and $\binom{n}{k}$. For convenience we will use the symbol $\binom{n}{k}$ in the next section. Note that although ABC and BCA are different permutations of the letters A, B, and C, they are the same combination. The following formula is stated without proof.

Combination Formula

> The number of combinations of n elements taken k at a time is given by
> $$ {}_nC_k = \binom{n}{k} = \frac{n!}{k!(n-k)!}. $$

Example 6. Use the Combination Formula to evaluate

(a) ${}_7C_3$. (b) ${}_{20}C_4$. (c) $\binom{12}{5}$.

Solution

(a) $${}_7C_3 = \frac{7!}{3!(7-3)!} = \frac{7!}{3!4!} = \frac{7\cdot 6\cdot 5\cdot 4\cdot 3\cdot 2\cdot 1}{(3\cdot 2\cdot 1)\cdot(4\cdot 3\cdot 2\cdot 1)} = \frac{7\cdot 6\cdot 5}{3\cdot 2\cdot 1} = 35.$$

(b) $${}_{20}C_4 = \frac{20!}{4!16!} = \frac{20\cdot 19\cdot 18\cdot 17\cdot(16!)}{4\cdot 3\cdot 2\cdot 1\cdot(16!)} = \frac{20\cdot 19\cdot 18\cdot 17}{4\cdot 3\cdot 2\cdot 1} = 4845.$$

(c) $$\binom{12}{5} = \frac{12!}{5!7!} = \frac{12\cdot 11\cdot 10\cdot 9\cdot 8\cdot(7!)}{5\cdot 4\cdot 3\cdot 2\cdot 1\cdot(7!)} = \frac{12\cdot 11\cdot 10\cdot 9\cdot 8}{5\cdot 4\cdot 3\cdot 2\cdot 1} = 792.$$

Although the computations involved with the formula for ${}_nC_k$ appear to be extensive, they can always be greatly simplified. For example, consider ${}_{25}C_7 = \dfrac{25!}{7!18!}$. Rather than perform all of the indicated multiplications, we may proceed as indicated in Example 6:

$${}_{25}C_7 = \frac{25\cdot \cancel{24}\cdot 23\cdot 22\cdot \cancel{21}\cdot \overset{2}{\cancel{20}}\cdot 19}{7\cdot 6\cdot 5\cdot 4\cdot 3\cdot 2\cdot 1}$$

$\cancel{24}$ $\cancel{10}$ $\cancel{21}$

$$= 25\cdot 23\cdot 22\cdot 2\cdot 19 = 480{,}700.$$

Thus we see that by carefully arranging our work, and by reducing as much as possible, we can greatly reduce the number of multiplications that need to be performed.

The following theorem is stated without proof.

Theorem 1

> If k is a nonnegative integer, $k > 0$ and $n \geq k$, then
> $$_nC_k = {}_nC_{n-k}.$$

One application of Theorem 1 is that it enables us to replace a combination formula with an equivalent combination formula that involves less computation. We illustrate this remark in the following example.

Example 7. Evaluate $_{10}C_7$

Solution: Direct application of Theorem 1 gives us

$$_{10}C_7 = {}_{10}C_3 = \frac{10 \cdot 9 \cdot 8}{3 \cdot 2 \cdot 1} = 10 \cdot 3 \cdot 4 = 120.$$

Example 8. E-Ling is going on a short vacation. She estimates that she will need 5 dresses for the vacation. If she owns 10 dresses, how many different sets of 5 dresses can she choose?

Solution: Because the order in which the dresses are chosen is not important, this is a combination problem. Since there are 10 dresses from which 5 are to be selected, the solution is given by

$$_{10}C_5 = \frac{10!}{5!5!} = 252.$$

EXERCISES 12.4

Solve Exercises 1–12. See Examples 1 and 2.

1. From a menu that offers 7 entrées and 9 desserts, how many meals consisting of an entrée and a dessert may be selected?
2. From a menu offering 6 entrées and 8 desserts, how many meals consisting of an entrée and a dessert are possible?
3. From a collection of 12 pairs of slacks and 15 blouses, how many ways can 1 pair of slacks and 1 blouse be selected?
4. From a selection of 8 pairs of pants and 11 shirts, how many ways can 1 pair of pants and 1 shirt be chosen?
5. A professor has a collection of 16 essay questions and 20 multiple-choice questions. In how many ways can she select 1 of each type of question for a two-question quiz?

6. In preparation for a new semester a student purchased 5 covers for term papers and 3 binders for computer programs. In how many ways can he select 1 cover and 1 binder?

7. From a group of 7 men and 5 women, how many committees can be formed consisting of 1 man and 1 woman?

8. From a group of 14 senators and 25 representatives, how many committees can be formed composed of 1 senator and 1 representative?

9. Linda has 10 novels, 7 volumes of poetry, and 4 science fiction books. In how many ways can she select 3 books consisting of 1 of each type of book?

10. A fruit stand has 6 types of apples, 3 types of oranges, and 4 types of grapes. How many selections consisting of 1 type of each fruit are possible?

11. How many license plates can be produced consisting of two different letters of the alphabet followed by three different digits?

12. How many license plates are possible consisting of two different nonzero digits followed by three different letters of the alphabet?

For Exercises 13–20, see Examples 3–5.

13. Evaluate.
(a) ${}_5P_3$ **(b)** ${}_8P_4$ **(c)** ${}_4P_4$

14. Evaluate.
(a) ${}_7P_2$ **(b)** ${}_9P_4$ **(c)** ${}_5P_5$

15. Evaluate.
(a) ${}_{12}P_5$ **(b)** ${}_{15}P_8$ **(c)** ${}_{13}P_7$

16. Evaluate.
(a) ${}_{12}P_6$ **(b)** ${}_{14}P_5$ **(c)** ${}_{15}P_6$

17. Using the letters of the word COMPUTER, how many different arrangements of distinct letters can be formed if each arrangement contains
(a) Three letters? **(b)** Five letters? **(c)** Seven letters?

18. Using the letters of the word HISTORY, how many different arrangements of distinct letters can be formed if each arrangement contains
(a) Three letters? **(b)** Four letters? **(c)** Five letters?

19. Each of the consecutive numbers 11, 12, 13, . . . , 48, 49 have been painted on a different Ping-Pong ball. In how many different ordered ways can 4 of the numbers be selected?

20. A professor has a collection of 25 essay questions. In how many different ways can he select and order a set of 3 of these questions?

For Exercises 21–30, see Examples 6–8.

21. Evaluate.
(a) ${}_5C_3$ **(b)** ${}_8C_4$ **(c)** ${}_4C_4$

22. Evaluate.
(a) ${}_7C_2$ **(b)** ${}_9C_4$ **(c)** ${}_5C_5$

23. Evaluate.
(a) $\binom{12}{5}$ **(b)** $\binom{15}{8}$ **(c)** $\binom{13}{7}$

24. Evaluate.
(a) $\binom{12}{6}$ **(b)** $\binom{14}{5}$ **(c)** $\binom{15}{6}$

25. How many committees consisting of 3 people can be formed from a group of 8 people?

26. From a group of 9 individuals, how many 4 member committees can be formed?

27. John has a collection of 15 hand-carved duck decoys. In how many ways can he select 4 of them for a local art show?

28. In how many ways can Allison select 5 of her 12 paintings to show at a local sidewalk art show?

Exercises 29 and 30 should be done with a calculator. Even if your calculator has a [n!] *key, you will have to be careful with Exercise 30.*

29. How many different unordered 5-card hands can be selected from a 52-card deck?

30. How many different unordered 13-card hands can be selected from a 52-card deck?

12.5 Expansion of a Binomial; Binomial Formula

From our previous work in Section 3.5 we know that

$$(a + b)^2 = a^2 + 2ab + b^2.$$

Similarly, we could verify that

$$(a + b)^3 = (a + b)(a + b)^2 = a^3 + 3a^2b + 3ab^2 + b^3.$$

In both of the equations above the expression on the right side is called the *expansion* of the binomial on the left side. In this section we discuss a method that will help us to quickly write the expansion of $(a + b)^n$ for any positive integer n.

For convenient reference we now list the expansion of $(a + b)^n$ for $n = 1, 2,$ and 3.

$$\begin{aligned}(a + b)^1 &= a + b\\ (a + b)^2 &= a^2 + 2ab + b^2\\ (a + b)^3 &= a^3 + 3a^2b + 3ab^2 + b^3\end{aligned}$$

Looking at these expansions for $n = 1, 2,$ and 3, we make the following observations about the expanded form of $(a + b)^n$.

1. It contains $n + 1$ terms.
2. The first term is a^n and the last term is b^n.
3. For each term, the sum of the exponents of a and b is n.

4. As we move to the right, in each succeeding term, the exponent on a decreases by 1 while the exponent on b increases by 1.

If we apply these observations to the expansion of $(a + b)^4$, we would conjecture that

$$(a + b)^4 = \underline{\ ?\ }\,a^4 + \underline{\ ?\ }\,a^3b + \underline{\ ?\ }\,a^2b^2 + \underline{\ ?\ }\,ab^3 + \underline{\ ?\ }\,b^4.$$

The blanks, (?), indicate the missing coefficients. To determine these missing coefficients, we now consider the multiplication of polynomials from a different point of view.

Recall from Section 3.5 that when we multiply two polynomials, each term of one polynomial is multiplied by each term of the other polynomial. This is the same as forming all possible products of two terms with exactly one term selected from each of the two polynomials. For example, consider the product of $(x + y + z)$ and $(x + 1)$. Since there are three terms in the first polynomial and two terms in the second polynomial, the multiplication principle of Section 12.4 tells us that there will be $3 \cdot 2 = 6$ terms in the product. The reader should verify this claim by performing the actual multiplication. We now apply this idea to the expansion of $(a + b)^4$. We begin by writing

$$(a + b)^4 = \underbrace{(a + b)}_{P_1}\underbrace{(a + b)}_{P_2}\underbrace{(a + b)}_{P_3}\underbrace{(a + b)}_{P_4},$$

where for reference we have labeled the four polynomial factors on the right side of the equation as P_1, P_2, P_3, and P_4. Extending the alternative method for multiplying two polynomials to four polynomials, we see that each term in the expansion of $(a + b)^4$ is formed by selecting exactly one term from each of the four factors P_1, P_2, P_3, and P_4. For example, if we select b from P_1, a from P_2, b from P_3, and b from P_4, the resulting product is ab^3. In general, each term will have the form $a^{4-k}b^k$, where $k = 0, 1, 2, 3$, or 4. For instance, when $k = 2$ we get a^2b^2. Notice that if we select exactly two b's, the remaining two factors must be a's. Also, since there are four polynomials from which to select the two b's and the order of their selection is not important, the term a^2b^2 can be obtained in exactly ${}_4C_2 = \binom{4}{2} = \dfrac{4!}{2!2!} = 6$ ways. Thus the coefficient of the a^2b^2 term is $\binom{4}{2} = 6$. In a similar way, we could show that the coefficient of each of the terms of the form $a^{4-k}b^k$ is $\binom{4}{k}$, $k = 0, 1, 2, 3, 4$. Thus

$$\begin{aligned}(a + b)^4 &= \binom{4}{0}a^4 + \binom{4}{1}a^3b + \binom{4}{2}a^2b^2 + \binom{4}{3}ab^3 + \binom{4}{4}b^4\\ &= a^4 + 4a^3b + 6a^2b^2 + 4ab^3 + b^4. \text{ Note that } \binom{4}{0} = \binom{4}{4} = 1.\end{aligned}$$

Generalizing the results of the preceding discussion to all positive integers, we now state the following two formulas (usually proved in more advanced mathematics courses).

Binomial Coefficient Formula

In the expansion of $(a + b)^n$, where n is a positive integer, the coefficient of the term $a^{n-k}b^k$ is

$$\binom{n}{k} = \frac{n!}{k!(n-k)!}. \qquad (k = 0, 1, 2, \ldots, n)$$

Example 1. In the expansion of $(a + b)^{10}$, what are the coefficients of the terms (a) a^6b^4? (b) a^8b^2? (c) a^3b^7?

Solution: Using the Binomial Coefficient Formula, we see that:
(a) Here $k = 4$; thus the coefficient is

$$\binom{10}{4} = \frac{10!}{4!6!} = \frac{10 \cdot 9 \cdot 8 \cdot 7}{4 \cdot 3 \cdot 2 \cdot 1} = 210.$$

(b) Here $k = 2$; thus the coefficient is

$$\binom{10}{2} = \frac{10!}{2!8!} = \frac{10 \cdot 9}{2 \cdot 1} = 45.$$

(c) Here $k = 7$; thus the coefficient is

$$\binom{10}{7} = \frac{10!}{7!3!} = \frac{10 \cdot 9 \cdot 8}{3 \cdot 2 \cdot 1} = 120.$$

Binomial Formula

For any positive integer n,

$$(a + b)^n = a^n + \binom{n}{1}a^{n-1}b + \binom{n}{2}a^{n-2}b^2 + \cdots + \binom{n}{k-1}a^{n-k+1}b^{k-1} + \cdots + \binom{n}{n-1}ab^{n-1} + b^n.$$

Example 2. Use the Binomial Formula to expand $(a + b)^7$.

Solution: Direct application of the Binomial Formula with $n = 7$ gives us the following equation.

$$(a + b)^7 = a^7 + \binom{7}{1}a^6b + \binom{7}{2}a^5b^2 + \binom{7}{3}a^4b^3 + \binom{7}{4}a^3b^4 + \binom{7}{5}a^2b^5 + \binom{7}{6}ab^6 + b^7$$

$$= a^7 + 7a^6b + 21a^5b^2 + 35a^4b^3 + 35a^3b^4 + 21a^2b^5 + 7ab^6 + b^7.$$

Notice by Theorem 1 that

$$\binom{7}{1} = \binom{7}{6}, \quad \binom{7}{2} = \binom{7}{5}, \quad \text{and} \quad \binom{7}{3} = \binom{7}{4}.$$

Example 3. Use the Binomial Formula to expand $(x - 2)^5$.

Solution: In this case $n = 5$. Now if we rewrite $(x - 2)$ as $[x + (-2)]$, and let $a = x$ and $b = -2$ in the Binomial Formula, we obtain

$$\begin{aligned}(x - 2)^5 \\ &= x^5 + \binom{5}{1}x^4(-2) + \binom{5}{2}x^3(-2)^2 + \binom{5}{3}x^2(-2)^3 + \binom{5}{4}x(-2)^4 + (-2)^5 \\ &= x^5 + 5x^4(-2) + 10x^3(4) + 10x^2(-8) + 5x(16) + (-32) \\ &= x^5 - 10x^4 + 40x^3 - 80x^2 + 80x - 32.\end{aligned}$$

Notice, in the Binomial Formula, that the exponent on b in the kth term is $k - 1$. Furthermore, notice that the coefficient of the kth term is

$$\binom{n}{k-1} = \frac{n!}{(k-1)!(n-k+1)!}.$$

Example 4. In the expansion of $(a + b)^{12}$, what is the coefficient of the
(a) Third term? (b) Sixth term? (c) Tenth term?

Solution: The coefficient of the kth term in the expansion of $(a + b)^n$ is $\binom{n}{k-1}$. For this example $n = 12$; thus

(a) Because $k = 3$ and therefore $k - 1 = 2$, $\binom{12}{3-1} = \binom{12}{2} = \frac{12!}{2!10!} = 66.$

(b) Because $k = 6$ and therefore $k - 1 = 5$, $\binom{12}{6-1} = \binom{12}{5} = \frac{12!}{5!7!} = 792.$

(c) Because $k = 10$ and therefore $k - 1 = 9$, $\binom{12}{10-1} = \binom{12}{9} = \frac{12!}{9!3!} = 220.$

Example 5. What is the fifth term in the expansion of $(2x - 1)^9$?

Solution: In this example $n = 9$ and $k = 5$; therefore, $k - 1 = 4$. Thus the coefficient is $\binom{9}{4}$ and the entire term is given by the expression

$$\binom{9}{4}(2x)^5(-1)^4 = \frac{9!}{4!5!}(2x)^5(-1)^4$$
$$= (126)(32x^5)(1)$$
$$= 4032x^5.$$

For small values of n, a convenient way of obtaining the coefficients of the terms in the expansion of $(a + b)^n$ is through the use of *Pascal's triangle*. The first six rows of Pascal's triangle are shown in Figure 12.2. Notice that each row of Pascal's triangle

	ROW
1 1	1
1 2 1	2
1 3 3 1	3
1 4 6 4 1	4
1 5 10 10 5 1	5
1 6 15 20 15 6 1	6

Figure 12.2

begins and ends with a 1. Furthermore, except for the first and last number in any row, each number is the sum of the numbers on either side of it in the row above. For example, we have written rows 3 and 4 of Pascal's triangle in Figure 12.3. Note that $4 = 1 + 3$, $6 = 3 + 3$, and $4 = 3 + 1$.

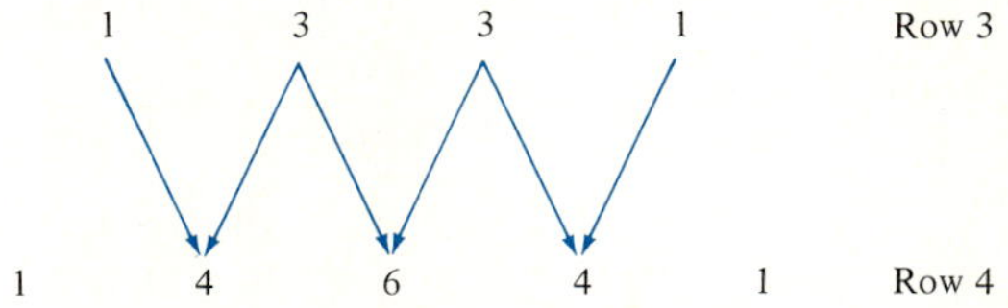

Figure 12.3

Also notice that the numbers in the third row of Pascal's triangle correspond to the coefficients in the expansion of $(a + b)^3$. That is,

$$(a + b)^3 = a^3 + 3a^2b + 3ab^2 + b^3$$
$$= 1a^3 + 3a^2b + 3ab^2 + 1b^3.$$

row 3 of Pascal's triangle $\quad 1 \quad 3 \quad 3 \quad 1$ (each ↕ corresponding to the coefficients above)

We note that the numbers in the nth row of Pascal's triangle correspond to the coefficients in the expansion of $(a + b)^n$.

Example 6. Determine the seventh row of Pascal's triangle and use it to write the expansion of $(x - y)^7$.

Solution: From Figure 12.2 we see that the *sixth* row of Pascal's triangle is

$$1 \quad 6 \quad 15 \quad 20 \quad 15 \quad 6 \quad 1$$

The first and last number of the seventh row are 1. The rest of the seventh row is obtained by adding elements in the sixth row as shown by the arrows in Figure 12.4.

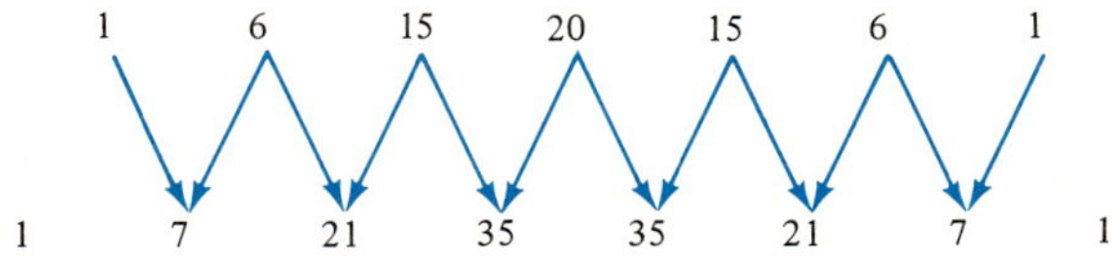

Figure 12.4

It follows that

$$\begin{aligned}(x - y)^7 &= x^7 + 7x^6(-y) + 21x^5(-y)^2 + 35x^4(-y)^3 \\ &\qquad + 35x^3(-y)^4 + 21x^2(-y)^5 + 7x(-y)^6 + (-y)^7. \\ &= x^7 - 7x^6y + 21x^5y^2 - 35x^4y^3 + 35x^3y^4 - 21x^2y^5 + 7xy^6 - y^7.\end{aligned}$$

We conclude this section with the following observation about the expansion of $(a + b)^n$, Pascal's triangle, and the formula for $\binom{n}{k}$.

Since the elements of the nth row of Pascal's triangle are the coefficients of $(a + b)^n$, we observe that:

(1) The kth element of the nth row of Pascal's triangle is $\binom{n}{k - 1}$.

Example 7. Verify that statement (1) is true for (a) $n = 3$ and $k = 2$; (b) $n = 5$ and $k = 3$.

Solution

(a) From Figure 12.2 we see that the second element from the left in row 3 is 3. Also, $\binom{3}{2 - 1} = \binom{3}{1} = \dfrac{3!}{1!2!} = 3.$

(b) From Figure 12.2 we see that starting from the left, the third element in row 5 is 10. Furthermore,

$$\binom{5}{3 - 1} = \binom{5}{2} = \frac{5!}{2!3!} = 10.$$

EXERCISES 12.5

For Exercises 1–6, determine the coefficient of the indicated term of the expansion. See Example 1.

1. For $(a + b)^6$, what is the coefficient of a^4b^2?

2. For $(a + b)^5$, what is the coefficient of a^2b^3?

3. For $(a + b)^{11}$, what is the coefficient of a^5b^6?

4. For $(a + b)^{11}$, what is the coefficient of a^7b^4?

5. For $(a + b)^{15}$, what is the coefficient of a^2b^{13}?

6. For $(a + b)^{15}$, what is the coefficient of $a^{13}b^2$?

For Exercises 7–18, expand the binomial. See Examples 2, 3, and 6.

7. $(a + b)^5$ **8.** $(a + b)^6$ **9.** $(a + b)^7$ **10.** $(a + b)^8$

11. $(a - b)^6$ **12.** $(a - b)^7$ **13.** $(x + 3)^4$ **14.** $(x + 2)^5$

15. $(2x - 1)^5$ **16.** $(2x - 1)^4$ **17.** $(3x - 4)^3$ **18.** $(3x - 2)^5$

For Exercises 19 and 20, see Example 4.

19. In the expansion of $(a + b)^9$ what is the coefficient of **(a)** The sixth term? **(b)** The eighth term?

20. In the expansion of $(a + b)^{20}$ what is the coefficient of **(a)** The third term? **(b)** The eighteenth term?

For Exercises 21–30, see Example 5.

21. What is the sixth term in the expansion of $(x + 2)^{10}$?

22. What is the fourth term in the expansion of $(y + 3)^{13}$?

23. What is the fifth term in the expansion of $(w - 2)^{15}$?

24. What is the third term in the expansion of $(z - 3)^{15}$?

25. Write the first three terms of the expansion of $(x - 1)^{14}$.

26. Write the first three terms of the expansion of $(x + 1)^{16}$.

27. Write the first four terms of the expansion of $(2x + 1)^7$.

28. Write the first two terms of the expansion of $(3x + 1)^4$.

29. Write the first two terms of the expansion of $(3y - 2)^4$.

30. Write the first two terms of the expansion of $(2z + 3)^4$.

For Exercises 31–34, use the formula $\binom{n}{k-1}$ *to find the* k*th element in the* n*th row of Pascal's Triangle. See Example 7.*

31. The sixth element in the 13th row.

32. The seventh element in the 17th row.

33. The third element in the 50th row.

34. The third element in the 45th row.

Key Words

Arithmetic sequence (progression)
  Common difference
Binomial coefficient
Binomial expansion
Binomial formula
Combination
Factorial ($n!$)
Geometric sequence (progression)
  Common ratio
Multiplication principle
Pascal's triangle
Permutation
Sequence
  Finite
  Infinite
Series
  Finite
  Infinite
Summation notation
  Index of summation
  Sigma
Terms of a sequence
  First
  General
  nth

CHAPTER 12 REVIEW

SECTION 12.1

Write out the first four terms of each sequence ($n > 1$) with the given nth term.

1. $a_n = n^4$ **2.** $b_n = \dfrac{1}{n+2}$ **3.** $c_n = (-2)^n$

Evaluate each sum.

4. $\sum_{i=1}^{3} 5i$ **5.** $\sum_{k=1}^{4} (2k+1)$ **6.** $\sum_{j=1}^{3} (\sqrt{3})^{2j}$

Write without summation notation.

7. $\sum_{i=1}^{n} i^2$ **8.** $\sum_{k=1}^{n} \dfrac{1}{k}$ **9.** $\sum_{k=2}^{5} x^k$

SECTION 12.2

For each arithmetic sequence, find the common difference d and the next term.

10. 3, 8, 13, 18, 23, . . . **11.** 10, 7, 4, 1, −2, . . . **12.** 6, 10, 14, 18, 22, . . .

For each arithmetic sequence with first term a_1 and common difference d, find the indicated term.

13. $a_1 = 5,\ d = 4;\ a_{11}$ **14.** $a_1 = -6,\ d = 3;\ a_{12}$ **15.** $a_1 = -4,\ d = -3;\ a_{21}$

16. Sum the arithmetic series

$$2 + 4 + 6 + 8 + \cdots + 60.$$

Solve the following word problem.

17. A woman executive is employed at a starting salary of $60,000 per year with an annual raise of $3000 per year. What will her total earnings in wages be in 10 years if she continues in this salary arrangement?

SECTION 12.3

Find the common ratio of each geometric sequence and the next term.

18. 5, 30, 180, . . . **19.** $\frac{1}{10}, \frac{1}{50}, \frac{1}{250}, \ldots$ **20.** $-4, 8, -16, \ldots$

For each geometric sequence with first term a and common ratio r, find the term indicated.

21. $a = 64,\ r = \frac{1}{4};\ a_6$ **22.** $a = \frac{2}{27},\ r = 3;\ a_7$

Sum each geometric series.

23. $1 + \frac{1}{3} + \frac{1}{27} + \frac{1}{81} + \frac{1}{243}$ **24.** $\sum_{n=0}^{\infty} (\frac{3}{5})^n$

Solve the following word problem.

25. A ball is dropped from a height of 128 feet and always rebounds one-half the height it falls. Find the height it reaches at the top of the seventh bounce.

SECTION 12.4

Solve each word problem. A calculator may prove useful.

26. From a collection of 8 shirts and 5 ties, how many ways can 1 shirt and 1 tie be selected?

27. At a neighborhood pet show there were 12 dogs, 7 cats, and 5 hamsters. In how many ways can the judges select 3 winners at the show with 1 winner from each type of pet?

28. How many license plates can be made consisting of two letters of the alphabet followed by two different nonzero digits? Assume that the letters may be the same.

29. Evaluate.
(a) ${}_7P_3$ **(b)** ${}_6P_2$ **(c)** ${}_{11}P_4$

30. Using the letters of the word EQUATION, how many different arrangements of distinct letters can be formed if each arrangement contains
(a) Four letters? **(b)** Six letters?

31. Evaluate.
(a) ${}_7C_3$ (b) ${}_6C_2$ (c) ${}_{11}C_4$

Solve each word problem.

32. How many committees of 2 people can be formed from a group of 8 people?

33. How many different unordered 7-card hands can be selected from a deck of 52 cards?

SECTION 12.5

34. What is the coefficient of a^3b^3 in the expansion of $(a + b)^6$?

35. Expand $(a + b)^5$.

36. Expand $(x - 3)^4$.

37. In the expansion of $(a + b)^8$, what is the coefficient of the seventh term?

38. What is the fifth term in the expansion of $(x - 1)^9$?

39. Write the first four terms of the expansion of $(x - 1)^{11}$.

40. Use the formula for $\binom{n}{k-1}$ to find the third element in the twenty-second row of Pascal's triangle.

CHAPTER 12 TEST

For Problems 1–5, each of the sequences is either an arithmetic or geometric sequence. Find the fifth term of the sequence.

1. 4, 8, 12, 16, . . . **2.** 4, −8, 16, −32, . . . **3.** −7, −12, −17, −22, . . .

4. $\frac{1}{2}, \frac{1}{8}, \frac{1}{32}, \frac{1}{128}$ **5.** $\frac{2}{3}, 1, \frac{4}{3}, \frac{5}{3}, \ldots$

For Problems 6 and 7, for each arithmetic sequence with first term a_1 and common difference d, find the indicated term.

6. $a_1 = 10, d = 3; a_{13}$ **7.** $a_1 = -7, d = -3; a_8$

8. Evaluate the following sum.

$$\sum_{k=1}^{40} (2k - 1)$$

For Problems 9 and 10, for each geometric sequence with first term a and common ratio r, find the term indicated.

9. $a = 6, r = \frac{1}{2}; a_{10}$ **10.** $a = 81, r = \frac{1}{3}; a_8$

For Problems 11 and 12, evaluate the sum.

11. $\sum_{k=1}^{6} (\frac{1}{3})^{k-1}$ **12.** $\sum_{n=0}^{\infty} (\frac{1}{6})^{n}$

Solve Problems 13 and 14.

13. In the first year of his retirement a man saves \$150 from his pension. In each succeeding year of his retirement he saves \$25 more than the year before. How much will he have saved at the end of 10 years?

14. A housewife puts \$.20 in a cookie jar at the beginning of the month and doubles the amount that she puts in every month thereafter. How much will she have in the cookie jar at the end of one year?

For Problems 15–26, a calculator may prove useful.

15. From a collection of 5 pairs of shoes and 14 pairs of socks, how many ways can 1 pair of socks and 1 pair of shoes be selected?

16. For promotional purposes a store is selecting some customers to win three prizes each. The three prizes must be selected as follows: one from category A, one from category B, and one from category C. In how many ways can the three prizes be awarded if there are 12 prizes in category A, 9 prizes in category B, and 4 prizes in category C?

17. Evaluate.
(a) ${}_8P_2$ **(b)** ${}_4P_4$ **(c)** ${}_{10}P_4$

18. Using the letters of the word MUSIC, how many different arrangements of different letters can be formed if each arrangement contains
(a) Two letters? **(b)** Four letters? **(c)** Five letters?

19. Evaluate.
(a) ${}_8C_2$ **(b)** ${}_4C_4$ **(c)** ${}_{10}C_4$

20. How many committees of 3 people can be formed from a group of 12 people?

21. What is the coefficient of a^4b^3 in the expansion of $(a + b)^7$?

22. Expand $(a + b)^6$.

23. Expand $(x - 2)^4$.

24. In the expansion of $(a + b)^9$, what is the coefficient of the sixth term?

25. Write the first four terms of the expansion of $(x - 1)^{13}$.

26. Use the formula for $\binom{n}{k-1}$ to find the fifth element in the thirtieth row of Pascal's triangle.

APPENDIX A
Synthetic Division

We now discuss an abbreviated procedure for dividing one polynomial by a polynomial of the form $x - r$. The procedure is called *synthetic division*. We begin by reviewing polynomial division as discussed in Section 4.2.

In order to divide $3x^4 - 50x^2 + 10x + 15$ by $x - 4$, we write the dividend as $3x^4 + 0x^3 - 50x^2 + 10x + 15$ to account for all the powers of x. Next we proceed as follows:

$$
\begin{array}{r l}
 & 3x^3 + 12x^2 - 2x + 2 \\
x - 4 \,\big) & 3x^4 + 0x^3 - 50x^2 + 10x + 15 \\
 & \underline{3x^4 - 12x^3} \\
 & \quad 12x^3 - 50x^2 \\
 & \quad \underline{12x^3 - 48x^2} \\
 & \qquad - 2x^2 + 10x \\
 & \qquad \underline{- 2x^2 + 8x} \\
 & \qquad\qquad 2x + 15 \\
 & \qquad\qquad \underline{2x - 8} \\
 & \text{remainder} \longrightarrow 23.
\end{array}
$$

Thus $(3x^4 - 50x^2 + 10x + 15) \div (x - 4)$ is $3x^3 + 12x^2 - 2x + 2$, with a remainder of 23.

We now repeat the division above, but this time we *do not write the variables*. Thus this step is similar to the way in which we solved systems of linear equations using the augmented matrix as discussed in Section 9.8. We have highlighted some numbers (coefficients) with squares, □, or parentheses, ( ), for the purpose of future reference.

$$
\begin{array}{rrr|rrrrr}
 & & & 3 & 12 & -2 & 2 & \\ \cline{4-8}
1 & -4 &) & 3 & 0 & -50 & 10 & 15 \\
 & & & \boxed{3} & -12 & & & \\ \cline{4-5}
 & & & & 12 & (-50) & & \\
 & & & & \boxed{12} & -48 & & \\ \cline{5-6}
 & & & & & -2 & (10) & \\
 & & & & & \boxed{-2} & 8 & \\ \cline{6-7}
 & & & & & & 2 & (15) \\
 & & & & & & \boxed{2} & -8 \\ \cline{8-8}
 & & & & & \multicolumn{2}{r}{\text{remainder} \longrightarrow} & 23.
\end{array}
$$

We now draw attention to the following three facts about the preceding division in which the variables were omitted.

1. The numbers in the squares are the same as the numbers immediately above them.
2. The numbers in the squares are the same as the top line of numbers. That is, the numbers in the squares are the coefficients of the quotient.
3. The numbers in parentheses are the same as the numbers in the dividend directly above them.

To avoid all this repetition, we rewrite the division problem without the numbers in the squares or parentheses.

$$
\begin{array}{rrr|rrrrr}
 & & & 3 & 12 & -2 & 2 & \\ \cline{4-8}
1 & -4 &) & 3 & 0 & -50 & 10 & 15 \\
 & & & & -12 & & & \\ \cline{4-5}
 & & & & 12 & & & \\
 & & & & & -48 & & \\ \cline{5-6}
 & & & & & -2 & & \\
 & & & & & & 8 & \\ \cline{6-7}
 & & & & & & 2 & \\
 & & & & & & & -8 \\ \cline{7-8}
 & & & & & \multicolumn{2}{r}{\text{remainder} \longrightarrow} & 23.
\end{array}
$$

In this form each line below the dividend contains exactly one term. Thus we could condense the problem to four lines, as follows:

$$
\begin{array}{lrrr|rrrrr}
(0) & & & & 3 & 12 & -2 & 2 & \\ \cline{5-9}
(1) & 1 & -4 &) & 3 & 0 & -50 & 10 & 15 \\
(2) & & & & & -12 & -48 & 8 & -8 \\ \cline{5-9}
(3) & & & & & 12 & -2 & 2 & 23
\end{array}
$$

When using synthetic division the divisor has a form such as $x - 4$. Thus, since the leading coefficient is always 1, we may delete it. Also note that if we introduce a 3 as the leading element in line (3), then line (3) with the exception of the last number, gives us the coefficients of the quotient. The last number on line (3) is the remainder. Therefore, we could delete line (0). Thus we may rewrite the problem in the following manner.

		column 1	column 2	column 3	column 4	column 5
(1)	$-4\rfloor$	3	0	−50	10	15
(2)			−12	−48	8	−8
(3)		3	12	− 2	2	23

In this form notice that the numbers in line (2) are obtained by multiplying the number in the preceding column of line (3) by −4. Also, the numbers in line (3) are obtained by subtracting the numbers in line (2) from the numbers in line (1). Finally, note that if we replaced the number −4 with 4, we would change the sign of each number in line (2). Thus the numbers in line (3) would be obtained by adding lines (1) and (2). If we make these changes, the problem is written as follows:

(1)	$4\rfloor$	3	0	−50	10	15
(2)			12	48	−8	8
(3)		3	12	−2	2	23

This shortened arrangement of the computations associated with polynomial division is called *synthetic division*.

The following steps summarize the synthetic division procedure for dividing a polynomial $P(x)$ by $x - r$. The polynomial $P(x)$ is the dividend, $x - r$ is the divisor, and r is called the *multiplier*. The work is arranged in three lines, numbered (1), (2), and (3) with the multiplier r located at the left of line (1).

Step 1. In line (1) write the coefficients of $P(x)$ in order of descending powers of the variable with the multiplier separated at the left. Remember to include a 0 coefficient for any missing powers.

Step 2. Write the first coefficient of $P(x)$ in line (3), directly below its position in line (1).

Step 3. Multiply r and the number in column 1 of line (3). Place this product in column 2 of line (2).

Step 4. Add the numbers in column 2 of lines (1) and (2). Place the sum in column 2 of line (3).

Step 5. Repeat Steps 3 and 4 for each of the remaining columns. Thus the next step is to multiply r and the number in column 2 of line (3) and place the result in column 3 of line (2).

Step 6. The *last number* in line (3) is the *remainder*. The preceding numbers in line (3) are the *coefficients of the quotient*. Note that the quotient will be a polynomial with degree one less than the degree of the dividend.

The following examples illustrate the process of synthetic division using the preceding steps.

Example 1. Use synthetic division to divide $2x^3 + 7x^2 - 5x + 7$ by $x + 2$.

Solution: To use synthetic division, the divisor must be in the form $x - r$. To get $x + 2$ in this form, we rewrite it as $x - (-2)$. Thus the multiplier $r = -2$. Now apply the six steps as listed above. We begin with steps 1 and 2. Thus we write the following:

$$\begin{array}{lr|rrrr}
(1) & -2 & 2 & 7 & -5 & 7 \\
(2) & & & & & \\ \hline
(3) & & 2 & & &
\end{array}$$

Next we place the product of the multiplier -2 and the number in column 1 of line (3) in column 2 of line (2). Thus we now have

$$\begin{array}{lr|rrrr}
(1) & -2 & 2 & 7 & -5 & 7 \\
(2) & & & -4 & & \\ \hline
(3) & & 2 & & &
\end{array}$$

Adding the numbers in column 2 (step 4), we get

$$\begin{array}{lr|rrrr}
(1) & -2 & 2 & 7 & -5 & 7 \\
(2) & & & -4 & & \\ \hline
(3) & & 2 & 3 & &
\end{array}$$

If we continue in this manner, the end result is

$$\begin{array}{lr|rrrr}
(1) & -2 & 2 & 7 & -5 & 7 \\
(2) & & & -4 & -6 & 22 \\ \hline
(3) & & 2 & 3 & -11 & 29
\end{array}$$

Thus the quotient is $2x^2 + 3x - 11$ and the remainder is 29.

Example 2. Use synthetic division to divide $x^5 + 3x^4 - 100x^2 - 540x + 210$ by $x - 5$.

Solution: The multiplier is 5. Since there is no x^3 term, we must include a zero coefficient for this missing term. Thus we have the following:

$$\begin{array}{r|rrrrrr}
5 & 1 & 3 & 0 & -100 & -540 & 210 \\
 & & 5 & 40 & 200 & 500 & -200 \\ \hline
 & 1 & 8 & 40 & 100 & -40 & 10
\end{array}$$

Therefore, the quotient is $x^4 + 8x^3 + 40x^2 + 100x - 40$ and the remainder is 10.

EXERCISES

For Exercises 1–20, use synthetic division to find the quotient and remainder. See Examples 1 and 2.

1. $(x^2 + 8x + 7) \div (x - 1)$

2. $(x^2 + 13x + 5) \div (x - 1)$

3. $(3x^2 - 5x + 12) \div (x - 2)$

4. $(2x^2 + 11x - 1) \div (x - 2)$

5. $(2x^2 + 5x - 47) \div (x - 4)$

6. $(3x^2 - 7x + 1) \div (x - 3)$

7. $(x^2 + 11x + 15) \div (x + 1)$

8. $(x^2 + 7x + 1) \div (x + 1)$

9. $(x^3 + 4x^2 + 7x + 6) \div (x - 2)$

10. $(x^3 + 5x^2 + x + 2) \div (x - 3)$

11. $(2x^3 - 10x^2 - 31x + 17) \div (x - 7)$

12. $(4x^3 - 23x^2 + 5x - 70) \div (x - 6)$

13. $(2x^3 - 21x + 10) \div (x - 3)$

14. $(5x^3 - 17x + 1) \div (x - 2)$

15. $(2x^4 - 13x^3 + 16x^2 + 1) \div (x - 5)$

16. $(3x^4 - 15x^3 + 47x + 7) \div (x - 4)$

17. $(2x^6 - 17x^4 + x^2 + 31x) \div (x + 3)$

18. $(3x^6 - 50x^4 + 35x^2 + 10x) \div (x + 4)$

19. $(x^6 + 2x - 1) \div (x + 1)$

20. $(x^6 + 2x^5 - 3x + 1) \div (x + 1)$

APPENDIX B

The Remainder Theorem and the Factor Theorem

When we divide the polynomial $P(x) = 3x^3 - 11x^2 - 5x + 6$ by $x - 4$, the quotient is $3x^2 + x - 1$ and the remainder is 2. That is, using either synthetic division or regular polynomial division as shown below, we have

$$
\begin{array}{r|rrrr}
 & 3x^2 & +\ x & -\ 1 & \\
\hline
x - 4 & 3x^3 & -\ 11x^2 & -\ 5x & +\ 6 \\
 & 3x^3 & -\ 12x^2 & & \\
\cline{2-3}
 & & x^2 & -\ 5x & \\
 & & x^2 & -\ 4x & \\
\cline{3-4}
 & & & -\ x & +\ 6 \\
 & & & -\ x & +\ 4 \\
\cline{4-5}
 & & & \text{remainder} \longrightarrow & 2.
\end{array}
$$

Notice also that

$$
\begin{aligned}
P(4) &= 3(4)^3 - 11(4)^2 - 5(4) + 6 \\
&= 192 - 176 - 20 + 6 \\
&= 2.
\end{aligned}
$$

Note that the remainder 2 is equal to $P(4)$.

The following theorem states that this equality is true in general when we divide a polynomial $P(x)$ by $(x - r)$.

The Remainder Theorem

> If the polynomial $P(x)$ is divided by $x - r$, the remainder R is equal to $P(r)$. That is, $R = P(r)$.

Proof. Let $Q(x)$ denote the resulting quotient when $P(x)$ is divided by $(x - r)$ and let R be the remainder. Thus

$$(1) \qquad P(x) = Q(x)(x - r) + R.$$

The equation given by (1) is true for all values of x. In particular, it is true when $x = r$. That is,

$$\begin{aligned} P(r) &= (r - r)Q(r) + R \\ &= 0 \cdot Q(r) + R \\ &= 0 + R \\ &= R. \end{aligned}$$

Thus $P(r) = R$.

Example 1. Use the remainder theorem to determine the remainder when

$$P(x) = 3x^3 - 5x - 8$$

is divided by
(a) $x - 2$; (b) $x + 1$.

Solution

(a) $R = P(2) = 3(2)^3 - 5(2) - 8$
$= 24 - 10 - 8 = 6$
Thus the remainder is 6.

(b) $R = P(-1) = 3(-1)^3 - 5(-1) - 8$
$= -3 + 5 - 8 = -6$
Thus the remainder is -6.

The next two theorems are very useful for finding factors of the form $(x - a)$ of a polynomial $P(x)$, where a is an integer. We state the first theorem without a proof.

Theorem 1

If a polynomial with leading coefficient 1,

$$P(x) = x^n + a_{n-1}x^{n-1} + \cdots + a_2x^2 + a_1x + a_0$$

has integer coefficients and $(x - a)$ is a factor of $P(x)$, where a is an integer, then a is an integer factor of the constant a_0.

Example 2. Use Theorem 1 to find all possible factors of the form $(x - a)$ for each polynomial $P(x)$.
(a) $P(x) = x^3 + 3x^2 - 4x + 3$. (b) $P(x) = x^5 - 7x^3 + x + 10$.

Solution

(a) The possible integer factors of 3 are 1, 3, -1, and -3. Thus the *possible* factors of $P(x)$ that have the form $(x - a)$ are $(x - 1)$, $(x - 3)$, $(x + 1)$, and $(x + 3)$.

(b) The possible factors of 10 are ± 1, ± 2, ± 5, and ± 10. Thus the possible factors of $P(x)$ that have the form $(x - a)$ are $(x - 1)$, $(x + 1)$, $(x - 2)$, $(x + 2)$, $(x - 5)$, $(x + 5)$, $(x - 10)$, and $(x + 10)$.

The next theorem follows immediately from the remainder theorem.

The Factor Theorem

> (a) If the polynomial $(x - r)$ is a factor of the polynomial $P(x)$, then $P(r) = 0$.
> (b) If $P(r) = 0$, then $(x - r)$ is a factor of the polynomial $P(x)$.

Proof. The remainder theorem states that when a polynomial $P(x)$ is divided by $(x - r)$, the remainder $R = P(r)$. That is, given that

$$P(x) = (x - r)Q(x) + R$$

we have

(1) $$P(x) = (x - r)Q(x) + P(r), \qquad \text{since } R = P(r).$$

(a) If $(x - r)$ is a factor of $P(x)$, then $P(x) = (x - r)Q(x)$ and the remainder $R = 0$; that is, $P(r) = 0$.

(b) If $P(r) = 0$, then since $P(r)$ is the remainder R, it follows from (1) that

$$P(x) = (x - r)Q(x).$$

Therefore, $(x - r)$ is a factor of $P(x)$.

Example 3. Use the factor theorem to show that $x - 3$ is a factor of

$$2x^3 - 7x^2 + 4x - 3.$$

Solution: Let $P(x) = 2x^3 - 7x^2 + 4x - 3$. Thus

$$\begin{aligned} P(3) &= 2(3)^3 - 7(3)^2 + 4(3) - 3 \\ &= 54 - 63 + 12 - 3 \\ &= 0. \end{aligned}$$

Since $P(3) = 0$, the factor theorem tells us that $(x - 3)$ is a factor of $P(x)$.

Example 4. Use the factor theorem to determine if $x + 1$ is a factor of

$$x^3 + 3x^2 + 4x + 5.$$

Solution: Let $P(x) = x^3 + 3x^2 + 4x + 5$. To check if $(x + 1)$ is a factor of $P(x)$, we need to evaluate $P(-1)$. Thus

$$\begin{aligned} P(-1) &= (-1)^3 + 3(-1)^2 + 4(-1) + 5 \\ &= -1 + 3 - 4 + 5 \\ &= 3. \end{aligned}$$

Since $P(-1) \neq 0$, $(x + 1)$ is not a factor of $P(x)$.

EXERCISES

For Exercises 1–10, use the remainder theorem to determine the remainder of the indicated division. See Example 1.

1. $(x^2 + 5x - 4) \div (x - 1)$
2. $(x^2 + 7x - 5) \div (x - 1)$
3. $(2x^2 - 3x - 4) \div (x + 1)$
4. $(3x^2 - x - 2) \div (x + 1)$
5. $(4x^3 + x^2 - 29) \div (x - 2)$
6. $(5x^3 - 4x^2 - 17) \div (x - 2)$
7. $(5x^4 - 2x^2 - 300) \div (x - 3)$
8. $(2x^4 - 25x - 80) \div (x - 3)$
9. $(14x^5 + 3x^2 - 15) \div (x - 1)$
10. $(19x^6 - 4x^3 - 13) \div (x - 1)$

For Exercises 11–20, use Theorem 1 to find all possible factors of the form $(x - a)$ for each polynomial. See Example 2.

11. $P(x) = x^2 + 6x + 5$
12. $P(x) = x^2 + 8x + 7$
13. $P(x) = x^2 - x - 2$
14. $P(x) = x^2 + 5x - 11$
15. $P(x) = x^3 - 1$
16. $P(x) = x^3 + 1$
17. $P(x) = x^3 + x^2 + 7$
18. $P(x) = x^3 - x + 13$
19. $P(x) = x^4 + x^3 - 2x^2 + 11$
20. $P(x) = x^4 - x^2 + 3x - 7$

For Exercises 21–30, use the factor theorem to show that the first polynomial is a factor of the second polynomial. See Example 3.

21. $x - 1;\ x^2 - 3x + 2$
22. $x - 1;\ x^2 + 5x - 6$
23. $x + 1;\ 2x^2 + 5x + 3$
24. $x + 1;\ 5x^2 + 7x + 2$
25. $x + 2;\ x^3 - x^2 - 7x - 2$
26. $x - 2;\ x^3 + x^2 - 5x - 2$
27. $x + 2;\ 2x^3 - 7x + 2$
28. $x + 3;\ 2x^3 - 15x + 9$
29. $x - 3;\ x^4 - 2x^3 - 8x - 3$
30. $x - 2;\ x^5 - x^3 - 11x - 2$

For Exercises 31–40, use factor theorem to determine if the first polynomial is a factor of the second polynomial. See Example 4.

31. $x - 2;\ -3x^2 + x + 10$
32. $x - 3;\ x^2 + x - 12$
33. $x + 1;\ x^3 + x^2 + x - 1$
34. $x + 1;\ x^3 + x^2 - x + 1$
35. $x - 1;\ x^4 + x^2 + 11$
36. $x - 2;\ x^5 + x^3 + 4$
37. $x - 3;\ 2x^2 - 5x - 3$
38. $x - 2;\ 5x^2 - 11x + 2$
39. $x + 2;\ x^3 + 2x^2 + 2x + 4$
40. $x + 3;\ x^3 + 3x^2 + 5x + 15$

APPENDIX C

Table of Common Logarithms

x	0	1	2	3	4	5	6	7	8	9
1.0	.0000	.0043	.0086	.0128	.0170	.0212	.0253	.0294	.0334	.0374
1.1	.0414	.0453	.0492	.0531	.0569	.0607	.0645	.0682	.0719	.0755
1.2	.0792	.0828	.0864	.0899	.0934	.0969	.1004	.1038	.1072	.1106
1.3	.1139	.1173	.1206	.1239	.1271	.1303	.1335	.1367	.1399	.1430
1.4	.1461	.1492	.1523	.1553	.1584	.1614	.1644	.1673	.1703	.1732
1.5	.1761	.1790	.1818	.1847	.1875	.1903	.1931	.1959	.1987	.2014
1.6	.2041	.2068	.2095	.2122	.2148	.2175	.2201	.2227	.2253	.2279
1.7	.2304	.2330	.2355	.2380	.2405	.2430	.2455	.2480	.2504	.2529
1.8	.2553	.2577	.2601	.2625	.2648	.2672	.2695	.2718	.2742	.2765
1.9	.2788	.2810	.2833	.2856	.2878	.2900	.2923	.2945	.2967	.2989
2.0	.3010	.3032	.3054	.3075	.3096	.3118	.3139	.3160	.3181	.3201
2.1	.3222	.3243	.3263	.3284	.3304	.3324	.3345	.3365	.3385	.3404
2.2	.3424	.3444	.3464	.3483	.3502	.3522	.3541	.3560	.3579	.3598
2.3	.3617	.3636	.3655	.3674	.3692	.3711	.3729	.3747	.3766	.3784
2.4	.3802	.3820	.3838	.3856	.3874	.3892	.3909	.3927	.3945	.3962
2.5	.3979	.3997	.4014	.4031	.4048	.4065	.4082	.4099	.4116	.4133
2.6	.4150	.4166	.4183	.4200	.4216	.4232	.4249	.4265	.4281	.4298
2.7	.4314	.4330	.4346	.4362	.4378	.4393	.4409	.4425	.4440	.4456
2.8	.4472	.4487	.4502	.4518	.4533	.4548	.4564	.4579	.4594	.4609
2.9	.4624	.4639	.4654	.4669	.4683	.4698	.4713	.4728	.4742	.4757
3.0	.4771	.4786	.4800	.4814	.4829	.4843	.4857	.4871	.4886	.4900
3.1	.4914	.4928	.4942	.4955	.4969	.4983	.4997	.5011	.5024	.5038
3.2	.5051	.5065	.5079	.5092	.5105	.5119	.5132	.5145	.5159	.5172
3.3	.5185	.5198	.5211	.5224	.5237	.5250	.5263	.5276	.5289	.5302
3.4	.5315	.5328	.5340	.5353	.5366	.5378	.5391	.5403	.5416	.5428
3.5	.5441	.5453	.5465	.5478	.5490	.5502	.5514	.5527	.5539	.5551
3.6	.5563	.5575	.5587	.5599	.5611	.5623	.5635	.5647	.5658	.5670
3.7	.5682	.5694	.5705	.5717	.5729	.5740	.5752	.5763	.5775	.5786
3.8	.5798	.5809	.5821	.5832	.5843	.5855	.5866	.5877	.5888	.5899
3.9	.5911	.5922	.5933	.5944	.5955	.5966	.5977	.5988	.5999	.6010
4.0	.6021	.6031	.6042	.6053	.6064	.6075	.6085	.6096	.6107	.6117
4.1	.6128	.6138	.6149	.6160	.6170	.6180	.6191	.6201	.6212	.6222
4.2	.6232	.6243	.6253	.6263	.6274	.6284	.6294	.6304	.6314	.6325
4.3	.6335	.6345	.6355	.6365	.6375	.6385	.6395	.6405	.6415	.6425
4.4	.6435	.6444	.6454	.6464	.6474	.6484	.6493	.6503	.6513	.6522
4.5	.6532	.6542	.6551	.6561	.6571	.6580	.6590	.6599	.6609	.6618
4.6	.6628	.6637	.6646	.6656	.6665	.6675	.6684	.6693	.6702	.6712
4.7	.6721	.6730	.6739	.6749	.6758	.6767	.6776	.6785	.6794	.6803
4.8	.6812	.6821	.6830	.6839	.6848	.6857	.6866	.6875	.6884	.6893
4.9	.6902	.6911	.6920	.6928	.6937	.6946	.6955	.6964	.6972	.6981
5.0	.6990	.6998	.7007	.7016	.7024	.7033	.7042	.7050	.7059	.7067
5.1	.7076	.7084	.7093	.7101	.7110	.7118	.7126	.7135	.7143	.7152
5.2	.7160	.7168	.7177	.7185	.7193	.7202	.7210	.7218	.7226	.7235
5.3	.7243	.7251	.7259	.7267	.7275	.7284	.7292	.7300	.7308	.7316
5.4	.7324	.7332	.7340	.7348	.7356	.7364	.7372	.7380	.7388	.7396
x	0	1	2	3	4	5	6	7	8	9

x	0	1	2	3	4	5	6	7	8	9
5.5	.7404	.7412	.7419	.7427	.7435	.7443	.7451	.7459	.7466	.7474
5.6	.7482	.7490	.7497	.7505	.7513	.7520	.7528	.7536	.7543	.7551
5.7	.7559	.7566	.7574	.7582	.7589	.7597	.7604	.7612	.7619	.7627
5.8	.7634	.7642	.7649	.7657	.7664	.7672	.7679	.7686	.7694	.7701
5.9	.7709	.7716	.7723	.7731	.7738	.7745	.7752	.7760	.7767	.7774
6.0	.7782	.7789	.7796	.7803	.7810	.7818	.7825	.7832	.7839	.7846
6.1	.7853	.7860	.7868	.7875	.7882	.7889	.7896	.7903	.7910	.7917
6.2	.7924	.7931	.7938	.7945	.7952	.7959	.7966	.7973	.7980	.7987
6.3	.7993	.8000	.8007	.8014	.8021	.8028	.8035	.8041	.8048	.8055
6.4	.8062	.8069	.8075	.8082	.8089	.8096	.8102	.8109	.8116	.8122
6.5	.8129	.8136	.8142	.8149	.8156	.8162	.8169	.8176	.8182	.8189
6.6	.8195	.8202	.8209	.8215	.8222	.8228	.8235	.8241	.8248	.8254
6.7	.8261	.8267	.8274	.8280	.8287	.8293	.8299	.8306	.8312	.8319
6.8	.8325	.8331	.8338	.8344	.8351	.8357	.8363	.8370	.8376	.8382
6.9	.8388	.8395	.8401	.8407	.8414	.8420	.8426	.8432	.8439	.8445
7.0	.8451	.8457	.8463	.8470	.8476	.8482	.8488	.8494	.8500	.8506
7.1	.8513	.8519	.8525	.8531	.8537	.8543	.8549	.8555	.8561	.8567
7.2	.8673	.8579	.8585	.8591	.8597	.8603	.8609	.8615	.8621	.8627
7.3	.8633	.8639	.8645	.8651	.8657	.8663	.8669	.8675	.8681	.8686
7.4	.8692	.8698	.8704	.8710	.8716	.8722	.8727	.8733	.8739	.8745
7.5	.8751	.8756	.8762	.8768	.8774	.8779	.8785	.8791	.8797	.8802
7.6	.8808	.8814	.8820	.8825	.8831	.8837	.8842	.8848	.8854	.8859
7.7	.8865	.8871	.8876	.8882	.8887	.8893	.8899	.8904	.8910	.8915
7.8	.8921	.8927	.8932	.8938	.8943	.8949	.8954	.8960	.8965	.8971
7.9	.8976	.8982	.8987	.8993	.8998	.9004	.9009	.9015	.9020	.9025
8.0	.9031	.9036	.9042	.9047	.9053	.9058	.9063	.9069	.9074	.9079
8.1	.9085	.9090	.9096	.9101	.9106	.9112	.9117	.9122	.9128	.9133
8.2	.9138	.9143	.9149	.9154	.9159	.9165	.9170	.9175	.9180	.9186
8.3	.9191	.9196	.9201	.9206	.9212	.9217	.9222	.9227	.9232	.9238
8.4	.9243	.9248	.9253	.9258	.9263	.9269	.9274	.9279	.9284	.9289
8.5	.9294	.9299	.9304	.9309	.9315	.9320	.9325	.9330	.9335	.9340
8.6	.9345	.9350	.9355	.9360	.9365	.9370	.9375	.9380	.9385	.9390
8.7	.9395	.9400	.9405	.9410	.9415	.9420	.9425	.9430	.9435	.9440
8.8	.9445	.9450	.9455	.9460	.9465	.9469	.9474	.9479	.9484	.9489
8.9	.9494	.9499	.9504	.9509	.9513	.9518	.9523	.9528	.9533	.9538
9.0	.9542	.9547	.9552	.9557	.9562	.9566	.9571	.9576	.9581	.9586
9.1	.9590	.9595	.9600	.9605	.9609	.9614	.9619	.9624	.9628	.9633
9.2	.9638	.9643	.9647	.9652	.9657	.9661	.9666	.9671	.9675	.9680
9.3	.9685	.9689	.9694	.9699	.9703	.9708	.9713	.9717	.9722	.9727
9.4	.9731	.9736	.9741	.9745	.9750	.9754	.9759	.9763	.9768	.9773
9.5	.9777	.9782	.9786	.9791	.9795	.9800	.9805	.9809	.9814	.9818
9.6	.9823	.9827	.9832	.9836	.9841	.9845	.9850	.9854	.9859	.9863
9.7	.9868	.9872	.9877	.9881	.9886	.9890	.9894	.9899	.9903	.9908
9.8	.9912	.9917	.9921	.9926	.9930	.9934	.9939	.9943	.9948	.9952
9.9	.9956	.9961	.9965	.9969	.9974	.9978	.9983	.9987	.9991	.9996
x	0	1	2	3	4	5	6	7	8	9

ANSWERS
ANSWERS
ANSWERS
ANSWERS
ANSWERS

ANSWERS
Odd-Numbered Exercises and Chapter Reviews and Tests

ANSWERS ANSWERS

CHAPTER 1

Exercises 1.1

1. {6, 8, 10, 12} **3.** {0, *e*} **5.** {1, 7}
7. {3, 5, 7, . . . , 87, 89} **9.** {12, 14, 16, 18, . . .}
11. $\{x \mid x$ is an even natural number between 2 and 400 inclusive$\}$
13. $\{x \mid x$ is one of the first three letters of the alphabet$\}$
15. $\{x \mid x$ is a natural number greater than 4$\}$ **17.** $\in$
19. $\notin$ **21.** $\subseteq$ **23.** $\not\subseteq$ **25.** $\subseteq$ **27.** $\subseteq$
29. $\varnothing$, {a} **31.** $\varnothing$, {0} **33.** $\varnothing$, $\{x\}$, $\{y\}$, $\{x, y\}$
35. $\varnothing$, {2}, {4}, {2, 4} **37.** $\varnothing$, $\{a\}$, $\{b\}$, $\{c\}$, $\{a, b\}$, $\{a, c\}$, $\{b, c\}$, $\{a, b, c\}$
39. $\varnothing$, {2}, {3}, {4}, {5}, {2, 3}, {2, 4}, {2, 5}, {3, 4}, {3, 5}, {4, 5}, {2, 3, 4}, {2, 3, 5}, {3, 4, 5}, {2, 4, 5}, {2, 3, 4, 5}. **41.** {2, 3, 4, 5, 6, 7, 8, 10}
43. {2} **45.** {2, 4, 6, 7, 8, 10, 13} **47.** $\varnothing$
49. {4, 6, 7, 8, 10} **51.** {2, 11, 13} **53.** Finite **55.** Finite
57. Finite **59.** Infinite **61.** True **63.** False
65. True **67.** False **69.** 4 **71.** 5

Exercises 1.2

1. **(a)** {3, 7} **(b)** {0, 3, 7} **(c)** {−6, −2} **(d)** {3, 7} **(e)** {−6, −2, 0, 3, 7}
(f) $\left\{\dfrac{-5}{2}, \dfrac{4}{3}\right\}$ **3.** **(a)** $\{-4, -.3\overline{5}, 1.414, \frac{9}{2}\}$ **(b)** $\{\sqrt{2}, \pi\}$
(c) $\{-4, -.3\overline{5}, \sqrt{2}, 1.414, \pi, \frac{9}{2}\}$ **5.** Rational **7.** Irrational
9. Rational **11.** Rational **13.** Rational **15.** True
17. False **19.** True **21.** True **23.** True **25.** True

27. 

29. 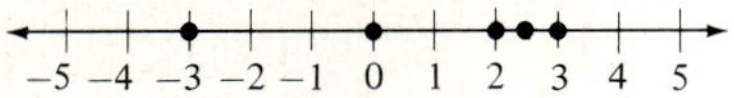

31. 10 33. −9.2 35. $\frac{1}{5}$ 37. $-\frac{1}{8}$ 39. 10 41. −4
43. −10 45. 8 47. −9 49. False 51. False

Exercises 1.3

1. The sum of x and 5.
3. The difference of x and 2.
5. Twice the sum of x and 4.
7. Nine times the difference of x and 2.
9. The sum of the product $3x$ and 1.
11. The difference of the product $7z$ and 3.
13. The quotient of the product xy by z.
15. The quotient of y by the sum of x and 2.
17. The product of the sum and difference of x and 2.

19. $2x - 3$ 21. $2y - 4$ 23. $2z + 10$ 25. $x - y + 1$ 27. $y - \dfrac{x}{2}$

29. 26 31. 24 33. 12 35. 16 37. 14 39. 56 41. 7 43. 30
45. 32 47. 9 49. 40 51. 14 53. 80 55. 12 57. 18 59. 28
61. 22 63. 3 65. 3 67. 3 69. 1 71. 2

Exercises 1.4

1. Commutative 3. Closure 5. Commutative 7. Additive inverse
9. Associative 11. Associative 13. Identity 15. Closure 17. Commutative
19. Identity 21. Inverse 23. Distributive 25. Distributive
27. Transitive or Substitution 29. Symmetric 31. Reflexive
33. Addition Property of Equality 35. Zero property 37. Substitution 39. $y + 10$
41. $8x$ 43. x 45. $21x$ 47. $z + 10$ 49. 0 51. $6x + 12$
53. $6x + 12$ 55. $8x + 5$ 57. $8x + 3$ 59. $12x + 9$ 61. $12x + 12$
63. $13x + 22$ 65. $18x$ 67. $20x$ 69. Closed 71. Not closed

Exercises 1.5

1. $<$ 3. $>$ 5. $>$ 7. $<$ 9. $<$ 11. $<$
13. x is less than or equal to 4.
15. z is greater than or equal to −2.
17. y is positive or y is greater than 0.
19. x is less than or equal to −3.

21. 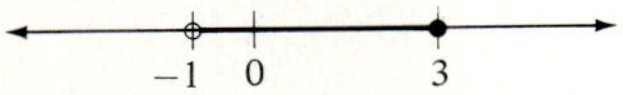

23. 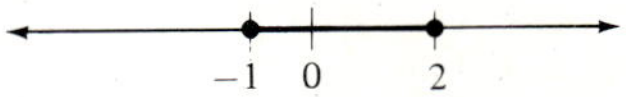

25. 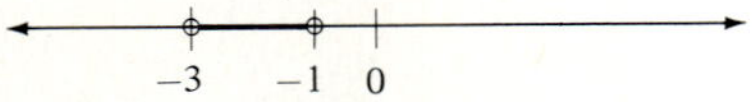

27. 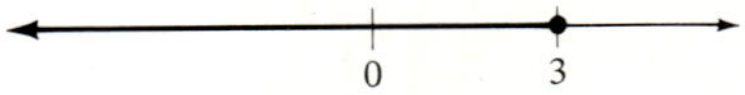

29. 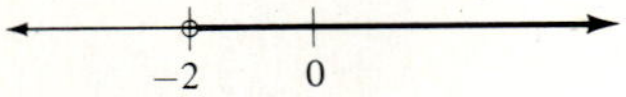

31. 

33. 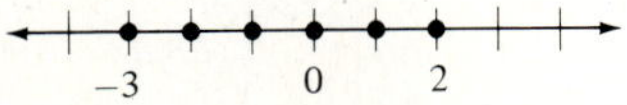

35. 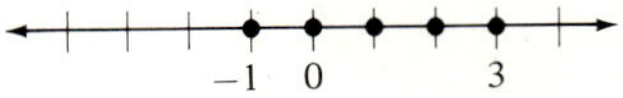

37. 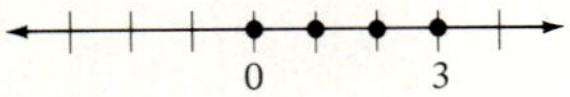

39. $>$ 41. $<$

Exercises 1.6

1. 10 3. −10 5. −4 7. −10 9. 8 11. −5 13. 4
15. −3 17. $\frac{3}{8}$ 19. −10 21. 12 23. 5 25. 1 27. 4
29. −2 31. 9 33. −6 35. 8 37. $-\frac{1}{4}$ 39. −5 41. 12
43. 20 45. −2 47. 4 49. −24 51. −25 53. $-\frac{2}{6} = -\frac{1}{3}$ 55. $\frac{1}{3}$
57. −5 59. 1 61. −4 63. −20 65. −12 67. 2 69. −2

71. 21 **73.** $\frac{4}{5}$ **75.** -17 **77.** 3 **79.** 3 **81.** 17 **83.** -17
85. True **87.** False **89.** True

Exercises 1.7

1. -16 **3.** -35 **5.** -1 **7.** -1 **9.** -30 **11.** -120 **13.** 20
15. 1 **17.** -24 **19.** 0 **21.** 18 **23.** $-\frac{1}{5}$ **25.** $-3x + 6$
27. $5z - 15$ **29.** $-x + y + z$ **31.** $-2s - 2t + 2u$ **33.** $-12t + 6u + 15v$
35. $9s - 6t + 3v$ **37.** 6 **39.** -4 **41.** -9 **43.** 4 **45.** -6
47. -14 **49.** -10 **51.** 10 **53.** -8 **55.** -1 **57.** -2
59. -4 **61.** $\frac{-3}{4}, \frac{3}{-4}$ **63.** $\frac{-4}{9}, -\frac{4}{9}$ **65.** $-\frac{5}{11}, \frac{5}{-11}$ **67.** $-\frac{7}{4}, \frac{7}{-4}$
69. 9 **71.** $\frac{5}{14}$ **73.** -1 **75.** -3 **77.** True **79.** True **81.** False

Chapter 1 Review

1. $\{5, 7, 9, 11\}$ **2.** $\{x \mid x$ is an even natural number between 6 and 90 inclusive$\}$
3. $\{p\}, \{q\}, \{r\}, \{p, q\}, \{p, r\}, \{q, r\}, \{p, q, r\}, \varnothing$ **4.** $\{-5, -3, 0, 2, 4\}$ **5.** $\{-5, 2, 4\}$
6. False **7.** True **8.** $\{\frac{3}{1}, 6\}$ **9.** $\{-3, -1\}$ **10.** $\{-3, -1, 0, \frac{3}{1}, 6\}$
11. $\{\frac{2}{3}, \frac{9}{4}\}$ **12.** Rational **13.** Irrational **14.** Rational **15.** 6 **16.** -9
17. -7.2 **18.** $\frac{1}{4}$ **19.** 8 **20.** -9 **21.** 3 **22.** $2x - 7$
23. $xy + 5$ **24.** 35 **25.** 18 **26.** 15 **27.** 30 **28.** 10 **29.** 4
30. Commutative **31.** Associative **32.** Distributive **33.** Identity
34. Symmetric **35.** Transitive or Substitution **36.** Substitution **37.** Zero property

38. $<$ **39.** $>$ **40.** $<$ **41.** $>$ **42.** 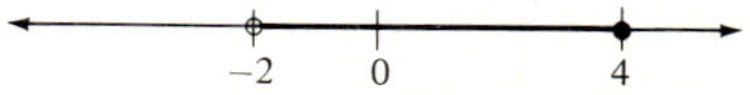

43. -4 **44.** 2 **45.** -10 **46.** 13 **47.** 10 **48.** 10 **49.** -12
50. 12 **51.** -40 **52.** $-5y + 15$ **53.** $-x + y + 4$ **54.** -9 **55.** 5
56. -4 **57.** 0 **58.** -3 **59.** -2 **60.** 2 **61.** 10

Chapter 1 Test

1. **(a)** $\{-2, -1, 0, 3, 4, 5, 6\}$ **(b)** $\{0, 3\}$

2. **(a)** $\{3, 4\}$ **(b)** $\{0, 3, 4\}$ **(c)** $\{-3, 0, 3, 4\}$ **(d)** $\left\{\frac{-5}{2}, \frac{22}{7}, 3.5, 4.\bar{2}\right\}$ **(e)** $\{-\pi, \sqrt{2}\}$

(f) $\left\{-\pi, -3, \frac{-5}{2}, 0, \sqrt{2}, 3, \frac{22}{7}, 3.5, 4.\bar{2}, 4\right\}$

3. **(a)** 5 **(b)** 5 **(c)** -7 **(d)** 3 **4.** **(a)** $x + y - 10$ **(b)** $10z + 5$ **5.** **(a)** 36 **(b)** 9 **(c)** 28
6. **(a)** Commutative **(b)** Associative **(c)** Identity **(d)** Identity **(e)** Distributive
7. **(a)** Transitive or substitution **(b)** Symmetric

8. 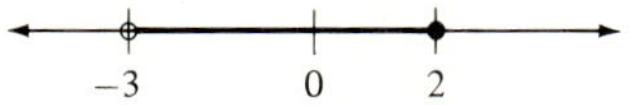

9. **(a)** -10 **(b)** -7 **(c)** 20 **(d)** 2 **(e)** 17 **(f)** -3
10. **(a)** 100 **(b)** -24 **(c)** -12 **(d)** -4 **(e)** 5 **(f)** 0 **11.** **(a)** -4 **(b)** 6 **12.** **(a)** -4 **(b)** 7
13. **(a)** $<$ **(b)** $<$

CHAPTER 2

Exercises 2.1

1. $5x$ **3.** $2z$ **5.** $3p$ **7.** $8r$ **9.** $4t$ **11.** $6t + 2$ **13.** $12v - 5$
15. $6x + 9$ **17.** $21z + 11$ **19.** $-2q - 17$ **21.** $5x + 14$ **23.** $w + 18$

25. $6p + 22$ **27.** $18r + 34$ **29.** $-27r + 21$ **31.** $-13x + 43$ **33.** $z + 1.2$
35. $.2x + 3$ **37.** $2w + 2$ **39.** $x + 1$

Exercises 2.2

1. {3} **3.** {4} **5.** {3} **7.** {4} **9.** {−2} **11.** {−5}
13. {−3} **15.** {−5} **17.** {4} **19.** {7} **21.** {3} **23.** {4}
25. {−3} **27.** {−4} **29.** {2} **31.** {20} **33.** {10} **35.** {7}
37. {−9} **39.** {−8} **41.** {4} **43.** {6} **45.** {17} **47.** {−2}
49. {5} **51.** {−6} **53.** {5} **55.** {4} **57.** {−3} **59.** {3}
61. $\mathbb{R}$ **63.** $\mathbb{R}$ **65.** { } **67.** $\mathbb{R}$ **69.** { } **71.** {−3} **73.** {.9}
75. {5} **77.** {−3.1} **79.** {−1.5}

Exercises 2.3

1. **(a)** $A = 78.5$ sq ft **(b)** $C = 31.4$ ft **3.** **(a)** $A = .50$ sq yd **(b)** $C = 2.51$ yd
5. **(a)** $A = 16.61$ sq ft **(b)** $C = 14.44$ ft **7.** 20°C **9.** 100°C **11.** 23°F
13. 95°F **15.** 140 miles **17.** 200 km **19.** $440
21. **(a)** $A = 96$ sq in., $P = 48$ in. **(b)** $A = 150$ sq in., $P = 60$ in. **23.** 324 sq ft
25. $V = 120$ cu ft, $S = 158$ sq ft **27.** $V = 87.92$ cu in., $S = 113.04$ sq in. **29.** $V = 32$ cu in.
31. **(a)** $105 **(b)** 19.2% **33.** **(a)** $197.50 **(b)** 6.1% **35.** $t = \frac{d}{r}$ **37.** $r = \frac{C}{2\pi}$
39. $b = \frac{2A}{h}$ **41.** $W = \frac{P - 2L}{2}$ **43.** $B = \frac{2A - hb}{h}$ **45.** 52 mph **47.** $3\frac{1}{2}$ hr
49. $1500 **51.** 10 ft **53.** 7 ft **55.** 21 in. **57.** 3 in. **59.** $78

Exercises 2.4

1. {−4, 4} **3.** {−13, 13} **5.** {−100, 100} **7.** { } **9.** {−5, 5}
11. $\{-\frac{1}{2}, \frac{1}{2}\}$ **13.** {−2.3, 2.3} **15.** {−6, 6} **17.** {−6, 4} **19.** {−8, 4}
21. {1, 3} **23.** {−3, 7} **25.** {−9, 1} **27.** {−5, 2} **29.** $\{-4, \frac{4}{3}\}$
31. $\{-2, \frac{12}{7}\}$ **33.** { } **35.** {−16, 4} **37.** {3, 9} **39.** $\{-10, -\frac{15}{2}\}$
41. $\{-4, -\frac{8}{3}\}$ **43.** {−2, 0} **45.** {−6, −2} **47.** {−4, 8} **49.** $\{-\frac{5}{2}, \frac{3}{2}\}$
51. $\{0, \frac{4}{5}\}$ **53.** {−5, −3} **55.** {−2, 12} **57.** $\{-\frac{4}{3}, \frac{2}{3}\}$ **59.** $\{-3, -\frac{7}{9}\}$
61. {1, 18} **63.** {−2} **65.** {2} **67.** $\{p \mid p \geq -3\}$ **69.** $\{v \mid v \geq 3\}$

EXERCISES 2.5

1. $\{x \mid x < 3\}$
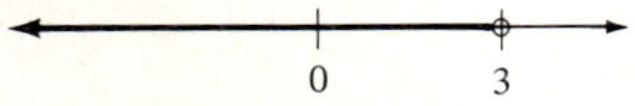

3. $\{w \mid w \geq 2\}$
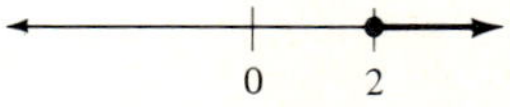

5. $\{p \mid p \leq 4\}$
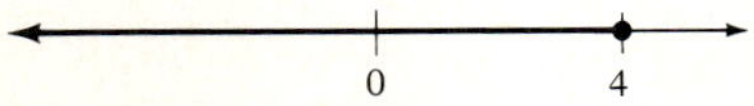

7. $\{r \mid r > -3\}$
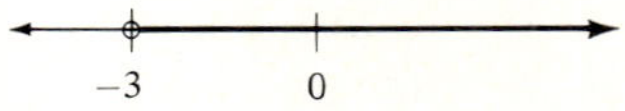

9. $\{x \mid x \geq -10\}$
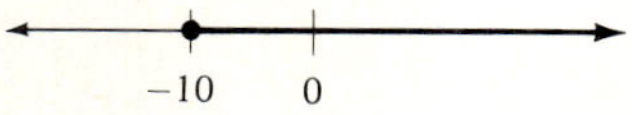

11. $\{u \mid u < \frac{2}{3}\}$
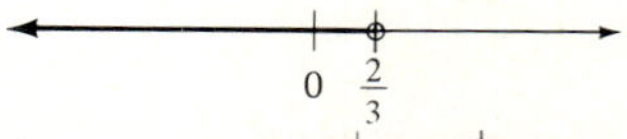

13. $\{p \mid p < 4\}$
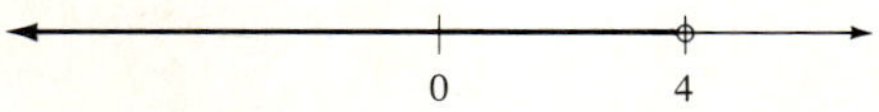

15. $\{z \mid z \leq 3\}$
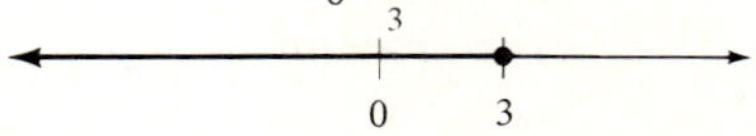

17. $\{t \mid t \leq 4\}$
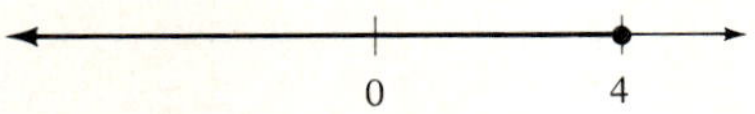

19. $\{u \mid u \geq \frac{3}{5}\}$
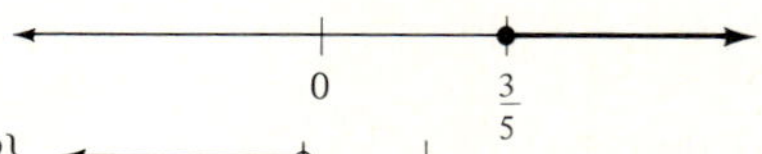

21. $\{w \mid w < 2\}$
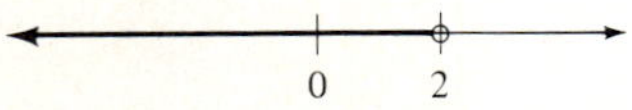

23. $\{x \mid x < -2\}$
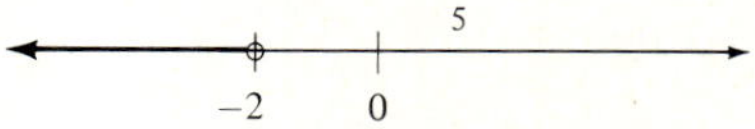

25. $\{p \mid p \geq 3\}$
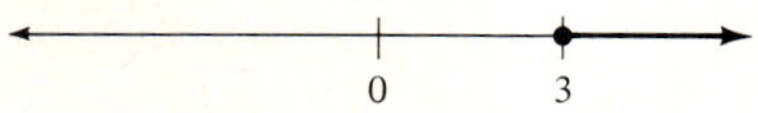

27. $\{s \mid s \geq 2\}$
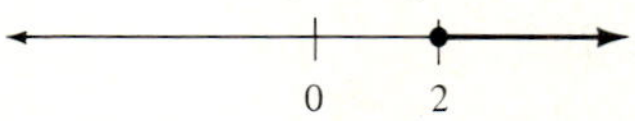

29. $\{u \mid u > 5\}$
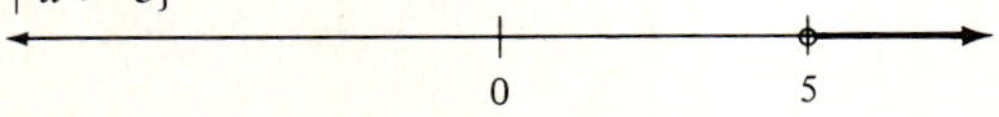

31. $\{p \mid p \leq 1\}$
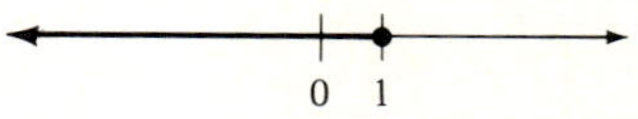

33. $\{u \mid u \geq 1\}$ 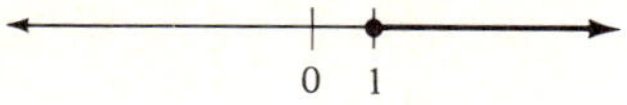

35. $\{w \mid w < 0\}$ 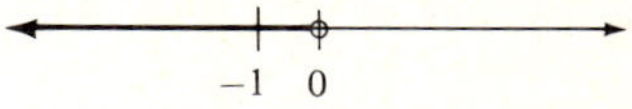

37. $\{r \mid r < 7\}$ 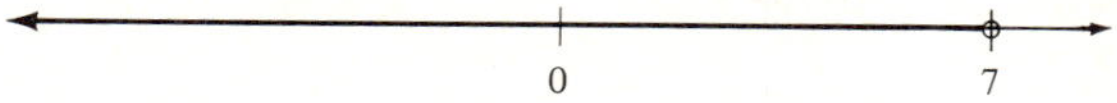

39. $\{x \mid x \leq 2\}$ 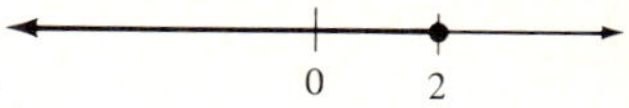

41. $\{u \mid u < 3\}$ 

43. $\{x \mid x \geq 7\}$ 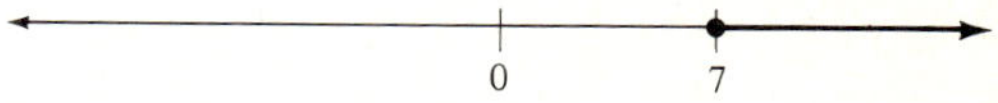

45. $\{y \mid y < 3\}$ 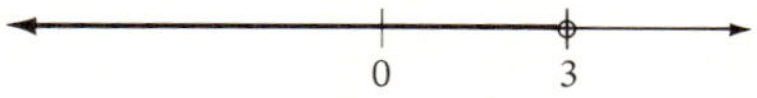

47. $\{t \mid t \leq -4\}$ 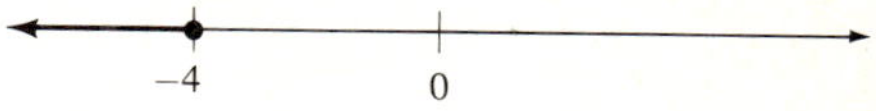

49. $\{v \mid v \leq 3\}$ 

51. $\{x \mid x \leq 4\} \cap \{x \mid x > 1\}$ 

53. $\{u \mid u > -2\} \cap \{u \mid u \leq 3\}$ 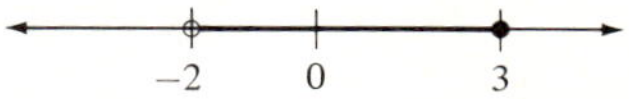

55. $\{w \mid w \geq 2\} \cap \{w \mid w \leq 4\}$ 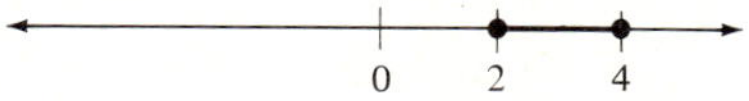

57. $\{x \mid x < 3\} \cup \{x \geq 5\}$ 

59. $\{w \mid w \geq 3\} \cup \{w \mid w \geq 1\} = \{w \mid w \geq 1\}$ 

61. $\{p \mid p > \frac{3}{2}\} \cup \{p \mid p > 0\} = \{p \mid p > 0\}$ 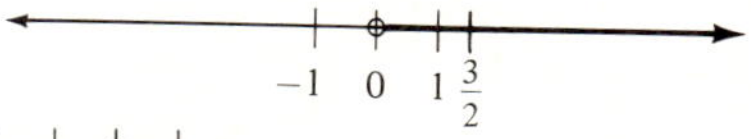

63. $\{x \mid 1 < x \leq 4\}$ 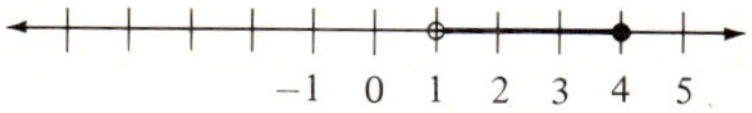

65. $\{w \mid -1 \leq w \leq 3\}$ 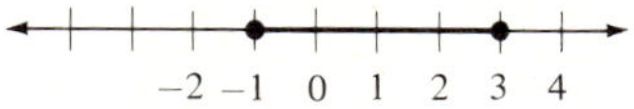

67. $\{p \mid -3 < p < -1\}$ 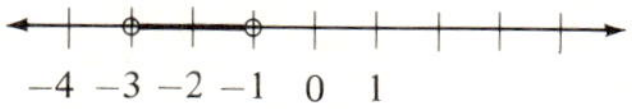

69. $\{x \mid -3 \leq x < 2\}$ 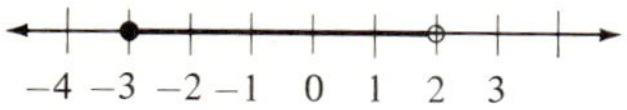

Exercises 2.6

1. $\{x \mid -5 \leq x \leq 5\}$ 

3. $\{w \mid -1 < w < 1\}$ 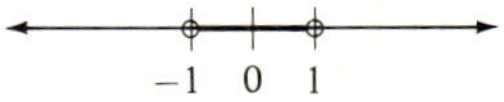

5. $\{p \mid p \leq -2 \text{ or } p \geq 2\}$ 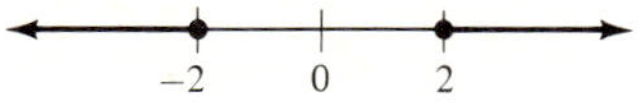

7. $\{r \mid r < -5 \text{ or } r > 5\}$ 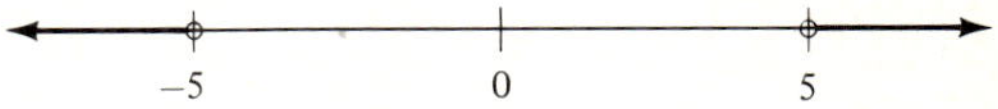

9. $\{x \mid -1 \leq x \leq 1\}$ 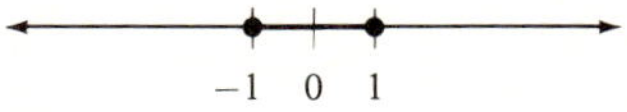

11. $\{w \mid w < -8 \text{ or } w > 8\}$ 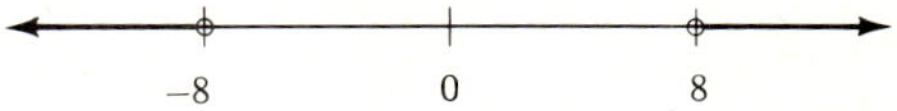

13. $\{x \mid 1 < x < 3\}$ 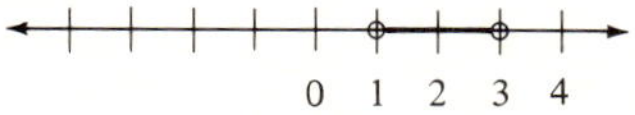

15. $\{z \mid -6 < z < 4\}$ 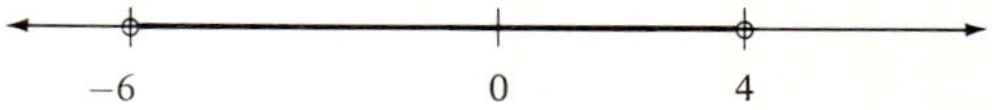

17. $\{p \mid -7 \le p \le 1\}$

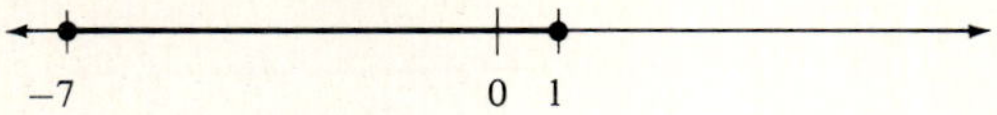

19. $\{r \mid -1 \le r \le 11\}$

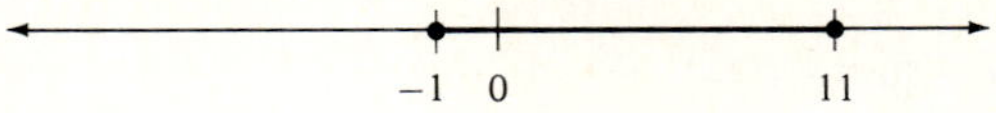

21. $\{t \mid t < -3 \text{ or } t > 1\}$

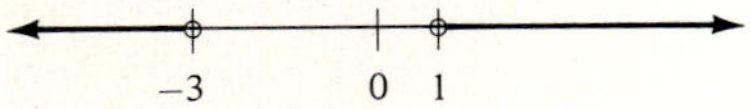

23. $\{v \mid v \le -2 \text{ or } v \ge 8\}$

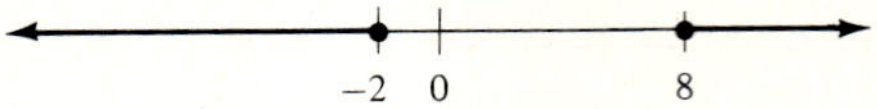

25. $\{x \mid -2 \le x \le 1\}$

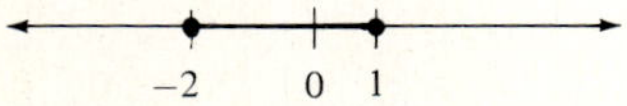

27. $\{x \mid -2 < x < \frac{8}{3}\}$

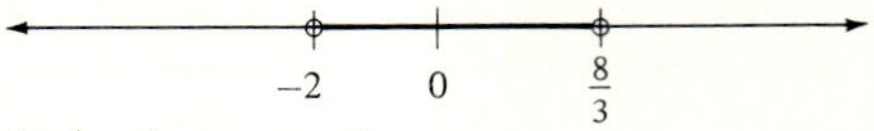

29. $\{p \mid -\frac{2}{5} < p < 0\}$

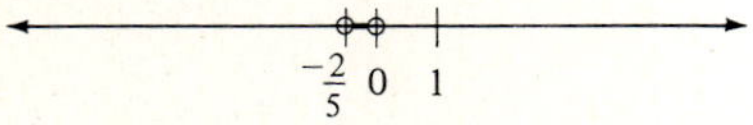

31. $\{w \mid -\frac{1}{3} \le w \le 2\}$

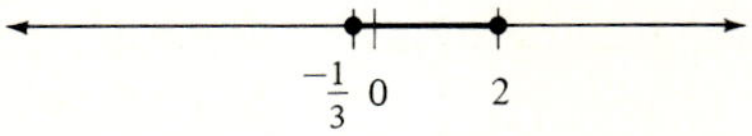

33. $\{x \mid x \le -3 \text{ or } x \ge \frac{3}{2}\}$

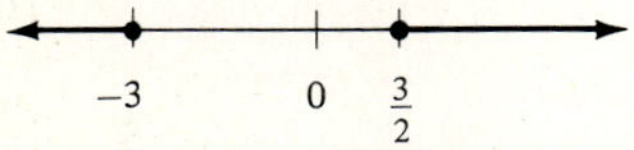

35. $\{w \mid w < -\frac{4}{7} \text{ or } w > 2\}$

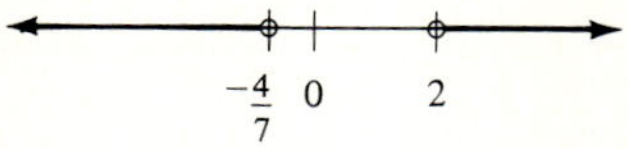

37. All numbers x such that the distance between x and 2 is less than or equal to 4.
39. All numbers w such that the distance between w and -1 is greater than 2.
41. All numbers w such that the distance between twice w and 5 is less than 1.
43. All numbers p such that the distance between $4p$ and -5 is greater than or equal to 7.
45. $|y - 4| < 3$ **47.** $|3x - 1| \ge 2$
49. $|x| \ge 0$ for all x; thus since $0 > -1$, $|x| > -1$ for all real numbers.
51. $|3w + 1| \ge 0$ for all w; thus there is no real number w such that $|3w + 1|$ is negative.

53. $\{x \mid -1 \le x \le 5\}$ 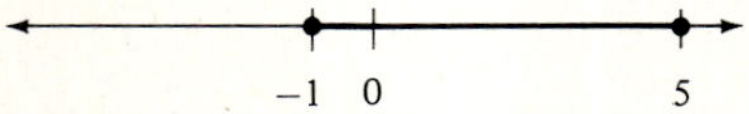

55. $\{y \mid 2 < y < 3\}$ 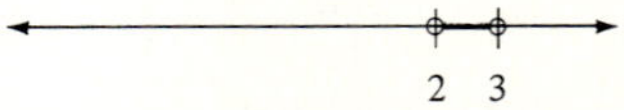

57. $\{w \mid w < 0 \text{ or } w > 2\}$ 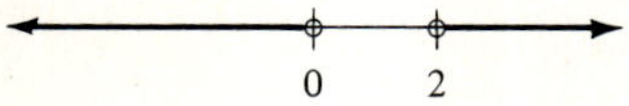

59. $\{p \mid p \le \frac{3}{5} \text{ or } p \ge 1\}$ 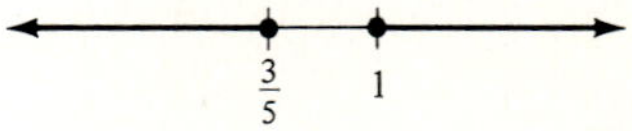

Exercises 2.7

1. $W = 9$ ft, $L = 41$ ft **3.** $W = 66$ m, $L = 134$ m **5.** 5 in., 11 in., and 13 in.
7. 10 cm, 14 cm, and 21 cm **9.** 7 ft **11.** 4 ft **13.** 11 **15.** 7 **17.** −7
19. 23, 24 **21.** 61, 62, 63 **23.** 17, 18, 19 **25.** \$8000 at 9%, \$4000 at 11%
27. \$10,000 at 8%, \$6000 at 10% **29.** \$4000 **31.** 8 miles **33.** 4 hr **35.** 55 mph
37. 50 in. **39.** 250 ml **41.** 4 qt **43.** 100 ml **45.** 11 **47.** 5 **49.** 40
51. 1.5 hr **53.** 2 miles **55.** 45 **57.** 145 lb **59.** 30 oz

Chapter 2 Review

1. $12w$ **2.** $10x + 4$ **3.** $y + 15$ **4.** $z + 1$ **5.** $\{2\}$ **6.** $\{10\}$
7. $\{3\}$ **8.** $\{1\}$ **9.** $\{3\}$ **10.** $\{4\}$ **11.** $A = 126$ sq in., $P = 50$ in.
12. $A = 49\pi = 153.86$ sq in., $C = 14\pi = 43.96$ in. **13.** 312 km
14. $A = 600$ sq in., $P = 120$ in. **15.** $t = \dfrac{i}{pr}$ **16.** $e = \dfrac{2s - nf}{n}$ **17.** $\{-11, 11\}$
18. $\{\ \}$ **19.** $\{0\}$ **20.** $\{-10, 4\}$ **21.** $\{-9, 4\}$ **22.** $\{-\frac{1}{2}, 4\}$ **23.** $\{2\}$
24. $\{-12, -\frac{2}{3}\}$ **25.** $\{x \mid x \ge 4\}$ 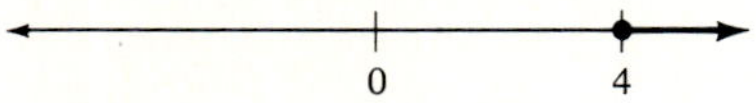

26. $\{y \mid y < -4\}$ 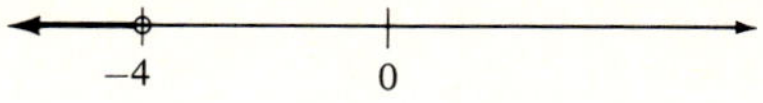

27. $\{w \mid w > -3\}$ 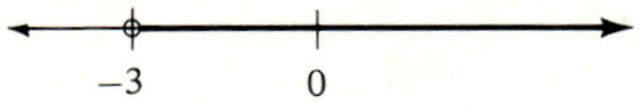

28. $\{z \mid z \leq 1\}$ 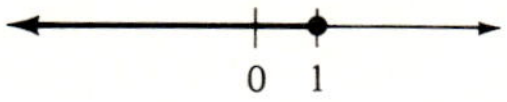

29. $\{p \mid -4 \leq p < 2\}$ 

30. $\{q \mid q \leq 5 \text{ or } q \geq 8\}$ 

31. $\{x \mid -7 < x < 3\}$ 

32. $\{-\frac{4}{3} \leq y \leq 0\}$ 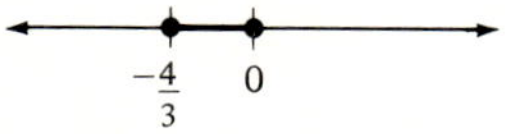

33. $\{z \mid z \leq -10 \text{ or } z \geq 0\}$ 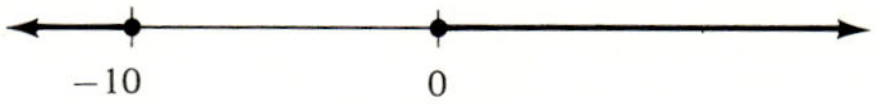

34. $\{w \mid w < -1 \text{ or } w > \frac{5}{3}\}$ 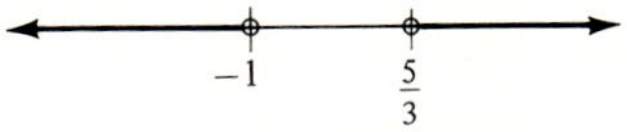

35. $W = 27$ in., $L = 36$ in. **36.** 73, 74, and 75
37. \$2000 **38.** 55 mph **39.** 3.2 qt **40.** $1\frac{1}{2}$ hr

Chapter 2 Test

1. $\{6\}$ **2.** $\{21\}$ **3.** $\{1\}$ **4.** $\{8\}$ **5.** $P = 68$ in., $A = 240$ sq in.

6. 3.5 hr **7.** $\{-11, 11\}$ **8.** $\{-1, 6\}$ **9.** $\{x \mid x \leq 3\}$ 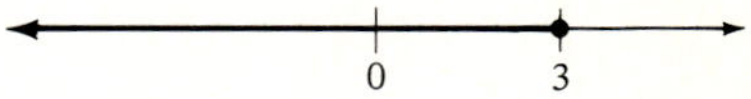

10. $\{y \mid y < -2\}$ 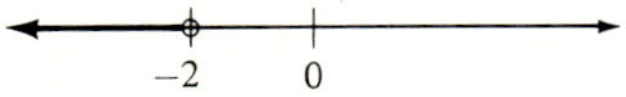

11. $\{z \mid z < 3\} \cap \{z \mid z \geq 1\}$ 

12. $\{w \mid w \leq 1\} \cup \{w \mid w > 4\}$ 

13. $\{x \mid -1 \leq x \leq -\frac{1}{2}\}$ 

14. $\{q \mid q < -\frac{1}{5}\} \cup \{q \mid q > \frac{3}{5}\}$ 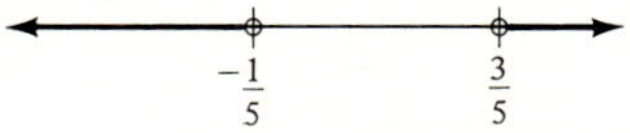

15. $W = 100$ in., $L = 150$ in. **16.** 91, 92, and 93 **17.** $\frac{8}{3} = 2\frac{2}{3}$ miles **18.** 100 ml

CHAPTER 3

Exercises 3.1

1. **(a)** 2 **(b)** 3 **(c)** 8 **3.** **(a)** -5 **(b)** 2 **(c)** 25 **5.** **(a)** 10 **(b)** 4 **(c)** -10000
7. **(a)** $\frac{1}{3}$ **(b)** 2 **(c)** $\frac{1}{9}$ **9.** **(a)** -1 **(b)** 4 **(c)** -1 **11.** 9^3 **13.** $(-7)^3$ **15.** x^2
17. z^4 **19.** r^2s^3 **21.** $-2a^2b^2$ **23.** $(-b)^2c^3$ or b^2c^3 **25.** $-m^3n^2$
27. $a^3 + a^2b$ **29.** $5m^2 - m^2n^3$ **31.** 7^{10} **33.** 5^9 **35.** $(-8)^5$ **37.** $(\frac{1}{4})^8$
39. x^5 **41.** x^7 **43.** x^8y^6 **45.** m^7n^5 **47.** $-6x^4y^7$ **49.** $a^5b^7c^9$

51. 1 **53.** $\frac{1}{125}$ **55.** 1 **57.** $\frac{1}{-32}$ **59.** y^3 **61.** $\frac{1}{t^5}$ **63.** $\frac{1}{x^2}$
65. $\frac{1}{r^6}$ **67.** $t^0 = 1$ **69.** xy **71.** $\frac{s}{r}$ **73.** $\frac{x^2}{y}$

Exercises 3.2

1. 400 **3.** 900 **5.** -1 **7.** 16 **9.** $9x^2$ **11.** $27r^3$ **13.** $8x^3y^3z^3$
15. $x^4y^4z^4t^4$ **17.** 3^4 **19.** $\frac{1}{5^4}$ **21.** y^{12} **23.** t^{12} **25.** 1 **27.** $\frac{1}{s^6}$
29. x^6y^{12} **31.** $\frac{y^6}{x^6}$ **33.** x^8y^{12} **35.** $3^4y^8 = 81y^8$ **37.** $\frac{16t^{12}}{z^8}$ **39.** $\frac{125y^3}{x^6}$
41. x^9y^7 **43.** $9r^{11}s^9$ **45.** 25 **47.** 3 **49.** $\frac{1}{4}$ **51.** $\frac{1}{9}$ **53.** $\frac{8}{27}$
55. $\frac{25}{9}$ **57.** 6 **59.** 8 **61.** z^6 **63.** $\frac{x^9}{z^6}$ **65.** b^6 **67.** $3a^{12}$
69. $\frac{a^{10}}{b^8}$ **71.** xy **73.** $\frac{a^5}{b^3c^2}$ **75.** $\frac{y^2}{z^2}$

Exercises 3.3

1. 3.7×10^2 **3.** 1×10^3 **5.** 2×10^{-3} **7.** 6.73×10^2 **9.** 6×10^5
11. 7.1×10^{-5} **13.** 4.87×10^7 **15.** 9.72×10^{-6} **17.** 860 **19.** 6810
21. .007 **23.** .000031 **25.** .00000572 **27.** .007531 **29.** 6.9×10^9
31. 8.4×10^7 **33.** 6.0×10^3 **35.** 4.8×10^{-5} **37.** 7.1×10^{-4} **39.** 4.0×10^4
41. 6.4×10^3 **43.** 8×10^4 **45.** 4×10^{14} **47.** 1.323×10^{25} lb **49.** 3×10^8 m/sec

Exercises 3.4

1. Polynomial **3.** Polynomial **5.** Polynomial **7.** Polynomial **9.** Not a polynomial
11. Polynomial **13.** Polynomial **15.** Not a polynomial **17.** Binomial; 1
19. Trinomial; 2 **21.** Monomial; 5 **23.** Binomial; 2 **25.** Polynomial; 4
27. Binomial; 3 **29.** Trinomial; 3 **31.** Polynomial; 2 **33.** $2x$
35. $10x$ **37.** $6x^2 + 7x$ **39.** $10t^2 - 4t$ **41.** $8x - 3y$ **43.** $-11r^2 + 3t$
45. $12xy^3 - 5xy$ **47.** $4p^3q - 7p^2q$ **49.** $8t - 7s$ **51.** $4x^2 - 7x + 5$
53. $-4y^4 - 4y^3 + y^2$ **55.** $x^3 + 4x$ **57.** $6x^3 + x^2 + 8x - 2$ **59.** $6xy^2 + xy$
61. $4x^2 + 4x$ **63.** $6t^2 - 9t + 1$ **65.** $5x^4 + x^3 + 2x^2 - x + 6$ **67.** $-6x^3 + 2x^2 + 6$
69. $4x^2 - 5x + 6$ **71.** $-4t^2 - 4t - 6$

Exercises 3.5

1. $5x^2 + 5xy$ **3.** $s^2t - st^2$ **5.** $3y^4 - 15y^2 + 6y$ **7.** $-6z^4 + 12z^3y - 6z^2y$
9. $6x^2y^3 - 3x^3y^2$ **11.** $x^5y^3 - 5xy^5 + x^2y^4 - 3xy^3$ **13.** $2a^2 + 7ab + 3b^2$
15. $6x^2 + 7xy - 20y^2$ **17.** $w^3 + 3w^2 - 2w + 8$ **19.** $3x^3 + 4x^2 + 5x - 6$
21. $2t^3 - 7t^2 - 7t + 12$ **23.** $x^3 + 4x^2y + 4xy^2 + y^3$ **25.** $6x^3 + 5x^2y + 5xy^2 + 2y^3$
27. $y^3 - 2y^2 - 5y + 6$ **29.** $6x^3 + 11x^2y + 6xy^2 + y^3$ **31.** $x^4 + 2x^2 + x + 2$
33. $2z^4 - 3z^3 + 7z^2 - 3z + 9$ **35.** $-6t^4 + 11t^3 - 2t^2 - 5t + 2$ **37.** $x^2 + 5x + 6$
39. $z^2 - 3z + 2$ **41.** $s^2 - 1$ **43.** $6x^2 + 11x + 3$ **45.** $9y^2 - 18y + 8$
47. $4x^2 - 9$ **49.** $-x^2 - x + 2$ **51.** $x^2 - 5x + 6$ **53.** $6r^2 + 5rs + s^2$
55. $6x^2 - 13xy + 6y^2$ **57.** $4x^2 + 16x + 16$ **59.** $9x^2 - 12x + 4$ **61.** $4r^2 - s^2$
63. $25x^2 + 30xy + 9y^2$ **65.** $4r^2 - 12rs + 9s^2$ **67.** $4p^2 - 9p^2$ **69.** $4x^4 - 4x^2 + 1$
71. $9x^4 - y^2$ **73.** $p^6 - q^6$ **75.** $x^{3n} - x^{2n+1}$ **77.** $z^{2n} + z^n$ **79.** $x^{2m} - 9$
81. $y^{2n} + 2y^n + 1$ **83.** $x^{2n} - y^{2n}$ **85.** $x^{2n} + 2x^ny^n + y^{2n}$

Exercises 3.6

1. $2 \cdot 3$ **3.** $2^3 \cdot 3$ **5.** $2 \cdot 11$ **7.** 5^2 **9.** $2^2 \cdot 5^2$ **11.** $(-1)2^2 \cdot 3^2$
13. $5 \cdot 7$ **15.** $2 \cdot 5 \cdot 7$ **17.** $2^2 \cdot 7$ **19.** Prime **21.** $5(x + 2)$ **23.** $6x(x + 3)$
25. $-4x(y - 4)$ **27.** $2xy(x - 9)$ **29.** $9x^2(x^2 - 3x + 2)$ **31.** $16x^2y(x + 2)$
33. $9ab^2(2ab + 3c)$ **35.** $6a(2a^2 - ab - 2b^2)$ **37.** $7a^2(3ab^2 - 2b + 1)$
39. $(z - 3)(x + 3t)$ **41.** $(a - b)(7s - 3t)$ **43.** $(a - b)(2 + 5c)$ **45.** $(m + 5)(m + b)$
47. $(2x - 3)(y + 5)$ **49.** $(q + 2)(q^2 - 3)$ **51.** $(y - 2)(3y^2 - 2)$

Exercises 3.7

1. $(x + 1)(x + 3)$ **3.** $(x + 1)(x + 7)$ **5.** $(x + 1)(x + 4)$ **7.** $(x - 5)(x + 1)$
9. $(y + 2)(y - 4)$ **11.** $(x + 2)(x - 9)$ **13.** $(x - 2)(x - 4)$ **15.** $(t - 2)(t - 7)$
17. $(x - 3)(x - 6)$ **19.** $(x - 1)(x + 9)$ **21.** $(w - 1)(w + 11)$ **23.** $(m - 2)(m + 6)$
25. $(x + 3y)(x + 6y)$ **27.** $(u - 4v)(u - 9v)$ **29.** $3(x - 2y)(x - 11y)$ **31.** $2x(y + 4)(y - 5)$
33. $(2x + 1)(x + 7)$ **35.** $(2x + 1)(x + 5)$ **37.** $(3y - 5)(y - 1)$ **39.** $y(3x - 1)(x - 5)$
41. $(7t - 5)(t + 1)$ **43.** $(5t - 3)(t + 3)$ **45.** $10(3t - 5)(2t + 3)$ **47.** $(3x + 2)(2x - 3)$
49. $y(2x + 5)(3x - 2)$ **51.** $(3x + 4y)(3x + 2y)$ **53.** $(3x + 2)(2x - 5)$ **55.** $(9x - 8)(x - 1)$
57. $10(2y + 1)(y - 3)$ **59.** $(5t - 2)(2t - 3)$ **61.** $z^2(2x - 1)(x - 2)$ **63.** $b^3(b + 2)(b + 3)$
65. $8xy(x + 2)(x + 1)$ **67.** $4t^2(3t + 4)(2t - 3)$ **69.** $3x(3x - 7y)(2x + 5y)$

Exercises 3.8

1. Perfect square trinomial **3.** Neither **5.** Difference of two squares
7. Perfect square trinomial **9.** Neither **11.** Neither **13.** $(x - 2)^2$
15. $(y + 7)^2$ **17.** $(y - 7)^2$ **19.** $(t - 1)^2$ **21.** $(2u - v)(2u + v)$
23. $(3x - 7y)(3x + 7y)$ **25.** $(2x + 8y)(2x - 8y)$ **27.** $(x - 2)(x + 2)(x^2 + 1)$
29. $(x - 3)(x + 3)(x^2 + 2)$ **31.** $(y - 1)(y + 1)(y - 2)(y + 2)$ **33.** $2y(x - 11)^2$
35. $a(a + 10)^2$ **37.** $(3y - 8)^2$ **39.** $4(a + 3b)^2$ **41.** $(3x - 4y)^2$
43. $(5x - 4y)^2$ **45.** $(x - 1)(x + 5)$ **47.** $(4 - a - b)(4 + a + b)$ **49.** $z(z - 2)$
51. $(x - 1)(x^2 + x + 1)$ **53.** $(m - 3)(m^2 + 3m + 9)$ **55.** $(u - v)(u^2 + uv + v^2)$
57. $(x + 4)(x^2 - 4x + 16)$ **59.** $(a - 6)(a^2 + 6a + 36)$ **61.** $(2a - b)(4a^2 + 2ab + b^2)$
63. $8(2m + n)(4m^2 - 2mn + n^2)$ **65.** $(z - 7)(z^2 + 7z + 49)$ **67.** $(y^2 + 1)(y^4 - y^2 + 1)$
69. $2(m + 2)(m^2 - 2m + 4)$ **71.** $x(x^2 - 3x + 3)$ **73.** $x(x^2 + 4x + 12)$

Chapter 3 Review

1. 49 **2.** -81 **3.** -64 **4.** $\frac{1}{9}$ **5.** 1 **6.** 3^4 **7.** b^3
8. a^3b^2 **9.** $-r^2s^4$ **10.** x^5 **11.** $y^0 = 1$ **12.** $\dfrac{a}{b}$ **13.** $64x^3$ **14.** z^8
15. $r^{12}s^8$ **16.** $\dfrac{1}{t^6}$ **17.** $\dfrac{y^9}{x^6}$ **18.** x^8y^3 **19.** $\dfrac{t^6}{s^9}$ **20.** $\dfrac{b}{a}$ **21.** $\dfrac{r^3}{s^2}$
22. ab **23.** 5.3×10^9 **24.** 2.3×10^{-7} **25.** .0072 **26.** 32,800
27. 2×10^5 or 200,000 **28.** $12x^2 - 3x$ **29.** $5a^3b - 8a$ **30.** $12t^2 - 2t - 3$
31. $4y^4 + 2y^3$ **32.** $x^3y^5 - xy^3$ **33.** $8x^2 + 2xy - 15y^2$ **34.** $9r^2 + 6rs + s^2$
35. $y^3 - 1$ **36.** $9r^2 - 4s^2$ **37.** $2 \cdot 7$ **38.** $2 \cdot 3 \cdot 3 \cdot 5$ **39.** $7x(x + 4)$
40. $3x^2y(1 - 9y)$ **41.** $(x - 6)(x + 1)$ **42.** $(3x + 1)(x + 2)$ **43.** $x(y + 1)(y + 2)$
44. $(2a - b)^2$ **45.** $(2x - 5y)(2x + 5y)$ **46.** $(3t + 4s)^2$ **47.** $(s - 2)(s^2 + 2s + 4)$
48. $(2t + 3)(4t^2 - 6t + 9)$

Chapter 3 Test

1. 100 **2.** -125 **3.** 1 **4.** $\dfrac{1}{-8}$ **5.** x^9y^4 **6.** $r^2s^2t^5$ **7.** t^8
8. $\dfrac{x^3y^2}{z^2}$ **9.** $\dfrac{b^4}{a^6}$ **10.** 8.9×10^{-6} **11.** 9.3×10^7 **12.** $8x^2 + 4x + 4$

13. $7y^5 - 7z^2 + 6$ **14.** $x^3y^2 - x^2y^3$ **15.** $10x^2 + 17x + 3$ **16.** $6x^2 - 23xy - 18y^2$
17. $16t^2 + 24t + 9$ **18.** $a^{2m} - b^{2n}$ **19.** $x^3 + 1$ **20.** $(y + 1)(x - 1)(x + 1)$
21. $(x - 1)(x^2 + 1)$ **22.** $(2x + 1)(3x + 4)$ **23.** $(y - 7)^2$ **24.** $2(2x - 5y)(2x + 5y)$
25. $(4a - 5b)(16a^2 + 20ab + 25b^2)$ **26.** $(3y^2 + 2)(9y^4 - 6y^2 + 4)$

CHAPTER 4

Exercises 4.1

1. 0 **3.** Defined for all real numbers **5.** 0 **7.** 2 **9.** -6 **11.** 5
13. $\frac{3}{5}$ **15.** $\frac{3x}{2}$ **17.** $\frac{7x}{10y}$ **19.** $\frac{2ac^2}{3b}$ **21.** $\frac{5r}{s}$ **23.** $\frac{1}{4x}$ **25.** $\frac{w + 1}{w - 1}$
27. $\frac{4}{z + 3}$ **29.** $\frac{p}{p + 5}$ **31.** $2(y - 2)$ **33.** -1 **35.** $-\frac{1}{2}$ **37.** $\frac{24}{32}$
39. $\frac{12x^3}{15x^2}$ **41.** $\frac{28p^2q}{12pq^2}$ **43.** $\frac{x^2 - 6x + 5}{x^2 - 1}$ **45.** $\frac{w^2 + 9w + 14}{w^2 + 4w + 4}$
47. $\frac{2p^2 + p - 3}{2p^2 + 17p + 21}$ **49.** True **51.** False

Exercises 4.2

1. $6x + 2$ **3.** $5w + 2$ **5.** $8p^2 + 6p$ **7.** $3u^4 + 2u^2 - u$ **9.** $3y^3 - 2y^2 + y$
11. $6w^5 + 4w^4 - 3w^3 - 2w^2$ **13.** $2x^2 + x + 3$ **15.** $8v^3 + 4v^2 - 2$
17. $3y^7 - 4y^3 + 1$ **19.** $-3a^2 + 6a - 2$ **21.** $2x + 1$ **23.** $3y + 2$
25. $4w + 6$ **27.** $3z + 2$ **29.** $x^2 + 1$ **31.** $4p^2 + 2p + 1$
33. $2q^2 - q - 3$ with remainder 2 **35.** $x^2 + x - 2$ **37.** $3y^2 + 2y - 1$ with remainder 4
39. $2u - 2$ with remainder $5u + 4$

Exercises 4.3

1. $\frac{7}{9}$ **3.** $\frac{3}{x^2}$ **5.** $\frac{3z}{w^2}$ **7.** $\frac{8b^3}{a}$ **9.** $\frac{p + 3}{p - 1}$ **11.** $\frac{(r + 4)^3}{r - 3}$
13. $\frac{(2w - 5)^2}{(3w + 1)^2}$ **15.** $\frac{2}{x + 3}$ **17.** $\frac{41}{15a}$ **19.** $-2(x - 4)$ **21.** $\frac{2}{3(w - 11)}$
23. $\frac{-5(p - 1)}{2(p + 1)}$ **25.** $-\frac{2(r + 3)}{3(r - 3)}$ **27.** $x - 3$ **29.** $\frac{a + 5}{a + 6}$ **31.** $\frac{4x}{3y}$
33. $\frac{3p^2r}{5qs}$ **35.** $\frac{w^2t^2v}{30}$ **37.** $\frac{-12x}{5py^4}$ **39.** $\frac{a}{5(a + 3)}$ **41.** $\frac{3(c - 2)}{4(c + 3)}$
43. $\frac{1}{20}$ **45.** $\frac{7}{2}$ **47.** $\frac{5(t - 4)}{6(t + 1)}$ **49.** $\frac{p + 7}{p + 1}$ **51.** $\frac{3(b + 2)}{5(b - 2)}$ **53.** $\frac{a - 1}{3a + 2}$
55. $\frac{10b^2}{a}$ **57.** $\frac{a}{7(a - 3)}$ **59.** $\frac{2q^2}{(q + 3)}$

Exercises 4.4

1. $\frac{7}{10}$ **3.** $\frac{1}{6}$ **5.** $\frac{4}{5}$ **7.** $\frac{3}{z}$ **9.** $\frac{13}{2w + 7}$ **11.** $\frac{8a}{5a - 3}$ **13.** 2
15. $\frac{2}{x + 3}$ **17.** $\frac{41}{15a}$ **19.** $\frac{-1}{3p}$ **21.** $\frac{5r + 2}{2r^2}$ **23.** $\frac{y - x}{xy}$
25. $\frac{20b + 6a}{75a^2b^2}$ **27.** $\frac{x - 5}{x(x + 5)}$ **29.** $\frac{4}{w - 2}$ **31.** $\frac{11b - 9}{b^2 - 9}$ **33.** $\frac{2y^2 - y - 3}{(y - 3)(y + 3)}$

35. $\dfrac{-x^2}{(x+1)(x+2)}$ **37.** $\dfrac{w}{(3w+1)(2w+1)}$ **39.** $\dfrac{2y^2+2y-2}{(y-2)(y+2)(y+3)}$

41. $\dfrac{-2b}{(b-4)(b+4)(b+4)}$ **43.** $\dfrac{3(c+6)}{(c-3)(c+3)}$ **45.** $\dfrac{q+5}{(q-1)(q+1)}$

47. $\dfrac{2x}{(x-2)(x+2)}$ **49.** $\dfrac{2(w+1)}{(w+2)(w+5)}$ **51.** $\dfrac{6ab+2b^2}{(a-b)(a+b)(a+b)}$

53. $\dfrac{-p^2-5pq}{(p-q)(p+q)(p+q)}$ **55.** $\dfrac{2y^2+42y-18}{7(y-3)(y+3)}$ **57.** $\dfrac{2(x-16)}{x^2-16}$ **59.** $\dfrac{w-4}{w+5}$

Exercises 4.5

1. $\frac{8}{10}=\frac{4}{5}$ **3.** $\frac{7}{11}$ **5.** $\frac{420}{300}=\frac{7}{5}$ **7.** $\frac{4200}{4500}=\frac{14}{15}$ **9.** $\frac{4200}{5400}=\frac{7}{9}$ **11.** $\frac{5400}{8400}=\frac{9}{14}$

13. The states in question are Montana, Minnesota, Missouri, Mississippi, Michigan, Maryland, Massachusetts, and Maine. Thus the ratio is $\frac{8}{42}=\frac{4}{21}$.

15. $\frac{56}{77}=\frac{8}{11}$ **17.** $\frac{5}{7}$ **19.** $\frac{4}{7}$ **21.** 21 **23.** 7 **25.** 2 **27.** 16

29. $\frac{5}{2}$ **31.** 63 **33.** 12 **35.** -1 **37.** 14 **39.** -6

41. 225 miles **43.** 13.5 in. **45.** 315 ml **47.** 960 **49.** 12.5 lb

Exercises 4.6

1. 1 **3.** 15 **5.** $\frac{6}{5}$ **7.** $-\frac{3}{7}$ **9.** $13\frac{1}{2}$ **11.** 15 **13.** 11

15. 1 **17.** $\frac{1}{4}$ **19.** 41 **21.** 5, extraneous **23.** -4, extraneous

25. -5, extraneous **27.** -6, extraneous **29.** 4 **31.** -2, extraneous **33.** 6

35. 4, extraneous **37.** 3 **39.** 2, extraneous **41.** 2 **43.** -3, extraneous

45. -3 **47.** 0 **49.** -2, extraneous

Exercises 4.7

1. 5 **3.** 9 **5.** 6 **7.** $\frac{7}{12}$ **9.** $\frac{17}{23}$ **11.** 3 **13.** 10 **15.** 13

17. 4 **19.** 7 **21.** 3 mph **23.** 5 mph **25.** 20 mph **27.** 25 mph

29. **(a)** 2.1 hr or 2 hr and 6 min **(b)** 420 **31.** **(a)** 18 min **(b)** 600

33. **(a)** 6 hours **(b)** $\frac{1}{6}$ **(c)** $\frac{3}{5}$ **35.** **(a)** 30 min **(b)** $\frac{1}{3}$ **(c)** $\frac{5}{8}$ **37.** 15 min

39. $\frac{45}{8}$ ohms **41.** $\frac{10}{3}$ ohms

Exercises 4.8

1. $\frac{35}{18}$ **3.** $\frac{1}{6}$ **5.** $\frac{10}{9}$ **7.** $\dfrac{x}{6}$ **9.** $\dfrac{b^2}{a}$ **11.** $\dfrac{p^3q^2}{4}$ **13.** $\dfrac{a+1}{2(a-1)}$

15. $\dfrac{2d}{c}$ **17.** $\dfrac{p-q}{q}$ **19.** $\dfrac{9y}{12y-4}$ **21.** $\dfrac{12p^2}{10pq^2-5q^2}$ **23.** $\dfrac{p}{q}$

25. $\frac{2}{3}$ **27.** $\dfrac{10x}{3}$ **29.** $\dfrac{18b^2-3}{6b^2+2}$ **31.** $v-u$ **33.** $\dfrac{y-x}{1-y}$ **35.** $\dfrac{2x(xy-9)}{3y(4+xy)}$

37. -1 **39.** $\dfrac{5w+2}{5w+8}$ **41.** $-\dfrac{2}{a+1}$ **43.** $\dfrac{b}{a}$ **45.** $\dfrac{q+p}{q-p}$ **47.** $\dfrac{z^2}{z^2-w^2}$

49. t **51.** $\dfrac{ab}{b^2+a^2}$ **53.** $\dfrac{xy}{y+x}$

Chapter 4 Review

1. $\frac{2}{3}$ **2.** $\dfrac{5y^2}{3}$ **3.** $\dfrac{2w^4}{5x^2}$ **4.** $\dfrac{q}{q+1}$ **5.** $\dfrac{3(z-2)}{7}$ **6.** -1

7. $4v^3-3v+1$ **8.** $6x^2+5x$ with remainder 3 **9.** $\dfrac{3y}{2x^3}$ **10.** $\dfrac{bd}{6ac}$

11. $\frac{w+3}{w+1}$ **12.** $\frac{2}{3(x+2)}$ **13.** $\frac{w+3}{w+1}$ **14.** $-\frac{2(z+3)}{3(z-3)}$ **15.** $\frac{1}{y-5}$
16. 1 **17.** $\frac{z+2}{2z^2}$ **18.** $\frac{4}{w+3}$ **19.** $\frac{-a}{(a+1)(a+2)}$ **20.** 1 **21.** $\frac{2}{1}$
22. $\frac{1}{2}$ **23.** 5 **24.** -5 **25.** 360 miles **26.** -11 **27.** 17 **28.** -3
29. -2 **30.** 11 **31.** 2 mph **32.** **(a)** 2.1 hr = 2 hr and 6 min **(b)** $.7 = \frac{7}{10}$
33. $\frac{2}{3}$ **34.** $\frac{a+b}{b}$ **35.** $\frac{x^2}{3x^2-6}$ **36.** $\frac{-8}{3a+4}$

Chapter 4 Test

1. $\frac{9a^2}{4}$ **2.** $\frac{4y}{3xz}$ **3.** $\frac{u-3}{u+2}$ **4.** $\frac{wt}{8}$ **5.** $v+2$ **6.** 18 **7.** $\frac{x-4}{x+4}$
8. 1 **9.** $\frac{5y+4}{4y^2}$ **10.** $\frac{2x-2}{x^2+2x} = \frac{2(x-1)}{x(x+2)}$ **11.** $\frac{2w^2+w+2}{w^2-4}$
12. $\frac{x^2+6x+12}{(x-2)(x+2)^2}$ **13.** $\frac{5p+11}{(p+1)(p-1)(p+3)}$ **14.** $\frac{1}{5}$ **15.** $\frac{5}{3}$ **16.** 288
17. -7 **18.** -14 **19.** -7, extraneous **20.** $\frac{a^2+1}{a^2-1}$ **21.** $\frac{2x-1}{2x+1}$
22. $8z^2-6z+2$ **23.** $5x^2-12x-1$ with remainder $3x+1$ **24.** $\frac{20}{27}$ **25.** 24 min

CHAPTER 5

Exercises 5.1

1. 9 **3.** -12 **5.** -1 **7.** 5 **9.** 2 **11.** Not real **13.** Not real
15. 9 **17.** $5z$ **19.** $-\frac{1}{2}$ **21.** $3x$ **23.** $\frac{3}{5}$ **25.** xy **27.** $-5ab$
29. 5 **31.** x^2 **33.** y^2 **35.** $5^2 = 25$ **37.** Not real **39.** $x+y$

Exercises 5.2

1. 6 **3.** -5 **5.** Not real **7.** -2 **9.** 64 **11.** 16 **13.** $\frac{1}{10}$
15. Not real **17.** 729 **19.** $\frac{8}{27}$ **21.** Not real **23.** $\frac{1}{64}$ **25.** $\sqrt[4]{x^3}$
27. $\sqrt[3]{x^2y^2}$ **29.** $\sqrt[3]{4y^2}$ **31.** $\sqrt{x^2+y^2}$ **33.** 27 **35.** 2 **37.** 7
39. 5 **41.** $6x^{3/4}$ **43.** $a^1 = a$ **45.** y **47.** x^2y^4 **49.** $\frac{x^2}{y^4}$
51. $\frac{b^{3/2}}{a}$ **53.** $8x^6y^9$ **55.** $\frac{2y^2}{xz^3}$

Exercises 5.3

1. 9 **3.** $3y$ **5.** $\frac{1}{3}$ **7.** x **9.** ab **11.** 5 **13.** $\frac{4}{3}$ **15.** a
17. $\frac{a}{b}$ **19.** $\frac{3}{a}$ **21.** $7\sqrt{2}$ **23.** $\frac{\sqrt{6}}{6}$ **25.** $\frac{\sqrt[3]{15}}{3}$ **27.** 25
29. $\frac{\sqrt{2x}}{2x}$ **31.** $\frac{\sqrt{3a}}{a}$ **33.** $x\sqrt{y}$ **35.** $3a\sqrt[3]{2a}$ **37.** $ab\sqrt[4]{ab^3}$ **39.** x
41. $\frac{5x\sqrt{2x}}{y}$ **43.** $\frac{\sqrt[3]{a^2}}{a}$ **45.** $3xy^2\sqrt{3x}$ **47.** $a^2b^3\sqrt{ab}$ **49.** $2xy^2\sqrt[3]{2x^2y}$
51. $\frac{a\sqrt[4]{2b^2}}{2b}$ **53.** $\frac{\sqrt[3]{4x^2}}{2x}$ **55.** $\frac{\sqrt[3]{9y^2}}{3y^2}$ **57.** $\frac{\sqrt[3]{5x}}{x}$
59. $\sqrt{3^2+4^2} = \sqrt{9+16} = \sqrt{25} = 5$; but $\sqrt{3^2} + \sqrt{4^2} = 3 + 4 = 7$.

Exercises 5.4

1. $12\sqrt{3}$ **3.** $3\sqrt{6}$ **5.** $11\sqrt{5}$ **7.** $10\sqrt[3]{4}$ **9.** $6\sqrt{x}$ **11.** $2\sqrt[3]{ab}$
13. $2\sqrt{x} + 3\sqrt{y}$ **15.** $5\sqrt[3]{a} + \sqrt{a}$ **17.** $5a\sqrt{b} + \sqrt{b}$ or $(5a + 1)\sqrt{b}$
19. $2u\sqrt[3]{v} - 2\sqrt[3]{v}$ or $(2u - 2)\sqrt[3]{v}$ **21.** $3\sqrt{5}$ **23.** $-4\sqrt{2}$ **25.** $11\sqrt{5}$
27. $8y\sqrt{2y}$ **29.** $18s\sqrt{5s}$ **31.** $15x\sqrt{2y}$ **33.** $5a\sqrt[3]{a}$ **35.** $5x^2\sqrt[3]{x}$
37. $9x^2y^2\sqrt[3]{x}$ **39.** $\sqrt{3}$ **41.** $3\sqrt{2}$ **43.** $\frac{\sqrt{x}}{x}$

Exercises 5.5

1. $2\sqrt{2} + 2$ **3.** 9 **5.** $14 - 3\sqrt{7}$ **7.** -18 **9.** -12 **11.** $a^2 - a^3$
13. 6 **15.** $-2\sqrt[4]{x^3}$ **17.** $7 + 3\sqrt{5}$ **19.** $7 + 7\sqrt{2}$ **21.** $x + 4 + 4\sqrt{x}$
23. $a - 2\sqrt{ab} + b$ **25.** 18 **27.** 12 **29.** 4 **31.** -2 **33.** -1
35. $x - 2$ **37.** $2 - \sqrt{3}$ **39.** $\sqrt{10} + 3$ **41.** $2(4 - \sqrt{5})$ **43.** $2(\sqrt{6} - \sqrt{5})$
45. $\sqrt{7} + \sqrt{5}$ **47.** $2\sqrt{2} + \sqrt{5}$ **49.** $(\sqrt{3} + \sqrt{2})^2$ **51.** $2(\sqrt{x+3} + \sqrt{x})$
53. $\sqrt{x} - 1$ **55.** $\sqrt{x} + \sqrt{y}$ **57.** $\sqrt{2x} + \sqrt{x}$ **59.** $\sqrt{y} - \sqrt{z}$ **61.** $\sqrt{z}(z + 1)$

Exercises 5.6

1. 16 **3.** 36 **5.** 11 **7.** 5 **9.** 4 **11.** 5 **13.** 6
15. No solution **17.** 4 **19.** No solution **21.** 10 **23.** No solution
25. 64 **27.** 2 **29.** 6 **31.** -5 **33.** No solution **35.** 3 **37.** 4
39. 10 **41.** 3

Exercises 5.7

1. $5 + 3i$ **3.** $5 - i$ **5.** $-5 + 10i$ **7.** $4 + 5i$ **9.** $3 - 12i$ **11.** $4i$
13. $7 + 7i$ **15.** 6 **17.** 2 **19.** $2i$ **21.** $-9 + 6i$ **23.** $2 + i$
25. $10i$ **27.** $6 + 17i$ **29.** $8 - 6i$ **31.** $25i$ **33.** **(a)** $1 - i$ **(b)** 2
35. **(a)** $2 + 4i$ **(b)** 20 **37.** **(a)** $-3i$ **(b)** 9 **39.** **(a)** $5 - 2i$ **(b)** 29 **41.** $-4 - 2i$
43. $1 - 3i$ **45.** $-4 + 3i$ **47.** i **49.** $1 - i$ **51.** $3 - i$ **53.** $4i$
55. $8i$ **57.** $-2i$ **59.** $i\sqrt{7}$ **61.** $6 - i$ **63.** $2 + 10i$ **65.** $i4\sqrt{2}$
67. $-i3\sqrt{2}$ **69.** -4 **71.** $i3$ or $3i$ **73.** -8 **75.** $5i$ **77.** $-8i$
79. $i^4 = 1,\ i^5 = i,\ i^6 = -1,\ i^7 = -i,\ i^8 = 1$

Chapter 5 Review

1. 9 **2.** -3 **3.** x^2 **4.** $2y^2$ **5.** 9 **6.** -2 **7.** 27 **8.** $\frac{1}{4}$
9. $b^{1/4}$ **10.** a **11.** $6xy^3$ **12.** $4xy^{1/3}$ **13.** $\frac{x^9}{y^8}$ **14.** $\frac{t^{1/2}}{s^{1/4}}$ **15.** 4
16. $2z$ **17.** $2x$ **18.** $10\sqrt{3}$ **19.** $\frac{\sqrt{7}}{7}$ **20.** $2y\sqrt[3]{2y}$ **21.** $7\sqrt{3}$
22. $19\sqrt{3}$ **23.** $3y\sqrt[3]{y}$ **24.** $-7 + 3\sqrt{3}$ **25.** $8 + 5\sqrt{3}$ **26.** $3 - \sqrt{2}$
27. 3 **28.** 5 **29.** -4 **30.** 9 **31.** 2 **32.** $9 + 3i$ **33.** $3 - 9i$
34. $-14 + 23i$ **35.** 17 **36.** $6 + 8i$ **37.** $2 - 3i$ **38.** $6i$ **39.** $-2i$
40. -4

Chapter 5 Test

1. -10 **2.** $3x$ **3.** x^3 **4.** 4 **5.** 8 **6.** 25 **7.** 4
8. $x^{2/3}$ **9.** $2x^2y$ **10.** $4x^{2/3}y^{1/2}$ **11.** $\frac{x^5}{y^2}$ **12.** $4x$ **13.** y **14.** $\frac{\sqrt[4]{y}}{y}$

15. $\frac{x}{y}$ **16.** $5x\sqrt{3x}$ **17.** $2xy\sqrt[3]{3xy^2}$ **18.** $16\sqrt{5}$ **19.** $4n\sqrt[3]{n}$
20. $-3 - \sqrt{3}$ **21.** -54 **22.** 3 **23.** -5 **24.** $8 + 5i$ **25.** $5 - 3i$
26. $11 + 13i$ **27.** $1 - i$ **28.** $8i$ **29.** $-5i$ **30.** -10

CHAPTER 6

Exercises 6.1

1. $\{3, 7\}$ **3.** $\{-2, 3\}$ **5.** $\{-9, -1\}$ **7.** $\{-\frac{5}{2}, \frac{10}{3}\}$ **9.** $\{-\frac{7}{2}, 0, 1\}$
11. $\{0, 1\}$ **13.** $\{-3, 0\}$ **15.** $\{0, 3\}$ **17.** $\{-4, -2\}$ **19.** $\{-7, 4\}$
21. $\{-7, 6\}$ **23.** $\{-14, -2\}$ **25.** $\{\frac{1}{3}, \frac{1}{2}\}$ **27.** $\{-5, \frac{3}{2}\}$ **29.** $\{\frac{1}{4}, 6\}$
31. $\{-\frac{3}{2}, 0, \frac{5}{6}\}$ **33.** $\{0, \frac{2}{3}\}$ **35.** $\{-4, 0, 4\}$ **37.** $\{-12, 0, 12\}$ **39.** $\{-3, 5\}$
41. $\{-3, \frac{1}{2}\}$ **43.** $\{-\frac{1}{6}, 1\}$ **45.** $\{0, 1, 5\}$ **47.** $\{-3, 4\}$ **49.** $\{-\frac{1}{2}, 2\}$
51. $\{-1, \frac{1}{2}\}$ **53.** $\{-5, -2\}$ **55.** $\{-7, 0\}$ **57.** $\{-5, 5\}$
59. $\{5\}$, -3 is extraneous **61.** $\{-\frac{1}{3}, 2\}$ **63.** $\{-4, 4\}$ **65.** $\{-5, 5\}$
67. $\{-3, 1\}$ **69.** $\{0, 2\}$ **71.** $\{\frac{1}{4}\}$, 2 is extraneous **73.** $\{-7, 8\}$ **75.** $\{1, 3\}$
77. $\{2, 4\}$

Exercises 6.2

1. $\{\pm 8\}$ **3.** $\{\pm\frac{3}{2}\}$ **5.** $\{\pm .2\}$ **7.** $\{\pm 5\sqrt{3}\}$ **9.** $\{\pm\sqrt{13}\}$ **11.** $\{\pm 10\sqrt{3}\}$
13. $\{\pm 3i\}$ **15.** $\{\pm 5i\sqrt{2}\}$ **17.** $\{-3, 11\}$ **19.** $\{-12, 6\}$ **21.** $\{-2, 8\}$
23. $\{-1, 2\}$ **25.** $\{-7, 3\}$ **27.** $\{3 \pm 2i\}$ **29.** $\left\{\frac{-3 \pm 5i}{2}\right\}$ **31.** $\{-5 \pm 2\sqrt{7}\}$
33. $\{4 \pm 2\sqrt{5}\}$ **35.** $\{-10 \pm 5\sqrt{5}\}$ **37.** $\{7 \pm 5\sqrt{2}\}$ **39.** $\{-9 \pm 5\sqrt{3}\}$
41. $\{4 \pm \sqrt{15}\}$ **43.** $\{12 \pm 5\sqrt{6}\}$ **45.** $\left\{\frac{-3 \pm \sqrt{13}}{2}\right\}$ **47.** $\{\frac{1}{2}, -1\}$
49. $\{-3, -\frac{1}{2}\}$ **51.** $\{-1, \frac{4}{3}\}$ **53.** $\left\{\frac{-3 \pm \sqrt{57}}{4}\right\}$ **55.** $\left\{\frac{3 \pm \sqrt{89}}{10}\right\}$
57. $\{-1 \pm 5i\}$ **59.** $\{2 \pm 3i\sqrt{3}\}$ **61.** $\{2 \pm 3i\}$ **63.** $\left\{\frac{-5 \pm 2i\sqrt{11}}{3}\right\}$
65. $\{3 \pm 2\sqrt{3}\}$ **67.** $\{-2 \pm \sqrt{6}\}$ **69.** $\{3 \pm \sqrt{26}\}$

Exercises 6.3

1. $\{-\frac{1}{4}, 1\}$ **3.** $\{-\frac{1}{2}, \frac{5}{2}\}$ **5.** $\{-2 \pm \sqrt{3}\}$ **7.** $\{-1, 3\}$ **9.** $\{-\frac{1}{2}, 2\}$
11. $\{1 \pm 2i\}$ **13.** $\{-1 \pm \sqrt{6}\}$ **15.** $\{-5 \pm 10i\}$ **17.** $\left\{\frac{3 \pm \sqrt{29}}{10}\right\}$
19. $\{-1 \pm i\sqrt{5}\}$ **21.** $\{-4 \pm \sqrt{10}\}$ **23.** $\left\{\frac{-1 \pm i\sqrt{15}}{2}\right\}$ **25.** $\left\{\frac{3 \pm \sqrt{21}}{2}\right\}$
27. $\left\{\frac{-1 \pm i\sqrt{23}}{6}\right\}$ **29.** $\left\{\frac{4 \pm \sqrt{11}}{5}\right\}$ **31.** $\left\{\frac{3 \pm \sqrt{3}}{2}\right\}$ **33.** $\{3 \pm \sqrt{13}\}$
35. $\left\{\frac{5 \pm \sqrt{5}}{2}\right\}$ **37.** $\{-4 \pm 7i\}$ **39.** $\{3 \pm 4i\}$ **41.** $\{-1 \pm \sqrt{5}\}$
43. $\left\{\frac{-2 \pm \sqrt{6}}{2}\right\}$ **45.** $\left\{\frac{-7 \pm \sqrt{57}}{2}\right\}$ **47.** $\{-1 \pm i\sqrt{5}\}$ **49.** $\left\{\frac{3 \pm i\sqrt{7}}{2}\right\}$
51. Two distinct irrational roots **53.** Two complex conjugate roots
55. Two distinct rational roots **57.** One rational root

Exercises 6.4

1. 13 in. **3.** 15 ft **5.** $2\sqrt{5} = 4.47$ in. **7.** $\dfrac{15 - \sqrt{113}}{2} = 2.18$ ft

9. 8 and 9 or -9 and -8 **11.** 5, 6, and 7 or -7, -6, and -5

13. 14 and 15 or -15 and -14 **15.** 13 and 15 or -15 and -13 **17.** In 6 sec

19. In 50 sec **21.** In $\dfrac{5 + 3\sqrt{5}}{4} = 2.93$ sec **23.** In $\dfrac{15 + 5\sqrt{10}}{4} = 7.70$ sec

25. 6 mph **27.** 7 mph **29.** 18 mph

Exercises 6.5

1. $\{x \mid x \le -2 \text{ or } x \ge 4\}$ 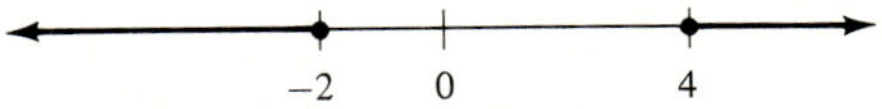

3. $\{x \mid x < -1 \text{ or } x > 1\}$ 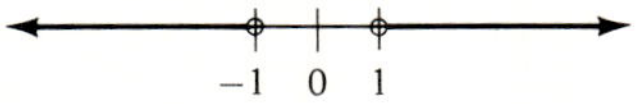

5. $\{x \mid -1 \le x \le 4\}$ 

7. $\{x \mid -3 < x < -1\}$ 

9. $\{x \mid -4 < x < \frac{1}{2}\}$ 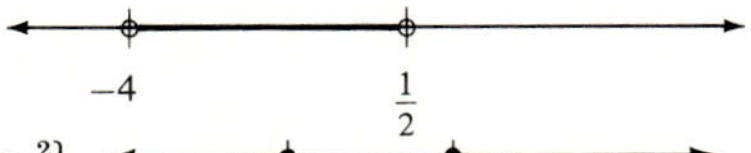

11. $\{x \mid x \le -2 \text{ or } x \ge \frac{2}{3}\}$ 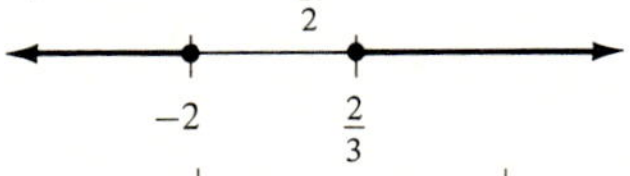

13. $\{x \mid x < -5 \text{ or } x > 0\}$ 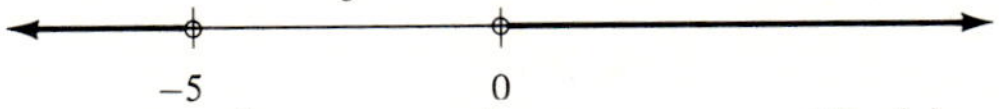

15. $\{x \mid 0 \le x \le 4\}$  **17.** $\{\ \}$ **19.** $\mathbb{R}$

21. $\{x \mid -4 < x < -1 \text{ or } x > 3\}$ **23.** $\{x \mid x < -2 \text{ or } -1 < x < 1\}$ **25.** $\{x \mid 2 < x < 7\}$

27. $\{x \mid -7 < x < -3\}$ **29.** $\{x \mid -2 < x < -\frac{5}{3}\}$ **31.** $\{x \mid x < -1 \text{ or } x > -\frac{1}{2}\}$

33. $\{x \mid -1 < x < 3 \text{ or } x > 5\}$ **35.** $\{x \mid x < -6 \text{ or } -2 < x < -1\}$

37. $\{x \mid -5 < x < 1 \text{ or } x > 10\}$ **39.** $\{x \mid x < -2 \text{ or } 2 < x < 6\}$

Chapter 6 Review

1. $\{-2, 4\}$ **2.** $\{-3, 0\}$ **3.** $\{-4, 7\}$ **4.** $\{-5, -3\}$ **5.** $\{2, 9\}$

6. $\{7\}$; -5 extraneous **7.** $\{-6, 6\}$ **8.** $\{0, 1\}$ **9.** $\{\pm 7\}$ **10.** $\{\pm 6\sqrt{2}\}$

11. $\{-7, 3\}$ **12.** $\{-4, 3\}$ **13.** $\{-5 \pm 5\sqrt{2}\}$ **14.** $\{-4 \pm 2\sqrt{5}\}$

15. $\{-2 \pm \sqrt{3}\}$ **16.** $\left\{\dfrac{1 \pm \sqrt{13}}{2}\right\}$ **17.** $\{-\frac{2}{3}, 1\}$ **18.** $\left\{\dfrac{-1 \pm i\sqrt{7}}{2}\right\}$

19. $\{-2 \pm i\}$ **20.** $\left\{\dfrac{7 \pm \sqrt{17}}{8}\right\}$ **21.** $3\sqrt{10}$ in. **22.** 12, 13 or -13, -12

23. 11, 13, or -13, -11 **24.** In 4 sec **25.** 4.5 mph

26. $\{x \mid x \le -1 \text{ or } x \ge 5\}$ 

27. $\{x \mid -6 < x < -5\}$ 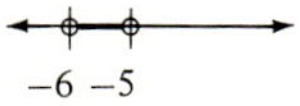

28. $\{x \mid x < -\frac{1}{2} \text{ or } x > 5\}$ 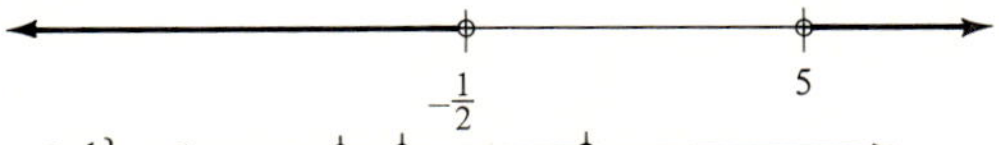 **29.** $\{\ \}$

30. $\{x \mid x \le -3 \text{ or } -2 \le x \le 1\}$ 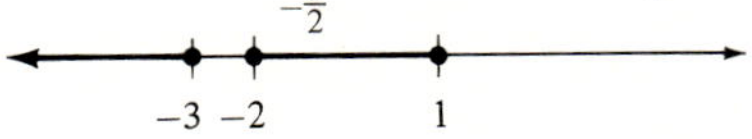

31. $\{x \mid -1 < x < \frac{1}{2}\}$ 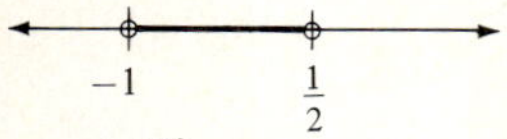

32. $\{x \mid x < -4 \text{ or } 3 < x < 10\}$ 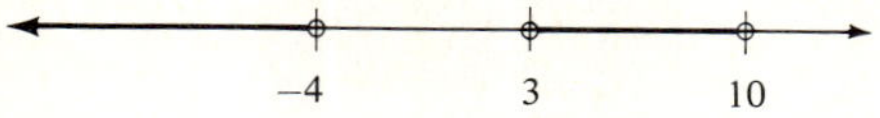

Chapter 6 Test

1. $\{-2, 3\}$ **2.** $\{-3, -2\}$ **3.** $\{-3, 0, 3\}$ **4.** $\{-2, 6\}$ **5.** $\{-1, 4\}$
6. $\{-4, 1\}$ **7.** $\{\pm 6\sqrt{2}\}$ **8.** $\{1 \pm 2i\}$ **9.** $\{-2 \pm \sqrt{5}\}$ **10.** $\left\{\frac{-3 \pm \sqrt{19}}{2}\right\}$
11. $\left\{\frac{3 \pm \sqrt{5}}{2}\right\}$ **12.** $\left\{\frac{-5 \pm \sqrt{13}}{6}\right\}$ **13.** $\left\{\frac{-1 \pm i\sqrt{3}}{2}\right\}$ **14.** $\{2\}$
15. $-8 + \sqrt{30}$ in. **16.** 4.5 mph **17.** $\{x \mid x < 2 \text{ or } x > 3\}$ **18.** $\{x \mid -2 \le x \le 4\}$
19. $\{x \mid x < -2 \text{ or } 1 < x < 3\}$ **20.** $\{x \mid -3 < x < 2\}$

CHAPTER 7

Exercises 7.1

Exercises 1–29 (odd-numbered):

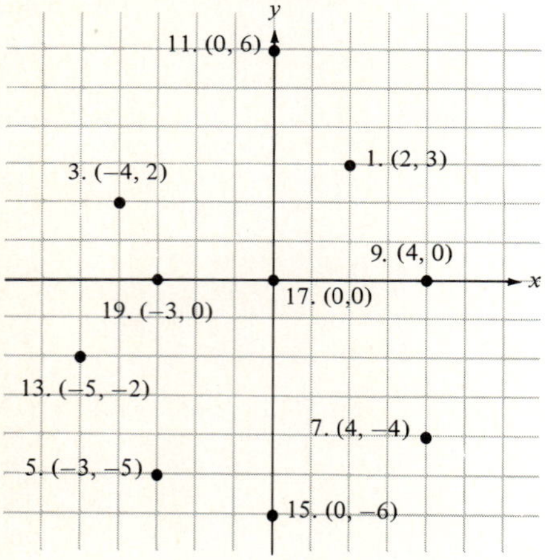

(a)

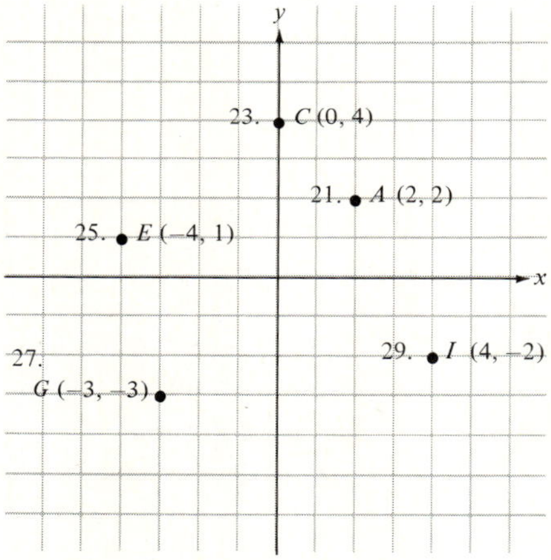

(b)

31. **(a)** III **(b)** I **(c)** I, III **33.** **(a)** 0 **(b)** 0 **35.** B(7, 2), C(7, 5), D(3, 5)

Exercises 7.2

1. (1, 2), (2, 1), (−1, 4) **3.** (0, −2), (6, 0), (9, 1) **5.** $(\frac{1}{3}, 2)$, $(2, -\frac{1}{2})$, (3, −2)
7. (0, 1), (1, 4), (−1, −2) **9.** (3, 2), (3, 4), (3, −3) **11.** (0, −3), (2, 0), (4, 3)
13. (0, 3), (2, 4), (−2, 2)

15.

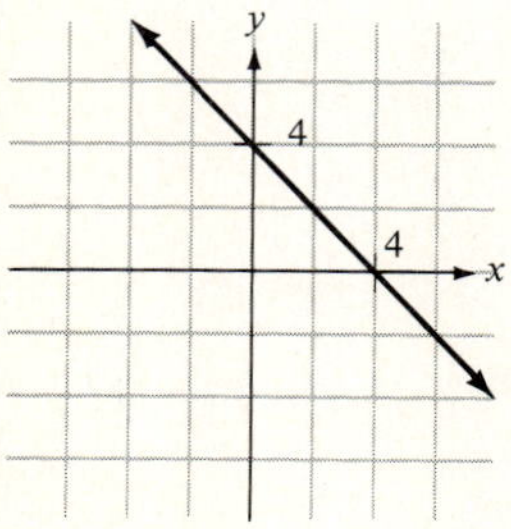

17.

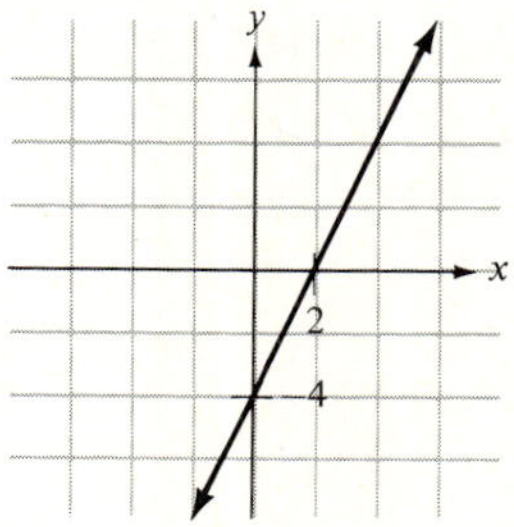

19.

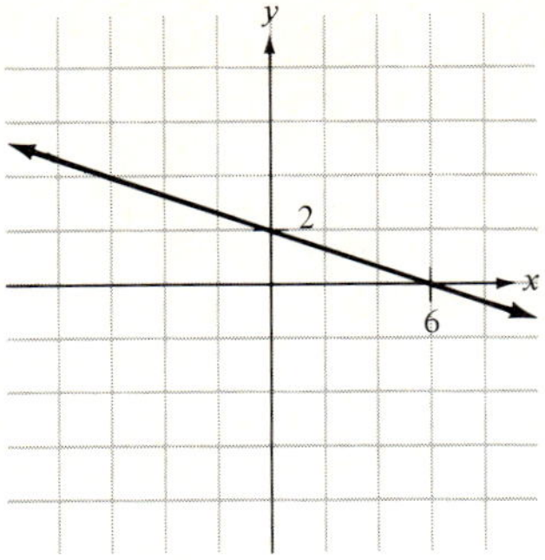

21.

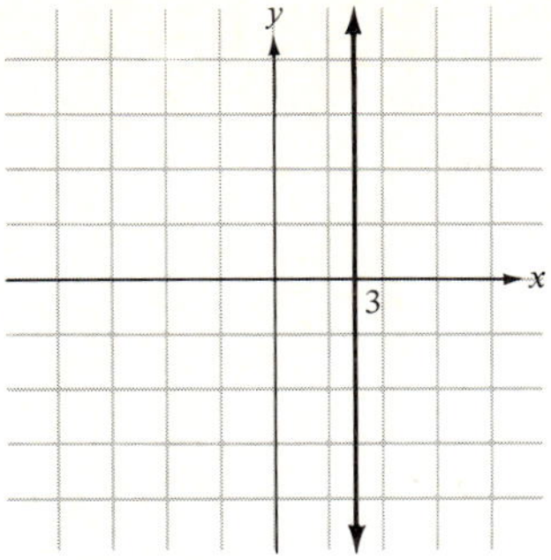

23.

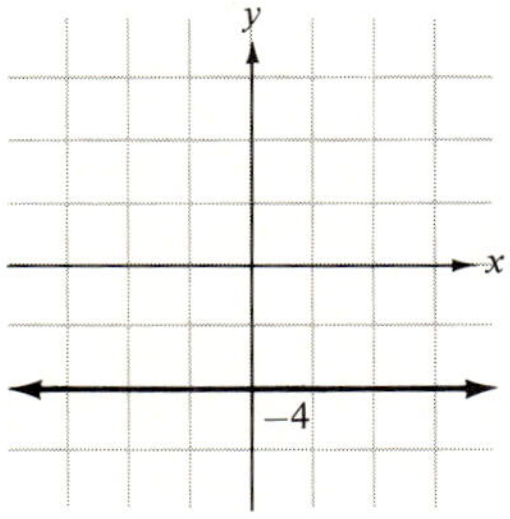

25.

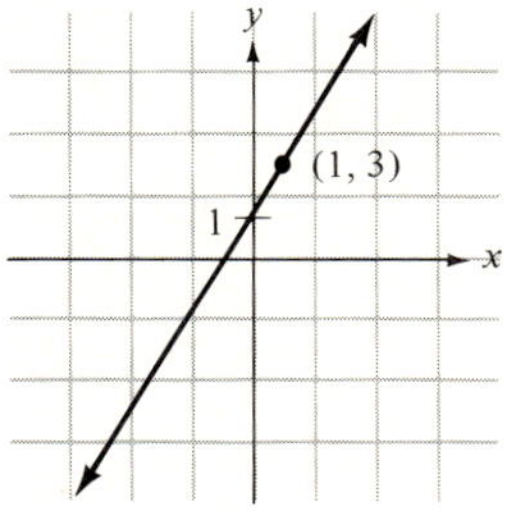

27.

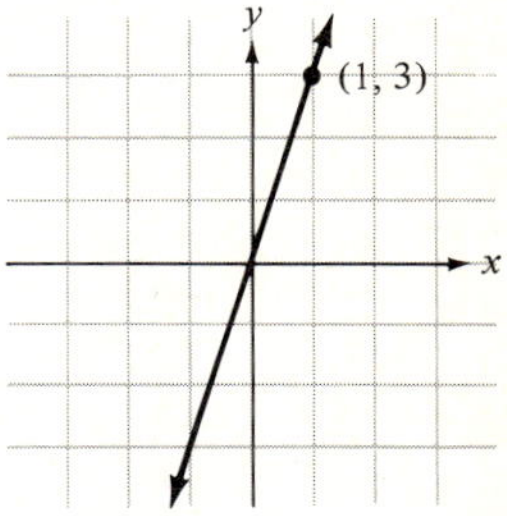

29.

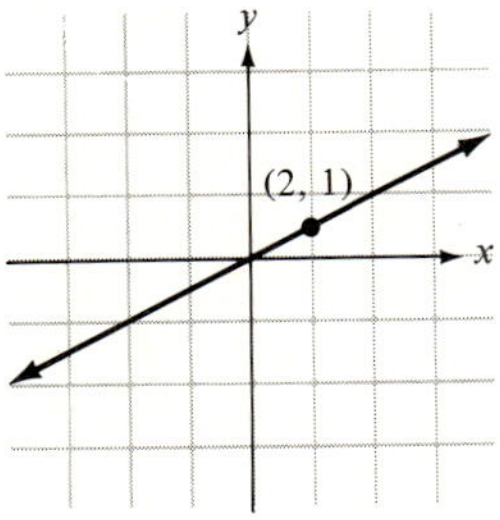

31.

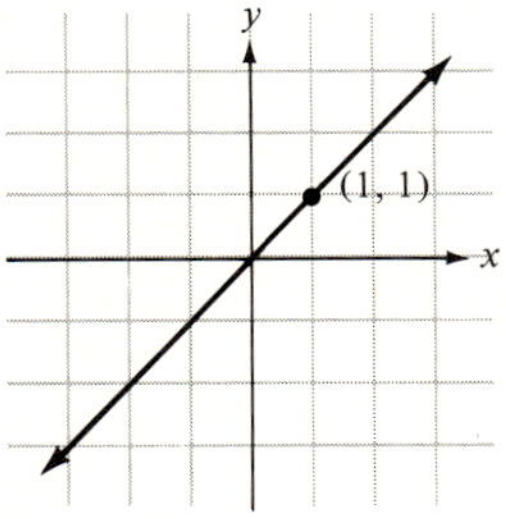

33.

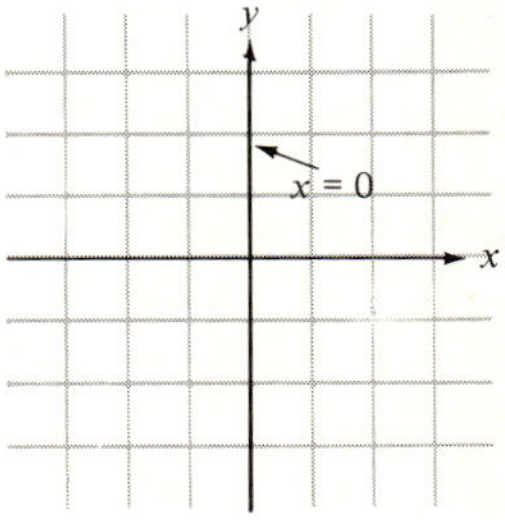

35. $y = 4x$ **37.** $y = 2x + 3$ **39.** $x + y = 4$ **41.** $3y = 4x + 2$

Exercises 7.3

1. 1 **3.** $\frac{3}{2}$ **5.** $-\frac{3}{4}$ **7.** -1 **9.** 0 **11.** Not defined **13.** $-\frac{2}{5}$
15. **(a)** 3 **(b)** $-\frac{1}{3}$ **17.** **(a)** -2 **(b)** $\frac{1}{2}$ **19.** **(a)** $\frac{3}{5}$ **(b)** $-\frac{5}{3}$
21. **(a)** 0 **(b)** Not defined **23.** **(a)** Not defined **(b)** 0 **25.** $\dfrac{-2}{3}$ **27.** 1
29. $\dfrac{-3}{5}$ **31.** 2 **33.** -4 **35.** $\frac{3}{2}$ **37.** $\frac{2}{3}$

Exercises 7.4

1. $y = 5x - 3$ **3.** $y = -3x + 2$ **5.** $y = -x - 3$ **7.** $y = -\frac{1}{3}x + 1$
9. $y = -\frac{2}{3}x + \frac{1}{3}$ **11.** $y = 3x + \frac{2}{5}$ **13.** $y = 2x + 3$ **15.** $y = x - 3$
17. $y = 8$ **19.** $y = -\frac{3}{5}x - 2$ **21.** $y = .25x + .3$ **23.** $y = -.72x + .8$
25. -4 **27.** 2 **29.** -3 **31.** $\frac{5}{2}$ **33.** 0 **35.** -1
37. $y = 3x - 1$ **39.** $y = -5x + 6$ **41.** $y = x - 3$ **43.** $y = -\frac{3}{5}x + 3$
45. $x = 5$ **47.** $y = -8$ **49.** $y = 0$ **51.** $y = 3x + 2$ **53.** $y = \frac{2}{5}x + 3$
55. $y = 2x - 4$

57.
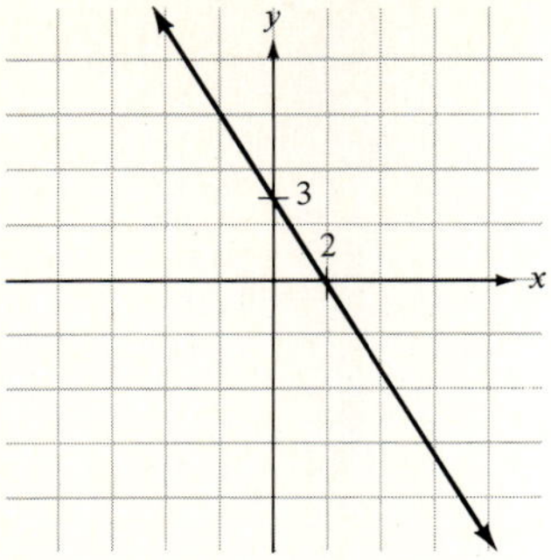

59.
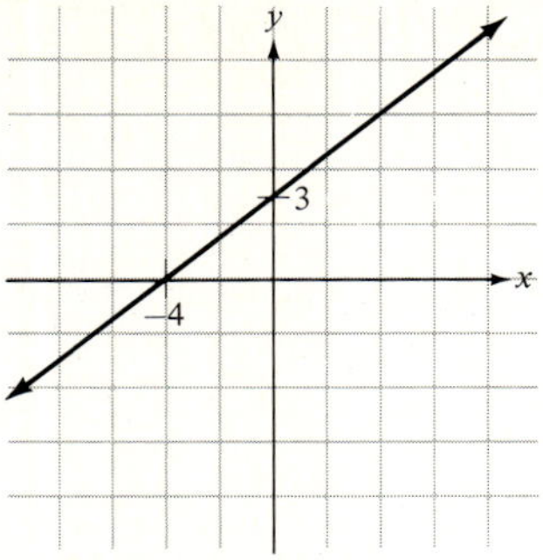

61.
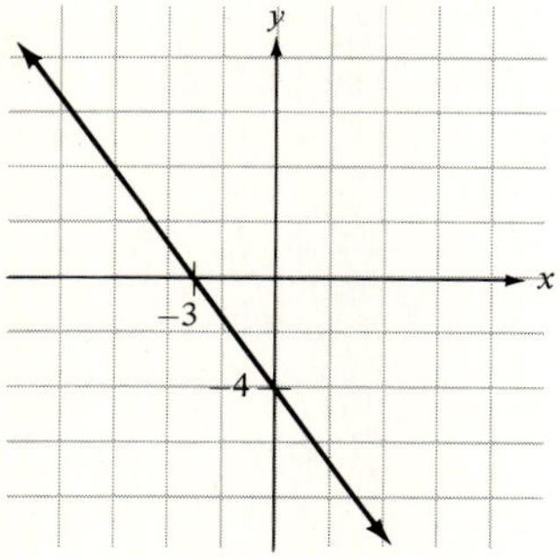

63.
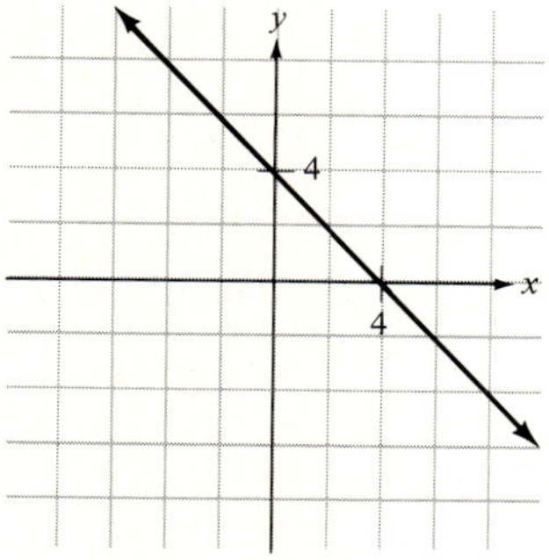

65. 9° Fahrenheit

Exercises 7.5

1.
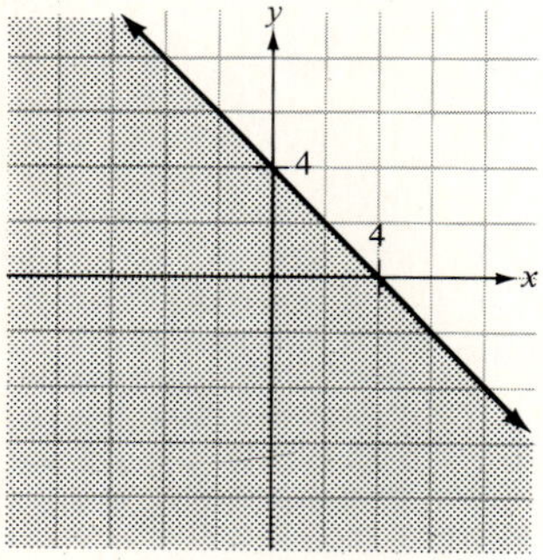

3.
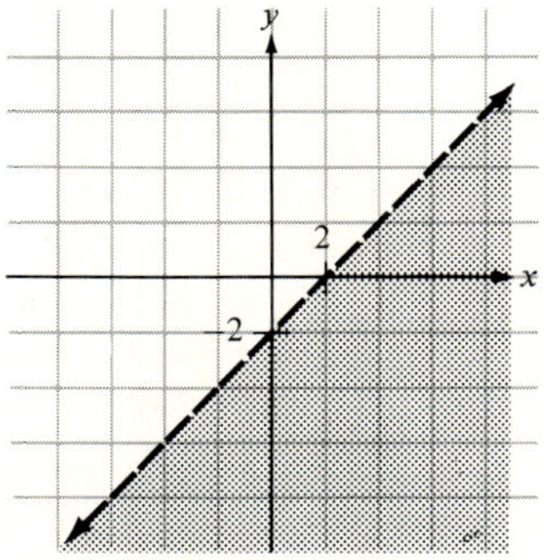

5.
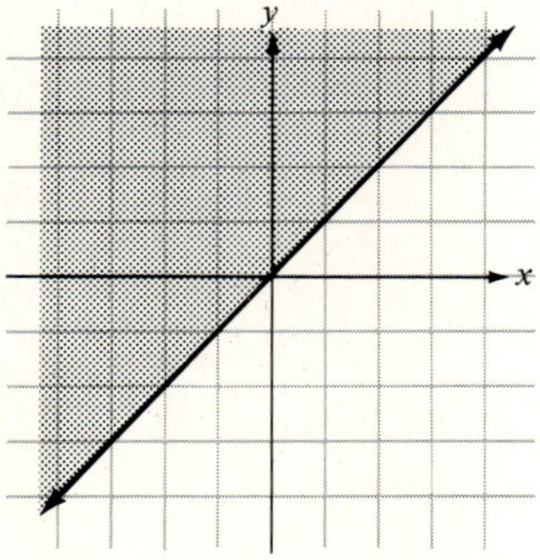

7.
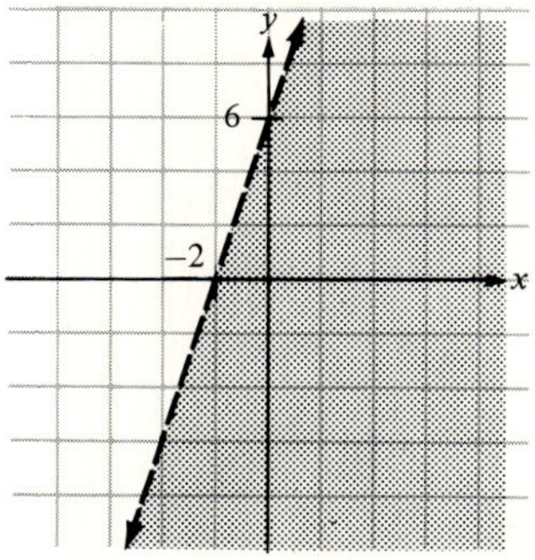

9.

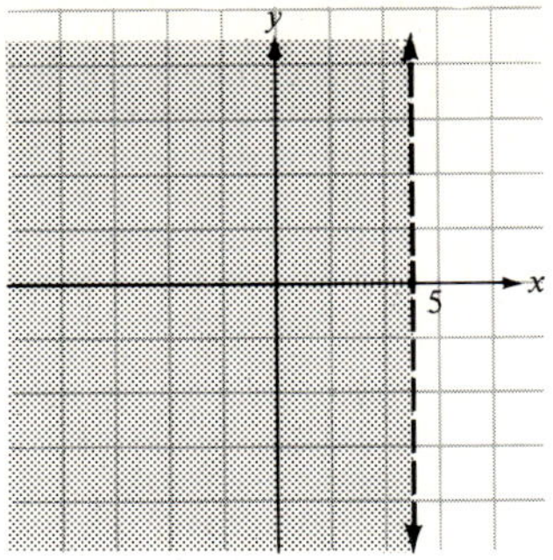

11.

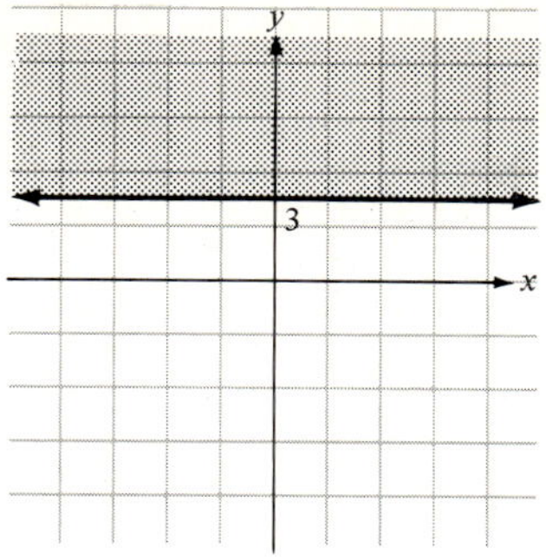

13.

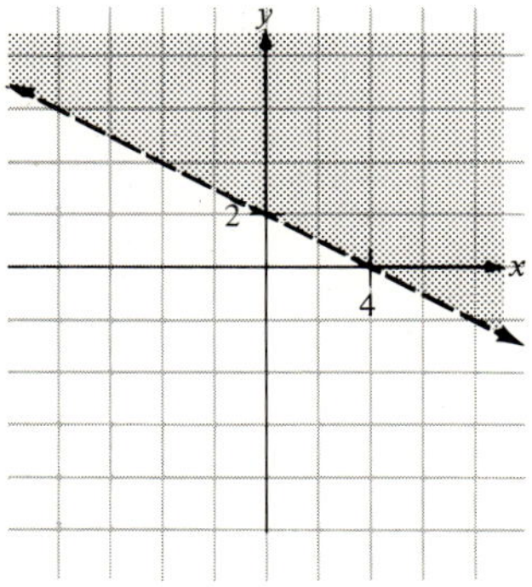

15.

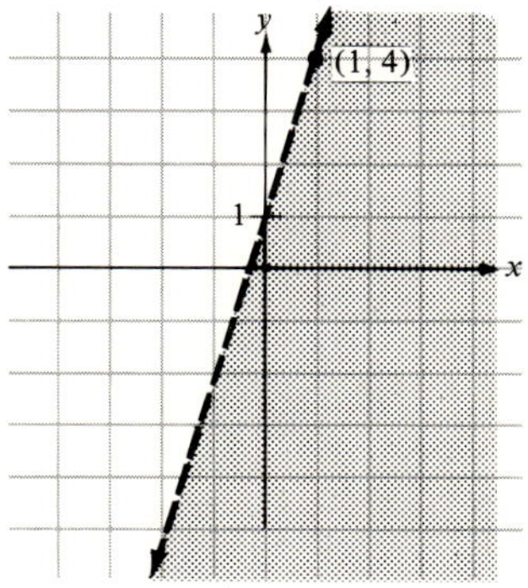

17.

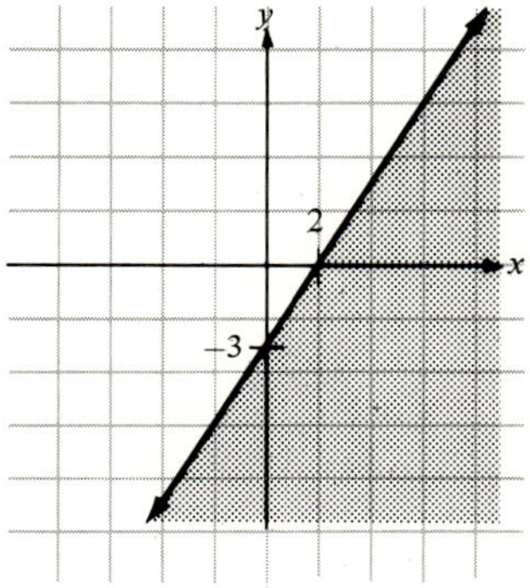

19.

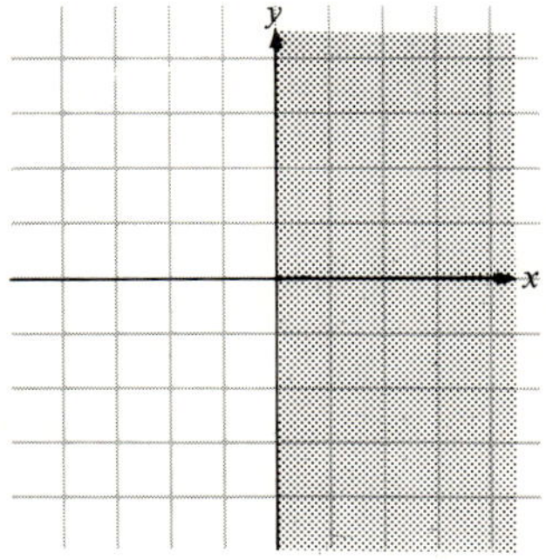

21.

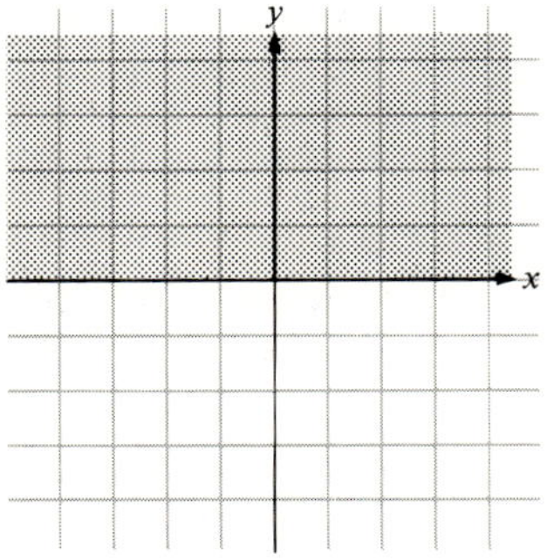

23.

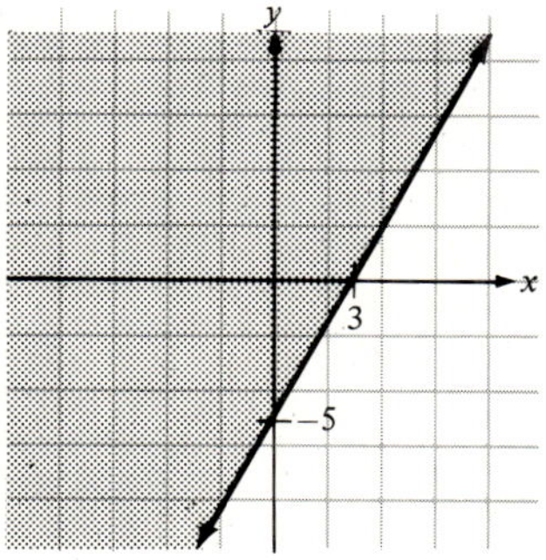

25.

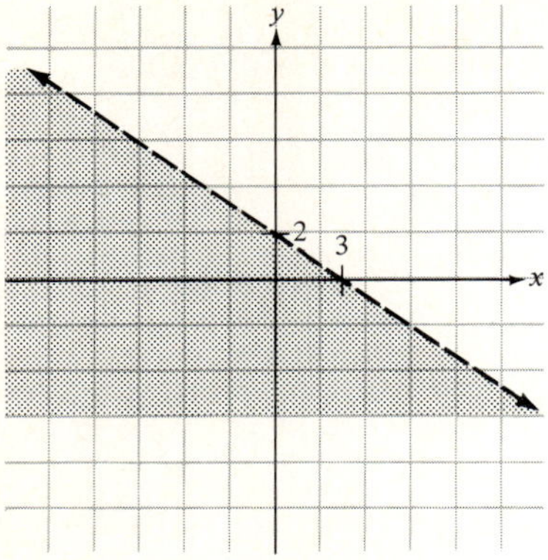

27.

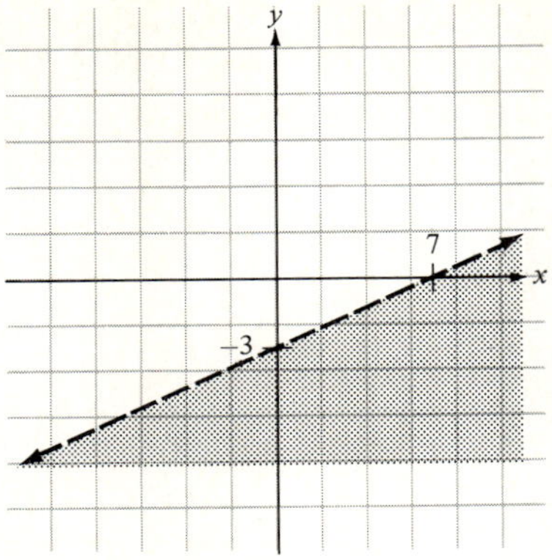

29.

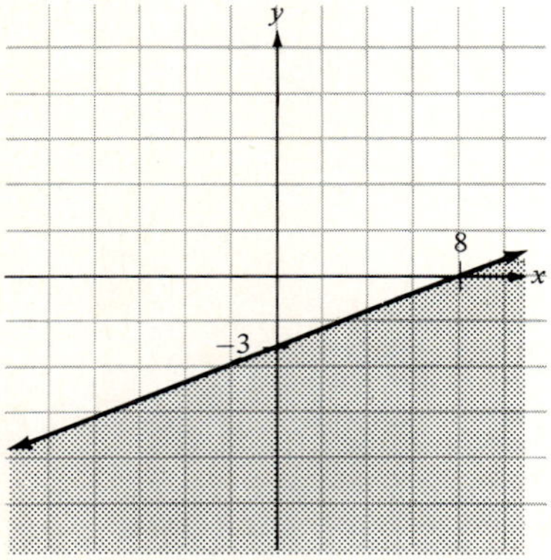

31.

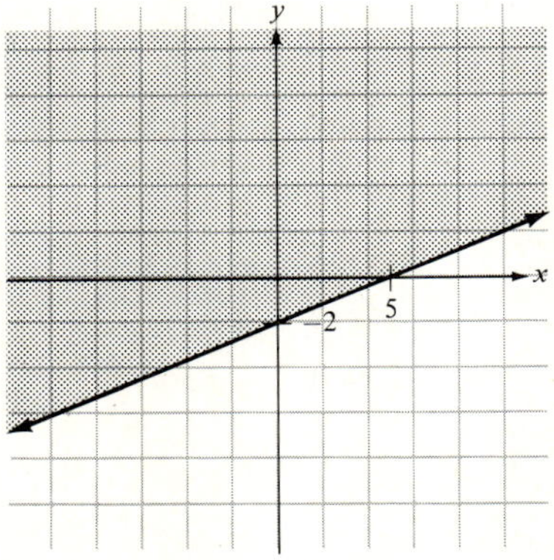

33.

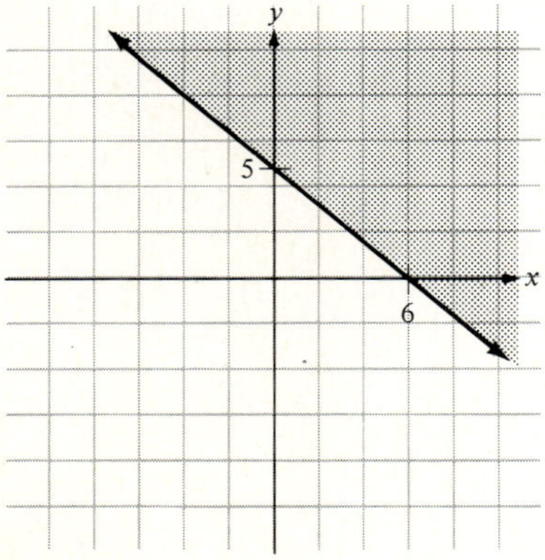

35.

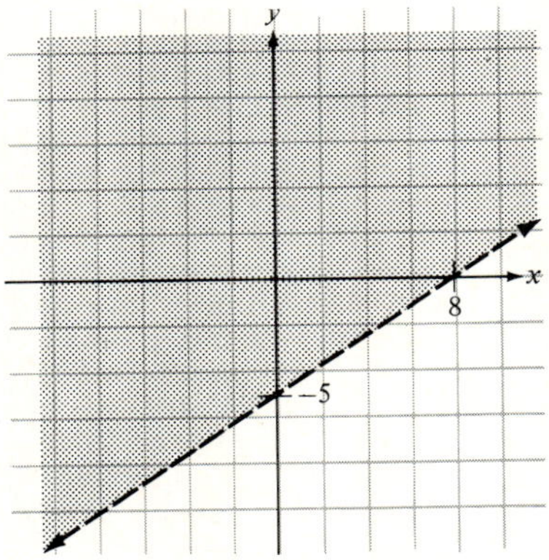

37.

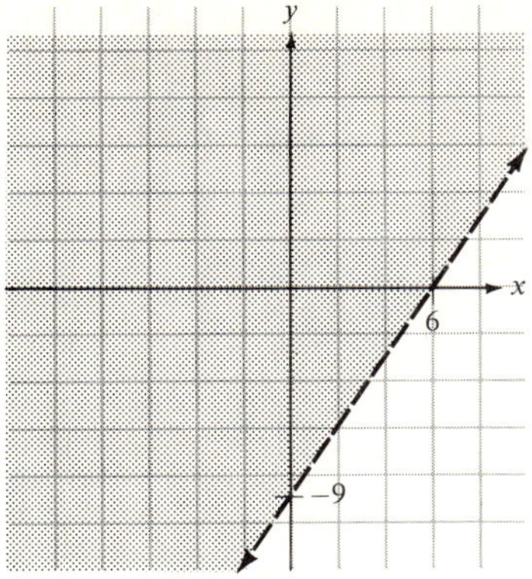

39.

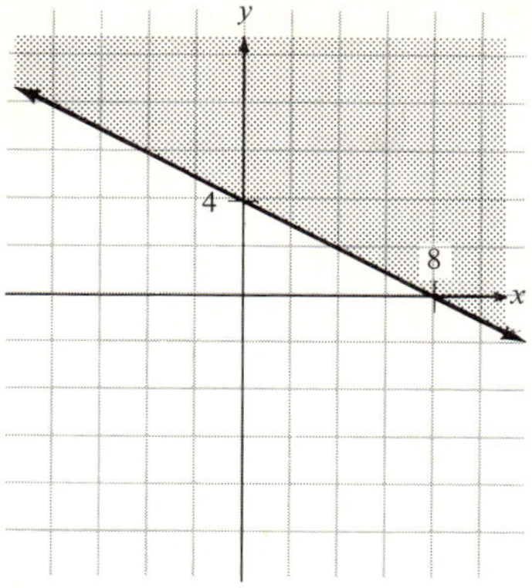

Chapter 7 Review
Exercises 1–5:

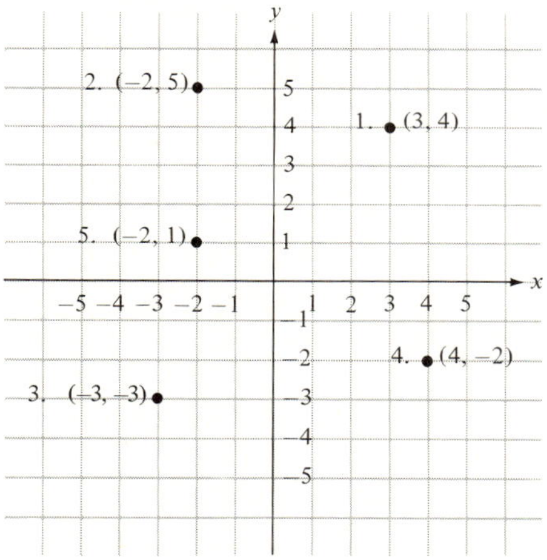

6. (5, 0) **7.** (0, 0) **8.** (−1, 5) **9.** (0, −3) **10.** (1, −6)
11. $A(2, 1)$, $C(5, 4)$ **12.** $A(2, 3)$, $C(9, 7)$ **13.** (0, −6), ($\frac{7}{2}$, 1), (4, 2)
14. (4, −9), (−4, −3), (−12, 3)

15.

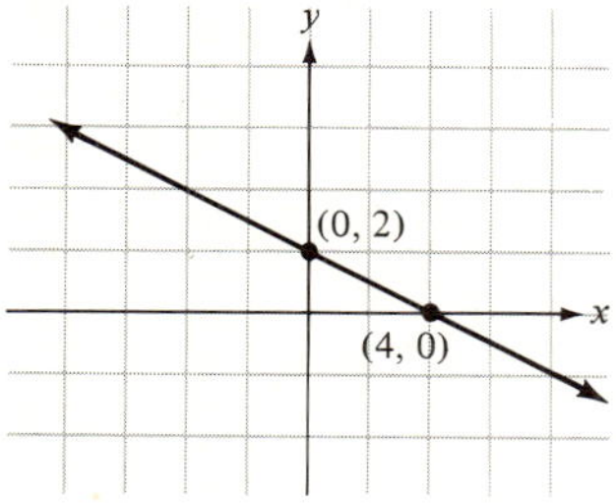

16.

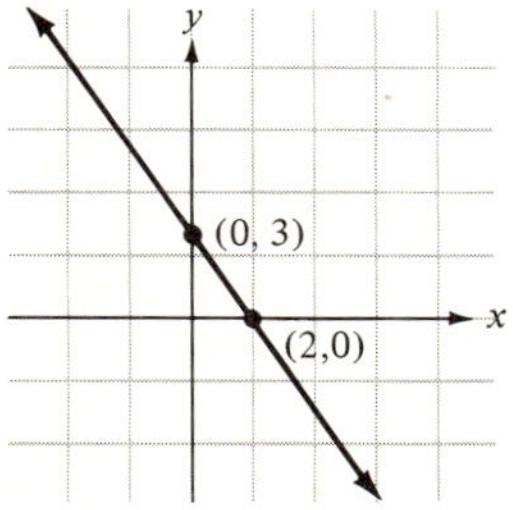

17.

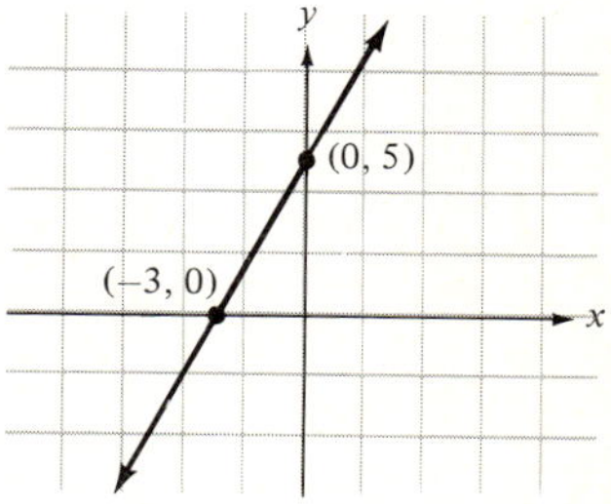

18. $3x + 2y = 14$ **19.** $\frac{3}{2}$ **20.** $-\frac{7}{4}$ **21.** $-\frac{3}{4}$ **22.** −2 **23.** −1
24. **(a)** $-\frac{1}{2}$ **(b)** 2 **25.** −1 **26.** $\frac{3}{5}$ **27.** −5 **28.** $\frac{4}{5}$ **29.** Slope $m = 3$.
30. $y = 2x - 5$ **31.** $y = \frac{4}{3}x + 6$ **32.** $y = -4x + 6$ **33.** $y = .65x + .31$
34. $y = 5x + 1$ **35.** $y = -\frac{4}{3}x + 9$ **36.** x-intercept, −2; y-intercept, 5
37. x-intercept, 5; y-intercept, 4 **38.** 14 cu cm **39.** 5 in.

40.

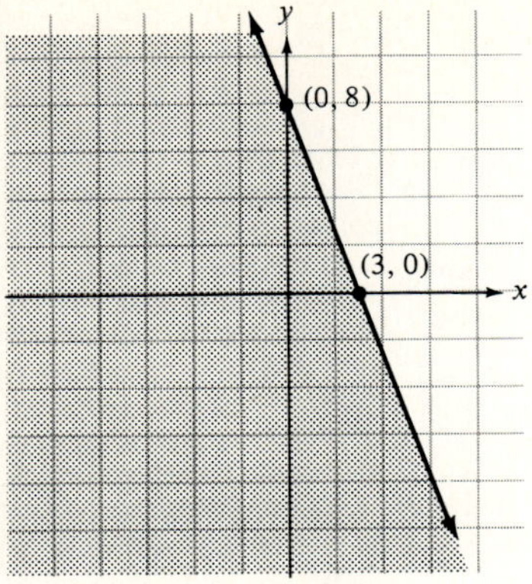

41.

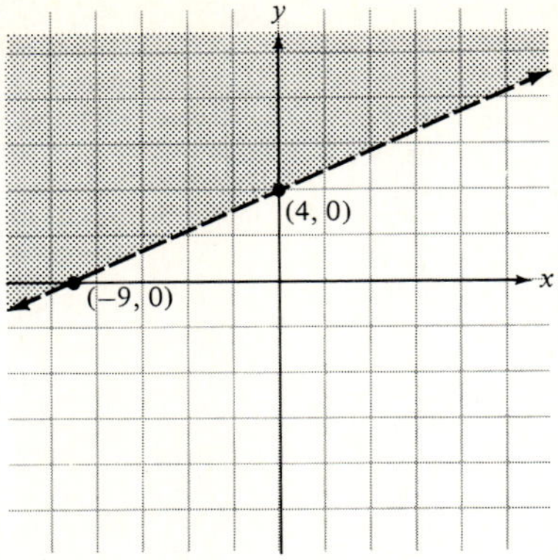

Chapter 7 Test

1.

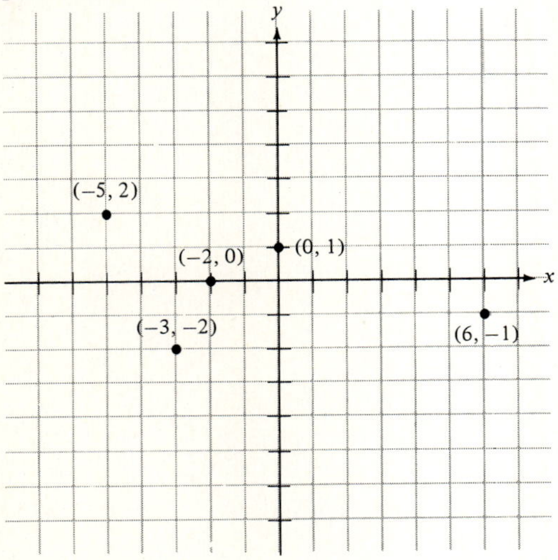

2. A, (2, 3); C, (8, 8); D, (2, 8)
3. (3, 5), (4, 10), (2, 0), (0, −10)
4. $(0, \frac{10}{3})$, (10, 0), $(-4, \frac{14}{3})$, (1, 3)

5.

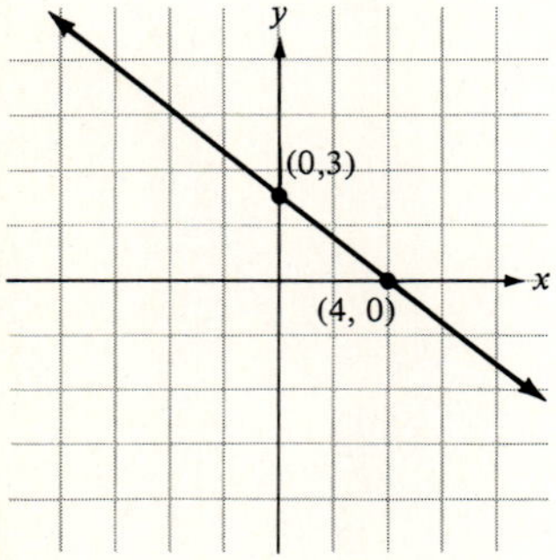

6.

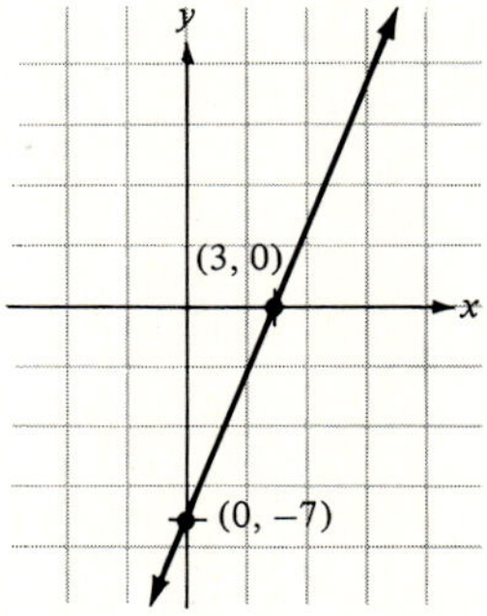

7. −2 **8.** 0 **9.** $\frac{1}{3}$ **10.** $5x + 2y = 17$ **11.** $y = -4x + 14$
12. $y = -5x + 2$ **13.** $y = \frac{3}{4}x + \frac{7}{5}$ **14.** $x = 7$ **15.** $y = -2x + 7$
16. $y = -3x + 4$ **17.** $y = \frac{4}{7}x + 3$ **18.** $y = 4x + 5$ **19.** 3 **20.** 6
21. x-intercept, 4; y-intercept, −5 **22.** x-intercept, −7; y-intercept, 2 **23.** 7 **24.** 2

25.

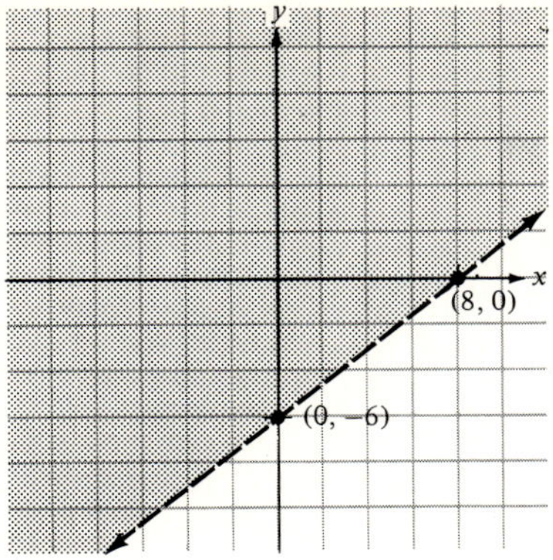

26.

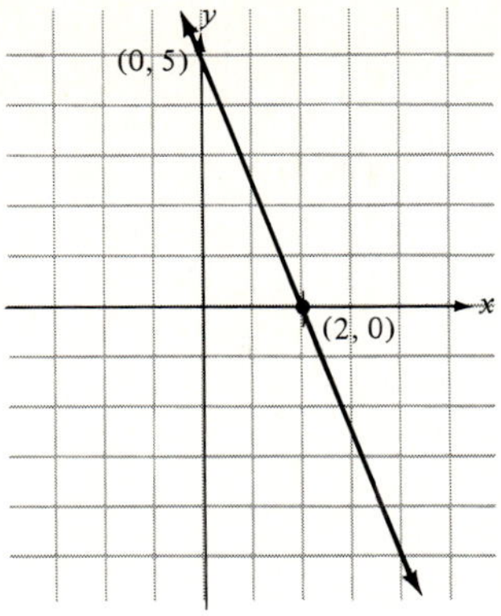

CHAPTER 8

Exercises 8.1

1. 10 **3.** 3 **5.** 5 **7.** 4 **9.** 10 **11.** 13
13. $(x - 3)^2 + (y - 4)^2 = 9$ **15.** $(x + 5)^2 + (y - 1)^2 = 3$ **17.** $x^2 + y^2 = 25$
19. $x^2 + (y - 3)^2 = 13$ **21.** $(x + 4)^2 + (y + 4)^2 = 49$ **23.** $(x - \sqrt{2})^2 + (y - 1)^2 = 16$
25. $x^2 + y^2 = 16$ **27.** $x^2 + y^2 = 4$

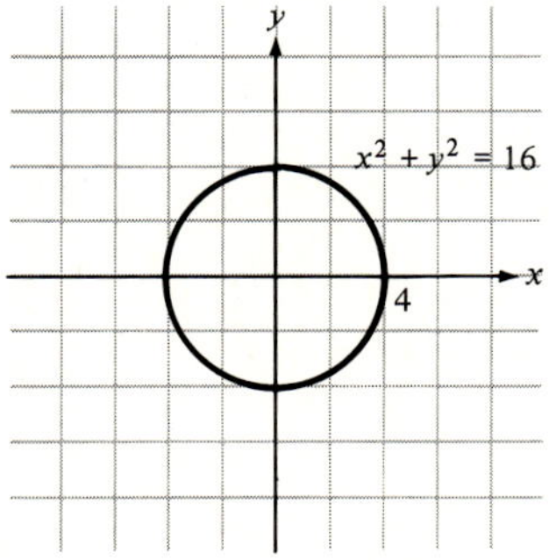

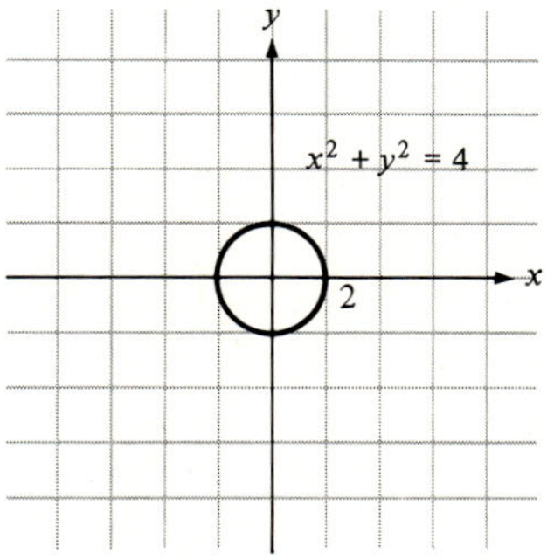

29. $x^2 + y^2 = 2$ **31.** $x^2 + y^2 = 36$

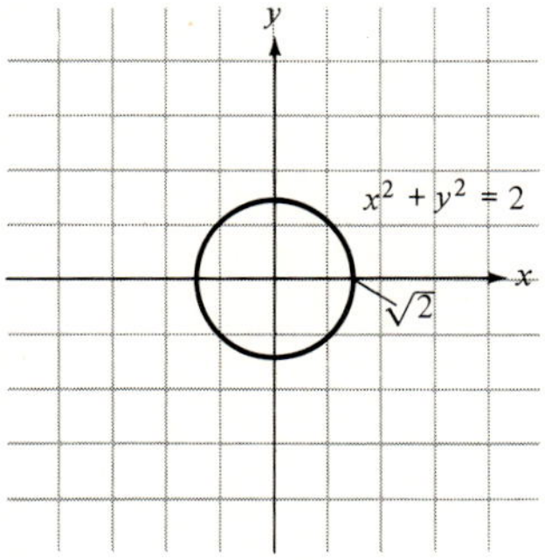

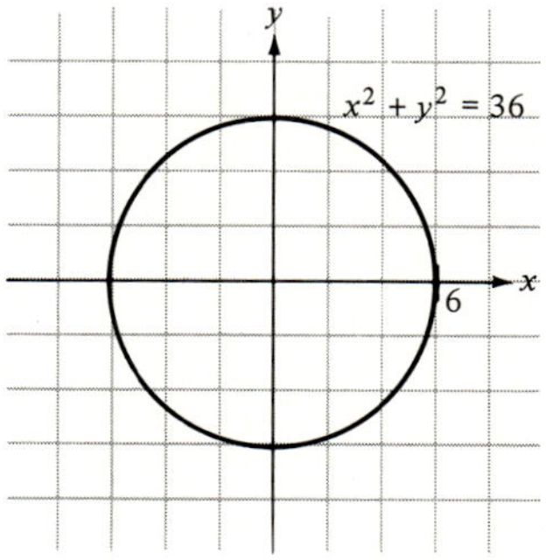

33. $(2, 1)$; $r = 5$ **35.** $(0, 0)$; $r = \sqrt{10}$ **37.** $(-3, 1)$; $r = \sqrt{5}$ **39.** $(1, 2)$; $r = 3$
41. $(4, -1)$; $r = 5$ **43.** $(0, 0)$; $r = 2$ **45.** $(6, 0)$; $r = 6$ **47.** $(\frac{3}{2}, 0)$; $r = \frac{3}{2}$
49. $(\frac{1}{2}, \frac{1}{2})$; $r = 1$ **51.** $x^2 + y^2 = 100$ **53.** $(x - 2)^2 + (y - 1)^2 = 25$

Exercises 8.2

1.
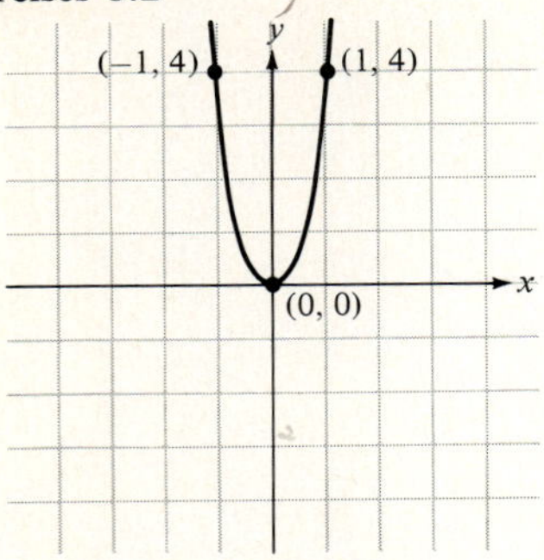

3.
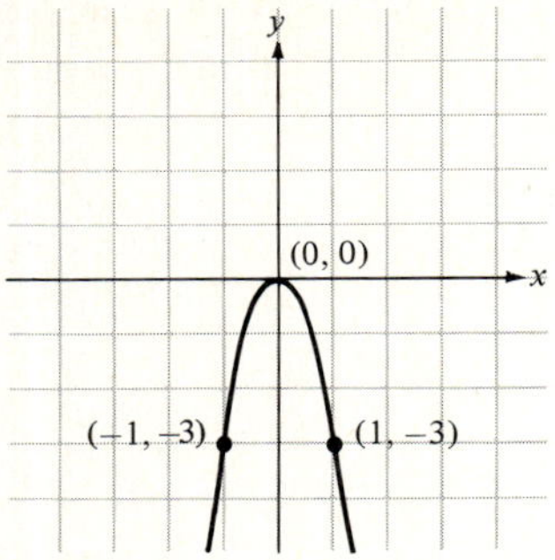

5.
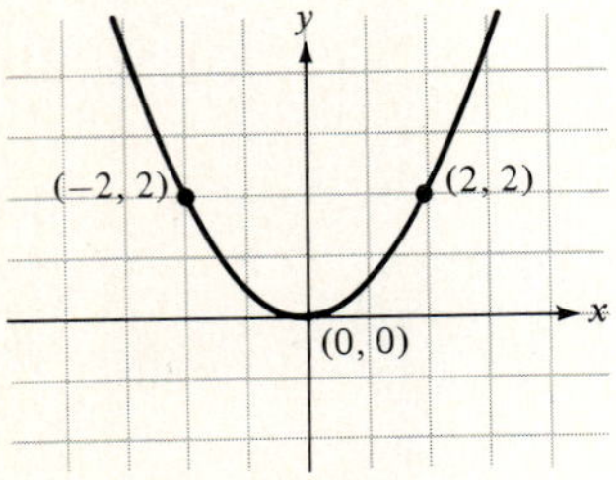

7.
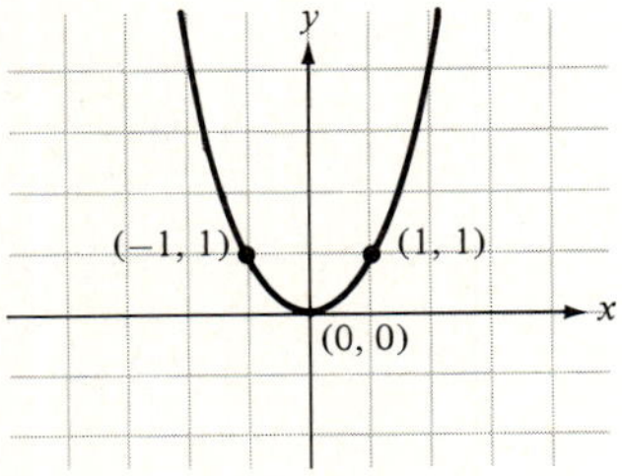

9.
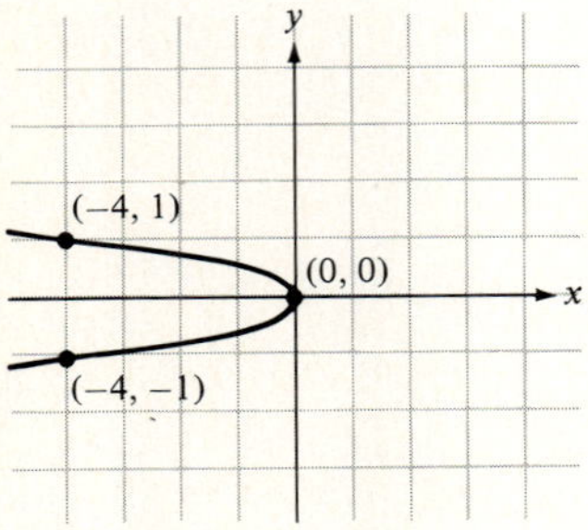

11.
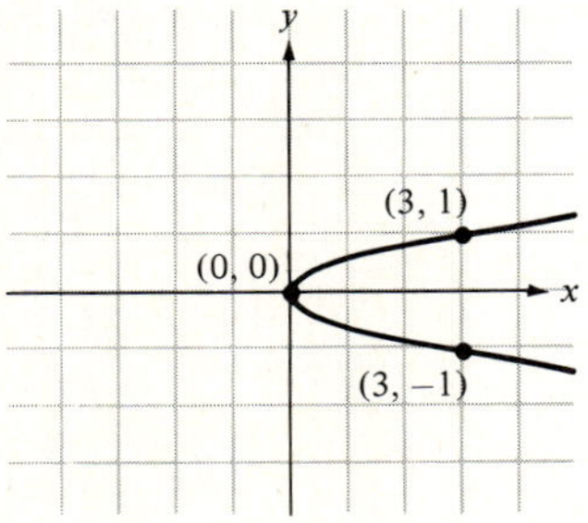

13.
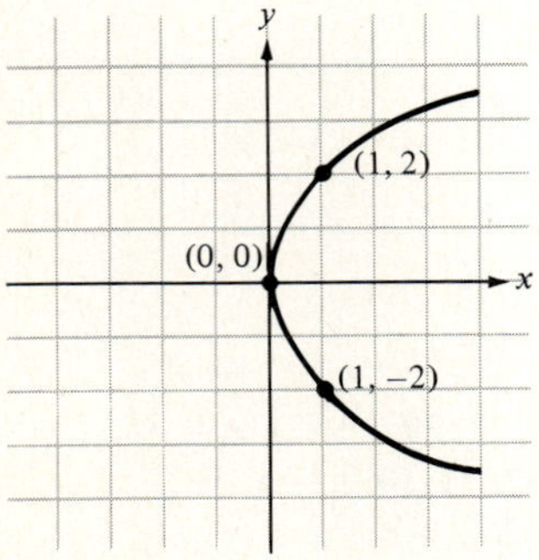

15.
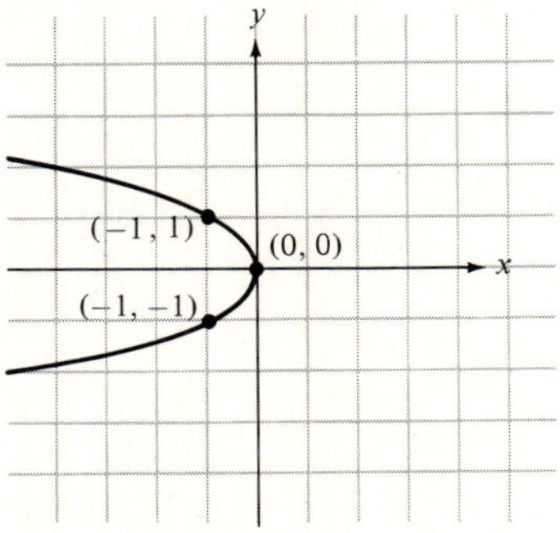

17.

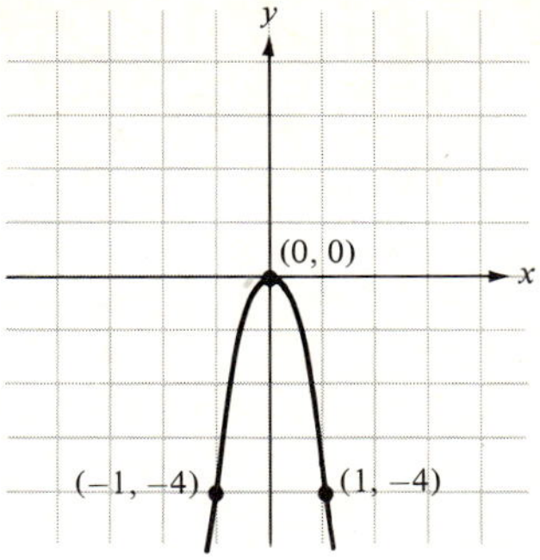

19.

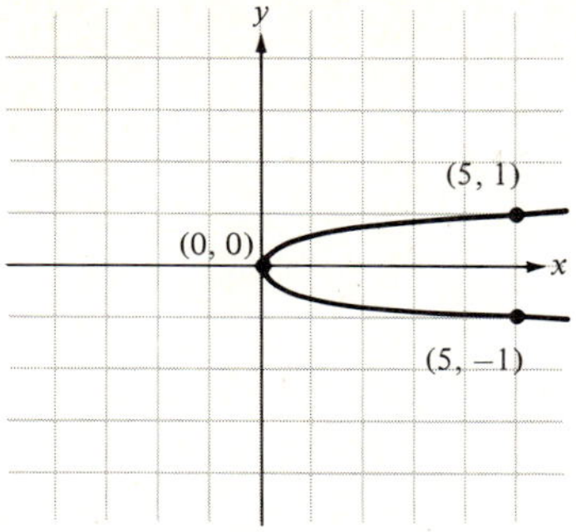

21. **(a)** (3, 1) **(b)** $x = 3$

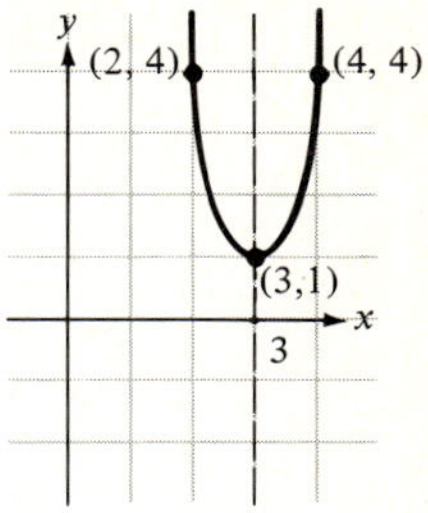

23. **(a)** (4, 2) **(b)** $x = 4$

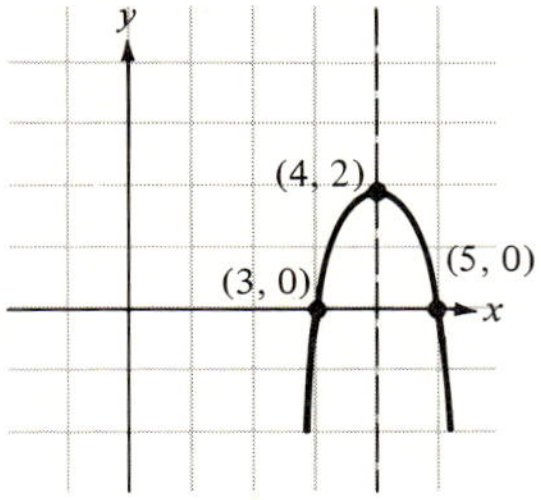

25. **(a)** $(-2, 3)$ **(b)** $x = -2$

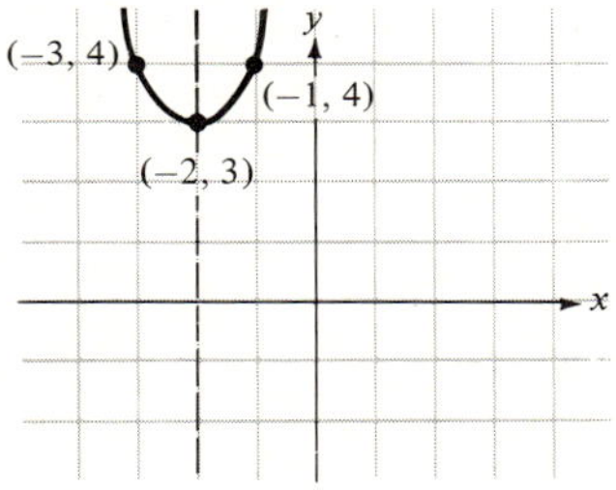

27. **(a)** (0, 1) **(b)** $x = 0$

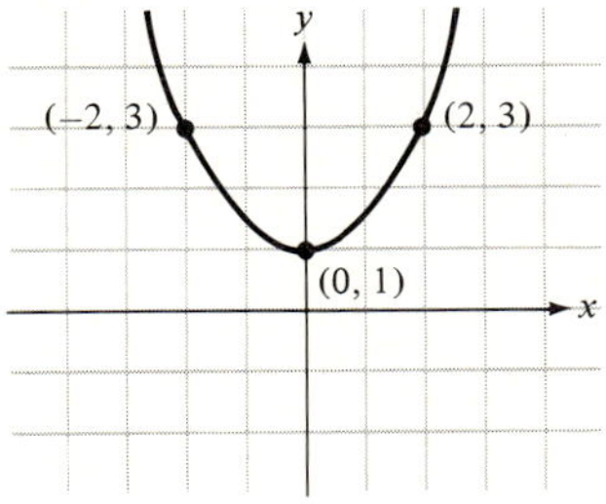

29. **(a)** $(-3, 0)$ **(b)** $x = -3$

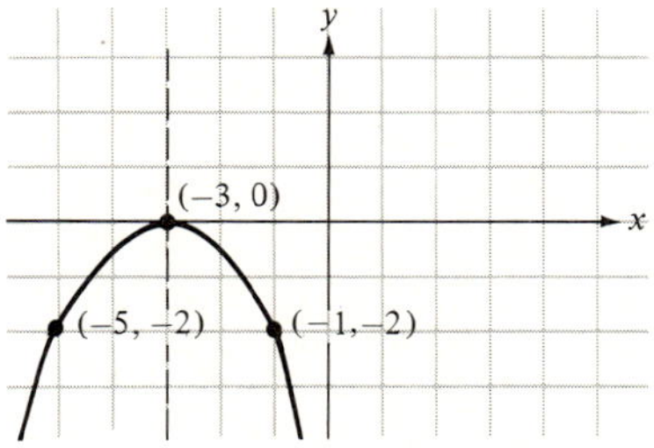

31. **(a)** $(-3, -2)$ **(b)** $x = -3$

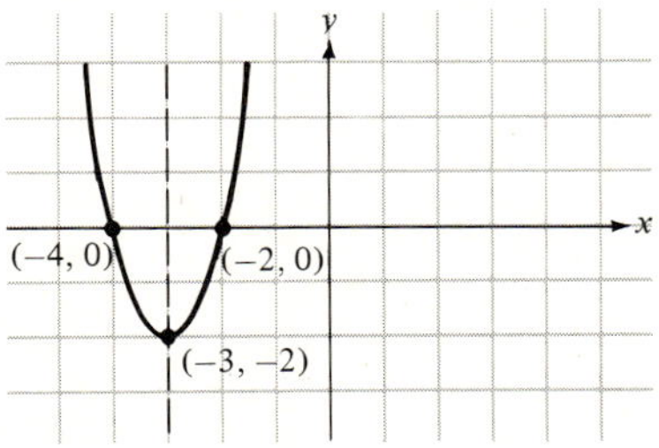

33. **(a)** (1, 2) **(b)** $x = 1$

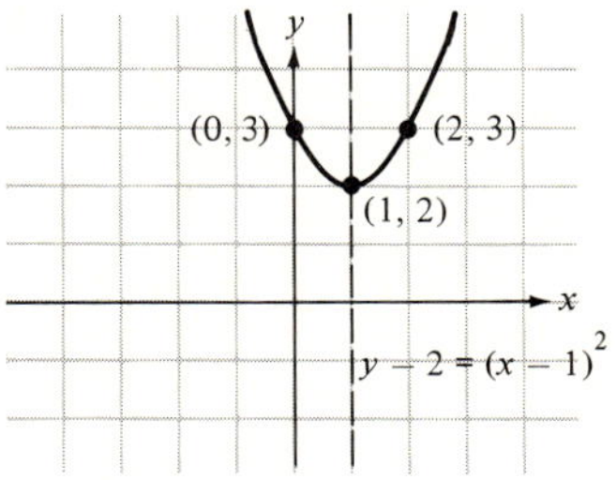

35. **(a)** $(-1, 1)$ **(b)** $x = -1$

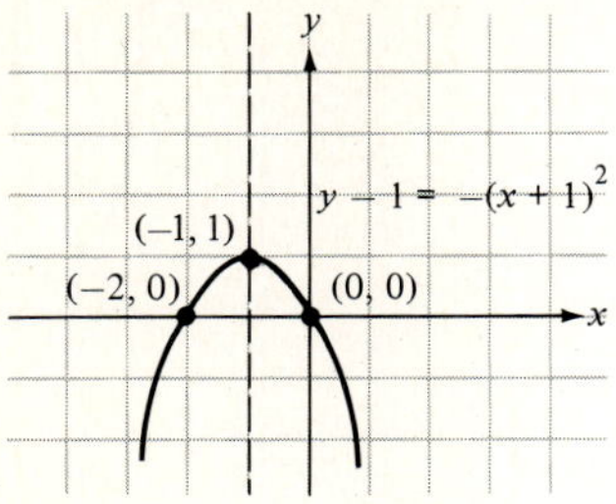

37. **(a)** $(0, 2)$ **(b)** $x = 0$

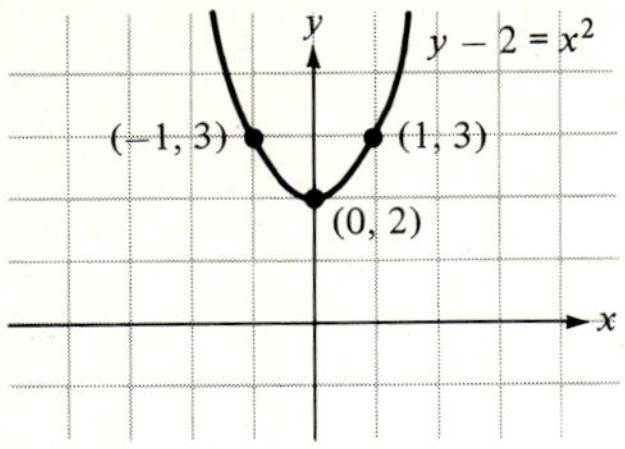

39. **(a)** $(-3, 2)$ **(b)** $x = -3$

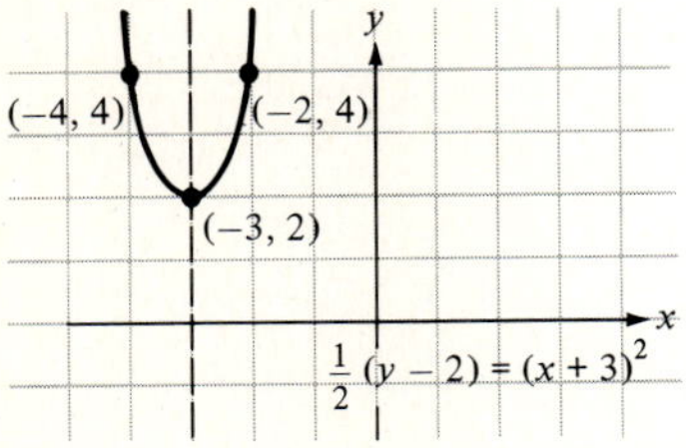

41. **(a)** $(3, -2)$ **(b)** $x = 3$

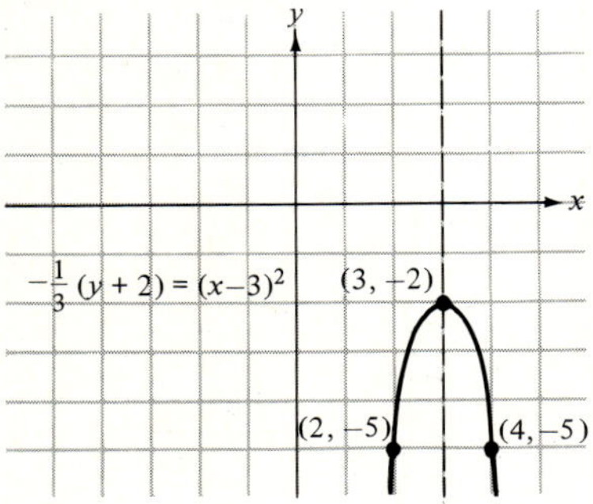

Exercises 8.3

1.

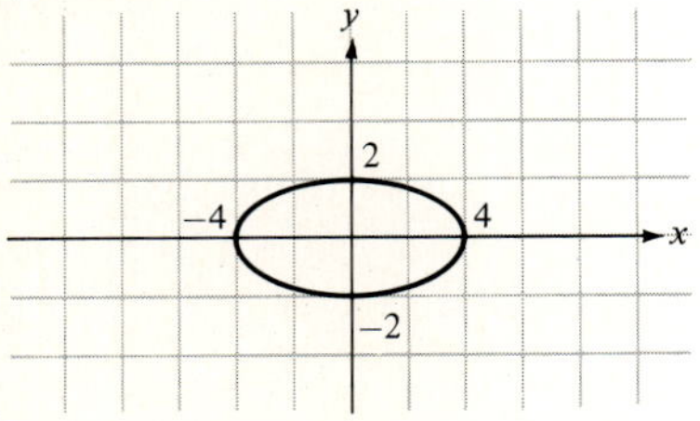

3.

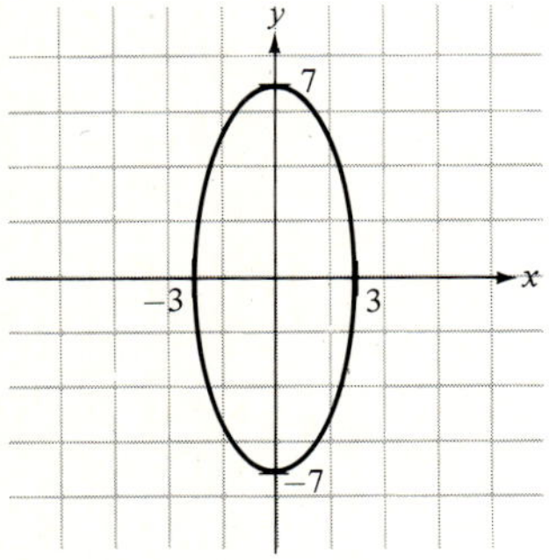

5.

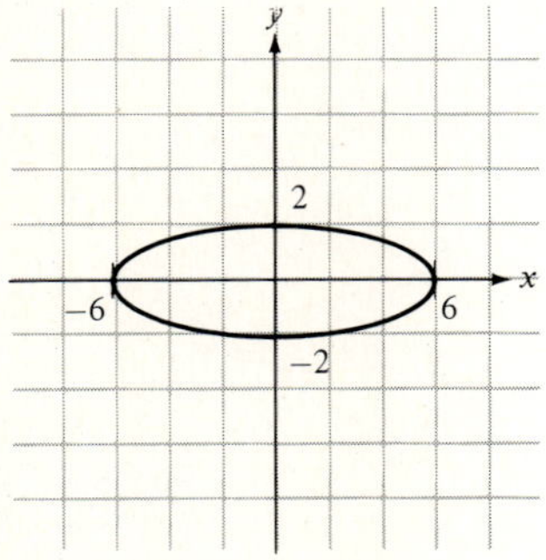

7.

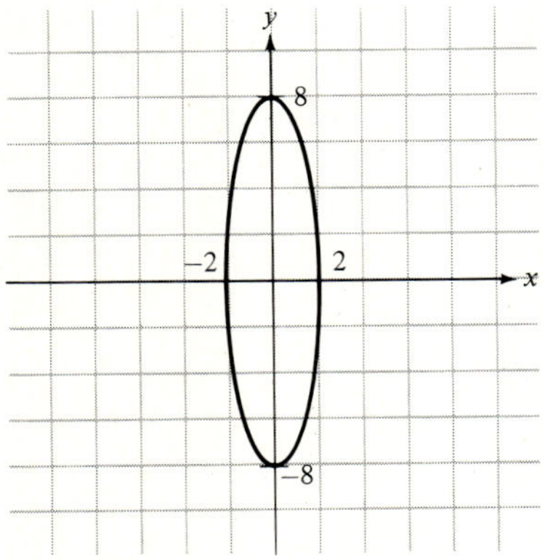

9.

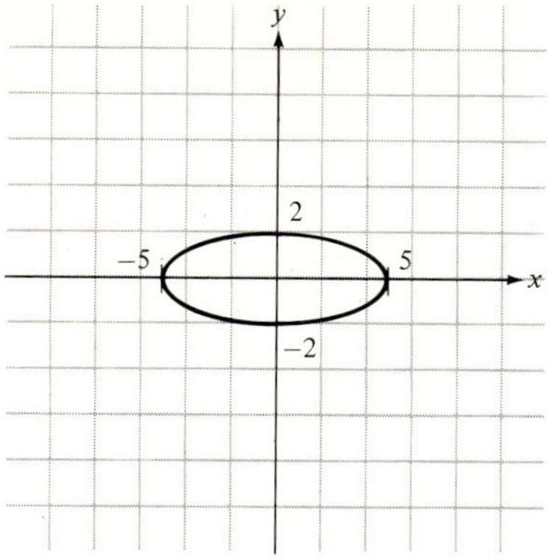

11.

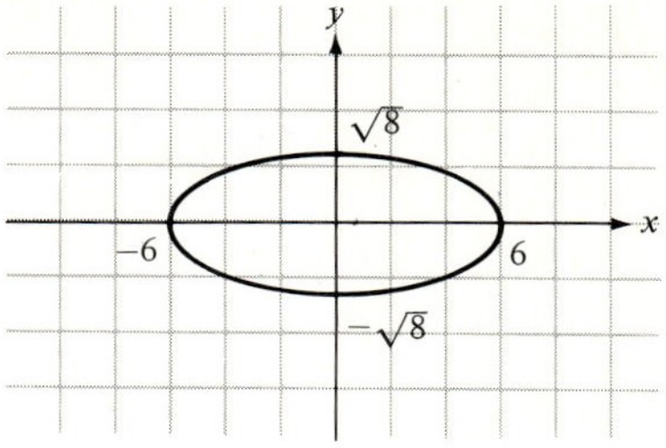

13.

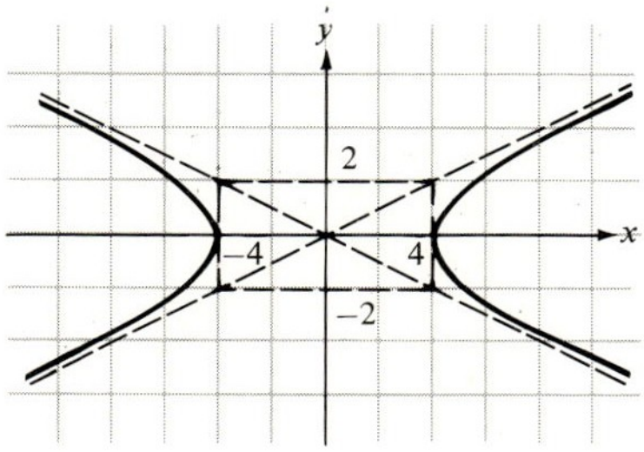

15.

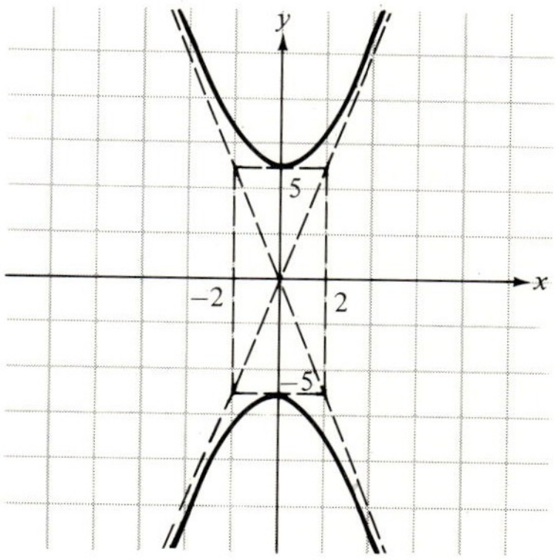

17.

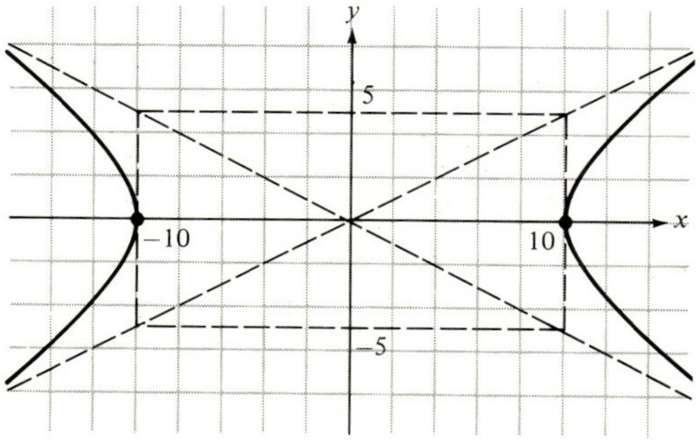

19.

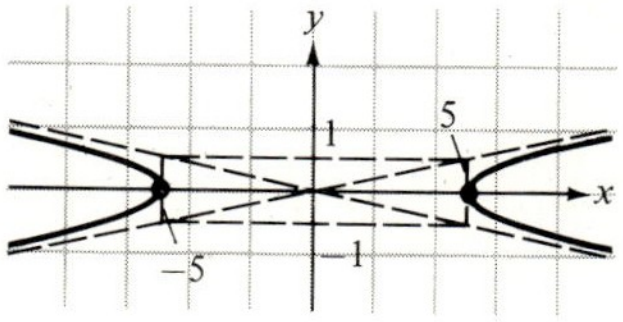

21.

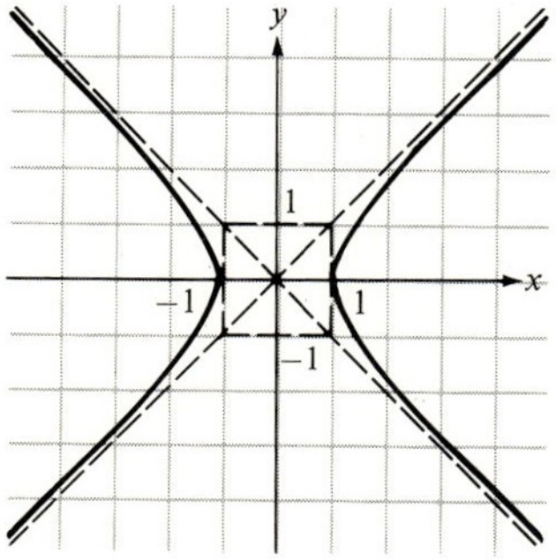

23.

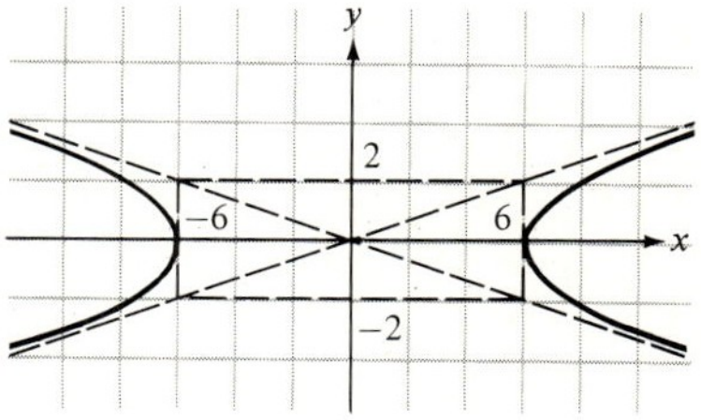

25.

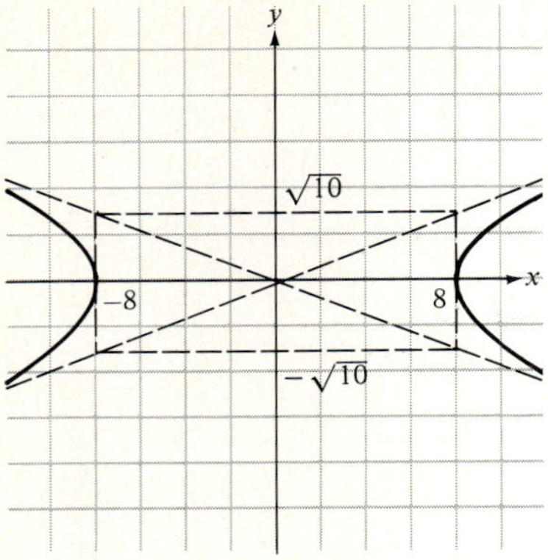

27. Circle **29.** Ellipse **31.** Hyperbola **33.** Ellipse **35.** Hyperbola **37.** Circle

39. **(a)** $x^2 + y^2 = a^2$, $x^2 + y^2 = b^2$ **(b)** a or b

Exercises 8.4

1.

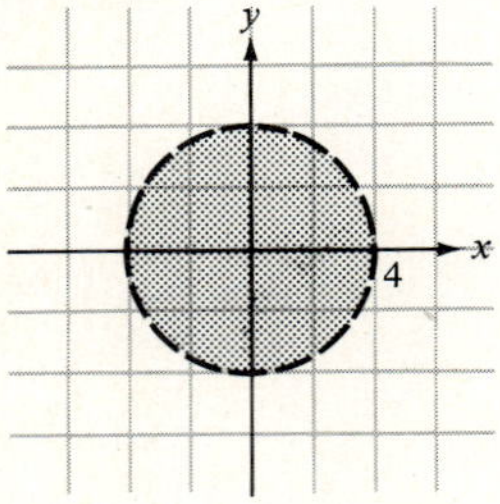

3.

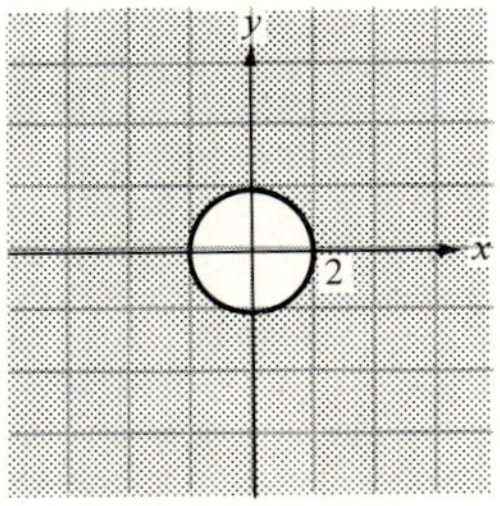

5.

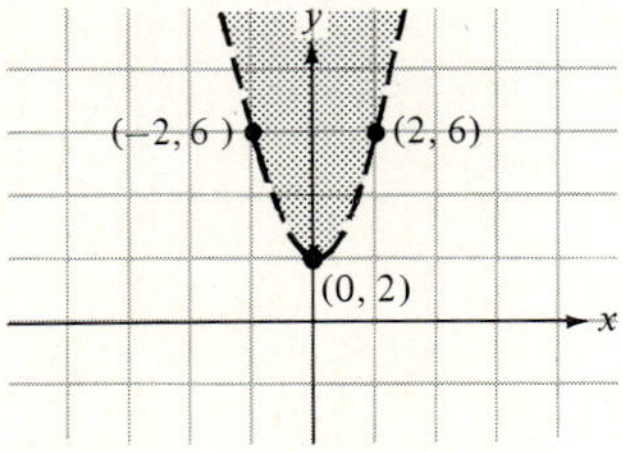

7.

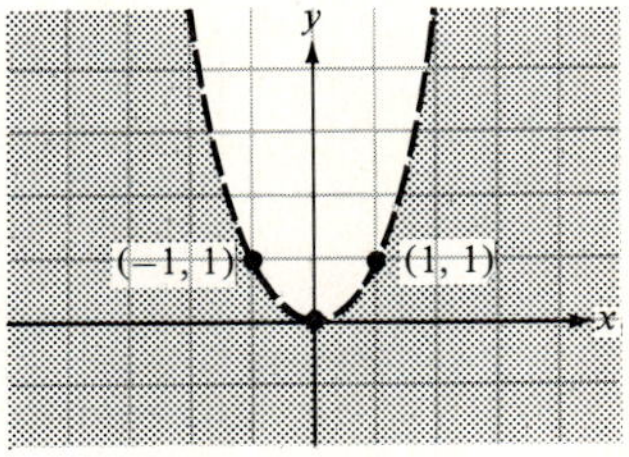

9.

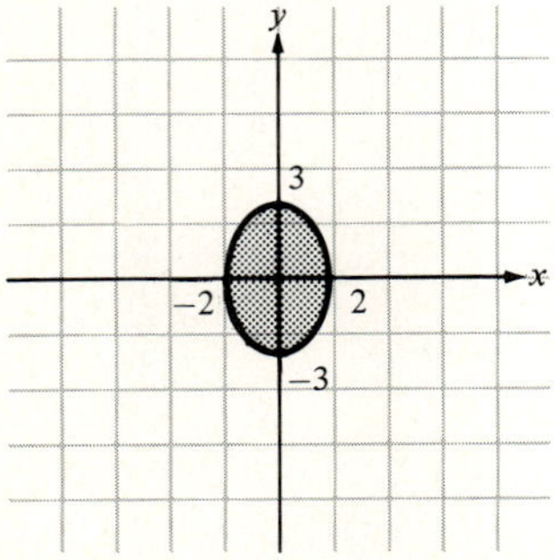

11.

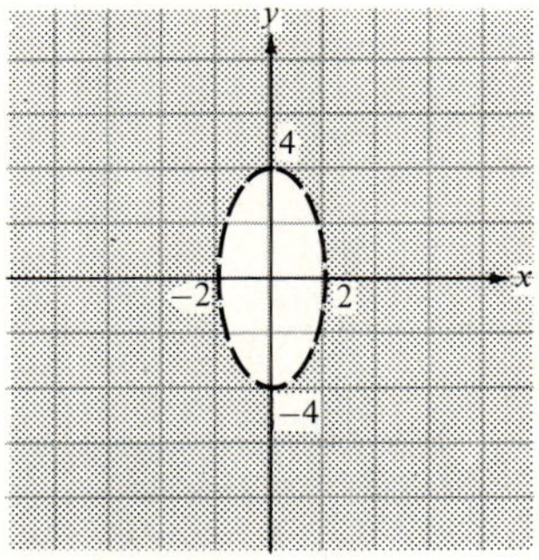

13.
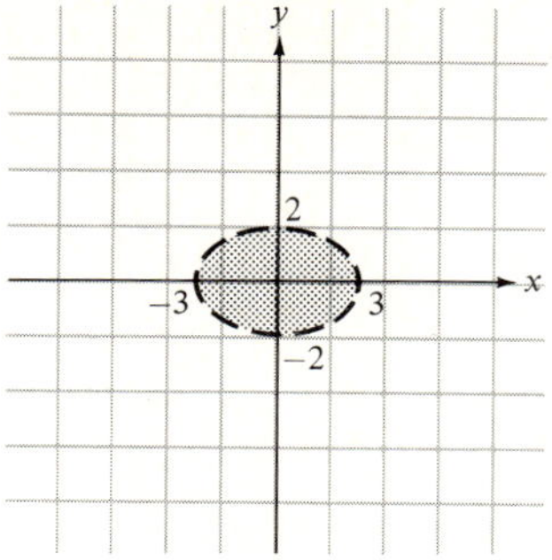

15.
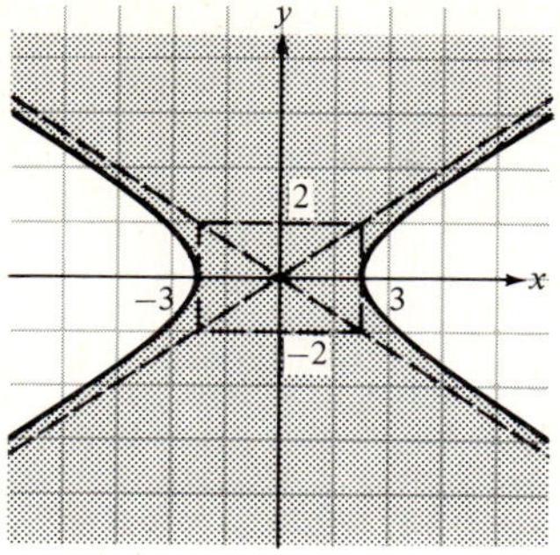

17.
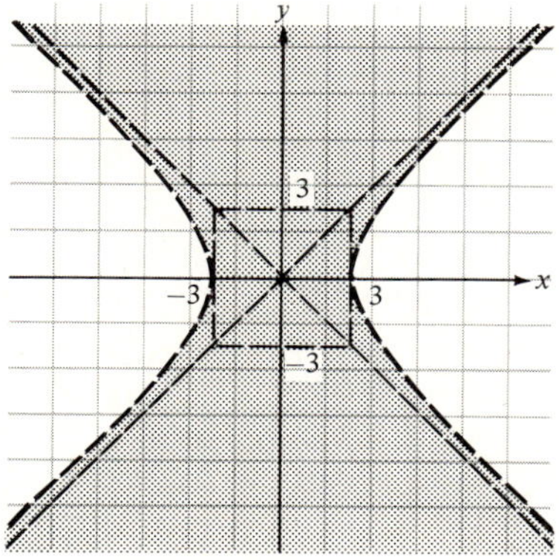

19.
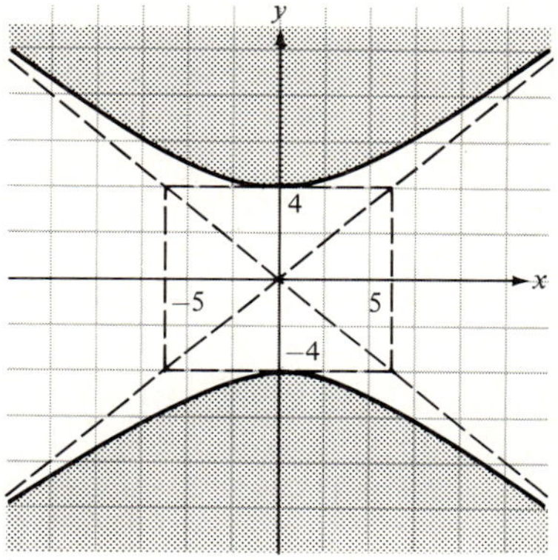

21.
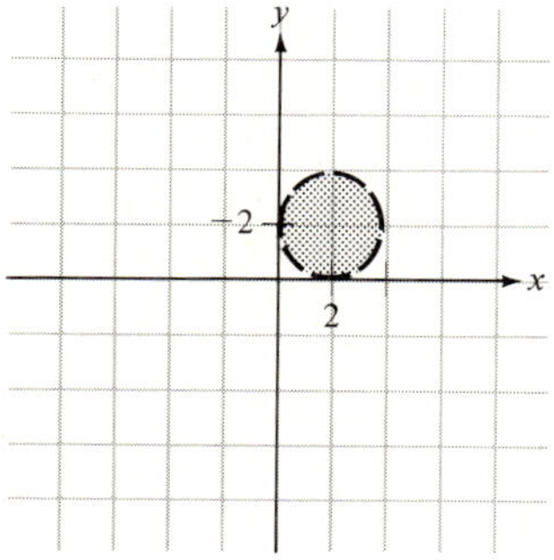

23.
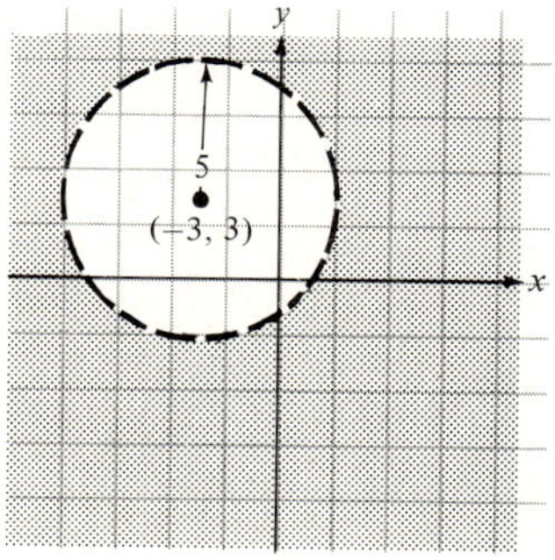

25.
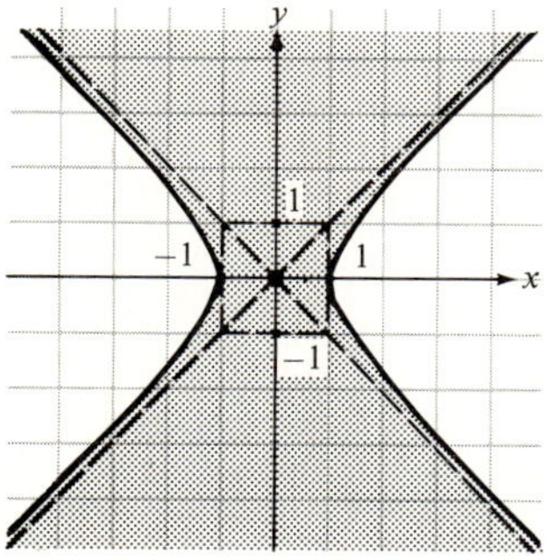

27.
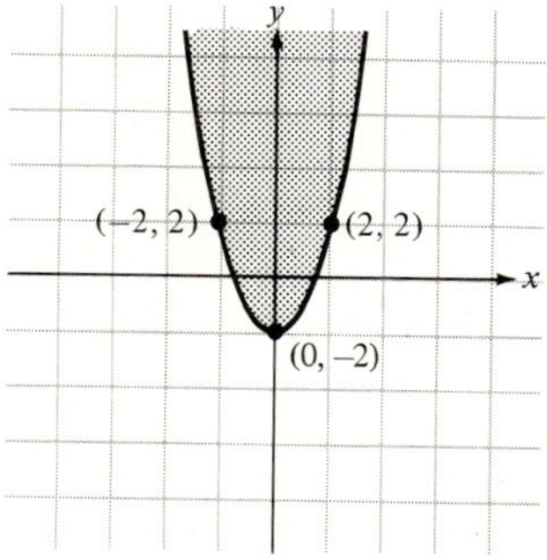

29.

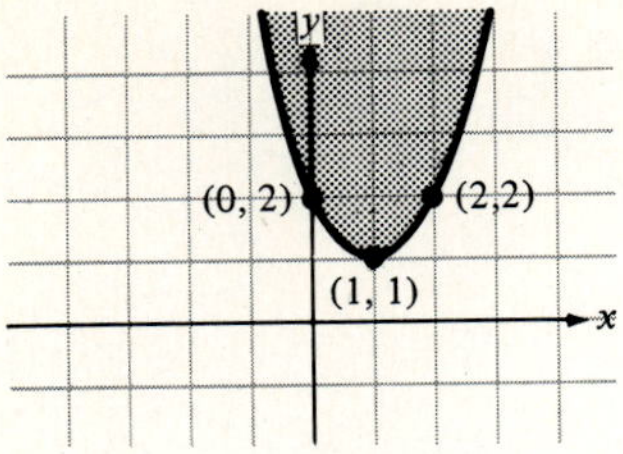

Chapter 8 Review

1. 5 **2.** 5 **3.** 10 **4.** 13 **5.** $(x - 5)^2 + (y - 2)^2 = 4.$

6. $(x + 3)^2 + (y - 4)^2 = 16$ **7.** $(x + 8)^2 + (y + 2)^2 = 25.$

8. $x^2 + y^2 = 9$

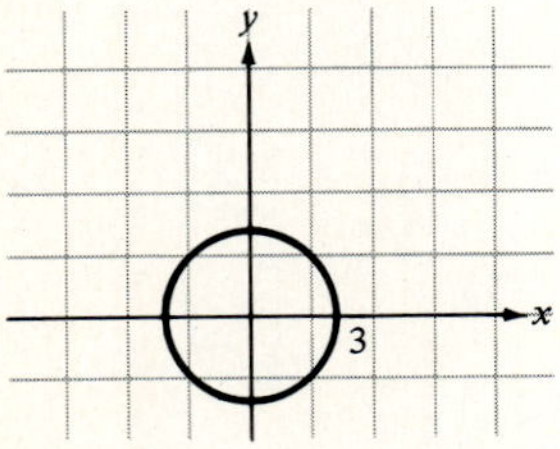

9. $x^2 + y^2 = 16$

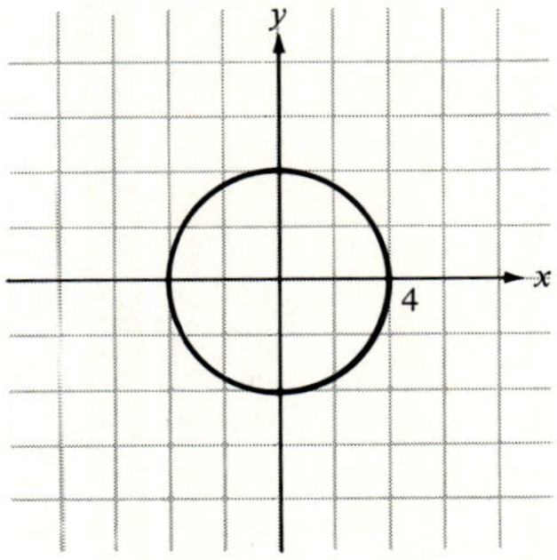

10. $x^2 + y^2 = 3$

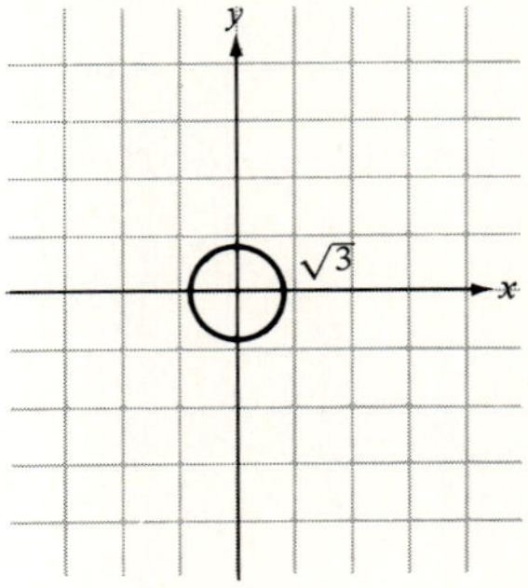

11. $(3, 1); r = 4$ **12.** $(-4, 2); r = 3$ **13.** $(-6, 3); r = 2$ **14.** $(0, 0); r = \sqrt{5}$

15. $(2, 3); r = 2$ **16.** $(-1, 4); r = 3$ **17.** $(0, 5); r = 4$ **18.** $(0, 0); r = 2$

19. $x^2 + y^2 = 25$ **20.** $(x - 5)^2 + (y + 2)^2 = 25$

21.

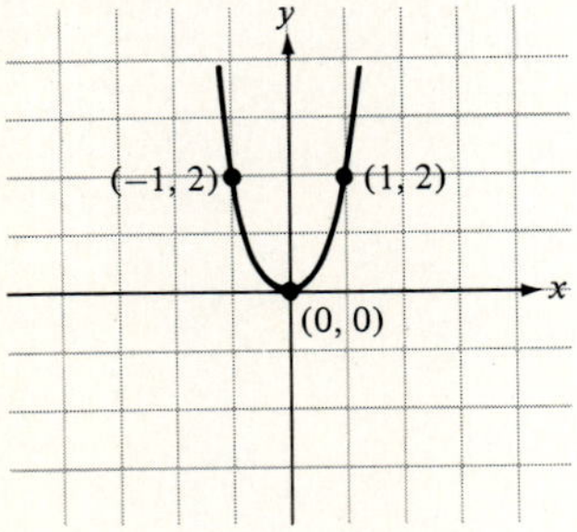

22.

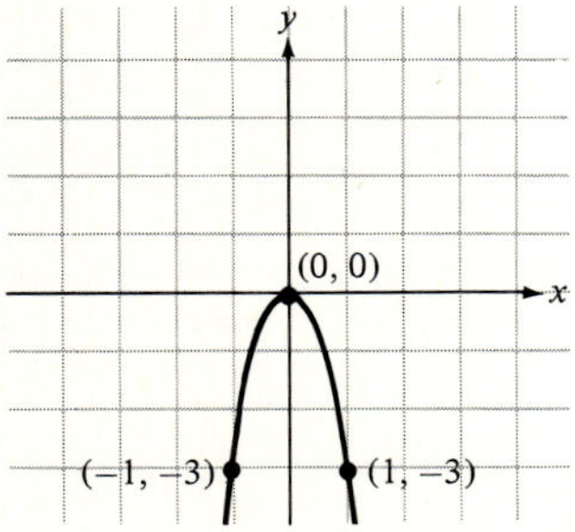

23.

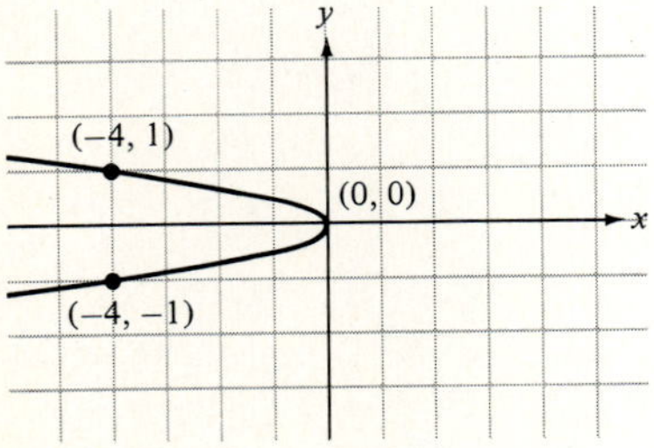

24.

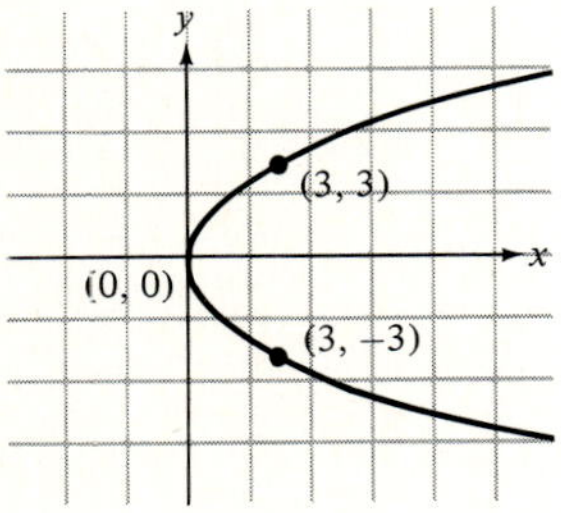

25. **(a)** $(2, -2)$ **(b)** $x = 2$
(c)

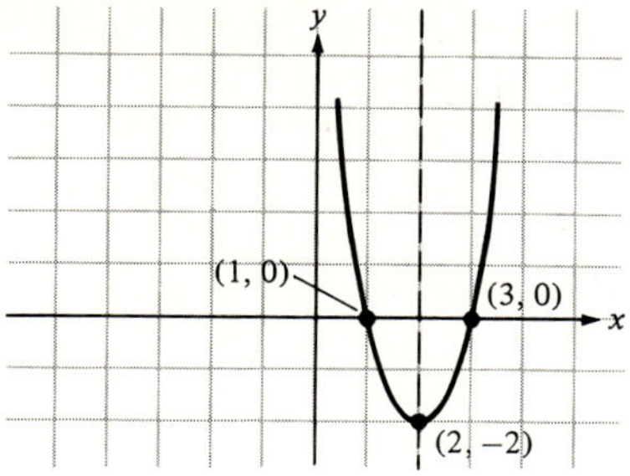

26. **(a)** $(-3, 4)$ **(b)** $x = -3$
(c)

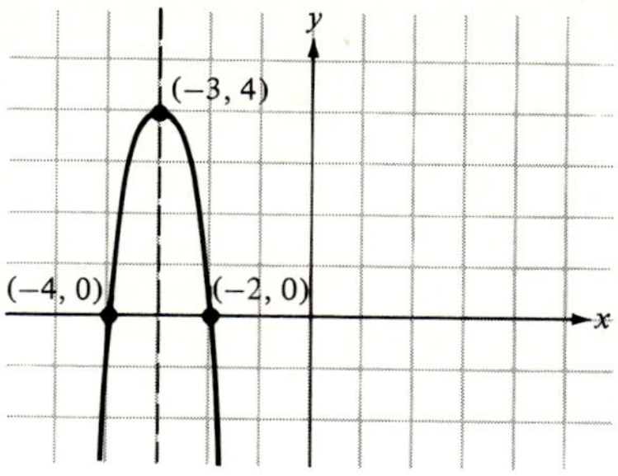

27. **(a)** $(-3, 0)$ **(b)** $x = -3$
(c)

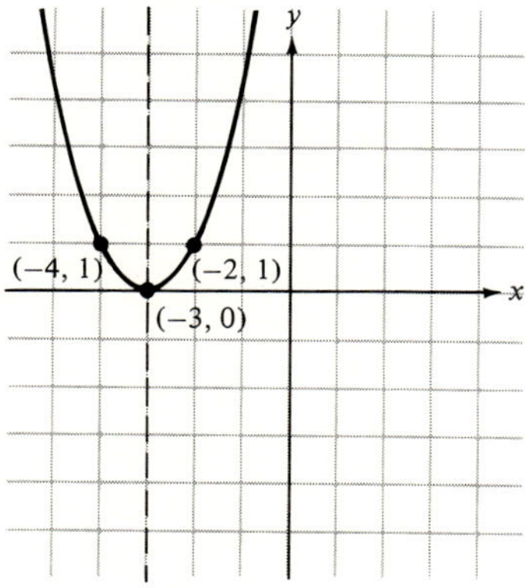

28. **(a)** $(2, 0)$ **(b)** $x = 2$
(c)

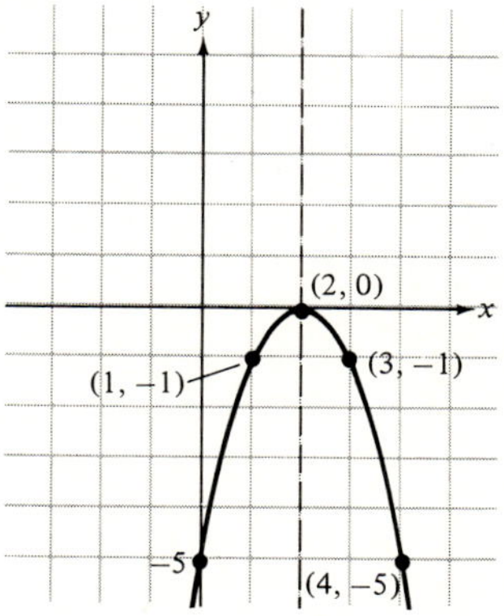

29. **(a)** $(3, 2)$ **(b)** $x = 3$
(c)

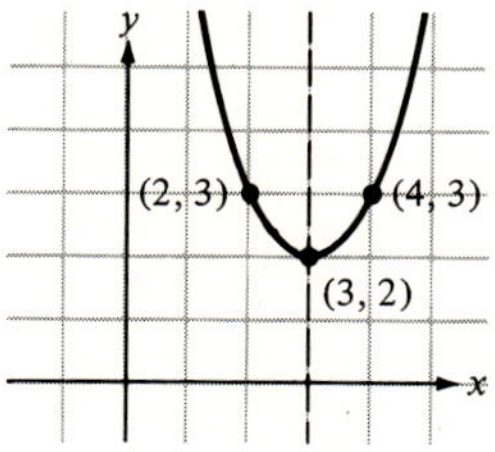

30. **(a)** $(-1, -4)$ **(b)** $x = -1$
(c)

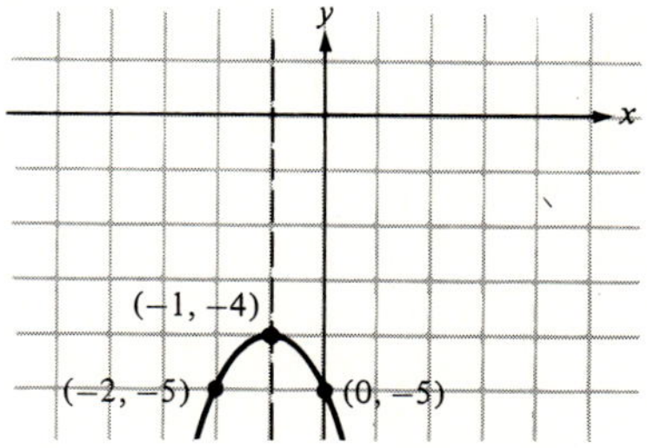

31.

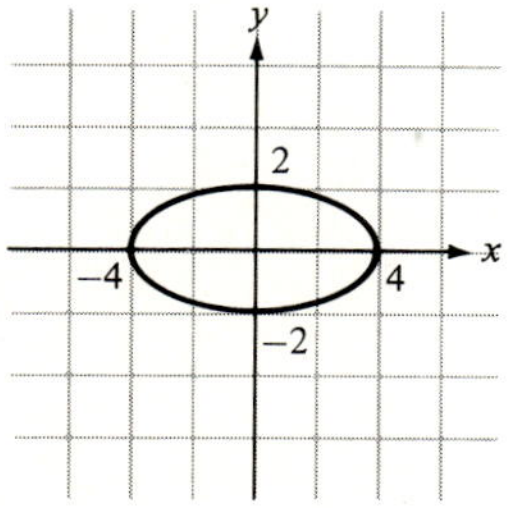

32.

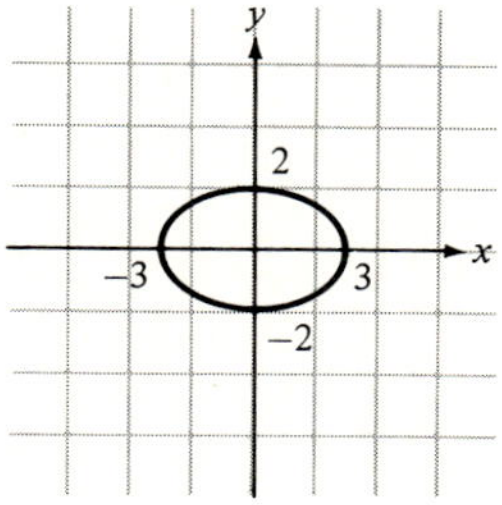

33.

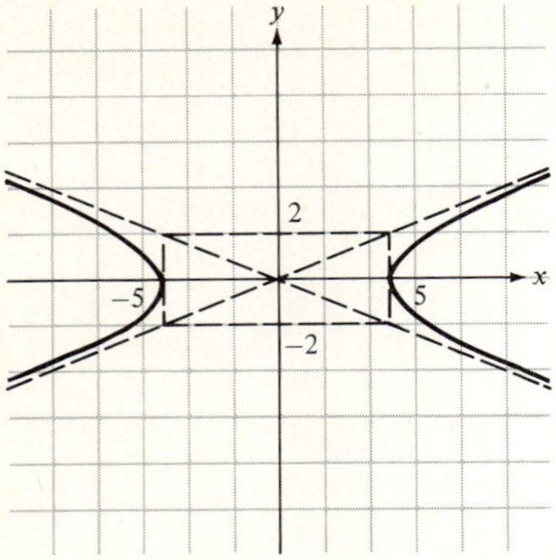

34.

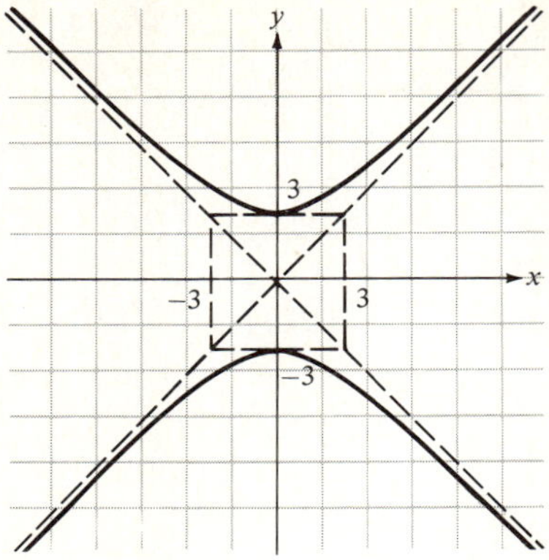

35. Circle **36.** Hyperbola **37.** Ellipse **38.** Circle **39.** Hyperbola **40.** Ellipse

41.

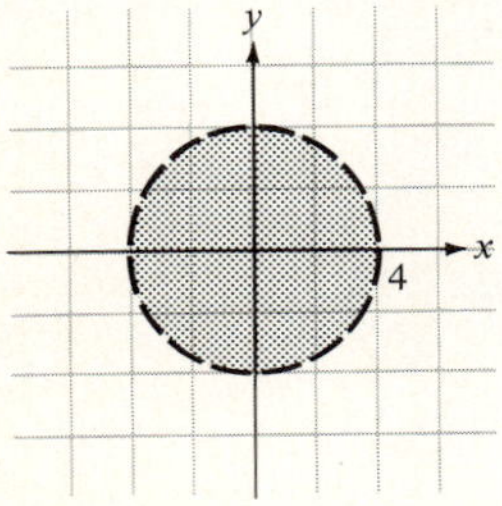

42.

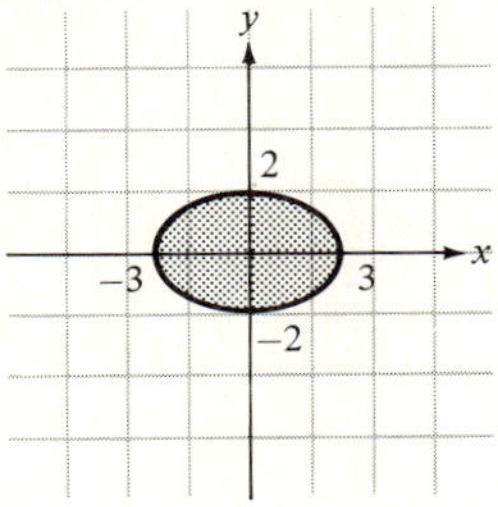

43.

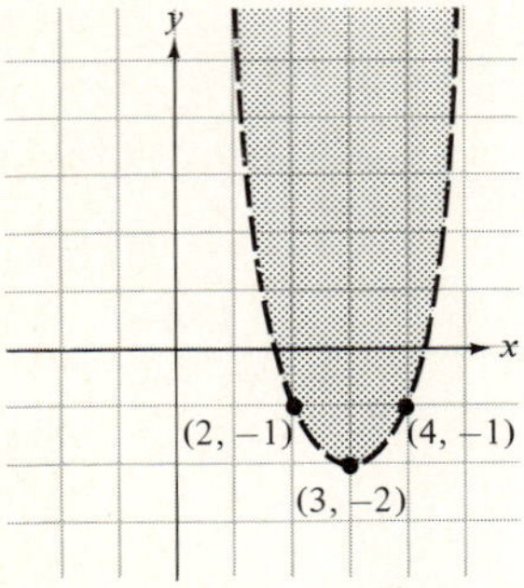

44.

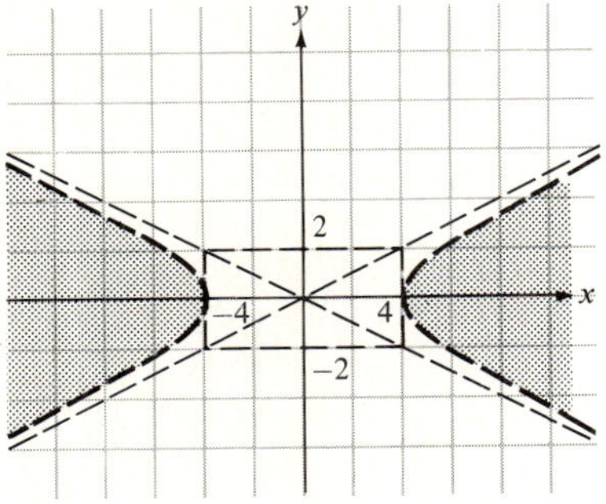

45.

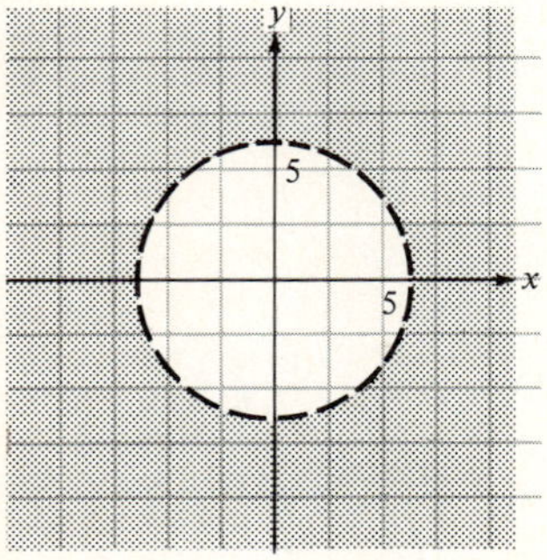

46.

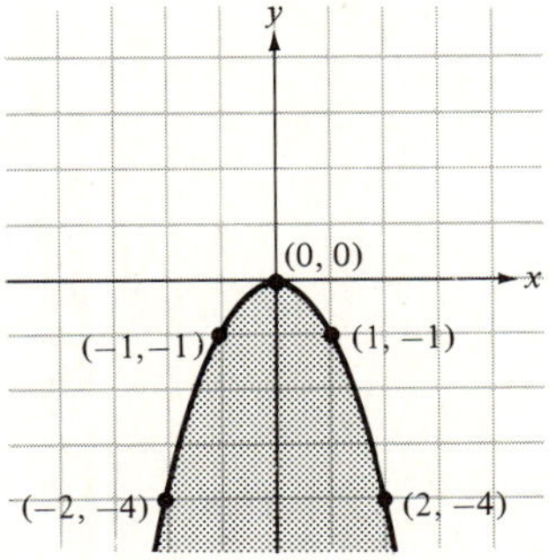

Chapter 8 Test

1. 5 **2.** $(x - 3)^2 + (y + 5)^2 = 16$ **3.** $x^2 + y^2 = 64$ **4.** **(a)** $(-2, 1)$ **(b)** $r = 2$

5. $(x - 1)^2 + (y - 2)^2 = 25$

6.

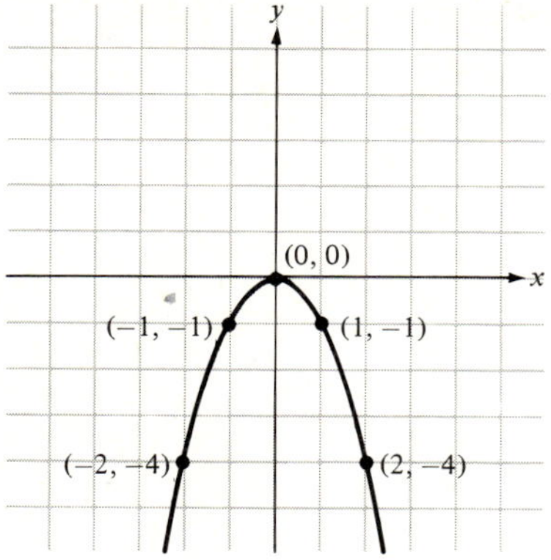

7.

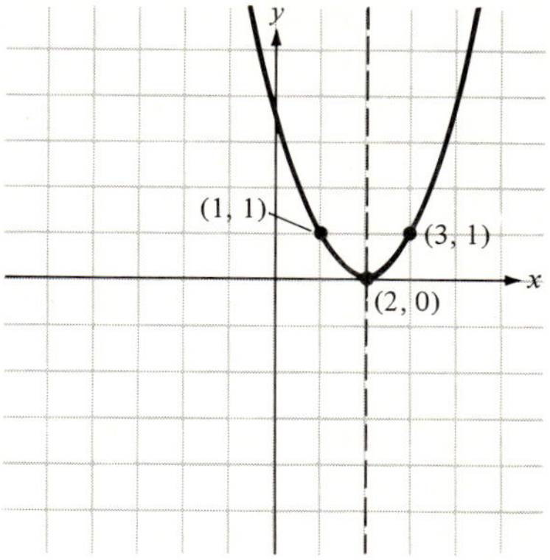

8. **(a)** (3, 1) **(b)** $x = 3$

(c)

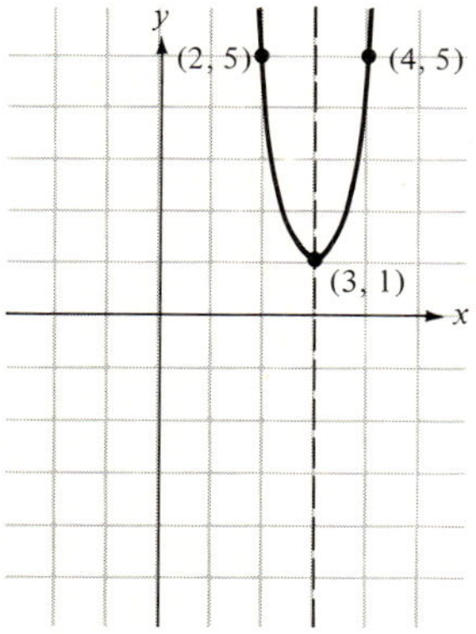

9. **(a)** $(-2, -1)$ **(b)** $x = -2$

(c)

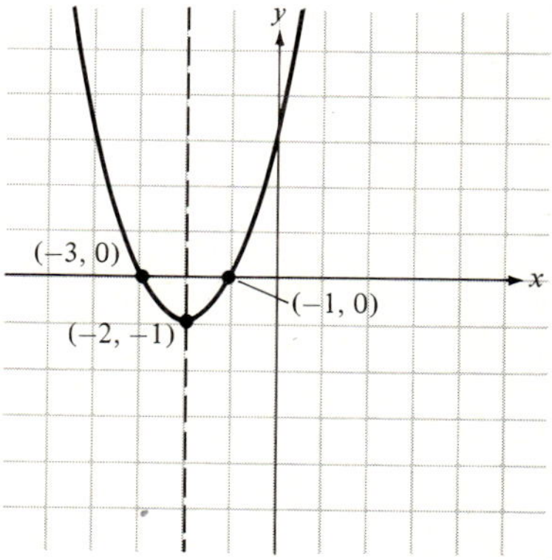

10.

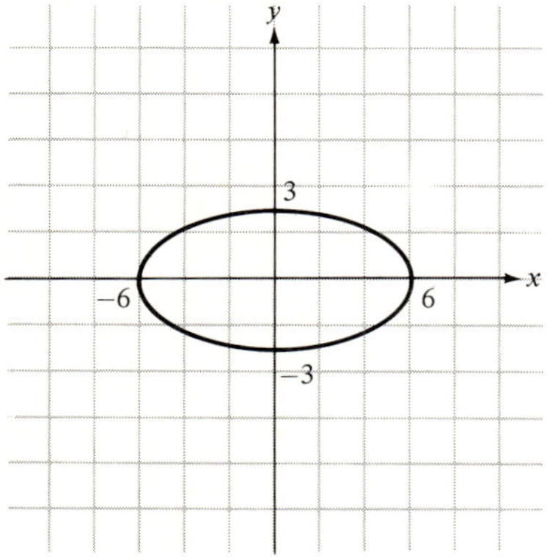

11.

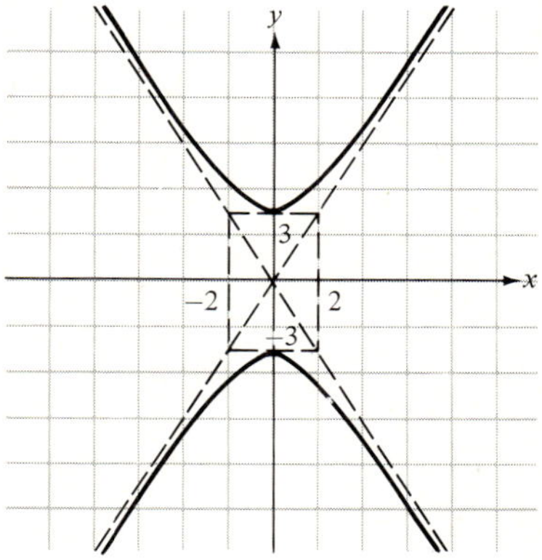

12. Parabola **13.** Hyperbola **14.** Circle **15.** Parabola **16.** Ellipse
17. Circle **18.** Hyperbola **19.** Ellipse **20.** Parabola

21.

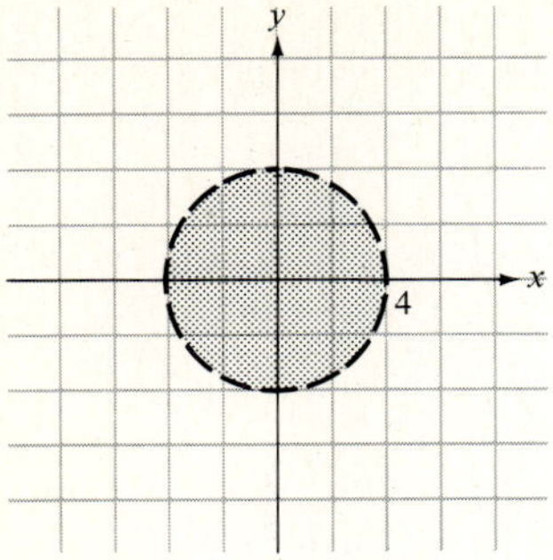

22.

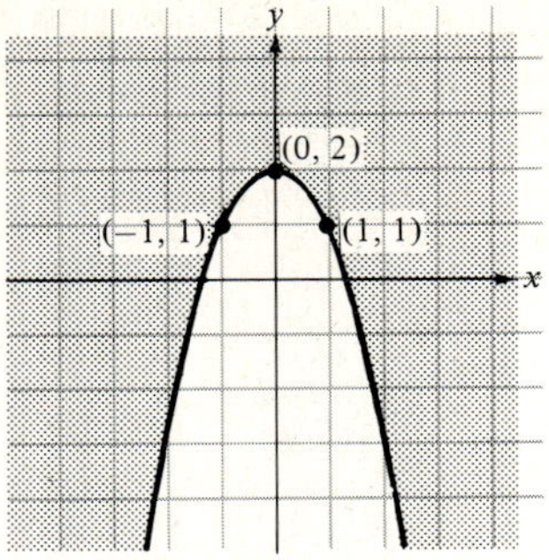

23.

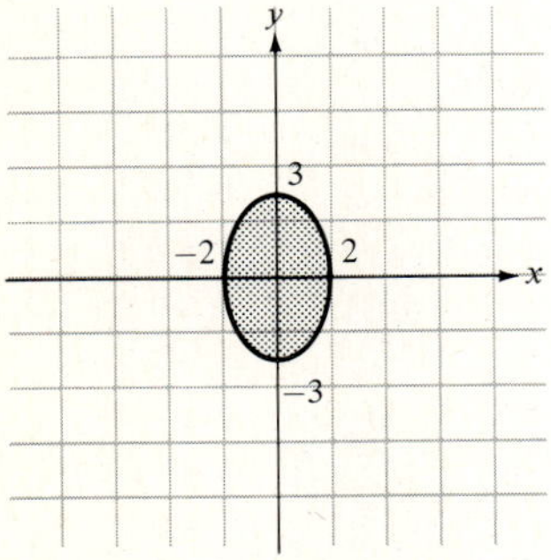

24.

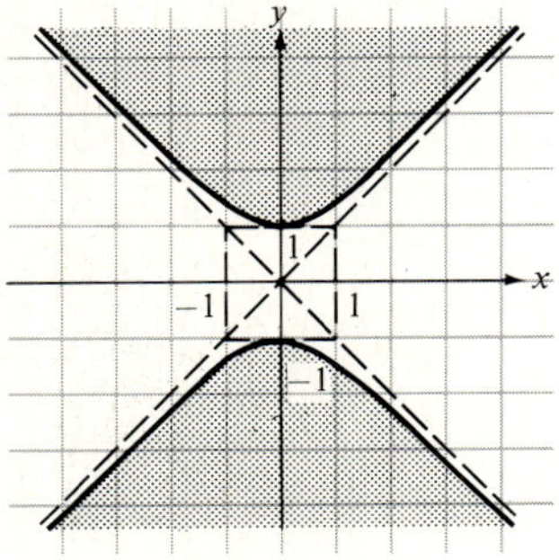

25.

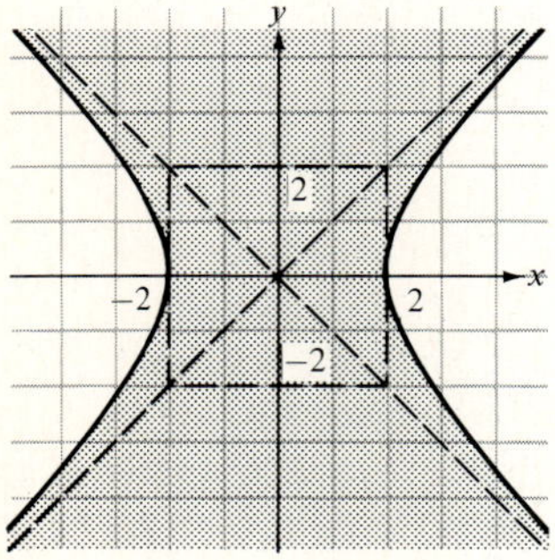

CHAPTER 9

Exercises 9.1

1. (1, 4) **3.** (2, 0) **5.** (4, 6) **7.** Dependent, $\{y \mid y = \frac{3}{5}x - 3\}$
9. Inconsistent **11.** (3, 2) **13.** (2, −2) **15.** (−2, 3) **17.** $(\frac{5}{2}, \frac{1}{2})$
19. (4, 2) **21.** $(\frac{7}{2}, 5)$ **23.** (2, −3) **25.** (29, −40) **27.** Inconsistent
29. Dependent, $\{y \mid y = \frac{3}{4}x - \frac{7}{4}\}$ **31.** (7, 2) **33.** (4, 4) **35.** (0, 1)
37. Inconsistent **39.** Dependent, $\{y \mid y = \frac{2}{3}x - \frac{7}{3}\}$ **41.** (2, −5) **43.** (3, 0)
45. (2, 2) **47.** Inconsistent **49.** (2, −5)

Exercises 9.2

1. (10, −6, 1) **3.** (2, 1, 3) **5.** (0, −2, 1) **7.** (1, 4, 2) **9.** (−1, 2, −3)
11. $(\frac{1}{2}, 1, \frac{1}{2})$ **13.** $(\frac{1}{3}, 1, -1)$ **15.** (−2, 1, 2) **17.** Dependent, $\{z \mid z = 4x + 2y - 8\}$
19. Dependent, $\{z \mid z = x + 2y - 6\}$ **21.** Inconsistent **23.** Inconsistent
25. (3, 2, −1) **27.** (−2, 1, 4) **29.** $(-\frac{4}{3}, 1, 5)$

Exercises 9.3

1. 13 and 29 **3.** 17 and 57 **5.** 124 and 204 **7.** 14, 17, and 23
9. 16, 30, and 62 **11.** 37 nickels, 23 dimes **13.** 36 nickels, 27 dimes
15. 50 at $.05, 20 at $.40 **17.** 6 at $.25, 14 at $.75 **19.** 20 oz of each solution
21. 24 g of 65% alloy, 36 g of 15% alloy **23.** 13 **25.** 18 yellowfins, 13 red spots, 3 catfish
27. 5 lb of Arabian, 12 lb of Colombian, 3 lb of Tanzanian **29.** $a = 3$, $b = -2$, and $c = -1$
31. Boat, 9 mph; current, 3 mph **33.** Plane, 450 mph; train, 70 mph **35.** 45 mph and 55 mph

Exercises 9.4

1.

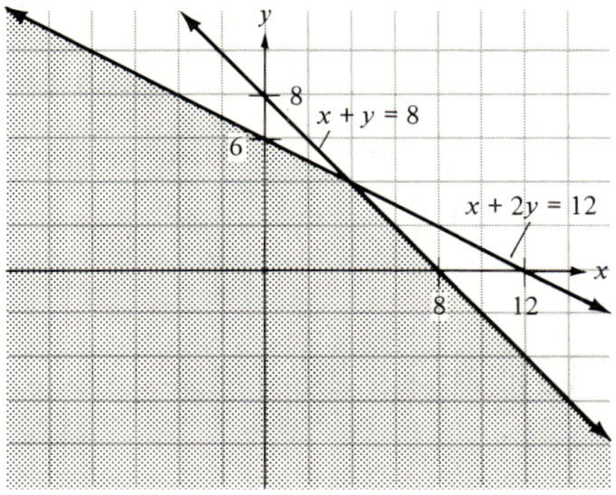

3.

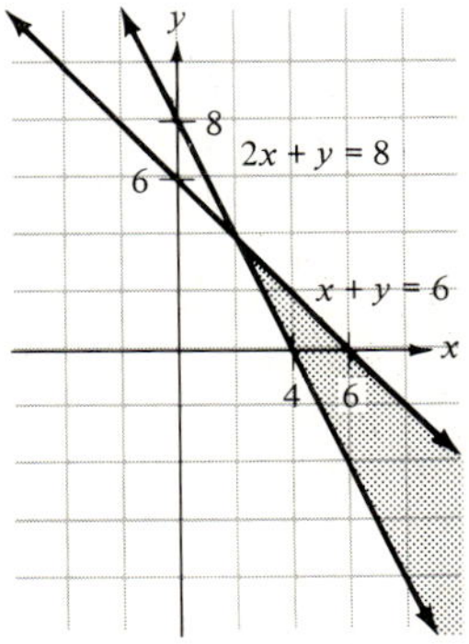

5.

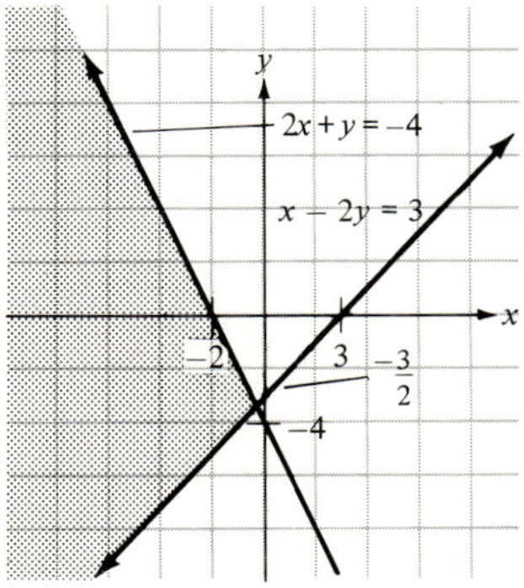

7.

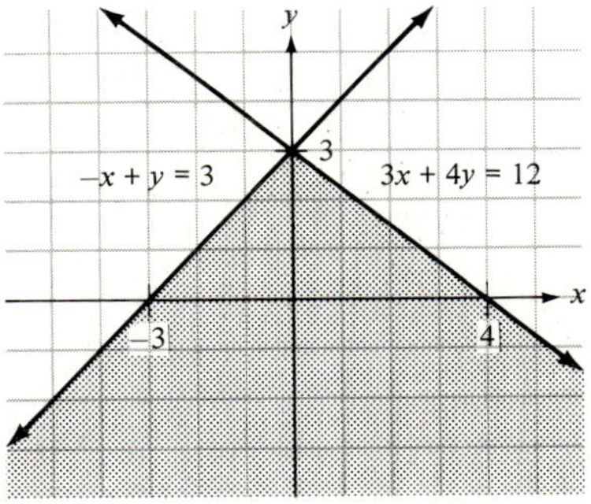

9.

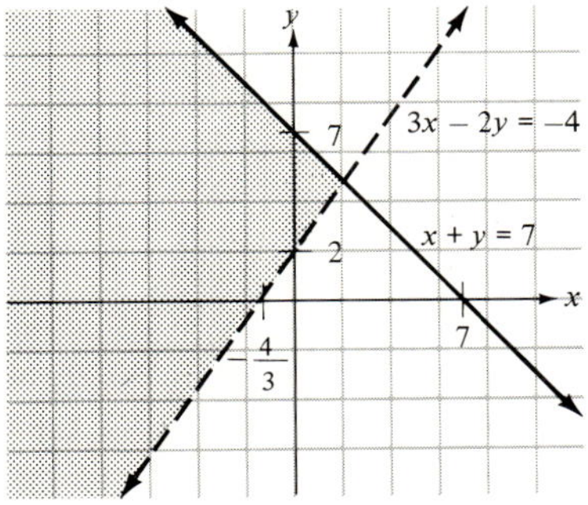

11.

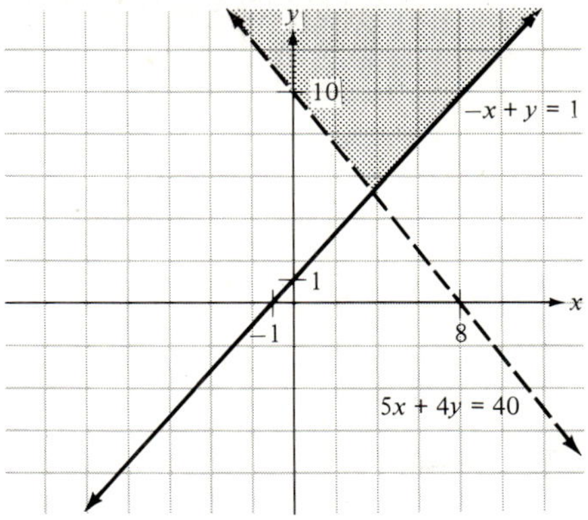

13.

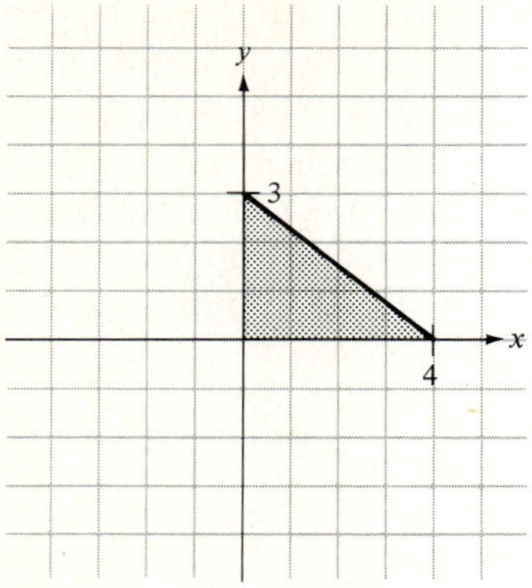

15.

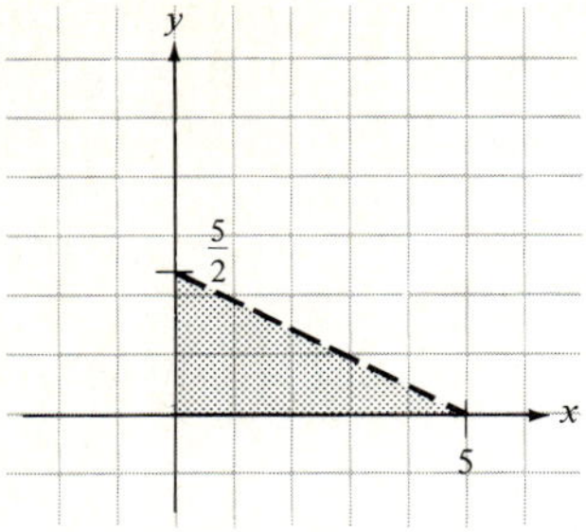

17.

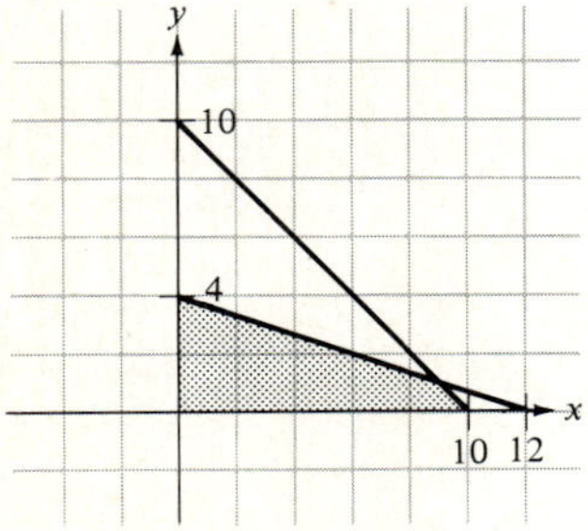

19.

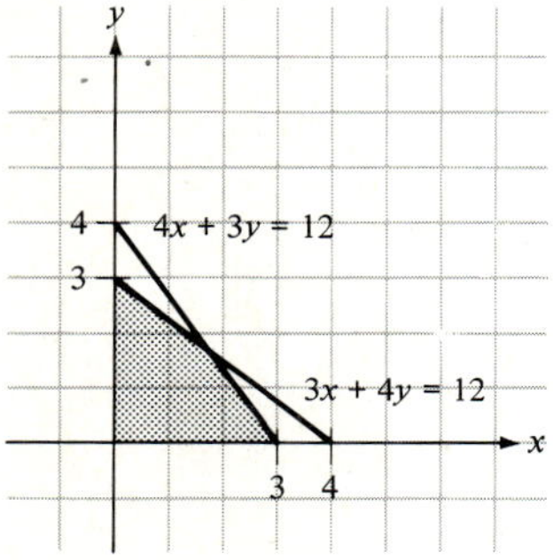

21.

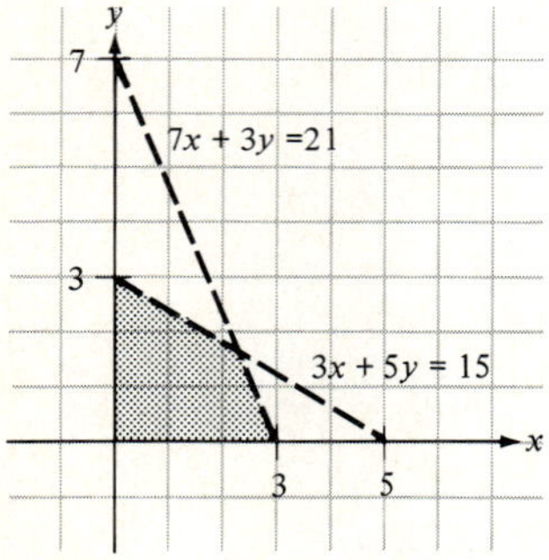

23.

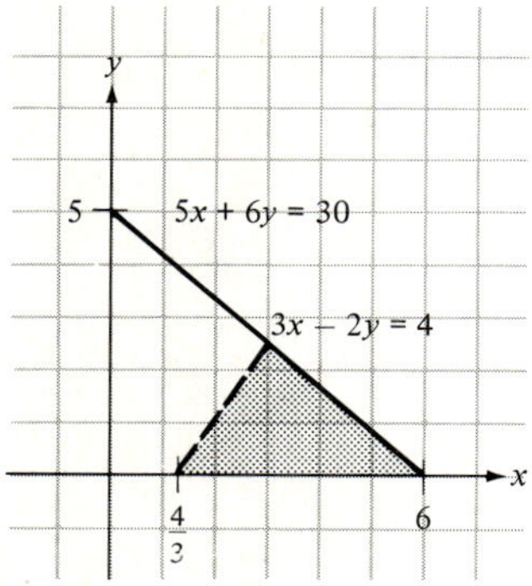

25.

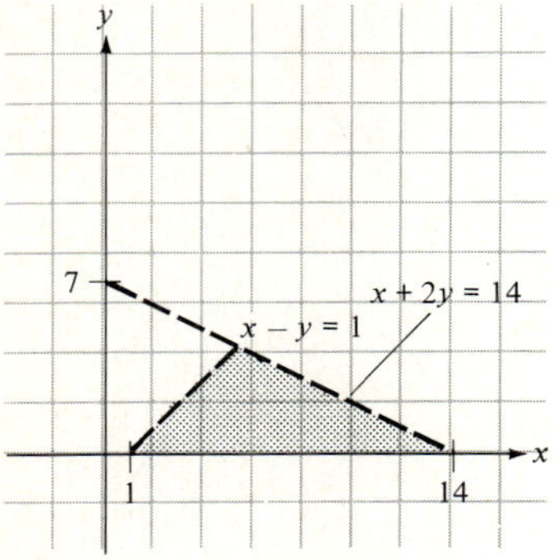

27.

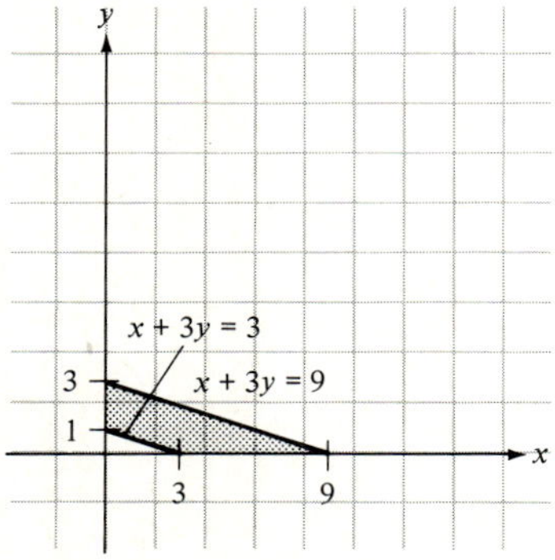

29.

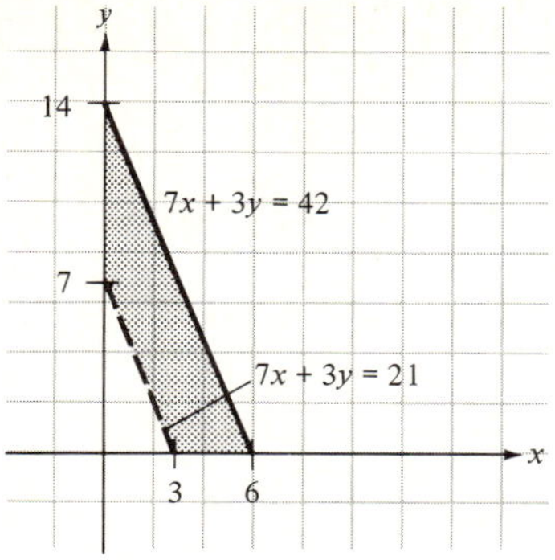

Exercises 9.5

1. $\{(-1, 0), (2, 3)\}$ **3.** $\{(1, -4), (-4, 11)\}$ **5.** $\{(-1, 5)\}$ **7.** $\{(0, 0), (-3, -6)\}$
9. $\{(-2, 0), (-1, 1)\}$ **11.** $\{(1, 0), (0, -1)\}$ **13.** $\{(2, 1)\}$ **15.** $\{(2, 10), (-4, 4)\}$
17. $\{(\sqrt{11}, 3), (-\sqrt{11}, 3), (\sqrt{11}, -3), (-\sqrt{11}, -3)\}$
19. $\{(\sqrt{2}, \sqrt{6}), (\sqrt{2}, -\sqrt{6}), (-\sqrt{2}, \sqrt{6}), (-\sqrt{2}, -\sqrt{6})\}$
21. $\left\{\left(\frac{6\sqrt{13}}{13}, \frac{6\sqrt{13}}{13}\right), \left(\frac{6\sqrt{13}}{13}, \frac{-6\sqrt{13}}{13}\right), \left(\frac{-6\sqrt{13}}{13}, \frac{6\sqrt{13}}{13}\right), \left(\frac{-6\sqrt{13}}{13}, \frac{-6\sqrt{13}}{13}\right)\right\}$
23. $\{(2, \sqrt{2}), (2, -\sqrt{2}), (-2, \sqrt{2}), (-2, -\sqrt{2})\}$ **25.** 4, 6 or -4, -6
27. 4, 7 or -4, -7 or -4, 7 or 4, -7 **29.** 3, 11 or -4, 18

Exercises 9.6

1. 2 **3.** -3 **5.** 0 **7.** 1 **9.** $-.5$ **11.** $\{(-2, 5)\}$
13. $\{(4, -2)\}$ **15.** $\{(-1, 4)\}$ **17.** $\{(-26, 46)\}$ **19.** $\{(\frac{6}{13}, -\frac{17}{13})\}$ **21.** $\{(9, 4)\}$
23. $\{(0, \frac{3}{2})\}$ **25.** $\{(-\frac{30}{7}, \frac{85}{7})\}$ **27.** $\{(\frac{235}{53}, \frac{118}{53})\}$ **29.** $\{(2, -3)\}$

Exercises 9.7

1. -186 **3.** 28 **5.** -2 **7.** -76 **9.** 90 **11.** 303
13. $\{(\frac{8}{7}, \frac{8}{7}, -\frac{3}{7})\}$ **15.** $\{(10, -\frac{55}{2}, -20)\}$ **17.** $\{(2, 3, 1)\}$ **19.** $\{(-2, 1, 3)\}$
21. $\{(\frac{5}{7}, -\frac{20}{21}, \frac{25}{21})\}$

Exercises 9.8

1. $\left[\begin{array}{rr|r} 4 & 5 & 3 \\ 7 & -8 & 15 \end{array}\right]$ **3.** $\left[\begin{array}{rr|r} 3 & -1 & -2 \\ -1 & 2 & 1 \end{array}\right]$ **5.** $\left[\begin{array}{rrr|r} 1 & 2 & -3 & 1 \\ 2 & -3 & 4 & 5 \\ 3 & 5 & -1 & 7 \end{array}\right]$

7. $\begin{aligned} 3x + 4y &= 10 \\ 2x - y &= 20 \end{aligned}$ **9.** $\begin{aligned} 3x + 5y + 7z &= 6 \\ 2x + z &= 4 \\ -2x + 3y + 6z &= 4 \end{aligned}$

11. $\{(1, 6)\}$ **13.** $\{(-5, 2)\}$ **15.** $\{(-4, 1)\}$ **17.** $\{(\frac{1}{3}, \frac{1}{2})\}$ **19.** $\{(9, -9, 8)\}$
21. $\{(2, -1, 1)\}$ **23.** $\{(-\frac{1}{5}, \frac{3}{5}, \frac{1}{5})\}$

Chapter 9 Review

1.

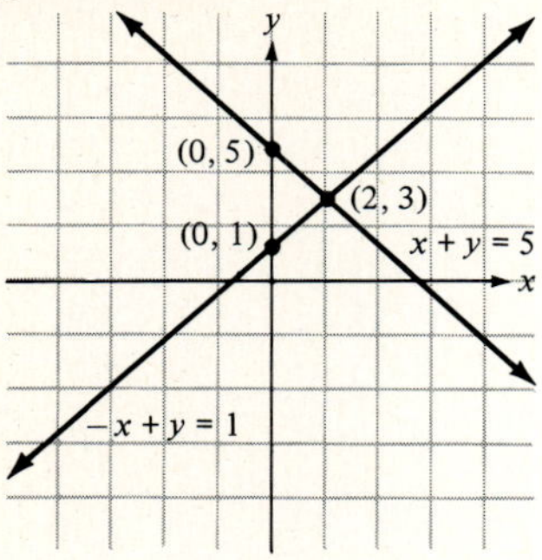

2.

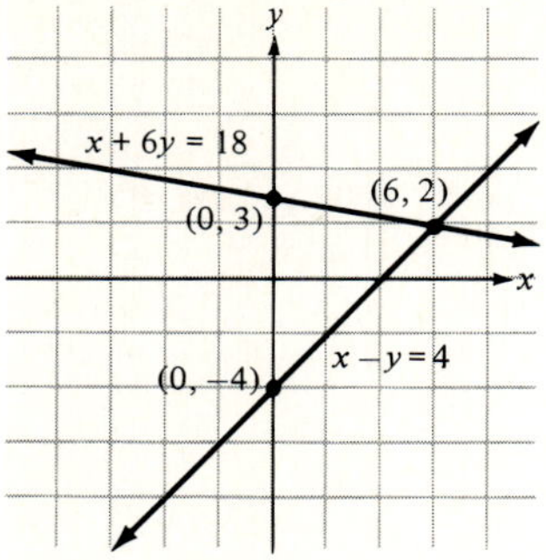

3. (5, 2) **4.** (1, 7) **5.** Inconsistent **6.** $(\frac{3}{2}, -\frac{1}{2})$
7. Dependent, $\{(x, y) \mid y = \frac{3}{2}x - \frac{5}{2}\}$ **8.** $(\frac{2}{3}, \frac{1}{4})$ **9.** (1, −2, −1) **10.** (2, −3, 1)
11. (−1, 3, 2) **12.** Dependent, $\{(x, y, z) \mid z = 2x + 3y + 4\}$ **13.** $(\frac{1}{3}, \frac{3}{4}, \frac{1}{2})$
14. Inconsistent **15.** 37 and 43 **16.** 11, 29, and 43
17. 16 oz of the 40% solution and 4 oz of the 90% solution
18. 84 of the 2.5-oz marbles and 76 of the 1.5-oz marbles
19. The speed of the boat in still water is 12.5 mph; the speed of the current is 2.5 mph.
20. $a = -2$, $b = 3$, and $c = 5$

21.

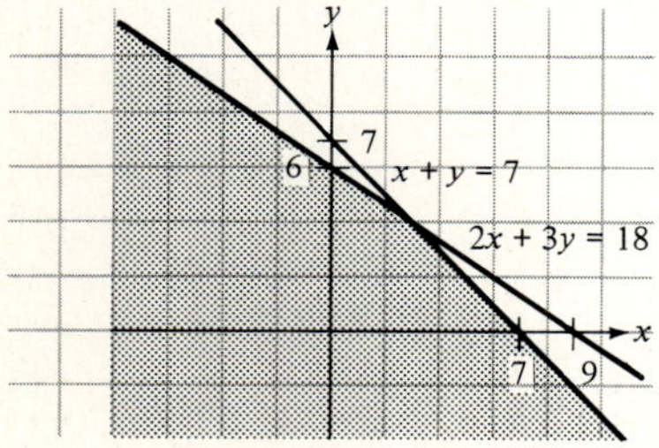

22.

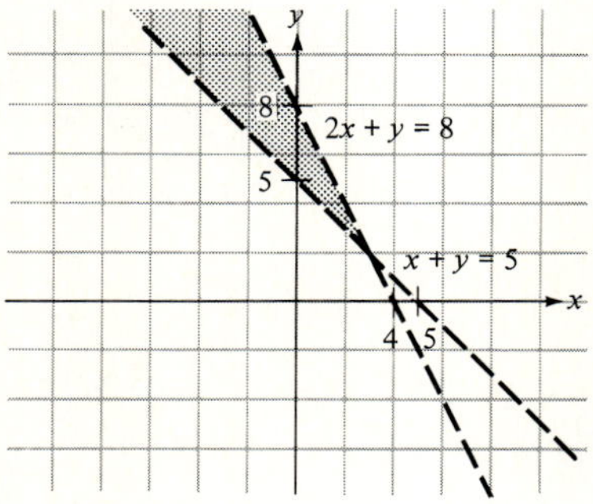

23.

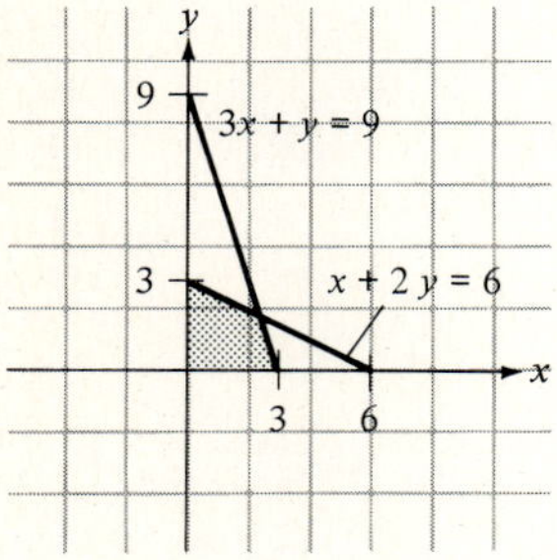

24. {(−1, 1), (3, 7)} **25.** $\left\{\left(\dfrac{\pm 6\sqrt{13}}{13}, \dfrac{\pm 6\sqrt{13}}{13}\right)\right\}$ **26.** 11 and 14 **27.** 3
28. −1 **29.** −17 **30.** 6 **31.** {(1, −2)} **32.** {(−2, 3)} **33.** $\{(\frac{3}{2}, -1)\}$
34. $\{(-\frac{30}{13}, \frac{40}{13})\}$ **35.** −29 **36.** 10 **37.** −27 **38.** 75
39. {(7, −2, −1)} **40.** $\{(-1, \frac{1}{2}, \frac{2}{3})\}$ **41.** {(−1, 1)} **42.** {(5, −2)}
43. {(31, −10, 4)} **44.** {(−7, −2, 2)}

Chapter 9 Test

1. $\{(2, 4)\}$ **2.** $\{(5, 2)\}$ **3.** $\{(-2, 5)\}$ **4.** Inconsistent **5.** $\{(1, -1, \frac{1}{3})\}$

6. Dependent, $\{z \mid z = \frac{3}{4}x + \frac{1}{2}y - 3\}$ **7.** 23 and 39 **8.** 22 at \$.25, 14 at \$.35

9.

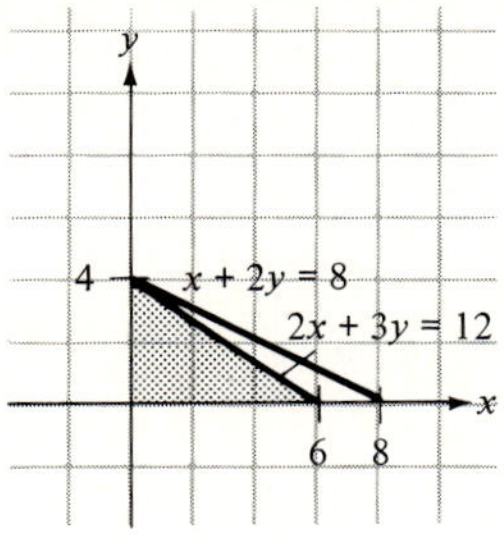

10. $\{(4, 28), (-1, -2)\}$ **11. (a)** 18 **(b)** -71

12. (a) $\{(\frac{7}{6}, -\frac{5}{6})\}$ **(b)** $\{(3, -5, -1)\}$ **13.** $\{(-2, 3, 5)\}$

CHAPTER 10

Exercises 10.1

1. $D = \{-2, -1, 1\}$; $R = \{-4, -2, 2\}$ **3.** $D = \{-3, 3\}$; $R = \{-4, 4\}$

5. $D = \{0, 2, 4\}$; $R = \{0, 3\}$ **7.** $D = \{-2, 2, 3\}$; $R = \{4, 9\}$ **9.** $D = \{0, 10\}$; $R = \{-8, 6, 8\}$

11. (a)

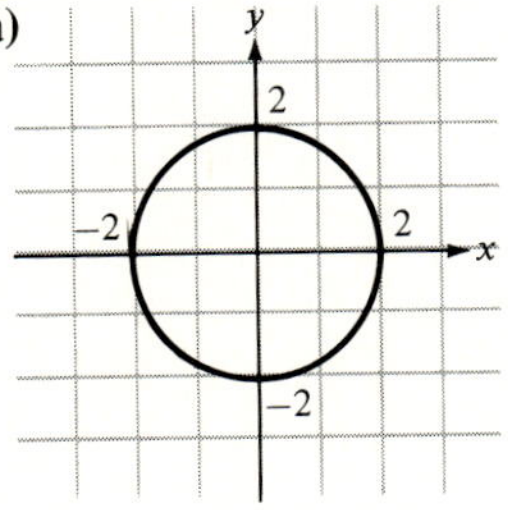

(b) $D = \{x \mid -2 \le x \le 2\}$ **(c)** $R = \{y \mid --2 \le y \le 2\}$

13. (a)

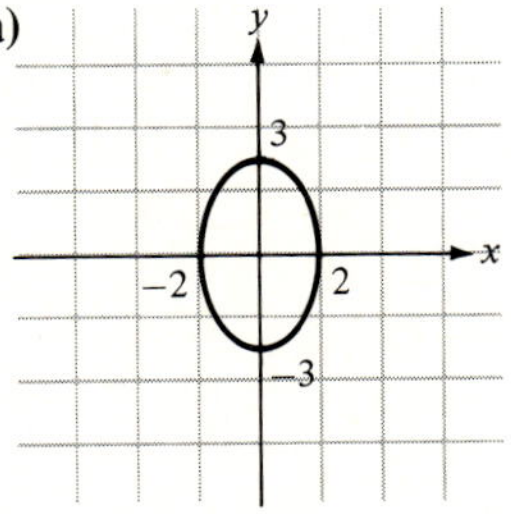

(b) $D = \{x \mid -2 \le x \le 2\}$ **(c)** $R = \{y \mid -3 \le y \le 3\}$

15. (a)

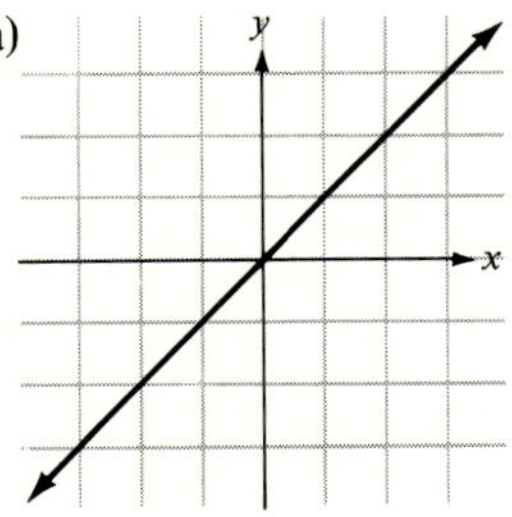

(b) $D = \{x \mid x \text{ is any real number}\}$ **(c)** $R = \{y \mid y \text{ is any real number}\}$

17. **(a)**

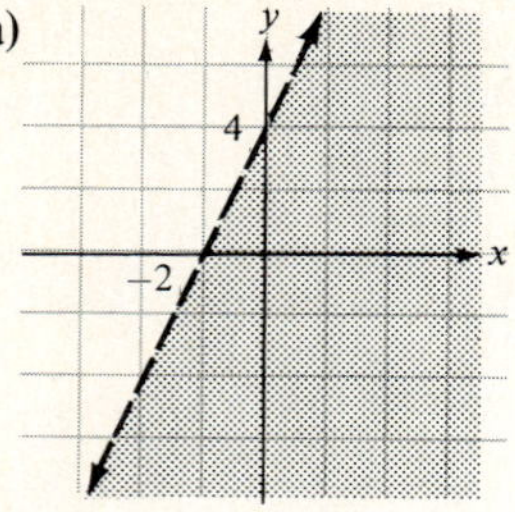

(b) $D = \{x \mid x \text{ is any real number}\}$ **(c)** $R = \{y \mid y \text{ is any real number}\}$

19. **(a)**

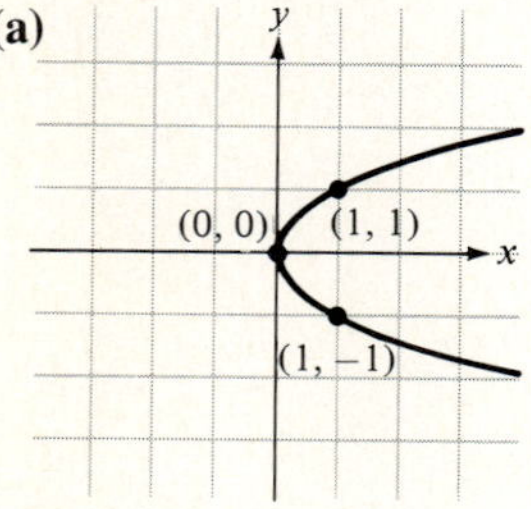

(b) $D = \{x \mid x \geq 0\}$ **(c)** $R = \{y \mid y \text{ is any real number}\}$

21. **(a)**

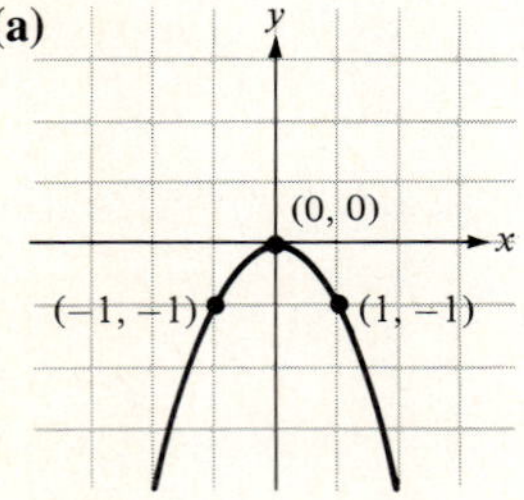

(b) $D = \{x \mid x \text{ is any real number}\}$ **(c)** $R = \{y \mid y \leq 0\}$

23. **(a)**

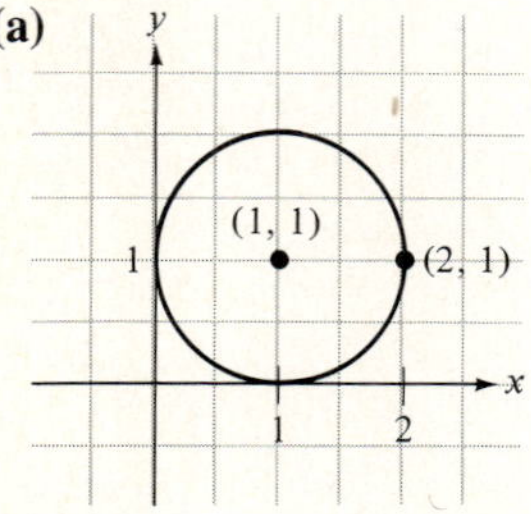

(b) $D = \{x \mid 0 \leq x \leq 2\}$ **(c)** $R = \{y \mid 0 \leq y \leq 2\}$

25. Equal to **27.** Less than **29.** Square root of

Exercises 10.2

1. Function; $R = \{0, 4, 8, 12\}$ **3.** Function; $R = \{-7, -1, 5\}$ **5.** Function; $R = \{1, 10\}$ **7.** Not a function **9.** Function; $R = \{0, 2, 3, 5\}$ **11.** Function; $R = \{4\}$ **13.** Function; $R = \{-2, -1, 0, 1, 2\}$ **15.** Not a function **17.** **(a)** -1 **(b)** -4 **(c)** 5 **19.** **(a)** 3 **(b)** 0 **(c)** 12 **21.** **(a)** 5 **(b)** 4 **(c)** 3 **23.** **(a)** $3x + 3h + 1$ **(b)** $3h$ **(c)** 3 **25.** **(a)** $3x^2 + 6hx + 3h^2$ **(b)** $6hx + 3h^2$ **(c)** $6x + 3h$ **27.** **(a)** $x^2 + 2hx + h^2 + x + h$ **(b)** $2hx + h^2 + h$ **(c)** $2x + h + 1$ **29.** $\{x \mid x \text{ is real and } x \neq 0\}$ **31.** $\{x \mid x \text{ is real and } x \geq -3\}$ **33.** $\{x \mid x \text{ is real and } x \neq -1, x \neq 5\}$ **35.** $\{x \mid x \text{ is any real number}\}$ **37.** $\{x \mid x \text{ is real and } x \neq \pm 2\}$ **39.** $\{x \mid x \text{ is real and } x \geq \frac{1}{3}\}$ **41.** $\{x \mid x \text{ is real and } x \neq \pm 2\}$ **43.** Function **45.** Not a function **47.** Function **49.** Not a function **51.** Function

Exercises 10.3

1. **(a)** (1, 2) **(b)** Minimum 2 **3.** **(a)** (−3, 6) **(b)** Maximum 6
5. **(a)** $(0, \frac{7}{2})$ **(b)** Minimum $\frac{7}{2}$ **7.** **(a)** (0, 0) **(b)** Maximum 0
9. **(a)** (−2, −28) **(b)** Minimum −28 **11.** **(a)** (.1, .5) **(b)** Minimum .5

13. **(a)**

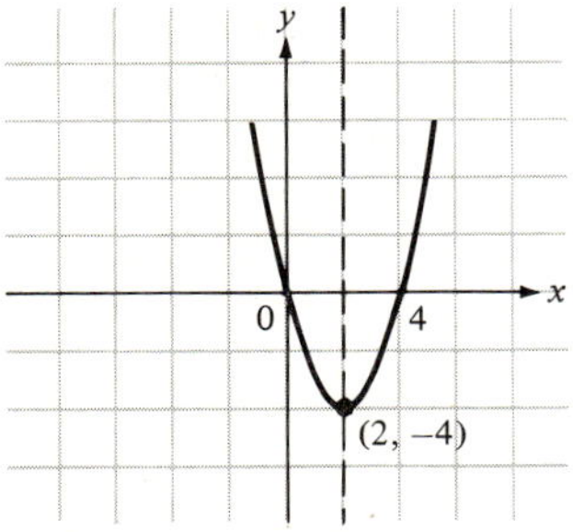

(b) Minimum; −4

15. **(a)**

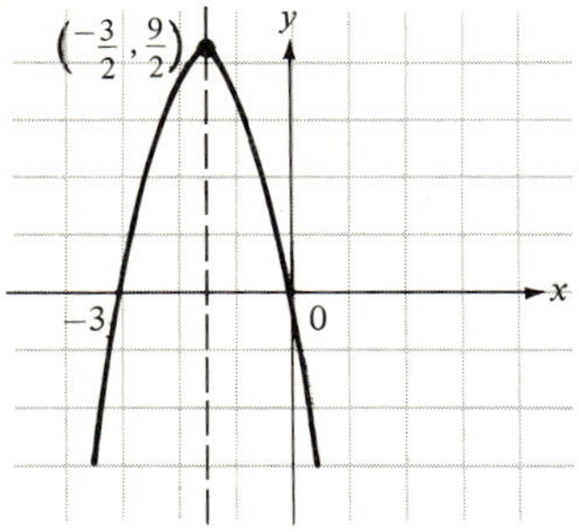

(b) Maximum; $\frac{9}{2}$

17. **(a)**

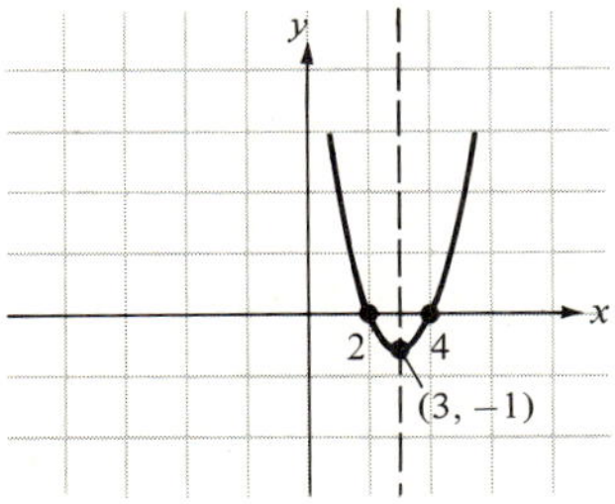

(b) Minimum; −1

19. **(a)**

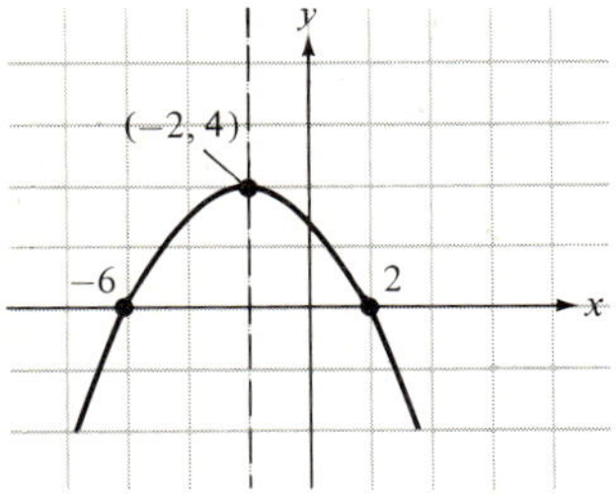

(b) Maximum; 4

21. **(a)**

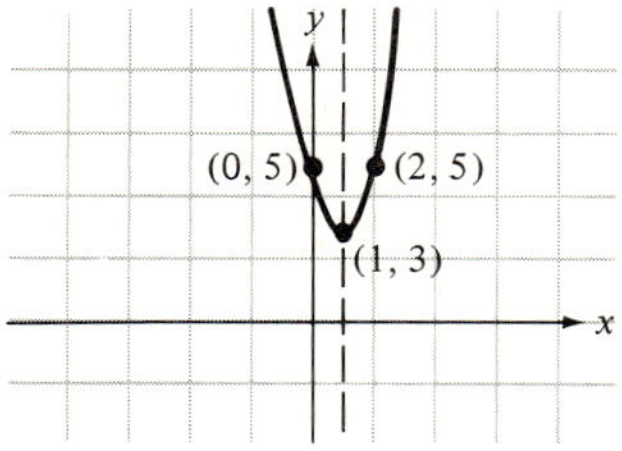

(b) Minimum; 3

23. **(a)**

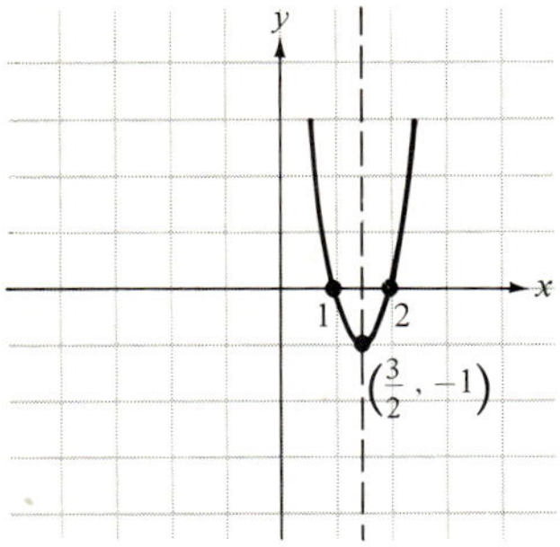

(b) Minimum; −1

25. 10 and 10 **27.** 1250 sq ft **29.** **(a)** 50 **(b)** $150

Exercises 10.4

1. Linear **3.** Constant **5.** Polynomial **7.** Constant **9.** Polynomial
11. Linear **13.** 7 **15.** −4 **17.** 5 **19.** 7 **21.** −40 **23.** 7
25. **(a)** $f(0) = 0, f(2) = 4, f(-2) = 4$ **(b)**

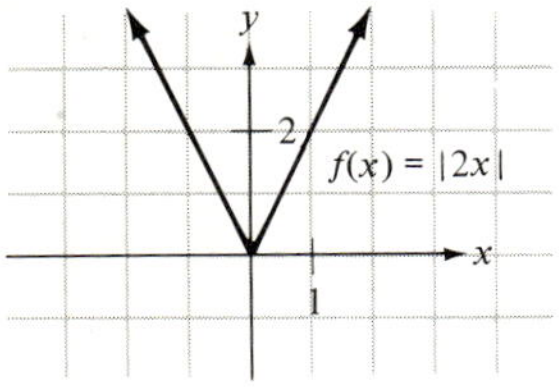

27. **(a)** $f(0) = 2, f(2) = 4, f(-2) = 4$ **(b)**

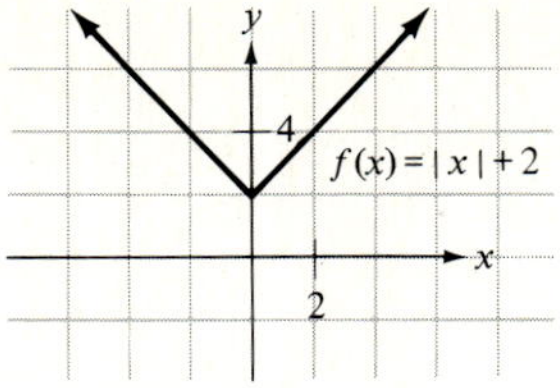

29. **(a)** $f(0) = 0, f(2) = -4, f(-2) = -4$ **(b)**

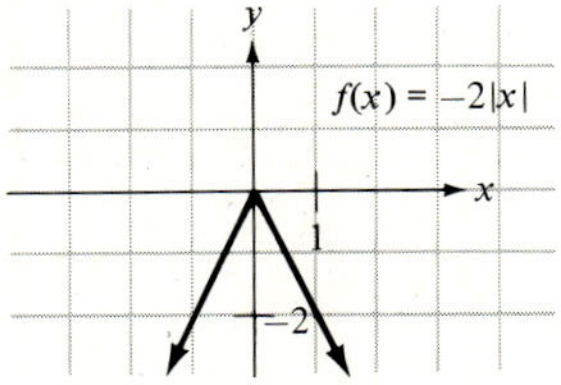

31. **(a)** $f(0) = 2, f(2) = 0, f(-2) = 4$ **(b)**

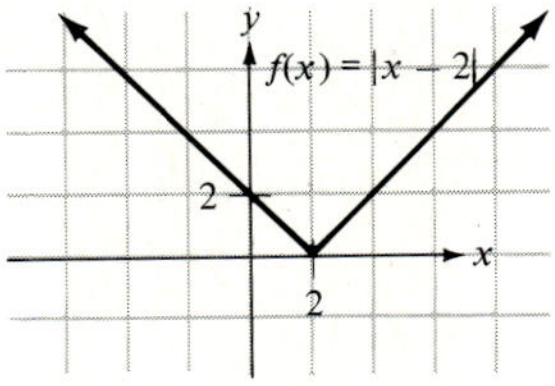

33. **(a)** $g(2) = 2,\ g(8) = 4,\ g(18) = 6$ **(b)**

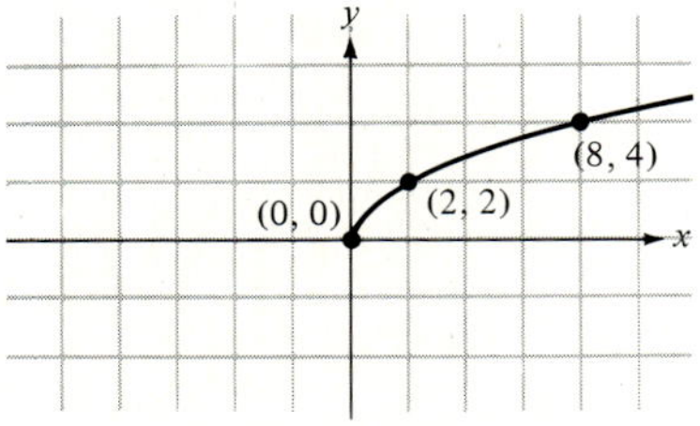

35. **(a)** $g(2) = -2,\ g(8) = -4,\ g(18) = -6$ **(b)**

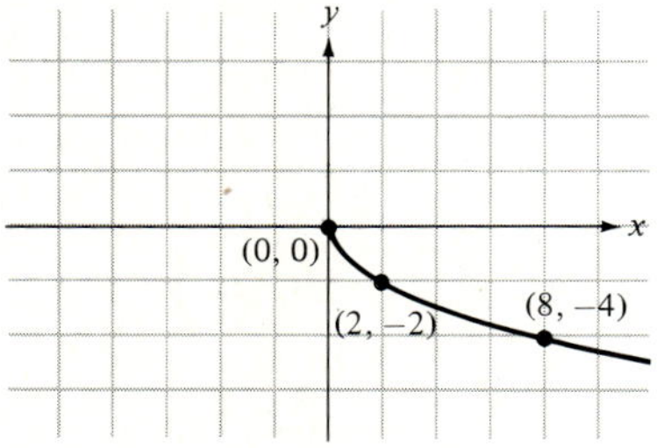

37. **(a)** $g(2) = 0,\ g(8) = \sqrt{6},\ g(18) = 4$ **(b)**

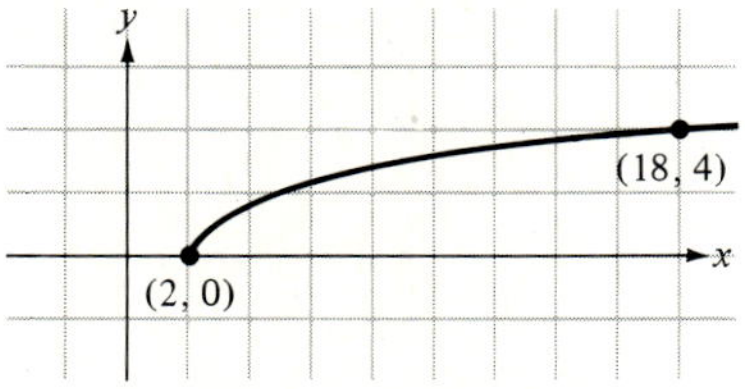

39.

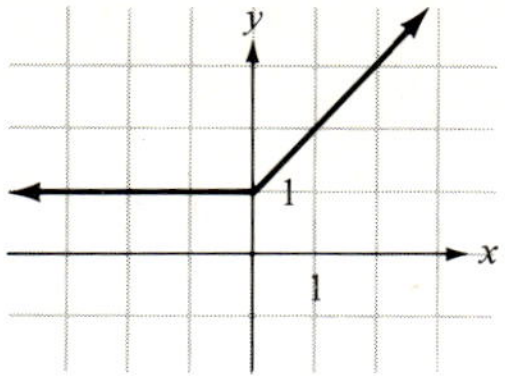

41.

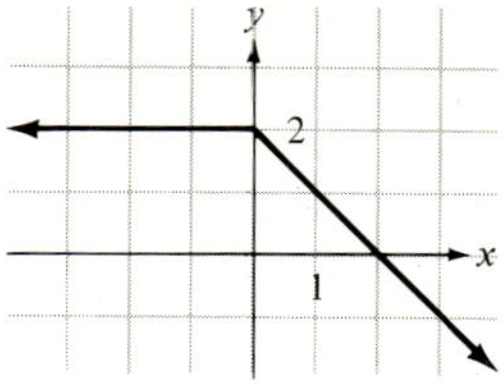

43. $f(1) = 3$; $f(0) = 2$; $f(-2) = -3$

45. $f(1) = 9$; $f(0) = 3$; $f(-2) = 5$

Exercises 10.5

1. $(f + g)(x) = 2x$; $(f - g)(x) = 2$; $(f \cdot g)(x) = x^2 - 1$; $\left(\dfrac{f}{g}\right)(x) = \dfrac{x + 1}{x - 1}$

3. $(f + g)(x) = x^2 + x + 5$; $(f - g)(x) = x^2 - x - 5$; $(f \cdot g)(x) = x^3 + 5x^2$; $\left(\dfrac{f}{g}\right)(x) = \dfrac{x^2}{x + 5}$

5. $(f + g)(x) = \dfrac{1}{x} + x$; $(f - g)(x) = \dfrac{1}{x} - x$; $(f \cdot g)(x) = 1$; $\left(\dfrac{f}{g}\right)(x) = \dfrac{1}{x^2}$

7. $(f + g)(x) = \sqrt{x} + |x|$; $(f - g)(x) = \sqrt{x} - |x|$; $(f \cdot g)(x) = |x|\sqrt{x}$; $\left(\dfrac{f}{g}\right)(x) = \dfrac{\sqrt{x}}{|x|}$

9. $(f + g)(x) = 2x^2$; $(f - g)(x) = -2$; $(f \cdot g)(x) = x^4 - 1$; $\left(\dfrac{f}{g}\right)(x) = \dfrac{x^2 - 1}{x^2 + 1}$

11. 9 **13.** -8 **15.** 35 **17.** -1 **19.** 0 **21.** 49

23. **(a)** $(f \circ g)(x) = 3x + 5$ **(b)** 17 **25.** **(a)** $(f \circ g)(x) = (x + 4)^2$ **(b)** 64

27. **(a)** $(f \circ g)(x) = \sqrt{2x + 1}$ **(b)** 3 **29.** **(a)** $(f \circ g)(x) = 3$ **(b)** 3

31. **(a)** $(f \circ g)(x) = (x + 2)^2 = x^2 + 4x + 4$ **(b)** $(g \circ f)(x) = x^2 + 2$

33. **(a)** $(f \circ g)(x) = |x|$ **(b)** $(g \circ f)(x) = x$ **35.** **(a)** $(f \circ g)(x) = x$ **(b)** $(g \circ f)(x) = x$

37. **(a)** $(f \circ g)(x) = x^2 + 3$ **(b)** $(g \circ f)(x) = x^2 + 3$

Exercises 10.6

1. One-to-one **3.** Not one-to-one **5.** One-to-one **7.** Not one-to-one **9.** One-to-one

11. Not one-to-one **13.** $f^{-1} = \{(1, -2), (3, 0), (4, 1)\}$ **15.** $f^{-1} = \{(1, 1), (4, 2), (9, 3)\}$

17. $f^{-1} = \{(2, 4), (3, 9), (5, 25)\}$ **19.** $f^{-1} = \{(2, -2), (3, -3), (5, -5)\}$ **21.** One-to-one

23. Not one-to-one **25.** Not one-to-one **27.** Not one-to-one

29. $f^{-1}(x) = x - 3$

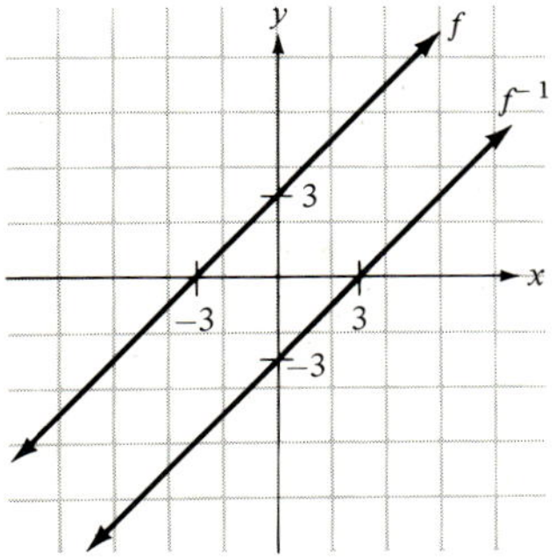

31. $f^{-1}(x) = \frac{1}{3}x + 2$

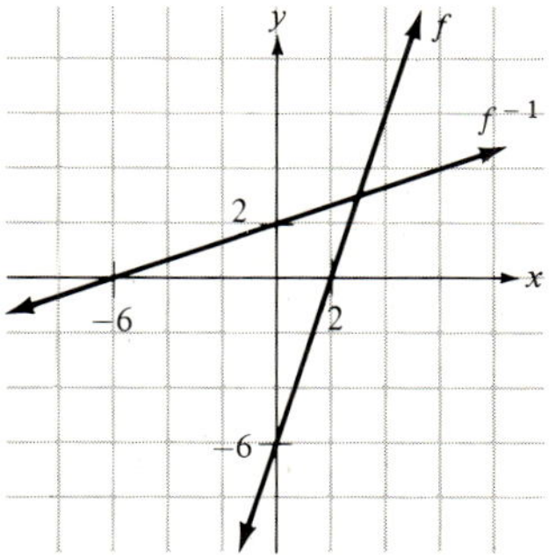

33. $f^{-1}(x) = \sqrt{x - 1};\ x \geq 1$

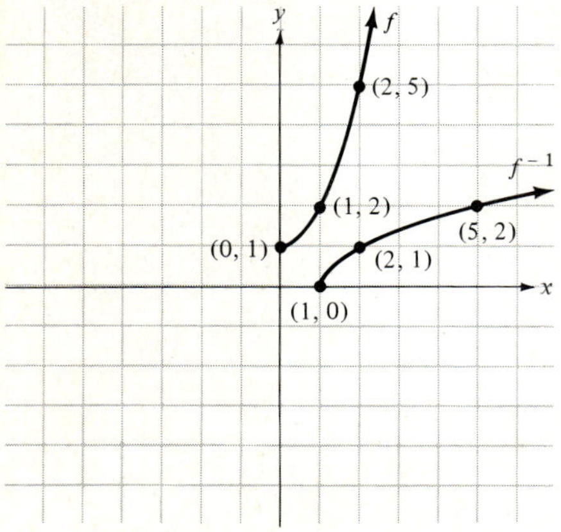

35. $f^{-1}(x) = -\frac{1}{2}x;\ x \geq 0$

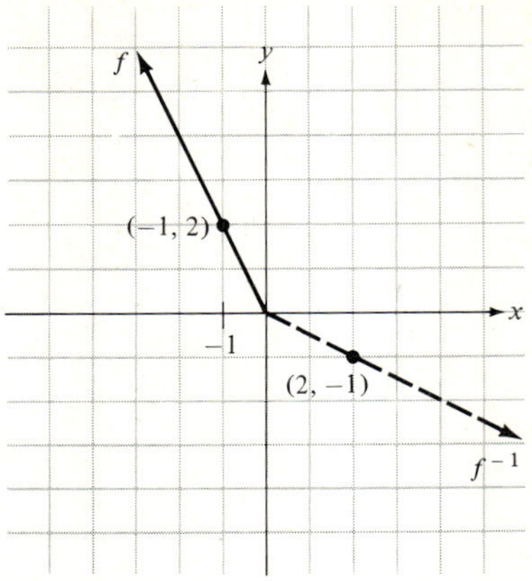

37. $f^{-1}(x) = \sqrt[3]{x}$

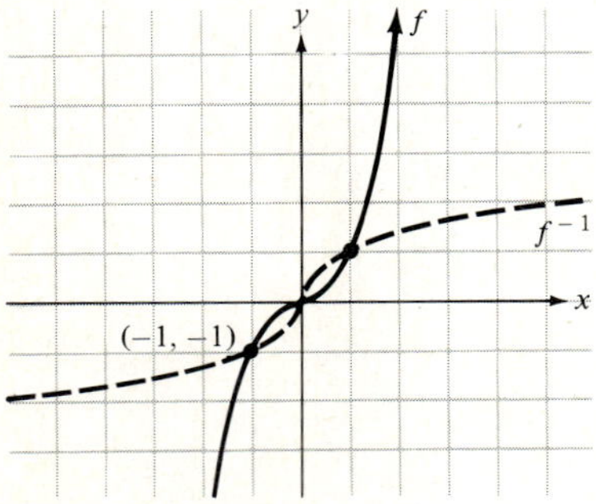

39. $f^{-1}(x) = x^2;\ x \geq 0$

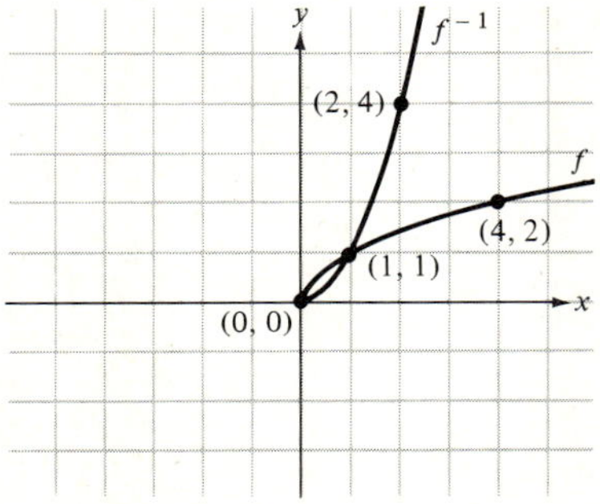

41. $f^{-1}(x) = \sqrt{9 - x^2};\ 0 \leq x \leq 3$ **43.** $f^{-1}(6) = -4$ **45.** $(f^{-1} \circ f)(5) = 5$

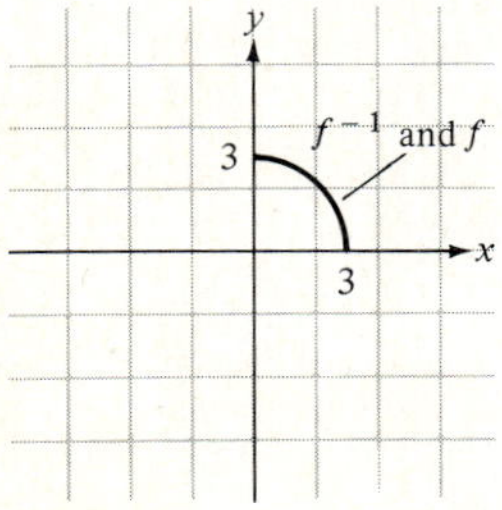

Exercises 10.7

1. $y = 3x;\ 27$ **3.** $y = 2t^2;\ 98$ **5.** $s = 3t^3;\ s = 81$ **7.** $w = 5\sqrt{z};\ 40$

9. $xy = 16;\ 2$ **11.** $s = \dfrac{32}{\sqrt{t}};\ 8$ **13.** $d = \dfrac{100}{s^2};\ 1$ **15.** $w = 4xy;\ 56$

17. $z = 5x^2y;\ 90$ **19.** \$1650 **21.** 400 ft **23.** $v = 8\sqrt{h}$ **25.** 128 lb

27. 605 watts

Chapter 10 Review

1. $D = \{-1, 0, 2\};\ R = \{-3, 0, 6\}$ **2.** $D = \{-3, -1, 3\};\ R = \{18, 2\}$

3. $D = \{x \mid -4 \leq x \leq 4\};\ R = \{y \mid -4 \leq y \leq 4\}$

4. $D = \{x \mid -2 \le x \le 2\}$; $R = \{y \mid -3 \le y \le 3\}$
5. $D = \{x \mid x \text{ is a real number}\}$; $R = \{y \mid y \text{ is a real number}\}$
6. $D = \{x \mid x \ge 0\}$; $R = \{y \mid y \text{ is a real number}\}$ **7.** $D = \{x \mid x \text{ is a real number}\}$; $R = \{y \mid y \ge 0\}$
8. $D = \{x \mid x \text{ is a real number}\}$; $R = \{y \mid y \text{ is a real number}\}$ **9.** Function; $\{0, 5, 10, 15\}$
10. Function; $\{3, 0, 15\}$ **11.** Function; $\{-1, 0, 1, 8\}$ **12.** Not a function
13. Function; $\{3\}$ **14.** Not a function **15. (a)** 3 **(b)** -7 **(c)** 8
16. (a) -1 **(b)** 3 **(c)** 0 **17. (a)** 3 **(b)** 18 **(c)** 18 **18.** All reals x, $x \ne 3$
19. $\{x \mid x \ge -2\}$ **20.** All reals x; $x \ne 5$, $x \ne 7$ **21.** All reals
22. All reals x; $x \ne \pm 1$ **23.** All reals x; $x \ne 1$, $x \ne 2$ **24.** Function
25. Not a function **26. (a)** $(1, -4)$ **(b)** Maximum value $= -4$
27. (a) 3 **(b)** Minimum value $= 1$ **28.** 15, 15 **29.** (a) 5 sec **(b)** 620 ft

30.

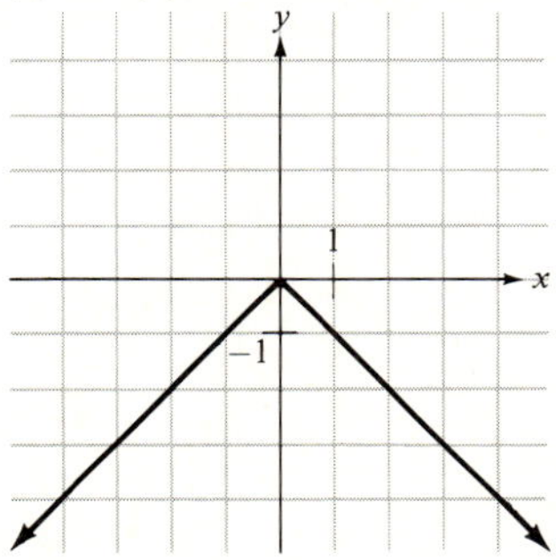

31.

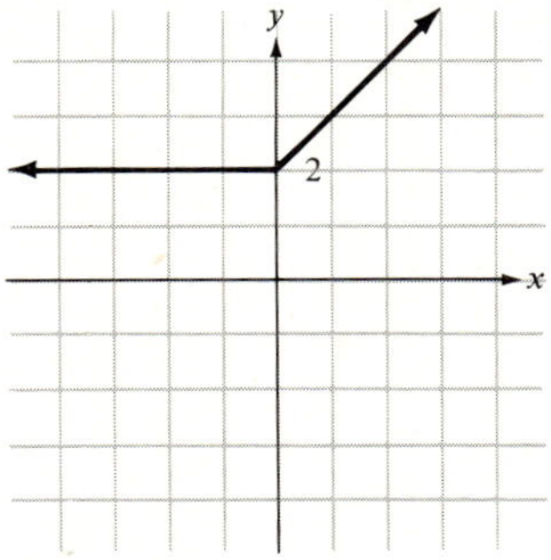

32. 6 **33.** 12 **34.** 87 **35.** 12.5
36. (a) $(f \circ g)(x) = x^2 + 4x + 3$ **(b)** $(g \circ f)(x) = x^2 + 1$
37. (a) $(f \circ g)(x) = (x^3 + 1)^2 = x^6 + 2x^3 + 1$ **(b)** $(g \circ f)(x) = x^6 + 1$
38. $f^{-1} = \{(-10, -1), (-3, 0), (4, 1)\}$ **39.** $f^{-1} = \{(4, 2), (9, 3), (16, 4)\}$
40. $f^{-1}(x) = \frac{1}{3}x + 3$ **41.** $f^{-1}(x) = 2\sqrt{x}$; $x \ge 0$

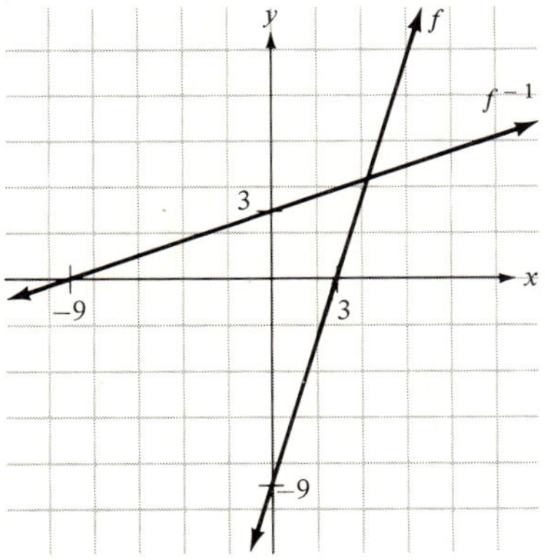

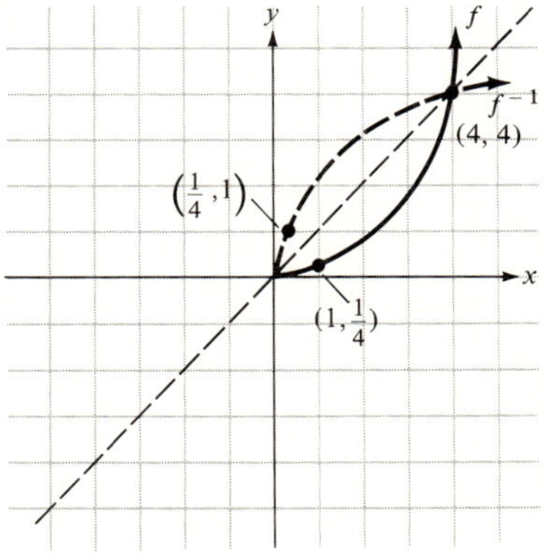

42. $y = 5x$; $y = 35$ **43.** $y = \dfrac{36}{x^2}$; $y = 1$ **44.** $w = 3xy$; $w = 45$

45. $I = \dfrac{V}{R}$; 4 amperes **46.** $T = \sqrt{2L}$; 4 sec

Chapter 10 Test

1. $D = \{-5, 3, 5, 6\}$; $R = \{2, 7, 8\}$ **2. (a)** Not a function **(b)** Function; $R = \{3, 6, 51\}$
3. (a) 11 **(b)** 2 **(c)** -4 **4.** $x^2 + 2hx + h^2$

5. **(a)** $\{x \mid x \text{ is real and } x \neq -4, x \neq 2\}$ **(b)** $\{x \mid x \text{ is real and } x > 4\}$
6. **(a)** Function **(b)** Not a function **7.** **(a)** Maximum value **(b)** 5 **8.** 20, 20
9. **(a)** 4 **(b)** 1 **(c)** 5 **10.** **(a)** 16 **(b)** 8 **(c)** 48 **(d)** 3 **11.** **(a)** 16 **(b)** 10
12. $f^{-1}(x) = \frac{1}{3}x - 2$ **13.** $z = 32$ **14.** **(a)** $k = \frac{1}{250}$ **(b)** $L = \dfrac{Av^2}{250}$

CHAPTER 11

Exercises 11.1

1.

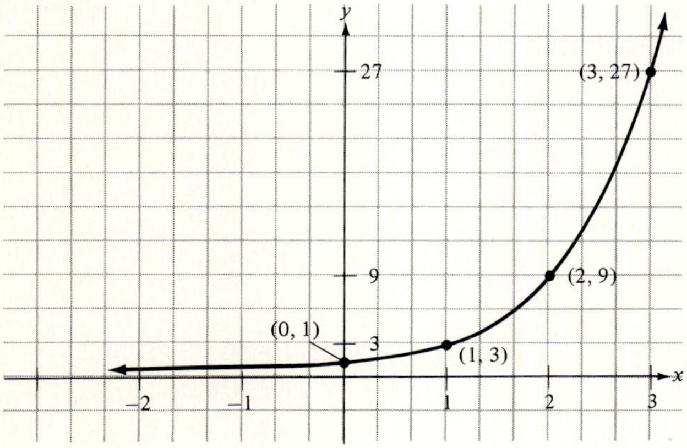

3.

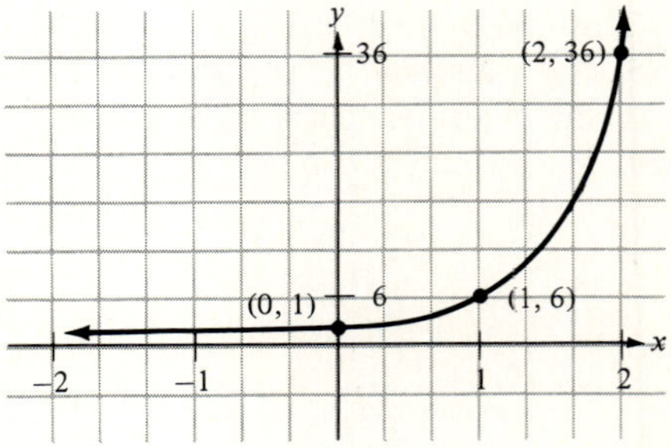

5.

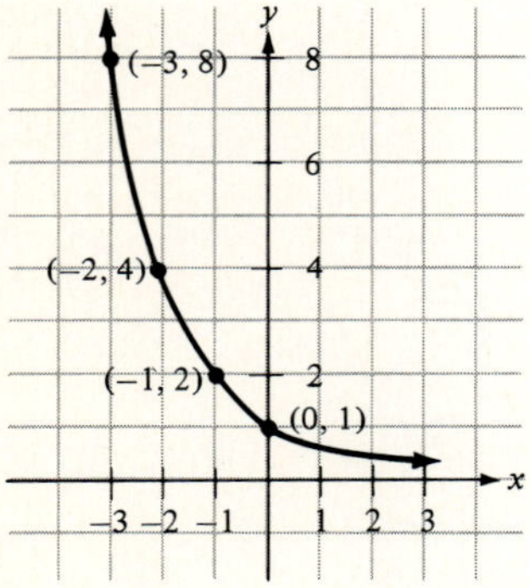

7.

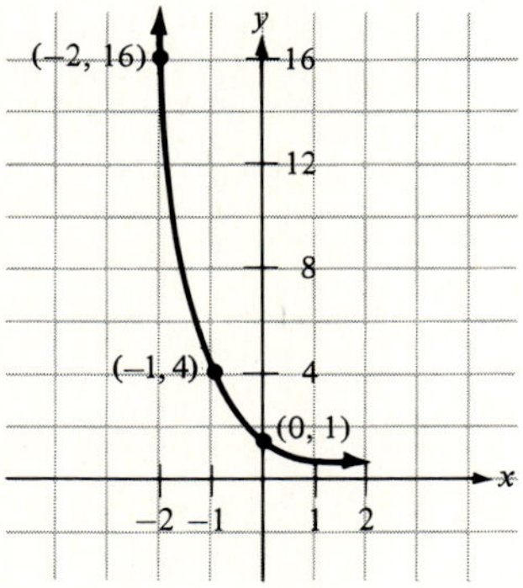

9.

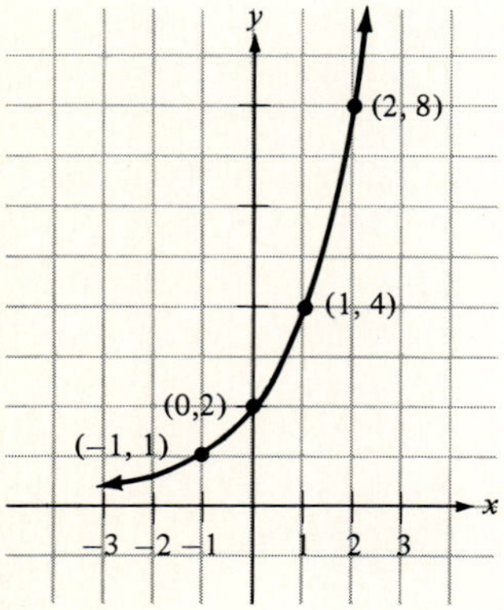

11.

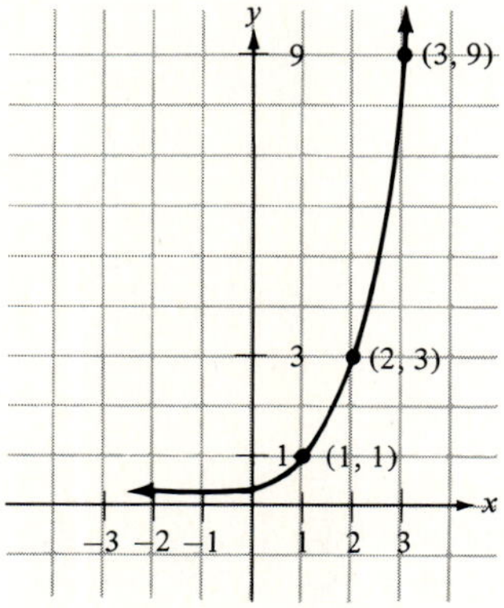

13.

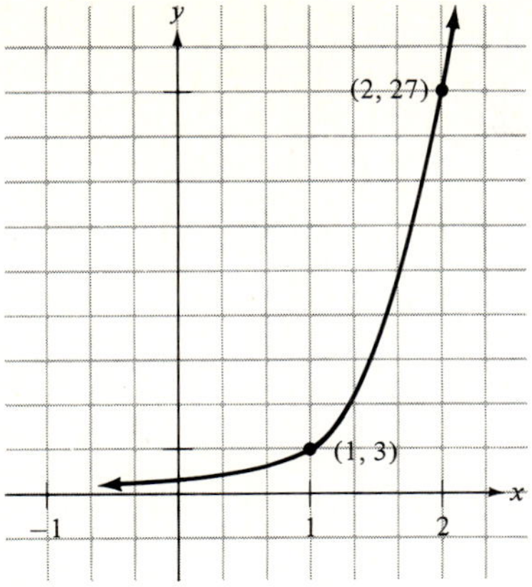

15.

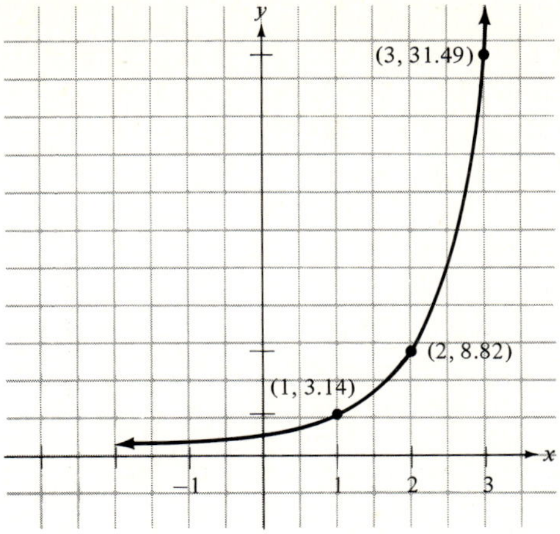

17.

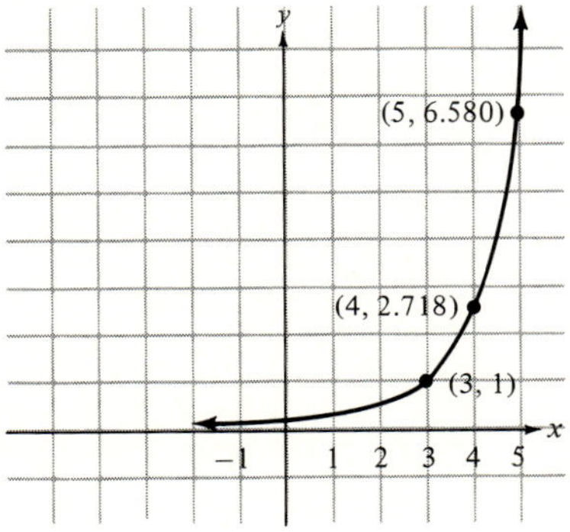

19.

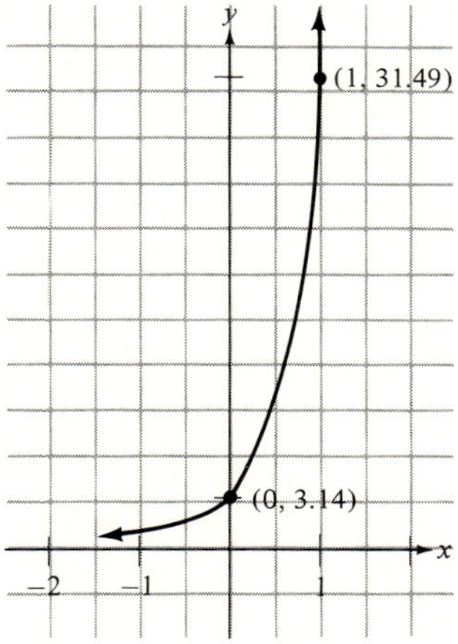

21. 2 **23.** $\frac{3}{2}$

25. −2.5 **27.** −1

29. −1.5

Exercises 11.2

1. $\log_2 16 = 4$ **3.** $\log_{10} 100 = 2$ **5.** $\log_5 625 = 4$ **7.** $\log_4 \frac{1}{16} = -2$

9. $\log_{2/5} \left(\frac{8}{125}\right) = 3$ **11.** $\log_{3/4} \left(\frac{16}{9}\right) = -2$ **13.** $3^2 = 9$ **15.** $10^2 = 100$

17. $5^3 = 125$ **19.** $2^{-2} = \frac{1}{4}$ **21.** $\left(\frac{1}{2}\right)^3 = \frac{1}{8}$ **23.** $\left(\frac{1}{2}\right)^{-4} = 16$

25.

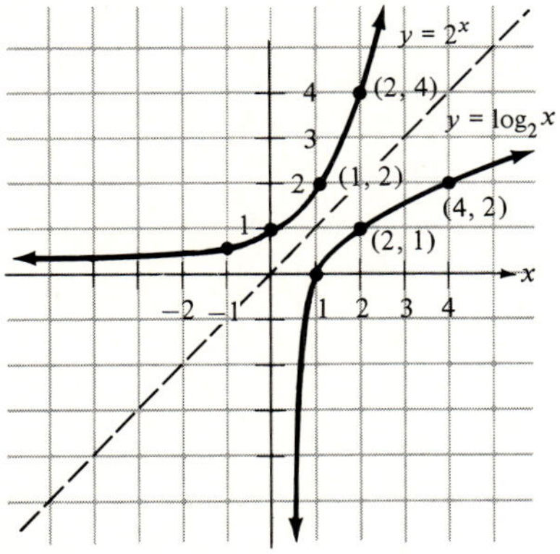

27.

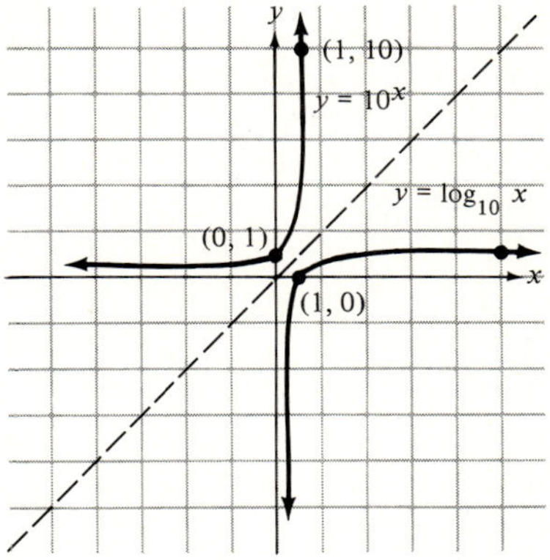

29.

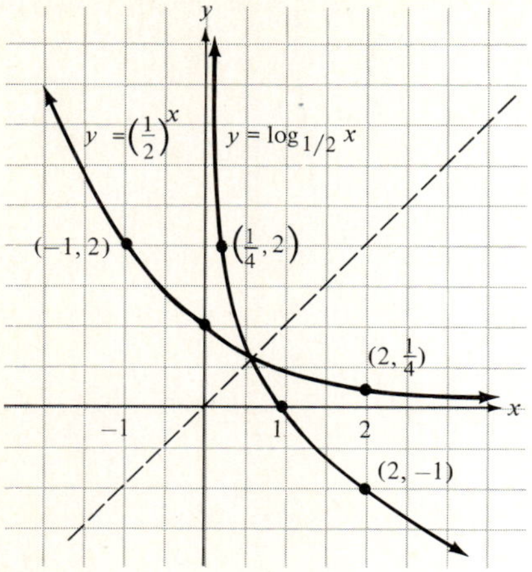

31. 3 **33.** 4 **35.** −2 **37.** 2 **39.** 4 **41.** $\frac{1}{2}$ **43.** $\frac{7}{2}$
45. $\frac{3}{2}$ **47.** $\frac{3}{2}$ **49.** 32 **51.** 9 **53.** $\frac{1}{10}$ **55.** $\frac{1}{64}$ **57.** 32 **59.** 64

Exercses 11.3

1. $\log_5 11 + \log_5 3$ **3.** $\log_{10} 17 + \log_{10} 2$ **5.** $\log_6 5 + \log_6 x$ **7.** $2.5 + \log_3 2$
9. $\frac{10}{3} + \log_2 3$ **11.** $\log_{10} 3 - \log_{10} 7$ **13.** $\log_2 5 - \log_2 x$ **15.** $2 - \log_7 w$
17. $4 \log_{10} 3$ **19.** $2 \log_5 7$ **21.** $(\frac{1}{2}) \log_5 2$ **23.** $(\frac{3}{4}) \log_7 w$
25. $\log_{10} 7 + \log_{10} w - \log_{10} 13$ **27.** $4 + \log_3 y - (\frac{1}{2}) \log_3 2$ **29.** $\log_2 3x$
31. $\log_5 \frac{11}{w}$ **33.** $\log_5 16w$ **35.** $\log_{10} \frac{12}{\sqrt{3}}$ **37.** $\log_{10} \frac{4\sqrt{w}}{8}$ **39.** $\log_{10} \frac{14x^2}{\sqrt{7}}$
41. $\frac{16}{3}$ **43.** $\frac{1}{5}$ **45.** 30 **47.** 25 **49.** 2

Exercises 11.4

1. .6911 **3.** 2.9063 **5.** 4.9595 **7.** 7.2504 − 10 **9.** 9.8513 − 10
11. 7.9425 **13.** 3.46 **15.** 74.4 **17.** 63100 **19.** .155 **21.** .000658
23. .120 **25.** 150,000,000 **27.** 7910 **29.** .00510 **31.** .000157
33. .000000742 **35.** .00000000523 **37.** 5.64 **39.** 4.53 **41.** .434
43. 118 **45.** .000000000295 **47.** 5,910 **49.** .00470 **51.** 89.2 **53.** 359

Exercises 11.5

For Exercises 1–19, if you use a calculator, your answer may differ in the fourth decimal place.

1. 1.4306 **3.** 3.170 **5.** 2.4403 **7.** 3.7713 **9.** 2.1962 **11.** 3.2099
13. 1.7918 **15.** 2.3026 **17.** 6.0064 **19.** 2.6878 **21.** 2.10 **23.** 4.25
25. .42 **27.** −1.50 **29.** 12.62 **31.** 1.74

Exercises 11.6

For Exercises 11.6, depending on your method of rounding numbers, your answers may vary slightly from those given.

1. 4.68 mg **3.** 2.62 mg **5.** 1473 years **7.** 19.5 mg **9.** 1179 years
11. In 45.5 min **13.** 2128 **15.** \$7780 **17.** \$7280 **19.** 8.04 years
21. 12.7 years

Chapter 11 Review

1.

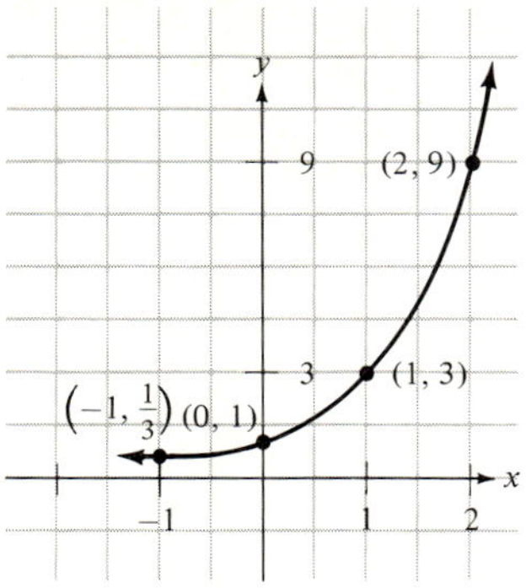

2.

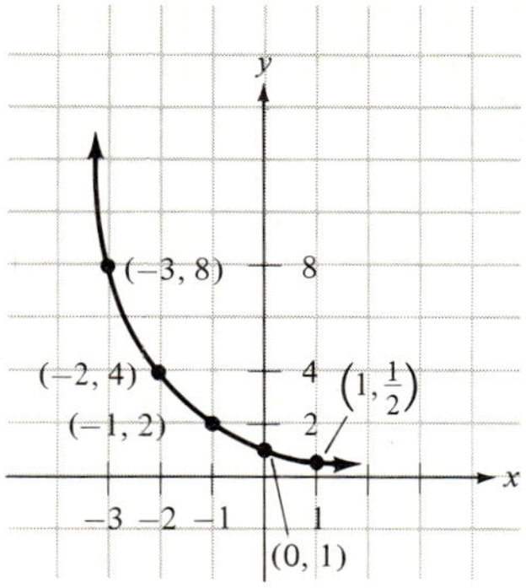

3.

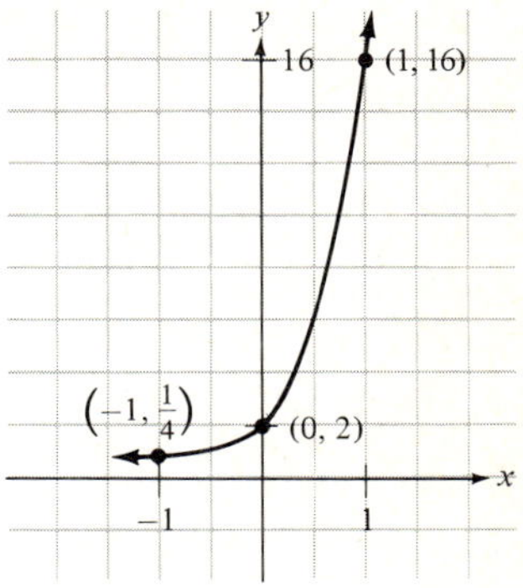

4. 2 **5.** 4 **6.** $-\frac{3}{2}$ **7.** $\log_5 625 = 4$ **8.** $\log_4 \frac{1}{16} = -2$
9. $\log_{10} 1000 = 3$ **10.** $9^2 = 81$ **11.** $5^{-2} = \frac{1}{25}$ **12.** $(\frac{1}{3})^2 = \frac{1}{9}$

13.

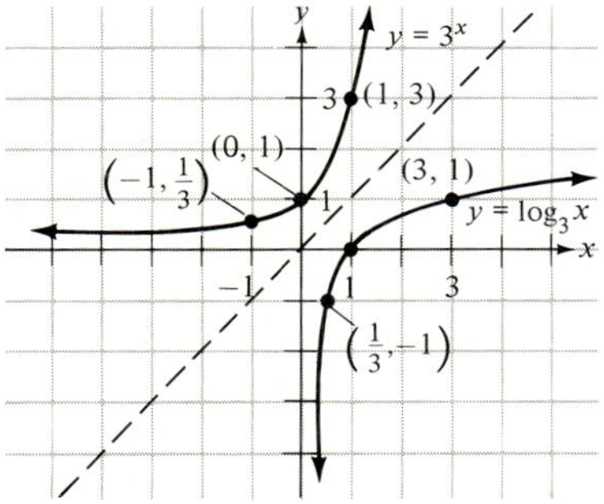

14.

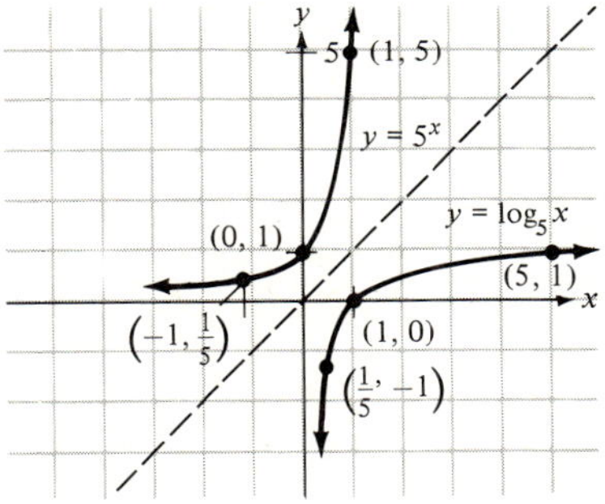

15. $\frac{1}{2}$ **16.** $\frac{1}{2}$ **17.** $\frac{3}{2}$ **18.** 16 **19.** 81 **20.** $\frac{1}{100}$
21. $\log_4 7 + \log_4 11$ **22.** $\log 3 + \log 13 + \log 19$ **23.** $\log 3 + \log x - \log 5$
24. $\log 29 - \log 7 - \log x$ **25.** $\frac{25}{9}$ **26.** 9, −4 is not a solution **27.** .7135
28. 3.0828 **29.** 7.4914 − 10 **30.** 737 **31.** .00832 **32.** 219,000,000
33. 409 **34.** 25,600 **35.** .261 **36.** 2.2146 **37.** 3.3223 **38.** 2.9957
39. 3.2188 **40.** 3.32 **41.** .59 **42.** 2.75 mg **43.** 8782 **44.** 12.7 years

Chapter 11 Test

1.

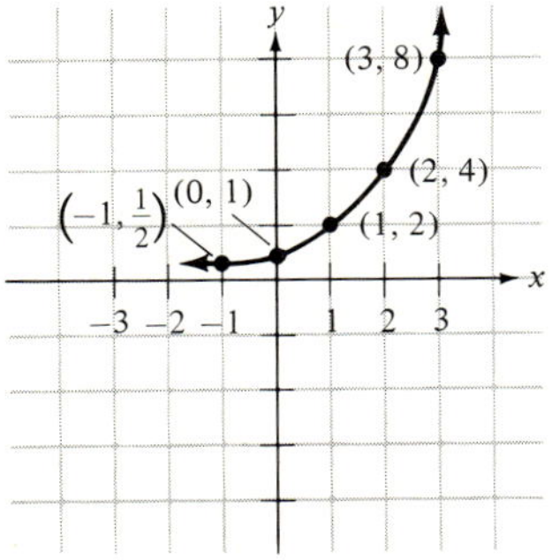

2.

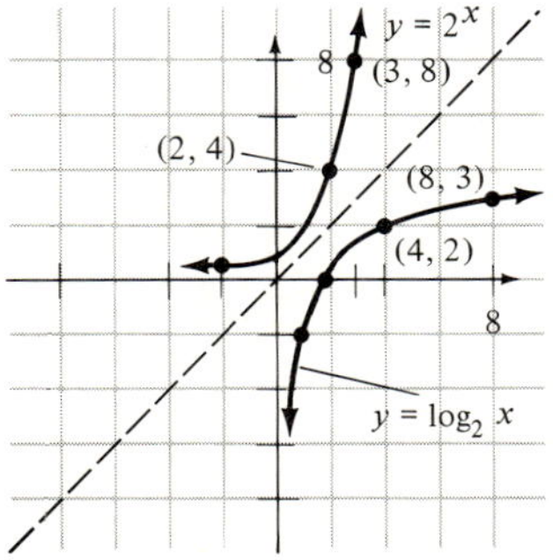

3. 2 **4.** 4 **5.** −2 **6.** −4 **7.** $\frac{1}{2}$ **8.** $-\frac{1}{2}$ **9.** 81
10. $\frac{1}{16}$ **11.** $\frac{9}{5}$ **12.** 625 **13.** .5694 **14.** 7.4533 − 10 **15.** 5.6739
16. 3.9513 **17.** 422 **18.** 6.59 **19.** 3.48 **20.** 8.60 **21.** 2.18
22. 2.99 **23.** 12.1 **24.** 2.65 mg **25.** 8.04 years

CHAPTER 12

Exercises 12.1

1. 1, 8, 27, 64 **3.** 2, 4, 6, 8 **5.** $\frac{1}{3}, \frac{1}{9}, \frac{1}{27}, \frac{1}{81}$ **7.** −1, 1, −1, 1 **9.** $\frac{1}{2}, \frac{1}{3}, \frac{1}{4}, \frac{1}{5}$
11. 5, 5, 5, 5 **13.** $1, \frac{1}{\sqrt{2}}, \frac{1}{\sqrt{3}}, \frac{1}{2}$ **15.** .1, .01, .001, .0001 **17.** 12
19. 38 **21.** 0 **23.** 4 **25.** $\frac{7}{8}$ **27.** 30 **29.** $1 + 2 + 3 + \cdots + n$
31. $\frac{1}{1 \cdot 2} + \frac{1}{2 \cdot 3} + \cdots + \frac{1}{n(n+1)}$ **33.** $1 + \frac{1}{8} + \frac{1}{27} + \frac{1}{64} + \cdots$
35. $1 + x + x^2 + x^3 + \cdots$ **37.** $x^2 + x^3 + x^4 + x^5 + \cdots$

Exercises 12.2

1. $d = 2$; 12, 14 **3.** $d = -3$; −4, −7 **5.** Not an arithmetic sequence
7. $d = 6$; 12, 18 **9.** 74 **11.** 33 **13.** −92 **15.** −184 **17.** 11
19. 242 **21.** 210 **23.** 629 **25.** 420 **27.** 125 **29.** −144
31. 120 **33.** $7625 **35. (a)** 240 ft **(b)** 1024 ft

Exercises 12.3

1. $r = 4$; 256 **3.** $r = \frac{1}{4}$; $\frac{1}{256}$ **5.** $r = .1$; .00001 **7.** $r = -2$; 32 **9.** $\frac{1}{27}$
11. −1 **13.** $\frac{32}{81}$ **15.** 63 **17.** 4096 **19.** 3645 **21.** $\frac{31}{16}$ **23.** $\frac{121}{81}$
25. $\frac{1}{3}$ **27.** $\frac{5}{4}$ **29.** 4 **31.** $\frac{5}{3}$ **33.** $\frac{2}{3}$ **35.** $\frac{1}{4}$ foot **37.** $\frac{2}{9}$ **39.** $\frac{64}{99}$

Exercises 12.4

1. 63 **3.** 180 **5.** 320 **7.** 35 **9.** 280 **11.** 468,000
13. (a) 60 **(b)** 1680 **(c)** 24 **15. (a)** 95,040 **(b)** 259,459,200 **(c)** 8,648,640
17. (a) 336 **(b)** 6720 **(c)** 40,320 **19.** 1,974,024 **21. (a)** 10 **(b)** 70 **(c)** 1
23. (a) 792 **(b)** 6435 **(c)** 1716 **25.** 56 **27.** 1365 **29.** 2,598,960

Exercises 12.5

1. 15 **3.** 462 **5.** 105 **7.** $a^5 + 5a^4b + 10a^3b^2 + 10a^2b^3 + 5ab^4 + b^5$
9. $a^7 + 7a^6b + 21a^5b^2 + 35a^4b^3 + 35a^3b^4 + 21a^2b^5 + 7ab^6 + b^7$
11. $a^6 - 6a^5b + 15a^4b^2 - 20a^3b^3 + 15a^2b^4 - 6ab^5 + b^6$ **13.** $x^4 + 12x^3 + 54x^2 + 108x + 81$
15. $32x^5 - 80x^4 + 80x^3 - 40x^2 + 10x - 1$ **17.** $27x^3 - 108x^2 + 144x - 64$
19. (a) 126 **(b)** 36 **21.** $8064x^5$ **23.** $21{,}840w^{11}$ **25.** $x^{14} - 14x^{13} + 91x^{12}$
27. $128x^7 + 448x^6 + 672x^5 + 560x^4$ **29.** $81y^4 - 216y^3$ **31.** 1287 **33.** 1225

Chapter 12 Review

1. 1, 16, 81, 256 **2.** $\frac{1}{3}, \frac{1}{4}, \frac{1}{5}, \frac{1}{6}$ **3.** −2, 4, −8, 16 **4.** 30 **5.** 24
6. 39 **7.** $1 + 4 + 9 + \cdots + n^2$ **8.** $1 + \frac{1}{2} + \frac{1}{3} + \cdots + \frac{1}{n}$
9. $x^2 + x^3 + x^4 + x^5$ **10.** $d = 5$; 28 **11.** $d = -3$; −5 **12.** $d = 4$; 26
13. 45 **14.** 27 **15.** −64 **16.** 930 **17.** $735,000 **18.** $r = 6$; 1080
19. $r = \frac{1}{5}$; $\frac{1}{1250}$ **20.** $r = -2$; 32 **21.** $\frac{1}{16}$ **22.** 54 **23.** $\frac{121}{81}$ **24.** $\frac{5}{2}$
25. 1 foot **26.** 40 **27.** 420 **28.** 48,672 **29. (a)** 210 **(b)** 30 **(c)** 7920

30. **(a)** 1680 **(b)** 20,160 **31.** **(a)** 35 **(b)** 15 **(c)** 330 **32.** 28 **33.** 133,784,560
34. 20 **35.** $a^5 + 5a^4b + 10a^3b^2 + 10a^2b^3 + 5ab^4 + b^5$
36. $x^4 - 12x^3 + 54x^2 - 108x + 81$ **37.** 28 **38.** $126x^5$
39. $x^{11} - 11x^{10} + 55x^9 - 165x^8$ **40.** 231

Chapter 12 Test

1. 20 **2.** 64 **3.** −27 **4.** $\frac{1}{512}$ **5.** 2 **6.** 46 **7.** −28
8. 1600 **9.** $\frac{1}{32}$ **10.** $\frac{1}{27}$ **11.** $\frac{728}{486} = \frac{364}{243}$ **12.** $\frac{6}{5}$ **13.** $2625
14. $819 **15.** 70 **16.** 432 **17.** **(a)** 56 **(b)** 24 **(c)** 5040
18. **(a)** 20 **(b)** 120 **(c)** 120 **19.** **(a)** 28 **(b)** 1 **(c)** 210 **20.** 220 **21.** 35
22. $a^6 + 6a^5b + 15a^4b^2 + 20a^3b^3 + 15a^2b^4 + 6ab^5 + b^6$ **23.** $x^4 - 8x^3 + 24x^2 - 32x + 16$
24. 126 **25.** $x^{13} - 13x^{12} + 78x^{11} - 286x^{10}$ **26.** 27,405

APPENDIX A

Exercises

1. $x + 9$, remainder = 16 **3.** $3x + 1$, remainder = 14 **5.** $2x + 13$, remainder = 5
7. $x + 10$, remainder = 5 **9.** $x^2 + 6x + 19$, remainder = 44
11. $2x^2 + 4x - 3$, remainder = −4 **13.** $2x^2 + 6x - 3$, remainder = 1
15. $2x^3 - 3x^2 + x + 5$, remainder = 26 **17.** $2x^5 - 6x^4 + x^3 - 3x^2 + 10x + 1$, remainder = −3
19. $x^5 - x^4 + x^3 - x^2 + x + 1$, remainder = −2

APPENDIX B

Exercises

1. 2 **3.** 1 **5.** 7 **7.** 87 **9.** 2
11. $(x + 1), (x - 1), (x + 5), (x - 5)$ **13.** $(x + 1), (x - 1), (x + 2), (x - 2)$
15. $(x + 1), (x - 1)$ **17.** $(x + 1), (x - 1), (x + 7), (x - 7)$
19. $(x + 1), (x - 1), (x + 11), (x - 11)$ **21.** $P(1) = 0$ **23.** $P(-1) = 0$
25. $P(-2) = 0$ **27.** $P(-2) = 0$ **29.** $P(3) = 0$ **31.** $P(2) = 0$, yes
33. $P(-1) = -2$, no **35.** $P(1) = 13$, no **37.** $P(3) = 0$, yes **39.** $P(-2) = 0$, yes

INDEX

INDEX INDEX

D

E

F